FUNDAMENTAL PRINCIPLES OF CLASSICAL MECHANICS

A Geometrical Perspective

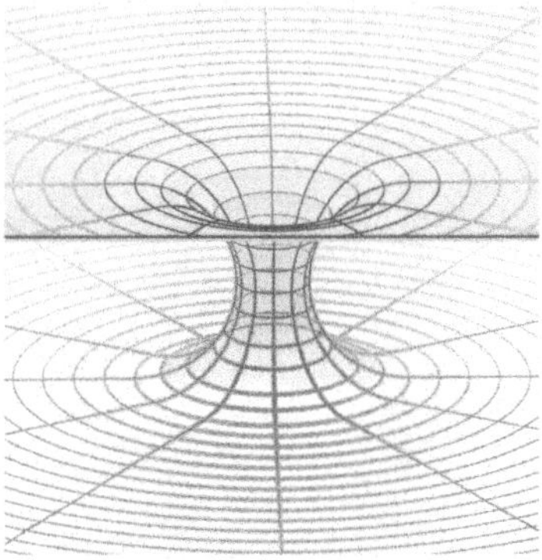

FUNDAMENTAL PRINCIPLES OF CLASSICAL MECHANICS

A Geometrical Perspective

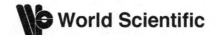

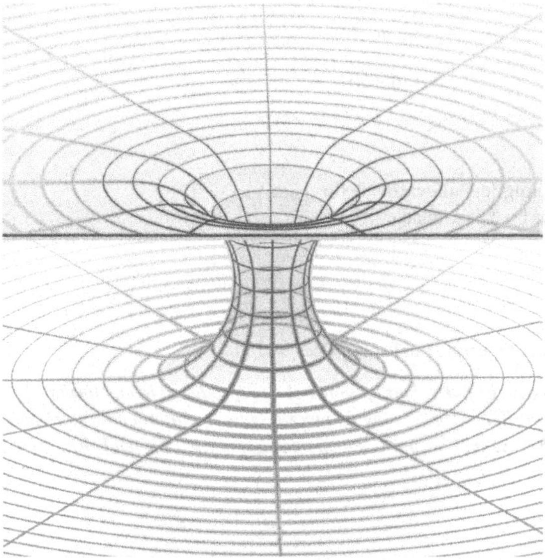

Kai S. Lam

California State Polytechnic University, Pomona, USA

World Scientific

NEW JERSEY • LONDON • SINGAPORE • BEIJING • SHANGHAI • HONG KONG • TAIPEI • CHENNAI

Published by

World Scientific Publishing Co. Pte. Ltd.
5 Toh Tuck Link, Singapore 596224
USA office: 27 Warren Street, Suite 401-402, Hackensack, NJ 07601
UK office: 57 Shelton Street, Covent Garden, London WC2H 9HE

Library of Congress Cataloging-in-Publication Data
Lam, Kai S. (Kai Shue), 1949– author.
Fundamental principles of classical mechanics : a geometrical perspective / by Kai S. Lam, California State Polytechnic University, USA.
pages cm
Includes bibliographical references and index.
ISBN 978-981-4551-48-9 (hardcover : alk. paper)
1. Mechanics. I. Title.
QA805.L245 2014
531.01'515--dc23
2014005901

British Library Cataloguing-in-Publication Data
A catalogue record for this book is available from the British Library.

Printed in Singapore

to Bonnie,
Nathan, Reuben, Aaron,
and
my Parents

Preface

This book is written with the belief that classical mechanics, as a theoretical discipline, possesses an inherent beauty, depth, and richness that far transcend its immediate applications in mechanical systems, although these are no doubt important in their own right. These properties are manifested, by and large, through the coherence and elegance of the mathematical structure underlying the discipline, and, at least in the opinion of this author, are eminently worthy of being communicated to physics students at the earliest stage possible. The present text is therefore addressed mainly to advanced undergraduate and beginning graduate physics students who are interested in an appreciation of the relevance of modern mathematical methods in classical mechanics, in particular, those derived from the much intertwined fields of topology and differential geometry, and also to the occasional mathematics student who is interested in important physics applications of these areas of mathematics. Its chief purpose is to offer an introductory and broad glimpse of the majestic edifice of the mathematical theory of classical dynamics, not only in the time-honored analytical tradition of Newton, Laplace, Lagrange, Hamilton, Jacobi, and Whittaker, but also the more topological/geometrical one established by Poincaré, and enriched by Birkhoff, Lyapunov, Smale, Siegel, Kolmogorov, Arnold, and Moser (as well as many others). The latter tradition has been somewhat inexplicably and politely ignored for many decades in the 20th century within the realm of physics instruction, and has only relatively recently regained favor in some physics textbooks under the guise of the more fashionable topics of chaos and complexity in dynamical systems theory. This unfortunate circumstance may perhaps in hindsight be attributable to related historical events: the rise of quantum mechanics just as Poincaré's contributions in celestial mechanics were coming to the fore at the dawn of the 20th century, and the subsequent competition for limited "curricular space" in physics instruction between classical and quantum mechanics. The irony from a historical perspective, of course, is that it was precisely the Hamilton-Jacobi theory within the Hamiltonian formulation of classical mechanics and its relationship to wave optics that precipitated the development of non-relativistic quantum theory, and that the formalism of action-angle variables in Hamiltonian mechanics, through the Bohr-Sommerfeld quantization rules, proved to be the royal road to the "old quantum theory". In addition, it was the Lagrangian formulation of classical mechanics that provided the groundwork for the Feynman path approach in quantum field theory.

We hope that the present text will make a modest contribution, along with many other excellent ones that have already appeared, towards the rehabilitation of Poincaré's tradition in the mainstream physics curricula. Because of the pervasive topological and geometrical character of this tradition, it is inevitable that a coherent, if somewhat spotty, introductory exposition (without straying too far into the broader field of dynamical systems) of a key group of concepts and tools of topology and differential geometry will have to be a central feature of the presentation, as well as an earnest attempt to convince the reader of the fundamental relevance of these mathematical tools in classical mechanics. These objectives give rise to the adoption of a "dual-track" character of the book, alternating between physics and mathematics. Our hope is to strike a delicate balance between the intuitive/physically specific and the abstract/mathematically general. The presentation will be mainly at a heuristic level, largely unburdened by a rigorous, systematic, and lengthy presentation of the mathematical background. The inevitable loss of generality, rigor, and even accuracy entailed by such a style of "physical" presentation will hopefully be compensated for by a sound degree of continuity from the physics to the mathematics (and vice versa), and by ample explicit calculations based on carefully chosen applications of the abstract mathematical machinery.

To achieve the aforementioned goals, the text seeks to build upon the conventional analytical treatment of the subject (based on vector calculus, multivariable calculus, and the solutions of differential equations) as presented in almost all undergraduate texts and most graduate ones, and substantially enrich the exposition with a healthy dose of the modern language and tools of topology and differential geometry, especially those related to the basic notions of *differentiable manifolds, tangent and cotangent spaces, the exterior differential calculus, homotopy, homology and cohomology, connections on fiber bundles, Riemannian geometry, symplectic geometry, and Lie groups and algebras.* The main pedagogical route to the introduction and elucidation of most of these mathematical topics will be the use of *Cartan's Method of Moving Frames* in the classical mechanical context of rigid-body dynamics, as promulgated by the geometer S. S. Chern. Historically, the mathematical development of the differential geometric theory of moving frames was in fact originally motivated by this classical mechanics problem. We seek to erect a pedagogical bridge, as it were, between the time-honored texts of, for example, Goldstein's "*Classical Mechanics*" on the one hand and Abraham and Marsden's "*Foundations of Mechanics*" on the other – allbeit at a more elementary and much less rigorous level. In this vein, the classic work "*Mathematical Methods of Classical Mechanics*" by Arnold immediately comes to mind as a sort of "gold standard"; but again, our aim is more modest: not only do we seek to build a bridge between the physical/analytical and the mathematical/geometrical, but also one between the undergraduate and graduate curricula. Due to the immense richness and depth of the field, however, we cannot lay claim to any degree of completeness or originality of treatment.

At the risk of alienating the more mathematically oriented reader – this is, after all, primarily a physics text – I have laid down the following guidelines for the writing of this book in regard to the sometimes tortured relationship between the physics and the mathematics contained therein.

- Except on occasion, when clarity, conciseness, and precision trump intuitive appeal, the usual abrupt definition–theorem–proof sequence encountered in the mathematical literature will be replaced by a more casual build-up with physical motivation and justification leading to the relevant mathematical concepts and facts. For example, the discussion of the mathematical equivalence of gauge fields and connections on fiber bundles is essentially built up from the elementary mechanical context of rigid bodies moving in Euclidean space. As another example, the discussion of the relevance of symplectic structures on manifolds and the crucial importance of Darboux's Theorem regarding symplectic manifolds will be preceded by a good amount of justification in more familiar analytical language. Once the informal build-up has achieved its purpose, however, we will not shy away from exploiting the logical sharpness, elegance and beauty of abstract concepts to arrive at results quickly, for example, in demonstrating the existence of certain integral invariants under canonical transformations (symplectomorphisms) in phase space.

- The analytical mode of development (because of its familiarity and concreteness) will almost always be presented first, as a precursor to and motivation for the more abstract and unfamiliar geometrical development. For example, Poisson brackets will be introduced first in the traditional "physics" manner (in terms of partial derivatives with respect to canonical coordinates and momenta) before being defined in terms of the value of the symplectic form on two vector fields on phase space. Once the geometrical notions have been firmly grounded in analytical representations, however, the strong interplay between the analytical (local) and the geometrical (global) viewpoints will be stressed.

- Within the geometrical exposition, (local) coordinates will almost always be used first in favor of the coordinate-free approach usually preferred by mathematicians. Darboux's Theorem guarantees that this can be done with impunity in the case of Hamiltonian mechanics, but in many other situations, such as the use of moving frames for rigid-body dynamics as a precursor to the introduction of frame bundles and the introduction of connections on principal bundles (gauge fields), the description in terms of local coordinates is usually much more intuitive. In addition, it yields easily interpretable analytical formulas for calculations, even though they may only be locally valid.

- When precise mathematical definitions are unavoidable, we strive to avoid layering unfamiliar, abstract, and overly technical ones in quick succession. If possible, specific examples will be used to motivate definitions of abstract concepts. For example, the general notion of connections on vector bundles is introduced through the much more familiar kinematical notion of angular velocities. As another example, the intuitive picture of two-dimensional KAM tori interlaced in three-dimensional energy hypersurfaces of four-dimensional phase space is used to motivate the notion of foliated spaces in general.

- Whole chapters serving as mathematical interludes dealing with specialized topics are placed at strategic positions within the text, rather than being lumped together entirely at the beginning of the book, or relegated to a set of appendices at the end. (We plead guilty – partially – to violating this rule at the beginning of the text, where we feel that it is necessary to engage in a higher concentration of mathematical presentation; but even there, important physics is linked to the mathematics whenever possible.) We hope that this practice will avoid having the patience of the non-mathematical reader taxed excessively, and preserve our stylistic goal of interweaving the physics with the mathematics. Examples are chapters dealing with the following topics: Exterior Calculus, Vector Calculus by Differential Forms, Lie Groups and Moving Frames, Connections on Fiber Bundles, Riemannian Curvature, and Symplectic Geometry.

- Full proofs of mathematical theorems are usually not presented, especially those involving a large amount of technical details. Proofs are given only when they are relatively direct, simple and illustrative of useful calculational techniques.

The physics prerequisites for this book can be simply stated: a solid lower-division course in classical mechanics. The mathematical prerequisites, however, are more nebulous, and, as can be expected, a bit more demanding. We hope the reader will have a good working knowledge of multivariable calculus, vector calculus (div, grad, curl and all that), some linear algebra (matrices and eigenvalue problems), some complex variable theory, elements of ordinary and partial differential equations, and finally, some exposure to introductory tensor analysis. Beyond these, everything is essentially developed from the ground up in the text, although, as we stipulated before, not comprehensively and rigorously. A certain degree of innate curiosity and what is often loosely referred to as mathematical maturity will also be useful.

The material contained in this book, its organization, and even the pedagogical strategies adopted in the delicate dance between the physics and the mathematics presentations, might be gleaned from the Chapter titles and their ordering. There is probably more than enough material for a year-long syllabus, but students and instructors should be able to adopt select portions according to their own interests and preferences without too much difficulty. The individual chapters are intended to be as thematically, and even on many occasions as technically, self-contained as possible. But no less important, we have also endeavored to render the entire sequencing into a coherent and logical flow. Numerous exercise problems, many of them supplied with generous and systematic hints, are strategically located within the running narrative of the text. It goes without saying that these form an integral part of the text, and the reader is urged to attempt as many of them as possible.

Special care has been devoted to cross-referencing of material for easy lookup by the reader, especially with regard to possibly unfamiliar physical or mathematical concepts. Within the text this will be done by thorough and careful references (forward and backward) through specific equation numbers, theorem and definition numbers, or direction of the reader to specific locations in the text. Outside of the text, every effort has been made to make the index as serviceable as possible.

About 40% of the material in this book, in less developed and organized form, were used at one time or another as class notes made available to students over the past decade or so when I taught the upper-division Classical Mechanics sequence at Cal Poly Pomona. Even though these tended to comprise the more physical parts of the text, the approach taken was perhaps more mathematical than customary and at times may have been a bit idiosyncratic. But throughout this period and for the most part, my students have been kind enough to put up with it and indeed provided honest and constructive feedback, and even occasional encouragement. It is to them that I would like to first and foremost express my gratitude. No less important to the task of writing was the collegial and congenial atmosphere afforded by my home department that I am fortunate enough to count myself a member of for the better part of the last three decades. Dr. Stephen McCauley, my current Department Chair in particular, was helpful and encouraging in every way possible. Amidst the inevitable heavy teaching duties at times, the Faculty Sabbatical Program of the California State University provided much needed focused periods of time to be devoted to this book project. In the matter of intellectual debt, I would like to single out and express my sincere gratitude to the late eminent differential geometer Professor S. S. Chern, who graciously and generously included me in a differential geometry book translation and expansion project some 15 years ago, and thus paved the way for my own modest book projects aimed at familiarizing physics students with that subject, of which the present text is the latest one. Much of the mathematical material on differential geometry in this book will bear the imprint of the valuable teachings and insights of Professor Chern, while the errors and inaccuracies remain of course entirely my own. Ms. E. H. Chionh of World Scientific Publishing Company, my able editor for my last published text, has again shepherded me with nothing but the highest standards of professionalism through the present one. To her I again extend my appreciation. Finally, my dear wife, Dr. Bonnie Buratti, and our three sons, Nathan, Reuben, and Arron, have all been, in their different ways, the bedrock as well as the inspirational muse of my life through the many ups and downs while this book was in progress over the past several years. To them I would like to say, as always and simply, thank you.

Kai S. Lam
Department of Physics and Astronomy,
California State Polytechnic University, Pomona *January 2014*

Contents

Chapter 1

Vectors, Tensors, and Linear Transformations

Classical mechanics is the quantitative study of the laws of motion for macroscopic physical systems with mass. The fundamental laws of this subject, known as *Newton's Laws of Motion*, are expressed in terms of second-order differential equations governing the time evolution of vectors in a so-called **configuration space** of a system (see Chapter 12). In an elementary setting, these are usually vectors in 3-dimensional Euclidean space, such as position vectors of point particles; but typically they can be vectors in higher dimensional and more abstract spaces. A general knowledge of the mathematical properties of vectors, not only in their most intuitive incarnations as directed arrows in physical space but as elements of abstract linear vector spaces, and those of linear operators (transformations) on vector spaces as well, is then indispensable in laying the groundwork for both the physical and the more advanced mathematical – more precisely topological and geometrical – concepts that will prove to be vital in our subject. In this beginning chapter we will review these properties, and introduce the all-important related notions of *dual spaces* and *tensor products* of vector spaces. The notational convention for vectorial and tensorial indices used for the rest of this book (except when otherwise specified) will also be established.

We start with a familiar situation. Consider a position vector $\boldsymbol{x}$ of a point particle in the two-dimensional plane $\mathbb{R}^2$ written in terms of its components x^1 and x^2 :

$$\boldsymbol{x} = x^1\,\boldsymbol{e}_1 + x^2\,\boldsymbol{e}_2 = x^i\,\boldsymbol{e}_i \quad . \tag{1.1}$$

The vectors $\boldsymbol{e}_1$ and $\boldsymbol{e}_2$ in (1.1) form what is called a **basis** of the **linear vector space** $\mathbb{R}^2$, and the components of a vector $\boldsymbol{x}$ are determined by the choice of the basis. In Fig. 1.1, we have chosen $\{\boldsymbol{e}_1, \boldsymbol{e}_2\}$ to be an **orthonormal basis**. This term simply means that $\boldsymbol{e}_1$ and $\boldsymbol{e}_2$ are each of unit length and are orthogonal (perpendicular) to each other. We note that the notions of length and orthogonality arise only after the introduction of an **inner (scalar) product** – an additional *metrical* structure – on the linear space [see (1.41) below)].

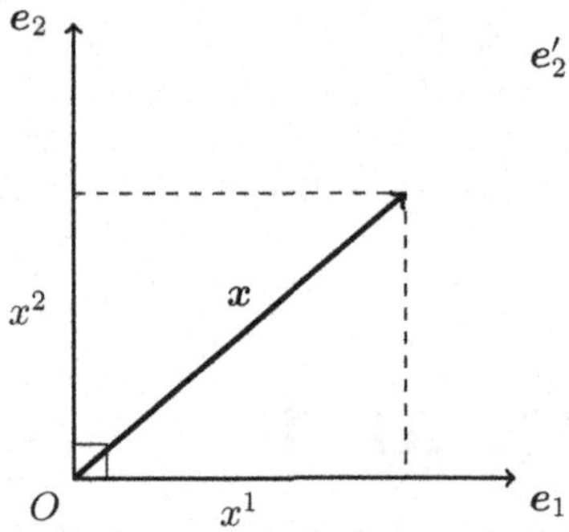

Fig. 1.1

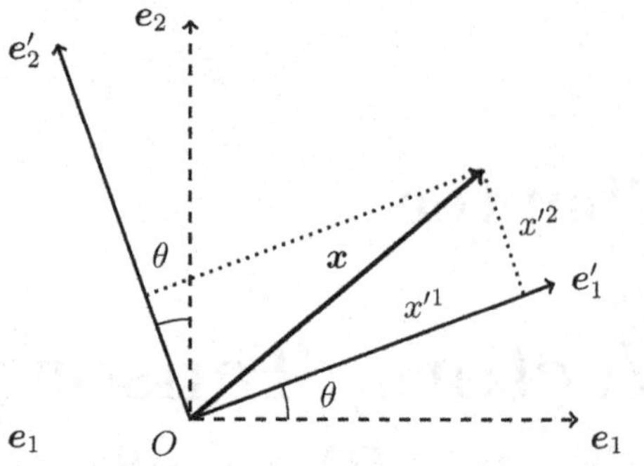

Fig. 1.2

Notice that in (1.1) we have used a superscript (upper index) for components and a subscript (lower index) for basis vectors, and repeated indices (one upper and one lower) are understood to be summed over (from 1 to the **dimension** of the vector space under consideration, which is two in the present case). This convenient notational device, called the *Einstein summation convention*, will be used routinely in this book from now on, and its elegance will hopefully become apparent as we proceed. (Repeated indices that are not meant to be summed over will be explicitly noted, while repeated summation indices that are both up or both down will be accompanied by an explicit summation sign.) The orthonormal basis $\{\boldsymbol{e}_1, \boldsymbol{e}_2\}$ is usually called a *reference* (*coordinate*) *frame* in physics.

Now consider the *same* vector $\boldsymbol{x}$ with *different* components x'^i with respect to a rotated orthonormal frame $\{\boldsymbol{e}'_1, \boldsymbol{e}'_2\}$, rotated from $\{\boldsymbol{e}_1, \boldsymbol{e}_2\}$ by an angle θ in the positive (anti-clockwise) sense (as shown in Fig. 1.2):

$$\boldsymbol{x} = x'^i \, \boldsymbol{e}'_i \quad . \tag{1.2}$$

It is simple to show that

$$(x')^1 = (\cos\theta)\, x^1 + (\sin\theta)\, x^2 \;, \quad (x')^2 = (-\sin\theta)\, x^1 + (\cos\theta)\, x^2 \quad . \tag{1.3}$$

Problem 1.1 Verify (1.3) by considering the geometry of Fig. 1.2.

Equation (1.3) can also be obtained by working with the basis vectors, instead of the components, directly. It is evident from Fig. 1.2 that

$$\boldsymbol{e}_1 = (\cos\theta)\, \boldsymbol{e}'_1 - (\sin\theta)\, \boldsymbol{e}'_2 \;, \quad \boldsymbol{e}_2 = (\sin\theta)\, \boldsymbol{e}'_1 + (\cos\theta)\, \boldsymbol{e}'_2 \quad . \tag{1.4}$$

Thus, (1.1) and (1.2) together, that is,

$$\boldsymbol{x} = x^i\, \boldsymbol{e}_i = x'^i\, \boldsymbol{e}'_i \ ,$$

imply that

$$x^1 \cos\theta\, \boldsymbol{e}'_1 - x^1 \sin\theta\, \boldsymbol{e}'_2 + x^2 \sin\theta\, \boldsymbol{e}'_1 + x^2 \cos\theta\, \boldsymbol{e}'_2 = x'^1\, \boldsymbol{e}'_1 + x'^2\, \boldsymbol{e}'_2 \quad . \tag{1.5}$$

Comparison of the coefficients of $\boldsymbol{e}'_1$ and $\boldsymbol{e}'_2$ on both sides of (1.5) immediately yields (1.3).

Let us now write (1.3) and (1.4) in matrix notation. Equations (1.3) can be written as the single matrix equation

$$(x'^1, x'^2) = (x^1, x^2) \begin{pmatrix} \cos\theta & -\sin\theta \\ \sin\theta & \cos\theta \end{pmatrix} \ , \tag{1.6}$$

while Eqs. (1.4) can be written as

$$\begin{pmatrix} \boldsymbol{e}_1 \\ \boldsymbol{e}_2 \end{pmatrix} = \begin{pmatrix} \cos\theta & -\sin\theta \\ \sin\theta & \cos\theta \end{pmatrix} \begin{pmatrix} \boldsymbol{e}'_1 \\ \boldsymbol{e}'_2 \end{pmatrix} \ . \tag{1.7}$$

Denote the 2×2 matrix in both (1.6) and (1.7) by

$$\left(a_i^j\right) = \begin{pmatrix} a_1^1 & a_1^2 \\ a_2^1 & a_2^2 \end{pmatrix} = \begin{pmatrix} \cos\theta & -\sin\theta \\ \sin\theta & \cos\theta \end{pmatrix} \ , \tag{1.8}$$

where *the lower index is the row index and the upper index is the column index.* (This index convention for matrices will be adopted in this book unless otherwise specified, and we warn the reader that, occasionally, it may lead to confusion.) Equations (1.6) and (1.7) can then be compactly written using the Einstein summation convention as

$$x'^i = a_j^i\, x^j \ , \tag{1.9}$$

$$\boldsymbol{e}_i = a_i^j\, \boldsymbol{e}'_j \ . \tag{1.10}$$

Note again that repeated pairs of indices, one upper and one lower, are summed over.

Equations (1.9) and (1.10) are equivalent, in the sense that either one is a consequence of the other. By way of illustrating the usefulness of the index notation, we again derive (1.9) from (1.10) *algebraically* as follows:

$$\boldsymbol{x} = x^j\, \boldsymbol{e}_j = x^j a_j^i\, \boldsymbol{e}'_i = x'^i\, \boldsymbol{e}'_i \quad . \tag{1.11}$$

The last equality implies (1.9).

In matrix notation, (1.9) and (1.10) can be written respectively as (note the order in which the matrices occur in the matrix products):

$$x' = x\,A\,, \tag{1.12}$$

$$\mathbf{e} = A\,\mathbf{e}'\,, \tag{1.13}$$

where A is the matrix in (1.8), $\mathbf{e}, \mathbf{e}'$ are the column matrices in (1.7), and x, x' are the row matrices in (1.6). The above equations imply that

$$x = x' A^{-1}\,, \tag{1.14}$$

$$\mathbf{e}' = A^{-1}\,\mathbf{e}\,, \tag{1.15}$$

where A^{-1} is the inverse matrix of A. The above transformation properties for basis vectors and the components of vectors are satisfied for any invertible matrix. These are summarized in the following table for a general vector $\boldsymbol{v} = v^i\,\boldsymbol{e}_i = v'^{\,i}\,\boldsymbol{e}'_i$.

$\boldsymbol{v} = v^i\,\boldsymbol{e}_i = v'^{\,i}\,\boldsymbol{e}'_i$	
$\boldsymbol{e}_i = a_i^j\,\boldsymbol{e}'_j$	$\boldsymbol{e}'_i = (a^{-1})_i^j\,\boldsymbol{e}_j$
$v'^{\,i} = a_j^i\,v^j$	$v^i = (a^{-1})_j^i\,v'^{\,j}$

Table 1.1

In all the indexed quantities introduced above, upper indices are called **contravariant indices** and lower indices are called **covariant indices**. These are distinguished because tensorial objects [see the discussion in the paragraph following Eq. (1.31) below] involving upper and lower indices transform differently under a change of basis, as evident from Table 1.1 for the case of vectors.

At this point it will be useful to recall the abstract concept of linear vector spaces. One speaks of a vector space V over a field F. In physics applications, the algebraic field is usually the field of real numbers $\mathbb{R}$ or the field of complex numbers $\mathbb{C}$, corresponding to so-called **real vector spaces** and **complex vector spaces**, respectively.

Definition 1.1. *A* ***vector space*** V *over an algebraic field* F *is a set of objects* $\boldsymbol{v} \in V$ *for which an internal commutative binary operation, called addition (denoted by* $+$*), and an external binary operation, called scalar multiplication, are defined such that*

(a) $\boldsymbol{v} + \boldsymbol{w} \in V$, *for all* $\boldsymbol{v}, \boldsymbol{w} \in V$ *(vector addition is closed);*

(b) $\boldsymbol{v} + \boldsymbol{w} = \boldsymbol{w} + \boldsymbol{v}$, *for all* $\boldsymbol{v}, \boldsymbol{w} \in V$ *(commutativity of vector addition);*

(c) $a\,\boldsymbol{v} \in V$, *for all* $a \in F$ *and all* $\boldsymbol{v} \in V$ *(scalar multiplication is closed), in particular* $1.\boldsymbol{v} = \boldsymbol{v}$ *for all* $\boldsymbol{v} \in V$;

(d) $a(\boldsymbol{v}+\boldsymbol{w}) = a\boldsymbol{v} + a\boldsymbol{w}$, *for all* $a \in F$ *and all* $\boldsymbol{v}, \boldsymbol{w} \in V$;

(e) $(ab)\,\boldsymbol{v} = a\,(b\boldsymbol{v})$, *for all* $a, b \in F$ *and all* $\boldsymbol{v} \in V$;

(f) *there exists a unique element* $\boldsymbol{0} \in V$, *called the* ***zero-vector*** *of* V *[to be distinguished from the zero of* F $(0 \in F)$*], such that* $\boldsymbol{0}+\boldsymbol{v} = \boldsymbol{v}$ *and* $0.\boldsymbol{v} = \boldsymbol{0}$ *for all* $\boldsymbol{v} \in V$.

The above properties are said to endow the set of objects in a vector space with a **linear structure**. As mentioned earlier and discussed below [following (1.41)], the familiar scalar product, more properly termed a *bilinear, nondegenerate, symmetric form* on the vector space, is an additional structure imposed on it, beyond the linear structure. It is easily verified that the set of "ordinary" vectors, pictured as directed arrows in one, two, or three-dimensional spaces, satisfies these properties. But we will have occasion to encounter more abstract vector spaces, such as function spaces, for example, where each function (satisfying a certain set of prescribed properties) is to be thought of as a vector (see, for example, our discussion of Fourier Analysis in Chapter 16). Abstract vector spaces will also be needed when we introduce the notion of tensor spaces later in this chapter.

The most important notion concerning linear vector spaces is that of linear independence. A set of non-zero vectors $\boldsymbol{v}_1, \ldots, \boldsymbol{v}_m \in V$ is said to be **linearly independent** if the vanishing of the *linear combination* $a_1\boldsymbol{v}_1 + \cdots + a_m\boldsymbol{v}_m = \boldsymbol{0}$ (where $a_1, \ldots, a_m \in F$) implies $a_1 = \cdots = a_m = 0$; otherwise the set of vectors is said to be *linearly dependent. The maximum number of linearly independent vectors in a vector space is called the* ***dimension*** *of the space. A maximal set of linearly independent vectors in a vector space is called a* ***basis*** *of the vector space.* It is quite obvious that, for a given vector space, the choice of a basis is not unique. Given a basis, any vector in the vector space can be expressed *uniquely* as a linear combination of the vectors in the basis. In physics, a change of basis is usually referred to as a change of **reference (coordinate) frame**. A subset $W \subset V$ of a vector space V which is in its own right a vector space is said to be a vector **subspace** of V. It is evident that the dimension of the subspace W cannot be greater than that of V, that $dim\, W = dim\, V$ if and only if $W = V$, and that W and V share the same zero-element. We note, without going into details concerning questions of convergence, that the dimension of a vector space may be (denumerably) infinite. Equations (1.9) through (1.15) in the above development are in fact valid for vector spaces of arbitrary dimension. In this book we will be concerned mainly with finite-dimensional vector spaces. A special kind of infinite dimensional vector spaces, called *Hilbert spaces*, is of primary importance in quantum mechanics.

Note that the matrix A given by (1.8) satisfies the following conditions:

$$A^{-1} = A^T \,, \tag{1.16}$$

$$\det(A) = 1 \,, \tag{1.17}$$

where A^T denotes the transpose of A. In general, an $n \times n$ matrix that satisfies (1.16) is called an **orthogonal matrix**; while one that satisfies both (1.16) and

(1.17) is called a **special orthogonal matrix**. Both of these types of matrices are very important in physics applications, especially in the dynamics of rigid bodies. Special orthogonal matrices represent pure rotations, while orthogonal matrices with determinant -1 represent inversions, or rotations plus inversions in n-dimensional real vector spaces. The set of all $n \times n$ orthogonal matrices forms a group, called the **orthogonal group** $O(n)$. The set of all $n \times n$ special orthogonal matrices forms a subgroup of $O(n)$, called the **special orthogonal group** $SO(n)$. Group theory plays a very crucial role in the understanding of the symmetries of the laws of physics, which lead to fundamental conservation laws. We will not delve into the details of this theory here except to give for reference the definition of a group, which is the simplest of *algebraic categories*: *A* ***group*** *G is a set of objects with an internal operation, called multiplication, that is both closed ($a, b \in G \Rightarrow ab \in G$) and associative [$a(bc) = (ab)c$], such that the set always possesses an element e called the identity ($ae = ea = a$ for every $a \in G$), and such that every element a always possesses an inverse element a^{-1} ($aa^{-1} = a^{-1}a = e$) which also belongs to the set.* [Among references on mathematical topics that are of interest to physicists, those on group theory occupy a unique position and are perhaps the most prevalent. The reader may consult, among many other excellent works, the monograph by Sternberg (Sternberg 1994).]

Rotations and inversions are specific examples of linear transformations on vector spaces (of arbitrary dimension). A (square) matrix A *represents* an abstract **linear operator (transformation)** (also denoted by A) on a vector space V. Mathematically we write $A : V \to V$. The property of **linearity** is specified by the following condition:

$$A(a\,\boldsymbol{x} + b\,\boldsymbol{y}) = a\,A(\boldsymbol{x}) + b\,A(\boldsymbol{y})\ , \quad \text{for all } a, b \in \mathbb{R}\,;\ \boldsymbol{x}, \boldsymbol{y} \in V\ , \tag{1.18}$$

where $\mathbb{R}$ denotes the field of real numbers.

Property (1.16) in fact follows from the following equivalent definition of an orthogonal transformation: *An orthogonal transformation is one which leaves the length (or* ***norm****) of a vector invariant.* This can be seen as follows. Suppose a linear transformation A sends a vector $\boldsymbol{v} \in V$ to another vector $\boldsymbol{v}' \in V$, that is, $\boldsymbol{v}' = A(\boldsymbol{v})$, or, in terms of components, $v'^i = a^i_j\, v^j$. On the other hand, the square of the length of v is given in terms of its components by [see (1.41) and (1.42) below]

$$\|\boldsymbol{v}\|^2 = \sum_j v^j v^j\ . \tag{1.19}$$

Then orthogonality of A implies that

$$\sum_j v^j v^j = \sum_i v'^i v'^i = \sum_i a^i_j a^i_k v^j v^k$$
$$= \sum_k a^i_j (A^T)^k_i v^j v^k = \sum_k (AA^T)^k_j v^j v^k\ . \tag{1.20}$$

Comparing the leftmost expression with the rightmost we see that $(AA^T)^k_j = \delta^k_j$, or $AA^T = 1$. Property (1.16) then follows.

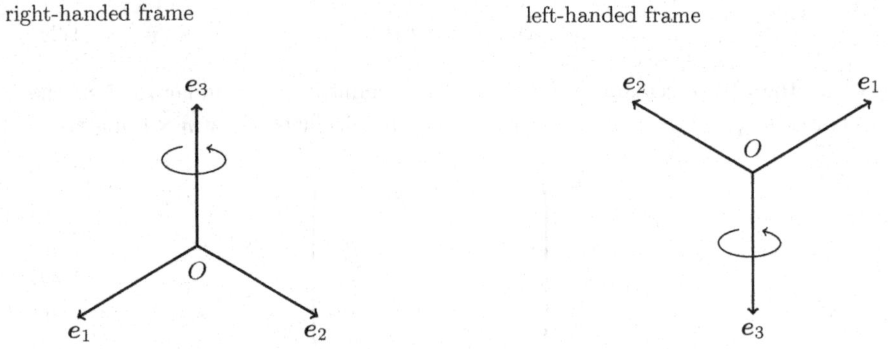

Fig. 1.3

We recall the following properties of determinants of matrices:

$$\det(AB) = \det A \det B \,, \tag{1.21}$$

$$\det(A^T) = \det(A) \,. \tag{1.22}$$

If A is orthogonal, we see that $(\det(A))^2 = 1$, which implies that $\det(A) = \pm 1$. Orthogonal matrices with determinant equal to -1 represent inversions or inversions plus rotations. An inversion changes the **orientation** (handedness) of a coordinate frame (see a more complete discussion of this concept in Chapter 2), which a pure rotation never does (see Fig. 1.3).

Problem 1.2 Write down explicitly the 3×3 orthogonal matrices representing rotations in 3-dimensional Euclidean space by 1) 45° about the e_1-axis, 2) 45° about the e_2-axis, and 3) 45° about the e_3-axis, all in the right-handed sense.

Problem 1.3 Show that $SO(2)$ is a **commutative (abelian) group**, that is, any two 2×2 matrices in $SO(2)$ commute with each other.

How does one obtain a concrete **matrix representation** of an abstract linear transformation $A : V \longrightarrow V$? We will see that a particular matrix representation results from a particular choice of basis $\{e_i\}$ for V. Consider an n-dimensional vector space V. For any $\boldsymbol{x} \in V$ given by $\boldsymbol{x} = x^i\, e_i$, the linearity condition (1.18) implies that

$$A(\boldsymbol{x}) = x^i\, A(e_i) \quad . \tag{1.23}$$

Thus the action of A on any vector $\boldsymbol{x} \in V$ is completely specified by the n actions $A(e_i)$, $i = 1, \ldots, n$. Since $A(e_i)$ is a vector in V, it can be expressed as a **linear combination** of the $\boldsymbol{e}_i$, that is

$$A(\boldsymbol{e}_i) = a_i^j\, \boldsymbol{e}_j \quad , \tag{1.24}$$

where the a_i^j are scalars in the field of real numbers $\mathbb{R}$. Similar to (1.8) the quantities a_i^j, $i = 1, \dots, n$, $j = 1, \dots, n$, can be displayed as an $n \times n$ matrix

$$(a_i^j) = \begin{pmatrix} a_1^1 & a_1^2 & \cdots & a_1^n \\ a_2^1 & a_2^2 & \cdots & a_2^n \\ \cdots & \cdots & \cdots & \cdots \\ a_n^1 & a_n^2 & \cdots & a_n^n \end{pmatrix} \quad . \tag{1.25}$$

Now suppose that, under the action of A, a vector $\boldsymbol{x} \in V$ is transformed into another vector $\boldsymbol{x}' \in V$. So

$$\boldsymbol{x}' = A(\boldsymbol{x}) = A(x^i\, \boldsymbol{e}_i) = x^i\, A(\boldsymbol{e}_i) = x^i a_i^j\, \boldsymbol{e}_j = x^j a_j^i\, \boldsymbol{e}_i \quad , \tag{1.26}$$

where in the last equality we have performed the interchange $(i \leftrightarrow j)$ since both i and j are dummy indices that are summed over. Equation (1.26) then implies

$$x'^{\,i} = a_j^i\, x^j \, , \tag{1.27}$$

which is formally the same as (1.9). Thus we have shown explicitly how *the matrix representation of a linear transformation depends on the choice of a basis set.*

Let us now investigate the transformation properties of the matrix representation of a linear transformation A under a change of basis:

$$\boldsymbol{e}_i = s_i^j\, \boldsymbol{e}'_j \; .$$

The required matrix $A' = (a_i'^{\,j})$ is given by [(cf. (1.24)]

$$A(\boldsymbol{e}'_i) = a_i'^{\,j}\, \boldsymbol{e}'_j \quad . \tag{1.28}$$

Using the equation for $\boldsymbol{e}'_i$ in terms of $\boldsymbol{e}_j$ in Table 1.1, we have

$$A(\boldsymbol{e}'_i) = A((s^{-1})_i^l\, e_l) = (s^{-1})_i^l\, A(e_l) = (s^{-1})_i^l\, a_l^k\, e_k = (s^{-1})_i^l\, a_l^k\, s_k^j\, \boldsymbol{e}'_j \, , \tag{1.29}$$

where in the second equality we have used the linearity property of A, in the third equality, Eq. (1.24), and in the fourth equality, Eq. (1.10). Comparison with (1.28) gives the desired result:

$$\boxed{a_i'^{\,j} = (s^{-1})_i^l\, a_l^k\, s_k^j} \quad . \tag{1.30}$$

In matrix notation (1.30) can be written

$$\boxed{A' = S^{-1}\, A\, S} \quad , \tag{1.31}$$

where $A' = (a'^{\,j}_i)$, $A = (a^j_i)$ and $S = (s^j_i)$ are the matrices with the indicated matrix elements. The transformation of matrices $A \mapsto A'$ given above is called a **similarity transformation.** These transformations are of great importance in physics. Two matrices related by an invertible matrix S as in (1.31) are said to be **similar.** Note that in (1.31), the matrix S itself may be viewed as representing a linear (invertible) transformation on V ($S : V \to V$, using the same symbol for the operator as for its matrix representation), and the equation as a relation between linear transformations. The equation (1.31) may also be represented succinctly by the following so-called **commutative diagram** (used frequently in the mathematics literature):

$$\begin{array}{ccc} V & \xrightarrow{A'} & V \\ {\scriptstyle S}\downarrow & & \downarrow{\scriptstyle S} \\ V & \xrightarrow{A} & V \end{array}$$

The way to read the diagram to arrive at the result (1.31) is as follows. Instead of using the operator A' to directly effect the linear transformation (from V to V) as represented by the top line (from left to right) of the diagram, one can instead start from the top left corner, use the operator S first on V (going down on the left line), followed by the operator A (going from left to right on the bottom line), then finally use the operator S^{-1} (going up on the right line).

According to (1.30) the upper index of a matrix (a^j_i) transforms like the upper index of a contravariant vector v^i (cf. Table 1.1), and the lower index of (a^j_i) transforms like the lower index of a basis vector e_i (cf. Table 1.1 also). In general one can consider a multi-indexed object

$$T^{i_1 \ldots i_r}_{j_1 \ldots j_s}$$

(with r upper indices and s lower indices) which transforms under a change of basis (or equivalently, a change of coordinate frames) $\boldsymbol{e}_i = s^j_i\, \boldsymbol{e}'_j$ (see Table 1.1) according to the following rule [compare carefully with (1.30)]

$$\boxed{(T')^{i_1 \ldots i_r}_{j_1 \ldots j_s} = s^{i_1}_{k_1} \ldots s^{i_r}_{k_r}\, (s^{-1})^{l_1}_{j_1} \ldots (s^{-1})^{l_s}_{j_s}\; T^{k_1 \ldots k_r}_{l_1 \ldots l_s}} \quad . \tag{1.32}$$

This multi-indexed object represents the *components* of a so-called (r, s)**-type tensor**, where r is called the **contravariant order** and s the **covariant order** of the tensor. Note carefully how the contravariant (upper) indices and the covariant (lower) indices remain in their respective upper and lower positions on both sides of the above equation, and how, on the right-hand side, the Einstein summation convention is used for the r sums over the repeated indices $k_1, \ldots, k_r$

and the s sums over the repeated indices $l_1, \ldots, l_s$. Thus a matrix (a_i^j) which transforms as (1.30) is a (1,1)-type tensor, and a vector v^i is a $(1,0)$-type tensor. (r,s)-type tensors with $r \neq 0$ *and* $s \neq 0$ are called **tensors of mixed type**. A (0,0)-type tensor is a scalar. The term "covariant" means that the transformation is the same as that of the basis vectors: $(\boldsymbol{e}')_i = (s^{-1})_i^j\, \boldsymbol{e}_j$ (see Table 1.1), while "contravariant" means that the indexed quantity transforms according to the inverse of the transformation of the basis vectors. Common examples of tensors occurring in physics are the *inertia tensor* I^{ij} in rigid body dynamics (see Chapter 20 and Problem 1.5 below), the *electromagnetic field tensor* $F_{\mu\nu}$ in Maxwell electrodynamics, and the *stress-energy tensor* $T_{\mu\nu}$ in the theory of relativity (the latter two behaving as tensors under *Lorentz transformations*, which preserve the lengths of 4-vectors in 4-dimensional spacetime).

Let us step back to look at the mathematical genesis of the concept of tensors more closely. We warn the reader that the somewhat abstract and seemingly convoluted development over the next several paragraphs [until Eq. (1.40)] may appear to be needlessly pedantic at this point for the purpose of physics applications. Our justification is that it will prepare the ground for developments of key concepts and miscellaneous applications in subsequent chapters.

Our objective is to construct a vector space out of two arbitrary vector spaces, by devising a *tensor product* between them, or "tensoring" them together, so to speak. The first step is to introduce the concept of *dual spaces*. Instead of linear operators $A: V \to V$, let us consider a linear function $f: V \to \mathbb{R}$. Recall that [cf. (1.18)] linearity means that, in this case, $f(a\boldsymbol{v} + b\boldsymbol{w}) = a\, f(\boldsymbol{v}) + b\, f(\boldsymbol{w})$, for all $\boldsymbol{v}, \boldsymbol{w} \in V$ and all $a, b \in \mathbb{R}$. Writing $\boldsymbol{v} = v^i\, \boldsymbol{e}_i$ (with respect to a particular basis), we have, by linearity, $f(v^i\, \boldsymbol{e}_i) = v^i\, f(\boldsymbol{e}_i)$. Let $f(\boldsymbol{e}_i) \equiv a_i \in \mathbb{R}$. (Notice the position of the covariant index i – a subscript.) Then we can write (verify it!) $f = a_i\, (\boldsymbol{e}^*)^i$, where $(\boldsymbol{e}^*)^i$ is a particular linear function on V such that its action is completely specified by $(\boldsymbol{e}^*)^i\, (\boldsymbol{e}_j) = \delta^i_j$ (the **Kronecker delta** symbol, defined to be equal to 1 if $i = j$ and equal to zero if $i \neq j$). If V is n-dimensional, the numbers a_i, $i = 1, \ldots, n$, can be considered as components of the "vector" f in an n-dimensional vector space V^*, called the dual space to V, with respect to the basis $\{(\boldsymbol{e}^*)^1, \ldots, (\boldsymbol{e}^*)^n\}$. Thus we have the following definition.

Definition 1.2. *The* **dual space** *to a vector space V, denoted V^*, is the vector space of all linear functions $f: V \to \mathbb{R}$ on V.*

The basis $\{(\boldsymbol{e}^*)^i\}$ of V^* is called the **dual basis** to the basis $\{\boldsymbol{e}_i\}$ of V, and vice versa. Since $dim\,(V) = dim\,(V^*)$, V is isomorphic to V^*, as follows from the general fact that *all finite dimensional vector spaces of the same dimension are isomorphic to each other*. Thus the vector spaces V and V^* are in fact dual to each other, or we can write $(V^*)^* = V$. Under the change of basis $\boldsymbol{e}_i = s_i^j\, \boldsymbol{e}'_j$ we have

$$f(\boldsymbol{e}'_i) \equiv a'_i = f((s^{-1})_i^j\, \boldsymbol{e}_j) = (s^{-1})_i^j\, f(\boldsymbol{e}_j) = (s^{-1})_i^j\, a_j\,. \tag{1.33}$$

This confirms the fact that the (lower) index in a_i is a covariant index. So vectors in V^* are $(0,1)$-type tensors.

Now let us observe that the set of all linear operators from V to V, denoted by $\mathcal{L}(V;V)$, is a vector space if we define the vector addition and scalar multiplication by $(A+B)(\boldsymbol{v}) = A(\boldsymbol{v}) + B(\boldsymbol{v})$ and $(aA)(\boldsymbol{v}) = aA(\boldsymbol{v})$, respectively, where $A, B \in \mathcal{L}(V;V)$, $\boldsymbol{v} \in V$ and $a \in \mathbb{R}$. As we demonstrated earlier, if V is n-dimensional, each $A \in \mathcal{L}(V;V)$ is represented by an $n \times n$ real matrix (a_i^j). So there is an isomorphic relationship between $\mathcal{L}(V;V)$ and the set of all $n \times n$ real matrices $M(n,\mathbb{R})$. On the other hand, $M(n,\mathbb{R})$ is also isomorphic to the vector space of all **bilinear maps** from $V^* \times V$ to $\mathbb{R}$, denoted by $\mathcal{L}(V^*,V;\mathbb{R})$, where bilinearity means linear in each argument. Indeed, given any $A \in \mathcal{L}(V^*,V;\mathbb{R})$, $\boldsymbol{w}^* = w_i\,(\boldsymbol{e}^*)^i \in V^*$ and $\boldsymbol{v} = v^i\,\boldsymbol{e}_i \in V$, we have, by bilinearity,

$$A(\boldsymbol{w}^*, \boldsymbol{v}) = w_j v^i\, A((\boldsymbol{e}^*)^j, \boldsymbol{e}_i)\,. \tag{1.34}$$

So the action of any $A \in \mathcal{L}(V^*, V\ \mathbb{R})$ is completely specified by the n^2 numbers $A((\boldsymbol{e}^*)^j, \boldsymbol{e}_i) \equiv a_i^j$, that is, the $n \times n$ matrix (a_i^j), and we have the following vector space isomorphisms

$$\mathcal{L}(V;\,V) \sim \mathcal{L}(V^*,V;\,\mathbb{R}) \sim M(n;\,\mathbb{R})\,. \tag{1.35}$$

There is yet another way to view the above vector space. For $\boldsymbol{v} \in V$ and $\boldsymbol{w}^* \in V^*$, consider an element $\boldsymbol{v} \otimes \boldsymbol{w}^* \in \mathcal{L}(V^*,V;\,\mathbb{R})$ defined by

$$\begin{aligned}(\boldsymbol{v} \otimes \boldsymbol{w}^*)(\boldsymbol{x}^*, \boldsymbol{y}) &= \boldsymbol{x}^*(\boldsymbol{v})\,.\,\boldsymbol{w}^*(\boldsymbol{y}) \\ &\equiv \langle\, \boldsymbol{x}^*, \boldsymbol{v}\,\rangle\langle\, \boldsymbol{w}^*, \boldsymbol{y}\,\rangle \qquad \text{for all } \boldsymbol{x}^* \in V^* \text{ and } \boldsymbol{y} \in V\,,\end{aligned} \tag{1.36}$$

where the notation $\boldsymbol{x}^*(\boldsymbol{v}) \equiv \langle\, \boldsymbol{x}^*, \boldsymbol{v}\,\rangle = \langle\, \boldsymbol{v}, \boldsymbol{x}^*\,\rangle$ indicates that the *pairing* between any $\boldsymbol{x}^* \in V^*$ and $\boldsymbol{v} \in V$ is symmetrical. Denote by $V \otimes V^*$ the vector space consisting of all finite sums of the form $\boldsymbol{v} \otimes \boldsymbol{w}^*$. It can be shown readily that the operation $\otimes$ is bilinear, that is, linear in both arguments. Thus $\boldsymbol{v} \otimes \boldsymbol{w}^* = v^i w_j\, \boldsymbol{e}_i \otimes (\boldsymbol{e}^*)^j$, and so $V \otimes V^*$ is spanned by the basis vectors $\boldsymbol{e}_i \otimes (\boldsymbol{e}^*)^j$. It follows that any $A \in V \otimes V^*$ can be written as the linear combination $A = a_i^j\, \boldsymbol{e}_j \otimes (\boldsymbol{e}^*)^i$. In other words, $V \otimes V^*$ is also isomorphic to the set of all $n \times n$ real matrices $M(n;\,\mathbb{R})$. Combining the above results we have the following strengthening of the isomorphism scheme of (1.35):

$$\mathcal{L}(V;\,V) \sim \mathcal{L}(V^*,V;\,\mathbb{R}) \sim V \otimes V^* \sim M(n;\,\mathbb{R})\,. \tag{1.37}$$

The vector space $V \otimes V^*$ is called the *tensor product* of the vector spaces V and V^*. Note that the ordering of the arguments in the binary operation $\otimes$ is important, and the complementary appearances of V and V^* in the second and third entries in the above equation.

It is readily seen that the notion of the tensor product as developed above can be generalized to that between *any* two vector spaces V and W (which may be of different dimensions). In fact, analogous to (1.37), we have the following definition:

Definition 1.3. *The* ***tensor product*** *of two vector spaces V and W, denoted $V \otimes W$, is the vector space of bilinear functions from $V^* \times W^*$ to $\mathbb{R}$, or real bilinear forms on $V^* \times W^*$:*

$$V \otimes W \equiv \mathcal{L}(V^*, W^*;\,\mathbb{R})\,. \tag{1.38}$$

Note again the duality of the vector spaces appearing on both sides of the equation. We note that $\mathcal{L}(V^*, W^*; \mathbb{R})$ can also be identified with the dual of the space of real bilinear forms on $V \times W$, that is, $\mathcal{L}(V^*, W^*; \mathbb{R}) \sim \mathcal{L}^*(V, W; \mathbb{R})$ (the reader can verify this easily).

By extension we can define the tensor product between any number of (different) vector spaces. An (r, s)-tensor [with components that transform according to (1.32)] can then be considered as an element in the tensor product space

$$V^r_s \equiv \underbrace{V \otimes \cdots \otimes V}_{r \text{ times}} \otimes \underbrace{V^* \otimes \cdots \otimes V^*}_{s \text{ times}} . \tag{1.39}$$

Such a tensor $\boldsymbol{T} \in V^r_s$ can be written in terms of components as

$$\boldsymbol{T} = T^{i_1 \dots i_r}_{j_1 \dots j_s} \, \boldsymbol{e}_{i_1} \otimes \cdots \otimes \boldsymbol{e}_{i_r} \otimes (\boldsymbol{e}^*)^{j_1} \otimes \cdots \otimes (\boldsymbol{e}^*)^{j_s} , \tag{1.40}$$

where the components transform according to (1.32) under the change of basis in V given by $\boldsymbol{e}_i = s^j_i \, \boldsymbol{e}'_j$ [and the associated change in V^* given by $(\boldsymbol{e}^*)^i = (s^{-1})^i_j \, (\boldsymbol{e}^{*\prime})^j$]. Note again the use of the Einstein summation convention.

After the above flight of abstraction we will return to more familiar ground for most of the remainder of this chapter – the special case of physical 3-dimensional Euclidean space – and review the notions of scalar products and cross products in this space. We will see that, while the definition of the scalar product as furnished in (1.41) below is in fact valid for any dimension n, that of the cross product given by (1.45) is specific to $n = 3$ only (although in the next chapter we will explore the important generalization of the concept to higher dimensions). The **scalar (inner) product** of two vectors $\boldsymbol{u}$ and $\boldsymbol{v}$ (in the same vector space) expressed in terms of components (with respect to a certain choice of basis) is defined by

$$(\boldsymbol{u}, \boldsymbol{v}) = \boldsymbol{u} \cdot \boldsymbol{v} \equiv \delta_{ij} \, u^i v^j = \sum_i u^i v^i = u^i v_i , \tag{1.41}$$

where the Kronecker delta symbol δ_{ij} (defined similarly as δ^i_j above) is considered to be a $(0, 2)$-type tensor, and can be regarded as a **metric** in Euclidean space $\mathbb{R}^3$. Loosely speaking, a metric in a space is a $(0, 2)$-type *non-degenerate, symmetric tensor field* which gives a prescription for measuring lengths and angles in that space. [For a definition of non-degeneracy, see discussion immediately above Eq. (1.65).] For example, the **norm** (length) of a vector $\boldsymbol{v}$ is defined to be [recall (1.19)]

$$\|\boldsymbol{v}\| \equiv \sqrt{(\boldsymbol{v}, \boldsymbol{v})} = \sqrt{\sum_i (v^i)^2} . \tag{1.42}$$

It is readily shown, by steps similar to those in (1.20), that *the scalar product of two vectors is invariant under orthogonal transformations.*

Referring to Fig. 1.4, in which the basis vector $\boldsymbol{e}_1$ is chosen to lie along the direction of $\boldsymbol{v}$, the angle θ between two vectors $\boldsymbol{u}$ and $\boldsymbol{v}$ is given in terms of the

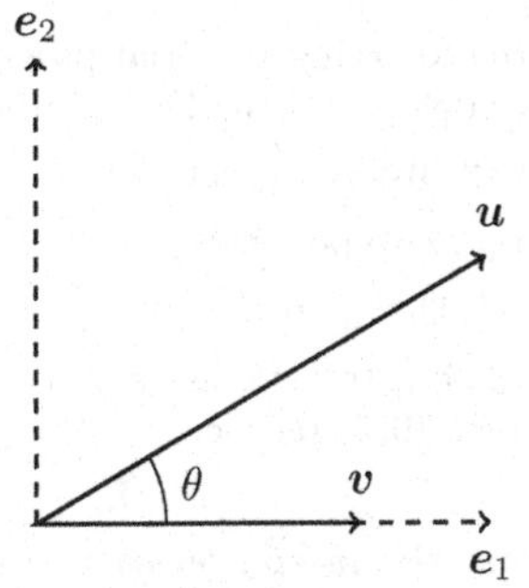

Fig. 1.4

scalar product by

$$u \cdot v = (u\cos\theta,\, u\sin\theta)\cdot(v,\, 0) = uv\,\cos\theta\,, \tag{1.43}$$

where u and v represent the magnitudes of the vectors $\boldsymbol{u}$ and $\boldsymbol{v}$, respectively.

The **cross product** (also called the **vector product**) can be defined in terms of the so-called **Levi-Civita tensor** (also known as the *completely anti-symmetric tensor*), which is a $(1,2)$-type tensor given as follows:

$$\varepsilon^i{}_{jk} = \begin{cases} 0 & , \; i,j,k \text{ not all distinct}\,; \\ +1 & , \; (ijk) \text{ is an even permutation of } (123)\,; \\ -1 & , \; (ijk) \text{ is an odd permutation of } (123)\,. \end{cases} \tag{1.44}$$

The cross product of two vectors $\boldsymbol{A}$ and $\boldsymbol{B}$ is then defined by

$$\boxed{\boldsymbol{A}\times\boldsymbol{B} = (\varepsilon^i{}_{jk}\, A^j B^k)\,\boldsymbol{e}_i}\quad . \tag{1.45}$$

The cross product gives a **Lie algebra structure** to the vector space $\mathbb{R}^3$, since this product satisfies the so-called **Jacobi identity**:

$$\boldsymbol{A}\times(\boldsymbol{B}\times\boldsymbol{C}) + \boldsymbol{B}\times(\boldsymbol{C}\times\boldsymbol{A}) + \boldsymbol{C}\times(\boldsymbol{A}\times\boldsymbol{B}) = 0\,. \tag{1.46}$$

The metric δ_{ij} and its inverse δ^{ij} can be used to raise and lower indices of tensors in $\mathbb{R}^3$. For example,

$$\varepsilon_{klm} = \delta_{kn}\varepsilon^n{}_{lm} = \varepsilon^k{}_{lm}\,. \tag{1.47}$$

In fact, because of the special properties of the Kronecker delta, one can use it to raise and lower indices "with impunity":

$$\varepsilon^i{}_{jk} = \varepsilon_{ijk} = \varepsilon_i{}^j{}_k = \dots\,. \tag{1.48}$$

Problem 1.4 Verify the following tensorial properties under orthogonal transformations for the Kronecker deltas and the Levi-Civita tensor:

(a) δ^{ij} transforms as a $(2,0)$-type tensor;

(b) δ_{ij} transforms as a $(0,2)$-type tensor;

(c) δ_i^j transforms as a $(1,1)$-type tensor;

(d) $\varepsilon^i{}_{jk}$ [as defined in (1.44)] transforms as a $(1,2)$-type tensor; ε_{ijk} [as defined in (1.47)] transforms as a $(0,3)$-tensor.

Problem 1.5 Show that the *inertia tensor density* $\mathcal{I}^{ij} \equiv \delta^{ij}\, x^l x_l - x^i x^j$, where the x^i are Cartesian coordinates and $i,j = 1,2,3$, transforms as a $(2,0)$-tensor under orthogonal transformations of the coordinates.

Problem 1.6 Use the definition of the cross product [(1.45)] to show that

$$\boldsymbol{A} \times \boldsymbol{B} = -\boldsymbol{B} \times \boldsymbol{A}\,. \tag{1.49}$$

Problem 1.7 Use (1.45) to show that

$$\boldsymbol{e}_1 \times \boldsymbol{e}_2 = \boldsymbol{e}_3\,, \quad \boldsymbol{e}_2 \times \boldsymbol{e}_3 = \boldsymbol{e}_1\,, \quad \boldsymbol{e}_3 \times \boldsymbol{e}_1 = \boldsymbol{e}_2\,. \tag{1.50}$$

The Levi-Civita tensor satisfies the following very useful **contraction property**:

$$\boxed{\varepsilon^k{}_{ij}\, \varepsilon_{klm} = \delta_{il}\delta_{jm} - \delta_{im}\delta_{jl}}\;. \tag{1.51}$$

This tensor is intimately related to the **determinant** of a matrix. One can easily see that

$$\det(a_i^j) = \begin{vmatrix} a_1^1 & a_1^2 & a_1^3 \\ a_2^1 & a_2^2 & a_2^3 \\ a_3^1 & a_3^2 & a_3^3 \end{vmatrix} = \varepsilon_{ijk}\, a_1^i a_2^j a_3^k\,. \tag{1.52}$$

Generalizing to an $n \times n$ matrix one has

$$\begin{aligned} \det(a_i^j) &= \sum_{\sigma \in S_n} (sgn\,\sigma)\, a_1^{\sigma(1)} \dots a_n^{\sigma(n)} = \sum_{\sigma \in S_n} (sgn\,\sigma)\, a^1_{\sigma(1)} \dots a^n_{\sigma(n)} \\ &= \varepsilon_{i_1 \dots i_n} a_1^{i_1} \dots a_n^{i_n}\,, \end{aligned} \tag{1.53}$$

where S_n is the **permutaton group** of n objects, $sgn\,\sigma = +1(-1)$ if σ is an even(odd) permutation of $(123\dots n)$, and $\varepsilon_{i_1 \dots i_n}$ is the **generalized Levi-Civita tensor** and is defined similarly to (1.44).

Equations (1.45) and (1.52) imply that

$$\boldsymbol{A}\times\boldsymbol{B}=\varepsilon^{i}{}_{jk}\,\boldsymbol{e}_i\,A^jB^k=\begin{vmatrix}\boldsymbol{e}_1 & \boldsymbol{e}_2 & \boldsymbol{e}_3\\ A^1 & A^2 & A^3\\ B^1 & B^2 & B^3\end{vmatrix}\ . \tag{1.54}$$

Problem 1.8 Verify the contraction property of the Levi-Civita tensor (1.51) by using the defining properties of the tensor given by (1.44) and the properties of the Kronecker delta.

We can use the contraction property of the Levi-Civita tensor (1.51) and the definition of the scalar product (1.41) to calculate the magnitude of the cross product as follows:

$$\begin{aligned}(\boldsymbol{A}\times\boldsymbol{B})\cdot(\boldsymbol{A}\times\boldsymbol{B})&=\sum_i(\boldsymbol{A}\times\boldsymbol{B})^i(\boldsymbol{A}\times\boldsymbol{B})^i\\ &=\sum_i\varepsilon^{i}{}_{jk}A^jB^k\,\varepsilon^{i}{}_{mn}A^mB^n=\varepsilon^{i}{}_{jk}\varepsilon_{imn}A^jB^kA^mB^n\\ &=(\delta_{jm}\delta_{kn}-\delta_{jn}\delta_{km})\,A^jB^kA^mB^n\\ &=A_jA^jB_nB^n-(A^jB_j)(A_kB^k)=A^2B^2-(\boldsymbol{A}\cdot\boldsymbol{B})^2\\ &=A^2B^2-A^2B^2\cos^2\theta=A^2B^2\sin^2\theta\ .\end{aligned} \tag{1.55}$$

Hence

$$|\boldsymbol{A}\times\boldsymbol{B}|=AB\sin\theta\ , \tag{1.56}$$

where θ is the angle between the vectors $\boldsymbol{A}$ and $\boldsymbol{B}$. Furthermore, using the properties of the Levi-Civita tensor (1.44), we see that

$$(\boldsymbol{A}\times\boldsymbol{B})\cdot\boldsymbol{A}=\sum_i\varepsilon^{i}{}_{jk}A^jB^kA^i=\varepsilon_{ijk}A^iA^jB^k=0\ . \tag{1.57}$$

Similarly,

$$(\boldsymbol{A}\times\boldsymbol{B})\cdot\boldsymbol{B}=0\ . \tag{1.58}$$

Thus the cross product of two vectors is always perpendicular to the plane defined by the two vectors. The **right-hand rule** for the direction of the cross product follows from (1.50).

As a further example in the use of the contraction property of the Levi-Civita tensor we will prove the following very useful identity, known as the "*bac minus cab*" *identity* in vector algebra.

$$\boxed{\boldsymbol{A}\times(\boldsymbol{B}\times\boldsymbol{C})=\boldsymbol{B}\,(\boldsymbol{A}\cdot\boldsymbol{C})-\boldsymbol{C}\,(\boldsymbol{A}\cdot\boldsymbol{B})}\ . \tag{1.59}$$

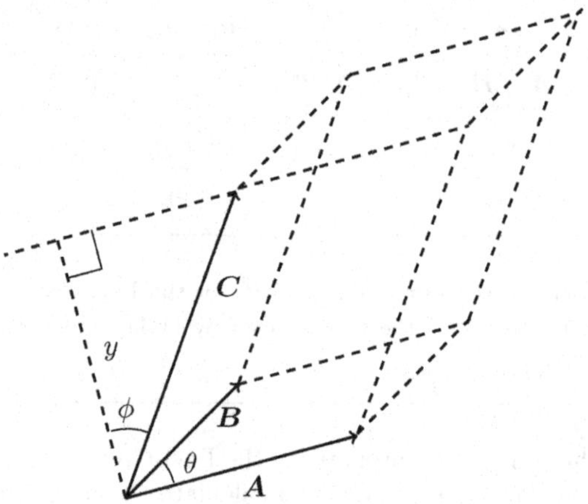

Fig. 1.5

Indeed, we have

$$\begin{aligned}(\boldsymbol{A}\times(\boldsymbol{B}\times\boldsymbol{C}))^i &= \varepsilon^i{}_{jk}A^j(\boldsymbol{B}\times\boldsymbol{C})^k \\ &= \varepsilon^i{}_{jk}A^j\varepsilon^k{}_{lm}B^lC^m = \varepsilon^i{}_{jk}\varepsilon^k{}_{lm}\,A^jB^lC^m = \varepsilon^k{}_{ij}\varepsilon_{klm}\,A^jB^lC^m \\ &= (\delta_{il}\delta_{jm}-\delta_{im}\delta_{jl})\,A^jB^lC^m \\ &= (\delta_{il}B^l)(\delta_{jm}A^jC^m)-(\delta_{im}C^m)(\delta_{jl}B^lA^j) \\ &= B^iA_mC^m - C^iA_lB^l = B^i(\boldsymbol{A}\cdot\boldsymbol{C})-C^i(\boldsymbol{A}\cdot\boldsymbol{B})\,.\end{aligned} \tag{1.60}$$

Problem 1.9 Use the properties of the Levi-Civita tensor to show that

$$\boldsymbol{A}\cdot(\boldsymbol{B}\times\boldsymbol{C}) = \boldsymbol{B}\cdot(\boldsymbol{C}\times\boldsymbol{A}) = \boldsymbol{C}\cdot(\boldsymbol{A}\times\boldsymbol{B})\,. \tag{1.61}$$

The absolute value of the quantity in (1.61) can be seen to be equal to the volume of the parallelopiped formed by the vectors $\boldsymbol{A}$, $\boldsymbol{B}$ and $\boldsymbol{C}$. Indeed, referring to Fig. 1.5,

$$\begin{aligned}\textit{volume} &= y\times\ \textit{area of parallelogram formed by } \boldsymbol{A} \textit{ and } \boldsymbol{B} \\ &= |\,y\cdot AB\sin\theta\,| = |\,C\cos\phi\, AB\sin\theta\,| = |\,\boldsymbol{C}\cdot(\boldsymbol{A}\times\boldsymbol{B})\,|\,.\end{aligned} \tag{1.62}$$

Problem 1.10 Use the "bac minus cab" identity (1.59) to show that

$$\begin{aligned}(\boldsymbol{A}\times\boldsymbol{B})\times(\boldsymbol{C}\times\boldsymbol{D}) &= \boldsymbol{C}\,(\boldsymbol{D}\cdot(\boldsymbol{A}\times\boldsymbol{B})) - \boldsymbol{D}\,(\boldsymbol{C}\cdot(\boldsymbol{A}\times\boldsymbol{B}))\\ &= \boldsymbol{B}\,(\boldsymbol{A}\cdot(\boldsymbol{C}\times\boldsymbol{D})) - \boldsymbol{A}\,(\boldsymbol{B}\cdot(\boldsymbol{C}\times\boldsymbol{D}))\,.\end{aligned} \tag{1.63}$$

Problem 1.11 Use the "bac minus cab" identity and (1.61) to show that

$$(\boldsymbol{A}\times\boldsymbol{B})\cdot(\boldsymbol{C}\times\boldsymbol{D}) = (\boldsymbol{A}\cdot\boldsymbol{C})(\boldsymbol{B}\cdot\boldsymbol{D}) - (\boldsymbol{A}\cdot\boldsymbol{D})(\boldsymbol{B}\cdot\boldsymbol{C})\,. \tag{1.64}$$

We conclude this chapter with some more mathematical digression on the usefulness of inner (scalar) products of vectors. The inner product, other than providing a metric structure for the vector space, serves another important mathematical purpose. It establishes a *canonical isomorphism* between a vector space V and its dual space V^*. This fact plays a crucial role in the mathematical appreciation of the relationship between the Lagrangian and the Hamiltonian formulations of classical mechanics (see Chapters 31 and 33), and actually in quantum mechanics also. Recall that, in general, *an inner product* $(\cdot\,,\,\cdot)$ *in a real vector space is a symmetric, nondegenerate, bilinear, real 2-form on the vector space.* This means that, given any ordered pair of vectors $\boldsymbol{v}$ and $\boldsymbol{w}$ in V, the scalar product $(\boldsymbol{v},\boldsymbol{w})$ is a real number. Symmetry means that $(\boldsymbol{v},\boldsymbol{w}) = (\boldsymbol{w},\boldsymbol{v})$. Nondegeneracy means that the condition $(\boldsymbol{v},\boldsymbol{w}) = 0$ for all $\boldsymbol{w}\in V$ implies $\boldsymbol{v} = \boldsymbol{0}$. Bilinearity means that the 2-form is linear with respect to both arguments: more specifically, for any $a, b\in\mathbb{R}$ and any $\boldsymbol{v},\boldsymbol{w},\boldsymbol{x}\in V$,

$$\begin{aligned}(a\boldsymbol{v}+b\boldsymbol{w}\,,\,\boldsymbol{x}) &= a\,(\boldsymbol{v},\boldsymbol{x}) + b\,(\boldsymbol{w},\boldsymbol{x})\,,\\ (\boldsymbol{v}\,,\,a\boldsymbol{w}+b\boldsymbol{x}) &= a\,(\boldsymbol{v},\boldsymbol{w}) + b\,(\boldsymbol{v},\boldsymbol{x})\,.\end{aligned} \tag{1.65}$$

We now assert that *the condition of nondegeneracy of the inner product is equivalent to the fact that the inner product induces a special (canonical) bijective [one-to-one (injective) and onto (surjective)] mapping between V and V^*.* Define a linear map $G: V\to V^*$ by

$$G(\boldsymbol{v})(\boldsymbol{w}) = (\boldsymbol{v},\boldsymbol{w})\,, \tag{1.66}$$

for all $\boldsymbol{v},\boldsymbol{w}\in V$. Note that $G(\boldsymbol{v})\in V^*$, as a linear function on V, is defined by its action on any $\boldsymbol{w}\in V$, and the linearity of G follows from the bilinearity of the inner product. We will show that G is the sought-for canonical isomorphism. Since the inner product is nondegenerate, it follows from (1.66) and the above definition of nondegeneracy that $G(\boldsymbol{v})(\boldsymbol{w}) = 0$ for all $\boldsymbol{w}\in V$ implies $\boldsymbol{v} = \boldsymbol{0}$, that is, $G(\boldsymbol{v})\in V^*$ is the zero-map on V (mapping all vectors in V to the zero-vector $\boldsymbol{0}$ in V^*). So $G(\boldsymbol{v}) = \boldsymbol{0}$ implies $\boldsymbol{v} = \boldsymbol{0}$. We say that the **kernel** of the linear map $G : V \to V^*$ is the zero-vector in V: $Ker\, G = \{\mathbf{0}\}$. Take any two vectors $\boldsymbol{v}_1, \boldsymbol{v}_2 \in V$. The vanishing of $Ker\, G$ means that

$$G(\boldsymbol{v}_1) = G(\boldsymbol{v}_2) \Longrightarrow G(\boldsymbol{v}_1 - \boldsymbol{v}_2) = \mathbf{0} \Longrightarrow \boldsymbol{v}_1 = \boldsymbol{v}_2 \; . \tag{1.67}$$

This says that the map G is injective. The fact that it is also surjective follows from the fact that V and V^*, as finite-dimensional vector spaces of the same dimension, are isomorphic. So G is a bijection. Conversely, suppose G is bijective, so it is certainly injective. This means that $Ker\, G = \mathbf{0}$; in other words, $G(\boldsymbol{v}) = \mathbf{0} \Longrightarrow \boldsymbol{v} = \mathbf{0}$. It follows from (1.66) that $G(\boldsymbol{v})(\boldsymbol{w}) = (\boldsymbol{v}, \boldsymbol{w}) = 0$ for all $\boldsymbol{w} \in V$ implies $\boldsymbol{v} = \mathbf{0}$; in other words, the inner product is nondegenerate. This establishes the equivalence referred to in the sentence following Eq. (1.65). The linear bijective map $G : V \to V^*$ defined by (1.66) is called the **conjugate isomorphism** between V and V^*.

In Hamiltonian mechanics, we will need *antisymmetric*, nondegenerate, bilinear 2-forms instead of symmetric ones. These are called **symplectic forms**. They will play a crucial role from Chapter 36 onwards.

Problem 1.12 Let $A : V \longrightarrow W$ be a linear operator, where both V and W are vector spaces over the same field. Prove the following relationship between the dimensions of the various vector spaces.

$$dim(V) = dim(Ker\, A) + dim(Im\, A) \; , \tag{1.68}$$

where $Ker\, A$ is the kernel, and $Im\, A \subset W$ is the image, of the operator A.
(*Hint*: First show that $Ker\, A$ and $Im\, A$ are vector subspaces of V and W, respectively.)

Problem 1.13 Let $W \subset V$ be a vector subspace of the vector space V. Define the **quotient space** V/W to be the space of **equivalence classes** $\{\boldsymbol{v}\}$, with the equivalence relationship $\sim$ defined by $\boldsymbol{v}_1 \sim \boldsymbol{v}_2$ if $\boldsymbol{v}_1 - \boldsymbol{v}_2 \in W$. Consider the projection map $\pi : V \longrightarrow V/W$ defined by $\pi(\boldsymbol{v}) = \{\boldsymbol{v}\}$. Use the result in the last problem to show that

$$dim\left(\frac{V}{W}\right) = dim(V) - dim(W) \; . \tag{1.69}$$

(*Hint*: $Ker\, \pi = W$.)

Chapter 2

Exterior Algebra: Determinants, Oriented Frames and Oriented Volumes

The intuitive notions of areas and volumes of parallelopipeds bounded by vectors in the Euclidean spaces $\mathbb{R}^2$ and $\mathbb{R}^3$ [see (1.62)] can be generalized to higher dimensions. In this chapter we will develop the mathematical formalism of *exterior algebra* – the algebra of antisymmetric tensors – which is the proper formalism required for the calculation of such volumes in $\mathbb{R}^n$, where n is any positive integer ≥ 2. The key operation in this algebra is a product called the *exterior* or *wedge product* (denoted by $\wedge$, hence the name), a generalization of the vector (cross) product for 3-dimensional vectors [cf. (1.45)] which will also be shown to be intimately connected to the notion of determinants of matrices. For our future work exterior algebra is absolutely essential in the understanding of the analytical concept of *differential forms*, which in turn is at the foundation of many differential geometric concepts explored in this book. As an introduction to the mathematics of differential forms and its applications in physics, we will see later (Chapter 7) that it generalizes and unifies in a most elegant and powerful fashion the familiar differential operators of 3-dimensional vector calculus. In this chapter we will focus on the algebraic aspects.

Our development will begin with the notion of the tensor product introduced in Chapter 1. Consider the rank-$(r,0)$ and rank-$(0,r)$ tensor product spaces

$$T^r(V) \equiv V_0^r = \underbrace{V \otimes \cdots \otimes V}_{r \text{ times}} \quad \text{and} \quad T^r(V^*) \equiv V_r^0 = \underbrace{V^* \otimes \cdots \otimes V^*}_{r \text{ times}} \,. \tag{2.1}$$

If $dim\,(V) = n$, the dimensions of $T^r(V)$ and $T^r(V^*)$, as vector spaces, are both n^r. They are, in fact, dual spaces of each other [$T^r(V^*)$ can be considered as the space of linear functions on $T^r(V)$ and vice versa] if we require, for any $\boldsymbol{v}_1, \dots, \boldsymbol{v}_r \in V$ and $(\boldsymbol{v}^*)^1, \dots, (\boldsymbol{v}^*)^r \in V^*$, that the action of $(\boldsymbol{v}^*)^1 \otimes \cdots \otimes (\boldsymbol{v}^*)^r \in T^r(V^*)$ on $\boldsymbol{v}_1 \otimes \cdots \otimes \boldsymbol{v}_r \in T^r(V)$ to yield a real number is given by

$$\begin{aligned} &((\boldsymbol{v}^*)^1 \otimes \cdots \otimes (\boldsymbol{v}^*)^r)(\boldsymbol{v}_1 \otimes \cdots \otimes \boldsymbol{v}_r) = (\boldsymbol{v}^*)^1(\boldsymbol{v}_1) \dots (\boldsymbol{v}^*)^r(\boldsymbol{v}_r) \\ &= \langle (\boldsymbol{v}^*)^1 , \boldsymbol{v}_1 \rangle \dots \langle (\boldsymbol{v}^*)^r , \boldsymbol{v}_r \rangle , \end{aligned} \tag{2.2}$$

where the *pairing* bracket notation $\langle (\boldsymbol{v}^*)^i , \boldsymbol{v}_i \rangle (= (\boldsymbol{v}^*)^i(\boldsymbol{v}_i))$ means the action of $(\boldsymbol{v}^*)^i$ on $\boldsymbol{v}_i$. Note that in the above equation there is no special relationship between $\boldsymbol{v}_i$ and $(\boldsymbol{v}^*)^i$. (There is no canonical relationship between vectors in V and vectors in V^*, even though these spaces are isomorphic, until V is endowed with a scalar product, as discussed at the end of the last chapter.) Particular elements $\boldsymbol{t} \in T^r(V)$ and $\boldsymbol{u} \in T^r(V^*)$ of these spaces can be written in terms of components as [cf. (1.40)]

$$\boldsymbol{t} = t^{i_1 \dots i_r}\, \boldsymbol{e}_{i_1} \otimes \cdots \otimes \boldsymbol{e}_{i_r} \,, \quad \boldsymbol{u} = u_{i_1 \dots i_r}\, (\boldsymbol{e}^*)^{i_1} \otimes \cdots \otimes (\boldsymbol{e}^*)^{i_r} \quad (i_1, \dots, i_r = 1, \dots, n) \,, \tag{2.3}$$

with respect to a particular choice of basis $\{\boldsymbol{e}_1, \dots \boldsymbol{e}_n\}$ for V and its dual basis $\{(\boldsymbol{e}^*)^1, \dots (\boldsymbol{e}^*)^n\}$ for V^*.

Within each of $T^r(V)$ and $T^r(V^*)$ we can construct subspaces of *symmetric tensors* $P^r(V) \subset T^r(V)$ and $P^r(V^*) \subset T^r(V^*)$, or subspaces of *antisymmetric tensors* $\Lambda^r(V) \subset T^r(V)$ and $\Lambda^r(V^*) \subset T^r(V^*)$; the latter are called **exterior spaces** of rank r and their elements are called **exterior vectors** [vectors in $\Lambda^r(V)$] or **exterior forms** [vectors in $\Lambda^r(V^*)$]. To identify these subspaces first recall the notion of the permutation group $\mathcal{S}_r$ of r objects and the $r!$ permutations $\sigma \in \mathcal{S}_r$ introduced in (1.53). Each of these permutations acts as a linear map on a tensor $\boldsymbol{t} \in T^r(V)$ according to the following rule, which specifies the action of the permuted tensor $\sigma\, \boldsymbol{t}$ on r arbitrary vectors $(\boldsymbol{v}^*)^1, \dots, (\boldsymbol{v}^*)^r \in V^*$:

$$(\sigma\, \boldsymbol{t})\,((\boldsymbol{v}^*)^1, \dots, (\boldsymbol{v}^*)^r) = \boldsymbol{t}\,((\boldsymbol{v}^*)^{\sigma(1)}, \dots, (\boldsymbol{v}^*)^{\sigma(r)}) \,. \tag{2.4}$$

For example, under the permutation $\sigma \in \mathcal{S}_3 : \sigma(1) = 2,\ \sigma(2) = 3,\ \sigma(3) = 1$, we have, for $\boldsymbol{t} \in T^3(V)$, $(\sigma\, \boldsymbol{t})\,((\boldsymbol{v}^*)^1, (\boldsymbol{v}^*)^2, (\boldsymbol{v}^*)^3) = \boldsymbol{t}\,((\boldsymbol{v}^*)^2, (\boldsymbol{v}^*)^3, (\boldsymbol{v}^*)^1)$. While the above definition is a "coordinate-free" one usually preferred by mathematicians, physicists may prefer the following equivalent one, which works on the coordinates:

$$\sigma\, \boldsymbol{t} = t^{i_{\sigma(1)} \dots i_{\sigma(r)}}\, \boldsymbol{e}_{i_1} \otimes \cdots \otimes \boldsymbol{e}_{i_r} \,. \tag{2.5}$$

Using the above example for σ, Eqs. (2.3) and (2.5) imply $\sigma \boldsymbol{t} = t^{i_2 i_3 i_1}\, \boldsymbol{e}_{i_1} \otimes \boldsymbol{e}_{i_2} \otimes \boldsymbol{e}_{i_3}$.

The permutations $\sigma \in \mathcal{S}_r$ act on $T^r(V^*)$ in a manner entirely analogous to (2.4) and (2.5):

$$(\sigma\, \boldsymbol{u})\,(\boldsymbol{v}_1, \dots, \boldsymbol{v}_r) = \boldsymbol{u}\,(\boldsymbol{v}_{\sigma(1)}, \dots, \boldsymbol{v}_{\sigma(r)}) \tag{2.6}$$

$$\sigma \boldsymbol{u} = u_{i_{\sigma(1)} \dots i_{\sigma(r)}}\, (\boldsymbol{e}^*)^{i_1} \otimes \cdots \otimes (\boldsymbol{e}^*)^{i_r} \,, \tag{2.7}$$

where $\boldsymbol{u} \in T^r(V^*)$ is given as in (2.3), and $\boldsymbol{v}_1, \dots, \boldsymbol{v}_r$ are any r vectors in V.

Problem 2.1 Show that the definitions given by (2.4) and (2.5) [similarly (2.6) and (2.7)] for the action of a permutation on a tensor are equivalent.

Hint: From (2.4) show that if $\boldsymbol{v}_1, \dots \boldsymbol{v}_r \in V$, then

$$\sigma(\boldsymbol{v}_1 \otimes \cdots \otimes \boldsymbol{v}_r) = \boldsymbol{v}_{\sigma^{-1}(1)} \otimes \cdots \otimes \boldsymbol{v}_{\sigma^{-1}(r)} \,. \tag{2.8}$$

In other words, σ permutes the *positions* of the vectors in a *tensor product monomial.*

We now define the properties of *symmetry* and *antisymmetry* for tensors.

Definition 2.1. *Let $\boldsymbol{t} \in T^r(V)$ or $T^r(V^*)$ and σ be any permutation in S_r. The tensor $\boldsymbol{t}$ is said to be **symmetric** if $\sigma\boldsymbol{t} = \boldsymbol{t}$. It is said to be **antisymmetric** if $\sigma\boldsymbol{t} = (sgn\,\sigma)\,\boldsymbol{t}$, where $sgn\,\sigma = +1\,(-1)$ if σ is an even (odd) permutation.*

The above "coordinate free" definition is equivalent to the following one given in terms of components:

Definition 2.1(a) *A tensor $\boldsymbol{t} \in T^r(V)$ given by $\boldsymbol{t} = t^{i_1 \dots i_r}\, \boldsymbol{e}_{i_1} \otimes \cdots \otimes \boldsymbol{e}_{i_r}$ with respect to a basis $\{\boldsymbol{e}_1, \dots, \boldsymbol{e}_n\}$ of V is said to be **symmetric** if $t^{i_{\sigma(1)} \dots i_{\sigma(r)}} = t^{i_1 \dots i_r}$, that is, its components are symmetric under any permutation of the indices. It is said to be **antisymmetric** if $t^{i_{\sigma(1)} \dots i_{\sigma(r)}} = (sgn\,\sigma)\, t^{i_1 \dots i_r}$, that is, its components are antisymmetric under any permutation of the indices. (The conditions are given by entirely analogous expressions for $\boldsymbol{t} \in T^r(V^*)$, where the tensorial components are quantities with lower indices.)*

It is clear that if $\boldsymbol{t}$ is antisymmetric, then $t^{i_1 \dots i_r} = 0$ whenever any pair of indices have the same value.

Problem 2.2 Show that the above two definitions for the symmetry (antisymmetry) of tensors are equivalent.

Out of any tensor $\boldsymbol{t}$ in $T^r(V)$ or $T^r(V^*)$ we can construct a symmetric tensor or antisymmetric tensor of the same rank by means of the **symmetrizing map** Sym_r or **antisymmetrizing map** $Asym_r$, defined as follows:

$$Sym_r(\boldsymbol{t}) = \frac{1}{r!} \sum_{\sigma \in S_r} \sigma\,\boldsymbol{t}\,, \quad Asym_r(\boldsymbol{t}) = \frac{1}{r!} \sum_{\sigma \in S_r} (sgn\,\sigma)\,\sigma\,\boldsymbol{t}\,. \tag{2.9}$$

We will state without proof the following fact: *the subspace of symmetric (antisymmetric) tensors of rank r is obtained by applying the symmetrizing (antisymmetrizing) map to the full space of tensors of rank r*, that is

$$P^r(V) = Sym_r(T^r(V))\,, \qquad \Lambda^r(V) = Asym_r(T^r(V))\,, \tag{2.10}$$

similarly for $P^r(V^*)$ and $\Lambda^r(V^*)$. It is understood that

$$P^0(V) = P^0(V^*) = \Lambda^0(V) = \Lambda^0(V^*) = \mathbb{R} , \tag{2.11}$$

$$P^1(V) = \Lambda^1(V) = V , \qquad P^1(V^*) = \Lambda^1(V^*) = V^* . \tag{2.12}$$

As an example of the use of (2.9), consider the case $dim\, V = 2$ and $r = 2$. So $dim\, T^2(V) = 2^2 = 4$; and given a basis $\{\mathbf{e}_1, \mathbf{e}_2\}$ for V, a basis for $T^2(V)$ can be chosen to be $\{\mathbf{e}_1 \otimes \mathbf{e}_1,\, \mathbf{e}_2 \otimes \mathbf{e}_2,\, \mathbf{e}_1 \otimes \mathbf{e}_2,\, \mathbf{e}_2 \otimes \mathbf{e}_1\}$. We then have

$$Sym_2(\mathbf{e}_1 \otimes \mathbf{e}_1) = \frac{1}{2}\,(\mathbf{e}_1 \otimes \mathbf{e}_1 + \mathbf{e}_1 \otimes \mathbf{e}_1) = \mathbf{e}_1 \otimes \mathbf{e}_1 , \tag{2.13}$$

$$Sym_2(\mathbf{e}_2 \otimes \mathbf{e}_2) = \mathbf{e}_2 \otimes \mathbf{e}_2 , \tag{2.14}$$

$$Sym_2(\mathbf{e}_1 \otimes \mathbf{e}_2) = Sym_2(\mathbf{e}_2 \otimes \mathbf{e}_1) = \frac{1}{2}\,(\mathbf{e}_1 \otimes \mathbf{e}_2 + \mathbf{e}_2 \otimes \mathbf{e}_1) , \tag{2.15}$$

$$Asym_2(\mathbf{e}_1 \otimes \mathbf{e}_1) = \frac{1}{2}\,(\mathbf{e}_1 \otimes \mathbf{e}_1 - \mathbf{e}_1 \otimes \mathbf{e}_1) = \mathbf{0} = Asym_2(\mathbf{e}_2 \otimes \mathbf{e}_2) , \tag{2.16}$$

$$Asym_2(\mathbf{e}_1 \otimes \mathbf{e}_2) = -Asym_2(\mathbf{e}_2 \otimes \mathbf{e}_1) = \frac{1}{2}\,(\mathbf{e}_1 \otimes \mathbf{e}_2 - \mathbf{e}_2 \otimes \mathbf{e}_1) . \tag{2.17}$$

For a general tensor $\boldsymbol{t} \in T^2(V)$ given in terms of components by $\boldsymbol{t} = t^{ij}\, \boldsymbol{e}_i \otimes \boldsymbol{e}_j$, we then have

$$Sym_2(\boldsymbol{t}) = \left(\frac{t^{ij} + t^{ji}}{2}\right) \boldsymbol{e}_i \otimes \boldsymbol{e}_j , \quad Asym_2(\boldsymbol{t}) = \left(\frac{t^{ij} - t^{ji}}{2}\right) \boldsymbol{e}_i \otimes \boldsymbol{e}_j . \tag{2.18}$$

We see that $P^2(V)$ is 3-dimensional, with a possible basis given by $\{\mathbf{e}_1 \otimes \mathbf{e}_1,\, \mathbf{e}_2 \otimes \mathbf{e}_2,\, \mathbf{e}_1 \otimes \mathbf{e}_2 + \mathbf{e}_2 \otimes \mathbf{e}_1\}$, while $\Lambda^2(V)$ is 1-dimensional, with a possible basis given by $\{\mathbf{e}_1 \otimes \mathbf{e}_2 - \mathbf{e}_2 \otimes \mathbf{e}_1\}$. In fact, since $T^2(V)$ is 4-dimensional, we have the *direct sum* relationship $T^2(V) = P^2(V) \oplus \Lambda^2(V)$. It is important to note, however, that $T^r(V)$ cannot be decomposed simply into a direct sum of just symmetric and antisymmetric subspaces when $r > 2$: there exist tensors of mixed symmetry in this case.

It is easily seen that [refer to (2.26) below]

$$dim\, \Lambda^r(V) = dim\, \Lambda^r(V^*) = \binom{n}{r} \quad \text{if } dim\,(V) = n . \tag{2.19}$$

The formula for $dim\, P^r(V)$ is more complicated and will not be given here. (The interested reader may consult, for example, Chapter 29 of Lam 2009.) In fact, for the purposes of the remainder of this chapter, we will not be considering the symmetric subspaces $P^r(V)$ any further.

Problem 2.3 Show that $P^r(V)$ and $\Lambda^r(V)$ are indeed vector subspaces of $T^r(V)$ [similarly for $P^r(V^*)$ and $\Lambda^r(V^*)$] by showing that

(a) the sum of any two symmetric (antisymmetric) tensors are still symmetric (antisymmetric);

(b) scalar multiplication of $\boldsymbol{t}$ in $T^r(V)$ or $T^r(V^*)$ by a real number $a \in \mathbb{R}$ does not change the symmetry character of the tensor;

(c) the zero tensor $\mathbf{0}$ in $T^r(V)$ or $T^r(V^*)$ is both symmetric and antisymmetric.

Problem 2.4 Construct the subspaces $P^3(V)$ and $\Lambda^3(V)$ of $T^3(V)$ using (2.9) and (2.10) when $dim\,(V) = 3$. What is the dimension of the symmetric subspace $P^3(V)$ in this case? Give an explicit basis for $P^3(V)$. Convince yourself that $T^3(V)$ is bigger than $P^3(V) \oplus \Lambda^3(V)$. Can you think of examples in physics, not necessarily within the realm of classical mechanics, where such tensor spaces may be useful?

Finally we are ready to introduce the notion of an exterior (wedge) product, which is a product defined *only* for exterior (antisymmetric) vectors such that the result is again an exterior vector.

Definition 2.2. *Let* $\boldsymbol{\xi} \in \Lambda^r(V)$ *and* $\boldsymbol{\eta} \in \Lambda^s(V)$ *be an exterior r-vector and an exterior s-vector, respectively. The* **exterior (wedge) product** *of* $\boldsymbol{\xi}$ *and* $\boldsymbol{\eta}$, *written* $\boldsymbol{\xi} \wedge \boldsymbol{\eta}$, *is the exterior* $(r+s)$*-vector given by*

$$\boldsymbol{\xi} \wedge \boldsymbol{\eta} = Asym_{r+s}\,(\boldsymbol{\xi} \otimes \boldsymbol{\eta}) \,. \tag{2.20}$$

Note that even though $\boldsymbol{\xi}$ and $\boldsymbol{\eta}$ are separately antisymmetric, their tensor product is not necessarily antisymmetric, and the antisymmetrizing map $Asym_{r+s}$ is required to make the exterior product as defined antisymmetric.

The most basic properties of exterior products will be stated as the following theorem (without proof).

Theorem 2.1. *Let* $\boldsymbol{\xi}, \boldsymbol{\xi}_1, \boldsymbol{\xi}_2 \in \Lambda^k(V)$, $\boldsymbol{\eta}, \boldsymbol{\eta}_1, \boldsymbol{\eta}_2 \in \Lambda^l(V)$, *and* $\boldsymbol{\zeta} \in \Lambda^m(V)$. *Then the following laws hold:*

(1) Distributive Law:

$$(\boldsymbol{\xi}_1 + \boldsymbol{\xi}_2) \wedge \boldsymbol{\eta} = \boldsymbol{\xi}_1 \wedge \boldsymbol{\eta} + \boldsymbol{\xi}_2 \wedge \boldsymbol{\eta} \,, \tag{2.21a}$$

$$\boldsymbol{\xi} \wedge (\boldsymbol{\eta}_1 + \boldsymbol{\eta}_2) = \boldsymbol{\xi} \wedge \boldsymbol{\eta}_1 + \boldsymbol{\xi} \wedge \boldsymbol{\eta}_2 \,. \tag{2.21b}$$

(2) Anticommutative Law:

$$\boldsymbol{\xi} \wedge \boldsymbol{\eta} = (-1)^{kl}\, \boldsymbol{\eta} \wedge \boldsymbol{\xi} \,. \tag{2.22}$$

(3) Associative Law:

$$(\boldsymbol{\xi} \wedge \boldsymbol{\eta}) \wedge \boldsymbol{\zeta} = \boldsymbol{\xi} \wedge (\boldsymbol{\eta} \wedge \boldsymbol{\zeta}) \,. \tag{2.23}$$

Several useful facts (especially from the viewpoint of calculations involving exterior products) are direct consequences of the above theorem and the definition of the exterior product. They will be collected in the list below for ease of reference. (Their proofs are not difficult but will not be given here; the reader is encouraged to supply them as exercises, or at least to verify them with specific examples.)

(a) For any $\boldsymbol{\xi}, \boldsymbol{\eta} \in \Lambda^1(V) = V$, the Anticommutative Law (2.22) implies

$$\boldsymbol{\xi} \wedge \boldsymbol{\eta} = -\boldsymbol{\eta} \wedge \boldsymbol{\xi}\,, \qquad \boldsymbol{\xi} \wedge \boldsymbol{\xi} = \mathbf{0}\,. \tag{2.24}$$

(b) If $dim\,(V) = n$, then

$$\Lambda^r(V) = \Lambda^r(V^*) = \{\mathbf{0}\} \quad \text{for } r > n\,. \tag{2.25}$$

(c) Suppose $\{\boldsymbol{e}_1, \dots, \boldsymbol{e}_n\}$ is a basis of V. Then a basis for $\Lambda^r(V)$ is given by

$$\{\boldsymbol{e}_{i_1} \wedge \dots \wedge \boldsymbol{e}_{i_r} \,|\, 1 \le i_1 < \dots < i_r \le n\}\,. \tag{2.26}$$

The dimension of $\Lambda^r(V)$ is given by (2.19), and any $\boldsymbol{\xi} \in \Lambda^r(V)$ can be expressed as

$$\boldsymbol{\xi} = \sum_{i_1 < \dots < i_r} \xi^{i_1 \dots i_r}\, \boldsymbol{e}_{i_1} \wedge \dots \wedge \boldsymbol{e}_{i_r}\,, \tag{2.27}$$

where each summation index runs from 1 to n, and the components $\xi^{i_1 \dots i_r}$ are antisymmetric under any permutation of the r indices, that is

$$\xi^{i_{\sigma(1)} \dots i_{\sigma(r)}} = (sgn\,\sigma)\, \xi^{i_1 \dots i_r}\,, \tag{2.28}$$

for any permutation $\sigma \in \mathcal{S}_r$.

(d) Let $\{\boldsymbol{e}_1, \dots, \boldsymbol{e}_n\}$ be a basis of V, and $(\boldsymbol{v}^*)^1, \dots, (\boldsymbol{v}^*)^r$ be r arbitrary vectors in V^*. We have the following **evaluation formula** for the exterior r-vector $\boldsymbol{e}_{i_1} \wedge \dots \wedge \boldsymbol{e}_{i_r}$:

$$\begin{aligned}
&(\boldsymbol{e}_{i_1} \wedge \dots \wedge \boldsymbol{e}_{i_r})((\boldsymbol{v}^*)^1, \dots, (\boldsymbol{v}_r)^r) \\
&= \frac{1}{r!} \sum_{\sigma \in \mathcal{S}_r} (sgn\,\sigma)\, \sigma(\boldsymbol{e}_{i_1} \otimes \dots \otimes \boldsymbol{e}_{i_r})((\boldsymbol{v}^*)^1, \dots, (\boldsymbol{v}^*)^r) \\
&= \frac{1}{r!} \sum_{\sigma \in \mathcal{S}_r} (sgn\,\sigma)(\boldsymbol{e}_{i_1} \otimes \dots \otimes \boldsymbol{e}_{i_r})((\boldsymbol{v}^*)^{\sigma(1)}, \dots, (\boldsymbol{v}^*)^{\sigma(r)}) \\
&= \frac{1}{r!} \sum_{\sigma \in \mathcal{S}_r} (sgn\,\sigma)\, \langle\, \boldsymbol{e}_{i_1}, (\boldsymbol{v}^*)^{\sigma(1)} \,\rangle \dots \langle\, \boldsymbol{e}_{i_r}, (\boldsymbol{v}^*)^{\sigma(r)} \rangle \\
&= \frac{1}{r!} \begin{vmatrix} \langle\, \boldsymbol{e}_{i_1}, (\boldsymbol{v}^*)^1 \,\rangle & \dots & \dots & \langle\, \boldsymbol{e}_{i_1}, (\boldsymbol{v}^*)^r \,\rangle \\ \langle\, \boldsymbol{e}_{i_2}, (\boldsymbol{v}^*)^1 \,\rangle & \dots & \dots & \langle\, \boldsymbol{e}_{i_2}, (\boldsymbol{v}^*)^r \,\rangle \\ \vdots & & & \vdots \\ \langle\, \boldsymbol{e}_{i_r}, (\boldsymbol{v}^*)^1 \,\rangle & \dots & \dots & \langle\, \boldsymbol{e}_{i_r}, (\boldsymbol{v}^*)^r \,\rangle \end{vmatrix}\,,
\end{aligned} \tag{2.29}$$

where in the last equality we have made use of the formula for the determinant given by (1.53).

(e) The exterior spaces $\Lambda^r(V) \subset T^r(V)$ and $\Lambda^r(V^*) \subset T^r(V^*)$ are dual to each other and inherit the pairing from $T^r(V)$ and $T^r(V^*)$ [as given by (2.2)]. The "normalized" inherited pairing between the exterior spaces (differing from the actual one only by a multiplicative factor of $1/r!$) is given by

$$\boxed{\langle \boldsymbol{v}_1 \wedge \cdots \wedge \boldsymbol{v}_r \,,\, (\boldsymbol{v}^*)^1 \wedge \cdots \wedge (\boldsymbol{v}^*)^r \rangle = det\,(\langle \boldsymbol{v}_i \,,\, (\boldsymbol{v}^*)^j \rangle)} \quad , \tag{2.30}$$

where $\boldsymbol{v}_1, \ldots, \boldsymbol{v}_r$ are arbitrary vectors in V and $(\boldsymbol{v}^*)^1, \ldots, (\boldsymbol{v}^*)^r$ are arbitrary vectors in V^*, and the right-hand side denotes the determinant of the $r \times r$ matrix whose (ij)-th element is as indicated.

(f) A set of vectors $\boldsymbol{v}_1, \ldots, \boldsymbol{v}_r \in V$ is linearly dependent if and only if their exterior product vanishes, that is, if

$$\boldsymbol{v}_1 \wedge \cdots \wedge \boldsymbol{v}_r = \boldsymbol{0} \,. \tag{2.31}$$

(g) Let $\boldsymbol{\omega} \in \Lambda^r(V)$, and $\boldsymbol{v}_1, \ldots, \boldsymbol{v}_r$ be r linearly independent vectors in V. Then $\boldsymbol{\omega}$ can be expressed as a linear combination of the $\boldsymbol{v}_i$'s with exterior vectors as "coefficients", that is, as

$$\boldsymbol{\omega} = \boldsymbol{v}_1 \wedge \boldsymbol{\psi}_1 + \cdots + \boldsymbol{v}_r \wedge \boldsymbol{\psi}_r \,, \tag{2.32}$$

where $\boldsymbol{\psi}_1, \ldots, \boldsymbol{\psi}_r \in \Lambda^{r-1}(V)$, if and only if

$$\boldsymbol{v}_1 \wedge \cdots \wedge \boldsymbol{v}_r \wedge \boldsymbol{\omega} = \boldsymbol{0} \,. \tag{2.33}$$

This is a generalization of the previous statement.

(h) Suppose $\boldsymbol{v}_1, \ldots, \boldsymbol{v}_r \in V$, and $\boldsymbol{w}_1, \ldots, \boldsymbol{w}_r$ are linear combinations of them given by $\boldsymbol{w}_i = a_i^j\, \boldsymbol{v}_j$. Then

$$\boldsymbol{w}_1 \wedge \cdots \wedge \boldsymbol{w}_r = det(a_i^j)\, \boldsymbol{v}_1 \wedge \cdots \wedge \boldsymbol{v}_r \,. \tag{2.34}$$

(i) Let $\{\boldsymbol{v}_1, \ldots, \boldsymbol{v}_r\}$ and $\{\boldsymbol{w}_1, \ldots, \boldsymbol{w}_r\}$ be two sets of vectors in V such that

$$\sum_{i=1}^{r} \boldsymbol{v}_i \wedge \boldsymbol{w}_i = \boldsymbol{0} \,. \tag{2.35}$$

If the vectors of one set are linearly independent, then each vector in the other set can be expressed as a linear combination of vectors of the first set, with the expansion coefficients forming a symmetric matrix. Thus, if $\boldsymbol{v}_1, \ldots, \boldsymbol{v}_r$ are linearly independent, then each $\boldsymbol{w}_i$ can be written $\boldsymbol{w}_i = (a_i^j)\, \boldsymbol{v}_j$, with $a_i^j = a_j^i$. This statement is known as **Cartan's Lemma**.

(j) Let $\{\boldsymbol{v}_1, \ldots, \boldsymbol{v}_r, \boldsymbol{w}_1, \ldots, \boldsymbol{w}_r\}$ and $\{\boldsymbol{v}'_1, \ldots, \boldsymbol{v}'_r, \boldsymbol{w}'_1, \ldots, \boldsymbol{w}'_r\}$ be two sets of vectors in V such that one of them is linearly independent. If

$$\sum_{i=1}^{r} \boldsymbol{v}_i \wedge \boldsymbol{w}_i = \sum_{i=1}^{r} \boldsymbol{v}'_i \wedge \boldsymbol{w}'_i \,, \tag{2.36}$$

then the other set is also linearly independent and each vector in it can be expressed as a linear combination of the vectors of the first set.

Problem 2.5 Let $\{\boldsymbol{e}_1, \boldsymbol{e}_2, \boldsymbol{e}_3, \boldsymbol{e}_4\}$ be a basis of a 4-dimensional vector space V. Write down explicitly a basis for $\Lambda^3(V)$ in terms of the $\boldsymbol{e}_i$'s.

Problem 2.6 Let $\{\boldsymbol{e}_1, \ldots, \boldsymbol{e}_n\}$ be a basis of an n-dimensional vector space V, and $\{(\boldsymbol{e}^*)^1, \ldots, (\boldsymbol{e}^*)^n\}$ be the dual basis in V^*. Show that, for $r \leq n$,

$$\langle\, \boldsymbol{e}_{i_1} \wedge \cdots \wedge \boldsymbol{e}_{i_r} \,,\, (\boldsymbol{e}^*)^{j_1} \wedge \cdots \wedge (\boldsymbol{e}^*)^{j_r} \,\rangle = det\,(\langle \boldsymbol{e}_{i_\alpha} \,,\, (\boldsymbol{e}^*)^{j_\beta} \rangle) = \delta^{j_1 \ldots j_r}_{i_1 \ldots i_r} \,, \tag{2.37}$$

where $\delta^{j_1 \ldots j_r}_{i_1 \ldots i_r}$ is the **generalized Kronecker delta symbol** given by

$$\delta^{j_1 \ldots j_r}_{i_1 \ldots i_r} = \begin{cases} 1 & \text{if } i_1, \ldots, i_r \text{ are distinct, and } \{j_1, \ldots, j_r\} \\ & \text{is an even permutation of } \{i_1, \ldots, i_r\}\,; \\ -1 & \text{if } i_1, \ldots, i_r \text{ are distinct, and } \{j_1, \ldots, j_r\} \\ & \text{is an odd permutation of } \{i_1, \ldots, i_r\}\,; \\ 0 & \text{otherwise.} \end{cases} \tag{2.38}$$

From the exterior spaces $\Lambda^r(V)$ we can build the so-called **exterior** (or **Grassmann**) **algebra**, defined on the space given by the *direct sum* (of vector spaces)

$$\Lambda(V) = \oplus \sum_{r=0}^{n} \Lambda^r(V) \,, \qquad dim\,(V) = n \,. \tag{2.39}$$

(Recall that $\Lambda^r(V) = \{\boldsymbol{0}\}$ for $r > n$.) A general element $\boldsymbol{\xi} \in \Lambda(V)$ can be written as

$$\boldsymbol{\xi} = \sum_{r=0}^{n} \boldsymbol{\xi}^r \,, \qquad \boldsymbol{\xi}^r \in \Lambda^r(V)\,; \tag{2.40}$$

and the exterior product of $\boldsymbol{\xi}$ and $\boldsymbol{\eta}$ in $\Lambda(V)$ is then defined by

$$\boldsymbol{\xi} \wedge \boldsymbol{\eta} = \sum_{r,s=0}^{n} \boldsymbol{\xi}^r \wedge \boldsymbol{\eta}^s \,. \tag{2.41}$$

Similarly, we have the exterior algebra

$$\Lambda(V^*) = \oplus \sum_{r=0}^{n} \Lambda^r(V^*) \,. \tag{2.42}$$

From Statement (c) and Eq. (2.26) above we see that the total number of elements in a basis for $\Lambda(V)$ [or $\Lambda(V^*)$], if $dim\,(V) = n$, is $\sum_{r=0}^{n} \binom{n}{r}$, which, on setting $x = 1$ in the binomial theorem

$$(1+x)^n = \sum_{r=0}^{n} \binom{n}{r} x^r \,, \tag{2.43}$$

is seen to be equal to 2^n. Thus

$$dim\,\Lambda(V) = dim\,\Lambda(V^*) = 2^n \,, \qquad dim\,(V) = n \,, \tag{2.44}$$

when the exterior algebras are considered as vector spaces.

The algebraic structure of an exterior algebra arising from the wedge product is preserved by linear maps. Suppose $A : V \to W$ is a linear map from a vector space V to another vector space W (which may have a different dimension from that of V). We introduce the important notion of a *pullback map* by the following "coordinate-free" definition.

Definition 2.3. *The* **pullback map** $A^* : \Lambda^r(W^*) \to \Lambda^r(V^*)$, *uniquely induced by a linear map* $A : V \to W$, *is the linear map that acts on an element* $\varphi \in \Lambda^r(W^*)$ *to produce an element* $A^*\varphi \in \Lambda^r(V^*)$ *defined by*

$$(A^*\varphi)(\boldsymbol{v}_1, \ldots, \boldsymbol{v}_r) = \varphi(A(\boldsymbol{v}_1), \ldots, A(\boldsymbol{v}_r)) \,, \tag{2.45}$$

for any $\boldsymbol{v}_1, \ldots, \boldsymbol{v}_r \in V$.

The pullback map has the important property that it commutes with the exterior product. This fact is stated more precisely as the following theorem:

Theorem 2.2. *Suppose* $A : V \to W$ *is a linear map from a vector space* V *to another vector space* W. *Then, for any* $\varphi \in \Lambda^r(W^*)$ *and* $\psi \in \Lambda^s(W^*)$,

$$A^*(\varphi \wedge \psi) = (A^*\varphi) \wedge (A^*\psi) \,. \tag{2.46}$$

Another way of stating this theorem is that the exterior product, as an algebraic structure, is *preserved* under the linear map A^* between the exterior algebras $\Lambda(W^*)$ and $\Lambda(V^*)$. We say that the pullback map A^* is an **algebra homomorphism** between the exterior algebras.

Problem 2.7 Prove that the pullback map A^* as defined above is in fact linear, that is, for any $a, b \in \mathbb{R}$ and any $\varphi_1, \varphi_2 \in \Lambda^r(W^*)$,

$$A^*(a\varphi_1 + b\varphi_2) = a\,A^*(\varphi_1) + b\,A^*(\varphi_2) \,.$$

Problem 2.8 Prove Theorem 2.2 by directly computing the action of the left-hand side of (2.46) on an ordered $(r+s)$-tuplet of vectors $\boldsymbol{v}_1, \dots, \boldsymbol{v}_{r+s}$ in V.

It is interesting to see that the pullback map induced by a linear map $A : V \to V$ actually provides the most natural definition of the concept of the determinant of a linear operator. If $dim\,(V) = n$, then A gives rise to the unique linear map $A^* : \Lambda^n(V^*) \to \Lambda^n(V^*)$ such that, for any $\varphi \in \Lambda^n(V^*)$ and any $\boldsymbol{v}_1, \dots, \boldsymbol{v}_n \in V$ [cf. (2.45)],

$$(A^*\varphi)(\boldsymbol{v}_1, \dots, \boldsymbol{v}_n) = \varphi(A(\boldsymbol{v}_1), \dots, A(\boldsymbol{v}_n)) \,. \tag{2.47}$$

Now (2.26) implies that $\Lambda^n(V^*)$ is a 1-dimensional vector space, so that any linear operator acting on it is just a scalar multiplication by a real number, where the scalar depends *solely* on the identity of the (abstract) linear operator, and not on the particular matrix representation of the operator. This number is called the **determinant** of the linear operator. So we have, for any $\varphi \in \Lambda^n(V^*)$,

$$A^*\varphi = det\,(A)\,\varphi \,. \tag{2.48}$$

Since different matrix representations of the same linear operator are related by similarity transformations [cf. (1.31)], we recover the familiar fact that *the determinant of a square matrix is invariant under similarity transformations.*

Problem 2.9 Use (2.47) and (2.48) to prove the following familiar fact for determinants: If $A : V \to V$ and $B : V \to V$ are linear operators on a vector space V, then

$$det\,(AB) = det\,(A) \cdot det\,(B) \,. \tag{2.49}$$

Problem 2.10 Establish the equivalence between the abstract "coordinate free" definition of the determinant of a linear operator A given by (2.48) and the elementary (computational) one given by (1.53), which depends on a particular matrix representation $A(\boldsymbol{e}_i) = a_i^j\, \boldsymbol{e}_j$ of the linear operator A.

Hint: Evaluate $(A^*\varphi)(\boldsymbol{e}_1, \dots, \boldsymbol{e}_n)$ using (2.47) and the multilinear property of $\varphi \in \Lambda^n(V^*)$, and then equate the result to $det\,(A) \cdot \varphi(\boldsymbol{e}_1, \dots, \boldsymbol{e}_n)$.

We will conclude this chapter with two elementary applications of exterior algebra: the orientation of reference frames and the calculation of oriented areas and volumes. The first has been alluded to in Chapter 1 already [see discussion following (1.22)], we will see here that the mathematical formalism of exterior algebra provides the most rigorous and efficient means of its description.

The notion of **orientation** (or **handedness**) arises intuitively from the realization that the fingers of our right hand or left hand can be used to represent two essentially different kinds of coordinate frames. In physics we call these the **right-handed** and **left-handed** frames; and no amount of "reorientation" (rotation) in physical space can change one to the other. This realization immediately leads to the idea that certain physical structures in space (such as molecules and polarized photons), and certain physical processes whose description depends intrinsically on spatial orientations (such as beta decay in nuclear physics and an electrical current generating a magnetic field in Maxwell electrodynamics), have to be characterized by their handedness, or **parity** (as physicists term the concept). Parity, in fact, was recognized as a fundamental attribute of physical systems; and the understanding of the fundamental laws of nature entails an understanding of whether it is conserved or not in fundamental interactions. It then becomes important to develop a mathematical way to unambiguously code the idea of orientation in such a way that it can be mathematically manipulated in calculations, not only for the most familiar situation of three spatial dimensions, but for all other dimensions as well. This can be done precisely by an exterior product of vectors in $\mathbb{R}^n$.

Consider first an orthonormal basis $\{\boldsymbol{e}_1, \boldsymbol{e}_2, \boldsymbol{e}_3\}$ in $V = \mathbb{R}^3$, and the top-ranked exterior 3-vector $\boldsymbol{e}_1 \wedge \boldsymbol{e}_2 \wedge \boldsymbol{e}_3 \in \Lambda^3(V)$. We say that $\boldsymbol{e}_1 \wedge \boldsymbol{e}_2 \wedge \boldsymbol{e}_3$ represents an *oriented frame* in $\mathbb{R}^3$. If we choose any other orthonormal basis $\{\boldsymbol{e}'_1, \boldsymbol{e}'_2, \boldsymbol{e}'_3\}$, Eq. (2.34), and the fact that the two bases must be related to each other by an orthogonal transformation (whose determinant must be either $+1$ or -1), imply that

$$\boldsymbol{e}'_1 \wedge \boldsymbol{e}'_2 \wedge \boldsymbol{e}'_3 = (\pm 1)\, \boldsymbol{e}_1 \wedge \boldsymbol{e}_2 \wedge \boldsymbol{e}_3 \,. \tag{2.50}$$

Thus we see that there are only two possible orientations as defined by the exterior products of orthonormal frames in space; if we label one of them as right-handed, then the other is left-handed. This confirms the fact that $\Lambda^3(\mathbb{R}^3)$ is 1-dimensional [as implied by (2.19)], and the two orientations just correspond to choosing the basis vector in $\Lambda^3(\mathbb{R}^3)$ as $\boldsymbol{e}_1 \wedge \boldsymbol{e}_2 \wedge \boldsymbol{e}_3$ or $-\boldsymbol{e}_1 \wedge \boldsymbol{e}_2 \wedge \boldsymbol{e}_3$. We can in fact generalize the concept of orientation of reference frames as referring to arbitrary frames formed by three linearly independent vectors in $\mathbb{R}^3$, whether orthonormal or not. Let $\{\boldsymbol{v}_1, \boldsymbol{v}_2, \boldsymbol{v}_3\}$ be a basis in $\mathbb{R}^3$ composed of the three indicated linearly independent vectors. With respect to the formerly chosen orthonormal basis $\{\boldsymbol{e}_i\}$, these vectors can be expressed as $\boldsymbol{v}_i = v_i^j\, \boldsymbol{e}_j$. We then have, again by (2.34),

$$\boldsymbol{v}_1 \wedge \boldsymbol{v}_2 \wedge \boldsymbol{v}_3 = det\,(v_i^j)\, \boldsymbol{e}_1 \wedge \boldsymbol{e}_2 \wedge \boldsymbol{e}_3 \,. \tag{2.51}$$

If $\boldsymbol{v}_1 \wedge \boldsymbol{v}_2 \wedge \boldsymbol{v}_3$, as a vector in the 1-dimensional space $\Lambda^3(\mathbb{R}^3)$, points in the same direction as $\boldsymbol{e}_1 \wedge \boldsymbol{e}_2 \wedge \boldsymbol{e}_3$, that is, $det\,(v_i^j) > 0$, we say that $\boldsymbol{v}_1 \wedge \boldsymbol{v}_2 \wedge \boldsymbol{v}_3$ is an oriented frame that has the same orientation as $\boldsymbol{e}_1 \wedge \boldsymbol{e}_2 \wedge \boldsymbol{e}_3$; otherwise it has the opposite orientation.

The above discussion clearly generalizes to an arbitrary finite-dimensional vector space. We thus have the following general definition of oriented frames:

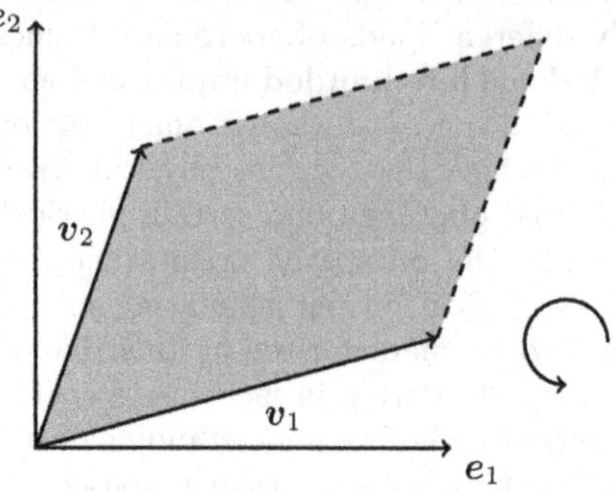

Fig. 2.1

Definition 2.4. *An* **oriented frame** *in an n-dimensional vector space V is an exterior n-vector $\boldsymbol{v}_1 \wedge \cdots \wedge \boldsymbol{v}_n \in \Lambda^n(V)$, where $\{\boldsymbol{v}_1, \dots, \boldsymbol{v}_n\}$ is any basis of V. Two oriented frames are said to have the same orientation if they are related by a positive multiplicative constant.*

We see that the totality of vectors in the 1-dimensional space $\Lambda^n(V)$ is partitioned into two **equivalence classes** having two different orientations: if one is labeled right-handed, then the other is left-handed.

Now consider two arbitrary vectors $\boldsymbol{v}_1, \boldsymbol{v}_2 \in \mathbb{R}^2$ and an oriented frame $\boldsymbol{e}_1 \wedge \boldsymbol{e}_2$ in the same space (see Fig. 2.1). We know from elementary vector algebra that the area of the parallelogram bounded by $\boldsymbol{v}_1$ and $\boldsymbol{v}_2$ and the two opposite sides (the shaded area in Fig. 2.1) is given by the absolute value of a cross product $\boldsymbol{v}_1 \times \boldsymbol{v}_2$ [cf. (1.54)], when the given 2-dimensional space $\mathbb{R}^2$ is imagined to be embedded in $\mathbb{R}^3$, so that each of the vectors $\boldsymbol{v}_1$ and $\boldsymbol{v}_2$ has a vanishing third component. If we express $\boldsymbol{v}_i$ in components as

$$\boldsymbol{v}_1 = v_1^1\, \boldsymbol{e}_1 + v_1^2\, \boldsymbol{e}_2 \,, \qquad \boldsymbol{v}_2 = v_2^1\, \boldsymbol{e}_1 + v_2^2\, \boldsymbol{e}_2 \,, \tag{2.52}$$

we have

$$\begin{aligned} &\text{area of parallelogram bounded by } \boldsymbol{v}_1 \text{ and } \boldsymbol{v}_2 \\ &= |\, \boldsymbol{v}_1 \times \boldsymbol{v}_2 \,| = \text{absolute value of } \begin{vmatrix} v_1^1 & v_1^2 \\ v_2^1 & v_2^2 \end{vmatrix} \,. \end{aligned} \tag{2.53}$$

Note that the determinant itself may be either positive or negative, and by (2.30), is precisely equal to the pairing expression $\langle\, \boldsymbol{v}_1 \wedge \boldsymbol{v}_2 \,,\, (\boldsymbol{e}^*)^1 \wedge (\boldsymbol{e}^*)^2 \,\rangle$. We say that this pairing gives the **oriented area** $A_{or}(\boldsymbol{v}_1 \wedge \boldsymbol{v}_2 \,;\, (\boldsymbol{e}^*)^1 \wedge (\boldsymbol{e}^*)^2)$ of the parallelogram bounded by the oriented frame $\boldsymbol{v}_1 \wedge \boldsymbol{v}_2$ with respect to the **area**

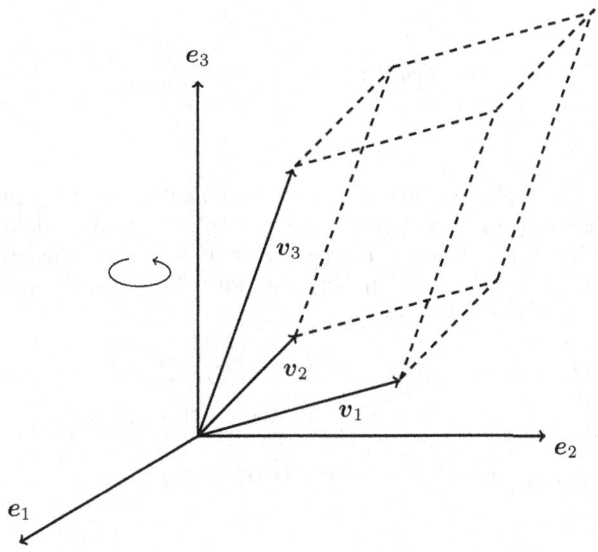

Fig. 2.2

form $(e^*)^1 \wedge (e^*)^2$:

$$A_{or}(v_1 \wedge v_2\,;\,(e^*)^1 \wedge (e^*)^2) = \langle\, v_1 \wedge v_2\,,\,(e^*)^1 \wedge (e^*)^2\,\rangle = \begin{vmatrix} v_1^1 & v_1^2 \\ v_2^1 & v_2^2 \end{vmatrix} \,. \tag{2.54}$$

Note that if the order of the factors in the exterior product in either entry of the pairing bracket is reversed (equivalent to interchanging either the two rows or the two columns of the determinant), then the oriented area would change by a sign. Also, if the two vectors v_1 and v_2 are linearly dependent, then, by (2.31), $v_1 \wedge v_2 = \mathbf{0}$, and the oriented area vanishes. This is obvious geometrically, since in this case v_1 would be parallel to v_2.

Next consider three arbitrary vectors $v_1, v_2, v_3 \in \mathbb{R}^3$ and an oriented frame $e_1 \wedge e_2 \wedge e_3$ in the same space formed by an orthonormal basis $\{e_1, e_2, e_3\}$ (see Fig. 2.2). Recall from (1.62) that the volume of the parallelopiped formed by the vectors v_1, v_2, v_3 is given by $|\,v_1 \cdot (v_2 \times v_3)\,|$. If we express v_i in terms of components:

$$v_1 = v_1^1\,e_1 + v_1^2\,e_2 + v_1^3\,e_3\,, \quad v_2 = v_2^1\,e_1 + v_2^2\,e_2 + v_2^3\,e_3\,, \quad v_3 = v_3^1\,e_1 + v_3^2\,e_2 + v_3^3\,e_3\,, \tag{2.55}$$

we find that, by the rules of 3-dimensional vector algebra for the scalar product and cross product [(1.41) and (1.45) respectively],

$$\boldsymbol{v}_1 \cdot (\boldsymbol{v}_2 \times \boldsymbol{v}_3) = \begin{vmatrix} v_1^1 & v_1^2 & v_1^3 \\ v_2^1 & v_2^2 & v_2^3 \\ v_3^1 & v_3^2 & v_3^3 \end{vmatrix} . \tag{2.56}$$

Equation (2.30) implies again that the determinant is precisely given by the pairing bracket $\langle\, \boldsymbol{v}_1 \wedge \boldsymbol{v}_2 \wedge \boldsymbol{v}_3 \,,\, (\mathbf{e}^*)^1 \wedge (\mathbf{e}^*)^2 \wedge (\mathbf{e}^*)^3 \,\rangle$. Analogous to (2.54), the **oriented volume** $V_{or}(\boldsymbol{v}_1 \wedge \boldsymbol{v}_2 \wedge \boldsymbol{v}_3 \,;\, (\mathbf{e}^*)^1 \wedge (\mathbf{e}^*)^2 \wedge (\mathbf{e}^*)^3)$ of the parallelopiped bounded by the oriented frame $\boldsymbol{v}_1 \wedge \boldsymbol{v}_2 \wedge \boldsymbol{v}_3$ with respect to the **volume form** $(\mathbf{e}^*)^1 \wedge (\mathbf{e}^*)^2 \wedge (\mathbf{e}^*)^3$ is given by

$$\begin{aligned} & V_{or}(\boldsymbol{v}_1 \wedge \boldsymbol{v}_2 \wedge \boldsymbol{v}_3 \,;\, (\mathbf{e}^*)^1 \wedge (\mathbf{e}^*)^2 \wedge (\mathbf{e}^*)^3) \\ & = \langle\, \boldsymbol{v}_1 \wedge \boldsymbol{v}_2 \wedge \boldsymbol{v}_3 \,,\, (\mathbf{e}^*)^1 \wedge (\mathbf{e}^*)^2 \wedge (\mathbf{e}^*)^3 \,\rangle = \begin{vmatrix} v_1^1 & v_1^2 & v_1^3 \\ v_2^1 & v_2^2 & v_2^3 \\ v_3^1 & v_3^2 & v_3^3 \end{vmatrix} . \end{aligned} \tag{2.57}$$

It is straightforward to generalize (2.54) and (2.57) to give the oriented volume $V_{or}(\boldsymbol{v}_1 \wedge \cdots \wedge \boldsymbol{v}_n \,;\, (\mathbf{e}^*)^1 \wedge \cdots \wedge (\mathbf{e}^*)^n)$ for a parallelopiped in the Euclidean space $\mathbb{R}^n$ bounded by the oriented frame $\boldsymbol{v}_1 \wedge \cdots \wedge \boldsymbol{v}_n$ with respect to the volume form $(\mathbf{e}^*)^1 \wedge \cdots \wedge (\mathbf{e}^*)^n$:

$$\begin{aligned} & V_{or}(\boldsymbol{v}_1 \wedge \cdots \wedge \boldsymbol{v}_n \,;\, (\mathbf{e}^*)^1 \wedge \cdots \wedge (\mathbf{e}^*)^n) \\ & = \langle\, \boldsymbol{v}_1 \wedge \cdots \wedge \mathbf{e}_n \,,\, (\mathbf{e}^*)^1 \wedge \cdots \wedge (\mathbf{e}^*)^n \,\rangle = \begin{vmatrix} v_1^1 & \dots & v_1^n \\ \vdots & \dots & \vdots \\ v_n^1 & \dots & v_n^n \end{vmatrix} . \end{aligned} \tag{2.58}$$

We mention in passing that since the above formula is valid for all positive integer values of n, the generic term "volume" can be used regardless of the dimension n, in lieu of "length", "area" and "volume" (which are the terms traditionally reserved for the special cases $n = 1, 2$ and 3 respectively).

The general notion of oriented volumes will play an important role in the formulation of the Stokes Theorem, a most important theorem concerned with the integrals of differential forms (see Chapter 8).

Chapter 3

The Hodge-Star Operator and the Vector Cross Product

The vector (cross) product as defined in Chapter 1 is supposed to be a map $\times : V \times V \to V$, where V is a 3-dimensional real vector space. While this product is of great use and even indispensable in many applications in physics (and we will indeed use it whenever it is expedient in this book), it does lead to mathematical ambiguities, and sometimes physical ones also. In this chapter we will highlight some of these issues, as they are closely related to the rules of exterior products presented in the previous chapter.

The "physics" definition of a vector quantity presented in Chapter 1 (cf. Table 1.1) is that under a change of reference frame $\boldsymbol{e}_i = a_i^j\, \boldsymbol{e}'_j$, the components of a vector $\boldsymbol{v} = v_i^j\, \boldsymbol{e}_j$ transform according to $v'^i = a_j^i\, v^j$. In the case of the pure inversion transformation $\boldsymbol{e}_i = -\delta_i^j\, \boldsymbol{e}'_j = -\boldsymbol{e}'_i$, then, all the components of a vector should simply change sign, regardless of the dimension of the vector space. In the definition of the cross product $\boldsymbol{C} = \boldsymbol{A} \times \boldsymbol{B}$ for 3-dimensional vectors given by (1.54), however, an inversion, while leading to sign changes for all components of both $\boldsymbol{A}$ and $\boldsymbol{B}$ (assuming that both vectors transform correctly under an inversion), leaves all components of the cross product vector $\boldsymbol{C}$ invariant. For this reason, any 3-dimensional vector arising from a cross product is sometimes referred to as a **pseudo-vector** (or an **axial vector**) in the physics literature, as opposed to a **polar** (or "true) vector. For example, the *angular momentum* $\boldsymbol{L}$ of a point particle of mass m is usually defined as $\boldsymbol{L} = \boldsymbol{r} \times \boldsymbol{p}$, where $\boldsymbol{r}$ is the position vector of the particle and $\boldsymbol{p} = m\boldsymbol{v}$ its linear momentum, with $\boldsymbol{v} = d\boldsymbol{r}/dt$ being its velocity. This qualifies the angular momentum as being a pseudo-vector; but the situation presents somewhat of a quandary because there are a variety of other physical quantities that are described by "true" vectors. Examples like this abound in physics, and have in fact led to some confusion in the physics literature where some authors have resorted to arbitrary conventions on the concomitant use of either right-handed or left-handed frames when cross products are involved in order to avoid inconsistency. This is unsatisfactory from the viewpoint of the mathematical formulation of the laws of physics, because one of the fundamental principles of physics, the Principle of Relativity, requires that *the laws of physics should be invariant under coordinate transformations, or changes of reference frames.*

In what follows we will see that the more mathematically consistent way of describing physical quantities involving cross products of 3-dimensional vectors of the form $\boldsymbol{A} \times \boldsymbol{B}$ is through the use of the exterior 2-vector $\boldsymbol{A} \wedge \boldsymbol{B}$. We will refrain from using the terms polar vectors and axial vectors. Our viewpoint is that there are only vectors and tensor products of vectors (tensors). All vectors are "true" vectors and all tensors are "true" tensors (as determined by their transformation properties under linear transformations of coordinate frames), and *retain their identities as vectors or tensors regardless of whether their components are expressed with respect to a right-handed or a left-handed frame.* As to whether physics equations remain invariant under spatial inversion of a coordinate frame, that depends on the physics described by the equations: some physical processes conserve *parity* or possess *mirror symmetry* (such as electromagnetism, gravity), others do not (such as weak interaction). Some physical quantities, such as the linear momentum, are represented mathematically by vectors, while others, such as the angular momentum and the magnetic field strength, are more properly represented by antisymmetric tensors (or exterior vectors). The cross product of two vectors in 3-dimensional Euclidean space can be viewed as a 3-vector only by accident – due to the circumstance that the two vectors to be crossed are 3-dimensional: If we adopt an oriented coordinate frame $\{\boldsymbol{e}_1 \wedge \boldsymbol{e}_2 \wedge \boldsymbol{e}_3\}$ in a 3-dimensional vector space V, either right-handed or left-handed, we end up with three independent exterior 2-vectors in $\Lambda^2(V)$: $\boldsymbol{e}_1 \wedge \boldsymbol{e}_2, \boldsymbol{e}_2 \wedge \boldsymbol{e}_3$ and $\boldsymbol{e}_3 \wedge \boldsymbol{e}_1$. Thus $\boldsymbol{A} \wedge \boldsymbol{B}$ has three independent components. This is what entitles $\boldsymbol{A} \wedge \boldsymbol{B}$ to be identified as a 3-dimensional "vector". But, as we mentioned before, this so-called "vector" has funny properties: Under the inversion $\boldsymbol{e}_i \to \boldsymbol{e}_i' = -\boldsymbol{e}_i \, (i = 1, 2, 3)$, the components of this "vector" do *not* change sign, as do those of a "true" vector! *The confusion arises solely from the seemingly convenient habit of calling something which is actually an exterior 2-vector (an antisymmetric rank-two tensor) a vector.* If we were dealing with 4-dimensional vectors $\boldsymbol{A}$ and $\boldsymbol{B}$ to begin with, we would not have the luxury of identifying the exterior vector $\boldsymbol{A} \wedge \boldsymbol{B}$ with a 4-dimensional vector, since by (2.19) it would have six independent components instead of four, as in the case of the electromagnetic field tensor (see Problem 3.2 below).

Going back to the 3-dimensional case, if $\boldsymbol{A} = A^1 \boldsymbol{e}_1 + A^2 \boldsymbol{e}_2 + A^3 \boldsymbol{e}_3$ and $\boldsymbol{B} = B^1 \boldsymbol{e}_1 + B^2 \boldsymbol{e}_2 + B^3 \boldsymbol{e}_3$, then the rules of the exterior product as given by Theorem 2.1 immediately imply

$$\boldsymbol{A} \wedge \boldsymbol{B} = (A^1 B^2 - A^2 B^1)\, \boldsymbol{e}_1 \wedge \boldsymbol{e}_2 + (A^2 B^3 - A^3 B^2)\, \boldsymbol{e}_2 \wedge \boldsymbol{e}_3 + (A^3 B^1 - A^1 B^3)\, \boldsymbol{e}_3 \wedge \boldsymbol{e}_1 \,. \tag{3.1}$$

The components of this exterior vector are precisely those of the cross product as given by (1.54):

$$\boldsymbol{A}\times\boldsymbol{B}=(A^1B^2-A^2B^1)\,\boldsymbol{e}_3+(A^2B^3-A^3B^2)\,\boldsymbol{e}_1+(A^3B^1-A^1B^3)\,\boldsymbol{e}_2\ , \quad (3.2)$$

and do not change sign under the inversion $\boldsymbol{e}_i\to-\boldsymbol{e}_i$, since $(-\boldsymbol{e}_i)\wedge(-\boldsymbol{e}_j)=\boldsymbol{e}_i\wedge\boldsymbol{e}_j$. Comparing the above two equations, it appears that if we can make the correlations:

$$\boldsymbol{e}_1\wedge\boldsymbol{e}_2\mapsto\boldsymbol{e}_3\ ,\quad \boldsymbol{e}_2\wedge\boldsymbol{e}_3\mapsto\boldsymbol{e}_1\ ,\quad \boldsymbol{e}_3\wedge\boldsymbol{e}_1\mapsto\boldsymbol{e}_2\ , \quad (3.3)$$

then we can identify $\boldsymbol{A}\wedge\boldsymbol{B}$ with $\boldsymbol{A}\times\boldsymbol{B}$. The mapping producing this correlation (but only in the 3-dimensional case) is known as the *Hodge-Star operator*, whose definition depends on two ingredients: the choice of an oriented frame in V, and a scalar product $(\ ,\ )_V$ in V. Before proceeding to define this operator in general, let us illustrate how it works in the 3-dimensional case first.

Suppose $\{\boldsymbol{e}_1,\boldsymbol{e}_2,\boldsymbol{e}_3\}$ is an orthonormal basis of V endowed with the scalar product given by (1.41), so that $(\boldsymbol{e}_i\ ,\ \boldsymbol{e}_j)=\delta_{ij}$. Pick the oriented frame $\boldsymbol{e}_1\wedge\boldsymbol{e}_2\wedge\boldsymbol{e}_3$ and label it a right-handed frame: $\boldsymbol{\sigma}_R=\boldsymbol{e}_1\wedge\boldsymbol{e}_2\wedge\boldsymbol{e}_3$. Then a left-handed one is given by $\boldsymbol{\sigma}_L=-\boldsymbol{\sigma}_R$. Consider a specific exterior 2-vector $\boldsymbol{A}\wedge\boldsymbol{B}$, where $\boldsymbol{A}$ and $\boldsymbol{B}$ are two vectors in V. For any $\boldsymbol{v}\in V$, then, $\boldsymbol{A}\wedge\boldsymbol{B}\wedge\boldsymbol{v}$ is an exterior 3-vector in $\Lambda^3(V)$ (a top ranked tensor), so it must be proportional to both $\boldsymbol{\sigma}_R$ and $\boldsymbol{\sigma}_L$ [since either one of these oriented frames can be considered as a basis vector in the 1-dimensional space $\Lambda^3(V)$]. So, for any $\boldsymbol{v}\in V$, we can write

$$\boldsymbol{A}\wedge\boldsymbol{B}\wedge\boldsymbol{v}=a_{R\,(\boldsymbol{A}\wedge\boldsymbol{B})}(\boldsymbol{v})\,\boldsymbol{\sigma}_R=a_{L\,(\boldsymbol{A}\wedge\boldsymbol{B})}(\boldsymbol{v})\,\boldsymbol{\sigma}_L\ , \quad (3.4)$$

where $a_{R\,(\boldsymbol{A}\wedge\boldsymbol{B})}(\boldsymbol{v})$ [and similarly $a_{L\,(\boldsymbol{A}\wedge\boldsymbol{B})}(\boldsymbol{v})$] is a scalar quantity resulting from a linear function characterized by $\boldsymbol{A}\wedge\boldsymbol{B}$ acting on the vector $\boldsymbol{v}\in V$. Hence both $a_{R\,(\boldsymbol{A}\wedge\boldsymbol{B})}(\boldsymbol{v})$ and $a_{L\,(\boldsymbol{A}\wedge\boldsymbol{B})}(\boldsymbol{v})$ belong to V^*, the dual space of V. Now recall that, according to the discussion at the end of Chapter 1, a scalar product in V induces a canonical (conjugate) isomorphism $G:V\to V^*$ between V and V^*, so that corresponding to each of $a_{R\,(\boldsymbol{A}\wedge\boldsymbol{B})}\in V^*$ and $a_{L\,(\boldsymbol{A}\wedge\boldsymbol{B})}\in V^*$ there is a unique vector $G^{-1}\left(a_{\substack{R\\L}\,(\boldsymbol{A}\wedge\boldsymbol{B})}\right)\in V$. These will be labeled $*_R(\boldsymbol{A}\wedge\boldsymbol{B})$ and $*_L(\boldsymbol{A}\wedge\boldsymbol{B})$, such that $\langle\, a_{R\,(\boldsymbol{A}\wedge\boldsymbol{B})}\ ,\ \boldsymbol{v}\,\rangle=(*_R(\boldsymbol{A}\wedge\boldsymbol{B})\ ,\ \boldsymbol{v})$ and $\langle\, a_{L\,(\boldsymbol{A}\wedge\boldsymbol{B})}\ ,\ \boldsymbol{v}\,\rangle=(*_L(\boldsymbol{A}\wedge\boldsymbol{B})\ ,\ \boldsymbol{v})$. In other words, corresponding to the two orientations, there are two distinct so-called Hodge-star operators: $*_{\substack{R\\L}}:\Lambda^2(V)\to\Lambda^1(V)=V$, whose actions on a general exterior 2-vector $\boldsymbol{A}\wedge\boldsymbol{B}$ (rendering it into two distinct vectors in V) are defined by

$$\boldsymbol{A}\wedge\boldsymbol{B}\wedge\boldsymbol{v}=(*_R(\boldsymbol{A}\wedge\boldsymbol{B})\ ,\ \boldsymbol{v})\,\boldsymbol{\sigma}_R=(*_L(\boldsymbol{A}\wedge\boldsymbol{B})\ ,\ \boldsymbol{v})\,\boldsymbol{\sigma}_L\ , \quad (3.5)$$

where $\boldsymbol{v}$ is *any* vector in the 3-dimensional vector space V. Since $\boldsymbol{\sigma}_L=-\boldsymbol{\sigma}_R$, we immediately obtain

$$*_R=-\,*_L\ . \quad (3.6)$$

The correlations (3.3) can then be rewritten as

$$*_R(\boldsymbol{e}_1\wedge\boldsymbol{e}_2)=\boldsymbol{e}_3\ ,\quad *_R(\boldsymbol{e}_2\wedge\boldsymbol{e}_3)=\boldsymbol{e}_1\ ,\quad *_R(\boldsymbol{e}_3\wedge\boldsymbol{e}_1)=\boldsymbol{e}_2\ , \quad (3.7)$$

and in general

$$*_R(\boldsymbol{A} \wedge \boldsymbol{B}) = - *_L (\boldsymbol{A} \wedge \boldsymbol{B}) = \boldsymbol{A} \times \boldsymbol{B} \,. \tag{3.8}$$

We can now generalize the notion of the Hodge-star operator with respect to a vector space V of arbitrary dimension. As a preliminary step we see that an inner product $(\;,\;)_V$ in V induces an inner product $(\;,\;)_{\Lambda^r(V)}$ in the exterior space $\Lambda^r(V)$ according to

$$\begin{aligned}(\boldsymbol{v}_1 \wedge \cdots \wedge \boldsymbol{v}_r\,,\, \boldsymbol{w}_1 \wedge \cdots \wedge \boldsymbol{w}_r)_{\Lambda^r(V)} &\equiv \langle\, \boldsymbol{v}_1 \wedge \cdots \wedge \boldsymbol{v}_r\,,\, (\boldsymbol{w}^*)^1 \wedge \cdots \wedge (\boldsymbol{w}^*)^r \,\rangle \\ &= det\,(\langle\, \boldsymbol{v}_i\,,\, (\boldsymbol{w}^*)^j \,\rangle)\,,\end{aligned} \tag{3.9}$$

where $\boldsymbol{v}_1, \ldots, \boldsymbol{v}_r, \boldsymbol{w}_1, \ldots, \boldsymbol{w}_r$ are arbitrary vectors in V and $(\boldsymbol{w}^*)^i = G(\boldsymbol{w}_i)$, with $G : V \to V^*$ being the canonical (conjugate) isomorphism induced by the inner product $(\;,\;)_V$ (cf. discussion at the end of Chapter 1). The general definition of the Hodge-star operator is then given below.

Definition 3.1. *Let V be a vector space of dimension n endowed with an inner product $(\;,\;)_V$. For an integer r satisfying $0 \le r \le n$ and an oriented frame $\boldsymbol{\sigma} \in \Lambda^n(V)$ in V, we define a* ***Hodge-star map (operator)*** $*_{\boldsymbol{\sigma}} : \Lambda^r(V) \to \Lambda^{n-r}(V)$ *as a map which satisfies the following property: For any $\boldsymbol{\lambda} \in \Lambda^r(V)$ and any $\boldsymbol{\omega} \in \Lambda^{n-r}(V)$,*

$$\boldsymbol{\lambda} \wedge \boldsymbol{\omega} = (*_{\boldsymbol{\sigma}}\boldsymbol{\lambda}\,,\, \boldsymbol{\omega})_{\Lambda^{n-r}(V)}\, \boldsymbol{\sigma}\,, \tag{3.10}$$

where $(\;,\;)_{\Lambda^{n-r}(V)}$ denotes the inner product induced in the exterior space $\Lambda^{n-r}(V)$ by the inner product $(\;,\;)_V$.

Problem 3.1 Consider $V = \mathbb{R}^3$ with the Euclidean metric (scalar product): $(\boldsymbol{e}_i\,,\, \boldsymbol{e}_j) = \delta_{ij}$, where $\{\boldsymbol{e}_1, \boldsymbol{e}_2, \boldsymbol{e}_3\}$ is an orthonormal basis of V. Choose the oriented frame $\boldsymbol{\sigma} = \boldsymbol{e}_1 \wedge \boldsymbol{e}_2 \wedge \boldsymbol{e}_3$. Verify the following:

(a) Eq. (3.7), that is,

$$*_{\boldsymbol{\sigma}}(\boldsymbol{e}_1 \wedge \boldsymbol{e}_2) = \boldsymbol{e}_3\,, \quad *_{\boldsymbol{\sigma}}(\boldsymbol{e}_2 \wedge \boldsymbol{e}_3) = \boldsymbol{e}_1\,, \quad *_{\boldsymbol{\sigma}}(\boldsymbol{e}_3 \wedge \boldsymbol{e}_1) = \boldsymbol{e}_2\,; \tag{3.11}$$

(b)

$$*_{\boldsymbol{\sigma}}\boldsymbol{e}_1 = \boldsymbol{e}_2 \wedge \boldsymbol{e}_3\,, \qquad *_{\boldsymbol{\sigma}}\boldsymbol{e}_2 = \boldsymbol{e}_3 \wedge \boldsymbol{e}_1\,, \qquad *_{\boldsymbol{\sigma}}\boldsymbol{e}_3 = \boldsymbol{e}_1 \wedge \boldsymbol{e}_2\,; \tag{3.12}$$

(c)

$$*_{\boldsymbol{\sigma}}\boldsymbol{\sigma} = *_{\boldsymbol{\sigma}}(\boldsymbol{e}_1 \wedge \boldsymbol{e}_2 \wedge \boldsymbol{e}_3) = 1\,, \tag{3.13}$$

where the quantity on the right-hand side of the last equality is the real number $1 \in \mathbb{R}$;

(d)

$$*_{\boldsymbol{\sigma}}1 = \boldsymbol{e}_1 \wedge \boldsymbol{e}_2 \wedge \boldsymbol{e}_3 = \boldsymbol{\sigma}\,, \tag{3.14}$$

where the quantity on the left-hand side of the first equality is the real number $1 \in \mathbb{R}$.

Problem 3.2 Consider $V = \mathbb{R}^4$ with the *non positive-definite* **Lorentz metric** (scalar product) given by

$$\begin{cases} (\mathbf{e}_1, \mathbf{e}_1) = -1\,, & \\ (\mathbf{e}_i, \mathbf{e}_i) = +1\,, & i = 2, 3, 4\,, \\ (\mathbf{e}_i, \mathbf{e}_j) = 0\,, & i \neq j\ , \end{cases} \tag{3.15}$$

where $\{\mathbf{e}_1, \mathbf{e}_2, \mathbf{e}_3, \mathbf{e}_4\}$ is an orthonormal basis of V. Choose the oriented frame $\boldsymbol{\sigma} = \mathbf{e}_1 \wedge \mathbf{e}_2 \wedge \mathbf{e}_3 \wedge \mathbf{e}_4$. Show that

(a) $$*_\sigma (\mathbf{e}_1 \wedge \mathbf{e}_2 \wedge \mathbf{e}_3) = \mathbf{e}_4\,, \qquad *_\sigma (\mathbf{e}_1 \wedge \mathbf{e}_2 \wedge \mathbf{e}_4) = -\mathbf{e}_3\,; \tag{3.16}$$

(b) $$*_\sigma (\mathbf{e}_1 \wedge \mathbf{e}_2) = \mathbf{e}_3 \wedge \mathbf{e}_4\,, \qquad *_\sigma (\mathbf{e}_3 \wedge \mathbf{e}_4) = -\mathbf{e}_1 \wedge \mathbf{e}_2\,; \tag{3.17}$$

(c) $$*_\sigma \mathbf{e}_1 = \mathbf{e}_2 \wedge \mathbf{e}_3 \wedge \mathbf{e}_4\,, \quad *_\sigma \mathbf{e}_2 = \mathbf{e}_1 \wedge \mathbf{e}_3 \wedge \mathbf{e}_4\,, \quad *_\sigma \mathbf{e}_3 = -\mathbf{e}_1 \wedge \mathbf{e}_2 \wedge \mathbf{e}_4\,; \tag{3.18}$$

(d) $$*_\sigma (\mathbf{e}_1 \wedge \mathbf{e}_2 \wedge \mathbf{e}_3 \wedge \mathbf{e}_4) = 1\,, \qquad *_\sigma 1 = -\mathbf{e}_1 \wedge \mathbf{e}_2 \wedge \mathbf{e}_3 \wedge \mathbf{e}_4\,, \tag{3.19}$$

where the quantity 1 is the real number in $\mathbb{R}$.

The vector space V in this problem with the given metric is precisely the 4-dimensional **Minkowski spacetime** in the theory of *special relativity*, where $\mathbf{e}_1$ represents the time axis, and $\mathbf{e}_2, \mathbf{e}_3, \mathbf{e}_4$ represent the spatial x, y, z axes, respectively.

Suppose $\{\mathbf{e}_1, \dots, \mathbf{e}_n\}$ is an orthonormal basis of a real vector space V endowed with an inner product $(\ ,\ )_V$ which may be non positive-definite (such as the Lorentz metric in Problem 3.2), so that

$$(\mathbf{e}_i, \mathbf{e}_j) = \pm\delta_{ij}\,, \qquad i, j = 1, \dots, n\,. \tag{3.20}$$

If $+1$'s appear P times and the -1's appear N times in the above group of equations, the integer $S \equiv P - N$ is called the **signature** of the inner product. Now suppose V is an n-dimensional real vector space endowed with an inner product $(\ ,\ )_V$ (which may be non positive-definite) with signature S. Let $\boldsymbol{\sigma} \in \Lambda^n(V)$ be an oriented frame in V, $\boldsymbol{\lambda}$ and $\boldsymbol{\mu}$ be arbitrary exterior vectors in $\Lambda^r(V)$ with $1 \leq r \leq n$, and $(\ ,\ )_{\Lambda^r(V)}$ be the induced inner product in $\Lambda^r(V)$. We collect below a list of useful formulas for the manipulations of expressions involving Hodge-star operators. These will be given without proof.

$$*_\sigma \boldsymbol{\sigma} = 1 \in \mathbb{R}\,; \tag{3.21}$$

$$*_\sigma 1 = (-1)^{(n-S)/2}\, \boldsymbol{\sigma}\,; \tag{3.22}$$

$$(\boldsymbol{\sigma}, \boldsymbol{\sigma})_{\Lambda^n(V)} = (-1)^{(n-S)/2}\,; \tag{3.23}$$

$$*_\sigma *_\sigma \boldsymbol{\lambda} = (-1)^{r(n-r)+(n-S)/2}\, \boldsymbol{\lambda}\,; \tag{3.24}$$

$$(\boldsymbol{\lambda}, \boldsymbol{\mu})_{\Lambda^r(V)} = *_\sigma *_\sigma *_\sigma (\boldsymbol{\mu} \wedge *_\sigma \boldsymbol{\lambda}) = *_\sigma *_\sigma *_\sigma (\boldsymbol{\lambda} \wedge *_\sigma \boldsymbol{\mu})\,; \tag{3.25}$$

$$\boldsymbol{\mu} \wedge *_\sigma \boldsymbol{\lambda} = \boldsymbol{\lambda} \wedge *_\sigma \boldsymbol{\mu} = (-1)^{(n-S)/2}\, (\boldsymbol{\lambda}, \boldsymbol{\mu})_{\Lambda^r(V)}\, \boldsymbol{\sigma}\,; \tag{3.26}$$

$$(*_\sigma \boldsymbol{\lambda}, *_\sigma \boldsymbol{\mu})_{\Lambda^{n-r}(V)} = *_\sigma (\boldsymbol{\lambda} \wedge *_\sigma \boldsymbol{\mu}) = *_\sigma (\boldsymbol{\mu} \wedge *_\sigma \boldsymbol{\lambda})\,. \tag{3.27}$$

Problem 3.3 Show that *the signature of an inner product is independent of the choice of the orthonormal basis.*

Problem 3.4 Verify formulas (3.21) to (3.27) for the specific cases of Problems 3.1 and 3.2.

We will now discuss some physics applications of the mathematical formalism presented above, for the 3-dimensional Euclidean case. Consider the equation of motion [see (12.26) and (12.27)]

$$\boldsymbol{r} \times \boldsymbol{F} = \frac{d\boldsymbol{L}}{dt}\,, \qquad \boldsymbol{L} = \boldsymbol{r} \times \frac{d\,(m\boldsymbol{r})}{dt} \tag{3.28}$$

arising from Newton's Laws for the description of rotational motion, where $\boldsymbol{r}$ is the position vector of a particle of mass m and $\boldsymbol{L}$ is its angular momentum. How do we transform this equation to one involving exterior vectors and products? We know that $\boldsymbol{r}$ is a "true" vector. So the answer depends on the mathematical nature of the force $\boldsymbol{F}$. If it is a "true" vector also it follows from (3.8) that the above equation is equivalent to

$$*_{\substack{R\\L}}(\boldsymbol{r} \wedge \boldsymbol{F}) = \frac{d}{dt}\left[*_{\substack{R\\L}}\left(\boldsymbol{r} \wedge \frac{d\,(m\boldsymbol{r})}{dt}\right)\right]\,. \tag{3.29}$$

Whether we use R on both sides or L on both sides, the above represents the *same* equation, since $*_R = -*_L$. It can be expressed in terms of components with respect to either a right-handed frame or a left-handed frame, that is, this equation is invariant under a (passive) spatial inversion of the coordinate frame. Notice that it is clearly time-reversal invariant also. By applying $*_R$ or $*_L$ on both sides of (3.29) again we have, from (3.24),

$$(\boldsymbol{r} \wedge \boldsymbol{F}) = \frac{d}{dt}\left(\boldsymbol{r} \wedge \frac{d\,(m\boldsymbol{r})}{dt}\right)\,. \tag{3.30}$$

In this form, the question of whether the right-hand rule or the left-hand rule should be used for a cross product does not even arise, and one can use either a right-handed or left-handed frame to equate the corresponding components on both sides of the equation. There is no ambiguity whatsoever.

Now suppose $\boldsymbol{F}$ is a *Lorentz force*: $\boldsymbol{F} = \frac{e}{c}\,\boldsymbol{v} \times \boldsymbol{B}$. This situation is somewhat more complicated since the magnetic field $\boldsymbol{B}$ is mathematically not a vector but an exterior 2-vector, as it is usually regarded in relativistic formulations (the antisymmetric electromagnetic field tensor). It can come in two varieties, the right-handed and the left-handed. We will denote these by $\boldsymbol{B}^{(R)}$ and $\boldsymbol{B}^{(L)}$, respectively. The right-handed kind, which is the kind encountered physically, behaves like a right-handed screw: if a current loop flows in the sense that the screw is rotated, the resulting magnetic field (viewed as a vector) is in the direction of the advancement of the screw. Similarly, a left-handed magnetic field behaves like a left-handed screw. We have, as tensors,

$$\boldsymbol{B}^{(L)} = -\boldsymbol{B}^{(R)} \,, \tag{3.31}$$

with

$$\boldsymbol{B}^{(R)} = \;\; B^1\, \boldsymbol{e}_2 \wedge \boldsymbol{e}_3 + B^2\, \boldsymbol{e}_3 \wedge \boldsymbol{e}_1 + B^3\, \boldsymbol{e}_1 \wedge \boldsymbol{e}_2 \,, \tag{3.32}$$

$$\boldsymbol{B}^{(L)} = -B^1\, \boldsymbol{e}_2 \wedge \boldsymbol{e}_3 - B^2\, \boldsymbol{e}_3 \wedge \boldsymbol{e}_1 - B^3\, \boldsymbol{e}_1 \wedge \boldsymbol{e}_2 \,, \tag{3.33}$$

$\boldsymbol{B}^{(R)}$ and $\boldsymbol{B}^{(L)}$ as defined above are actually the projections of the corresponding electromagnetic field strength tensor (with six independent components) from Minkowski spacetime onto Euclidean 3-space. Note that under the inversion $\boldsymbol{e}_i \to -\boldsymbol{e}_i$ (in Euclidean 3-space) the above equations remain the same, so they are valid expressions for both right-handed and left-handed frames. The customary way of writing Newton's equation of motion when the force is a Lorentz force, namely,

$$\boldsymbol{F} = \frac{e}{c}\, \boldsymbol{v} \times \boldsymbol{B} \,, \tag{3.34}$$

with the understanding that the magnetic field $\boldsymbol{B}$ is *physically* of the right-handed kind and that the right-handed rule is to be used in the cross product, should then be re-expressed tensorially as

$$\boldsymbol{F} = \frac{e}{c} \left[*_R \left\{ \boldsymbol{v} \wedge \left(*_R \boldsymbol{B}^{(R)} \right) \right\} \right] , \tag{3.35a}$$

where $\boldsymbol{F}$ and $\boldsymbol{v}$ are vectors and $\boldsymbol{B}^{(R)}$ is an exterior 2-vector. Despite the profusion of brackets (which can admittedly be bothersome, but which can also be abbreviated with more familiarity with the notation) the equation leaves no doubt as to what kind of mathematical objects one is dealing with, and stands as a correct tensor equation, with the correct transformation properties under any orthogonal transformation of coordinate frames, *whether a right-handed frame or a left-handed frame is used to express the components on both sides.* Now, because of (3.31) and the property of the Hodge star operator (3.6), the above equation is equivalent to

$$\boldsymbol{F} = \frac{e}{c} \left[*_L \left\{ \boldsymbol{v} \wedge \left(*_L \boldsymbol{B}^{(R)} \right) \right\} \right] , \tag{3.35b}$$

or

$$\boldsymbol{F} = \frac{e}{c} \left[*_L \left\{ \boldsymbol{v} \wedge \left(*_R \boldsymbol{B}^{(L)} \right) \right\} \right] , \tag{3.35c}$$

or

$$\boldsymbol{F} = \frac{e}{c} \left[*_R \left\{ \boldsymbol{v} \wedge \left(*_L \boldsymbol{B}^{(L)} \right) \right\} \right] . \tag{3.35d}$$

Each of the above four equations represents mathematically the *same* tensor equation, and so would yield the same results, *regardless of whether a right-handed or left-handed coordinate frame is used when one wishes to express the equation in component form.* So there is a certain freedom in the choice of the orientation indices R or L, even if, physically, the magnetic field is of the right-handed kind. On the other hand, the equation

$$\boldsymbol{F} = \frac{e}{c}\left[*_R\left\{\boldsymbol{v}\wedge\left(*_R\boldsymbol{B}^{(L)}\right)\right\}\right], \tag{3.36}$$

for example, would yield the wrong result, since it would lead to a force vector opposite in direction to that given by the correct equation.

We summarize the procedure and rules for dealing with any equation of motion written traditionally in terms of 3-dimensional cross products and "axial vectors" as follows:

(1) Determine which quantities are traditionally regarded as "axial vectors", namely, those possessing orientational properties (right-handed or left-handed, such as the magnetic field), and determine whether *physically* it is the right-handed kind or left-handed kind we want to use. Express these quantities as exterior 2-vectors (antisymmetric rank-two tensors) with the proper orientation index R or L (right or left).

(2) Write the equation of motion as a *tensor equation* in terms of vectors, exterior 2-vectors, and Hodge stars $*_R$ and $*_L$ replacing cross products so that the result agrees with the original "vector" equation [such as (3.34)] when, in the latter, proper choices had been made for the handedness of the cross product and the "axial vector".

(3) Assign a value of $+1$ to each orientation index R and a value of -1 to each orientation index L in the tensor equation. Then the equation remains invariant (the same equation) under arbitrary switches of the orientation indices as long as the product of the values of these indices in each side remains unchanged.

(4) Either a right-handed or a left-handed frame can be used for the purpose of expressing the tensor equation in component form. In other words, *the "physics" cannot depend on the choice of frames for the description of a physical process.*

Chapter 4

Kinematics and Moving Frames: From the Angular Velocity to Gauge Fields

Kinematics is the combined framework of physical concepts and mathematical formalism for the description of the motion of physical systems in Euclidean 3-dimensional physical space, or some other types of spaces in the case of *constrained motions*, known loosely as *differentiable manifolds* mathematically, that may be embedded in physical space. Since all motion takes place *in* space, a proper appreciation and understanding of the geometrical properties of manifolds beyond elementary Euclidean geometry is indispensable in setting up the formalism and manipulating with it. This will be done in stages in this book. The present chapter will set up the basic framework, and introduce the reader immediately to a deep structure in the geometry of *fiber bundles*, known as the *connection*, that will be shown to be very relevant even in our elementary setting, namely, the development of the concept of *angular velocity*. Without going into any details at this point, we would like to point out that the geometric notion of connections on fiber bundles also happens to be of basic importance in a variety of physics topics, from *geometric phases* in many areas of classical and quantal physics to *gauge field theories* – the quantized version of which form the core of our current understanding of the fundamental forces of nature. These important linkages between differential geometry and physics will be elaborated on further later in this chapter and in later chapters. But our focus will of course be on classical mechanics.

The fundamental physical concepts in kinematics are *the trajectory* (or path) of a point particle in space, its *velocity*, and its *acceleration*. These are very simply expressed mathematically, using elementary concepts in geometry and analysis, as follows.

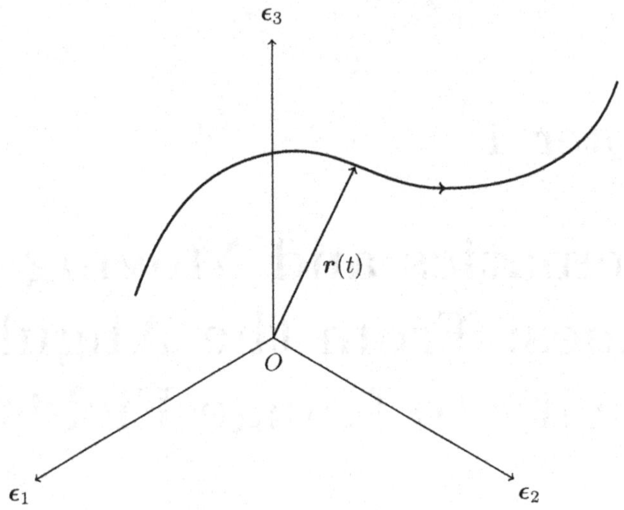

Fig. 4.1

Definition 4.1. *The* ***classical trajectory*** *of a point mass is a parametrized curve* $C : \mathbb{R} \to \mathbb{R}^3\,[\,t \mapsto \boldsymbol{r}(t)\,]$ *in physical space (Euclidean* $\mathbb{R}^3$*), for some interval* $a < t < b$ *in* $\mathbb{R}$*, where the parameter* t *is usually interpreted as the (physical) time and* $\boldsymbol{r}(t)$ *the* ***position vector*** *of the particle at the time* t *(see Fig. 4.1). Given a classical trajectory* $\boldsymbol{r}(t)$*, the (instantaneous)* ***velocity*** *at time* t*,* $\boldsymbol{v}(t)$*, and the (instantaneous)* ***acceleration*** *at time* t*,* $\boldsymbol{a}(t)$*, are defined to be*

$$\boldsymbol{v}(t) \equiv \frac{d\boldsymbol{r}}{dt} = \dot{\boldsymbol{r}}(t) \tag{4.1}$$

and

$$\boldsymbol{a}(t) \equiv \frac{d\boldsymbol{v}}{dt} = \frac{d^2\boldsymbol{r}}{dt^2} = \ddot{\boldsymbol{r}}(t)\,, \tag{4.2}$$

respectively. In the above equations, the single dot and double dot above a quantity mean the first and second derivatives with respect to time, respectively, of that quantity.

In computing the time derivatives in the above definition, it is assumed that a fixed basis $\{\epsilon_1, \epsilon_2, \epsilon_3\}$ in $\mathbb{R}^3$ has been chosen, usually referred to as a **space-fixed frame** in physics (see Fig. 4.1). The position vector, the velocity, and the acceleration in terms of components with respect to this frame can then be

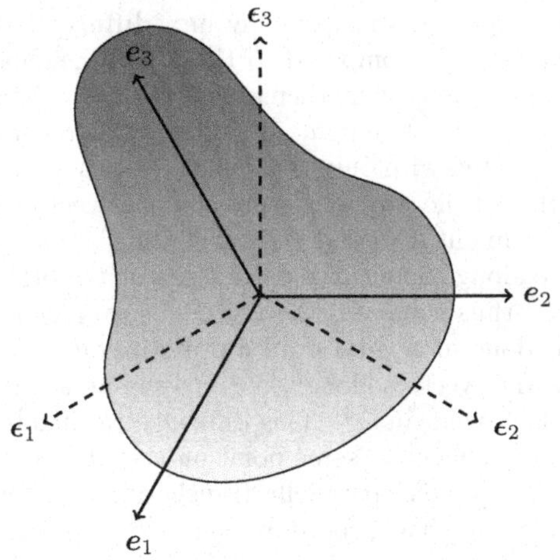

Fig. 4.2

written as

$$r(t) = x^i(t)\,\epsilon_i\ , \tag{4.3}$$

$$v(t) = v^i(t)\,\epsilon_i = \dot{x}^i\,\epsilon_i\ , \tag{4.4}$$

$$a(t) = a^i(t)\,\epsilon_i = \ddot{x}^i\,\epsilon_i\ . \tag{4.5}$$

To describe *rigid body motions*, in particular, rotations, it is often more convenient to use **body-fixed frames** (a body-fixed frame is one that is fixed with respect to the moving rigid body), usually referred to in mathematics as *moving frames* (see Fig. 4.2). From now on we will always denote body-fixed frames by $\{e_i\}$ and space-fixed frames by $\{\epsilon_i\}$. With respect to a moving frame, the position vector $r(t)$ of a point in the rigid body is given by

$$r(t) = x^i\,e_i\ , \tag{4.6}$$

where the components x^i remains constant in time as the rigid body moves in space. We then have formally, by (4.1),

$$v(t) = x^i\,\frac{de_i}{dt}\ . \tag{4.7}$$

The local geometry (but not the dynamics) of the rotational motion is apparently completely described by the information contained in the quantities de_i in the above equation. How is one to compute these "differentials" of basis vectors? In fact, what do these "differentials" mean? Naively, they are expected to tell us something about how the basis vectors of the moving frame $\{e_i\}$ at one point in physical space (more generally in a **differentiable manifold** in which a particle "lives") are compared to those at a neighboring point. But here we encounter the following mathematical difficulty: Vectors constituting the axes of a frame erected at one point in a differentiable manifold live in a linear vector space associated with only that point (known as the **tangent space** at the point), and there is no way a priori to compare vectors in tangent spaces at different points. It might be asked: Why not translate a vector at one point "parallelly" in space along some curve until it sits at the other point, and then add or subtract (and thus compare) the vectors as we please? The problem is that this cannot be done in general until a prescription is given on how to do **parallel translation** of vectors along some curve in an arbitrary differentiable manifold, not just in Euclidean $\mathbb{R}^3$. This difficulty can readily be appreciated if we imagine a vector tangent at some point on the surface of a 2-dimensional sphere S^2 and ask how we can parallelly translate the vector from that point to some neighboring point. The procedure seems to be obvious only if we view the translation from the vantage point of the 3-dimensional Euclidean space in which the 2-dimensional sphere is embedded. But then we are doing parallel translation in 3-dimensional Euclidean space, not in the 2-dimensional sphere. From the vantage point of an ant living on the surface of the sphere, who does not know that the sphere is embeddable in a higher dimensional space, it is not obvious at all how to tanslate the vector. Indeed, the notion of **parallelism** of vectors is not defined in an arbitrary differentiable manifold (of which the 3-dimensional Euclidean space in which a rigid body moves is only a very special example) until a structure known as an **affine connection** is imposed on the manifold. All these important differential geometrical notions – differential manifolds, tangent spaces, affine connections, parallel translations – and more, will be explored in more detail in subsequent chapters, beginning with the next one. For the moment, we will just state that *a given affine connection is encoded in the prescription for parallel translation in terms of the so-called* **covariant derivative**, denoted by D (as opposed to the symbol d used for ordinary differentials) in the following way:

$$\boxed{De_i = \omega_i^j e_j} \quad . \tag{4.8}$$

A **tangent vector** v is said to be *parallelly translated* (transported, or displaced) along a parametrized curve (parametrized by t) in the manifold if $Dv/dt = 0$ along the curve.

The very important equation (4.8) needs to be decoded in some detail. First we assume that we have made a *smooth* assignment of a frame $\{e_i\}$ at each point of the manifold in which the particle is moving (also called the **base manifold**). (For the exact definition of smoothness of a tangent vector field, see Def. 5.1.) This assignment is called a **frame field**. In general, *the covariant derivative is used to differentiate vector fields on a manifold*, and only makes sense when a connection is given. It is given *locally*, that is, within a particular coordinate neighborhood of the base manifold, by the **connection matrix of one-forms** (ω_i^j), or connection matrix of differential expressions. The notion of a *differential form*, in particular a 1-form, will be explained in more detail in Chapter 6. Here, as before, the subscript i will be the row index and the superscript j the column index, if the $\boldsymbol{e}_i$'s are interpreted as forming a column vector. (The reader should bear in mind, however, that elsewhere in this book, the other convention – superscript for row index and subscript for column index of a matrix – may be adopted instead; but equations with indices displayed explicitly are always unambiguous.) If the base manifold is n-dimensional, then the connection matrix is $n \times n$. It is important to realize that *a connection matrix is always given with respect to a particular frame field*, just as the matrix representation of a linear operator is always given with respect to a particular basis.

On the base manifold $\mathbb{R}^3$, (ω_i^j) is a 3×3 matrix of (differential) 1-forms. The velocity of a point $\boldsymbol{r} = x^i\,\boldsymbol{e}_i$ in the rigid body is then given by

$$\boldsymbol{v}(t) = \frac{D\boldsymbol{r}}{dt} = \frac{D}{dt}\,x^i\,\boldsymbol{e}_i = x^i\,\frac{D\boldsymbol{e}_i}{dt} = x^i\varphi_i^j\,\boldsymbol{e}_j\ , \tag{4.9}$$

where φ_i^j is a matrix of derivatives (with respect to time) defined by

$$\varphi_i^j \equiv \frac{\omega_i^j}{dt}\ . \tag{4.10}$$

Note that *with respect to a moving frame the velocity is more properly defined in terms of a covariant derivative rather than an ordinary derivative*, because one has to differentiate the frame fields $\boldsymbol{e}_i$ also, in addition to the components x^i. Hence we have written $\boldsymbol{v} = D\boldsymbol{r}/dt$ rather than $\boldsymbol{v} = d\boldsymbol{r}/dt$. On $\mathbb{R}^3$ (and only in a space of dimension 3), then, we can define a quantity ω^k with 3 components with respect to an orthonormal frame field $\{\boldsymbol{e}_i\}$ as follows:

$$\varphi_i^j \equiv \varepsilon_i{}^j{}_k\,\omega^k\ , \tag{4.11}$$

where $\varepsilon_i{}^j{}_k$ is the Levi-Civita tensor defined in (1.44). Then

$$\boldsymbol{v}(t) = x^i\varepsilon_i{}^j{}_k\omega^k\,\boldsymbol{e}_j = \varepsilon^j{}_{ki}\omega^k x^i\,\boldsymbol{e}_j\ . \tag{4.12}$$

By the definition of the cross product (1.45) this is recognized to be the familiar equation for $\boldsymbol{v}$ in terms of the angular velocity "vector" $\boldsymbol{\omega}$ with components ω^k:

$$\boxed{\boldsymbol{v} = \boldsymbol{\omega} \times \boldsymbol{r}}\ . \tag{4.13}$$

It is important to understand that, mathematically, $\boldsymbol{\omega} = (\omega^1, \omega^2, \omega^3)$ is not really a vector quantity (hence the quotes above), because it does not transform like one. As defined in terms of φ_i^j [(4.11)], which in turn is defined in terms of the connection matrix ω_i^j, it transforms like a vector only if the connection matrix transforms like a rank $(1,1)$-tensor. But *a connection matrix does not transform like a tensor* [according to the transformation rule of (1.32)] under a local change of frame fields. This can be demonstrated easily as follows. Suppose the frame $\{e_i\}$ is related to another $\{e_i'\}$ by (see Table 1.1)

$$e_i' = g_i^j \, e_j \,, \tag{4.14}$$

where (g_i^j) is an element in a matrix group of transformations. We will rewrite the above equation in matrix notation as

$$e' = g e \,, \tag{4.15}$$

where e and e' are understood to be column matrices of basis vectors and g is a square matrix of appropriate dimension. The connection matrix of one-forms is given with respect to the frames e and e' by the following matrix equations:

$$De = \omega e \,, \qquad De' = \omega' e' = \omega' g e \,. \tag{4.16}$$

On the other hand, by using a "product rule" for covariant differentiation (which will not be proved here), we have

$$De' = D(ge) = (dg)\, e + g\, De = (dg)\, e + g \omega e \,. \tag{4.17}$$

Thus

$$\omega' g = dg + g\omega \,, \tag{4.18}$$

which implies the following transformation rule for a connection matrix under a change of frames (known as a **gauge transformation** in physics).

$$\boxed{\omega' = (dg)\, g^{-1} + g\omega g^{-1}} \quad . \tag{4.19}$$

The presence of the first term on the right-hand side (in addition to the second one – a similarity transformation) makes ω non-tensorial [compare with (1.31)]. Note that since this term involves dg, it gives the effect of the dependence of the transformation matrix g on the particular location in configuration space. Equation (4.19) is an extremely important one in physics. Even though it may not be apparent at this point, it in fact gives the general transformation rule under a local change of frame fields (in *internal space*) for the so-called **gauge fields** describing the interactions between the fundamental constituents of matter – both for the *abelian* case, such as electromagnetic force, and the non-abelian case, such as the strong force. This fascinating link will be displayed more clearly in Chapter 22.

Suppose $\{e_i\}$ is an orthonormal frame. Then

$$e_i \cdot e_j = \delta_{ij} \,. \tag{4.20}$$

On covariant differentiation, we obtain

$$De_i \cdot e_j + e_i \cdot De_j = 0 \,. \tag{4.21}$$

It follows from (4.8) that

$$\omega_i^k\, e_k \cdot e_j + e_i \cdot \omega_j^l\, e_l = 0\,, \tag{4.22}$$

or

$$\omega_i^k\, \delta_{kj} + \omega_j^l\, \delta_{il} = 0\,. \tag{4.23}$$

Thus *the connection matrix* ω *with respect to an orthonormal frame field is antisymmetric:*

$$\omega_i^j + \omega_j^i = 0\,. \tag{4.24}$$

It follows from (4.11) that φ_i^j is also antisymmetric, and we can display it as a matrix in terms of the "components" of the angular velocity ω^i as follows:

$$(\varphi_i^j) = \begin{pmatrix} 0 & \omega^3 & -\omega^2 \\ -\omega^3 & 0 & \omega^1 \\ \omega^2 & -\omega^1 & 0 \end{pmatrix}\,. \tag{4.25}$$

We will now illustrate the use of the gauge transformation equation (4.19) by an elementary example: the calculation of the velocity and acceleration of a moving point in spherical polar coordinates, which by definition are coordinates with respect to moving frames. Suppose $g(\boldsymbol{x})$ is the space-dependent transformation matrix relating the body-fixed frame $\{\boldsymbol{e}_i\}$ to the space-fixed frame $\{\boldsymbol{\epsilon}_j\}$ at each point in space (Fig. 4.3). Then, in matrix notation, $\boldsymbol{e} = g\,\boldsymbol{\epsilon}$, and thus $\boldsymbol{\epsilon} = g^{-1}\,\boldsymbol{e}$. Since $D\boldsymbol{\epsilon} = 0$ ($\{\boldsymbol{\epsilon}_i\}$ is a space-fixed frame), we have

$$D\,\boldsymbol{e} = dg\cdot\boldsymbol{\epsilon} + g\,D\boldsymbol{\epsilon} = dg\cdot\boldsymbol{\epsilon} = (dg\cdot g^{-1})\,\boldsymbol{e}\,. \tag{4.26}$$

Comparison with (4.8) shows that, with respect to the moving frame, the connection matrix is given by

$$(\omega_i^j) = dg\cdot g^{-1}\,. \tag{4.27}$$

This result is also in accord with (4.19) if, in that equation, we interpret ω and ω' to be the connection matrices with respect to the space-fixed and moving frames, respectively.

Equation (4.27) will be used to calculate ω_i^j in terms of the spherical polar coordinates (r,θ,ϕ) in $\mathbb{R}^3$. The frame field ($\{\boldsymbol{e}_i\}$ as a function of spatial position) appropriate to this system of coordinates is shown in Fig. 4.4. We have, from the geometry of this figure,

$$\boldsymbol{e}_1 \equiv \boldsymbol{e}_r = (\sin\theta\cos\phi)\,\boldsymbol{\epsilon}_1 + (\sin\theta\sin\phi)\,\boldsymbol{\epsilon}_2 + \cos\theta\,\boldsymbol{\epsilon}_3\,, \tag{4.28}$$

$$\boldsymbol{e}_2 \equiv \boldsymbol{e}_\theta = (\cos\theta\cos\phi)\,\boldsymbol{\epsilon}_1 + (\cos\theta\sin\phi)\,\boldsymbol{\epsilon}_2 - \sin\theta\,\boldsymbol{\epsilon}_3\,, \tag{4.29}$$

$$\boldsymbol{e}_3 \equiv \boldsymbol{e}_\phi = -\sin\phi\,\boldsymbol{\epsilon}_1 + \cos\phi\,\boldsymbol{\epsilon}_2\,, \tag{4.30}$$

where $\boldsymbol{e}_r, \boldsymbol{e}_\theta, \boldsymbol{e}_\phi$ are three mutually orthogonal unit vectors along the radial, polar, and azimuthal directions, respectively, which originate from an arbitrary

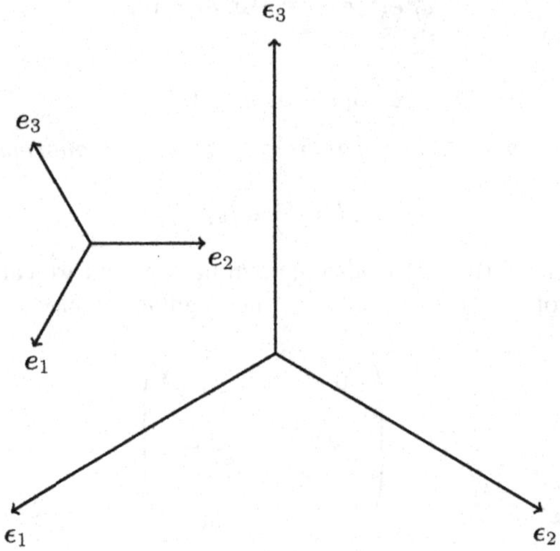
ϵ_3
e_3
e_2
e_1
ϵ_1
ϵ_2

Fig. 4.3

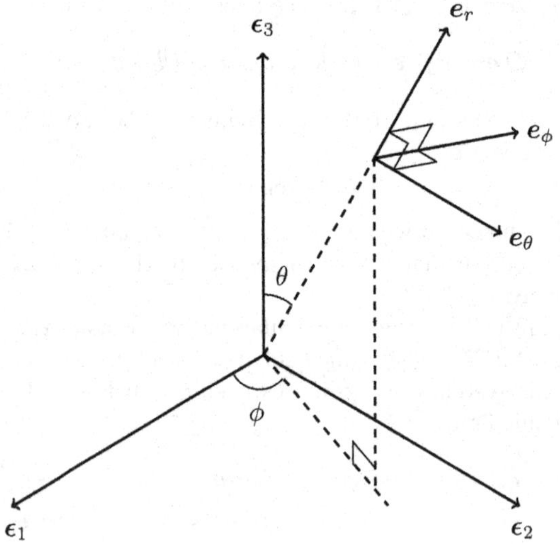
e_r
ϵ_3
e_ϕ
e_θ
θ
ϕ
ϵ_1
ϵ_2

Fig. 4.4

point $\boldsymbol{r}$ in Euclidean $\mathbb{R}^3$. They form a right-handed (oriented) frame at $\boldsymbol{r}$. Then the matrix g relating $\boldsymbol{e}$ to $\boldsymbol{\epsilon}$ is given by

$$g = \begin{pmatrix} \sin\theta\cos\phi & \sin\theta\sin\phi & \cos\theta \\ \cos\theta\cos\phi & \cos\theta\sin\phi & -\sin\theta \\ -\sin\phi & \cos\phi & 0 \end{pmatrix} . \tag{4.31}$$

It is easily verified that g as given above is an orthogonal matrix: $g^{-1} = g^T$. It follows from (4.27) that

$$(\omega_i^j) = \begin{pmatrix} 0 & d\theta & \sin\theta\, d\phi \\ -d\theta & 0 & \cos\theta\, d\phi \\ -\sin\theta\, d\phi & -\cos\theta\, d\phi & 0 \end{pmatrix} . \tag{4.32}$$

Equation (4.10) then gives

$$(\varphi_i^j) = \begin{pmatrix} 0 & \dfrac{d\theta}{dt} & \sin\theta\,\dfrac{d\phi}{dt} \\ -\dfrac{d\theta}{dt} & 0 & \cos\theta\,\dfrac{d\phi}{dt} \\ -\sin\theta\,\dfrac{d\phi}{dt} & -\cos\theta\,\dfrac{d\phi}{dt} & 0 \end{pmatrix} . \tag{4.33}$$

Notice that whereas the ω_i^j's are differential 1-forms, the φ_i^j's are ordinary derivatives. On comparison with (4.25), we have the following expressions for the components of the angular velocity in spherical coordinates:

$$\omega^1 = \cos\theta\,\dot{\phi}\,, \quad \omega^2 = -\sin\theta\,\dot{\phi}\,, \quad \omega^3 = \dot{\theta}\,. \tag{4.34}$$

Problem 4.1 Derive Eqs. (4.28) to (4.30) from the geometry shown in Fig. 4.4.

Problem 4.2 Verify that the transformation matrix g given by (4.31) is orthogonal.

Problem 4.3 Verify (4.32) by calculating the right-hand side of (4.27). Use the matrix for g given by (4.31).

Problem 4.4 For polar coordinates (r, θ) in the plane $\mathbb{R}^2$ the unit vectors of the moving frame are $\boldsymbol{e}_r$ and $\boldsymbol{e}_\theta$. Use (4.27), or simply by truncating the matrix in (4.32) (deleting the third row and third column), to show that $D\boldsymbol{e}_r = d\theta\,\boldsymbol{e}_\theta$ and

$De_\theta = -d\theta\, e_r$. Obtain the same results by expressing e_r and e_θ in terms of a space-fixed (rectangular) frame $\{\epsilon_1, \epsilon_2\}$ and directly computing De_r/dt and De_θ/dt.

We are now in a position to give general expressions for the velocity $\boldsymbol{v}$ and the acceleration $\boldsymbol{a}$ in terms of spherical coordinates in Euclidean $\mathbb{R}^3$. Writing

$$\boldsymbol{r} = r\, e_r \,, \tag{4.35}$$

we have [cf. (4.9)]

$$\boldsymbol{v} = \frac{D\boldsymbol{r}}{dt} = \dot{r}\, e_r + r\, \frac{De_r}{dt} = \dot{r}\, e_1 + r\, \frac{De_1}{dt} \,. \tag{4.36}$$

Dividing (4.8) by dt we obtain

$$\frac{De_r}{dt} = \frac{De_1}{dt} = \frac{\omega_1^2}{dt}\, e_2 + \frac{\omega_1^3}{dt}\, e_3 \,. \tag{4.37}$$

Using the explicit expressions for ω_i^j given by (4.32) we have

$$\boxed{\frac{De_r}{dt} = \dot{\theta}\, e_\theta + \dot{\phi}\sin\theta\, e_\phi} \,. \tag{4.38}$$

Equation (4.36) then gives

$$\boxed{\boldsymbol{v} = \dot{r}\, e_r + r\dot{\theta}\, e_\theta + r\dot{\phi}\sin\theta\, e_\phi} \,. \tag{4.39}$$

Note that the sum of last two terms on the right-hand side of the above equation is exactly equal to $\boldsymbol{\omega} \times \boldsymbol{r}$:

$$\boldsymbol{\omega} \times \boldsymbol{r} = \begin{vmatrix} e_r & e_\theta & e_\phi \\ \dot{\phi}\cos\theta & -\dot{\phi}\sin\theta & \dot{\theta} \\ r & 0 & 0 \end{vmatrix} = r\dot{\theta}\, e_\theta + r\dot{\phi}\sin\theta\, e_\phi \,. \tag{4.40}$$

This expression describes **tangential motion** (it gives the velocity tangent to the surface of a sphere), while the term involving $\dot{r}$ describes **radial motion**.

Differentiation of (4.39) with respect to time yields

$$\begin{aligned} \boldsymbol{a} = \frac{D\boldsymbol{v}}{dt} &= \ddot{r}\, e_r + \dot{r}\, \frac{De_r}{dt} + (\dot{r}\dot{\theta} + r\ddot{\theta})\, e_\theta + r\dot{\theta}\, \frac{De_\theta}{dt} \\ &\quad + (\dot{r}\dot{\phi}\sin\theta + r\dot{\theta}\dot{\phi}\cos\theta + r\ddot{\phi}\sin\theta)\, e_\phi + r\dot{\phi}\sin\theta\, \frac{De_\phi}{dt} \,. \end{aligned} \tag{4.41}$$

Similarly to (4.38), we can use (4.8) and (4.32) to obtain

$$\boxed{\frac{De_\theta}{dt} = -\dot{\theta}\, e_r + \dot{\phi}\cos\theta\, e_\phi\,, \qquad \frac{De_\phi}{dt} = -\dot{\phi}\sin\theta\, e_r - \dot{\phi}\cos\theta\, e_\theta} \quad . \tag{4.42}$$

Finally, the acceleration $\boldsymbol{a}$ is given in terms of spherical polar coordinates by

$$\begin{aligned} \boldsymbol{a} = \;& e_r\,(\ddot{r} - r\dot{\theta}^2 - r\dot{\phi}^2\sin^2\theta) \\ &+ e_\theta\,(2\dot{r}\dot{\theta} + r\ddot{\theta} - r\dot{\phi}^2\sin\theta\cos\theta) \\ &+ e_\phi\,(2\dot{r}\dot{\phi}\sin\theta + 2r\dot{\theta}\dot{\phi}\cos\theta + r\ddot{\phi}\sin\theta)\,. \end{aligned} \tag{4.43}$$

In the above equation the term $-e_r\,(r\dot{\theta}^2 + r\dot{\phi}^2\sin^2\theta)$ is called the **centripetal acceleration** (it is always directed along the radial direction and towards the origin of the coordinate system); while the sum of the terms proportional to $\dot{r}$ is called the **Coriolis acceleration**. The sum of the rest of the terms, excluding the one proportional to $\ddot{r}$, is called the **tangential acceleration**.

Problem 4.5 Verify (4.43) for the acceleration in spherical polar coordinates.

Chapter 5

Differentiable Manifolds: The Tangent and Cotangent Bundles

In an idealized fashion we can treat our everyday motions on the surface of the earth as motions restricted, or *constrained*, to a two-dimensional spherical surface S^2 embedded in 3-dimensional Euclidean (physical) space. In a similar fashion one can conceive of the motion of a particle on any other two-dimensional surface, for example, a torus, embedded in 3-dimensional space. Such motions are called constrained motions in general. We may not be aware of the details of the forces of constraint that keep the particle on the surface, and in many cases they may be impossible to determine, but we will see later how the dynamical effects of these forces can be encoded in the geometrical properties of the surface, more specifically, in the curvature of the surface. These situations prompt us to engage in a more detailed study of the mathematics of such surfaces, or their more general counterparts in higher dimensions, *differentiable (or smooth) manifolds.* We will refrain from treating this very basic topic in any complete or mathematically rigorous manner, and instead appeal to an intuitive understanding whenever possible (the interested reader may acquire more details as required from many basic texts on Differential Geometry – see, for example, Chern, Chen and Lam 1999). In this chapter we will focus on two main mathematical constructions: the tangent and cotangent bundles associated with a differentiable manifold. These will prove to be of great use in our later developments.

A finite-dimensional **differentiable manifold** M of dimension n looks locally like a neighborhood of a finite-dimensional Euclidean space (of the same dimension), so that locally, a point $x \in M$ can be characterized by a set of n **local coordinates** $\{x^1, \ldots, x^n\}$. Such a characterization, however, cannot in general be applied globally, and different "patches" of Euclidean space are required to cover a manifold. This is already evident from the example of the

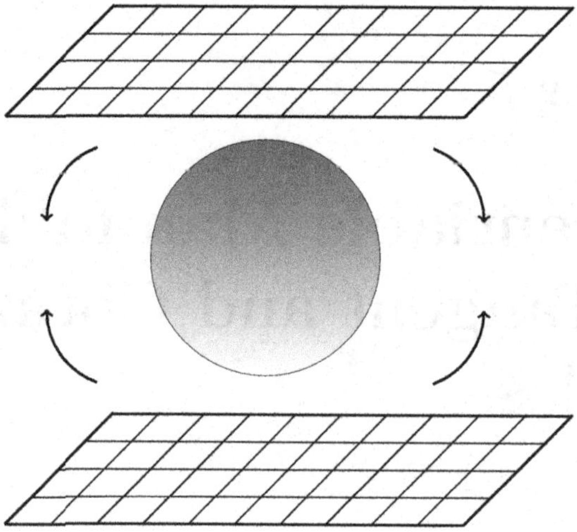

Fig. 5.1

2-sphere S^2 mentioned above, where a single set of two coordinates (latitude and longitude, for example), cannot be used throughout the whole surface. One requires at least two patches, each *homeomorphic* (topologically equivalent) to an open region of the Euclidean plane $\mathbb{R}^2$, one covering the northern hemisphere, and the other the southern hemisphere, for example (see Fig. 5.1). In obvious reference to map-making, we speak of a differentiable manifold as being covered by **coordinate charts**, each of which, denoted by (U, u^i), consists of a neighborhood $U \subset M$ and a set of local coordinates $u^1, \dots, u^n$ valid only in U. If two charts (U, u^i) and (W, w^i) are such that U and W overlap, that is, if $U \cap W \neq \emptyset$, then all the w's must be *smooth* (infinitely many times differentiable) functions of all the u's and vice versa. We then say that the two coordinate charts are **compatible** with each other. Thus one and the same point in M may have different coordinates for its description. A set of mutually compatible charts $(U, u^i), (V, v^i), (W, w^i), \dots$ such that $\{U, V, W, \dots\}$ covers M is called an **atlas** of the differentiable manifold M, and uniquely determines a so-called **differentiable structure** of the manifold.

We will now define the concept of a *tangent vector* at a point $x \in M$. This can be most intuitively done using generalizations of the kinematic concepts introduced in the last chapter. Consider a classical trajectory of a point particle (a parametrized curve) $\gamma : \mathbb{R} \to M$ [also written as $\gamma(t)$, where t is the

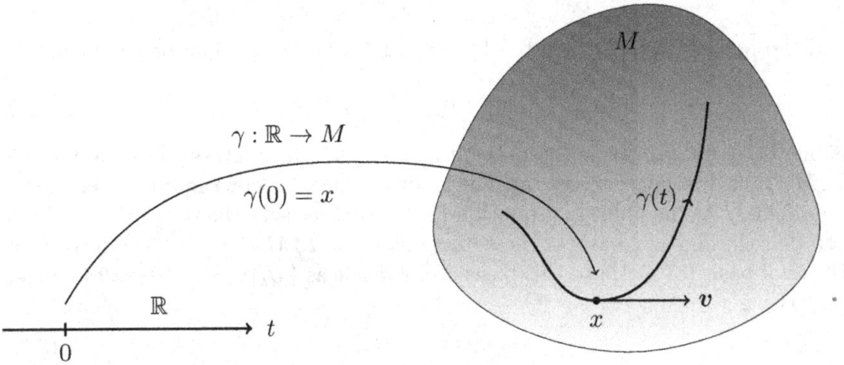

Fig. 5.2

parameter] passing through x such that $\gamma(0) = x$ (see Fig. 5.2). The **tangent vector** $\boldsymbol{v}$ to the curve $\gamma(t)$ at the point x is the contravariant vector whose i-th component v^i describes how "fast" (when t is interpreted as the time) the i-th coordinate of the point x, x^i, changes along the curve $\gamma(t)$ at $t = 0$:

$$v^i \equiv \frac{d}{dt}(x^i \circ \gamma)\bigg|_{t=0} = \frac{d}{dt}x^i(t)\bigg|_{t=0} = \dot{x}(0) , \tag{5.1}$$

where $\dot{x}(t)$ means differentiation of x with respect to the parameter t. On comparison with (4.4) this is seen to be the generalization of the velocity vector (to the case of motion in an arbitrary differentiable manifold instead of Euclidean 3-space) that we have introduced in the last chapter.

The set of all tangent vectors at a point $x \in M$ constitutes a linear vector space T_xM, called the **tangent space** at the point x. Now consider a function $f : M \to \mathbb{R}$. We may ask: How "fast" does f change along a tangent vector $\boldsymbol{v}$ at x? The answer is

$$\frac{\partial f}{\partial x^i} \frac{dx^i(t)}{dt}\bigg|_{t=0} = (\partial_i f)\, v^i , \tag{5.2}$$

where we have written $\partial_i f$ for $\partial f/\partial x^i$ to highlight the tensorial nature of the quantity, and used the Einstein summation convention. This quantity is called the **directional derivative** of the function f along the tangent vector $\boldsymbol{v}$ at x, and is usually written as $\boldsymbol{v}(f)$. The right-hand side of the above equation suggests that the directional derivative of f is the pairing between a contravariant vector with components v^i and a covariant vector with components $\partial_i f$. In fact, this covariant vector is the ordinary differential df of the function f:

$$df = \frac{\partial f}{\partial x^1}\, dx^1 + \cdots + \frac{\partial f}{\partial x^n}\, dx^n = \partial_i f\, dx^i \,, \tag{5.3}$$

and the pairing in (5.2), denoted by [cf. (2.2) and the explanation following]

$$\langle \boldsymbol{v} \,,\, df \rangle = \langle df \,,\, \boldsymbol{v} \rangle = (\partial_i f)\, v^i = \boldsymbol{v}(f) \,, \tag{5.4}$$

is one between vectors in the tangent space $T_x M$ and vectors in its dual space $T_x^* M$. This dual space to the tangent space is called the **cotangent space** of the manifold M at the point x. Equation (5.3) indicates that the set $\{dx^1, \ldots, dx^n\}$ can be regarded as a basis of the cotangent space $T_x^* M$. We can then introduce the dual basis (of $T_x M$) to this basis and write it as $\{\partial/\partial x^1, \ldots, \partial/\partial x^n\}$, since, on writing

$$\boldsymbol{v} = v^i \frac{\partial}{\partial x^i} = v^i\, \partial_i \quad , \tag{5.5}$$

and using this expression on the left-hand side of (5.4) we obtain the duality requirement

$$\boxed{\left\langle \frac{\partial}{\partial x^i} \,,\, dx^j \right\rangle = \frac{\partial x^j}{\partial x^i} = \delta_i^j} \quad . \tag{5.6}$$

For obvious reasons, $\{\partial_i\}$ and $\{dx^j\}$ are called the **coordinate bases** (or **natural bases**) of $T_x M$ and $T_x^* M$, respectively. The fact that the coordinate basis vectors in $T_x M$ are expressed in terms of differential operators may appear strange at first. But this is perfectly reasonable if we remember that $T_x M$, being dual to $T_x^* M$, is in fact the space of linear functions on $T_x^* M$ (see Chapter 1). Thus $\dfrac{\partial}{\partial x^i} \in T_x M$ acts on $\mathrm{d}f \in T_x^* M$ to yield the scalar $\dfrac{\partial f}{\partial x^i}$.

Under a general coordinate transformation

$$x^i \longrightarrow x'^i = x'^i(x^1, \ldots, x^n) \,, \quad i = 1, \ldots, n \,,$$

(which in general is not linear) we have, by the chain rule,

$$\partial_i = \frac{\partial}{\partial x^i} = \frac{\partial}{\partial x'^j} \frac{\partial x'^j}{\partial x^i} = \frac{\partial x'^j}{\partial x^i}\, \partial'_j \tag{5.7}$$

and

$$\mathrm{d}x^i = \frac{\partial x^i}{\partial x'^j}\, \mathrm{d}x'^j \quad . \tag{5.8}$$

We can rewrite the above two equations as

$$\partial_i = a_i^j\, \partial'_j \,, \qquad \mathrm{d}x^i = (a^{-1})_j^i\, \mathrm{d}x'^j \quad , \tag{5.9}$$

where

$$a_i^j \equiv \frac{\partial x'^j}{\partial x^i} \quad , \tag{5.10}$$

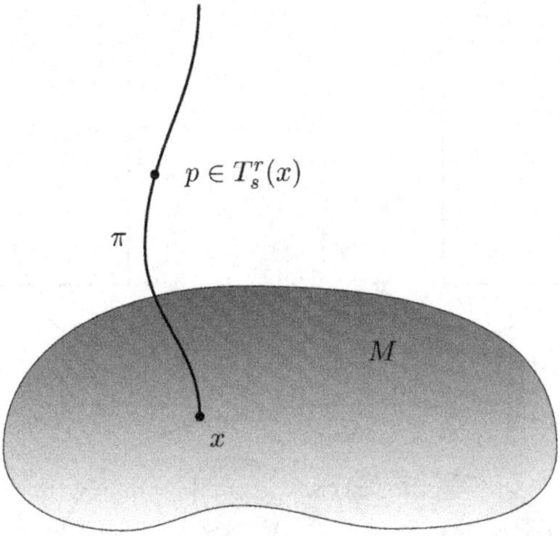

Fig. 5.3

is the **Jacobian matrix** for a local change of coordinates. The transformation rules (5.9) are in conformity with those given in Table 1.1 in Chapter 1 for the covariant (lower indexed) objects $\boldsymbol{e}_i$ and contravariant (upper indexed) objects $(\boldsymbol{e}^*)^j$, respectively.

Based on the notion of a tangent space T_xM and a cotangent space T^*_xM at a particular point $x \in M$, we can construct the (r, s)-type tensor space [cf. (1.39)]

$$T^r_s(x) = \underbrace{T_x \otimes \cdots \otimes T_x}_{\text{r times}} \otimes \underbrace{T^*_x \otimes \cdots \otimes T^*_x}_{\text{s times}} \tag{5.11}$$

at $x \in M$, and then the (r, s)-type **tensor bundle**

$$T^r_s = \bigcup_{x \in M} T^r_s(x) \quad . \tag{5.12}$$

Loosely speaking, T^r_s is a "bundling together" of (r, s)-type tensor spaces at all points of the manifold M. Each $T^r_s(x)$ is called the **fiber** of the bundle T^r_s at the point x. The natural projection

$$\pi : T^r_s \longrightarrow M \ , \tag{5.13}$$

which maps all points in $T^r_s(x)$ to the point $x \in M$, is a smooth surjective map. It is called the **bundle projection** (Fig. 5.3).

The construction of the tensor bundle T^r_s involves more than forming a union of sets [as given by (5.12)] and requires some additional structure. A detailed

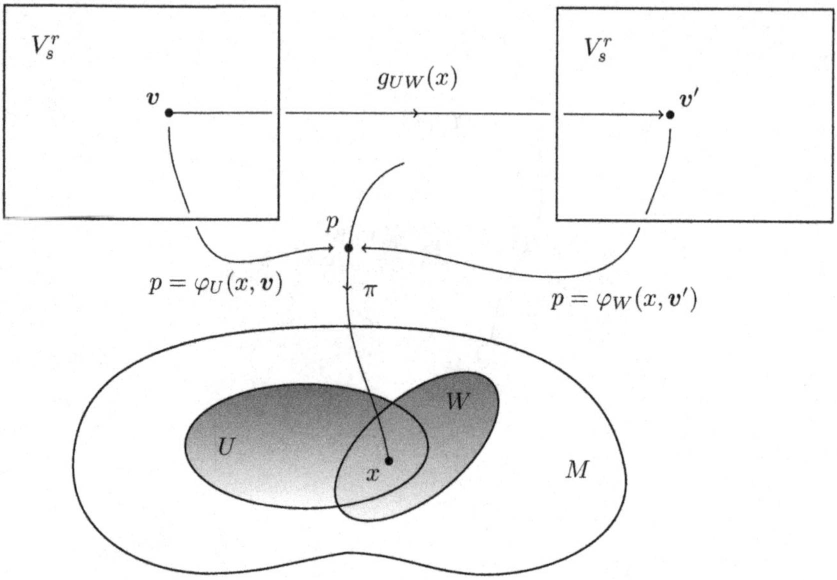

Fig. 5.4

treatment of the procedure is beyond the scope of this book (see Chern, Chen and Lam 1999, for example), and we will only provide a very brief and intuitive description here. It consists mainly of the following two steps (refer to Fig. 5.4):

1) specify a topological structure and a differentiable structure for T_s^r by local charts φ_U such that *locally the tensor bundle is a product*:

$$\varphi_U : U \times V_s^r \longrightarrow \bigcup_{x \in U} T_s^r(x) \quad ; \tag{5.14}$$

where (U, u^i) is a chart of M and V_s^r is a real linear vector space of the same dimension as $T_s^r(x)$;

2) specify a way whereby fibers "sticking out" of points in overlapping charts of M are "glued" together so that *the linear structure of the fiber is preserved.*

Step 2) above is understood with reference to Fig. 5.4 as follows. The vector space V_s^r is called the **typical fiber** of the tensor bundle T_s^r. This means that each fiber "above" the point x, that is, the set $\pi^{-1}(x) \subset T^r_s$, is isomorphic to V^r_s. The functions

$$g_{UW} : U \cap W \longrightarrow GL(V^r_s) \,, \tag{5.15}$$

where $GL(V)$ is the **general linear group** on V (the group of all linear transformations on the vector space V), called the **transition functions** ("glueing" functions) of the tensor bundle, are defined for any $\boldsymbol{v} \in V^r_s$ by (see Fig. 5.4)

$$[g_{UW}(x)](\boldsymbol{v}) \equiv (\varphi_W)^{-1}(\varphi_U(x, \boldsymbol{v})) \,. \tag{5.16}$$

The transition functions determine the so-called **bundle structure**. Preservation of the linear structure of the fiber (by the "glueing" process) means that the value of a transition function at each point x is a linear map, as can be verified readily from the above definition.

Among the T^r_s (for different values of r and s) we have the **tangent bundle** T^1_0 [also denoted $T(M)$] and the **cotangent bundle** T^0_1 [also denoted $T^*(M)$]. If $dim\,(M) = n$, then $dim\, T(M) = dim\, T^*(N) = 2n$. Furthermore, we can construct the so-called **exterior vector bundles** and **exterior form bundles** on M, given respectively by

$$\Lambda^r(M) \equiv \bigcup_{x \in M} \Lambda^r(T_x M) \quad \text{and} \quad \Lambda^r(M^*) \equiv \bigcup_{x \in M} \Lambda^r(T^*_x M) \,, \tag{5.17}$$

$T^r_s, \Lambda^r(M)$ and $\Lambda^r(M^*)$ are special cases of **vector bundles** (fiber bundles whose fibers are vector spaces). These are usually denoted $\pi : E \longrightarrow M$, where π is the natural projection [cf. (5.13)], E (the bundle itself) is called the **total space** (or **bundle space**), and M is called the **base manifold**.

Note that the total space E of a fiber bundle should be thought of as a differentiable manifold in its own right (with its own differentiable structure), and not necessarily as a product of the base manifold and the fiber. In fact, a fiber bundle is only locally a product [see (5.14)]; and in general cannot be represented as such globally. A bundle which is globally a product is called a **trivial bundle**, and much of the theory of fiber bundles is devoted to characterizing various kinds of non-trivialities. Non-trivial bundles play a most important role in modern theoretical physics: they form the mathematical basis for the description of many interesting physical objects, such as magnetic monopoles, instantons, and even molecular collision systems.

A smooth map $s : M \longrightarrow E$ such that

$$\pi \circ s = id : M \longrightarrow M \quad , \tag{5.18}$$

where id is the identity map, is called a **smooth section** of the vector bundle $\pi : E \longrightarrow M$. The set of all smooth sections of $\pi : E \longrightarrow M$, denoted $\Gamma(E)$, is an important vector space. It is also a $C^\infty(M)$ **module**, that is, a module over the ring of smooth functions on M. (Roughly, this algebraic concept means that two sections in $\Gamma(E)$ can be added together to yield another section in $\Gamma(E)$, a section can be multiplied by a smooth function to yield another section, and the usual distributive and associative rules for the addition and scalar multiplication are obeyed.) A section of a vector bundle $\pi : E \longrightarrow M$ is also called a **vector field** on the manifold M in the physics literature. Finally, a *smooth* assignment of a basis set in each tangent space $T_x(M)$ (for each $x \in M$) is called a **frame field** on the differentiable manifold M. We have used this last concept repeatedly under the guise of moving frames in the previous chapter in the special case when $M = \mathbb{R}^3$. In this special case, a single coordinate chart $(U = \mathbb{R}^3; x^1, x^2, x^3)$ suffices and the differentiable structure is quite simple. Also, for $M = \mathbb{R}^n$ more generally, the same linear structure can be imposed on the base manifold as each tangent space, even though *in general a differentiable manifold does not have a linear structure.*

The tangent and cotangent bundles associated with configuration spaces (as base manifolds) are the primary mathematical ingredients in the formulation of Lagrangian and Hamiltonian mechanics, respectively (see Chapters 31 and 33).

We will end this chapter with an elaboration of the notion of smoothness of a tangent vector field beyond the intuitive one we have been relying on so far, and hence by extension the notion of smoothness of any frame field. Consider any **tangent vector field** $\boldsymbol{X}$ on a smooth manifold M, or an assignment of a tangent vector at each point $x \in M$, denoted by $\boldsymbol{X}_x$. By the definition of a tangent vector (5.2), $\boldsymbol{X}_x$ acts on a smooth function on M, $f \in C^\infty(M)$, to yield the directional derivative of f at x, a real number. Hence it can be considered as a function (map) $\boldsymbol{X}_x : C^\infty(M) \to \mathbb{R}$. Now, corresponding to a smooth function f, we can associate another function, denoted by $\boldsymbol{X}f$, defined by

$$(\boldsymbol{X}f)(x) = \boldsymbol{X}_x f\ . \tag{5.19}$$

In other words, a tangent vector field $\boldsymbol{X}$ can be considered as a map from the set of smooth functions on M to the whole set of functions on M, not necessarily smooth. The technical definition of smoothness of a tangent vector field is then given as follows.

Definition 5.1. *Suppose $\boldsymbol{X}$ is a tangent vector field on a smooth manifold M. If, for any smooth function $f \in C^\infty(M)$, $\boldsymbol{X}f$ is also a smooth function on M, that is, $\boldsymbol{X}f \in C^\infty(M)$, then $\boldsymbol{X}$ is said to be a* **smooth tangent vector field** *on M.*

Note that this definition of smoothness of a vector field does not rely on the operation of taking the difference of the values (as vectors) of the vector field at two neighboring points in M, as one normally expects to do when the existence of derivatives (and hence smoothness) is required to be established. As we stressed in Chapter 4 [see discussion surrounding (4.8)], this procedure is undefined until the added structure of an *affine connection* (or a rule for taking *covariant derivatives*, or equivalently, a rule for *parallelly translating* vectors) has been imposed on the tangent bundle. This subtlety is not apparent in Euclidean $\mathbb{R}^3$, since in that case the so-called *metric compatible connection* [the particular affine connection that is compatible (in a sense to be defined later in Chapter 27) with the Euclidean metric] can always be set equal to zero locally, which in turn leads to zero *curvature*. On the other hand, there is another kind of derivative of a vector field in differential geometry relevant in the study of fluid flows, called the *Lie derivative*, which does not require a connection. This type of derivatives will be studied in a later chapter (Chapter 11).

There are two useful theorems concerning smooth tangent vector fields which we will state without proofs.

Theorem 5.1. *A tangent vector field* $\boldsymbol{X}$ *on a smooth manifold* n*-dimensional* M *is smooth if and only if, for any point* $x \in M$*, there exists a local coordinate system* $(U; x^1, \dots, x^n)$ *such that the restriction of* $\boldsymbol{X}$ *on* U*,* $\boldsymbol{X}|_U$*, can be expressed as*

$$\boldsymbol{X}|_U = \xi^i(x^1, \dots, x^n)\,\frac{\partial}{\partial x^i}\,, \tag{5.20}$$

where ξ^i*,* $1 \le i \le n$ *are smooth functions on* M*.*

Theorem 5.2. *Let* $\boldsymbol{X}$ *be a smooth tangent vector field on a smooth manifold* M*. If* $\boldsymbol{X}_y \neq \boldsymbol{0}$ *at a point* $y \in M$*, then there exists a local coordinate system* $(W; y^i)$ *such that*

$$\boldsymbol{X}|_W = \frac{\partial}{\partial y^1}\,. \tag{5.21}$$

Theorem 5.2 follows from Theorem 5.1. We will see later (Chapter 10) that it has important consequences for the integrability conditions for submanifolds embodied in the Frobenius Theorem.

Problem 5.1 Prove that the transition functions $g_{UW}(x)$ of a vector bundle as defined in (5.16) are linear maps.

Chapter 6

Exterior Calculus: Differential Forms

The exterior calculus, or the calculus of exterior differential forms, is an extremely basic, important, and powerful analytical tool that permeates many areas in differential geometry and differential topology; and given the relevance of these subjects in theoretical physics, it finds wide applications in physics also. In fact, we will show in the next chapter that, to begin with, it subsumes conventional vector calculus in three dimensions and generalizes it to an arbitrary number of dimensions. Its importance in classical mechanics derives from the fact that the latter, being the study of motions in arbitrary differential manifolds, is intimately connected with the geometry of manifolds. For an excellent introductory survey on the physical and mathematical applications of differential forms the reader can consult, for example, the classical text by Flanders (Flanders 1963).

To introduce the notion of differential forms we start with an n-dimensional differentiable manifold M, and construct the tensor bundle of exterior r-forms on M [cf. (5.17)]:

$$\Lambda^r(M^*) = \bigcup_{x \in M} \Lambda^r(T_x^* M) \,. \tag{6.1}$$

This is a vector bundle on M [the typical fiber, being isomorphic to $\Lambda^r(T_x^* M)$, has the structure of a vector space] and so it is meaningful to speak of the vector space of smooth sections on it. We denote this vector space by $A^r(M)$ ("A" for antisymmetric):

$$A^r(M) \equiv \Gamma(\Lambda^r(M^*)) \quad . \tag{6.2}$$

The above construction is summarized below for easy reference:

$$M \xrightarrow[\text{at } x \in M]{\text{cotangent space}} T_x^*(M) \xrightarrow[\text{exterior } r\text{-forms}]{\text{bundle of}} \Lambda^r(M^*) \xrightarrow[\text{on } \Lambda^r(M^*)]{\text{smooth sections}} A^r(M) \,.$$

As mentioned earlier, $A^r(M)$ is also a $C^\infty(M)$ module [elements in it can be multiplied by C^∞ (smooth) functions on M to yield other elements]. We can now give the definition of exterior differential forms.

Definition 6.1. *The elements of the $C^\infty(M)$ module $A^r(M)$ are called* ***exterior differential** r-**forms*** *on M. In other words, an exterior differential r-form on M is a smooth antisymmetric covariant tensor field of order r [a $(0,r)$-type tensor field] on M.*

Let us immediately illustrate this with a very simple example. Suppose $M = \mathbb{R}^3$ with local coordinates x, y, z for some point $p \in M$. Then the natural bases for $\Lambda^1(T_p^*M) = T_p^*M$, $\Lambda^2(T_p^*M)$ and $\Lambda^3(T_p^*M)$ would be $\{dx, dy, dz\}$, $\{dx\wedge dy, dy\wedge dz, dx\wedge dz\}$, and $\{dx\wedge dy\wedge dz\}$, respectively [cf. (2.26)]. General expressions for exterior 1-forms, 2-forms and 3-forms on M, denoted below by $\omega^1, \omega^2, \omega^3$, respectively, would then be written as

$$\omega^1 = f_1(x,y,z)\,dx + f_2(x,y,z)\,dy + f_3(x,y,z)\,dz\ ,$$
$$\omega^2 = g_1(x,y,z)\,dx\wedge dy + g_2(x,y,z)\,dy\wedge dz + g_3(x,y,z)\,dx\wedge dz\ ,$$
$$\omega^3 = h(x,y,z)\,dx\wedge dy\wedge dz\ ,$$

where $f_i(x,y,z), g_i(x,y,z)\,(i = 1,2,3)$, and $h(x,y,z)$ are all smooth functions on M. Note that the quantities $dx, dx\wedge dy$, etc., in the above equations are to be interpreted as vector fields (as the point p in the base manifold M varies over the neighborhood with the local coordinates x, y, z).

From $A^r(M)$ [$dim\,(M) = n$] we can construct the direct sum

$$A(M) = \oplus\sum_{r=0}^{n} A^r(M) = A^0(M)\oplus\cdots\oplus A^n(M)\quad , \tag{6.3}$$

which is in fact the space of all smooth sections of the exterior form bundle [cf. (2.42)]

$$\Lambda(M^*) = \bigcup_{x\in M} \Lambda(T_x^*M)\quad . \tag{6.4}$$

Elements of $A(M)$ are called **exterior differential forms** on M. Thus every exterior differential form $\omega \in A(M)$ can be written

$$\omega = \omega^0 + \omega^1 + \cdots + \omega^n\quad , \tag{6.5}$$

where $\omega^r \in A^r(M)$, $0 \le r \le n$. (Technically a differential form need not be an exterior form. The term "exterior" refers to the antisymmetry of the form in question. When no confusion arises, or when it is clear from the context, we will frequently from now on speak of just differential forms instead of exterior differential forms.)

We can impose a multiplication rule on $A(M)$ using the exterior (wedge) product for exterior vectors (or exterior forms) introduced in Chapter 2 (see Def .2.2). The wedge product at a single point $x \in M$ can be easily extended to the space of exterior differential forms $A(M)$ by a pointwise association: for any $x \in M$, $\omega_1, \omega_2 \in A(M)$, define

$$(\omega_1 \wedge \omega_2)(x) = \omega_1(x) \wedge \omega_2(x) \quad , \tag{6.6}$$

where the right-hand side is a wedge product of two exterior forms. With this multiplication rule, the space $A(M)$ then becomes a **graded algebra** with respect to addition, scalar multiplication, and the wedge product. The wedge product defines a map

$$\wedge : A^r(M^*) \times A^s(M^*) \longrightarrow A^{r+s}(M^*) \quad , \tag{6.7}$$

where $A^{r+s}(M^*) = \{\mathbf{0}\}$ when $r + s > n$. Let (U, x^i) be a chart of M. An exterior differential r-form ω can be expressed *locally* in the neighborhood U in terms of the local coordinates $x^1, \dots, x^n$ of M as

$$\omega = a_{i_1 \dots i_r}(x^1, \dots, x^r)\, \mathrm{d}x^{i_1} \wedge \dots \wedge \mathrm{d}x^{i_r} \quad , \tag{6.8}$$

where each of the $a_{i_1 \dots i_r}$ is a smooth function on U (i.e., a smooth function of the local coordinates $x^1, \dots, x^n$), and is required to be antisymmetric with respect to any permutation of its indices [cf. Def. 2.1(a)].

Since the fibers of $\Lambda^r(M)$ and $\Lambda^r(M^*)$ at a point $x \in M$ are dual spaces to each other, we can introduce a pairing between them according to (2.30). In terms of the natural bases $\left\{ \frac{\partial}{\partial x^{i_1}} \wedge \dots \wedge \frac{\partial}{\partial x^{i_r}} \right\}$, $i_1 \leq \dots \leq i_r$, of $\Lambda^r(T_x)$ and $\{\mathrm{d}x^{j_1} \wedge \dots \wedge \mathrm{d}x^{j_r}\}$, $j_1 \leq \dots \leq j_r$, of $\Lambda^r(T^*_x)$, we have

$$\boxed{\left\langle \frac{\partial}{\partial x^{i_1}} \wedge \dots \wedge \frac{\partial}{\partial x^{i_r}},\, \mathrm{d}x^{j_1} \wedge \dots \wedge \mathrm{d}x^{j_r} \right\rangle = \delta^{j_1 \dots j_r}_{i_1 \dots i_r}} \quad , \tag{6.9}$$

where the Kronecker delta symbol has been defined in (2.38). Thus the components of ω in terms of local coordinates can be expressed as

$$a_{i_1 \dots i_r} = \frac{1}{r!} \left\langle \frac{\partial}{\partial x^{i_1}} \wedge \dots \wedge \frac{\partial}{\partial x^{i_r}},\, \omega \right\rangle \quad . \tag{6.10}$$

(The reader should verify the above statement.)

The space $A(M)$ of exterior differential forms plays a crucial role in the study of differentiable manifolds, due to the existence of a most useful operator, the so-called exterior derivative operator d on $A(M)$, with the property that $d^2 = 0$. We state without proof the following theorem.

Theorem 6.1. *Let M be an n-dimensional smooth manifold. Then there exists a unique map*

$$d : A(M) \longrightarrow A(M) , \quad d(A^r(M)) \subset A^{r+1}(M) ,$$

the ***exterior derivative****, which satisfies the following properties:*

1) For any $\omega_1, \omega_2 \in A(M)$, $d(\omega_1 + \omega_2) = d\omega_1 + d\omega_2$.

2) Suppose $\omega_1 \in A^r(M)$. *Then, for any* $\omega_2 \in A(M)$,

$$\boxed{d(\omega_1 \wedge \omega_2) = d\omega_1 \wedge \omega_2 + (-1)^r \, \omega_1 \wedge d\omega_2} \quad . \tag{6.11}$$

Property 2) [Eq. (6.11)] is the "product rule" of exterior differentiation. We note that d is a *local operator*, which means that, if $\omega_1, \omega_2 \in A(M)$ and their restrictions to a neighborhood U of M are equal ($\omega_1|_U = \omega_2|_U$), then $d\omega_1|_U = d\omega_2|_U$.

Problem 6.1 Let $\omega \in A^r(M)$ be given by (6.8). Use Property 2) of Theorem 6.1 to show that

$$d\omega = (da_{i_1 \dots i_r}) \wedge dx^{i_1} \wedge \dots \wedge dx^{i_r} \; . \tag{6.12}$$

Problem 6.2 Use (6.12) to verify Property 2) of Theorem 6.1 when both $\omega_1 \in A^r(M)$ and $\omega_2 \in A^s(M)$ are monomials, that is, when

$$\omega_1 = a \, dx^{i_1} \wedge \dots \wedge dx^{i_r} \; , \qquad \omega_2 = b \, dx^{i_1} \wedge \dots \wedge dx^{i_s} \; ,$$

where a and b are both smooth functions on M. The case for non-monomials are taken care of by the linearity property of the exterior derivative [Property 1) of Theorem 6.1].

Problem 6.3 Use (6.12) to show that, for $f \in A^0(M)$, $d^2 f = 0$.

One of the most important (and useful) properties of the exterior derivative is the following result:

Theorem 6.2. *The square of the exterior derivative* d *vanishes identically: for every differential form* $\omega \in A(M)$, $d^2\,\omega = d(d\omega) = 0$.

Proof. We need only prove the theorem for a monomial $\omega \in A^r(M)$ since d is linear by property 1) of Theorem 6.1. Also, since d is a local operator, we need only consider

$$\omega = a \, dx^1 \wedge \dots \wedge dx^r \; .$$

By (6.12),

$$d\omega = da \wedge dx^1 \wedge \dots \wedge dx^r \; .$$

Differentiating once more and applying property 2) of Theorem 6.1 and the result of Problem 6.3 (namely, that $d^2 f = \mathbf{0}$ for all smooth functions on M), we have

$$\begin{aligned} d(d\omega) &= d(da) \wedge dx^1 \wedge \cdots \wedge dx^r - da \wedge d(dx^1 \wedge \cdots \wedge dx^r) \\ &= -da \wedge d(dx^1) \wedge \cdots \wedge dx^r + \ldots = 0 \,, \end{aligned}$$

since each of the x^i is a local coordinate function on U (a neighborhood of M), and so, by the result of Problem 6.3 again, $d^2 x^i = 0$. □

The result of Problem 6.1 [Eq. (6.12)] leads to a form that is most often used in computations of exterior derivatives:

$$\boxed{\begin{array}{ll} \text{If} & \omega|_U = a_{i_1 \ldots i_r}\, dx^{i_1} \wedge \cdots \wedge dx^{i_r} \in A^r(M)\,, \\ \text{then} & d\omega|_U = \dfrac{\partial a_{i_1 \ldots i_r}}{\partial x^j}\, dx^j \wedge dx^{i_1} \wedge \cdots \wedge dx^{i_r}\,. \end{array}} \tag{6.13}$$

The pairing between the tangent bundle $T(M)$ and the cotangent bundle $T^*(M)$ given (in a pointwise fashion, that is , for every $x \in M$) by (5.4) can be generalized to the following useful statement.

Theorem 6.3. *Let ω be a differential 1-form on a smooth manifold M; and $\boldsymbol{X}$ and $\boldsymbol{Y}$ are smooth tangent vector fields on M [smooth sections on $T(M)$]. Then the following pairing between the section spaces $\Gamma(\Lambda^2(M))$ and $A^2(M) = \Gamma(\Lambda^2(M^*))$ holds:*

$$\boxed{\langle \boldsymbol{X} \wedge \boldsymbol{Y}, d\omega \rangle = \boldsymbol{X} \langle \boldsymbol{Y}, \omega \rangle - \boldsymbol{Y} \langle \boldsymbol{X}, \omega \rangle - \langle [\boldsymbol{X}, \boldsymbol{Y}], \omega \rangle} \quad , \tag{6.14}$$

*where $[\boldsymbol{X}, \boldsymbol{Y}]$, called the **Lie bracket** of the tangent vector fields $\boldsymbol{X}$ and $\boldsymbol{Y}$, or the **Lie derivative** $L_{\boldsymbol{X}}\boldsymbol{Y}$ of the tangent vector field $\boldsymbol{Y}$ with respect to the tangent vector field $\boldsymbol{X}$, is defined by*

$$[\boldsymbol{X}, \boldsymbol{Y}] = L_{\boldsymbol{X}}\boldsymbol{Y} \equiv \boldsymbol{X}\boldsymbol{Y} - \boldsymbol{Y}\boldsymbol{X}\,, \tag{6.15}$$

with $\boldsymbol{X}\boldsymbol{Y}$ meaning the composition of the actions of the tangent vector fields $\boldsymbol{Y}$ followed by $\boldsymbol{X}$ in succession on a smooth function on M, and $\boldsymbol{X} \langle \boldsymbol{Y}, \omega \rangle$ means the action of the smooth tangent vector field $\boldsymbol{X}$ on the smooth function $\langle \boldsymbol{Y}, \omega \rangle$.

Proof. Since any pairing $\langle\, , \rangle$ is a bilinear function, it is sufficient to prove the theorem for $\omega = g df$, where g and f are both smooth functions on M. Thus

$$d\omega = dg \wedge df\,.$$

By (2.30) and (5.4),

$$\langle \boldsymbol{X} \wedge \boldsymbol{Y}, dg \wedge df \rangle = \begin{vmatrix} \langle \boldsymbol{X}, dg \rangle & \langle \boldsymbol{X}, df \rangle \\ \langle \boldsymbol{Y}, dg \rangle & \langle \boldsymbol{Y}, df \rangle \end{vmatrix} = \boldsymbol{X}(g)\boldsymbol{Y}(f) - \boldsymbol{X}(f)\boldsymbol{Y}(g)\,. \tag{6.16}$$

Since

$$\langle \boldsymbol{X}, \omega \rangle = \langle \boldsymbol{X}, gdf \rangle = g \cdot \boldsymbol{X}(f) , \tag{6.17}$$

(note that $\boldsymbol{X}$ only acts on the df part in the expression gdf) we have

$$\boldsymbol{Y} \langle \boldsymbol{X}, \omega \rangle = \boldsymbol{Y}(g)\boldsymbol{X}(f) + g \cdot \boldsymbol{Y}(\boldsymbol{X}(f)) . \tag{6.18}$$

Similarly,

$$\boldsymbol{X} \langle \boldsymbol{Y}, \omega \rangle = \boldsymbol{X}(g)\boldsymbol{Y}(f) + g \cdot \boldsymbol{X}(\boldsymbol{Y}(f)) . \tag{6.19}$$

It follows that the right-hand side of (6.14) is

$$\begin{aligned} &\boldsymbol{X} \langle \boldsymbol{Y}, \omega \rangle - \boldsymbol{Y} \langle \boldsymbol{X}, \omega \rangle - \langle [\boldsymbol{X}, \boldsymbol{Y}], \omega \rangle \\ &= \boldsymbol{X}(g)\boldsymbol{Y}(f) - \boldsymbol{Y}(g)\boldsymbol{X}(f) + g[\boldsymbol{X}(\boldsymbol{Y}(f)) - \boldsymbol{Y}(\boldsymbol{X}(f))] - g \langle [\boldsymbol{X}, \boldsymbol{Y}], df \rangle \\ &= \boldsymbol{X}(g)\boldsymbol{Y}(f) - \boldsymbol{Y}(g)\boldsymbol{X}(f) , \end{aligned} \tag{6.20}$$

which is precisely the right-hand side of the second equality in (6.16). For a more complete discussion on the Lie bracket (and Lie derivatives in general) see Chapter 11. □

The above theorem can be generalized to higher degrees as follows. (We will leave the proof for the reader as Problem 6.6.)

Corollary 6.1. *Let $\omega \in A^r(M)$ be a smooth exterior differential r-form on M, and $\boldsymbol{X}_1, \ldots, \boldsymbol{X}_{r+1}$ be smooth tangent vector fields on M. Then*

$$\begin{aligned} \langle \boldsymbol{X}_1 \wedge \cdots \wedge \boldsymbol{X}_{r+1}, d\omega \rangle &= \sum_{i=1}^{r+1} (-1)^{i+1} \boldsymbol{X}_i \langle \boldsymbol{X}_1 \wedge \cdots \wedge \hat{\boldsymbol{X}}_i \wedge \cdots \wedge \boldsymbol{X}_{r+1}, \omega \rangle \\ &+ \sum_{1 \le i < j \le r+1} (-1)^{i+j} \langle [\boldsymbol{X}_i, \boldsymbol{X}_j] \wedge \cdots \wedge \hat{\boldsymbol{X}}_i \wedge \cdots \wedge \hat{\boldsymbol{X}}_j \wedge \cdots \wedge \boldsymbol{X}_{r+1}, \omega \rangle , \end{aligned} \tag{6.21}$$

where the $\hat{\boldsymbol{X}}_i$ designates an omitted term.

Problem 6.4 Show that (6.21) reduces to (6.14) when $r = 1$. Expand the formula (6.21) (write down all the terms in the sums explicitly) for the case $r = 2$.

Problem 6.5 Prove Corollary 6.1 by the method of *mathematical induction.*

Problem 6.6 Show that *the Lie bracket of two tangent vector fields (or the Lie derivative of a tangent vector field with respect to another) is also a tangent vector field.* If two tangent vector fields $\boldsymbol{X}$ and $\boldsymbol{Y}$ are given in terms of components by

$\boldsymbol{X} = X^i \partial_i$ and $\boldsymbol{Y} = Y^i \partial_i$ [cf. (5.5)], show that the components of the Lie bracket or Lie derivative (6.15) are given by

$$[\boldsymbol{X}, \boldsymbol{Y}]^i = (L_{\boldsymbol{X}} \boldsymbol{Y})^i = X^j \partial_j Y^i - Y^j \partial_j X^i \,. \tag{6.22}$$

The pullback map defined by (2.45) on exterior forms can be readily extended to pullback maps on exterior differential forms. Suppose $f : M \longrightarrow N$ is a smooth map from a smooth manifold M to a smooth manifold N. Then it induces a linear map between the corresponding spaces of exterior differential forms:

$$f^* : A(N) \longrightarrow A(M) \quad , \tag{6.23}$$

known as the **pullback** map induced by f, defined successively by the following steps.

(1): *Define the pullback of a cotangent vector.* Suppose $f : M \longrightarrow N$ is a smooth map between smooth manifolds M and N, $x \in M$, and $y = f(x) \in N$. The pullback map induced by f

$$f^* : T_y^*(N) \longrightarrow T_x^*(M) \tag{6.24}$$

is given by

$$\boxed{f^*(d\phi)_y = (d(\phi \circ f))_x} \quad , \tag{6.25}$$

where $(d\phi)_y \in T_y^*(N)$ and ϕ is a function on N. The map f^* in (6.24) is sometimes called the **differential** of the smooth map f.

(2): *Define the **pushforward** of a tangent vector.* Let $f : M \longrightarrow N$ be a smooth map as in step (1), $\boldsymbol{X} \in T_x(M)$, and $d\phi \in T_y^*(N)$. Then the **tangent map** (or the **derivative map**) induced by f

$$f_* : T_x(M) \longrightarrow T_y(N) \tag{6.26}$$

is given by

$$\boxed{\langle f_* \boldsymbol{X} \,,\, d\phi \rangle_y = \langle \boldsymbol{X} \,,\, f^*(d\phi) \rangle_x} \quad . \tag{6.27}$$

Notice the *adjoint relationship* between the "upper star" and the "lower star" maps f_* and f^* in the above equation. These maps are obviously linear. The action of the tangent map f_* is illustrated in Fig. 6.1, where $\boldsymbol{X}$ is a tangent vector to M at $x \in M$ and $f_* \boldsymbol{X}$ is a tangent vector to N at $y = f(x) \in N$. The unique curve element γ lying in M and tangent to $\boldsymbol{X}$ is mapped by f to the curve element $f(\gamma)$ in N.

(3): *Define the pullback of a differential r-form.*

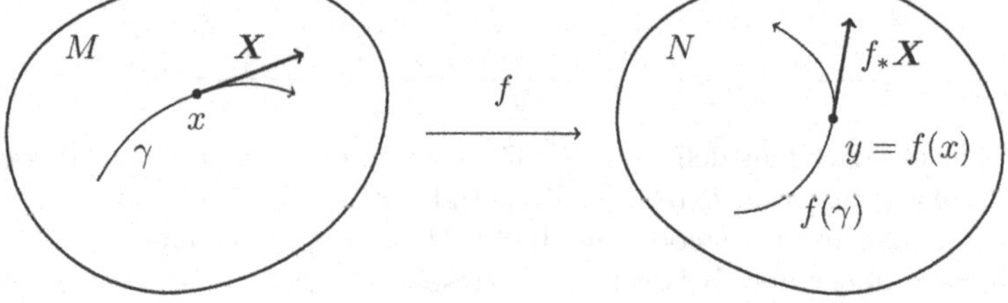

Fig. 6.1

Definition 6.2. *Suppose $f : M \longrightarrow N$ is a smooth map, $\boldsymbol{X}_1, \ldots, \boldsymbol{X}_r$ are r arbitrary smooth tangent vector fields on M, and $\beta \in A^r(N^*)$, $r \geq 1$, is a differential r-form on N. The* ***pullback*** *of β, $f^*\beta$, is given in a pointwise fashion [compare with (2.45)] by*

$$\langle \boldsymbol{X}_1 \wedge \cdots \wedge \boldsymbol{X}_r, f^*\beta \rangle_x = \langle f_*\boldsymbol{X}_1 \wedge \cdots \wedge f_*\boldsymbol{X}_r, \beta \rangle_{f(x)} \quad , \tag{6.28}$$

where x is an arbitrary point in M. If $\beta \in A^0(N^*)$, then

$$f^*\beta = \beta \circ f \in A^0(M^*) \quad . \tag{6.29}$$

The pairing in (6.28) can be computed by (6.9) when the differential forms and tangent vector fields are expressed in natural coordinates.

Problem 6.7 Suppose $dim(M) = m$ and $dim(N) = n$. Let $(x^1, \ldots, x^m)$ and $(y^1, \ldots, y^n)$ be local coordinates near $x \in M$ and $y \in N$, respectively. A map $f : M \longrightarrow N$ can be represented near the point x by the n functions

$$y^\alpha = f^\alpha(x^1, \ldots, x^m), \quad 1 \leq \alpha \leq n. \tag{6.30}$$

Show that the actions of f^* and f_* on the natural bases $\{dy^\alpha, 1 \leq \alpha \leq n\}$ and $\{\frac{\partial}{\partial x^i}, 1 \leq i \leq m\}$ are given by:

$$f^*(dy^\alpha) = \sum_{i=1}^m \left(\frac{\partial f^\alpha}{\partial x^i}\right)_x dx^i \quad , \tag{6.31}$$

$$f_*\left(\frac{\partial}{\partial x^i}\right) = \sum_{\beta=1}^n \left(\frac{\partial f^\beta}{\partial x^i}\right)_x \frac{\partial}{\partial y^\beta} \quad . \tag{6.32}$$

These results show that f_* *and* f^**, as linear maps, have the same matrix representation under the natural (coordinate) bases.*
Notice that Eqs. (6.31) and (6.32) are frequently written, with some abuse of notation, in the more transparent forms

$$dy^\alpha = \sum_{i=1}^{m} \frac{\partial y^\alpha}{\partial x^i}\, dx^i \,, \qquad \frac{\partial}{\partial x^i} = \sum_{\beta=1}^{n} \frac{\partial y^\beta}{\partial x^i} \frac{\partial}{\partial y^\beta} \,. \tag{6.33}$$

The inclusion of the symbols f^* and f_* makes it clear that linear transformations are involved between tangent (or cotangent) spaces at different points of manifolds.

Analogous to Theorem 2.2, the following theorem (stated without proof) asserts that the pullback map f^*, operating between the algebras of exterior differential forms, is an algebra homomorphism.

Theorem 6.4. *For any* $\omega, \eta \in A(N)$ *(exterior differential forms on a manifold* N*) and a given smooth function* $f : M \longrightarrow N$*,* f^* *commutes with the exterior product, that is,*

$$f^*(\omega \wedge \eta) = f^*\omega \wedge f^*\eta \quad . \tag{6.34}$$

In addition, the importance of the pullback map f^* also rests on the fact that it commutes with the exterior derivative. More specifically, we have the following useful theorem (again stated without proof).

Theorem 6.5. *Let* $f : M \longrightarrow N$ *be a smooth map. Then*

$$f^* \circ d = d \circ f^* : A(N) \longrightarrow A(M) \quad . \tag{6.35}$$

Equivalently, we have the following commuting diagram.

$$\begin{array}{ccc} A(N) & \xrightarrow{\;d\;} & A(N) \\ {\scriptstyle f^*}\downarrow & & \downarrow{\scriptstyle f^*} \\ A(M) & \xrightarrow{\;d\;} & A(M) \end{array} \quad .$$

Problem 6.8 Prove Theorem 6.5 for a monomial $\omega \in A^r(N)$ $[dim(N) = n]$, that is, for

$$\omega = b_{i_1 \dots i_r}(y_1, \dots, y_n)\, dy_{i_1} \wedge \dots \wedge dy_{i_r} \,,$$

and then justify that the theorem holds for all $\omega \in A(N)$.

Hints: Use (6.13) for exteriorly differentiating monomials, (6.31) for the pullback map, and Theorem 6.4.

Chapter 7

Vector Calculus by Differential Forms

In this chapter we will begin to demonstrate the power and elegance of differential forms and exterior differentiation by applying it to 3-dimensional vector calculus in Euclidean space. We will see that the exterior derivative operator d completely subsumes the vector calculus operators of "div, grad, curl and all that". On occasion, we will revert to writing the Cartesian coordinates as (x, y, z), rather than (x^1, x^2, x^3), in order to facilitate comparison with conventional notation used most frequently in physics and elementary calculus texts.

Let $f(x, y, z)$ be a smooth function (a zero-form) on $\mathbb{R}^3$. Then it yields the 1-form

$$df = \frac{\partial f}{\partial x}\, dx + \frac{\partial f}{\partial y}\, dy + \frac{\partial f}{\partial z}\, dz \equiv a_i(x, y, z)\, dx^i \tag{7.1}$$

on exterior differentiation. In vector calculus, we call the vector field with Cartesian components $(a_1, a_2, a_3) = (\partial f/\partial x, \partial f/\partial y, \partial f/\partial z)$ the **gradient** of the scalar function f, denoted by ∇f or *grad* f, and write

$$\nabla f = \frac{\partial f}{\partial x}\, \epsilon_1 + \frac{\partial f}{\partial y}\, \epsilon_2 + \frac{\partial f}{\partial z}\, \epsilon_3 \,, \tag{7.2}$$

where $\{\epsilon_1, \epsilon_2, \epsilon_3\}$ is a space-fixed orthonormal frame with respect to which the Cartesian components of a point in $\mathbb{R}^3$ are (x, y, z). Note, however, that at a particular point (x, y, z), the a_i are the components of a *covariant vector.* Indeed, under a general coordinate transformation $(x, y, z) \to (x', y', z')$, we have

$$df = \frac{\partial f}{\partial x^j}\, dx^j = \frac{\partial f}{\partial x^j}\, \frac{\partial x^j}{\partial x'^i}\, dx'^i = a_j\, \frac{\partial x^j}{\partial x'^i}\, dx'^i \equiv a'_i(x', y', z')\, dx'^i \,. \tag{7.3}$$

Hence the components of df transform as

$$a'_i = a_j\, \frac{\partial x^j}{\partial x'_i} \,. \tag{7.4}$$

On the other hand, for an arbitrary (tangent) vector field $\boldsymbol{A} = A^i(x,y,z)\,\partial/\partial x^i$, we have, under the above coordinate transformation,

$$\boldsymbol{A} = A^j \frac{\partial}{\partial x^j} = A^j \frac{\partial}{\partial x'^i}\frac{\partial x'^i}{\partial x^j} = \left(A^j \frac{\partial x'^i}{\partial x^j}\right)\frac{\partial}{\partial x'^i} \equiv A'^i \frac{\partial}{\partial x'^i} \,. \tag{7.5}$$

Thus

$$A'^i = A^j \frac{\partial x'^i}{\partial x^j} \,. \tag{7.6}$$

On comparison with (7.4), we see that the $A^i(x,y,z)$, being components of a contravariant vector at a point (x,y,z), transform inversely as the covariant vector components a_i, as expected (recall the transformation properties displayed in Table 1.1). We conclude that *the gradient of a scalar function is a covariant vector field.* Equation (7.2) is in fact somewhat misleading in the sense that it misrepresents the tensorial character of ∇f: not everything that has three components behaves like a contravariant vector!

Next consider a 1-form

$$\omega = A(x,y,z)\,dx + B(x,y,z)\,dy + C(x,y,z)\,dz \quad , \tag{7.7}$$

where A, B, C are smooth functions on $\mathbb{R}^3$. Deploying the rule for calculating exterior derivatives given by (6.13), we have

$$\begin{aligned} d\omega &= dA\wedge dx + dB\wedge dy + dC\wedge dz = \frac{\partial A}{\partial y}\,dy\wedge dx + \frac{\partial A}{\partial z}\,dz\wedge dx \\ &\quad + \frac{\partial B}{\partial x}\,dx\wedge dy + \frac{\partial B}{\partial z}\,dz\wedge dy + \frac{\partial C}{\partial x}\,dx\wedge dz + \frac{\partial C}{\partial y}\,dy\wedge dz \\ &= \left(\frac{\partial C}{\partial y} - \frac{\partial B}{\partial z}\right) dy\wedge dz + \left(\frac{\partial A}{\partial z} - \frac{\partial C}{\partial x}\right) dz\wedge dx + \left(\frac{\partial B}{\partial x} - \frac{\partial A}{\partial y}\right) dx\wedge dy \,. \end{aligned} \tag{7.8}$$

We recognize that the components of this 2-form are precisely those of the **curl** of a vector field $\boldsymbol{X}$ whose components are (A, B, C), denoted by $\nabla\times\boldsymbol{X}$ in vector calculus in reference to the cross product of two vectors [cf. (1.54)]:

$$\nabla\times\boldsymbol{X} = \begin{vmatrix} \epsilon_1 & \epsilon_2 & \epsilon_3 \\ \dfrac{\partial}{\partial x} & \dfrac{\partial}{\partial y} & \dfrac{\partial}{\partial z} \\ A & B & C \end{vmatrix} \,. \tag{7.9}$$

As in the case of the gradient of a scalar function we have to be careful about the tensorial character of this object. Equation (7.8) clearly implies that $d\omega$, being a two form, is a $(0,2)$-type (rank 2 covariant) tensor field [cf. Def. 6.1], and *not* a contravariant vector field [a $(1,0)$-type tensor field]. We can, however, render $d\omega$ into a 1-form, or a $(0,1)$-type (rank 1 covariant) tensor field, with the same components by using the Hodge star operator [see Def. 3.1 and Eq. (3.11)]:

$$*_{\boldsymbol{\sigma}}(d\omega) = \left(\frac{\partial C}{\partial y} - \frac{\partial B}{\partial z}\right) dx + \left(\frac{\partial A}{\partial z} - \frac{\partial C}{\partial x}\right) dy + \left(\frac{\partial B}{\partial x} - \frac{\partial A}{\partial y}\right) dz \,, \tag{7.10}$$

where $\boldsymbol{\sigma} = dx \wedge dy \wedge dz$ is the top (volume) form in $A(\mathbb{R}^3)$. As we discussed in Chapter 3, the fact that both the exterior spaces $\Lambda^2(T_p^*(\mathbb{R}^3))$ and $\Lambda^1(T_p^*(\mathbb{R}^3)) = T_p^*(\mathbb{R}^3)$ have the same dimension of 3 is an accident of the fact that the dimension of $\mathbb{R}^3$ is 3.

Finally consider the 2-form

$$\psi = A(x,y,z)\, dy \wedge dz + B(x,y,z)\, dz \wedge dx + C(x,y,z)\, dx \wedge dy \,, \tag{7.11}$$

where, again, A, B, C are smooth functions on $\mathbb{R}^3$. We have, on applying (6.13),

$$d\psi = \left(\frac{\partial A}{\partial x} + \frac{\partial B}{\partial y} + \frac{\partial C}{\partial z}\right) dx \wedge dy \wedge dz \,. \tag{7.12}$$

The quantity inside the parenthesis in the above equation is precisely the **divergence** of the vector field $\boldsymbol{X} = A\,\epsilon_1 + B\,\epsilon_2 + C\,\epsilon_3$, denoted in vector calculus by $\nabla \cdot \boldsymbol{X}$ or $div\, \boldsymbol{X}$:

$$\nabla \cdot \boldsymbol{X} = \frac{\partial A}{\partial x} + \frac{\partial B}{\partial y} + \frac{\partial C}{\partial z} \,. \tag{7.13}$$

In fact, using the Hodge-star operator we can also write [cf. (3.13)] (with $\boldsymbol{\sigma} = dx \wedge dy \wedge dz$)

$$*_{\boldsymbol{\sigma}}\,(d\psi) = \nabla \cdot \boldsymbol{X} \,, \tag{7.14}$$

which is a 0-form, or a function on $\mathbb{R}^3$.

Thus, the vector calculus operators *div*, *grad*, and *curl* are all different manifestations of one and the same operator, namely, the exterior derivative d, acting on a 2-form, a 0-form, and a 1-form, respectively.

On exteriorly differentiating (7.1) and (7.8), and using the fact that $d^2 = 0$ (cf. Theorem 6.2), we obtain

$$d^2 f = 0 \,, \qquad [\, f(x,,y,z) \in A^0(\mathbb{R}^3) \text{ is a smooth function on } \mathbb{R}^3 \,] \,, \tag{7.15}$$

$$d^2 \omega = 0 \,, \qquad [\, \omega = A(x,y,z)\, dx + B(x,y,z)\, dy + C(x,y,z)\, dz \in A^1(\mathbb{R}^3) \,] \,. \tag{7.16}$$

From the above discussion we see that these are, respectively, none other than the well-known vector calculus identities

$$\nabla \times (\nabla f) = 0 \,, \tag{7.17}$$

$$\nabla \cdot (\nabla \times \boldsymbol{X}) = 0 \,, \tag{7.18}$$

where $f(x,y,z)$ is a scalar function on $\mathbb{R}^3$ and $\boldsymbol{X}$ is a smooth vector field on $\mathbb{R}^3$.

Problem 7.1 Perform exterior differentiations on the appropriate differential forms to prove the following vector calculus identities:

$$\nabla(fg) = f\nabla g + g\nabla f \,, \tag{7.19}$$
$$\nabla \times (f\boldsymbol{X}) = \nabla f \times \boldsymbol{X} + f\nabla \times \boldsymbol{X} \,, \tag{7.20}$$
$$\nabla \cdot (f\boldsymbol{X}) = (\nabla f) \cdot \boldsymbol{X} + f\nabla \cdot \boldsymbol{X} \,, \tag{7.21}$$
$$\nabla \cdot (\boldsymbol{X} \times \boldsymbol{Y}) = \boldsymbol{Y} \cdot (\nabla \times \boldsymbol{X}) - \boldsymbol{X} \cdot (\nabla \times \boldsymbol{Y}) \,. \tag{7.22}$$

Hints: For (7.19) use

$$d(fg) = f\, dg + g\, df \,.$$

For (7.20) let $\boldsymbol{X} = (X^1(x,y,z), X^2(x,y,z), X^3(x,y,z))$ and use

$$d(f\omega) = df \wedge \omega + f\, d\omega \,,$$

where $\omega = X^1\, dx + X^2\, dy + X^3\, dz$. For (7.21) set $\psi = X^1\, dy\wedge dz + X^2\, dz\wedge dx + X^3\, dx\wedge dy$ and use

$$d(f\psi) = df \wedge \psi + f\, d\psi \,.$$

For (7.22) let $\boldsymbol{X}$ be as above and $\boldsymbol{Y} = (Y^1(x,y,z), Y^2(x,y,z), Y^3(x,y,z))$. Set

$$\theta = X^1\, dx + X^2\, dy + X^3\, dz \,, \quad \phi = Y^1\, dx + Y^2\, dy + Y^3\, dz \,, \quad \psi = \theta \wedge \phi \,.$$

Then use

$$d\psi = d\theta \wedge \phi - \theta \wedge d\phi \,.$$

Note the minus sign in the above equation, as follows from (6.11).

Problem 7.2 Prove the vector calculus identities (7.17) to (7.22) using the tensorial index notation introduced in Chapter 1. In this notation

$$\nabla f = \sum_i \partial_i f\, \boldsymbol{e}_i \,, \tag{7.23}$$
$$\nabla \cdot \boldsymbol{X} = \partial_i X^i \,, \tag{7.24}$$
$$\nabla \times \boldsymbol{X} = \varepsilon^{ij}{}_k\, \partial_j X^k\, \boldsymbol{e}_i \,, \tag{7.25}$$

where $\{\boldsymbol{e}_1, \boldsymbol{e}_2, \boldsymbol{e}_3\}$ is a space-fixed orthonormal frame in $\mathbb{R}^3$, $f(x^1, x^2, x^3)$ is a scalar field, $\boldsymbol{X} = X^i(x^1, x^2, x^3)\, \boldsymbol{e}_i$ is a vector field, $\partial_i = \partial/\partial x^i$, and $\varepsilon^{ij}{}_k$ is the Levi-Civita tensor.

Problem 7.3 Prove the following vector calculus identity, using either differential forms or the tensorial index method:

$$\nabla \times (\boldsymbol{X} \times \boldsymbol{Y}) = (\boldsymbol{Y} \cdot \nabla)\, \boldsymbol{X} - (\boldsymbol{X} \cdot \nabla)\, \boldsymbol{Y} + (\nabla \cdot \boldsymbol{Y})\, \boldsymbol{X} - (\nabla \cdot \boldsymbol{X})\, \boldsymbol{Y} \,. \tag{7.26}$$

Chapter 8

The Stokes Theorem

The usefulness of the vector calculus differential operators *div*, *grad* and *curl* rests on their intuitive physical and geometrical interpretations in 3-dimensional Euclidean space (as their names imply). We will first review these interpretations, give some physical examples, and then tie them to their generalizations in higher dimensional settings through the use of differential forms. The so-called *Stokes Theorem* for integrals of differential forms over regions in differentiable manifolds of arbitrary dimension will emerge to be the key fact that unifies many integral theorems in vector calculus.

Let $\boldsymbol{r} = x^i\, \boldsymbol{\epsilon}_i$ be the position vector of a particle, where x^i $(i = 1, 2, 3)$ are the Cartesian coordinates with respect to a space-fixed frame. For a scalar function $\phi(x^i)$ the gradient $\nabla\phi$ is then related to the 1-form $d\phi$ by

$$d\phi = \partial_i \phi \, dx^i = \nabla\phi \cdot d\boldsymbol{r} \, . \tag{8.1}$$

Thus, at each point in space (specified by the position vector $\boldsymbol{r}$), $\nabla\phi$ is normal (perpendicular) to the surface of constant ϕ (**equipotential surface**) passing through that point (see Fig. 8.1). Also, $\nabla\phi$ is along the direction in which ϕ increases the fastest.

In classical mechanics the gradient of a scalar field leads to the concept of a **conservative force field** (see Chapter 12). In fact, the conservative force field $\boldsymbol{F}(\boldsymbol{r})$ associated with a given scalar field $U(\boldsymbol{r})$, which is called the **potential function**, is identified as

$$\boxed{\boldsymbol{F} = -\nabla U} \, . \tag{8.2}$$

We then have, from (8.1),

$$dU = -\boldsymbol{F} \cdot d\boldsymbol{r} \, , \tag{8.3}$$

and thus the change in potential energy on going from point 1 to point 2 is given by a **line integral**:

$$\Delta U = U_2 - U_1 = -\int_1^2 \boldsymbol{F} \cdot d\boldsymbol{r} \, , \tag{8.4}$$

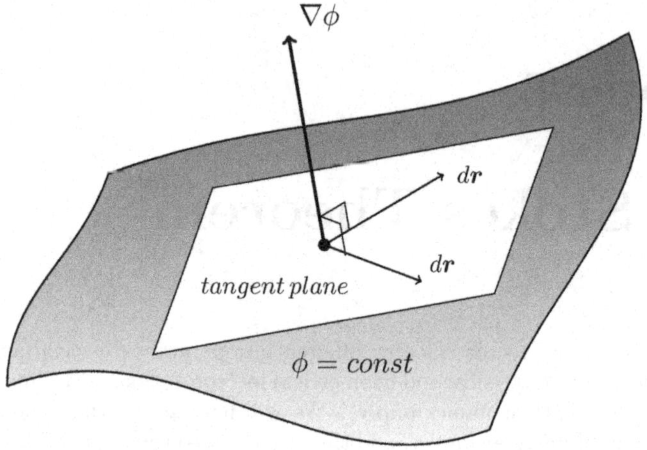

Fig. 8.1

where dr is a **line element** along a given path from 1 to 2. *The line integral above is independent of the path chosen*, since otherwise ΔU, which only depends on the two end points of the path, would not be a meaningful quantity. The line integral is called the **mechanical work** done by the force field $\boldsymbol{F}(\boldsymbol{r})$ from point 1 to point 2, and the 1-form $\boldsymbol{F} \cdot d\boldsymbol{r}$ will be referred to as the *work one-form.* We see immediately from (7.17) that *if the force field $\boldsymbol{F}$ is conservative, its curl vanishes.* It is interesting to ask whether the converse is true, that is, whether a curl-free field is always conservative. We will see in Chapter 21 that the answer depends on the topology of the manifold in which the force field is defined. For the moment we note that it is in the affirmative if $\boldsymbol{F}$ is defined on the Euclidean space $\mathbb{R}^3$. Thus, *if $\boldsymbol{F}$ is a force field defined on $\mathbb{R}^3$, it is conservative if and only if $\nabla \times \boldsymbol{F} = 0$.*

Problem 8.1 A two-dimensional force field is given in terms of its Cartesian components by $F_x = y$, $F_y = -x$ (see Fig. 8.2). Show that this force field is not conservative by showing that its curl does not vanish identically.

The divergence of a vector field $\boldsymbol{A}(\boldsymbol{r}) = A^i(\boldsymbol{r})\, \boldsymbol{\epsilon}_i$, defined in terms of Cartesian coordinates x^i by $\nabla \cdot \boldsymbol{A} \equiv \partial A^i / \partial x^i = \partial_i A^i$ [cf. (7.13)], satisfies **Gauss'**

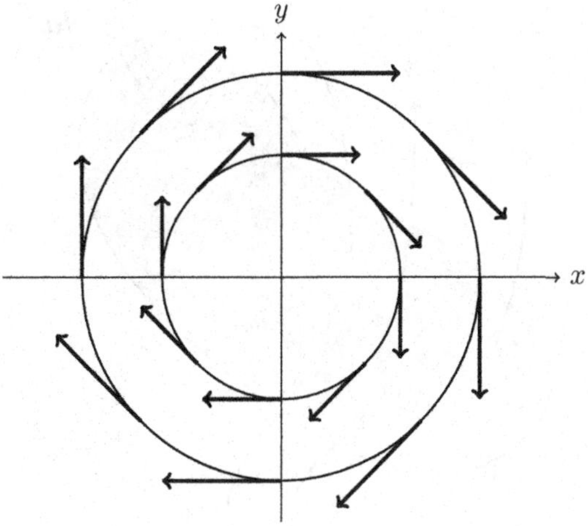

Fig. 8.2

Theorem (see Fig. 8.3):

$$\boxed{\int_D \nabla \cdot \boldsymbol{A}\, d^3r = \int_{\partial D} \boldsymbol{A} \cdot d\boldsymbol{a}} \quad , \tag{8.5}$$

where D represents a volume domain of integration in $\mathbb{R}^3$, ∂D represents the *boundary* of D, and $d\boldsymbol{a}$ is an infinitesimal *outward* normal vector area element.

The area integral on the right-hand side of (8.5) is called the **flux** of the vector field $\boldsymbol{A}$ through the area ∂D. Thus *the divergence of a vector field at a certain point in space is equal to the flux per unit volume, in the limit of vanishing volume, at that point.* Gauss' Theorem plays an important role in Newtonian gravitation and electrostatics. The corresponding laws governing the gravitational field $\boldsymbol{g}$ and the electrostatic field $\boldsymbol{E}$ are

$$\nabla \cdot \boldsymbol{g} = -4\pi G\rho \tag{8.6}$$

and

$$\nabla \cdot \boldsymbol{E} = 4\pi\rho\,, \tag{8.7}$$

respectively, where G is the Newtonian gravitational constant, and ρ stands for mass density in the gravitational context and charge density in the electrostatic

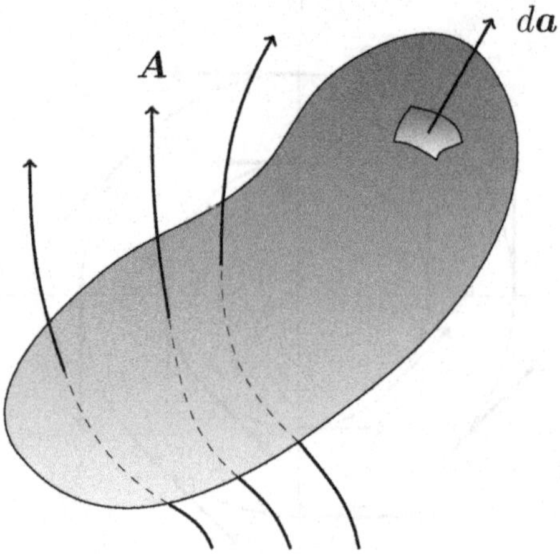

Fig. 8.3

context. Gauss' Theorem immediately leads to the following integral forms for the above laws:

$$\oint \boldsymbol{g} \cdot d\boldsymbol{a} = -4\pi G \int \rho \, d^3r = -4\pi G M \,, \tag{8.8}$$

$$\oint \boldsymbol{E} \cdot d\boldsymbol{a} = 4\pi \int \rho \, d^3r = 4\pi Q \,, \tag{8.9}$$

where M and Q are the total mass and total charge enclosed by the volume that is integrated over, respectively.

Gauss' Theorem is also used to give the integral form of the **equation of continuity**, which is a statement of conservation (no spontaneous generation or loss of material) obeyed by any fluid flow, for example, the flow of liquids and electrical currents through different regions in space. The differential form of this equation is given by

$$\nabla \cdot \boldsymbol{J} + \frac{\partial \rho}{\partial t} = 0 \,, \tag{8.10}$$

where $\boldsymbol{J}$ is the **current density**, measured by the amount of "stuff" (for example, mass or charge) flowing through a certain area perpendicular to the direction of flow per unit area per unit time, and ρ is the volume density of the "stuff". The integral of the divergence term over a certain volume D is, according to Gauss' Theorem,

$$\int_D \nabla \cdot \boldsymbol{J}\, d^3r = \oint_{\partial D} \boldsymbol{J} \cdot d\boldsymbol{a} \,. \tag{8.11}$$

Thus the integral form of the equation of continuity appears as, on integrating (8.10) over a volume D (see Fig. 8.2),

$$\oint_{\partial D} \boldsymbol{J} \cdot d\boldsymbol{a} = -\frac{\partial}{\partial t} \int_D \rho\, d^3r \,. \tag{8.12}$$

This equation has a very intuitive physical interpretation: the flux of "stuff" through a closed area is equal to the negative time rate of change of the amount of "stuff" contained in the volume that is bounded by that area.

The **curl** of a vector field $\boldsymbol{A}$, defined in terms of Cartesian coordinates by [cf. (7.9) and (7.25)]

$$\nabla \times \boldsymbol{A} = (\varepsilon^{ij}{}_k\, \partial_j A^k)\, \boldsymbol{e}_i = \begin{vmatrix} \boldsymbol{e}_1 & \boldsymbol{e}_2 & \boldsymbol{e}_3 \\ \dfrac{\partial}{\partial x^1} & \dfrac{\partial}{\partial x^2} & \dfrac{\partial}{\partial x^3} \\ A^1 & A^2 & A^3 \end{vmatrix} \,, \tag{8.13}$$

satisfies **Stokes' Theorem** of vector calculus (see Fig. 8.4):

$$\boxed{\int_D (\nabla \times \boldsymbol{A}) \cdot d\boldsymbol{a} = \int_{\partial D} \boldsymbol{A} \cdot d\boldsymbol{l}} \,, \tag{8.14}$$

where $d\boldsymbol{l}$ is a line element along the boundary ∂D of an open surface D, and has a direction consistent with the "right-hand rule", that is, the directions of $d\boldsymbol{l}$ and the vector area element $d\boldsymbol{a}$ are consistent with the orientation of a right-handed frame [see Fig. 8.4 and refer to the discussion of oriented frames in Chapter 3]. The integral on the right-hand side of the above equation is called the **circulation** of the vector field $\boldsymbol{A}$ around the (oriented) loop ∂D. Thus Stokes' Theorem implies that *the curl of a vector field at a certain point in space is equal to the circulation per unit area, in the limit of vanishing area, at that point.* As such, the curl of the velocity field describing a fluid flow is immediately useful in representing, for example, *vortices* in fluid dynamics.

All the integral theorems mentioned so far in this chapter relate integrals over a domain and integrals over its boundary. This relationship is at the heart of calculus. We will now cast these theorems in the language of differentiable forms, and thereby arrive at a general theorem valid for manifolds of any dimension – the Stokes Theorem – which subsumes all the vector calculus integral theorems. The primary concept is the integral of a differential r-form over a certain *oriented* domain of dimension r embedded in a differentiable manifold, of which all the integrals occurring so far in this chapter are special examples. Again, we will only present the conceptual outlines without any pretense to mathematical rigor.

First we recall the discussion of oriented frames in Chapter 2 to introduce the concept of an *orientable manifold* [cf. Def. 2.4].

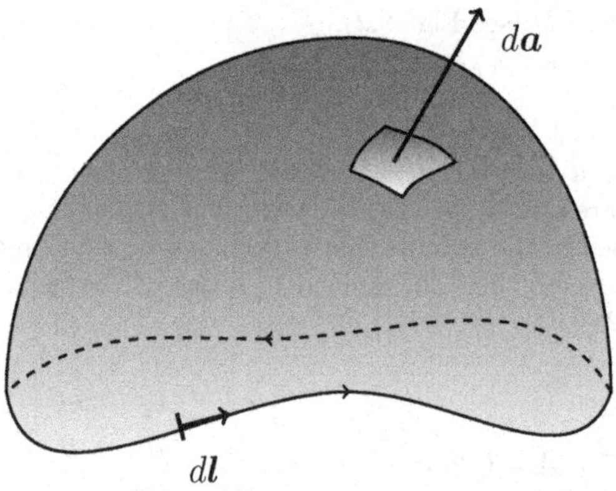

Fig. 8.4

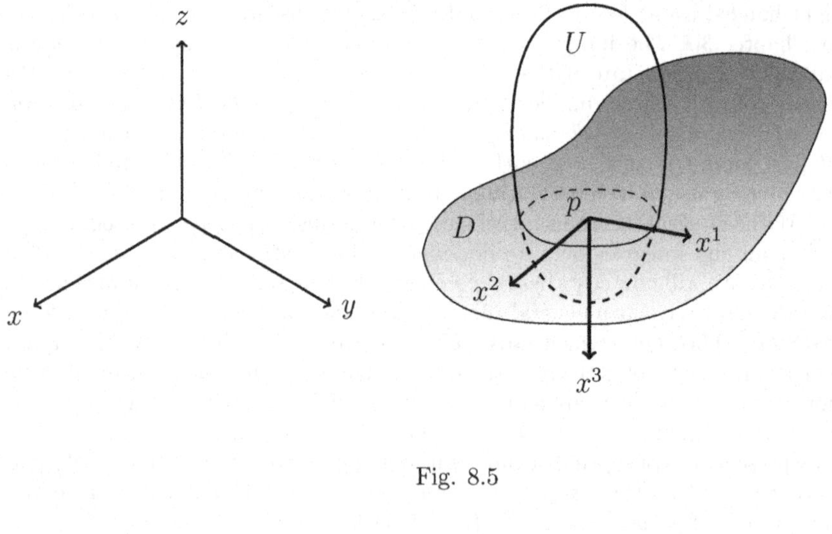

Fig. 8.5

Definition 8.1. *An n-dimensional smooth manifold M is said to be* ***orientable*** *if there exists a continuous and non-vanishing differential n-form (a top form) ω on M. Given such an ω, M is said to be* ***oriented*** *(by ω). Two such forms differing only by a positive function factor are said to assign the same* ***orientation*** *to M.*

Problem 8.2 Justify that if a connected manifold M is orientable, then there exists exactly two orientations on M; if one is labelled right-handed, the other is labelled left-handed.

As immediate and the most familiar examples $\mathbb{R}, \mathbb{R}^2$ and $\mathbb{R}^3$ are orientable, being oriented by the top-forms $dx, dx \wedge dy$, and $dx \wedge dy \wedge dz$, respectively. A 2-dimensional torus embedded in $\mathbb{R}^3$ is orientable, while a Möbius strip is not (there is no "inside" or "outside" to the strip). For manifolds requiring multiple coordinate charts one needs to consider how the top forms restricted to different charts are related to each other. Suppose an n-dimensional orientable manifold M is oriented by an exterior form ω, and $(U; x^i)$ is a local coordinate chart in M. Then $dx^1 \wedge \cdots \wedge dx^n$ and $\omega|_U$ (ω restricted to U) are the same up to a non-zero function factor. If the factor is positive, then we say that $(U; x^i)$ is a coordinate chart **compatible** (consistent) with the orientation of M. Obviously, for any oriented manifold, there exists a coordinate covering which is compatible with the orientation of the manifold: every coordinate chart in the covering is compatible with the orientation of M, and the Jacobian of the change of coordinates between any two coordinate neighborhoods with non-empty intersection is always positive (at every point in the intersection). Conversely, it is also true that if there exists a compatible coordinate covering such that the Jacobian of the change of coordinates in the non-empty intersection of any two coordinate neighborhoods in the covering is always positive, then M is orientable. The proof of this last statement will not be given here.

We will next make precise the notion of an *induced orientation* on the boundary ∂D of a domain D in an orientable manifold.

Definition 8.2. *Suppose M is an n-dimensional smooth manifold. A* ***domain with boundary*** *D is a subset of M with two kinds of points:*

1) ***interior points****, each of which has a neighborhood in M contained entirely in D;*

2) ***boundary points*** *p, for each of which there exists a coordinate chart (U, x^i) in M such that $x^i(p) = 0, i = 1, \ldots, n$, and*

$$U \cap D = \{q \in U \,|\, x^n(q) \geq 0\} \quad .$$

A coordinate system x^i with the above property is called an ***adapted coordinate system*** *for the boundary point p. The set of all boundary points of D is called the boundary of D, denoted by ∂D.*

The case $D \subset M = \mathbb{R}^3$ is illustrated in Fig. 8.5.

Let M be an oriented n-dimensional manifold, and D be a domain with boundary in M. At a point $p \in \partial D$, choose an adapted coordinate system x^i such that $dx^1 \wedge \cdots \wedge dx^n$ is compatible with the given orientation of M. Then $(x^1, \ldots, x^{n-1})$ is a local coordinate system of ∂D at p. The orientation of ∂D specified by

$$(-1)^n dx^1 \wedge \cdots \wedge dx^{n-1} \tag{8.15}$$

is called the **induced orientation** on the boundary ∂D. Note that for $n = 1$, D is a closed directed line segment and ∂D consists of just two points, the beginning point 1 and the endpoint 2. The induced orientation on ∂D would be a 0-form on the boundary, that is, a function f defined on the two points 1 and 2, such that $f(2) = 1$ and $f(1) = -1$. Now consider (8.4) concerning the gradient of a scalar function $\phi(x, y, z)$ again. Let the path of the line integral C be the parametrized curve $x(t), y(t), z(t)$. Then on C the potential function ϕ is implicitly a function of t. Suppose at the beginning point of the path $t = t_1$ and at the end point $t = t_2\ (> t_1)$. Equation (8.4) can thus be written as

$$\int_{t_1}^{t_2} d\phi = \phi(t_2) - \phi(t_1) \,. \tag{8.16}$$

This is precisely the **Fundamental Theorem of Calculus**. Let $M = \mathbb{R}$ and a domain with boundary $D \subset M$ be the closed interval $[t_1, t_2]$ with orientation dt. Thus $\partial D = \{t_1, t_2\}$ with the above described induced orientation, and we can write (8.16) in terms of differential forms as

$$\int_D d\phi = \int_{\partial D} \phi \,, \tag{8.17}$$

where ϕ is a 0-form and on the right-hand side it is integrated over the 0-dimensional domain ∂D (consisting of two points) with the induced orientation (from D).

With reference to Fig. 8.5 we see that for $M = \mathbb{R}^3$ and a 3-dimensional domain D with boundary, the induced orientation on ∂D [with respect to a local adapted coordinate system (x^1, x^2, x^3)] is, according to (8.15), $(-1)^3\, dx^1 \wedge dx^2 = dx^2 \wedge dx^1$. This can be associated through the Hodge star operator $*_{dx\wedge dy\wedge dz}$ to $-dx^3$, that is, the *outward* pointing normal direction, which is defined to be the direction of a vector area element $\boldsymbol{da}$ on ∂D. Consider a vector field $\boldsymbol{A}(x, y, z) = (A_1(x, y, z), A_2(x, y, z), A_3(x, y, z))$ on M and the 2-form ψ on M given by

$$\psi = A_1\, dy \wedge dz + A_2\, dz \wedge dx + A_3\, dx \wedge dy \,. \tag{8.18}$$

Then, as in (7.12),

$$d\psi = \nabla \cdot \boldsymbol{A}\, dx \wedge dy \wedge dz \,, \tag{8.19}$$

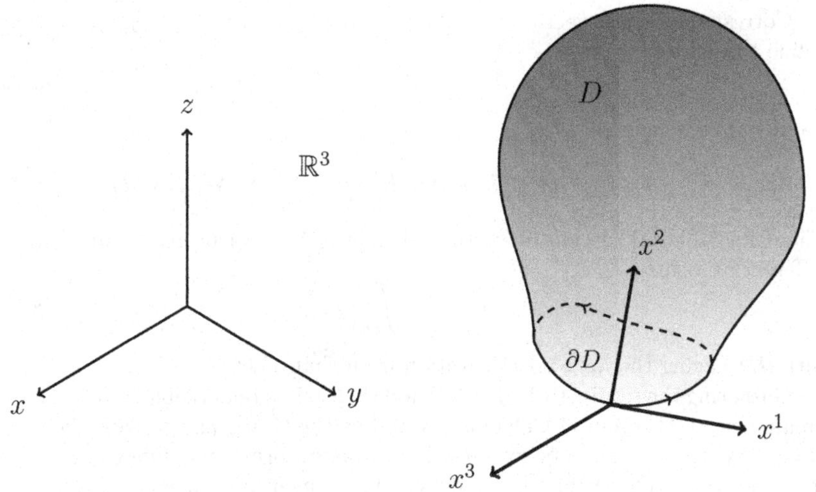

Fig. 8.6

and so Gauss' Theorem (8.5) can be written in terms of differential forms as

$$\int_D d\psi = \int_{\partial D} \psi \,, \tag{8.20}$$

with the boundary ∂D having the induced orientation from that of D.

Note that in integral expressions in elementary calculus presentations the wedge products are absent, so that Gauss' Theorem, for example, is stated as

$$\int_{\partial D} (A_1\, dydz + A_2\, dzdx + A_3\, dxdy) = \int_D \left(\frac{\partial A_1}{\partial x} + \frac{\partial A_2}{\partial y} + \frac{\partial A_3}{\partial z} \right) dxdydz \,. \tag{8.21}$$

The Riemann integrals in this equation assumes that the volume element $dx \wedge dy \wedge dz$ is compatible with the orientation of D; and the area elements $dy \wedge dz$, $dz \wedge dx$, and $dx \wedge dy$ are all compatible with the induced orientation on ∂D. The wedge products in the integrands of (8.20) makes this fact explicit.

Next let D be an oriented surface in $\mathbb{R}^3$ with oriented boundary ∂D. The orientations on D and ∂D are chosen as follows. First consider D as a subset of M, a 2-dimensional manifold embedded in $\mathbb{R}^3$. Let M have an orientation inherited from an appropriate one in $\mathbb{R}^3$. For example, choose an orientation $dx \wedge dy \wedge dz$ on $\mathbb{R}^3$, and an orientation $dx^1 \wedge dx^2$ on M so that $dx^1 \wedge dx^2 \wedge dx^3$ (in local coordinates) is compatible with the one chosen on $\mathbb{R}^3$. Further, choose x^1

and x^2 to be *adapted coordinates* on D so that dx^2 points towards the interior of D. The induced orientation on ∂D is then, by (8.15), $(-1)^2\,dx^1 = dx^1$ (see Fig. 8.6).

Corresponding to a vector field $\boldsymbol{A}(x,y,z) = (A_1(x,y,z), A_2(x,y,z), A_3(x,y,z))$ define the 1-form

$$\omega = A_1\,dx + A_2\,dy + A_3\,dz\ . \tag{8.22}$$

As in (7.8) we have

$$d\omega = (\nabla\times\boldsymbol{A})_x\,dy\wedge dz + (\nabla\times\boldsymbol{A})_y\,dz\wedge dx + (\nabla\times\boldsymbol{A})_z\,dx\wedge dy\ . \tag{8.23}$$

Thus the Stokes Theorem of vector calculus (8.14) can be recast in terms of differential forms as

$$\int_D d\omega = \int_{\partial D}\omega\ , \tag{8.24}$$

with ∂D having the induced orientation from that of D.

Gathering the results (8.17), (8.20) and (8.24), it is remarkable to see that the Fundamental Theorem of Calculus, as well as the Gauss and Stokes Theorems of vector calculus, can all be expressed in the same form using differential forms. It is even more remarkable that the formal statement appearing in (8.24) holds for manifolds of arbitrary dimension. This is the general Stokes Theorem on integrals of differential forms. Before stating the theorem we need to define more carefully the very concept of an integral of a differential form on arbitrary finite-dimensional differentiable manifolds. Again, we will only outline the procedure.

Consider a general differential form $\omega \in A(M)$. In order for its integral over a domain in M to exist we first need to impose the (topological) condition that it has a *compact support*. The **support** of ω, denoted by $supp\,(\omega)$, is the point set in M defined as follows:

$$supp\,(\omega) \equiv \overline{\{x\in M\,|\,\omega(x)\neq 0\}}\ , \tag{8.25}$$

where $\overline{\{\}}$ means the closure of the set $\{\}$, and $\omega(x)$ is the value of the differential form ω at the point $x \in M$ (remember that a differential form is a tensor field on M). This is an extension of the notion of the support of a function $f : M \to \mathbb{R}$: $supp\,(f) = \overline{\{x\in M\,|\,f(x)\neq 0\}}$. One of the complications in defining the integral of a differential form is that the differential manifold M on which it is defined may require for its covering multiple coordinate patches (charts) W_i, unlike the special familiar cases of $\mathbb{R}^n$. In this case one needs a technical device known as a **partition of unity** subordinate to an atlas of charts $\{W_i\}$ to patch together the local expressions of the differential form corresponding to the different charts as a *finite* sum, with an appropriate weight for each term in the sum so that all the weights add up to unity (this is the origin of the term "partition of unity"). Such a partition of unity, which consists of a set of functions $\{g_\alpha\}$ with compact supports and satisfying the following conditions:

(1) for each α, $0 \le g_\alpha(x) \le 1$ for all $x \in M$ and there exists an open set W_i in the family of charts such that $supp\,(g_\alpha) \subset W_i$;

(2) for each $x \in M$ there exists a neighborhood U of x that intersects $supp\,(g_\alpha)$ for only *finitely* many α;

(3) $\sum_\alpha g_\alpha(x) = 1$ for all $x \in M$,

can be proven to exist for any smooth manifold M. The following steps will lead up to the definition of the integral of a differential form ω on an oriented manifold M, with $dim\,(M) = n$.

(a) First define the integral of an n-form (a top form) on M. Choose a family of coordinate charts (an atlas) $\{W_i\}$ that is compatible with the orientation of M. If $supp\,(\omega)$ happens to be inside a coordinate neighborhood $U \subset W_i$ with local coordinates $x^1, \dots, x^n$, then ω can be expressed as

$$\omega = f(x^1, \dots, x^n)\, dx^1 \wedge \dots \wedge dx^n \,. \tag{8.26}$$

In this case we define

$$\int_M \omega \equiv \int_U f(x^1, \dots, x^n)\, dx^1 \dots dx^n \,, \tag{8.27}$$

where the integral on the right-hand side is the usual *Riemann integral*. Note that in (8.26) $dx^1 \wedge \dots \wedge dx^n$ is supposed to be compatible with the given orientation of M.

(b) If $supp\,(\omega)$ does not fall inside a particular W_i, we make use of a partition of unity $\{g_\alpha\}$ subordinate to $\{W_i\}$ to write

$$\omega = \left(\sum_\alpha g_\alpha \right) \omega = \sum_\alpha (g_\alpha \cdot \omega) \,. \tag{8.28}$$

Since $supp\,(g_\alpha \cdot \omega) \subset supp\,(g_\alpha)$, by condition (1) of a partition of unity there exists, for each α, some other coordinate neighborhood W_{j_α} such that $supp\,(g_\alpha \cdot \omega) \subset supp\,(\omega) \subset W_{j_\alpha}$. Note that j_α is not unique. Suppose there is another integer j'_α such that $supp\,(\omega) \subset W_{j'_\alpha}$. One can prove that

$$\int_{W_{j_\alpha}} g_\alpha \cdot \omega = \int_{W_{j'_\alpha}} g_\alpha \cdot \omega \,. \tag{8.29}$$

So one can define, for any α,

$$\int_M g_\alpha \cdot \omega \equiv \int_{W_j} g_\alpha \cdot \omega \,, \tag{8.30}$$

where W_j is any coordinate neighborhood that satisfies the condition $supp\,(\omega) \subset W_j$. The integral on the right-hand side is again understood to be the usual Riemann integral [see step (a) above].

(c) Since $supp\,(\omega)$ is compact by assumption, it intersects only finitely many $supp\,(g_\alpha)$ by condition (2) of the definition of a partition of unity. Hence the right-hand side of (8.28) is a sum (over α) of finitely many terms. One can also prove that the integral of this sum over M, as defined by (8.30), is independent of the choice of a partition of unity subordinate to $\{W_i\}$. So we can define the integral of an n-form ω over an n-dimensional manifold M by

$$\boxed{\int_M \omega \equiv \sum_\alpha \int_M g_\alpha \cdot \omega} \quad , \tag{8.31}$$

where $\{g_\alpha\}$ is *any* partition of unity subordinate to a coordinate covering of M.

(d) Having defined the integral of an n-form with compact support over an n-dimensional manifold M, we can now define the integral of an r-form with compact support, for $r < n$, over any r-dimensional submanifold N of M. For this step we use the device of the *pullback map* (introduced in Chapter 6) of an embedding map. Let

$$h : N \longrightarrow M$$

be an embedding of an r-dimensional manifold N into an n-dimensional manifold M, $r < n$; and ω be a differential r-form on M with compact support. Then, from Def. 6.2, the pullback $h^*\omega$ is a differential r-form on N with compact support. So the integral

$$\int_N h^*\omega$$

is well-defined according to steps (a) to (c) above [cf. (8.31)]. We then define the integral of φ on the submanifold $h(N)$ by

$$\boxed{\int_{h(N)} \omega \equiv \int_N h^*\omega} \quad . \tag{8.32}$$

This completes the definition of the integral of a differential form with compact support.

We are now ready to state without proof the Stokes Theorem. (For details on the proof, the reader may consult various standard texts on differential geometry, for example, Chern, Chen, and Lam 1999.)

Theorem 8.1 (Stokes Theorem). *Suppose D is a domain with boundary in an n-dimensional oriented manifold M, and ω is an exterior differential $(n-1)$-form on M with compact support. Then*

$$\boxed{\int_D d\omega = \int_{\partial D} \omega} \quad , \tag{8.33}$$

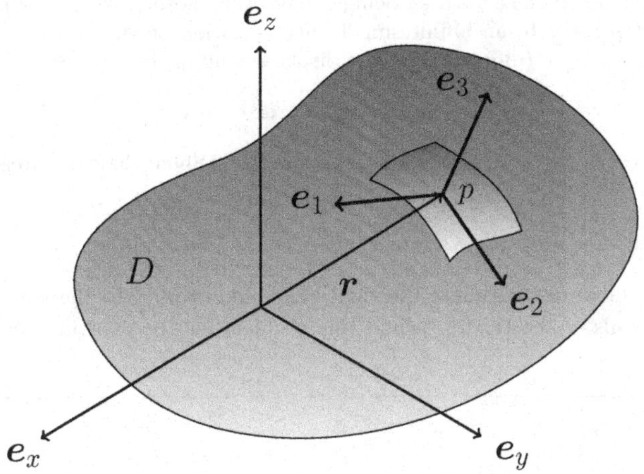

Fig. 8.7

where ∂D is the boundary of D with the induced orientation. If $\partial D = \emptyset$ (the empty set), then the integral on the right-hand side is zero.

Let $\omega, \omega_1, \omega_2$ be exterior differential n-forms on M (of dimension n) with compact support. Then $\omega_1 + \omega_2$ has compact support, as has $c\omega$, for any real number c. By the definition (8.31), it is obvious that

$$\int_M (\omega_1 + \omega_2) = \int_M \omega_1 + \int_M \omega_2 \quad , \tag{8.34}$$

$$\int_M c\omega = c \int_M \omega \quad . \tag{8.35}$$

Thus the integration $\int_M$ can be viewed as a **linear functional** on the set of all n-forms on M with compact support. Equation (8.31) is in fact a generalization of the Riemann (multiple) integral to the case of manifolds. Writing $\int_{\partial D} \omega = (\partial D, \omega)$ and $\int_D d\omega = (D, d\omega)$, where either entry in each bracket of the right-hand sides can be regarded as a linear functional acting on the other, we see that the Stokes Theorem [Eq. (8.33)] can be written as

$$(\partial D, \omega) = (D, d\omega) \; . \tag{8.36}$$

Thus this theorem describes a *duality* between the **boundary operator** ∂ and the exterior derivative operator d. The latter is also called the **coboundary operator**. ∂ and d can also be regarded as *adjoint operators* of each other.

Problem 8.3 Refer to Fig. 8.7, where $(\boldsymbol{e}_1, \boldsymbol{e}_2, \boldsymbol{e}_3)$ is a moving orthonormal frame at a boundary point p of D, with $\boldsymbol{e}_3$ being the outward normal to the boundary ∂D at p. If we let p move to an infinitesimally near position on ∂D, we can write the vector-valued 1-form $d\boldsymbol{r}$ (differential of the displacement vector on ∂D) as

$$d\boldsymbol{r} = dx\,\boldsymbol{e}_x + dy\,\boldsymbol{e}_y + dz\,\boldsymbol{e}_z = \omega^1 \boldsymbol{e}_1 + \omega^2 \boldsymbol{e}_2 \quad , \tag{8.37}$$

where $\omega^1 = \lambda_1 dx^1, \omega^2 = \lambda_2 dx^2$ are 1-forms on ∂D. Show that the area element $\omega^1 \wedge \omega^2\, \boldsymbol{e}_3$ is given by

$$\omega^1 \wedge \omega^2\, \boldsymbol{e}_3 = (dy \wedge dz)\,\boldsymbol{e}_x + (dz \wedge dx)\,\boldsymbol{e}_y + (dx \wedge dy)\,\boldsymbol{e}_z \quad . \tag{8.38}$$

This equation allows us to equate the right-hand side of (8.5) to the left-hand side of (8.21). (Details of the theory behind this problem will be explained in the next chapter.)

Chapter 9

Cartan's Method of Moving Frames: Curvilinear Coordinates in $\mathbb{R}^3$

We have mentioned briefly in Chapter 4 that the Method of Moving Frames provides the mathematical foundations for *gauge field theories* in modern physics. Historically, however, it arose from a consideration of the most basic of physics problems: the classical mechanical description of the motion of rigid bodies in three-dimensional Euclidean space $\mathbb{R}^3$. The key idea was to use a set of *parametrized frames* to explore the geometrical properties of the underlying manifold. It originated from the works of the 19th century geometers Darboux and Cotton. (For an authoritative account of the history and a mathematical review of the subject of moving frames, see Chern 1985.) But it was E. Cartan, a pioneer of differential geometry in the early 20th century, who, through the use of exterior differentiation, developed it to new heights. Cartan's work was subsequently further developed and used with great facility by the differential geometer, S. S. Chern, who exploited it, among other things, to give an intrinsic proof of the *Gauss-Bonnet-Chern Theorem* (see Chapter 26), a classical and far-reaching theorem in differential geometry. This proof in turn opened the door to Chern's discovery of the so-called *Chern classes*, topological invariants which have become prime objects of study in modern differential geometry, and of late, in physics also, where certain integrals of Chern classes, the *Chern numbers*, have been identified as *topological quantum numbers* in a variety of physical systems. (For an introductory account of a physical application of Chern numbers, see, for example, Lam 2009.) In this chapter we will present the elements of the method of moving frames and illustrate their use in studying the geometry of Euclidean $\mathbb{R}^3$. In particular, we will show how it leads naturally to a unified derivation for the useful expressions of the vector calculus differential operators in curvilinear coordinates. Although these results are usually derived in introductory physics and mathematics texts by more elementary means, the

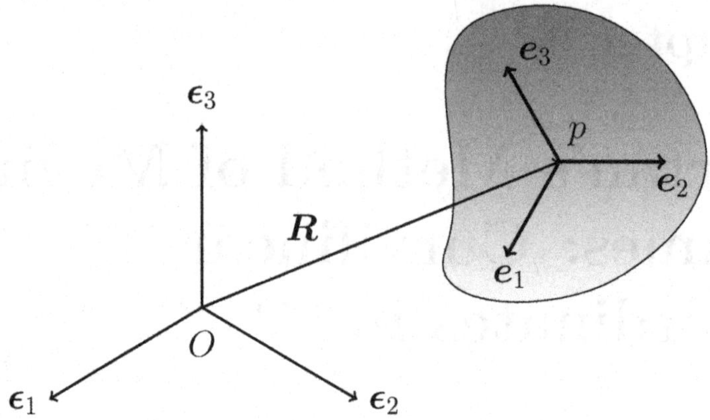

Fig. 9.1

approach presented here has the advantage of a general applicability to arbitrary manifolds.

Consider a rigid body moving in three-dimensional Euclidean space $\mathbb{R}^3$, with an orthonormal frame $\{\boldsymbol{e}_i\}$ $(i = 1, 2, 3)$ rigidly attached to it. Let $\boldsymbol{R}(t)$ be the position vector of the origin of this moving frame at time t. Then we can write the following equations describing the translational and rotational motions of the body:

$$\frac{d\boldsymbol{R}}{dt} = \varphi^i(t)\,\boldsymbol{e}_i\,, \tag{9.1}$$

$$\frac{d\boldsymbol{e}_i}{dt} = \varphi_i^j(t)\,\boldsymbol{e}_j\,, \quad 1 \leq i, j \leq 3\,, \tag{9.2}$$

with

$$\varphi_i^j(t) + \varphi_j^i(t) = 0\,. \tag{9.3}$$

The last equation is a consequence of the orthonormality of the frames [cf. (4.20)]. The functions $\varphi^i(t)$ and $\varphi_i^j(t)$ give a complete description of the motion of the orthonormal frame and hence of the rigid body. The above equations imply that the moving frames $\{\boldsymbol{e}_i\}$ (with origins at different points in space) are parametrized by the single variable t. There are great mathematical advantages, however, in considering families of orthonormal frames depending on multiple parameters. For this situation, Eqs. (9.1) and (9.2) will need to be replaced by partial differential equations whose coefficients (the analogs of the φ^i and the φ_i^j) satisfy certain *integrability conditions*. The most convenient way to represent these is by using exterior derivatives. This is done as follows.

Consider, in $\mathbb{R}^3$, a space-fixed orthonormal frame $\{\epsilon_1, \epsilon_2, \epsilon_3\}$ with origin located at O, denoted by $(O;\, \epsilon_1, \epsilon_2, \epsilon_3)$, and a moving orthonormal frame $\{\boldsymbol{e}_1, \boldsymbol{e}_2, \boldsymbol{e}_3\}$ with origin located at the point p, whose coordinates with respect to the space-fixed frame are (R^1, R^2, R^3), denoted by $(p;\, \boldsymbol{e}_1, \boldsymbol{e}_2, \boldsymbol{e}_3)$. Denote the position vector $\overrightarrow{Op}$ by $\boldsymbol{R}$ (see Fig. 9.1). We can write

$$\boldsymbol{R} = R^i\, \epsilon_i\ , \qquad \boldsymbol{e}_i = g_i^j(\boldsymbol{R})\, \epsilon_j\ , \tag{9.4}$$

where $(g_i^j(\boldsymbol{R}))$ is a special orthogonal matrix (of order 3) for any $\boldsymbol{R}$, that is, $(g_i^j(\boldsymbol{R})) \in SO(3)$ [cf. (4.14) and (4.15)]. For an infinitesimal displacement of the point p, we have

$$d\boldsymbol{R} = dR^i\, \epsilon_i\ , \qquad d\boldsymbol{e}_i = dg_i^j\, \epsilon_j\ , \tag{9.5}$$

where the symbol "d" can be viewed as an exterior derivative. Defining dR to be the row matrix (dR^1, dR^2, dR^3), ϵ and $\boldsymbol{e}$ to be the column matrices $(\epsilon_1, \epsilon_2, \epsilon_3)^T$ and $(\boldsymbol{e}_1, \boldsymbol{e}_2, \boldsymbol{e}_3)^T$, respectively ($T$ means transpose), g to be the 3×3 matrix (g_i^j) (with i being the row index and j the column index), and introducing the notation $dp \equiv d\boldsymbol{R}$, we can write the above equations in matrix form as

$$dp = dR \cdot \epsilon\ , \qquad d\boldsymbol{e} = dg \cdot \epsilon\ . \tag{9.6}$$

The second of Eqs. (9.4) implies $\epsilon = g^{-1}\, \boldsymbol{e}$. Hence the above equations can be rewritten as

$$dp = (dR \cdot g^{-1})\, \boldsymbol{e}\ , \qquad d\boldsymbol{e} = (dg \cdot g^{-1})\, \boldsymbol{e}\ . \tag{9.7}$$

Define the 1×3 matrix of 1-forms

$$\theta = (\omega^1, \omega^2, \omega^3) \equiv dR \cdot g^{-1}\ , \tag{9.8}$$

and the 3×3 matrix of 1-forms [cf. (4.27)]

$$\omega = (\omega_i^j) \equiv dg \cdot g^{-1}\ . \tag{9.9}$$

Equations (9.7) then become

$$dp = \theta \cdot \boldsymbol{e}\ , \qquad d\boldsymbol{e} = \omega \cdot \boldsymbol{e}\ . \tag{9.10}$$

The reader is cautioned not to confuse the (1-forms) ω^i defined above with the components of the angular velocity defined in Chapter 4 (also denoted by ω^i) [cf. (4.25) and (4.34)]. Referring to the first of Eqs. (9.5), $\{dR^i\}$ is called the **coframe field (dual frame field)** to the (natural) frame field $(p;\, \{\epsilon_i\})$, with the pairing $\langle\, dR^i, \epsilon_j\, \rangle = \delta_j^i$ [cf. (5.6), where $\partial/\partial R^j$ would be interpreted as ϵ_j]. The first of Eqs. (9.10) then implies that $\{\omega^i\}$ is the coframe field to the frame field $\{e_i\}$ such that the pairing between them also satisfies the duality condition $\langle e_i, \omega^j \rangle = \delta_i^j$. Exteriorly differentiating the equations (9.10), we obtain, on recalling the "product rule" for exterior derivatives (6.11) and the fact that $d^2 = 0$ (Theorem 6.2),

$$0 = d\theta \cdot e - \theta \wedge \omega\, e\,, \qquad 0 = d\omega \cdot e - \omega \wedge \omega\, e\,. \tag{9.11}$$

Canceling the e in both equations we have

$$d\theta = \theta \wedge \omega\,, \qquad d\omega = \omega \wedge \omega\,, \tag{9.12}$$

where $\wedge$ is interpreted as both a wedge product and matrix multiplication (which is why in these equations the wedge product of 1-forms is not identically zero). The above equations appear explicitly in terms of components as

$$d\omega^i = \omega^j \wedge \omega_j^i\,, \tag{9.13}$$

$$d\omega_i^j = \omega_i^k \wedge \omega_k^j\,, \qquad 1 \le i, j, k \le 3\,. \tag{9.14}$$

Exteriorly differentiating the matrix equation $g \cdot g^{-1} = 1$, we have

$$dg \cdot g^{-1} + g \cdot d(g^{-1}) = dg \cdot g^{-1} + (dg \cdot g^{-1})^T = 0\,, \tag{9.15}$$

where we have used the fact that $g^{-1} = g^T$ (since g is an orthogonal matrix). Equation (9.9) then implies that the matrix of 1-forms ω is antisymmetric:

$$\omega_i^j + \omega_j^i = 0\,. \tag{9.16}$$

This result has already been obtained as (4.24).

Equations (9.13) and (9.14) for the six linearly independent one-forms ω^i and ω_i^j are the integrability conditions referred to above, and are recognized to be the so-called **Maurer-Cartan structure equations** for the *motion group* $E(3)$ in Euclidean $\mathbb{R}^3$ (consisting of rigid-body rotations and translations), which is a six-dimensional Lie group. Elements of this group transform the orthonormal moving frames to each other. The structure equations, together with the antisymmetry condition (9.16), determine completely the geometry of Euclidean $\mathbb{R}^3$. In fact, ω_i^j is the so-called **Levi-Civita connection** matrix of one-forms on the tangent bundle $T(\mathbb{R}^3)$ corresponding to the standard Euclidean metric, introduced in Chapter 4 [cf. (4.8) and see Chapter 27]. One way to see this is to recall the gauge transformation rule (4.19), which implies that ω (the connection matrix expressed with respect to the frame field $\{e_i\}$) gives the same connection as the zero-matrix (expressed with respect to $\{\epsilon_i\}$). Note that for this connection, the covariant derivative is the same as the exterior derivative: $De_i = de_i$ [compare (4.8) with the second equation of (9.10)]. In geometric language, (9.13) states that this connection is **torsion free**, while (9.14) implies that it leads to zero **curvature**. Equation (9.16) is also a statement of the condition of **metric compatibility** of the connection. This group of intimately related and important differential geometric concepts will be explained more systematically in subsequent chapters and their relevance to classical mechanics demonstrated (see Chapters 27).

Our analysis above proceeded from the assumption that the orthonormal frame field $\{e_i\}$ exists in Euclidean $\mathbb{R}^3$. The integrability conditions (9.13) and (9.14) follow as a consequence. It is a significant result that the converse is true also: the validity of the integrability conditions related to the motion group in $\mathbb{R}^3$ implies the existence of orthonormal frame fields. In fact, the whole analysis applies to the group of rigid motions $E(N)$ in Euclidean $\mathbb{R}^N$ in general; and we will state (without proof) the more general result as the following theorem, known as the **Fundamental Theorem for Moving Frames in $\mathbb{R}^N$**.

Theorem 9.1. *Suppose ω^α and $\omega^\gamma_\beta = -\omega^\beta_\gamma$ $(1 \le \alpha, \beta, \gamma \le N)$ are differential 1-forms on an n-dimensional manifold M depending (locally) on n variables $(x^1, \dots, x^n)$. Then there exists an n-parameter family of orthonormal frames $\{e_1, \dots, e_N\}$ (or a frame field) on M with the given differential forms satisfying*

$$dp = \theta \cdot \boldsymbol{e} \,, \qquad d\boldsymbol{e} = \omega \cdot \boldsymbol{e} \,, \tag{9.17}$$

where $p \in M$, $\theta \equiv (\omega^1, \dots, \omega^N), \omega \equiv (\omega^\beta_\alpha)$, and $\boldsymbol{e} = (e_1, \dots, e_N)^T$, if and only if the differential forms satisfy the following ***integrability conditions****:*

$$d\omega^\alpha = \omega^\beta \wedge \omega^\alpha_\beta \,, \qquad d\omega^\beta_\alpha = \omega^\gamma_\alpha \wedge \omega^\beta_\gamma \,. \tag{9.18}$$

Moreover, any two such families of orthonormal frames are related by a rigid motion in $\mathbb{R}^N$ [an element in the motion group $E(N)$].

A few immediate comments on this theorem are in order. In our derivation of the integrability conditions (9.13) and (9.14), the Euclidean space $\mathbb{R}^3$ plays the role of M, and the motion group is $E(3)$, the group of rigid motions in $\mathbb{R}^3$, which as we mentioned before, is a 6-dimensional Lie group. In this case, since $dim\,\mathbb{R}^3 = 3$, we have a 3-parameter family of orthonormal frames. In the abstract context of the mathematical theorem, however, the manifold M (of arbitrary dimension) and the Lie group $E(N)$ (which is also a differentiable manifold in its own right, of dimension $N(N+1)/2$) are not necessarily related to each other. In fact, in the rigid-body dynamics context [described by (9.1) and (9.2)] $M = \mathbb{R}$ and we have a 1-parameter family of orthonormal frames (where the parameter t is usually interpreted to be the time). The integrability conditions (9.18) are so-called because, due to the *Frobenius Theorem* (an important theorem that we will discuss in the next chapter), they guarantee the existence of a function $f : M \to E(N)$ which defines an *integral submanifold* of the manifold $M \times E(N)$. The function f, of course, defines a frame field on M, because each group element $g \in E(N)$ corresponds to a rigid motion in $\mathbb{R}^N$ of a space-fixed frame $(O; \epsilon_1, \dots, \epsilon_N)$.

We will now refocus our attention on the case $M = \mathbb{R}^3$. The 3-form $dR^1 \wedge dR^2 \wedge dR^3$ is then a top-form in $\mathbb{R}^3$, called the *volume element.* From (9.8) we have

$$\omega^i = dR^j (g^{-1})^i_j \,. \tag{9.19}$$

Recalling (2.34) we then obtain

$$\omega^1 \wedge \omega^2 \wedge \omega^3 = det\,(g^{-1})\, dR^1 \wedge dR^2 \wedge dR^3 \,. \tag{9.20}$$

Since g is an orthogonal matrix [$g \in SO(3)$], its determinant, and also that of its inverse, is equal to one. Hence we have the following useful expression for the invariance of the volume form under rotations:

$$\omega^1 \wedge \omega^2 \wedge \omega^3 = dR^1 \wedge dR^2 \wedge dR^3 \,, \tag{9.21}$$

where the R^i are interpreted as the usual rectangular coordinates of an arbitrary point p in $\mathbb{R}^3$. We can also write the first of Eqs. (9.6) and (9.7) as the following expressions for *line elements* in $\mathbb{R}^3$ (on recalling that, at an arbitrary point p, $\epsilon_i = \partial/\partial R^i$):

$$\begin{aligned} dp = d\boldsymbol{R} &= dR^1\, \epsilon_1 + dR^2\, \epsilon_2 + dR^3\, \epsilon_3 \\ &= dR^1\, \frac{\partial}{\partial R^1} + dR^2\, \frac{\partial}{\partial R^2} + dR^3\, \frac{\partial}{\partial R^3} = \omega^1\, e_1 + \omega^2\, e_2 + \omega^3\, e_3 \,. \end{aligned} \tag{9.22}$$

The above two equations, together with the following duality conditions for pairings

$$\left\langle dR^i \,,\, \frac{\partial}{\partial R^j} \right\rangle = \langle\, \omega^i \,,\, e_j \,\rangle = \delta^i_j \qquad (i, j = 1, 2, 3) \,, \tag{9.23}$$

are the basis for a general procedure to derive the transformation rules for vector calculus differential operators between rectangular and curvilinear coordinates.

Problem 9.1 Consider spherical polar coordinates with respect to the orthonormal frame field $\{e_r, e_\theta, e_\phi\}$ [given by (4.28) to (4.30)]. The rectangular coordinates (x, y, z) and the spherical polar coordinates (r, θ, ϕ) of a point p are related by

$$x = r \sin\theta \cos\phi \,, \quad y = r \sin\theta \sin\phi \,, \quad z = r \cos\theta \,.$$

Setting $x = R^1, y = R^2, z = R^3$, where R^i are the quantities appearing in (9.4), use (9.8) and the explicit expression for the matrix g given by (4.31) to show that

$$\omega^1 = dr \,, \qquad \omega^2 = r d\theta \,, \qquad \omega^3 = r \sin\theta d\phi \,. \tag{9.24}$$

Hence verify that

$$dx \wedge dy \wedge dz = \omega^1 \wedge \omega^2 \wedge \omega^3 = r^2 \sin\theta\, dr \wedge d\theta \wedge d\phi \,, \tag{9.25}$$

which is the familiar result for the volume element in spherical coordinates (usually written without the wedge products).

Problem 9.2 Assume the duality condition $\langle\, dR^i, \epsilon_j \,\rangle = \delta^i_j$. Use (9.8) in the form

$$\omega^i = dR^j\, (g^{-1})^i_j \tag{9.26}$$

and the second of Eqs. (9.4), that is, $e_i = g_i^j \epsilon_j$ [with $g \in SO(N)$] to prove the duality condition $\langle \omega^i , e_j \rangle = \delta^i_j$. Verify this result for the spherical polar frame field $\{e_r, e_\theta, e_\phi\}$ with the ω^i given in Problem 9.1.

Problem 9.3 Use the results for ω^i given in Problem 9.1 and the explicit expression for the connection matrix of 1-forms (ω_i^j) given in (4.32) to verify the integrability conditions (9.13) and (9.14) for the spherical polar orthonormal frame field $\{e_r, e_\theta, e_\phi\}$.

The **curvilinear coordinates** (x^1, x^2, x^3) of a point $p \in \mathbb{R}^3$ [whose rectangular coordinates are (R^1, R^2, R^3)] associated with the frame field $\{e_1, e_2, e_3\}$ are defined by

$$\omega^1 = \lambda_1(x^1,x^2,x^3)\,dx^1\,,\quad \omega^2 = \lambda_2(x^1,x^2,x^3)\,dx^2\,,\quad \omega^3 = \lambda_3(x^1,x^2,x^3)\,dx^3\,, \tag{9.27}$$

where $\lambda_i(x^1,x^2,x^3)$ $(i = 1,2,3)$ are functions on $\mathbb{R}^3$. These coordinates can obviously be used as local coordinates in $\mathbb{R}^3$. So we have

$$\left\langle dx^i , \frac{\partial}{\partial x^j} \right\rangle = \delta^i_j = \langle \omega^i , e_j \rangle = \lambda_i \,\langle dx^i ,\, e_j \rangle\,, \tag{9.28}$$

where on the right-hand side of the last equlity, the repeated index "i" does not correspond to a sum. Thus

$$e_i = \frac{1}{\lambda_i}\frac{\partial}{\partial x^i} \qquad (i = 1,2,3)\,. \tag{9.29}$$

Note that, whereas the e_i's are unit vectors in the tangent space $T_p(\mathbb{R}^3)$, the vectors $\partial/\partial x^i$ are not (although they still span the tangent space).

Consider the **gradient** operator [cf. (7.2)]. We have, for a scalar function $f : \mathbb{R}^3 \to \mathbb{R}$,

$$\begin{aligned} df &= \nabla f \cdot d\boldsymbol{R} = (\nabla f)\cdot(\omega^1\, e_1 + \omega^2\, e_2 + \omega^3\, e_3) \\ &= \frac{\partial f}{\partial x^i}\, dx^i = \sum_i \frac{1}{\lambda_i}\frac{\partial f}{\partial x^i}\,\omega^i\,. \end{aligned} \tag{9.30}$$

Hence we obtain

$$\boxed{\nabla f = \frac{1}{\lambda_1}\frac{\partial f}{\partial x^1}\, e_1 + \frac{1}{\lambda_2}\frac{\partial f}{\partial x^2}\, e_2 + \frac{1}{\lambda_3}\frac{\partial f}{\partial x^3}\, e_3}\,. \tag{9.31}$$

Next consider the **Laplacian** operator acting on a scalar function f. In rectangular coordinates $(R^1, R^2, R^3) = (x, y, z)$ it is defined by

$$\nabla^2 f \equiv \frac{\partial^2 f}{\partial x^2} + \frac{\partial^2 f}{\partial y^2} + \frac{\partial^2 f}{\partial z^2}\,. \tag{9.32}$$

We will show that

$$\nabla^2 f\, dx \wedge dy \wedge dz = d(*\, df) \quad , \tag{9.33}$$

where in this equation (and in what follows) the Hodge-star operator is understood to be with respect to the orientation $dx \wedge dy \wedge dz$, that is, $*_{dx\wedge dy\wedge dz}$ [cf. Def. 3.1]. Starting with

$$df = \frac{\partial f}{\partial x}\, dx + \frac{\partial f}{\partial y}\, dy + \frac{\partial f}{\partial z}\, dz \quad , \tag{9.34}$$

we have, from (3.12),

$$*\, df = \frac{\partial f}{\partial x}\, dy \wedge dz + \frac{\partial f}{\partial y}\, dz \wedge dx + \frac{\partial f}{\partial z}\, dx \wedge dy \quad . \tag{9.35}$$

On exterior differentiation [using (6.13)] it follows that

$$d(*\, df) = \left(\frac{\partial^2 f}{\partial x^2} + \frac{\partial^2 f}{\partial y^2} + \frac{\partial^2 f}{\partial z^2} \right) dx \wedge dy \wedge dz \quad , \tag{9.36}$$

which is the result (9.33). It follows from (9.21) that

$$d(*df) = \nabla^2 f\, \omega^1 \wedge \omega^2 \wedge \omega^3 \, . \tag{9.37}$$

On the other hand, operating on the right-hand side of the last equality of (9.30) with the help of (3.12) again, and recalling $\omega^i = \lambda_i\, dx^i$ [see (9.27)], we obtain

$$\begin{aligned} *df &= \frac{1}{\lambda_1}\frac{\partial f}{\partial x^1}\,\omega^2 \wedge \omega^3 + \frac{1}{\lambda_2}\frac{\partial f}{\partial x^2}\,\omega^3 \wedge \omega^1 + \frac{1}{\lambda_3}\frac{\partial f}{\partial x^3}\,\omega^1 \wedge \omega^2 \\ &= \frac{\lambda_2\lambda_3}{\lambda_1}\left(\frac{\partial f}{\partial x^1}\right) dx^2 \wedge dx^3 + \frac{\lambda_1\lambda_3}{\lambda_2}\left(\frac{\partial f}{\partial x^2}\right) dx^3 \wedge dx^1 \\ &+ \frac{\lambda_1\lambda_2}{\lambda_3}\left(\frac{\partial f}{\partial x^3}\right) dx^1 \wedge dx^2 \quad . \end{aligned} \tag{9.38}$$

On exterior differentiation [using (6.13)] this equation yields

$$\begin{aligned} d(*df) = &\left\{ \frac{\partial}{\partial x^1}\left(\frac{\lambda_2\lambda_3}{\lambda_1}\frac{\partial f}{\partial x^1}\right) + \frac{\partial}{\partial x^2}\left(\frac{\lambda_1\lambda_3}{\lambda_2}\frac{\partial f}{\partial x^2}\right) + \frac{\partial}{\partial x^3}\left(\frac{\lambda_1\lambda_2}{\lambda_3}\frac{\partial f}{\partial x^3}\right) \right\} \\ &dx^1 \wedge dx^2 \wedge dx^3 \quad . \end{aligned} \tag{9.39}$$

But by (9.27),

$$dx^1 \wedge dx^2 \wedge dx^3 = \frac{\omega^1 \wedge \omega^2 \wedge \omega^3}{\lambda_1\lambda_2\lambda_3} \quad . \tag{9.40}$$

Using this expression in (9.39) and comparing with (9.37), we obtain the following formula for the Laplacian of a scalar function f in the curvilinear coordinates x^1, x^2, x^3:

$$\boxed{\nabla^2 f = \frac{1}{\lambda_1\lambda_2\lambda_3}\left\{ \frac{\partial}{\partial x^1}\left(\frac{\lambda_2\lambda_3}{\lambda_1}\frac{\partial f}{\partial x^1}\right) + \frac{\partial}{\partial x^2}\left(\frac{\lambda_1\lambda_3}{\lambda_2}\frac{\partial f}{\partial x^2}\right) + \frac{\partial}{\partial x^3}\left(\frac{\lambda_1\lambda_2}{\lambda_3}\frac{\partial f}{\partial x^3}\right) \right\}} \, . \tag{9.41}$$

Moving on to the **divergence** we let $\boldsymbol{A} = A^1\, \boldsymbol{e}_1 + A^2\, \boldsymbol{e}_2 + A^3\, \boldsymbol{e}_3$ be a vector field on $\mathbb{R}^3$, where $A^1(x^1, x^2, x^3)$, $A^2(x^1, x^2, x^3)$ and $A^3(x^1, x^2, x^3)$ are smooth functions on $\mathbb{R}^3$. With the components of $\boldsymbol{A}$, construct the 2-form

$$\psi = A^1\, \omega^2 \wedge \omega^3 + A^2\, \omega^3 \wedge \omega^1 + A^3\, \omega^1 \wedge \omega^2 \quad , \tag{9.42}$$

which, in terms of rectangular coordinates (x, y, z), can be expressed as

$$\psi = \mathcal{A}^1\, dy \wedge dz + \mathcal{A}^2\, dz \wedge dx + \mathcal{A}^3\, dx \wedge dy \;, \tag{9.43}$$

where $\mathcal{A}^i$ are the rectangular components of the vector field $\boldsymbol{A}$ and are related to the curvilinear components A^i by

$$\mathcal{A}^i = A^j\, g^i_j \;, \tag{9.44}$$

with the matrix g given by (9.4). We then have, from (7.12), (7.13) and (9.21) (in which we set $R^1 = x, R^2 = y, R^3 = z$),

$$d\psi = (\nabla \cdot \boldsymbol{A})\, \omega^1 \wedge \omega^2 \wedge \omega^3 \quad . \tag{9.45}$$

Using (9.27) to rewrite (9.42) as

$$\psi = \lambda_2\lambda_3 A^1 dx^2 \wedge dx^3 + \lambda_1\lambda_3 A^2 dx^3 \wedge dx^1 + \lambda_1\lambda_2 A^3 dx^1 \wedge dx^2 \quad , \tag{9.46}$$

we can compute $d\psi$ to obtain

$$d\psi = \left\{ \frac{\partial}{\partial x^1}\left(\lambda_2\lambda_3 A^1\right) + \frac{\partial}{\partial x^2}\left(\lambda_1\lambda_3 A^2\right) + \frac{\partial}{\partial x^3}\left(\lambda_1\lambda_2 A^3\right) \right\} dx^1 \wedge dx^2 \wedge dx^3 \;. \tag{9.47}$$

Using (9.40) again yields the following expression for the divergence of a vector field in curvilinear coordinates:

$$\boxed{\nabla \cdot \boldsymbol{A} = \frac{1}{\lambda_1\lambda_2\lambda_3} \left\{ \frac{\partial}{\partial x^1}\left(\lambda_2\lambda_3 A^1\right) + \frac{\partial}{\partial x^2}\left(\lambda_1\lambda_3 A^2\right) + \frac{\partial}{\partial x^3}\left(\lambda_1\lambda_2 A^3\right) \right\}} \quad . \tag{9.48}$$

Problem 9.4 Prove that the two 2-forms on the right-hand sides of (9.42) and (9.43) are indeed equal to each other. Proceed by using the Hodge star on each expression and show that the resulting expressions are the same. The one-to-one property of the Hodge star map then establishes the claim. *Hints*: Use (9.8) and (9.44).

Finally we consider the **curl**. Start with the vector field $\boldsymbol{A} = A^1\, \boldsymbol{e}_1 + A^2\, \boldsymbol{e}_2 + A^3\, \boldsymbol{e}_3 = \mathcal{A}^1\, \partial/\partial x + \mathcal{A}^2\, \partial/\partial y + \mathcal{A}^3\, \partial/\partial z$ again. Its dual 1-form is

$$\alpha = A^1\omega^1 + A^2\omega^2 + A^3\omega^3 = \mathcal{A}^1\, dx + \mathcal{A}^2\, dy + \mathcal{A}^3\, dz \;. \tag{9.49}$$

According to (7.10), $\nabla \times \boldsymbol{A}$ is obtained from α through the relation

$$*(d\alpha) = (\nabla \times \boldsymbol{A})^1\omega^1 + (\nabla \times \boldsymbol{A})^2\omega^2 + (\nabla \times \boldsymbol{A})^3\omega^3 \quad . \tag{9.50}$$

Writing

$$\alpha = \lambda_1 A^1 dx^1 + \lambda_2 A^2 dx^2 + \lambda_3 A^3 dx^3 \quad , \tag{9.51}$$

we can exteriorly differentiate to obtain

$$\begin{aligned} d\alpha = \; & \frac{\partial}{\partial x^2}(\lambda_1 A^1) dx^2 \wedge dx^1 + \frac{\partial}{\partial x^3}(\lambda_1 A^1) dx^3 \wedge dx^1 \\ & + \frac{\partial}{\partial x^1}(\lambda_2 A^2) dx^1 \wedge dx^2 + \frac{\partial}{\partial x^3}(\lambda_2 A^2) dx^3 \wedge dx^2 \\ & + \frac{\partial}{\partial x^1}(\lambda_3 A^3) dx^1 \wedge dx^3 + \frac{\partial}{\partial x^2}(\lambda_3 A^3) dx^2 \wedge dx^3 \\ = \; & \left\{ \frac{\partial}{\partial x^2}(\lambda_3 A^3) - \frac{\partial}{\partial x^3}(\lambda_2 A^2) \right\} dx^2 \wedge dx^3 \\ & + \left\{ \frac{\partial}{\partial x^3}(\lambda_1 A^1) - \frac{\partial}{\partial x^1}(\lambda_3 A^3) \right\} dx^3 \wedge dx^1 \\ & + \left\{ \frac{\partial}{\partial x^1}(\lambda_2 A^2) - \frac{\partial}{\partial x^2}(\lambda_1 A^1) \right\} dx^1 \wedge dx^2 \\ = \; & \frac{1}{\lambda_2\lambda_3} \left\{ \frac{\partial}{\partial x^2}(\lambda_3 A^3) - \frac{\partial}{\partial x^3}(\lambda_2 A^2) \right\} \omega^2 \wedge \omega^3 \\ & + \frac{1}{\lambda_1\lambda_3} \left\{ \frac{\partial}{\partial x^3}(\lambda_1 A^1) - \frac{\partial}{\partial x^1}(\lambda_3 A^3) \right\} \omega^3 \wedge \omega^1 \\ & + \frac{1}{\lambda_1\lambda_2} \left\{ \frac{\partial}{\partial x^1}(\lambda_2 A^2) - \frac{\partial}{\partial x^2}(\lambda_1 A^1) \right\} \omega^1 \wedge \omega^2 \quad . \end{aligned} \tag{9.52}$$

Noting that

$$*(\omega^2 \wedge \omega^3) = \omega^1 \; , \quad *(\omega^3 \wedge \omega^1) = \omega^2 \; , \quad *(\omega^1 \wedge \omega^2) = \omega^3 \; , \tag{9.53}$$

Eq. (9.50) then allows us to write

$$\boxed{\nabla \times \boldsymbol{A} = \frac{1}{\lambda_1\lambda_2\lambda_3} \begin{vmatrix} \lambda_1 \boldsymbol{e}_1 & \lambda_2 \boldsymbol{e}_2 & \lambda_3 \boldsymbol{e}_3 \\ \dfrac{\partial}{\partial x^1} & \dfrac{\partial}{\partial x^2} & \dfrac{\partial}{\partial x^3} \\ \lambda_1 A^1 & \lambda_2 A^2 & \lambda_3 A^3 \end{vmatrix}} \quad . \tag{9.54}$$

Problem 9.5 Show that for *cylindrical coordinates*, where $x^1 = r$ (radial distance), $x^2 = \phi$ (azimuthal angle on the xy-plane), $x^3 = z$,

$$\omega^1 = dr \; , \qquad \omega^2 = r\,d\phi \; , \qquad \omega^3 = dz \; . \tag{9.55}$$

Problem 9.6 Use the general formulas for the gradient, Laplacian, divergence and curl in curvilinear coordinates [(9.31), (9.41), (9.48) and (9.54), respectively] to show that for the spherical polar coordinates $x^1 = r$, $x^2 = \theta$, $x^3 = \phi$,

$$\nabla f = \frac{\partial f}{\partial r}\, \boldsymbol{e}_r + \frac{1}{r}\frac{\partial f}{\partial \theta}\, \boldsymbol{e}_\theta + \frac{1}{r\sin\theta}\frac{\partial f}{\partial \phi}\, \boldsymbol{e}_\phi \quad , \tag{9.56}$$

$$\nabla^2 f = \frac{1}{r^2 \sin\theta}\left\{ \frac{\partial}{\partial r}\left(r^2 \sin\theta \frac{\partial f}{\partial r}\right) + \frac{\partial}{\partial \theta}\left(\sin\theta \frac{\partial f}{\partial \theta}\right) + \frac{\partial}{\partial \phi}\left(\frac{1}{\sin\theta}\frac{\partial f}{\partial \phi}\right)\right\} , \tag{9.57}$$

$$\nabla \cdot \boldsymbol{A} = \frac{1}{r^2}\frac{\partial}{\partial r}\left(r^2 A_r\right) + \frac{1}{r\sin\theta}\frac{\partial}{\partial \theta}\left(\sin\theta\, A_\theta\right) + \frac{1}{r\sin\theta}\frac{\partial A_\phi}{\partial \phi} \quad , \tag{9.58}$$

$$\begin{aligned} \nabla \times \boldsymbol{A} = & \frac{1}{r\sin\theta}\left\{ \frac{\partial}{\partial \theta}\left(A_\phi \sin\theta\right) - \frac{\partial A_\theta}{\partial \phi}\right\} \boldsymbol{e}_r \\ & + \frac{1}{r\sin\theta}\left\{ \frac{\partial A_r}{\partial \phi} - \sin\theta \frac{\partial}{\partial r}\left(rA_\phi\right)\right\} \boldsymbol{e}_\theta + \frac{1}{r}\left\{ \frac{\partial}{\partial r}\left(rA_\theta\right) - \frac{\partial A_r}{\partial \theta}\right\} \boldsymbol{e}_\phi \quad , \end{aligned} \tag{9.59}$$

where, on $\mathbb{R}^3$, f is a smooth function and

$$\boldsymbol{A} = A_r\, \boldsymbol{e}_r + A_\theta\, \boldsymbol{e}_\theta + A_\phi\, \boldsymbol{e}_\phi \tag{9.60}$$

is a smooth vector field.

Chapter 10

Mechanical Constraints: The Frobenius Theorem

Examples of constrained motions in classical mechanics abound. Broadly speaking these are motions in which the variables of the *configuration space* of the classical system are not all independent. In this chapter we will only consider *time-independent constraints.* Simple examples are motions of a point particle in $\mathbb{R}^3$ confined to either a fixed curve (a 1-dimensional submanifold of $\mathbb{R}^3$) or a fixed surface (a 2-dimensional submanifold of $\mathbb{R}^3$). More generally we may have the trajectory of a point in an n-dimensional differentiable manifold confined to a fixed m-dimensional submanifold, where $m < n$. These are referred to as **holonomic constraints**. In these cases, the submanifold can be specified by a set of $r = n - m$ (time-independent) equations relating the n local coordinates $(x^1, \ldots, x^n)$ of a neighborhood of M:

$$f_1(x^1, \ldots, x^n) = 0 \, , \ldots\ldots \, , f_r(x^1, \ldots, x^n) = 0 \, , \tag{10.1}$$

where $f_1, \ldots, f_r$ are smooth functions on M. One can then introduce a set of m coordinates, $q^1(x^1, \ldots, x^n), \ldots, q^m(x^1, \ldots, x^n)$, which can be used as local coordinates on the m-dimensional submanifold. For example, motion confined to a two-dimensional spherical surface of radius R embedded in $\mathbb{R}^3$ is given by the single constraint equation $x^2+y^2+z^2 = R^2$, with the angle coordinates $\theta(x, y, z)$ and $\phi(x, y, z)$ serving as coordinates on the surface. For motion confined to the z-axis (in $\mathbb{R}^3$), the two constraint equations would be, trivially, $x = 0$ and $y = 0$, with the single coordinate z as the submanifold coordinate.

Situations where the relationship between the manifold coordinates cannot be expressed in the form of a set of equations as in (10.1) are said to exhibit **non-holonomc constraints**. A particularly interesting subset of which involves explicitly time-independent constraint equations relating the velocities dx^i/dt (instead of coordinates), or equivalently, after elimination of the differential dt, a set of $r < n$ equations relating $dx^1, \dots, dx^n$:

$$\begin{aligned} \omega^{m+1} &\equiv f_{11}(x^1,\dots,x^n)\,dx^1 + \dots + f_{1n}(x^1,\dots,x^n)\,dx^n = 0\,, \\ &\vdots \\ \omega^n &\equiv f_{r1}(x^1,\dots,x^n)\,dx^1 + \dots + f_{rn}(x^1,\dots,x^n)\,dx^n = 0\,, \end{aligned} \tag{10.2}$$

where $m = n - r$ and ω^s, $m+1 \le s \le n$ stand for the r 1-forms on M as indicated. A frequently cited example is that of a vertically oriented disk rolling without slipping on the horizontal plane. This case involves a 4-dimensional configuration space with, for example, (x, y) being the rectangular coordinates of the point of contact between the disk and the horizontal plane, θ the orientation angle (with respect to some fixed direction in space) of the (horizontal) rotational axis of the disk, and ϕ the rotational angle of a fixed radial line on the disk. The two constraint equations can be written as:

$$dx - R\sin\theta\,d\phi = 0\,, \qquad dy + R\cos\theta\,d\phi = 0\,, \tag{10.3}$$

where R is the radius of the disk. The general system of equations (10.2) is called a **Pfaffian** system. An interesting, and important, mathematical question is then the following: Under what conditions will this system of equations be *completely integrable*, that is, the solution be expressible as the set of equations (10.1)? When a solution does exist, the submanifold that it determines is known as the **integral submanifold** of the Pfaffan system, which then describes holonomic constraints. If the Pfaffian system is not completely integrable, it describes non-holonomic constraints. The answer to the above-posed question is given by the Frobenius Theorem.

Before we state the theorem, let us first assume integrability and investigate its consequences. First write the analytical conditions for complete integrability of the Pfaffian system (10.2). These can be expressed as the existence of $r\,(= n-m)$ functions $q^{m+1}(x^1,\dots,x^n),\dots,q^n(x^1,\dots,x^n)$ such that

$$dq^{m+1} = \omega^{m+1} = 0\,, \dots\dots\dots\,, dq^n = \omega^n = 0\,. \tag{10.4}$$

One then seeks another m functions $q^1(x^1,\dots,x^n),\dots q^m(x^1,\dots x^n)$ such that the entire set of n functions $\{q^1,\dots,q^m,q^{m+1}\dots,q^n\}$ can be used as local coordinates in M while $\{q^1,\dots,q^m\}$ can be used as local coordinates in the integral submanifold of the Pfaffian system. [To get a feel for the significance of (10.4), consider the trivial case of the submanifold – the xy-plane – defined by the equation $dz = 0$.] If such a set of coordinates exists, the tangent vector fields $\partial/\partial q^1,\dots,\partial/\partial q^m,\partial/\partial q^{m+1},\dots,\partial/\partial q^n$ would form a frame field on M, while $\partial/\partial q^1,\dots,\partial/\partial q^m$ would form a frame field on the integral submanifold. The r 1-forms $\omega^{m+1},\dots,\omega^n$ of the Pfaffian system (10.2) would then span an r-dimensional subspace of the n-dimensional one spanned by the coframe field $\omega^1,\dots,\omega^m,\omega^{m+1},\dots,\omega^n$ dual to the frame field $\partial/\partial q^1,\dots,\partial/\partial q^n$, which satisfies the duality pairing conditions

$$\left\langle \omega^i, \frac{\partial}{\partial q^j} \right\rangle = \delta^i_j \qquad (i,j = 1,\dots,n)\,. \tag{10.5}$$

The n-dimensional space spanned by the coframe field $\omega^1, \dots, \omega^n$ is precisely the cotangent space at a particular point $q \in M$, and correspondingly, the n-dimensional space spanned by the frame field $\partial/\partial q^1, \dots, \partial/\partial q^n$ is precisely the tangent space at the point q.

Denote the m-dimensional integral submanifold, if it exists, by S. The tangent space at a point $q \in S$ would then be spanned by $\partial/\partial q^1, \dots, \partial/\partial q^m$. Let another frame field on S be $\boldsymbol{X}_1, \dots, \boldsymbol{X}_m$. These tangent vector fields would then be related to $\partial/\partial q^1, \dots, \partial/\partial q^m$ by the linear transformations

$$\boldsymbol{X}_\alpha = a_\alpha^\beta(q)\,\frac{\partial}{\partial q^\beta} \qquad (\alpha, \beta = 1, \dots, m)\,, \tag{10.6}$$

where the coefficients a_α^β are functions on S. Let us calculate an arbitrary Lie bracket $[\boldsymbol{X}_\alpha, \boldsymbol{X}_\beta]$ among these tangent vector fields [cf. (6.15)]. We obtain after some algebra, since $[\partial/\partial q^\alpha\,, \partial/\partial q^\beta] = 0$,

$$[\boldsymbol{X}_\alpha, \boldsymbol{X}_\beta] = \sum_{\gamma=1}^{m} C_{\alpha\beta}^{\gamma}\,\boldsymbol{X}_\gamma\,, \tag{10.7}$$

where

$$C_{\alpha\beta}^{\gamma} = \sum_{\delta,\eta}\left(a_\alpha^\delta\,\frac{\partial a_\beta^\eta}{\partial q^\delta} - a_\beta^\delta\,\frac{\partial a_\alpha^\eta}{\partial q^\delta}\right)(a^{-1})_\eta^\gamma \tag{10.8}$$

and a^{-1} denotes the inverse matrix of the matrix (a_α^β). Equation (10.7) states that the Lie bracket of any two vector fields of the set $\{\boldsymbol{X}_1, \dots, \boldsymbol{X}_m\}$, which spans the tangent space $T_q(S)$ at any point q of the integral submanifold S, is a linear combination of the same set, and hence belongs to $T_q(S)$ also. Equation (10.7) satisfied by a set of tangent vector fields is the so-called **Frobenius condition** for the set, which is then seen to be a necessary condition for complete integrability of the Pfaffian set (10.2).

Problem 10.1 Use the result for the calculation of Lie brackets given by (6.22) to verify the Frobenius condition (10.7).

A set of m linearly independent smooth tangent vector fields $\{\boldsymbol{X}_1, \dots, \boldsymbol{X}_m\}$ on an n-dimensional manifold M (with $m < n$) spans an m-dimensional smooth *tangent subspace field* on M, also called a smooth **distribution** on M, denoted by L^m. We then write

$$L^m = span\,\{\boldsymbol{X}_1, \dots, \boldsymbol{X}_m\}\,. \tag{10.9}$$

We have thus seen that, for an arbitrary distribution L^m on an n-dimensional differential manifold M (with $m < n$), the necessary condition for the existence of a local coordinate chart $(Q;\, q^1, \dots, q^n)$ in M such that $L^m|_Q$ (the restriction of L^m to the neighborhood Q) is spanned by $\{\partial/\partial q^1, \dots, \partial/\partial q^m\}$, is that the Frobenius condition be satisfied by any two vector fields in L^m, that is, if $\boldsymbol{X}, \boldsymbol{Y} \in L^m$, then $[\boldsymbol{X}, \boldsymbol{Y}] \in L^m$. The Frobenius Theorem states that the Frobenius condition is also a sufficient condition. We will not prove the sufficiency part, and will now state the **Frobenius Theorem.**

Theorem 10.1. *Suppose the (everywhere) linearly independent tangent vector fields $\boldsymbol{X}_1, \dots, \boldsymbol{X}_m$ span an m-dimensional smooth distribution L^m on an n-dimensional differential manifold M with $m < n$. Then there exists a local coordinate chart $(Q;\, q^1, \dots, q^n)$ in M such that*

$$L^m|_Q = span\left\{\frac{\partial}{\partial q^1}, \dots, \frac{\partial}{\partial q^m}\right\} \tag{10.10}$$

if and only if $\boldsymbol{X}_1, \dots, \boldsymbol{X}_m$ satisfy the Frobenius condition [(10.7)].

This important theorem can also be conveniently, and usefully, stated in its dual form, using the 1-forms $\omega^{m+1}, \dots, \omega^n$ in the Pfaffian system (10.2). Consider again an m-dimensional smooth distribution on M given by $L^m = span\{\boldsymbol{X}_1, \dots, \boldsymbol{X}_m\}$. Then, for any $q \in M$, $L^m(q)$ is an m-dimensional linear subspace of the tangent space $T_q(M)$. Define the following linear subspace to the cotangent space $T_q^*(M)$:

$$(L^m(q))^\perp \equiv \{\omega \in T_q^*(M) \,|\, \langle \boldsymbol{X}_q, \omega \rangle = 0 \ \text{ for any } \ \boldsymbol{X}_q \in L^m(q)\}\,. \tag{10.11}$$

This subspace of $T_q^*(M)$ is called the *annihilator subspace* of $L^m(q)$. In a neighborhood of any point $q \in M$, there then exists $n - m$ linearly independent 1-forms $\omega^{m+1}, \dots, \omega^n$ that span the annihilator subspace $(L^m(q))^\perp$ at any point in the neighborhood, with their dual tangent vector fields denoted by $\boldsymbol{X}_{m+1}, \dots, \boldsymbol{X}_n$. These, together with the dual 1-forms $\omega^1, \dots, \omega^m$ to $\boldsymbol{X}_1, \dots, \boldsymbol{X}_m$, then form a coframe field in the neighborhood dual to the frame field $\boldsymbol{X}_1, \dots, \boldsymbol{X}_n$. Locally, the distribution L^m is then determined by the Pfaffian system (10.2):

$$\omega^s = 0 \qquad (m + 1 \le s \le n)\,. \tag{10.12}$$

Recalling (6.14) we have, for $1 \le \alpha, \beta \le m$, $m + 1 \le s \le n$,

$$\begin{aligned}\langle \boldsymbol{X}_\alpha \wedge \boldsymbol{X}_\beta, d\omega^s \rangle &= \boldsymbol{X}_\alpha \langle \boldsymbol{X}_\beta, \omega^s \rangle - \boldsymbol{X}_\beta \langle \boldsymbol{X}_\alpha, \omega^s \rangle - \langle [\boldsymbol{X}_\alpha, \boldsymbol{X}_\beta], \omega^s \rangle \\ &= -\langle [\boldsymbol{X}_\alpha, \boldsymbol{X}_\beta], \omega^s \rangle\,.\end{aligned} \tag{10.13}$$

Hence the set $\{\boldsymbol{X}_1, \dots, \boldsymbol{X}_m\}$ satisfies the Frobenius condition if and only if

$$\langle \boldsymbol{X}_\alpha \wedge \boldsymbol{X}_\beta, d\omega^s \rangle = 0 \qquad (1 \le \alpha, \beta \le m,\ m + 1 \le s \le n)\,. \tag{10.14}$$

Now $d\omega^s$ can be expanded in terms of *all* the dual 1-forms $\omega^1, \dots, \omega^n$ as

$$d\omega^s = \sum_{t=m+1}^{n} \psi_t^s \wedge \omega^t + \sum_{\alpha,\beta=1}^{m} a_{\alpha\beta}^s \omega^\alpha \wedge \omega^\beta\,, \tag{10.15}$$

where the ψ_t^s's are 1-forms on M and the smooth function coefficients $a^s_{\alpha\beta}$ are antisymmetric with respect to the lower indices. Plugging this expression into (10.14) we obtain, on recalling (2.30) with the duality conditions $\langle \boldsymbol{X}_j, \omega^i \rangle = \delta^i_j$,

$$a^s_{\alpha\beta} = 0 \qquad (1 \le \alpha, \beta \le m,\ m+1 \le s \le n)\,. \tag{10.16}$$

Thus (10.15) becomes

$$d\omega^s = \sum_{t=m+1}^{n} \psi_t^s \wedge \omega^s\,. \tag{10.17}$$

This equation is often expressed as

$$d\omega^s = 0\, mod\, (\omega^{m+1}, \dots, \omega^n) \qquad (m+1 \le s \le n)\,, \tag{10.18}$$

and also known as the Frobenius condition. It is the counterpart of the Frobenius condition given by (10.7). So the Frobenius Theorem 10.1 can be recast in its somewhat simpler dual form as

Theorem 10.2. *The Pfaffian system of r equations of 1-forms*

$$\omega^s = 0 \qquad (1 \le s \le r) \tag{10.19}$$

is completely integrable if and only if

$$d\omega^s = \sum_{t=1}^{r} \psi_t^s \wedge \omega^t\,, \tag{10.20}$$

where the ψ_t^s are 1-forms. Equivalently,

$$d\omega^s = 0\, mod\, (\omega^1 \dots, \omega^r)\,. \tag{10.21}$$

Problem 10.2 Use Theorem 10.2 to give a necessary and sufficient condition for the Pfaffian equation

$$A(x,y,z)\,dx + B(x,y,z)\,dy + C(x,y,z)\,dz = 0 \tag{10.22}$$

to be integrable in $\mathbb{R}^3$. Integrability in this case means the existence of a smooth function f satisfying $df(x,y,z) = Adx + Bdy + Cdz = 0$, so that a 2-dimensional integral submanifold is determined by the equation $f(x,y,z) = const.$ The function f is known as a **first integral** of the Pfaffian equation. Is the equation

$$x\,dx + y\,dy + z\,dz = 0 \tag{10.23}$$

integrable? If so, find a first integral for it.

Problem 10.3 Consider a mechanical system with three degrees of freedom x, y and ϕ $(-\infty < x, y < \infty,\ 0 \le \phi \le 2\pi)$ with one constraint relation

$$\sin\phi\,\dot{x} + \cos\phi\,\dot{y} = 0\,. \tag{10.24}$$

Show, by using Theorem 10.2 that the associated Pfaffian equation

$$\sin\phi\, dx + \cos\phi\, dy = 0 \tag{10.25}$$

is not integrable.

Chapter 11

Flows and Lie Derivatives

The intuitive kinematical picture of fluid flows in $\mathbb{R}^3$, in which the velocity field $\boldsymbol{v}(\boldsymbol{r},t)$ (t being the time) is the tangent vector field whose integral curves are the flow lines tangent to $\boldsymbol{v}$ at every point (at which the field is defined), lays the physical groundwork for an appreciation of flows in arbitrary differentiable manifolds. In general the velocity field satisfies a partial differential equation involving $\partial \boldsymbol{v}/\partial t$ [see Eq. (13.65)]. In this chapter, we will focus on the so-called *static fields*, which do not depend on the time t explicitly. The mathematical theory of static flows will play a crucial role in Hamiltonian dynamics (Chapter 39). The key concept in this theory is the so-called *one-parameter group of diffeomorphisms* on a differentiable manifold.

Consider a differentiable manifold M and a Lie group G that acts on elements in M, such as the group $SO(3)$ which acts on $\mathbb{R}^3$. Picture a point traversing a curve in M over some period of time (a flow line). Let this flow line be described by a map $\varphi : \mathbb{R} \times M \to M$, and write, for any $(t, x) \in \mathbb{R} \times M$,

$$\varphi_t(x) \equiv \varphi(t, x) \,. \tag{11.1}$$

If $\varphi_t \in G$ satisfies the conditions:

(1) $\varphi_0(x) = x$;

(2) $\varphi_s \circ \varphi_t = \varphi_{s+t}$ for any $s, t \in \mathbb{R}$,

then we call φ_t a **one-parameter group of diffeomorphisms** on M, the parameter being t. The above two properties guarantee that $\{\varphi_t\}$ is indeed a group (a subgroup of G in fact), with the identity being φ_0 and the inverse given by $\varphi_t^{-1} = \varphi_{-t}$. *The term **diffeomorphism** generally means a differentiable map whose inverse is also differentiable.* The flow line can be more appropriately expressed as the *parametrized curve* in M passing through the point $x \in M$ at $t = 0$:

$$\gamma_x(t) = \varphi_t(x) \,. \tag{11.2}$$

It is called the **orbit** of φ_t through the point x.

If, at each point $x \in M$, a tangent vector $\boldsymbol{X}_x \in T_x(M)$ tangent to the curve $\gamma_x(t)$ at $t = 0$ is assigned, we obtain the so-called **induced tangent vector field** $\boldsymbol{X}$ corresponding to the one-parameter group of diffeomorphisms φ_t on M. It is readily seen that $\boldsymbol{X}$ is a smooth tangent vector field. The parametrized curve $\gamma_x(t)$ is said to be an **integral curve** of the tangent vector field $\boldsymbol{X}$. It has the property that, if $y = \gamma_x(s)$ (that is, y is on the orbit of γ_x), then $\boldsymbol{X}_y$ is a tangent vector tangent to the curve $\gamma_x(t)$ at $t = s$. This is expressed symbolically as

$$(\varphi_s)_* \boldsymbol{X}_x = \boldsymbol{X}_{\gamma_x(s)} \,, \tag{11.3}$$

where $(\varphi_s)_*$ is the tangent map induced by φ_s [cf. (6.27)]. In local coordinates x^i, the induced tangent vector field can be expressed as

$$\boldsymbol{X}_x = X^i_x \left(\frac{\partial}{\partial x^i} \right)_x \,, \tag{11.4}$$

where the components X^i are given by [recall (5.1)]

$$X^i_x = \left. \frac{dx^i(\gamma_x(t))}{dt} \right|_{t=0} \,. \tag{11.5}$$

Equation (11.3) then implies that, if $y = \gamma_x(s)$, then

$$\boldsymbol{X}_y = X^i_y \left(\frac{\partial}{\partial x^i} \right)_y \,, \tag{11.6}$$

with

$$X^i_y = \left. \frac{dx^i(\gamma_p(t))}{dt} \right|_{t=s} \,. \tag{11.7}$$

If $\boldsymbol{X}_x \neq 0$ at $x \in M$, then by Theorem 5.2 there exist local coordinates u^i near x such that $\boldsymbol{X} = \partial / \partial u^1$. By (11.5) φ_t then acts on $x = (u^1, \dots, u^n)$ in the following simple way:

$$\varphi_t(u^1, \dots, u^n) = (u^1 + t, u^2, \dots, u^n) \,, \tag{11.8}$$

that is, simply as a displacement along the u^1-axis.

Problem 11.1 Use Def. 5.1 for a smooth tangent vector field to prove that the induced tangent vector field (induced by a one-parameter group of diffeomorphisms) as defined above is a smooth tangent vector field.

The above discussion shows that every one-parameter group of diffeomorphisms induces a tangent vector field. Conversely, one can ask the question:

Is every tangent vector field induced by a one-parameter group of diffeomorphisms? To answer this question for a given tangent vector field $\boldsymbol{X} = X^i\,\partial/\partial x^i$, one needs to examine the solutions to the following system of first-order ordinary differential equations:

$$\frac{dx^i(t)}{dt} = X^i(x^1(t), \ldots, x^n(t)) \qquad (1 < i < n)\,, \tag{11.9}$$

with given initial conditions, which defines a so-called *dynamical system* (see Chapter 18). A basic theorem in the theory of ordinary differential equations guaranteeing the uniqueness of solutions satisfying a certain set of initial conditions implies that the answer to the above question is in the affirmative, at least locally. This result is stated (without proof) as the following theorem.

Theorem 11.1. *Suppose $\boldsymbol{X}$ is a smooth tangent vector field on a differential manifold M. Then for every point $x \in M$ there exists a neighborhood U of x and a local one-parameter group φ_t of diffeomorphisms on U, with $|t| < \epsilon$ for some real number $\epsilon > 0$, such that $\boldsymbol{X}|_U$ (the restriction of $\boldsymbol{X}$ on U) is precisely the tangent vector field induced by φ_t.*

As an immediate consequence of this theorem we have the following slightly more general global statement for *compact* manifolds.

Corollary 11.1. *Every smooth tangent vector field on a smooth compact manifold M is induced by a one-parameter group of diffeomorphisms on M.*

A general remark on the singularities of tangent vector fields is in order here. A **singularity** of a tangent vector field $\boldsymbol{X}$ on a smooth manifold M is a point $x \in M$ at which $\boldsymbol{X}_x = \boldsymbol{0}$. The nature of smooth tangent vector fields near non-singular points is quite simple, as the result of Theorem 5.2 indicates. Near a singularity, however, the properties of the field can be quite complicated. In general *the singularities of a smooth tangent vector field are closely related to the topological properties of the underlying manifold M*. For example, there are no singularity-free smooth tangent vector fields on an even-dimensional sphere S^n (n even), but such a tangent vector field always exists on a torus. These very important issues will find applications in Hamiltonian dynamics and will be discussed in more detail in later chapters (Chapters 18 and 38).

Suppose $\psi : M \to M$ is a diffeomorphism, and φ_t is a one-parameter group of diffeomorphisms on M, with its induced tangent vector field $\boldsymbol{X}$. By (6.27) and (6.25), we have, for a point $x \in M$ and a smooth function f on M,

$$\begin{aligned}(\psi_* \boldsymbol{X}_x) f = \boldsymbol{X}_x (f \circ \psi) &= \frac{d}{dt} f \circ \psi(\varphi_t(x)) \Big|_{t=0} \\ &= \frac{d}{dt} f(\psi \circ \varphi_t \circ \psi^{-1}(\psi(x))) \Big|_{t=0}\,,\end{aligned} \tag{11.10}$$

which implies that $\psi_* \boldsymbol{X}_x$ is the tangent vector at $\psi(x)$ to the orbit of the one-parameter group of diffeomorphisms $\psi \circ \varphi_t \circ \psi^{-1}$ running through the point

$\psi(x) \in M$. Thus $\psi_* \boldsymbol{X}$ is the induced tangent vector field of the one-parameter group of diffeomorphisms $\psi \circ \varphi_t \circ \psi^{-1}$ on M.

We can now define the important concept of an *invariant tangent vector field.*

Definition 11.1. *Let* $\boldsymbol{X}$ *be a smooth tangent vector field on a smooth differentiable manifold* M*, and* $\psi : M \to M$ *is a diffeomorphism on* M*.* $\boldsymbol{X}$ *is said to be an* **invariant tangent vector field** *under* ψ *if*

$$\psi_* \boldsymbol{X} = \boldsymbol{X} . \tag{11.11}$$

From (11.10) we immediately have the following result.

Theorem 11.2. *Let* $\boldsymbol{X}$ *be the tangent vector field on a manifold* M *induced by the one-parameter group of diffeomorphisms* φ_t *on* M*. It is invariant under the diffeomorphism* $\psi : M \to M$ *if and only if*

$$\psi \circ \varphi_t = \varphi_t \circ \psi , \tag{11.12}$$

that is, φ_t *commutes with* ψ*.*

Let us now recall the notion of the Lie derivative of tangent vector fields first introduced in Chapter 6. The Lie derivative of a vector field $\boldsymbol{Y}$ with respect to a vector field $\boldsymbol{X}$ was defined in (6.15) as a Lie bracket: $L_{\boldsymbol{X}}\boldsymbol{Y} \equiv [\boldsymbol{X}, \boldsymbol{Y}] = \boldsymbol{XY} - \boldsymbol{YX}$. We will see that it can actually be more geometrically defined – as a derivative – using the (local) one-parameter group of diffeomorphisms corresponding to $\boldsymbol{X}$. Choose a point $x \in M$. Consider the integral curve $\gamma_x(t)$ of $\boldsymbol{X}$ passing through the point x at $t = 0$. Again, since there is no canonical relationship between vectors in tangent spaces at different points of M, we cannot directly take the vector difference $\boldsymbol{Y}_{\varphi_t(x)} - \boldsymbol{Y}_x$. But we can use the tangent map $(\varphi_t^{-1})_*$ to first "push" the vector $\boldsymbol{Y}_{\varphi_t(x)} \in T_{\varphi_t(x)}(M)$ back (in time) to the vector $(\varphi_t^{-1})_* \boldsymbol{Y}_{\varphi_t(x)} \in T_x(M)$, which can then be directly compared with $\boldsymbol{Y}_x$. The **Lie derivative** $L_{\boldsymbol{X}}\boldsymbol{Y}$ can then be defined as follows:

$$\boxed{(L_{\boldsymbol{X}}\boldsymbol{Y})_x \equiv \lim_{t\to 0} \frac{(\varphi_t^{-1})_* \boldsymbol{Y}_{\varphi_t(x)} - \boldsymbol{Y}_x}{t} = \lim_{t\to 0} \frac{\boldsymbol{Y}_x - (\varphi_t)_* \boldsymbol{Y}_{\varphi_t^{-1}(x)}}{t}} . \tag{11.13}$$

Superficially this definition bears little resemblance to that given by (6.15) in terms of Lie brackets, but they can be reconciled as follows. Let $f : M \to \mathbb{R}$ be a smooth function on M, and let

$$F(t) \equiv f(\varphi_t(x)) . \tag{11.14}$$

Then

$$F(t) - F(0) = \int_0^1 \frac{dF(st)}{ds}\, ds = t \int_0^1 F'(u)|_{u=st}\, ds . \tag{11.15}$$

It follows from (11.14) that, for a point $x \in M$,

$$f(\varphi_t(x)) = f(x) + t g_t(x) , \tag{11.16}$$

where

$$g_t(x) \equiv \int_0^1 F'(u)|_{u=st}\, ds = \int_0^1 \left.\frac{df(\varphi_u(x))}{du}\right|_{u=st} ds , \tag{11.17}$$

and thus

$$g_0(x) = \int_0^1 \left.\frac{df(\varphi_u(x))}{du}\right|_{u=0} ds = \left.\frac{df(\varphi_u(x))}{du}\right|_{u=0} = \boldsymbol{X}_x f . \tag{11.18}$$

Applying the right-hand side of the second equality of (11.13) to f, we have

$$\begin{aligned}
&\left(\lim_{t\to 0} \frac{\boldsymbol{Y}_x - (\varphi_t)_* \boldsymbol{Y}_{\varphi_t^{-1}(x)}}{t}\right) f = \lim_{t\to 0} \frac{\boldsymbol{Y}_x f - \boldsymbol{Y}_{\varphi_t^{-1}(x)}(f\circ\varphi_t)}{t} \\
&= \lim_{t\to 0} \frac{\boldsymbol{Y}_x f - \boldsymbol{Y}_{\varphi_t^{-1}(x)}(f + t g_t)}{t} = \lim_{t\to 0} \frac{\boldsymbol{Y}_x f - \boldsymbol{Y}_{\varphi_t^{-1}(x)} f}{t} - \lim_{t\to 0} \boldsymbol{Y}_{\varphi_t^{-1}(x)} g_t \\
&= \lim_{t\to 0} \frac{\boldsymbol{Y}_{\varphi_t(x)} f - \boldsymbol{Y}_x f}{t} - \boldsymbol{Y}_x g_0(x) \\
&= \lim_{t\to 0} \frac{[(\boldsymbol{Y}f)\circ\varphi_t](x) - [(\boldsymbol{Y}f)\circ\varphi_0](x)}{t} - \boldsymbol{Y}_x \boldsymbol{X}_x f \\
&= \frac{d}{dt}[(\boldsymbol{Y}f)\circ\varphi_t](x)\Big|_{t=0} - (\boldsymbol{Y}\boldsymbol{X})_x f = \boldsymbol{X}_x(\boldsymbol{Y}f) - (\boldsymbol{Y}\boldsymbol{X})_x f = [\,\boldsymbol{X},\boldsymbol{Y}\,]_x f ,
\end{aligned} \tag{11.19}$$

where $\boldsymbol{Y}f$ is considered as a function on M [according to (5.19)]. Since the function f is arbitrary, the equivalence of the two definitions of the Lie derivative (6.15) and (11.13) is established.

The concept of the Lie derivative of a tangent vector field on a manifold M can be immediately generalized to any tensor field on M. In fact, if $\varphi_t(x) = y$, the pullback map $(\varphi_t)^*$ gives a homomorphism $(\varphi_t)^* : T_y^*(M) \to T_x^*(M)$ between cotangent spaces. This pullback map together with the tangent map $(\varphi_t^{-1})_*$ [cf. (6.26)] then induce a homomorphism $\Phi_t : T_s^r(\varphi_t(x)) \to T_s^r(x)$ between tensor spaces as follows: For any $\boldsymbol{v}_1, \dots, \boldsymbol{v}_r \in T_{\varphi_t(x)}(M)$ and $(\boldsymbol{v}^*)^1, \dots (\boldsymbol{v}^*)^s \in T^*_{\varphi_t(x)}(M)$,

$$\begin{aligned}
&\Phi_t(\boldsymbol{v}_1 \otimes \dots \boldsymbol{v}_r \otimes (\boldsymbol{v}^*)^1 \otimes \dots \otimes (\boldsymbol{v}^*)^s) \\
&\equiv (\varphi_t^{-1})_* \boldsymbol{v}_1 \otimes \dots \otimes (\varphi_t^{-1})_* \boldsymbol{v}_r \otimes \varphi_t^*(\boldsymbol{v}^*)^1 \otimes \dots \otimes \varphi_t^*(\boldsymbol{v}^*)^s .
\end{aligned} \tag{11.20}$$

The Lie derivative $L_{\boldsymbol{X}}\boldsymbol{\xi}$ of an (r,s)-type tensor field $\boldsymbol{\xi}$ on M (with respect to the tangent vector field $\boldsymbol{X}$) can then be defined, in analogy to (11.13), as follows:

$$L_{\boldsymbol{X}}\boldsymbol{\xi} \equiv \lim_{t\to 0} \frac{\Phi_t(\boldsymbol{\xi}) - \boldsymbol{\xi}}{t} . \tag{11.21}$$

Obviously, $L_{\boldsymbol{X}}\boldsymbol{\xi}$ is also an (r, s)-type tensor field. The Lie derivative of a function f on M is defined by

$$L_{\boldsymbol{X}} f \equiv \boldsymbol{X} f = X^i \frac{\partial f}{\partial x^i} , \tag{11.22}$$

which, by (5.19), is another function on M. The Lie derivative of a differential form ω on M is of special interest. It is a special case of (11.21), given by

$$L_{\boldsymbol{X}}\omega = \lim_{t\to 0} \frac{\varphi_t^* \omega - \omega}{t} = \frac{d}{dt}(\varphi_t^* \omega)\bigg|_{t=0} . \tag{11.23}$$

One can show the following pairing relationship: For any r smooth tangent vector fields $\boldsymbol{Y}_1, \dots, \boldsymbol{Y}_r$, a smooth tangent vector field $\boldsymbol{X}$, and an r-form ω (all on M),

$$\begin{aligned} &\langle \boldsymbol{Y}_1 \wedge \cdots \wedge \boldsymbol{Y}_r \,,\, L_{\boldsymbol{X}}\omega \rangle \\ &= \boldsymbol{X} \langle \boldsymbol{Y}_1 \wedge \cdots \wedge \boldsymbol{Y}_r \,,\, \omega \rangle - \sum_{i=1}^{r} \langle \boldsymbol{Y}_1 \wedge \cdots \wedge L_{\boldsymbol{X}}\boldsymbol{Y}_i \wedge \cdots \wedge \boldsymbol{Y}_r \,,\, \omega \rangle . \end{aligned} \tag{11.24}$$

There is a group of very useful operational formulas, called the *Cartan formulas*, for the Lie derivative $L_{\boldsymbol{X}}$ applied to any differential form ω. These are most conveniently stated in terms of the so-called **contraction map**, $\boldsymbol{X} \lrcorner : A^r(M) \to A^{r-1}(M)$ acting on the space of r-forms, defined as follows:

(1) If $r = 0$, then $\boldsymbol{X} \lrcorner$ acts on $A^0(M)$ as the zero-map.

(2) If $r = 1$, then

$$\boldsymbol{X} \lrcorner\, \omega \equiv \langle \boldsymbol{X} \,,\, \omega \rangle . \tag{11.25}$$

(3) If $r > 1$, then for any $r-1$ smooth tangent vector fields $\boldsymbol{Y}_1, \dots \boldsymbol{Y}_{r-1}$, we have

$$\langle \boldsymbol{Y}_1 \wedge \cdots \wedge \boldsymbol{Y}_{r-1} \,,\, \boldsymbol{X} \lrcorner\, \omega \rangle = \langle \boldsymbol{X} \wedge \boldsymbol{Y}_1 \wedge \cdots \wedge \boldsymbol{Y}_{r-1} \,,\, \omega \rangle . \tag{11.26}$$

The **Cartan formulas** are then stated as follows:

$$L_{\boldsymbol{X}}\omega = d(\boldsymbol{X} \lrcorner\, \omega) + \boldsymbol{X} \lrcorner\, d\omega ; \tag{11.27}$$

$$L_{\boldsymbol{X}}(\boldsymbol{Y} \lrcorner\, \omega) - \boldsymbol{Y} \lrcorner\, (L_{\boldsymbol{X}}\omega) = [\boldsymbol{X}, \boldsymbol{Y}] \lrcorner\, \omega ; \tag{11.28}$$

$$L_{\boldsymbol{X}}(L_{\boldsymbol{Y}}\omega) - L_{\boldsymbol{Y}}(L_{\boldsymbol{X}}\omega) = L_{[\boldsymbol{X},\boldsymbol{Y}]}\omega ; \tag{11.29}$$

$$d(L_{\boldsymbol{X}}\omega) = L_{\boldsymbol{X}} d\omega . \tag{11.30}$$

Problem 11.2 For any $\alpha \in A^r(M)$, $\beta \in A^s(M)$, and $\boldsymbol{X}$ a tangent vector field on M, show that

$$\boldsymbol{X} \lrcorner (\alpha \wedge \beta) = (\boldsymbol{X} \lrcorner \alpha) \wedge \beta + (-1)^r \, \alpha \wedge (\boldsymbol{X} \lrcorner \beta) \,. \tag{11.31}$$

Problem 11.3 Prove (11.24) directly for the case $r = 1$, that is

$$\langle \boldsymbol{Y} \,,\, L_{\boldsymbol{X}} \omega \rangle = \boldsymbol{X} \langle \boldsymbol{Y} \,,\, \omega \rangle - \langle L_{\boldsymbol{X}} \boldsymbol{Y} \,,\, \omega \rangle \,, \tag{11.32}$$

and then prove the general formula by induction.

Problem 11.4 Use the definitions of the contraction map and the Lie derivative acting on differential forms to prove the Cartan formulas (11.27) to (11.30).

For some concrete examples of the use of Lie derivatives in fluid mechanics the reader is referred to the discussion at the end of Chapter 13. These derivatives are also used in connection with flows in symplectic manifolds in our later development in Chapter 40.

Chapter 12

Newton's Laws: Inertial and Non-inertial Frames

In this chapter we begin the study of **dynamics**, which traditionally is concerned with the understanding of why and how things move in *space* under the influence of *forces*. We will see, however, that the distinction between dynamics and kinematics is not always clearcut, due to certain ambiguities as to the objective nature of physical forces, depending on the nature of the reference frame used for the description of the motion. We will see that, when so-called *non-inertial (moving) frames* are used, *fictitious forces* appear which are entirely geometrical in character.

Classical dynamics is based *entirely* on *Newton's three laws of motion.* Before stating these laws we note the crucial fact that all the kinematical concepts introduced so far – the position, velocity and acceleration vectors – are all *reference frame-dependent* quantities. Newton's Laws can be stated as follows.

(N1) **Newton's First Law**. There exists in nature a special class of reference frames, called **inertial frames**, which is characterized by the following property: *With respect to an inertial frame, an object will always travel with a constant velocity unless acted upon by a* **force**.

(N2) **Newton's Second Law**. With respect to an inertial frame, an object's response to a given *net* force $\boldsymbol{F}$ is described by the **equation of motion**

$$\boldsymbol{F} = \frac{d\boldsymbol{p}}{dt} , \tag{12.1}$$

where $\boldsymbol{p}$, called the **linear momentum** of the object, is defined by

$$\boldsymbol{p} \equiv m\boldsymbol{v} , \tag{12.2}$$

with $\boldsymbol{v}$ being the velocity of the object and m, called the **inertial mass**, is a scalar quantity describing an intrinsic property of the object.

(N3) **Newton's Third Law**. When two objects in isolation exert forces on each other, these forces (called **action** and **reaction**) are equal in magnitude and opposite in direction (as vectors).

Several comments on some technical and philosophical issues regarding the above statements of Newton's Laws are in order at the outset.

The laws are easily stated but can be very difficult to apply in practice. One of the main difficulties is that in most situations, the forces are not known *a priori*. Another is that the mathematics involved (mainly that of differential equations and by extension differential geometry) can be very complicated technically for systems involving more than two objects, even if the forces are known. For example, the famous *Three-Body Problem* in celestial mechanics has occupied the attention of prominent physicists and mathematicians since the time of Newton for more than three hundred years (see Chapter 43). But, at least in principle, the three laws provide a *complete* theoretical basis for classical mechanics.

If the mass m of an object is constant, then Newton's Second Law can be re-stated as

$$\boldsymbol{F} = m\boldsymbol{a} \qquad (m \text{ constant}) , \tag{12.3}$$

where $\boldsymbol{a} = d\boldsymbol{v}/dt$ is the acceleration.

The Third Law is actually a consequence of the Second Law *and* the so-called Principle of **Conservation of Linear Momentum**, when the latter is applied to an isolated system of two objects mutually exerting forces on each other. Conversely, the Principle of Conservation of Linear Momentum follows from the Second and Third Laws. We will discuss this all-important principle and its applications in much more detail in a later chapter (Chapter 19). It states that, *in the absence of external forces, the total linear momentum of a system is conserved* (does not change in time). Consider an isolated two-object system with linear momenta $\boldsymbol{p}_1$ and $\boldsymbol{p}_2$. The total momentum is then $\boldsymbol{P} = \boldsymbol{p}_1 + \boldsymbol{p}_2$. Conservation of momentum then implies

$$\frac{d\boldsymbol{P}}{dt} = \frac{d\boldsymbol{p}_1}{dt} + \frac{d\boldsymbol{p}_2}{dt} = 0 . \tag{12.4}$$

By the Second Law, on the other hand,

$$\frac{d\boldsymbol{p}_1}{dt} = \boldsymbol{F}_{12} , \qquad \frac{d\boldsymbol{p}_2}{dt} = \boldsymbol{F}_{21} , \tag{12.5}$$

where $\boldsymbol{F}_{12}$ is the force that object 2 exerts on object 1 and $\boldsymbol{F}_{21}$ is the force that object 1 exerts on object 2. Equation (12.4) then implies $\boldsymbol{F}_{12} = -\boldsymbol{F}_{21}$, which is exactly what the Third Law states.

It appears that from a logical standpoint, the First Law is a consequence of the Second Law, since, from the version of the Second Law given by (12.3), $\boldsymbol{a} = 0$ implies that $\boldsymbol{v} =$ constant (in time). The First Law then appears to be logically redundant. This is actually a matter of some controversy. Our viewpoint is that the First Law in fact contains information beyond the Second Law. It is both an *empirical statement* – postulating the *physical* existence of inertial frames in nature – and a logical-mathematical one – stating mathematically the consequence of a force (non-constant velocity) on the motion of a particle, when that motion is described kinematically with respect to an inertial frame. In this interpretation, the First Law gives a logical-mathematical definition of the concept of an inertial frame in terms of the (mathematically ill-defined but physically familiar and immensely useful) concept of the force and asserts that such frames exist objectively in nature. In this book we will not be preoccupied with these subtle logico-philosophical issues. In fact, when there is no confusion, we will just assume that the reference frames that we use are inertial ones, unless otherwise stated. *The final arbiter on this matter will, as in any scientific theory, be the experimental validation of the theoretical predictions based on this assumption*, and historically, classical mechanics has proved to be remarkably successful within its legitimate realm of applications.

Now our *physical* understanding of forces always tends to attribute an *objectively existing* agent – something that is doing the push and pull – to a force, whose existence is independent of the choice of a reference frame for the description of motion. The following problem immediately arises. Suppose that with respect to a certain frame, an object is *observed* by a certain observer to be moving with constant velocity (that is, in a straight line and with constant speed), and the observer knows (by whatever means) that the forces (pushes and pulls) exerted on the object by all physically existing agents add up to zero (as vectors). Then the observer can conclude from the First Law that the frame that he has chosen is an inertial frame, and that the Second Law is also (trivially) satisfied. Suppose, then, that another observer chooses another frame that is accelerating with respect to the first (rotating with respect to it, for example) to describe the *same* motion. What appeared to be a straight path by the first observer may now appear to be a *curve* to the second, who would thus conclude that the object under observation is accelerating. If the second observer had no knowledge of the physical origin of the forces that the first observer had, he would not be able to say unambiguously whether it is actually *physical* forces that are causing the object to accelerate, or that he is using the wrong (non-inertial) kind of reference frame for the description of the motion. In fact, as we will see below, he has the option of attributing that part of the acceleration due to the use of non-inertial frames to the action of so-called "*fictitious*" forces, and still using Newton's Second Law for calculational and predictive purposes. Or, with a much more radical leap, he can, as Einstein did in the formulation of his *General Theory of Relativity*, attribute the physical origin of forces (such as gravity) to the intrinsic *curvature of space* (actually the four-dimnensional *spacetime*) itself. In any case it is quite remarkable that, more than three hundred years ago, Newton had the tremendous foresight of separating his First Law from the Second, and thus opened the door to path-breaking investigations by later physicists and mathematicians of the role of reference frames and their relationship to the *geometrical structure of the underlying space* in the formulation of the fundamental laws of nature. In a later chapter we will examine in more detail, after the necessary mathematical apparatus of Riemannian geom-

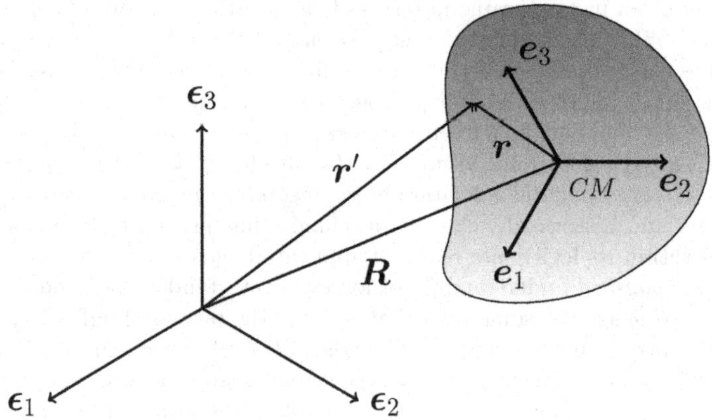

Fig. 12.1

etry has been developed, the relationship between the curvature of space and the use of non-inertial frames [see the discussion following (25.40) in Chapter 25 and also the development in Chapter 27]. Meanwhile, we will study the consequences that result from the use of non-inertial frames.

Let $\{\epsilon_1, \epsilon_2, \epsilon_3\}$ be an orthonormal inertial frame, and $\{\boldsymbol{e}_1, \boldsymbol{e}_2, \boldsymbol{e}_3\}$ be another orthonormal frame that is moving in an arbitrary manner with respect to it. Thus $\{\boldsymbol{e}_1, \boldsymbol{e}_2, \boldsymbol{e}_3\}$ is in general a **non-inertial frame.** For example, it may be a body-fixed frame with the origin located at the *center of mass* (CM) of a rigid body (Fig. 12.1). Consider a point mass m moving in space under the influence of some force. Suppose its position vector with respect to the $\{\boldsymbol{e}_i\}$ frame is $\boldsymbol{r}$, and with respect to the $\{\epsilon_i\}$ frame is $\boldsymbol{r}'$. From Fig. 12.1 we have

$$\boldsymbol{r}' = \boldsymbol{r} + \boldsymbol{R}\,. \tag{12.6}$$

Thus

$$\frac{d\boldsymbol{r}'}{dt} = \frac{D\boldsymbol{r}}{dt} + \boldsymbol{V}\,, \qquad \frac{d^2\boldsymbol{r}'}{dt^2} = \frac{D^2\boldsymbol{r}}{dt^2} + \frac{d\boldsymbol{V}}{dt}\,, \tag{12.7}$$

where

$$\boldsymbol{V} \equiv \frac{d\boldsymbol{R}}{dt}\,. \tag{12.8}$$

Note that we have used the covariant derivative for vectors expressed with reference to moving frames, which reduce to ordinary derivatives when inertial (space-fixed) frames are used (cf. Chapter 4). Write

$$\boldsymbol{r} = x^i\, \boldsymbol{e}_i \,. \tag{12.9}$$

Then, following the developments in Chapter 4, we have

$$\begin{aligned} \frac{D\boldsymbol{r}}{dt} &= \frac{dx^i}{dt}\, \boldsymbol{e}_i + x^i \frac{D\boldsymbol{e}_i}{dt} = \frac{dx^i}{dt}\, \boldsymbol{e}_i + x^i \varphi_i^j\, \boldsymbol{e}_j \\ &= \frac{dx^i}{dt}\, \boldsymbol{e}_i + x^i \varepsilon_i{}^j{}_k \omega^k\, \boldsymbol{e}_j = \frac{dx^i}{dt}\, \boldsymbol{e}_i + \varepsilon^j{}_{ki} \omega^k x^i\, \boldsymbol{e}_j \,, \end{aligned} \tag{12.10}$$

where we have used (4.8) and (4.11), and φ_i^j has been defined in (4.10). The above equation can be rewritten in ordinary vector notation as [compare with (4.13)]

$$\boxed{\frac{D\boldsymbol{r}}{dt} = \frac{dx^i}{dt}\, \boldsymbol{e}_i + \boldsymbol{\omega} \times \boldsymbol{r}} \,. \tag{12.11}$$

To calculate the acceleration it is most convenient to start with the following expression [cf. (12.10)]:

$$\frac{D\boldsymbol{r}}{dt} = (\dot{x}^j + x^i \varphi_i^j)\, \boldsymbol{e}_j \,. \tag{12.12}$$

Differentiating once with respect to t, we have

$$\frac{D^2\boldsymbol{r}}{dt^2} = (\dot{x}^j + x^i \varphi_i^j) \varphi_j^k\, \boldsymbol{e}_k + \left(\ddot{x}^j + x^i \frac{d\varphi_i^j}{dt} + \dot{x}^i \varphi_i^j \right) \boldsymbol{e}_j \,, \tag{12.13}$$

or, on collecting terms on the right-hand side,

$$\boxed{\frac{D^2\boldsymbol{r}}{dt^2} = \ddot{x}^j \boldsymbol{e}_j + 2\dot{x}^i \varphi_i^j \boldsymbol{e}_j + x^i \varphi_i^j \varphi_j^k \boldsymbol{e}_k + x^i \frac{d\varphi_i^j}{dt}\, \boldsymbol{e}_j} \,. \tag{12.14}$$

The above expression is valid for an arbitrary base manifold (the space in which the particle is moving, which is in general a *differentiable manifold* with a *Riemannian metric*, not necessarily a Euclidean one). In this general case the $\boldsymbol{e}_i$'s form an orthonormal basis of the *tangent space* at the instantaneous position of the particle. In the special case when the base manifold is three-dimensional Euclidean space (which is the most commonly considered case), it is customary to write (12.14) in terms of the angular velocity vector $\boldsymbol{\omega}$ [defined in (4.11)], the velocity $\boldsymbol{v}$ of the particle with respect to the moving (non-inertial) $\{\boldsymbol{e}_i\}$ frame:

$$\boldsymbol{v} \equiv \dot{x}^i\, \boldsymbol{e}_i \,, \tag{12.15}$$

and the position vector $\boldsymbol{r}$ of the particle, also with respect to the moving $\{\boldsymbol{e}_i\}$ frame [(15.4)]. We have

$$\varphi_i^j \dot{x}^i \boldsymbol{e}_j = \varepsilon_i{}^j{}_k \omega^k \dot{x}^i \boldsymbol{e}_j = \varepsilon^j{}_{ki} \omega^k \dot{x}^i \boldsymbol{e}_j = \boldsymbol{\omega} \times \boldsymbol{v} \,. \tag{12.16}$$

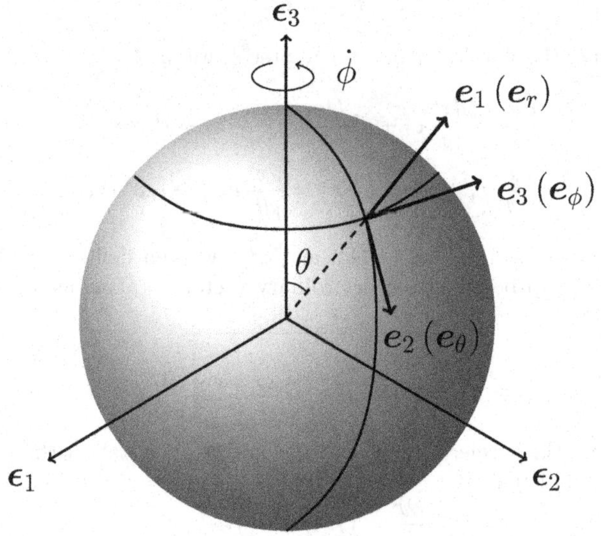

Fig. 12.2

Similarly,

$$x^i \frac{d\varphi_i^j}{dt} \boldsymbol{e}_j = x^i \dot{\varphi}_i^j \boldsymbol{e}_j = \dot{\boldsymbol{\omega}} \times \boldsymbol{r} \tag{12.17}$$

and

$$x^i \varphi_i^k \varphi_k^j \boldsymbol{e}_j = x^i \varepsilon_i{}^k{}_m \omega^m \varepsilon_k{}^j{}_n \omega^n \boldsymbol{e}_j = \varepsilon^j{}_{nk} \omega^n (\varepsilon^k{}_{mi} \omega^m x^i) \boldsymbol{e}_j = \boldsymbol{\omega} \times (\boldsymbol{\omega} \times \boldsymbol{r})\,. \tag{12.18}$$

Finally, using (12.14), the second equation of (12.7) can be written, for motion taking place in three-dimensional Euclidean space, as

$$\boxed{\ddot{x}^i \boldsymbol{e}_i = \frac{d^2 \boldsymbol{r}'}{dt^2} - \frac{d\boldsymbol{V}}{dt} - 2\boldsymbol{\omega} \times \boldsymbol{v} - \boldsymbol{\omega} \times (\boldsymbol{\omega} \times \boldsymbol{r}) - \dot{\boldsymbol{\omega}} \times \boldsymbol{r}}\quad . \tag{12.19}$$

In the above equation, the left-hand side represents the acceleration of the particle with respect to the non-inertial frame $\{\boldsymbol{e}_i\}$. The first term on the right-hand side is the acceleration of the particle with respect to the inertial frame $\{\boldsymbol{\epsilon}_i\}$, and is thus, by Newton's Second Law, equal to the external (non-fictitious) force on the particle divided by its mass m. The other terms on the right-hand side (the ones with negative signs) represent the so-called **fictitious forces**. These are distinguished as follows.

$$-m\frac{d\boldsymbol{V}}{dt} = \text{fictitious force due to the acceleration of the origin of the noninertial frame } \{\boldsymbol{e}_i\}\,, \tag{12.20}$$

$$-2m\boldsymbol{\omega}\times\boldsymbol{v} = \textbf{Coriolis force}\,, \tag{12.21}$$

$$-m\boldsymbol{\omega}\times(\boldsymbol{\omega}\times\boldsymbol{r}) = \textbf{centrifugal force}\,. \tag{12.22}$$

The last two are fictitious forces due to the rotational motion of the non-inertial frame.

Equation (12.19) implies that if $d\boldsymbol{V}/dt = \boldsymbol{0}$ [the origin of the moving frame is moving with a uniform velocity with respect to the (space-fixed) inertial frame] and $\boldsymbol{\omega} = \boldsymbol{0}$ [abscence of rotational motion of the moving frame] the accelerations of the particle are the same with respect to both frames. In this case the moving frame is said to be parallelly translated with a uniform velocity with respect to the inertial frame (**Galilean transformation**). According to Newton's Laws, the moving frame in this case is an inertial frame also. Hence the existence of one inertial frame implies that of infinitely many others, and Newton's Laws are said to be invariant under Galilean transformations.

Problem 12.1 Consider a non-inertial frame $\{\boldsymbol{e}_1, \boldsymbol{e}_2, \boldsymbol{e}_3\}$ attached to the surface of the earth as it rotates around its axis with an angular speed $\dot{\phi}$ (see Fig. 12.2), with $\boldsymbol{e}_1 = \boldsymbol{e}_r$, $\boldsymbol{e}_2 = \boldsymbol{e}_\theta$, $\boldsymbol{e}_3 = \boldsymbol{e}_\phi$. Use (4.33) to show that

$$(\varphi_i^j) = \begin{pmatrix} 0 & 0 & \sin\theta\,\dot{\phi} \\ 0 & 0 & \cos\theta\,\dot{\phi} \\ -\sin\theta\,\dot{\phi} & -\cos\theta\,\dot{\phi} & 0 \end{pmatrix}, \tag{12.23}$$

and thus, from (4.25), that with respect to the $\{\boldsymbol{e}_i\}$ frame, the angular velocity vector $\boldsymbol{\omega}$ is given by

$$\omega^1 = \cos\theta\,\dot{\phi}\,, \quad \omega^2 = -\sin\theta\,\dot{\phi}\,, \quad \omega^3 = 0\,. \tag{12.24}$$

Problem 12.2 Consider the non-inertial frame introduced in the last Problem. Let $\boldsymbol{R} = R\boldsymbol{e}_1$ and $\boldsymbol{V} = d\boldsymbol{R}/dt$, where R is the radius of the earth (Fig. 12.2). Use (12.14) to show that

$$\frac{d\boldsymbol{V}}{dt} = \boldsymbol{\omega}\times(\boldsymbol{\omega}\times\boldsymbol{R})\,, \tag{12.25}$$

where $\boldsymbol{\omega}$ is given by (12.24). From (12.19), $-md\boldsymbol{V}/dt$ is the fictitious centrifugal force on a particle of mass m at the surface of the earth that is associated with the rotation of the earth.

Problem 12.3 A bee flies with constant velocity $\boldsymbol{v}_0$ northwards over a circular turntable rotating anti-clockwise with constant angular speed ω. Joe is an observer standing on the edge of the turntable and rotating with it. At $t = 0$, the bee is right above the center of the turntable, and Joe is directly due east. The speed v_0 is such that Joe and the bee will meet when the turntable has rotated a quarter turn after $t = 0$. Consider an orthonormal moving frame $(\boldsymbol{e}_1 \boldsymbol{e}_2)$ that is attached to the turntable, with the origin at the place where Joe is standing and $\boldsymbol{e}_1$ pointing due south at $t = 0$. Write the position vector of the bee relative to the moving frame as $\boldsymbol{r} = x^i \boldsymbol{e}_i$.

(a) Find $x_1(t)$ and $x_2(t)$ as functions of t directly from the geometry of the above description.

(b) Show a qualitative sketch of the curved trajectory of the bee from Joe's viewpoint, based on the results in (a).

(c) Since the flight path of the bee appears to Joe to be a curve, he thinks there must be forces acting on it, according to Newton's laws. He then uses Eq. (12.19) (with $d^2\boldsymbol{r}'/dt^2 = 0$) to solve for $\boldsymbol{r}(t)$:

$$\ddot{x}^i \boldsymbol{e}_i = -\frac{d\boldsymbol{V}}{dt} - 2\boldsymbol{\omega} \times \boldsymbol{v} - \boldsymbol{\omega} \times (\boldsymbol{\omega} \times \boldsymbol{r}) - \dot{\boldsymbol{\omega}} \times \boldsymbol{r} \quad ,$$

where $\boldsymbol{V} = d\boldsymbol{R}/dt$. Show that this equation is equivalent to the following set of coupled equations for x_1 and x_2:

$$\ddot{x}_1 = 2\omega\dot{x}_2 + \omega^2 x_1 \,, \quad \ddot{x}_2 = \omega^2 R - 2\omega\dot{x}_1 + \omega^2 x_2 \,.$$

(d) Show that the results from (a) satisfy the above set of coupled equations.

We will conclude this chapter with some immediate and far-reaching consequences of Newton's Laws as they are applied to single-particle dynamics – the conservation principles of linear momentum, angular momentum, and energy.

From Newton's Second Law (12.1) the **Principle of Conservation of Linear Momentum** immediately follows: *The linear momentum of a particle is conserved (remains constant in time) if the net force acting on it is zero.*

The **angular momentum** $\boldsymbol{L}$ of a particle at a position $\boldsymbol{r}$ is given by

$$\boldsymbol{L} = \boldsymbol{r} \times \boldsymbol{p} \,, \tag{12.26}$$

and the **torque** $\boldsymbol{\tau}$ due to a force $\boldsymbol{F}$ is

$$\boldsymbol{\tau} = \boldsymbol{r} \times \boldsymbol{F} \,. \tag{12.27}$$

Then we have

$$\boldsymbol{\tau} = \frac{d\boldsymbol{L}}{dt} \,, \tag{12.28}$$

where $\boldsymbol{\tau}$ is the net torque (torque due to the net force) acting on the particle. Indeed, from (12.1), (12.26) and (12.27),

$$\frac{d\boldsymbol{L}}{dt} = \boldsymbol{r} \times \frac{d\boldsymbol{p}}{dt} + \frac{d\boldsymbol{r}}{dt} \times \boldsymbol{p} = \boldsymbol{r} \times \boldsymbol{F} + \frac{\boldsymbol{p}}{m} \times \boldsymbol{p} = \boldsymbol{\tau} \,. \tag{12.29}$$

Equation (12.29) implies the **Principle of the Conservation of Angular Momentum**: *The angular momentum of a particle is conserved if the net torque acting on it vanishes.*

The **kinetic energy** K of a particle of mass m moving with a velocity $\boldsymbol{v}$ is given by

$$K = \frac{1}{2} m \boldsymbol{v} \cdot \boldsymbol{v} = \frac{1}{2} m v^2 = \frac{p^2}{2m} . \tag{12.30}$$

A position-dependent force $\boldsymbol{F}(\boldsymbol{r})$ acting on a particle as it moves along a curve C in $\mathbb{R}^3$ from position $\boldsymbol{r}_1$ to $\boldsymbol{r}_2$ is said to do an amount of **mechanical work** W_{12} on the particle given by the line integral

$$W_{12} = \int_{\boldsymbol{r}_1\,(C)}^{\boldsymbol{r}_2} \boldsymbol{F}(\boldsymbol{r}) \cdot d\boldsymbol{r} . \tag{12.31}$$

This line integral in general depends on the curve C. Let $\boldsymbol{F}_{\boldsymbol{net}}$ be the net (position-dependent) force acting on a particle as it moves along some path C from point 1 to point 2 (with position vectors $\boldsymbol{r}_1$ and $\boldsymbol{r}_2$, respectively). Then, if the mass m of the particle remains constant as it moves, Newton's Second Law in the form (12.2) implies that

$$\begin{aligned} &\int_1^2 \boldsymbol{F}_{net} \cdot d\boldsymbol{r} = m \int_1^2 \frac{d\boldsymbol{v}}{dt} \cdot d\boldsymbol{r} = m \int d\boldsymbol{v} \cdot \frac{d\boldsymbol{r}}{dt} = m \int_{v_1}^{v_2} \boldsymbol{v} \cdot d\boldsymbol{v} \\ &= \frac{m}{2} \int_{v_1}^{v_2} d(\boldsymbol{v} \cdot \boldsymbol{v}) = \frac{m}{2} \int_{v_1}^{v_2} d(v^2) = K_2 - K_1 , \end{aligned} \tag{12.32}$$

where v_1 and v_2 are the speeds, and K_1 and K_2 are the kinetic energies, of the particle at points 1 and 2, respectively. Defining the change in kinetic energy ΔK by $\Delta K \equiv K_2 - K_1$, we have the **work-energy theorem**:

$$\int_1^2 \boldsymbol{F}_{net} \cdot d\boldsymbol{r} = \Delta K . \tag{12.33}$$

We will now introduce the important notions of a *conservative force field* and the associated *potential energy*:

Definition 12.1. *A force field $\boldsymbol{F}(\boldsymbol{r})$ on a region $R \subset \mathbb{R}^3$ is said to be **conservative** if the 1-form $\boldsymbol{F} \cdot d\boldsymbol{r}$ is integrable, that is, if there exists a function $\phi(\boldsymbol{r})$ on R such that*

$$d\phi = \boldsymbol{F} \cdot d\boldsymbol{r} . \tag{12.34}$$

Definition 12.2. *The **potential energy** function $U(\boldsymbol{r})$ associated with a conservative force field $\boldsymbol{F}_c(\boldsymbol{r})$ is defined to be*

$$U(\boldsymbol{r}) \equiv - \int_{\boldsymbol{r}_0}^{\boldsymbol{r}} \boldsymbol{F}_c \cdot d\boldsymbol{r} , \tag{12.35}$$

where $\boldsymbol{r}_0$, the zero of the potential energy function, is an arbitrarily chosen point in the region $R \subset \mathbb{R}^3$ in which $\boldsymbol{F}_c(\boldsymbol{r})$ is defined, and the integral can be carried out along an arbitrary curve in R joining the points $\boldsymbol{r}_0$ and $\boldsymbol{r}$.

Define the total **mechanical energy** E to be the sum of the kinetic and potential energies:

$$E \equiv K + U \,. \tag{12.36}$$

Then, by the Work-Energy Theorem (12.33) and the definition of the potential energy function (12.35),

$$\begin{aligned} \frac{dE}{dt} &= \frac{dK}{dt} + \frac{dU}{dt} = \boldsymbol{F}_{net} \cdot \frac{d\boldsymbol{r}}{dt} + \frac{dU}{dt} = \boldsymbol{F}_{net} \cdot \boldsymbol{v} + \frac{\partial U}{\partial x^i}\frac{dx^i}{dt} + \frac{\partial U}{\partial t} \\ &= \boldsymbol{F}_{net} \cdot \boldsymbol{v} + \nabla U \cdot \boldsymbol{v} + \frac{\partial U}{\partial t} = (\boldsymbol{F}_{net} - \boldsymbol{F}_c) \cdot \boldsymbol{v} + \frac{\partial U}{\partial t} \,, \end{aligned} \tag{12.37}$$

where $\boldsymbol{F}_c$ is the conservative force field associated with the potential energy function U. Thus we have the **Principle of Conservation of Mechanical Energy**: *The total mechanical energy of a particle is conserved if it is acted on only by a conservative force field and if the potential energy associated with the force field does not depend explicitly on time.*

Chapter 13

Simple Applications of Newton's Laws

The examples discussed in this chapter are chosen on the basis of their relevance in fundamental mechanics problems as well as the relative simplicity of the mathematical techniques required for their solutions. They are meant to convey the flavor of some of the elementary and standard approaches used in the direct applications of Newton's Laws and are, in the order presented, *the Kepler Problem, the motion of charged particles in electric and magnetic fields, tides, and fluid motion.* Except for the last example (fluid motion), they are all treated within the framework of single-particle dynamics. The following two chapters will continue with the illustration of elementary approaches: potential theory to treat forces due to extended objects (Chapter 14) and non-inertial forces (Chapter 15).

Let us first consider the **Kepler Problem**. It is defined by the equation of motion

$$\frac{d\boldsymbol{p}}{dt} = -\frac{\alpha \boldsymbol{r}}{r^3} \qquad (\alpha > 0) , \tag{13.1}$$

where $\boldsymbol{p}$ is the linear momentum of a particle of mass m, $\boldsymbol{r}$ is its position vector with respect to an inertial frame, and α is a positive constant. The problem of two point particles interacting through an *inverse square law*, for example, can be reduced to a problem of this form (see Problem 13.1). Equation (13.1) represents a set of three second order differential equations, whose solutions give the trajectory $\boldsymbol{r}(t)$ of the particle. We will not solve this equation explicitly here. Instead, the geometrical nature of the trajectory will be determined by the *constants of motion* of this problem.

Problem 13.1 For the two-body problem with only gravitational interaction, show

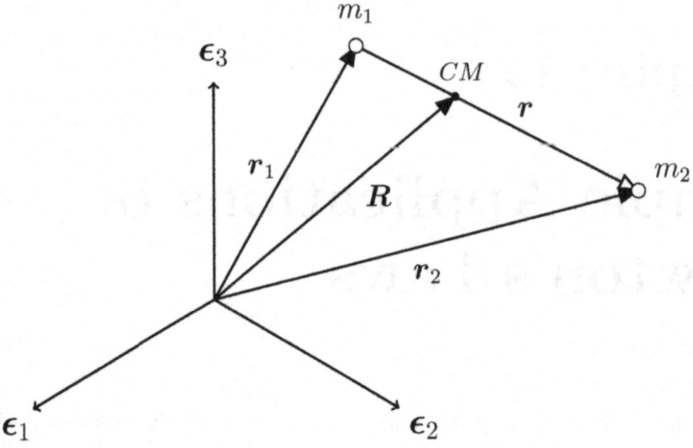

Fig. 13.1

that the equations of motion are

$$\dot{\boldsymbol{R}} = constant\ , \tag{13.2}$$

$$\mu \ddot{\boldsymbol{r}} = -Gm_1 m_2 \frac{\boldsymbol{r}}{r^3}\ , \tag{13.3}$$

where m_1, m_2 are the masses of the two bodies, G is Newton's constant of gravitation, $\boldsymbol{R}$ is the *center-of-mass* position vector of the two-body system (with respect to an inertial frame), $\boldsymbol{r}$ is the position vector of m_2 relative to m_1, and μ is the *reduced mass* of the two-body system, given by

$$\mu = \frac{m_1 m_2}{m_1 + m_2}\ . \tag{13.4}$$

The various position vectors are shown in Fig. 13.1.

The force given by (13.1) is a special example of a **central force** (one which is along the radial direction). Consequently, the torque due to that force (with respect to the origin of the coordinate system) vanishes [cf. (12.27)], and the angular momentum $\boldsymbol{L}$ is conserved:

$$\frac{d\boldsymbol{L}}{dt} = 0\ , \tag{13.5}$$

that is, $\boldsymbol{L}$ *is a constant of motion.* This fact can also be demonstrated explicitly as follows. From $\boldsymbol{L} = \boldsymbol{r} \times \boldsymbol{p}$ we have

$$\frac{d\boldsymbol{L}}{dt} = \frac{d\boldsymbol{r}}{dt} \times \boldsymbol{p} + \boldsymbol{r} \times \frac{d\boldsymbol{p}}{dt} = \frac{\boldsymbol{p}}{m} \times \boldsymbol{p} + \left(-\frac{\alpha}{r^3}\right) \boldsymbol{r} \times \boldsymbol{r} = 0 \,. \tag{13.6}$$

Define the **Runge-Lenz vector** $\boldsymbol{M}$ as follows:

$$\boxed{\boldsymbol{M} = \frac{\boldsymbol{p} \times \boldsymbol{L}}{m} - \frac{\alpha \boldsymbol{r}}{r}} \,. \tag{13.7}$$

We will show that $\boldsymbol{M}$ *is also a constant of motion.* Indeed,

$$\begin{aligned}
\frac{d\boldsymbol{M}}{dt} &= \frac{1}{m}\frac{d\boldsymbol{p}}{dt} \times \boldsymbol{L} + \frac{\boldsymbol{p}}{m} \times \frac{d\boldsymbol{L}}{dt} - \alpha \frac{d}{dt}\left(\frac{\boldsymbol{r}}{r}\right) \\
&= -\frac{\alpha}{mr^3}\, \boldsymbol{r} \times \boldsymbol{L} - \alpha \frac{d}{dt}\left(\frac{\boldsymbol{r}}{r}\right) \\
&= -\frac{\alpha}{mr^3}\, \boldsymbol{r} \times (\boldsymbol{r} \times \boldsymbol{p}) - \alpha \left(-\frac{\dot{r}\boldsymbol{r}}{r^2} + \frac{1}{r}\frac{d\boldsymbol{r}}{dt}\right) \\
&= -\frac{\alpha}{mr^3}\left(\boldsymbol{r}(\boldsymbol{r} \cdot \boldsymbol{p}) - r^2 \boldsymbol{p}\right) + \frac{\alpha \dot{r}}{r^2}\boldsymbol{r} - \frac{\alpha \boldsymbol{p}}{mr} \\
&= -\frac{\alpha \boldsymbol{r}}{r^3}\left(\frac{d\boldsymbol{r}}{dt} \cdot \boldsymbol{r}\right) + \alpha \frac{\dot{r}\boldsymbol{r}}{r^2} = -\frac{\alpha \boldsymbol{r}}{r^3} r\dot{r} + \alpha \frac{\dot{r}\boldsymbol{r}}{r^2} = 0 \,,
\end{aligned} \tag{13.8}$$

where the fourth equality follows from the "bac - cab" identity (1.59) and the next to last equality from (4.39), which implies that $\boldsymbol{v} \cdot \boldsymbol{r} = \dot{r}r$.

The constants of motion $\boldsymbol{L}$ and $\boldsymbol{M}$ satisfy the following conditions:

$$\boldsymbol{L} \cdot \boldsymbol{r} = \boldsymbol{L} \cdot \boldsymbol{M} = 0 \,, \tag{13.9}$$

which imply that $\boldsymbol{L}$ *is perpendicular to the plane of the orbit* and $\boldsymbol{M}$ *lies on the plane of the orbit.* Equation (13.9) can be demonstrated easily:

$$\boldsymbol{L} \cdot \boldsymbol{r} = \boldsymbol{r} \cdot (\boldsymbol{r} \times \boldsymbol{p}) = \boldsymbol{p} \cdot (\boldsymbol{r} \times \boldsymbol{r}) = 0 \,, \tag{13.10}$$

$$\boldsymbol{L} \cdot \boldsymbol{M} = \boldsymbol{L} \cdot \left(\frac{\boldsymbol{p} \times \boldsymbol{L}}{m} - \frac{\alpha \boldsymbol{r}}{r}\right) = \frac{1}{m}\boldsymbol{p} \cdot (\boldsymbol{L} \times \boldsymbol{L}) - \frac{\alpha}{r}\boldsymbol{r} \cdot \boldsymbol{L} = 0 \,. \tag{13.11}$$

To find the equation of the orbit we calculate $\boldsymbol{M} \cdot \boldsymbol{r}$:

$$\boldsymbol{M} \cdot \boldsymbol{r} = \frac{\boldsymbol{r} \cdot (\boldsymbol{p} \times \boldsymbol{L})}{m} - \alpha r = \frac{L^2}{m} - \alpha r \,. \tag{13.12}$$

Choosing the x-axis to be along the direction of $\boldsymbol{M}$, we can write the above equation in terms of polar coordinates (r, θ) as (Fig. 13.2)

$$Mr\cos\theta = \frac{L^2}{m} - \alpha r \,, \tag{13.13}$$

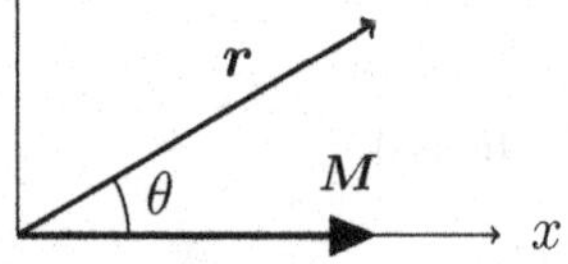

Fig. 13.2

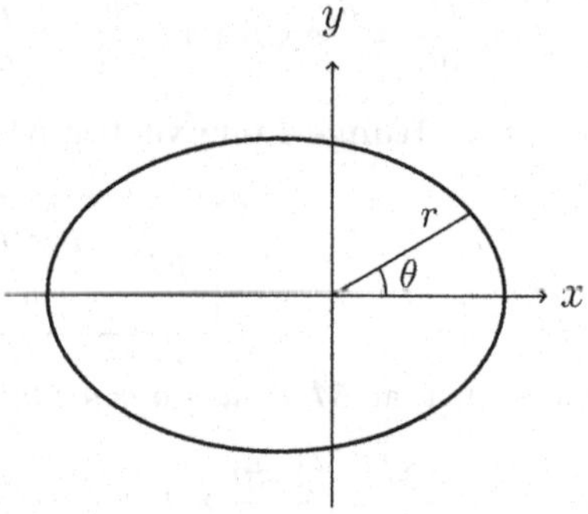

Fig. 13.3

or

$$\boxed{r = \frac{L^2}{\alpha m \left(1 + \frac{M}{\alpha}\cos\theta\right)}} \,. \tag{13.14}$$

This is the equation of a *conic section*, with the *eccentricity* equal to M/α (Fig. 13.3).

Problem 13.2 Show that the magnitude of the Runge-Lenz vector is given by

$$M = \sqrt{\alpha^2 + \frac{2EL^2}{m}}\,, \tag{13.15}$$

where the total energy E is given by

$$E = \frac{m\dot{r}^2}{2} + \frac{L^2}{2mr^2} - \frac{\alpha}{r}\,. \tag{13.16}$$

Next we consider the classical motion of a charged particle in combined electric and magnetic fields. For simplicity we will only treat the case of static (time-independent) fields. According to the **Lorentz Force Law** the equation of motion of a charged particle of charge e and mass m in an electric field $\boldsymbol{E}$ and magnetic field $\boldsymbol{H}$ is given by

$$m\ddot{\boldsymbol{r}} = e\boldsymbol{E} + \frac{e}{c}\,\dot{\boldsymbol{r}} \times \boldsymbol{H}\,, \tag{13.17}$$

where c is the speed of light in vacuum. Again, $\boldsymbol{r}$ is the position vector of the charged particle with respect to an inertial frame.

We first note that a change in the kinetic energy of the particle is produced by the electric field $\boldsymbol{E}$ alone. This is demonstrated as follows:

$$\begin{aligned} \frac{d}{dt}\left(\frac{1}{2}mv^2\right) &= \frac{m}{2}\,2\boldsymbol{v}\cdot\frac{d\boldsymbol{v}}{dt} = \boldsymbol{v}\cdot(m\ddot{\boldsymbol{r}}) \\ &= e\boldsymbol{E}\cdot\boldsymbol{v} + \frac{e}{c}\,(\boldsymbol{v}\times\boldsymbol{H})\cdot\boldsymbol{v} = e\boldsymbol{E}\cdot\boldsymbol{v}\,. \end{aligned} \tag{13.18}$$

If $\boldsymbol{E}$ is an electrostatic field, then

$$\boldsymbol{E} = -\nabla\phi\,, \tag{13.19}$$

where ϕ is the *electrostatic potential.* Equation (13.18) then reads

$$\frac{d}{dt}\left(\frac{1}{2}mv^2\right) = -e\nabla\phi\cdot\frac{d\boldsymbol{r}}{dt} = -e\frac{d\phi}{dt}\,, \tag{13.20}$$

or

$$\frac{d}{dt}\left(\frac{1}{2}mv^2 + e\phi\right) = 0\,. \tag{13.21}$$

The quantity

$$\mathcal{E} \equiv \frac{1}{2}mv^2 + e\phi \tag{13.22}$$

can be identified as the total energy of the particle; and (13.21) is just a statement of the conservation of energy.

Consider the special case of constant $\boldsymbol{E}$ and $\boldsymbol{H}$, with $\boldsymbol{E} = 0$ and $\boldsymbol{H} = H\,\boldsymbol{e}_3 = (0, 0, H)$. Then

$$\dot{\boldsymbol{r}} \times \boldsymbol{H} = \begin{vmatrix} \boldsymbol{e}_1 & \boldsymbol{e}_2 & \boldsymbol{e}_3 \\ \dot{x} & \dot{y} & \dot{z} \\ 0 & 0 & H \end{vmatrix} = \dot{y}H\,\boldsymbol{e}_1 - \dot{x}H\,\boldsymbol{e}_2\,. \tag{13.23}$$

The equation of motion (13.17) thus becomes

$$m\ddot{x} = \frac{e}{c}\,\dot{y}H\,, \qquad m\ddot{y} = -\frac{e}{c}\,\dot{x}H\,, \qquad m\ddot{z} = 0\,. \tag{13.24}$$

The last of these equations implies

$$z(t) = z(0) + \dot{z}(0)\,t\,. \tag{13.25}$$

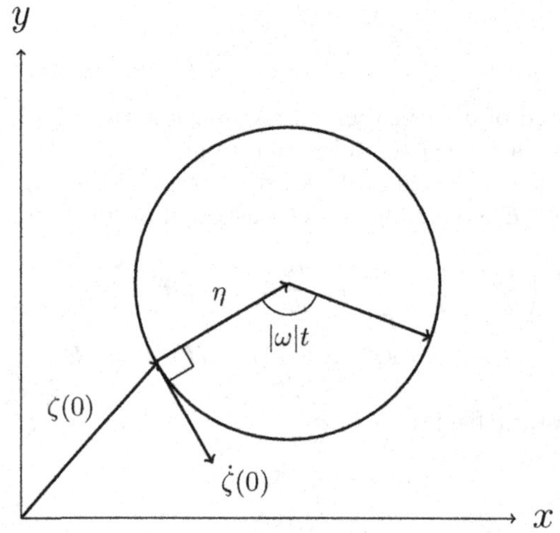

Fig. 13.4

So the charged particle moves along the z-direction with constant velocity $\dot{z}(0)$. To obtain the motion projected on the xy plane we have to solve the first two coupled equations in (13.24). Defining

$$\omega \equiv \frac{eH}{mc}, \tag{13.26}$$

the coupled equations can be written

$$\ddot{x} = \omega\dot{y}, \qquad \ddot{y} = -\omega\dot{x}. \tag{13.27}$$

These can be solved most readily by invoking the complex function $\zeta(t) : \mathbb{R} \to \mathbb{C}$ defined by

$$\zeta(t) \equiv x(t) + i\,y(t). \tag{13.28}$$

Then, from (13.27),

$$\ddot{\zeta} = \ddot{x} + i\,\ddot{y} = \omega\dot{y} + i\,(-\omega\dot{x}) = -i\,\omega(\dot{x} + i\,\dot{y}), \tag{13.29}$$

or

$$\ddot{\zeta} = -i\,\omega\dot{\zeta}. \tag{13.30}$$

This equation can be integrated once to yield

$$\dot{\zeta}(t) = \dot{\zeta}(0)\,e^{-i\omega t}, \tag{13.31}$$

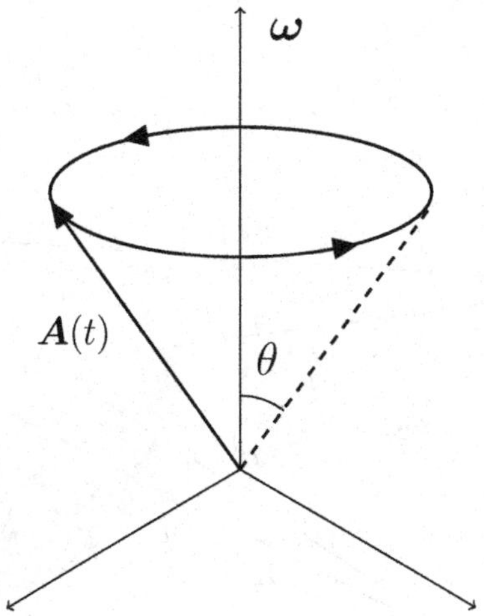

Fig. 13.5

and then another time to obtain

$$\zeta(t) = \zeta(0) + \int_0^t \dot{\zeta}(t)\, dt = \zeta(0) + \dot{\zeta}(0) \int_0^t e^{-i\omega t}\, dt \; . \tag{13.32}$$

The final result is

$$\zeta(t) = \zeta(0) + \frac{\dot{\zeta}(0)}{i\omega}(1 - e^{-i\omega t}) \; , \tag{13.33}$$

where

$$\zeta(0) = x(0) + i\, y(0) \; , \qquad \dot{\zeta}(0) = \dot{x}(0) + i\, \dot{y}(0) \; , \tag{13.34}$$

$$|\, \dot{\zeta}(0)\, | = v \quad \text{(constant speed)} \; . \tag{13.35}$$

Let us assume that the charged particle is an electron so that $e < 0$ and thus $\omega < 0$. The complex numbers in (13.33) can be viewed as two-dimensional vectors and the projected motion of the particle on the xy plane can be obtained diagrammatically in a straightforward manner. In fact, defining the constant complex number

$$\eta \equiv \frac{\dot{\zeta}(0)}{i\omega} \; , \tag{13.36}$$

so that

$$|\, \eta\, | = \left| \frac{\dot{\zeta}(0)}{\omega} \right| = v \, \left| \frac{mc}{eH} \right| \; , \tag{13.37}$$

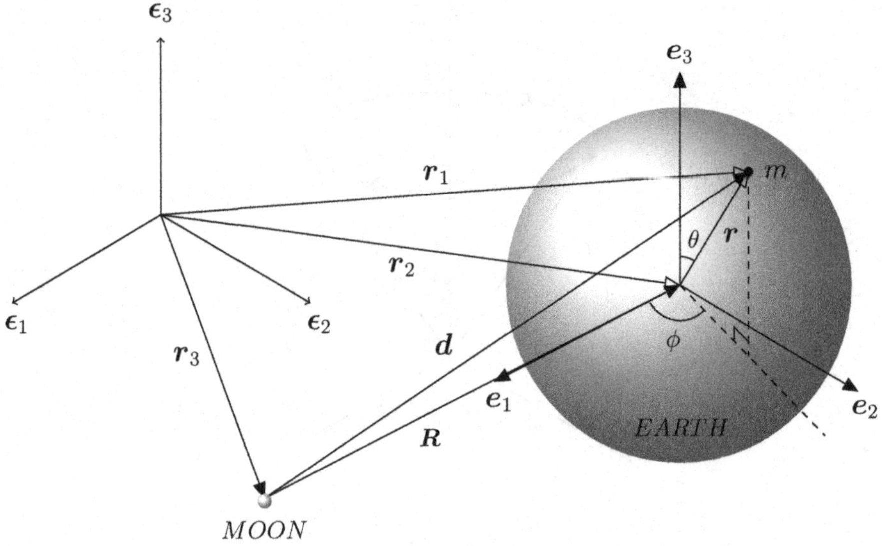

Fig. 13.6

we see that the projected motion is uniform circular motion with angular speed $|\omega| = (|e|H)/(mc)$ (see Fig. 13.4). The radius of the circular orbit is $|\eta| = v/|\omega|$.

Problem 13.3 Show that the time evolution of a vector $\boldsymbol{A}(t)$ whose motion is described by the differential equation

$$\frac{d\boldsymbol{A}}{dt} = \boldsymbol{\omega} \times \boldsymbol{A} \qquad (\boldsymbol{\omega} = \text{constant}) \tag{13.38}$$

is the *precessional motion* shown in Fig. 13.5 (with the indicated orientation), with precessional angular speed $|\boldsymbol{\omega}|$ and $\theta = \text{constant}$.

We will now turn to a simple model of tidal motion. Consider the earth-moon system as shown in Fig. 13.6. The quantity m is a point mass on the

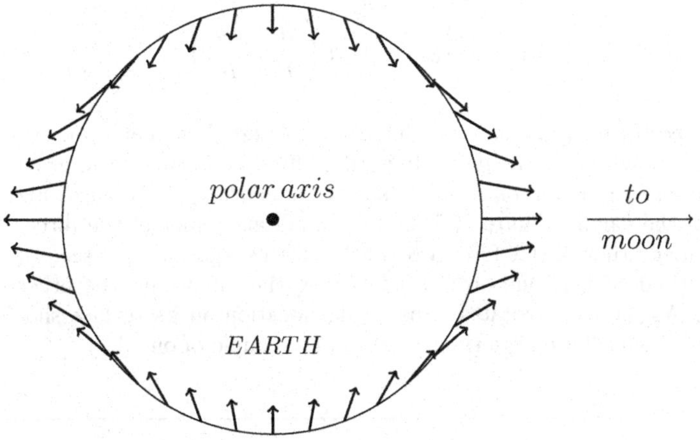

Fig. 13.7

earth's ocean surface, M_E is the mass of the earth, and M is the mass of the moon. $\{\boldsymbol{\epsilon}_1, \boldsymbol{\epsilon}_2, \boldsymbol{\epsilon}_3\}$ is a space-fixed orthonormal frame, while $\{\boldsymbol{e}_1, \boldsymbol{e}_2, \boldsymbol{e}_3\}$ is a moving orthonormal frame with origin at the center of the earth and $\boldsymbol{e}_1$ along the direction from the center of the earth to the moon (considered as a point mass). The $\boldsymbol{e}_1$ axis is also considered to lie on the equatorial plane of the earth. Thus θ is the colatitude of the point mass m. In our model, the surface of the earth is assumed to be completely covered by an ocean.

As seen from Fig. 13.6,

$$\boldsymbol{r} = \boldsymbol{r}_1 - \boldsymbol{r}_2 \,, \tag{13.39}$$

$$\boldsymbol{R} = \boldsymbol{r}_2 - \boldsymbol{r}_3 \,, \tag{13.40}$$

$$\boldsymbol{d} = \boldsymbol{r}_1 - \boldsymbol{r}_3 = \boldsymbol{R} + \boldsymbol{r} \,. \tag{13.41}$$

The motions of the mass point m and the earth (of mass M_E) due to gravitational forces are determined, according to Newton's second law and Newton's universal law of gravitation, by the following two equations:

$$m\ddot{\boldsymbol{r}}_1 = -\frac{GmM_E\hat{\boldsymbol{r}}}{r^2} - \frac{GmM\hat{\boldsymbol{d}}}{d^2} \,, \tag{13.42}$$

$$M_E\ddot{\boldsymbol{r}}_2 = -\frac{GM_EM\hat{\boldsymbol{R}}}{R^2} \,. \tag{13.43}$$

Dividing (13.42) by m and (13.43) by M_E, and then subtracting the second equation from the first, we obtain, on recalling (13.39), the following equation of motion for the relative position vector $\boldsymbol{r}$:

$$\ddot{\boldsymbol{r}} = -\frac{GM_E\hat{\boldsymbol{r}}}{r^2} - GM\left(\frac{\hat{\boldsymbol{d}}}{d^2} - \frac{\hat{\boldsymbol{R}}}{R^2}\right) . \tag{13.44}$$

The second term on the right-hand side of the above equation is the tide-generating force per unit mass. It is the difference between the gravitational forces due to the moon on a unit mass on the surface of the earth and at the center of the earth. Figure 13.7 shows a qualitative plot of the direction and relative magnitude of this force around the earth's equator. It is seen that there are two tidal bulges, one on the side facing the moon and the other on the far side. As the earth rotates, a particular location on its surface should then experience two tidal bulges (high tides) in the course of one day.

Problem 13.4 Use (13.44) to justify the qualitative features of Fig. 13.7.

From (13.41), $d^2 = R^2 + r^2 + 2\boldsymbol{R}\cdot\boldsymbol{r}$, and so

$$d = R\left(1 + \frac{2\boldsymbol{R}\cdot\boldsymbol{r}}{R^2} + \frac{r^2}{R^2}\right)^{1/2} . \tag{13.45}$$

Since $r \ll R$, we can expand in powers of r/R to obtain

$$d \sim R\left(1 + \frac{\boldsymbol{R}\cdot\boldsymbol{r}}{R^2} + \dots\right) . \tag{13.46}$$

It follows that

$$\frac{1}{d^3} \sim \frac{1}{R^3}\left(1 + \frac{\boldsymbol{R}\cdot\boldsymbol{r}}{R^2}\right)^{-3} \sim \frac{1}{R^3}\left(1 - \frac{3\boldsymbol{R}\cdot\boldsymbol{r}}{R^2}\right) = \frac{1}{R^3} - \frac{3\hat{\boldsymbol{R}}\cdot\boldsymbol{r}}{R^4} . \tag{13.47}$$

Using this result, we can write

$$\begin{aligned} &\frac{\hat{\boldsymbol{d}}}{d^2} - \frac{\hat{\boldsymbol{R}}}{R^2} = \frac{\boldsymbol{d}}{d^3} - \frac{\boldsymbol{R}}{R^3} = \frac{\boldsymbol{R}+\boldsymbol{r}}{d^3} - \frac{\boldsymbol{R}}{R^3} \\ &= \boldsymbol{R}\left(\frac{1}{d^3} - \frac{1}{R^3}\right) + \frac{\boldsymbol{r}}{d^3} \sim -\frac{3(\hat{\boldsymbol{R}}\cdot\boldsymbol{r})\hat{\boldsymbol{R}}}{R^3} + \frac{\boldsymbol{r}}{R^3} - \frac{3(\hat{\boldsymbol{R}}\cdot\boldsymbol{r})}{R^3}\frac{\boldsymbol{r}}{R} \\ &\sim \frac{1}{R^3}\{-3\hat{\boldsymbol{R}}(\hat{\boldsymbol{R}}\cdot\boldsymbol{r}) + \boldsymbol{r}\} , \end{aligned} \tag{13.48}$$

where in the right-hand side of the last equality we have neglected the last term on the right-hand side of the next to last equality (since it is small compared to the other two terms by a factor of $\boldsymbol{r}/R$). Since $\hat{\boldsymbol{R}} = -\boldsymbol{e}_1$, we can then write

$$\frac{\hat{\boldsymbol{d}}}{d^2} - \frac{\hat{\boldsymbol{R}}}{R^2} \sim \frac{1}{R^3}(-3x\,\boldsymbol{e}_1 + \boldsymbol{r})\,, \tag{13.49}$$

where $\boldsymbol{r} = x\,\boldsymbol{e}_1 + y\,\boldsymbol{e}_2 + z\,\boldsymbol{e}_3$. Equation (13.44) then implies

$$\ddot{\boldsymbol{r}} = -\frac{GM_E\,\hat{\boldsymbol{r}}}{r^2} + \frac{GM}{R^3}(3x\,\boldsymbol{e}_1 - \boldsymbol{r})\,. \tag{13.50}$$

We can express the right-hand side of the above equation as the gradient of a scalar field. Thus

$$\ddot{\boldsymbol{r}} = -\nabla\Phi(\boldsymbol{r})\,, \tag{13.51}$$

where

$$\Phi(\boldsymbol{r}) = -\frac{GM_E}{r} - \frac{GM}{R^3}\left(\frac{3x^2}{2} - \frac{r^2}{2}\right)\,. \tag{13.52}$$

Since $x = r\sin\theta\cos\phi$ (see Fig. 13.6), Eq. (13.52) can also be written

$$\Phi(\boldsymbol{r}) = -\frac{GM_E}{r} - \frac{GM}{r}\left(\frac{r}{R}\right)^3\left(\frac{3}{2}\sin^2\theta\cos^2\phi - \frac{1}{2}\right)\,. \tag{13.53}$$

Problem 13.5 Verify the validity of (13.51) with $\Phi(\boldsymbol{r})$ given by (13.53). [Use the formula for the calculation of the gradient in spherical coordinates given by (9.56).]

The surface of the ocean (assumed to cover the entire surface of the earth) must be an **equipotential surface** under the action of the gravitational force due to the earth itself and the tidal gravitational forces due to the moon. In other words, $\Phi(\boldsymbol{r})$ must be constant on the surface of the ocean (see Fig. 13.8). In the absence of tidal forces (equivalently $M = 0$), the surface of the ocean is represented by the dotted line in Fig. 13.8. The constant value of $\Phi(\boldsymbol{r})$ on this surface is approximately $-GM_E/R_E$, where R_E can be taken to be the mean radius of the earth. Thus the actual ocean (equipotential) surface satisfies the equation

$$-\frac{GM_E}{R_E} = -\frac{GM_E}{r} - \frac{GM}{r}\left(\frac{r}{R}\right)^3\left(\frac{3}{2}\sin^2\theta\cos^2\phi - \frac{1}{2}\right)\,. \tag{13.54}$$

Multiplying the above equation by $-R_E r/(GM_E)$, we have

$$r - R_E = \left(\frac{M}{M_E}\right)\left(\frac{r}{R}\right)^3 R_E\left(\frac{3}{2}\sin^2\theta\cos^2\phi - \frac{1}{2}\right)\,. \tag{13.55}$$

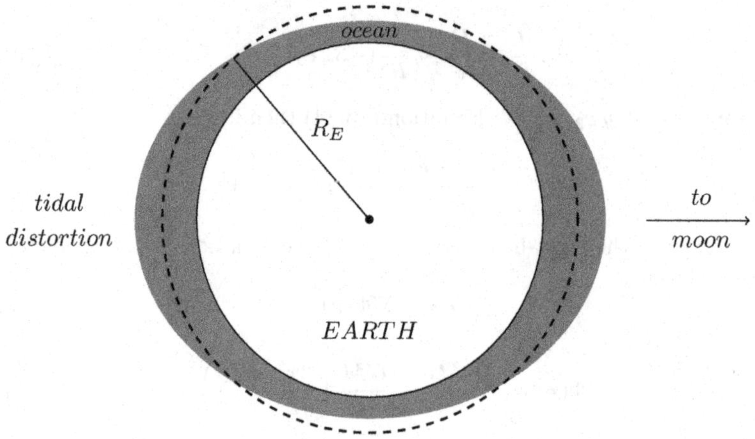

Fig. 13.8

We can now define the so-called *tidal distortion*

$$h(\theta, \phi) \equiv r - R_E \ , \tag{13.56}$$

which only depends on the angles θ and ϕ, by approximating r on the right-hand side of (13.55) by R_E. Thus

$$h(\theta, \phi) \sim \left(\frac{M}{M_E}\right) \frac{R_E^4}{R^3} \left(\frac{3}{2} \sin^2\theta \cos^2\phi - \frac{1}{2}\right) \ . \tag{13.57}$$

For a given colatitude θ, maximum h (high tide) positions occur at $\cos^2\phi = 1$, that is, $\phi = 0, \pi$; and minimum h (low tide) positions occur at $\cos^2\phi = 0$, that is, $\phi = \pi/2, 3\pi/2$. The *tidal range*, $\Delta h \equiv h_{max} - h_{min}$, for a given θ is given by

$$\Delta h = \frac{3MR_E^4 \sin^2\theta}{2M_E R^3} \ . \tag{13.58}$$

Maximum tidal distortion h_M occurs at the equator ($\theta = \pi/2$, $\phi = 0$ or π):

$$h_M \sim \frac{MR_E^4}{M_E R^3} \ . \tag{13.59}$$

The entire development in the above paragraphs on tidal forces would also have been valid if we replace the moon by the sun. Indeed it is interesting to

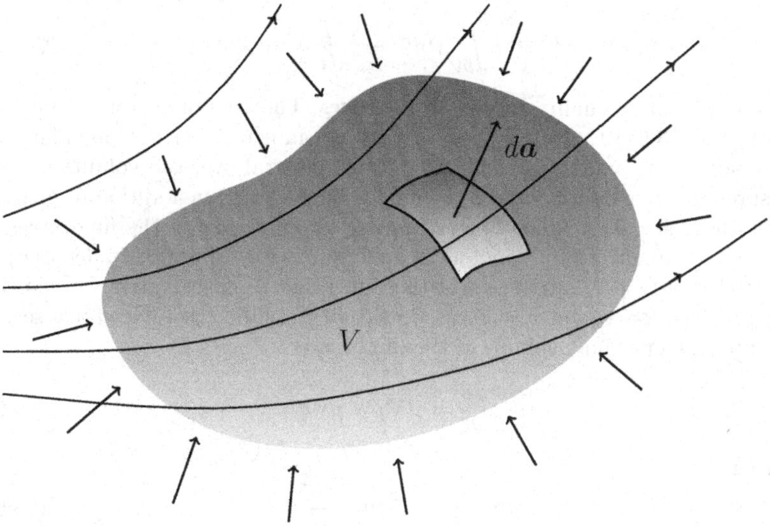

Fig. 13.9

use (13.59) to compare lunar and solar tidal effects. Using the following data:

$$
\begin{aligned}
M_{moon} &= \text{mass of moon} \sim 7.4 \times 10^{22}\, kg \,, \\
M_{sun} &= \text{mass of sun} \sim 2 \times 10^{30}\, kg \,, \\
R_{moon} &= \text{mean earth-to-moon distance} \sim 3.8 \times 10^{8}\, m \,, \\
R_{sun} &= \text{mean earth-to-sun distance} \sim 1.5 \times 10^{11}\, m,
\end{aligned}
$$

we estimate that

$$\frac{h_M(moon)}{h_M(sun)} \sim \left(\frac{M_{moon}}{M_{sun}}\right)\left(\frac{R_{sun}}{R_{moon}}\right)^3 = \left(\frac{7.4\times 10^{22}}{2\times 10^{30}}\right) \times \left(\frac{1.5\times 10^{11}}{3.8\times 10^{8}}\right)^3 \sim 2.2 \,. \tag{13.60}$$

We will finally study fluid motion. First we consider **ideal fluids**, in which there is no dissipation of energy. Figure 13.9 shows an arbitrary volume V inside a region of fluid flow, with arrows pointing (perpendicularly) towards the surface of V representing pressure forces on the fluid contained in V. The **streamlines** (marked by arrows) indicate the direction of the flow, and are *integral curves* to a **velocity field** $\boldsymbol{v}(\boldsymbol{r}, t)$, which is tangent to the streamlines at every point of the flow (cf. discussion in Chapter 11). The area element $d\boldsymbol{a}$ of the boundary to the volume points in the outward normal direction.

Suppose $p(\boldsymbol{r})$ represents the **pressure field**, which is assumed to be a scalar field. Then the net pressure force $\boldsymbol{F}$ acting on the volume is given by

$$\boldsymbol{F} = -\oint_{\partial V} p\, d\boldsymbol{a} = -\oint_{V} \nabla p\, d^3 r\,, \tag{13.61}$$

where the second equality follows from Stokes' Theorem of vector calculus [cf. (8.14)]. In (13.61) the first integral is an area integral over the boundary ∂V of the volume V and the second is a volume integral over the volume V. The pressure force per unit volume is then given by $-\nabla p$. If, in addition to pressure forces, there are other kinds of conservative forces acting on the fluid, we may represent their effects by a potential function $\phi(\boldsymbol{r})$. Specifically, let $\phi(\boldsymbol{r})$ = potential energy per unit mass (of the fluid) due to conservative forces other than pressure forces and ρ = mass density of the fluid, then Newton's second law applied to per unit volume of the fluid reads

$$\rho \frac{d\boldsymbol{v}}{dt} = -\nabla p - \rho \nabla \phi\,. \tag{13.62}$$

Writing

$$d\boldsymbol{v} = (d\boldsymbol{r} \cdot \nabla)\boldsymbol{v} + \frac{\partial \boldsymbol{v}}{\partial t} dt\,, \tag{13.63}$$

we have

$$\frac{d\boldsymbol{v}}{dt} = (\boldsymbol{v} \cdot \nabla)\boldsymbol{v} + \frac{\partial \boldsymbol{v}}{\partial t}\,. \tag{13.64}$$

In the physics literature the above derivative is sometimes referred to as the **material derivative**. Substituting the right-hand side of this equation for $d\boldsymbol{v}/dt$ in (13.62) and rearranging, we obtain **Euler's equation** for the motion of ideal fluids:

$$\boxed{\frac{\partial \boldsymbol{v}}{\partial t} + (\boldsymbol{v} \cdot \nabla)\boldsymbol{v} = -\frac{\nabla p}{\rho} - \nabla \phi}\,. \tag{13.65}$$

Using the vector identity

$$\frac{1}{2}\nabla v^2 = \boldsymbol{v} \times (\nabla \times \boldsymbol{v}) + (\boldsymbol{v} \cdot \nabla)\, \boldsymbol{v}\,, \tag{13.66}$$

Euler's equation can also be written in the useful form

$$\frac{\partial \boldsymbol{v}}{\partial t} + \frac{1}{2}\nabla v^2 - \boldsymbol{v} \times (\nabla \times \boldsymbol{v}) = -\frac{\nabla p}{\rho} - \nabla \phi\,. \tag{13.67}$$

Problem 13.6 Prove the vector identity (13.66) by using the index formalism for tensors introduced in Chapter 1.

Problem 13.7 Use Euler's Equation (13.65) to determine the shape of the surface of an incompressible fluid subject to the gravitational field near the surface of the earth, if the fluid is contained in a cylindrical vessel which rotates about its vertical axis with a constant angular speed ω. (See a different approach to the problem in Chapter 15.)

We will consider the case of the **steady flow** of **incompressible fluids**. The first condition means that $\partial \boldsymbol{v}/\partial t = 0$, while the second implies $\rho =$ constant. For steady flow the flow pattern is time-independent and can be represented by **streamlines**, which are integral curves to the *Pfaffian* system of differential equations [cf. (10.2) and the discussion following]:

$$\frac{dx}{v_x} = \frac{dy}{v_y} = \frac{dz}{v_z} \,. \tag{13.68}$$

[At a point in the fluid, $\boldsymbol{v} = (v_x, v_y, v_z)$ is along the same direction as the line element $d\boldsymbol{r} = (dx, dy, dz)$ along the flow line passing through that point.]

The tangential directions to the streamlines are given by the unit vector $\hat{\boldsymbol{s}} \equiv \boldsymbol{v}/v$ at each point in the fluid. Obviously

$$\hat{\boldsymbol{s}} \cdot (\boldsymbol{v} \times (\nabla \times \boldsymbol{v}) = 0 \,, \tag{13.69}$$

since $\boldsymbol{v} \times (\nabla \times \boldsymbol{v}) \perp \boldsymbol{v}$ (from the definition of the cross product). Also

$$\boldsymbol{s} \cdot \nabla f = \frac{\partial f}{\partial s} \,, \tag{13.70}$$

where $\boldsymbol{s} = s\,\hat{\boldsymbol{s}}$, and f is an arbitrary scalar field. On projecting (13.67) along $\boldsymbol{s}$ (taking the scalar product of that equation with $\boldsymbol{s}$) and remembering $\partial \boldsymbol{v}/\partial t = 0$ for steady flow, we then obtain

$$\frac{\partial}{\partial s}\left(\frac{v^2}{2} + \frac{p}{\rho} + \phi\right) = 0 \,. \tag{13.71}$$

Finally we arrive at **Bernoulli's equation** governing the steady flow of incompressible ideal fluids:

$$\boxed{\frac{p}{\rho} + \frac{v^2}{2} + \phi = \text{ constant along a streamline}} \,. \tag{13.72}$$

To study *viscous flow* with energy dissipation (non-ideal fluids), it will be useful to first rewrite Euler's equation with the help of the *equation of continuity* [cf. (8.10)]:

$$\nabla \cdot (\rho \boldsymbol{v}) + \frac{\partial \rho}{\partial t} = 0 \,, \tag{13.73a}$$

or, in tensorial notation

$$\frac{\partial \rho}{\partial t} = -\partial_k(\rho v^k) \,. \tag{13.73b}$$

It then follows from Euler's equation [(13.65)] (with $\phi = 0$), which in tensorial notation can be written as

$$\frac{\partial v^i}{\partial t} = -v^k\,\partial_k v^i - \frac{1}{\rho}\,\delta^{ik}\partial_k p\,, \tag{13.74}$$

and the equation of continuity (13.73) that

$$\frac{\partial}{\partial t}(\rho v^i) = v^i\frac{\partial \rho}{\partial t} + \rho\frac{\partial v^i}{\partial t} = -\delta^{ik}\,\partial_k p - \partial_k(\rho v^i v^k)\,. \tag{13.75}$$

We can then rewrite Euler's equation (in the absence of an external conservative field) in the following form

$$\frac{\partial}{\partial t}(\rho v^i) = -\partial_k \Pi^{ik}\,, \tag{13.76}$$

where

$$\Pi^{ik} \equiv \delta^{ik}p + \rho v^i v^k \tag{13.77}$$

is a $(2,0)$-type symmetric tensor called the **momentum flux density tensor**, so named because it gives the flux of the i-th component of the momentum per unit time across unit area perpendicular to the k-th spatial direction.

Problem 13.8 Show that the tensor Π^{ik} defined by (13.77) indeed represents a momentum flux density by integrating both sides of Euler's equation in the form (13.76) over a volume D in the body of the fluid, and then applying Gauss' Theorem [(8.5)] to the integral on the right-hand side to get

$$\int_D \partial_k \Pi^{ik}\, d^3r = \oint_{\partial D} \Pi^{ik} da_k\,, \tag{13.78}$$

where ∂D is the boundary of D and da_k ia an area element along the k-th direction.

In the presence of internal viscous forces, there must be irreversible momentum transfer from locations where the velocity is high to those where it is low. To account for this situation we alter the expression for the momentum flux density tensor given by (13.77) as follows:

$$\Pi^{ik} = \delta^{ik}p + \rho v^i v^k - \sigma'^{ik} = -\sigma^{ik} + \rho v^i v^k\,, \tag{13.79}$$

where

$$\sigma^{ik} \equiv -\delta^{ik}p + \sigma'^{ik} \tag{13.80}$$

is called the **viscosity stress tensor**. It is postulated that, when the velocity gradients are small, the irreversible terms in the viscosity stress tensor, that is, those making up σ'^{ik}, depend linearly on such gradients. The most general form for σ'^{ik}, as a symmetric $(2,0)$-type tensor that depends linearly on the velocity gradients, can then be written as

$$\sigma'^{ik} = a(\delta^{kl}\,\partial_l v^i + \delta^{il}\,\partial_l v^k) + b\delta^{ik}\,\partial_l v^l\,, \tag{13.81}$$

where a and b are constants. Note that this automatically vanishes when the whole fluid is in uniform rotational motion (see Problem 13.9).

Problem 13.9 Use (13.81) to show that the irreversible part of the viscosity stress tensor, that is, the tensor σ'^{ik}, vanishes when $\boldsymbol{v} = \boldsymbol{\omega} \times \boldsymbol{r}$, where $\boldsymbol{\omega}$ is a constant angular velocity vector .

It is convenient to rewrite σ'^{ik} as follows:

$$\sigma'^{ik} = \eta\left(\delta^{il}\,\partial_l v^k + \delta^{kl}\,\partial_l v^i - \frac{2}{3}\,\delta^{ik}\,\partial_l v^l\right) + \zeta\,\delta^{ik}\,\partial_l v^l\,. \tag{13.82}$$

This expression has the property that the tensorial quantity within the parentheses vanishes on contraction with respect to the indices i and k. (The reader should verify this.) The positive scalar quantities η and ζ are called **coefficients of viscosity**. In general they are functions of the temperature T and the pressure p, and are therefore functions of the position $\boldsymbol{r}$ in the fluid.

In view of (13.76), (13.79) and (13.80), Euler's equation [in the form (13.74)] then has to be generalized by adding the term $\partial_k\sigma'^{ik}$ to the right-hand side when internal viscous forces are present:

$$\rho\left(\frac{\partial v^i}{\partial t} + v^k\,\partial_k v^i\right) = -\delta^{ik}\,\partial_k p + \partial_k\sigma'^{ik}\,. \tag{13.83}$$

If we assume that the coefficients of viscosity η and ζ are constant in the fluid (independent of $\boldsymbol{r}$), then

$$\begin{aligned}\partial_k\sigma'^{ik} &= \eta\left(\delta^{kl}\,\partial_k\partial_l v^i + \frac{1}{3}\,\delta^{ik}\,\partial_k(\partial_l v^l)\right) + \zeta\,\delta^{ik}\,\partial_k(\partial_l v^l)\\ &= \eta\,\delta^{kl}\,\partial_k\partial_l v^i + \left(\zeta + \frac{\eta}{3}\right)\,\delta^{ik}\,\delta_k(\partial_l v^l)\\ &= \eta\,\nabla^2 v^i + \left(\zeta + \frac{\eta}{3}\right)\,\delta^{ik}\,\delta_k(\nabla\cdot\boldsymbol{v})\,,\end{aligned} \tag{13.84}$$

where in the last line we have used the ordinary vector calculus notation, with ∇^2 and $\nabla\cdot$ being the *Laplacian* and the *divergence* operators, respectively [cf. (9.32) and (7.13)]. For *incompressible fluids*, ρ is constant (independent of both $\boldsymbol{r}$ and t), and the equation of continuity (13.73) implies that $\nabla\cdot\boldsymbol{v} = 0$. The modified Euler's equation (13.83) with the viscous term given by (13.84) then becomes

$$\boxed{\frac{\partial\boldsymbol{v}}{\partial t} + (\boldsymbol{v}\cdot\nabla)\boldsymbol{v} = -\frac{\nabla p}{\rho} + \frac{\eta}{\rho}\,\nabla^2\boldsymbol{v}}\quad. \tag{13.85}$$

This is known as the **Navier-Stokes equation** in fluid mechanics (in the absence of external fields). [Compare with Euler's equation (13.65).] Its solution in general poses great mathematical difficulties. In fact, the following (stated roughly) is still an open mathematical problem: *For any given initial smooth and divergence-free velocity field $\boldsymbol{v}(\boldsymbol{r},0)$ and positive constant coefficient of viscosity η, do there exist smooth, divergence-free, spatially bounded (as $|\boldsymbol{r}| \to \infty$), and spatially normalizable vector velocity fields $\boldsymbol{v}(\boldsymbol{r},t)$ on $\mathbb{R}^3 \times [0,\infty)$ and smooth scalar pressure fields $p(\boldsymbol{r},t)$ on $\mathbb{R}^3 \times [0,\infty)$ that satisfy the Navier-Stokes equation (13.85)?* For various specific and physically relevant solutions, the reader can refer to, for example, Landau and Liftshitz 1979.

To conclude this chapter we will demonstrate how the Lie derivative (cf. Chapter 11) is naturally suited for the mathematical description of fluid flows. Still within the context of the steady flow of incompressible fluids, we assume in addition that the external potential vanishes: $\phi(\boldsymbol{r}) = 0$. Euler's equation (13.65) then reads

$$(\boldsymbol{v}\cdot\nabla)\,\boldsymbol{v} = -\nabla\left(\frac{p}{\rho}\right) . \tag{13.86}$$

Define the (vector) **vorticity** field

$$\boldsymbol{w} \equiv \nabla \times \boldsymbol{v} , \tag{13.87}$$

and the scalar field

$$\Phi(\boldsymbol{r}) \equiv \frac{1}{2}(\boldsymbol{v}\cdot\boldsymbol{v}) + \frac{p}{\rho} . \tag{13.88}$$

Then

$$\begin{aligned} \nabla\Phi &= \nabla\left(\frac{1}{2}\boldsymbol{v}\cdot\boldsymbol{v}\right) + \nabla\left(\frac{p}{\rho}\right) = \boldsymbol{v}\times(\nabla\times\boldsymbol{v}) + (\boldsymbol{v}\cdot\nabla)\,\boldsymbol{v} + \nabla\left(\frac{p}{\rho}\right) \\ &= \boldsymbol{v}\times\boldsymbol{w} + (\boldsymbol{v}\cdot\nabla)\,\boldsymbol{v} + \nabla\left(\frac{p}{\rho}\right) , \end{aligned} \tag{13.89}$$

where the second equality follows from (13.66). The simplified Euler's equation (13.86) then leads to

$$\nabla\Phi = \boldsymbol{v}\times\boldsymbol{w} . \tag{13.90}$$

On taking the scalar product on both sides of this equation with $\boldsymbol{v}$, we obtain

$$\boldsymbol{v}\cdot\nabla\Phi = \boldsymbol{v}\cdot(\boldsymbol{v}\times\boldsymbol{w}) = 0 . \tag{13.91}$$

The left-hand side of the first equality is precisely the Lie derivative $L_{\boldsymbol{v}}\Phi$ of the scalar field Φ [according to (11.22)]. Hence Euler's equation (13.86) can be expressed much more succinctly as

$$\boxed{L_{\boldsymbol{v}}\Phi = 0} . \tag{13.92}$$

Now (13.90) also implies

$$\begin{aligned} 0 &= \nabla\times\nabla\Phi = \nabla\times(\boldsymbol{v}\times\boldsymbol{w}) \\ &= (\boldsymbol{w}\cdot\nabla)\,\boldsymbol{v} - (\boldsymbol{v}\cdot\nabla)\,\boldsymbol{w} + (\nabla\cdot\boldsymbol{w})\,\boldsymbol{v} - (\nabla\cdot\boldsymbol{v})\,\boldsymbol{w} , \end{aligned} \tag{13.93}$$

where the first equality follows from (7.17) and the third from (7.26) (cf. Problem 7.3). On assuming that the velocity field is sourceless, so that $\nabla \cdot \boldsymbol{v} = 0$, and recognizing that $\nabla \cdot \boldsymbol{w} = \nabla \cdot (\nabla \times \boldsymbol{v}) = 0$ [by (7.18)], the above equation amounts to

$$(\boldsymbol{w} \cdot \nabla)\, \boldsymbol{v} - (\boldsymbol{v} \cdot \nabla)\, \boldsymbol{w} = 0 \,. \tag{13.94}$$

The left-hand side, by (6.22), is precisely the Lie derivative $L_{\boldsymbol{w}}$. So

$$\boxed{L_{\boldsymbol{w}}\boldsymbol{v} = L_{\boldsymbol{v}}\boldsymbol{w} = 0} \quad , \tag{13.95}$$

where the first equality also follows from (6.22). These equations express succinctly the relationship between a fluid velocity field and its vorticity.

Chapter 14

Potential Theory: Newtonian Gravitation

In this chapter we will illustrate the use of Newton's Laws on a most important topic, Newtonian gravitation due to a continuous distribution of mass, whose successful application to celestial mechanics in the seventeenth century historically established the validity of classical mechanics, and indeed, laid the foundations for the development of modern physics. We will see that the treatment of this problem makes essential use of some of the techniques of vector and tensor analysis introduced in Chapter 1.

Consider first the motion of a point particle of mass m in a conservative gravitational force field $\boldsymbol{F}(\boldsymbol{r})$. Newton's second law gives [cf. (8.2) and (12.3)]

$$m\ddot{\boldsymbol{r}} = \boldsymbol{F} = -\nabla U \,, \tag{14.1}$$

where U is the potential energy of the particle corresponding to $\boldsymbol{F}$. Let

$$\boldsymbol{F} = m\boldsymbol{g} \,, \qquad U = m\phi \,. \tag{14.2}$$

Then the **gravitational field** $\boldsymbol{g}$ is given by

$$\boldsymbol{g} = -\nabla\phi \,, \tag{14.3}$$

where ϕ is called the **gravitational potential**.

Newton's Universal Law of Gravitation is equivalent to *Gauss' Law* for the gravitational field, which can be stated as follows:

$$\boxed{\nabla \cdot \boldsymbol{g} = -4\pi G\rho} \quad , \tag{14.4}$$

where

$$G = 6.67 \times 10^{-11}\, N \cdot m^2/kg^2$$

is *Newton's constant of gravitation*, and ρ is the local mass density. For the gravitational field due to a point mass M located at the origin of the coordinate

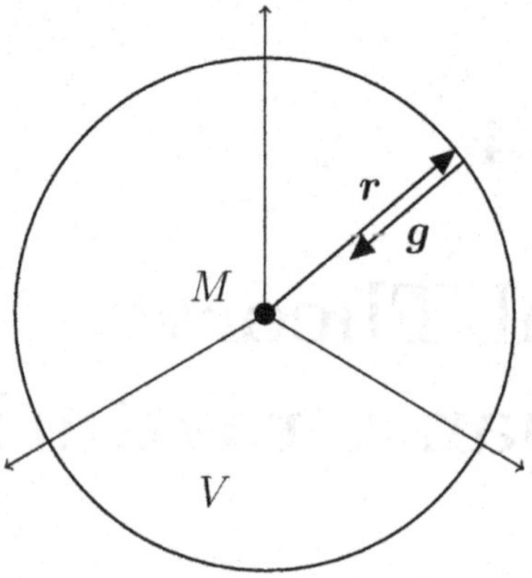

Fig. 14.1

system (Fig. 14.1) and remembering that in this case $\boldsymbol{g}$ is radially inward (the gravitational force is always attractive), we can apply Gauss' Theorem of vector calculus [cf. (8.5)] to obtain,

$$\int_V \nabla \cdot \boldsymbol{g}\, d^3r = \oint_{\partial V} \boldsymbol{g} \cdot d\boldsymbol{a} = -4\pi r^2 g \,, \tag{14.5}$$

where the first integral is a volume integral over a 3-dimensional sphere V of radius r centered at the origin, and the second is a surface integral over the boundary ∂V of V. Equation (14.4) then gives $-4\pi g r^2 = -4\pi GM$, or

$$g = \frac{GM}{r^2} \,, \tag{14.6}$$

which is the familiar form of Newton's law of gravitation giving the magnitude of the gravitational field due to a point mass.

Using (14.3), (14.4) becomes

$$\boxed{\nabla^2 \phi = 4\pi G \rho} \,, \tag{14.7}$$

which is a second-order partial differential equation called **Poisson's equation** for the gravitational potential. For a point mass M at the origin, $\rho(\boldsymbol{r}) = M\,\delta^3(\boldsymbol{r})$, and

$$\phi(\boldsymbol{r}) = -\frac{GM}{r} \,. \tag{14.8}$$

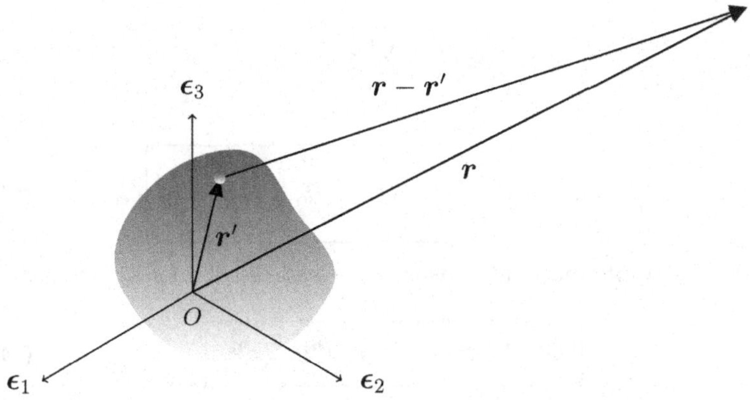

Fig. 14.2

For a continuous mass distribution (Fig. 14.2), the formal solution to Poisson's equation is given by

$$\phi(\boldsymbol{r}) = -G \int d^3r' \, \frac{\rho(\boldsymbol{r}')}{|\,\boldsymbol{r} - \boldsymbol{r}'\,|} \,, \tag{14.9}$$

where $\boldsymbol{r}'$ is the position vector of an infinitesimal volume element of the mass distribution and the integral is over the volume of the mass distribution.

This formal solution is the basis of the **multipole expansion** of $\phi(\boldsymbol{r})$, which we will now demonstrate. First we use Taylor's expansion (expanding around the point $\boldsymbol{r}' = 0$) to write

$$\frac{1}{|\,\boldsymbol{r} - \boldsymbol{r}'\,|} = \left\{ 1 + (-\boldsymbol{r}' \cdot \nabla) + \frac{1}{2!}(-\boldsymbol{r}' \cdot \nabla)^2 + \dots \right\} \frac{1}{r} \,. \tag{14.10}$$

Then it follows from (14.9) that

$$\phi(\boldsymbol{r}) = -G \sum_{n=0}^{\infty} \frac{1}{n!} \int d^3r' \, \rho(\boldsymbol{r}')(-\boldsymbol{r}' \cdot \nabla)^n \left(\frac{1}{r} \right) \,. \tag{14.11}$$

Write the operator $(-\boldsymbol{r}' \cdot \nabla)^n$ in terms of Cartesian coordinates as

$$(-\boldsymbol{r}' \cdot \nabla)^n = (-1)^n x'^{i_1} \dots x'^{i_n} \frac{\partial}{\partial x^{i_1}} \cdots \frac{\partial}{\partial x^{i_n}} \,, \tag{14.12}$$

where repeated indices are summed over from 1 to 3 (Einstein's summation convention). Note that the gradient is with respect to the unprimed coordinates.

Then the *multipole expansion* of the gravitational potential due to a continuous mass distribution is given by

$$\phi(\boldsymbol{r}) = \sum_{n=0}^{\infty} \phi_{(n)}(\boldsymbol{r}) \,, \tag{14.13}$$

where

$$\boxed{\phi_{(n)}(\boldsymbol{r}) \equiv G\frac{(-1)^{n+1}}{n!}\, Q^{i_1\dots i_n}\, \partial_{i_1}\dots\partial_{i_n}\left(\frac{1}{r}\right)} \,, \tag{14.14}$$

and the **multipole moment tensor** of the mass distribution $Q^{i_1\dots i_n}$ is defined by

$$\boxed{Q^{i_1\dots i_n} \equiv \int d^3r'\, \rho(\boldsymbol{r}') x'^{i_1}\dots x'^{i_n}} \,. \tag{14.15}$$

Note that $Q^{i_1\dots i_n}$ is a contravariant tensor of rank n, or a $(n, 0)$-type tensor.

As examples, we will calculate the first three lowest order contributions to the multipole expansion for the gravitational potential $\phi(\boldsymbol{r})$ in (14.13): $\phi_{(0)}, \phi_{(1)}$ and $\phi_{(2)}$. The term $\phi_{(0)}$ is called the **monopole** term; it is simply given by

$$\phi_{(0)}(\boldsymbol{r}) = -G\,\frac{\int d^3r \rho(\boldsymbol{r}')}{r} = -\frac{GM}{r} \,. \tag{14.16}$$

Comparing with (14.8), we see that this term gives the potential due to a point mass, whose mass is equal to the entire mass of the extended object under consideration, located at the origin of the coordinate system.

Next we calculate $\phi_{(1)}$, the so-called **dipole** term. From (14.14) we have

$$\phi_{(1)}(\boldsymbol{r}) = GQ^i\,\partial_i\left(\frac{1}{r}\right) , \tag{14.17}$$

where the **dipole moment tensor** Q^i is given, according to (14.15), by

$$Q^i = \int d^3r'\, \rho(\boldsymbol{r}')\, x'^i = MR^i \,. \tag{14.18}$$

In the above equation, the second equality follows from the definition of the *center of mass position vector* (see Chapter 19), with R^i being its i-th coordinate. Writing $x^1 = x$, $x^2 = y$ and $x^3 = z$. Equation (14.18) is equivalent to

$$Q^x = \int d^3r\rho(\boldsymbol{r})\, x \,, \quad Q^y = \int d^3r\rho(\boldsymbol{r})\, y \,, \quad Q^z = \int d^3r\rho(\boldsymbol{r})\, z \,. \tag{14.19}$$

The partial derivatives in (14.17) can be calculated easily. We have

$$\frac{\partial}{\partial x}\left(\frac{1}{r}\right) = -\frac{x}{r^3} \,, \quad \frac{\partial}{\partial y}\left(\frac{1}{r}\right) = -\frac{y}{r^3} \,, \quad \frac{\partial}{\partial z}\left(\frac{1}{r}\right) = -\frac{z}{r^3} \,. \tag{14.20}$$

Problem 14.1 Verify Eqs. (14.20) using the chain rule of calculus and the fact that $r^2 = x^2 + y^2 + z^2$.

Thus

$$Q^i \frac{\partial}{\partial x^i}\left(\frac{1}{r}\right) = Q^x\left(-\frac{x}{r^3}\right) + Q^y\left(-\frac{y}{r^3}\right) + Q^z\left(-\frac{z}{r^3}\right) . \tag{14.21}$$

Defining the **dipole moment vector** $\boldsymbol{d}$ by

$$\boldsymbol{d} = Q^x\, \boldsymbol{\epsilon}_1 + Q^y\, \boldsymbol{\epsilon}_2 + Q^z\, \boldsymbol{\epsilon}_3 , \tag{14.22}$$

where $\{\boldsymbol{\epsilon}_1, \boldsymbol{\epsilon}_2, \boldsymbol{\epsilon}_3\}$ is an inertial orthonormal frame [cf. Fig. 14.2], it follows from (14.17) that the dipole term can be written as

$$\phi_{(1)}(\boldsymbol{r}) = -G\frac{\boldsymbol{d}\cdot\boldsymbol{r}}{r^3} . \tag{14.23}$$

Note that according to the second equality of (14.18), $\boldsymbol{d} = 0$ if the center of mass lies at the origin of the coordinate system. *Thus the dipole term can always be made to vanish by choosing a reference frame whose origin coincides with the position of the center of mass.*

Next we calculate the quadrupole term $\phi_{(2)}$. By (14.14) we have

$$\phi_{(2)}(\boldsymbol{r}) = -\frac{G}{2}\, Q^{ij} \frac{\partial}{\partial x^i}\frac{\partial}{\partial x^j}\left(\frac{1}{r}\right) , \tag{14.24}$$

where the **quadrupole moment tensor** Q^{ij} is given by

$$Q^{ij} = \int d^3r'\, \rho(\boldsymbol{r}') x'^i x'^j . \tag{14.25}$$

The second order partial derivatives can be calculated as follows:

$$\begin{aligned} \frac{\partial}{\partial x^i}\frac{\partial}{\partial x^j}\left(\frac{1}{r}\right) &= \frac{\partial}{\partial x^i}\left(-\frac{1}{r^2}\frac{\partial r}{\partial x^j}\right) = -\frac{\partial}{\partial x^i}\left(\frac{x^j}{r^3}\right) \\ &= -\left(-\frac{3x^j x^i}{r^5} + \frac{\delta^{ij}}{r^3}\right) = \frac{3x^i x^j}{r^5} - \frac{\delta^{ij}}{r^3} . \end{aligned} \tag{14.26}$$

The quadrupole term of $\phi(\boldsymbol{r})$ can then be written as

$$\phi_{(2)}(\boldsymbol{r}) = -\frac{G}{2r^5}\int d^3r'\, \rho(\boldsymbol{r}')\{3(\boldsymbol{r}'\cdot\boldsymbol{r})^2 - r^2 r'^2\} . \tag{14.27}$$

Equation (14.25) shows that Q^{ij} is a symmetric matrix. Hence it can be diagonalized by a suitable choice of coordinate frame. Assuming this is done, we can write

$$Q^{ij} = \begin{pmatrix} Q^{xx} & 0 & 0 \\ 0 & Q^{yy} & 0 \\ 0 & 0 & Q^{zz} \end{pmatrix} . \tag{14.28}$$

Then (14.24) yields

$$\phi_{(2)}(\boldsymbol{r}) = -\frac{G}{2}\left\{ Q^{xx}\left(\frac{3x^2}{r^5} - \frac{1}{r^3}\right) + Q^{yy}\left(\frac{3y^2}{r^5} - \frac{1}{r^3}\right) + Q^{zz}\left(\frac{3z^2}{r^5} - \frac{1}{r^3}\right)\right\} . \tag{14.29}$$

This expression simplifies more if the mass distribution has *axial* or *azimuthal symmetry* (symmetry along the ϕ direction). In this case, $Q^{xx} = Q^{yy}$ (see Fig. 14.3), and (14.29) implies

$$\begin{aligned} \phi_{(2)}(\boldsymbol{r}) &= -\frac{G}{2}\left[Q^{xx}\left\{\frac{3(x^2+y^2)}{r^5} - \frac{2}{r^3}\right\} + Q^{zz}\left(\frac{3z^2}{r^5} - \frac{1}{r^3}\right)\right] \\ &= -\frac{G}{2}\left[Q^{xx}\left\{\frac{3(r^2-z^2)}{r^5} - \frac{2}{r^3}\right\} + Q^{zz}\left(\frac{3z^2}{r^5} - \frac{1}{r^3}\right)\right] \\ &= -\frac{G}{2r^3}\left[Q^{xx}\left(1 - \frac{3z^2}{r^2}\right) + Q^{zz}\left(\frac{3z^2}{r^2} - 1\right)\right] . \end{aligned} \tag{14.30}$$

If we define in general the scalar **quadrupole moment** Q by

$$Q \equiv 2Q^{zz} - (Q^{xx} + Q^{yy}) , \tag{14.31}$$

so that, with axial symmetry,

$$Q = 2(Q^{zz} - Q^{xx}) , \tag{14.32}$$

then we have the following useful expression for the quadrupole term of the gravitational potential due to a mass distribution with axial symmetry:

$$\boxed{\phi_{(2)}(\boldsymbol{r}) = -\frac{GQ}{4r^3}(3\cos^2\theta - 1)} , \tag{14.33}$$

where θ is the polar angle of the field point. The scalar quadrupole moment Q gives a measure of the "flattening" of the mass distribution along the z-axis. A mass distribution is known as **prolate** when $Q > 0$ (cigar-shaped) [Fig. 14.4(a)] and **oblate** (pancake-shaped) when $Q < 0$ [Fig. 14.4(b)].

For an *ellipsoid of revolution* (with axial symmetry and uniform mass distribution) (Fig. 14.5) it can be shown that

$$Q = \frac{2}{5}M(c^2 - a^2) , \tag{14.34}$$

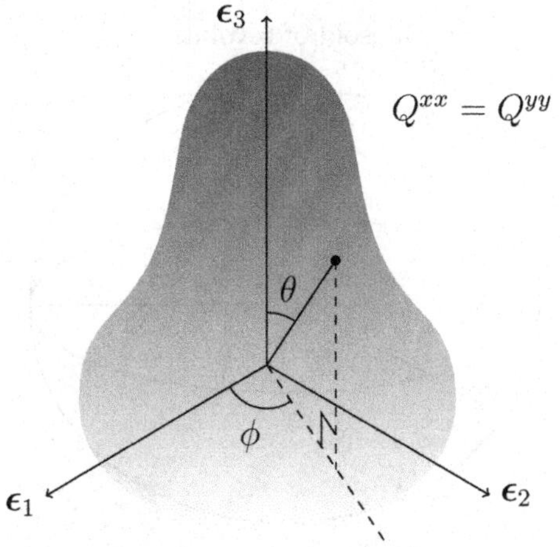

Fig. 14.3

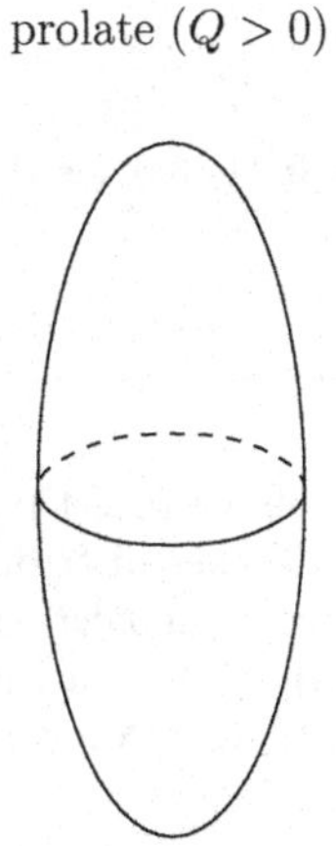

Fig. 14.4(a)

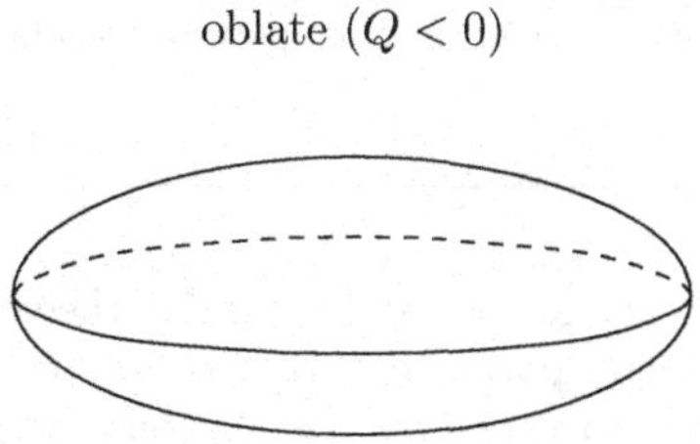

Fig. 14.4(b)

ellipsoid of revolution

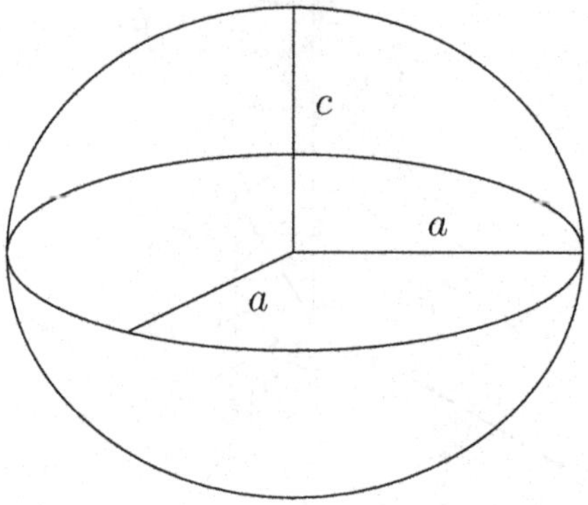

Fig. 14.5

where a and c are the lengths of the semiaxes of the ellipsoid (as shown in Fig. 14.5) and M is the total mass of the distribution.

Problem 14.2 Use (14.25) and (14.32) to prove the result (14.34) for the scalar quadrupole moment of an ellipsoid of revolution.

We will now use the potential theory as developed above to estimate the values of two parameters that characterize the shape and mass distribution of the earth. To a good approximation the earth has the shape of an *oblate ellipsoid of revolution* flattened slightly at the poles (see Fig. 14.5). The departure from perfect sphericity is described by a dimensionless parameter ε known as the *oblateness*:

$$\varepsilon \equiv \frac{a-c}{a}\,, \tag{14.35}$$

where a is the equatorial radius and c is the distance from the center of the earth to either pole. For the earth, $a - c \sim 21.5\,km$. Using $a \sim 6.37 \times 10^3\,km$, one obtains $\varepsilon \sim 1/300$. Assume for the moment that the mass distribution of the earth is uniform (which is not the case), its scalar quadrupole moment Q is given by (14.34). It can be rewritten in terms of the oblateness ε as follows:

$$Q = -\frac{2}{5}M(a^2 - c^2) = -\frac{2}{5}M\frac{(a+c)(a-c)a}{a} \sim -\frac{2}{5}M\frac{2a(a-c)a}{a} = -\frac{4}{5}Ma^2\varepsilon \,. \tag{14.36}$$

But the mass density ρ of the earth is not constant; in fact it increases towards the center. To account for this fact, we introduce a dimensionless parameter λ in place of ε in the expression for Q and write

$$Q = -\frac{4}{5}Ma^2\lambda \,, \qquad \lambda < \varepsilon \,. \tag{14.37}$$

On recalling (14.16) and (14.33), we then have the following expression for the gravitational potential due to the earth up to the quadrupole term [with respect to a CM (center-of-mass) frame whose z-axis points from the center of the earth to the north pole]:

$$\boxed{\phi(\boldsymbol{r}) = -\frac{GM}{r} + \frac{GMa^2\lambda}{5r^3}(3\cos^2\theta - 1)} \,. \tag{14.38}$$

We will work with a body-fixed CM frame, that is, one that is rotating with the earth. With respect to this non-inertial frame, every part of the earth is in equilibrium under the *combined* gravitational and centrifugal forces. In particular the surface of the earth must be an equipotential surface under these combined forces. The centrifugal potential ϕ_c corresponding to the centrifugal force [cf. (12.2)] can be written as:

$$\phi_c = -\frac{1}{2}\omega^2(x^2 + y^2) = -\frac{1}{2}\omega^2 r^2 \sin^2\theta \,, \tag{14.39}$$

where ω is the angular speed of rotation of the earth about its axis. Using (14.38) for the gravitational potential ϕ and (14.39) for the centrifugal potential ϕ_c, and equating the values of the total potential $\phi + \phi_c$ at the north pole and the equator, we have

$$-\frac{GM}{c} + \frac{2GMa^2\lambda}{5c^3} = -\frac{GM}{a} - \frac{GM\lambda}{5a} - \frac{1}{2}\omega^2 a^2 \,. \tag{14.40}$$

Now, (14.35) and the fact that the oblateness $\varepsilon \ll 1$ for the earth imply that

$$\frac{1}{c} \sim \frac{1}{a}(1 + \varepsilon) \,. \tag{14.41}$$

Using the above equation, (14.40) reduces to

$$-\frac{GM\varepsilon}{a} + \frac{3GM\lambda}{5a} = -\frac{1}{2}\omega^2 a^2 \,. \tag{14.42}$$

Define the quantity

$$g_0 \equiv \frac{GM}{a^2} \,, \tag{14.43}$$

which is an approximate value for the acceleration due to gravity anywhere on the earth's surface. Then (14.42) can be written as

$$2\varepsilon - \frac{6}{5}\lambda = \frac{\omega^2 a}{g_0} . \tag{14.44}$$

We will obtain an independent equation relating ε and λ in order to solve for these parameters. The gravitational field $\boldsymbol{g}(\boldsymbol{r})$ can be obtained from the gravitational potential $\phi(\boldsymbol{r})$ by $\boldsymbol{g} = -\nabla\phi$. Thus, on using (14.38) for $\phi(\boldsymbol{r})$ and (9.56) (the expression for the gradient in spherical coordinates), we have

$$g_r = -\frac{\partial\phi}{\partial r} = -\frac{GM}{r^2} + \frac{3GMa^2\lambda}{5r^4}(3\cos^2\theta - 1) , \tag{14.45}$$

$$g_\theta = -\frac{1}{r}\frac{\partial\phi}{\partial\theta} = \frac{6GMa^2\lambda\cos\theta\sin\theta}{5r^4} . \tag{14.46}$$

Note that $g_\theta = 0$ at both poles ($\theta = 0, \pi$) and the equator ($\theta = \pi/2$). We then have

$$\begin{aligned} |\, g_{pole} \,| &= \frac{GM}{c^2} - \frac{6GMa^2\lambda}{5c^4} \sim \frac{GM}{a^2}(1+2\varepsilon) - \frac{6GM\lambda}{5a^2} \\ &= \frac{GM}{a^2}(1 + 2\varepsilon - \frac{6\lambda}{5}) = g_0(1 + 2\varepsilon - \frac{6\lambda}{5}) , \end{aligned} \tag{14.47}$$

$$|\, g_{eq} \,| = \frac{GM}{a^2} + \frac{3GM\lambda}{5a^2} \sim g_0(1 + \frac{3\lambda}{5}) , \tag{14.48}$$

where $|\, g_{pole} \,|$ and $|\, g_{eq} \,|$ are the magnitudes of the gravitational field $\boldsymbol{g}$ at the poles and the equator, respectively.

From (14.39), the centrifugal force field $\boldsymbol{g}_c$ at the poles vanishes, while at the equator it is in the outward radial direction and has a magnitude given by $\omega^2 a$ [both of these facts can be easily verified by calculating the gradient of ϕ_c, with ϕ_c given by (14.39)]. We then obtain the following expression for the difference in acceleration (due to both gravitational and centrifugal forces) at either pole (north or south) and the equator:

$$\Delta g \equiv |\, g_{pole} \,| - |\, g_{eq} \,| + \omega^2 a . \tag{14.49}$$

Using (14.44), (14.47) and (14.48) in the above equation we have

$$\frac{\Delta g}{g_0} = 4\varepsilon - 3\lambda . \tag{14.50}$$

One can then use the approximate empirical values:

$$\Delta g = 5.2\ cm/sec^2 , \qquad \omega^2 a = 3.4\ cm/sec^2 , \qquad g_0 = 980\ cm/sec^2 , \tag{14.51}$$

to solve the simultaneous equations (14.44) and (14.50) and finally obtain the following estimates for ε and λ:

$$\varepsilon = 3.37 \times 10^{-3} , \qquad \lambda = 2.72 \times 10^{-3} . \tag{14.52}$$

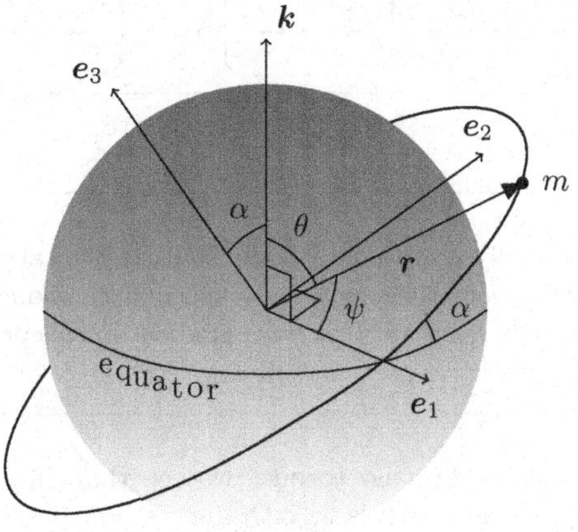

Fig. 14.6

Centrifugal forces at an arbitrary position on the surface of the earth will be studied in more detail in the next chapter.

Problem 14.3 Confirm the above numerical results for ε and λ by solving the simultaneous algebraic equations (14.44) and (14.50) with the given numerical values in (14.51).

Let us now consider the effects of the quadrupole moment of the earth on an orbiting satellite above its surface. Consider a satellite of mass m in a circular orbit of radius r over the earth, with the orbital plane inclined at an angle α to the equator (see Fig. 14.6). Using (14.38) for the gravitational potential of the satellite (up to the quadrupole term) we can obtain the gravitational field $\boldsymbol{g}$ acting on it as follows:

$$\begin{aligned}\boldsymbol{g}(\boldsymbol{r}) = -\nabla\phi(\boldsymbol{r}) &= -\frac{GM}{r^2}\,\boldsymbol{e}_r - \frac{GMa^2\lambda}{5}\nabla\left(\frac{3\cos^2\theta - 1}{r^3}\right)\\ &= -\frac{GM}{r^2}\,\boldsymbol{e}_r - \frac{GMa^2\lambda}{5}\left\{-\frac{3(3\cos^2\theta - 1)}{r^4}\,\boldsymbol{e}_r + \frac{(-6\cos\theta\sin\theta)}{r^4}\,\boldsymbol{e}_\theta\right\}.\end{aligned} \tag{14.53}$$

Collecting terms we can write

$$\boxed{g(r) = -\left\{\frac{GM}{r^2} - \frac{3GMa^2\lambda(3\cos^2\theta - 1)}{5r^4}\right\} e_r + \left\{\frac{6GMa^2\lambda \sin\theta\cos\theta}{5r^4}\right\} e_\theta} \; . \tag{14.54}$$

Problem 14.4 Carefully go through all the steps in the calculation of the gradient of the gravitational potential $\phi(r)$ [given by (14.38)] leading to (14.54) for the gravitational field $g(r)$ by using (9.56) for the gradient in spherical coordinates.

Since a radial force exerts no torque, we see that the main effect of the quadrupole moment of the earth is to introduce a non-vanishing torque due to the θ-component of g. As seen from (14.54) the quadrupole moment also leads to a perturbative radial force which depends on the polar angle θ and varies with r according to r^{-4}. Writing $r = r\,e_r$, the net torque τ is given by

$$\tau = mr \times g = \frac{6GMa^2\lambda m \sin\theta\cos\theta}{5r^3}\, e_r \times e_\theta \, . \tag{14.55}$$

Referring to Fig. 14.6, we see that

$$\cos\theta = \hat{k}\cdot e_r \, , \quad \sin\theta\, e_r \times e_\theta = \hat{k}\times e_r \, . \tag{14.56}$$

Thus

$$\tau = \frac{6GMa^2\lambda m}{5r^3}(\hat{k}\cdot e_r)(\hat{k}\times e_r) \, . \tag{14.57}$$

We will choose a moving frame $\{e_1, e_2, e_3\}$ so that e_1 lies on the intersection of the orbital plane of the satellite and the equatorial plane of the earth (refer to Fig. 14.6). Then $\hat{k}\cdot e_1 = 0$ and

$$\hat{k} = \sin\alpha\, e_2 + \cos\alpha\, e_3; \, , \tag{14.58}$$

$$e_r = \cos\psi\, e_1 + \sin\psi\, e_2 \, . \tag{14.59}$$

It follows from these equations that

$$\hat{k}\cdot e_r = \sin\alpha\,\sin\psi \, , \tag{14.60}$$

$$\hat{k}\times e_r = \begin{vmatrix} e_1 & e_2 & e_3 \\ 0 & \sin\alpha & \cos\alpha \\ \cos\psi & \sin\psi & 0 \end{vmatrix} \tag{14.61}$$

$$= (-\sin\psi\,\cos\alpha)\, e_1 + (\cos\psi\,\cos\alpha)\, e_2 + (-\cos\psi\,\sin\alpha)\, e_3 \, .$$

As will be justified at the end of our calculations, the orbital motion of the satellite is much faster than the precessional motion of the axis of the orbit. Hence, as a first approximation, we can average the torque over one period of the orbital motion. Defining the orbital average of a function $f(\psi)$ by

$$\langle f(\psi) \rangle \equiv \frac{1}{2\pi} \int_0^{2\pi} d\psi \, f(\psi) \,, \tag{14.62}$$

we have

$$\begin{aligned} \langle \boldsymbol{\tau} \rangle = \frac{6GMma^2\lambda}{5r^3} \{ & \sin\alpha\cos\alpha \, \langle -\sin^2\psi \rangle \, \boldsymbol{e}_1 \\ & + \sin\alpha\cos\alpha \, \langle \sin\psi\cos\psi \rangle \, \boldsymbol{e}_2 - \sin^2\alpha \, \langle \sin\psi\cos\psi \rangle \, \boldsymbol{e}_3 \} \,. \end{aligned} \tag{14.63}$$

From (14.62) we have

$$\langle \sin^2\psi \rangle = \frac{1}{2} \,, \qquad \langle \sin\psi\cos\psi \rangle = 0 \,. \tag{14.64}$$

Thus

$$\langle \boldsymbol{\tau} \rangle = -\frac{3GMma^2\lambda}{5r^3} \sin\alpha\cos\alpha \, \boldsymbol{e}_1 \,, \tag{14.65}$$

which can also be written as

$$\langle \boldsymbol{\tau} \rangle = -\frac{3GMma^2\lambda\cos\alpha}{5r^3} (\hat{\boldsymbol{k}} \times \boldsymbol{e}_3) \,, \tag{14.66}$$

since $\sin\alpha \, \boldsymbol{e}_1 = \hat{\boldsymbol{k}} \times \boldsymbol{e}_3$ [as can be readily seen from (14.58)].

To simplify our calculations further we assume that the satellite orbit is circular. In this case we can write the orbital angular momentum $\boldsymbol{L}$ as

$$\boldsymbol{L} = m\boldsymbol{r} \times \boldsymbol{v} = mr^2\omega \, \boldsymbol{e}_3 \,, \tag{14.67}$$

where ω is the orbital angular speed. Thus

$$\boldsymbol{e}_3 = \left(\frac{1}{mr^2\omega} \right) \boldsymbol{L} \,. \tag{14.68}$$

Equations (12.29) and (14.66) then imply that

$$\frac{d\boldsymbol{L}}{dt} = \langle \boldsymbol{\tau} \rangle = \left\{ -\frac{3}{5} \left(\frac{GM}{r^3} \right) \left(\frac{ma^2\lambda}{mr^2\omega} \right) \cos\alpha \, \hat{\boldsymbol{k}} \right\} \times \boldsymbol{L} \,. \tag{14.69}$$

Using the approximation $GM/r^3 \sim \omega^2$ (for uniform circular motion of the satellite), the above equation can be written as

$$\frac{d\boldsymbol{L}}{dt} = \left\{ -\frac{3\lambda\omega}{5} \left(\frac{a}{r} \right)^2 \cos\alpha \, \hat{\boldsymbol{k}} \right\} \times \boldsymbol{L} \,. \tag{14.70}$$

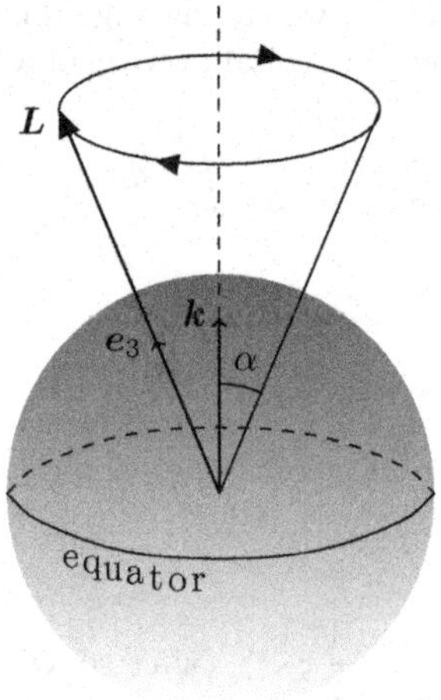

Fig. 14.7

This is of the form

$$\boxed{\frac{d\boldsymbol{L}}{dt} = \boldsymbol{\Omega} \times \boldsymbol{L}} \quad , \tag{14.71}$$

where

$$\boldsymbol{\Omega} \equiv -\frac{3\lambda\omega}{5}\left(\frac{a}{r}\right)^2 \cos\alpha\, \hat{\boldsymbol{k}} \, , \tag{14.72}$$

is approximately time-independent.

The physical interpretation of (14.71) is that $\boldsymbol{L}$, and hence the orbital axis of the satellite, *precesses* around the direction of the earth's rotational axis $\hat{\boldsymbol{k}}$ [but opposite in sense to the rotation of the earth because of the negative sign in (14.72) for $\boldsymbol{\Omega}$]. This precession takes place at the (approximately) constant rate $\Omega = (3\lambda a^2\omega\cos\alpha)/(5r^2)$ (Fig. 14.7).

It is useful to give a numerical estimate of the precessional velocity Ω for a typical satellite orbit. Assume that

$$
\begin{aligned}
a &= \text{ mean radius of earth } \sim 6.37 \times 10^6\, m \,, \\
r &= a + 400\, Km = (6.37 + 0.4) \times 10^6\, m = 6.77 \times 10^6\, m \,, \\
\lambda &= 2.72 \times 10^{-3} \quad [\text{from } (14.52)] \,, \\
\omega &= \sqrt{\frac{GM}{r^3}} = \sqrt{\frac{6.67 \times 10^{-11} \times 5.98 \times 10^{24}}{(6.77 \times 10^6)^3}}\, sec^{-1} = 1.13 \times 10^{-3}\, sec^{-1} \,, \\
\alpha &= 30^\circ \,, \text{ thus } \cos\alpha = 0.866 \,.
\end{aligned}
$$

Then, from (14.72),

$$
\Omega = \frac{3 \times 2.72 \times 10^{-3} \times 1.13 \times 10^{-3}}{5} \left(\frac{6.37}{6.77} \right)^2 \times 0.866\, sec^{-1} \sim 1.41 \times 10^{-6}\, sec^{-1} \,. \tag{14.73}
$$

This value of Ω indeed satisfies the condition $\Omega \ll \omega$ claimed earlier, and justifies the averaging of $\boldsymbol{\tau}$ over one cycle of the orbit in (14.63). The precessional period T_p is seen to be

$$
T_p = \frac{2\pi}{\Omega} \sim 4.46 \times 10^6\, sec \sim 51.6\,\text{days} \,, \tag{14.74}
$$

while the orbital period of the satellite T is

$$
T = \frac{2\pi}{\omega} \sim 5.56 \times 10^3\, sec \sim 1.5\,\text{hrs} \,. \tag{14.75}
$$

Chapter 15

Centrifugal and Coriolis Forces

In this chapter we study two familiar examples of fictitious forces – centrifugal and Coriolis forces, a general discussion of which was given in Chapter 12.

Our first example will be the centrifugal force correction to $\boldsymbol{g}_0$, the acceleration due to the earth's gravity on the surface of the earth, as a result of the rotation of the earth. (The other main corrections, due to the slight flattening of the earth at the poles and its non-uniform density, were discussed in the last chapter.) The noninertial frame we are using will be the moving frame introduced in Problem 12.1 (see Fig. 12.2). Consider a point mass m located at the origin ($\boldsymbol{r} = 0$) of this frame, and hence rotating with the earth. Since $\boldsymbol{r} = \dot{\boldsymbol{\omega}} = \boldsymbol{v} = \boldsymbol{0}$ in this situation, it follows from (12.19) and (12.25) that the corrected acceleration due to the earth's gravity, $\boldsymbol{g}$, taking into account the centrifugal force due to the rotation of the earth, is given by

$$\boldsymbol{g} = \boldsymbol{g}_0 - \boldsymbol{\omega} \times (\boldsymbol{\omega} \times \boldsymbol{R}) \,, \tag{15.1}$$

where $\boldsymbol{g}_0 = -g_0\, \boldsymbol{e}_1$ ($g_0 \sim 9.8\, m/sec^2$), and $R =$ mean radius of the earth. In the above equation, the angular velocity vector $\boldsymbol{\omega}$ is given by (12.24). Figures 15.1 and 15.2 show the directions of the centrifugal force and $\boldsymbol{g}$, respectively.

From Fig. 15.1, it is easily seen that the magnitude of the centrifugal acceleration is given by

$$|\, \boldsymbol{\omega} \times (\boldsymbol{\omega} \times \boldsymbol{R})\,| = \dot{\phi}^2 R \sin\theta \,, \tag{15.2}$$

while from Fig. 15.2, the horizontal and vertical components of $\boldsymbol{g}$ are given by

$$g_h = \dot{\phi}^2 R \sin\theta \cos\theta \,, \qquad g_v = g_0 - \dot{\phi}^2 R \sin^2\theta \,. \tag{15.3}$$

Some numerical estimates are given as follows. Take

$$R \sim 6.37 \times 10^6\, m \,, \tag{15.4}$$

$$\dot{\phi} \sim \frac{2\pi}{86,164}\, sec^{-1} \sim 7.29 \times 10^{-5}\, sec^{-1} \,. \tag{15.5}$$

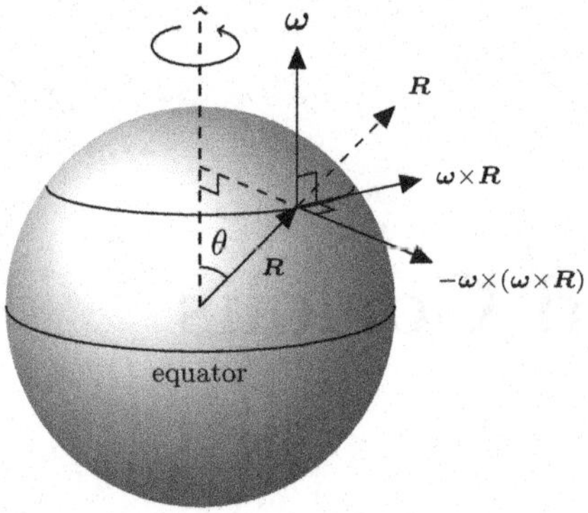

Fig. 15.1

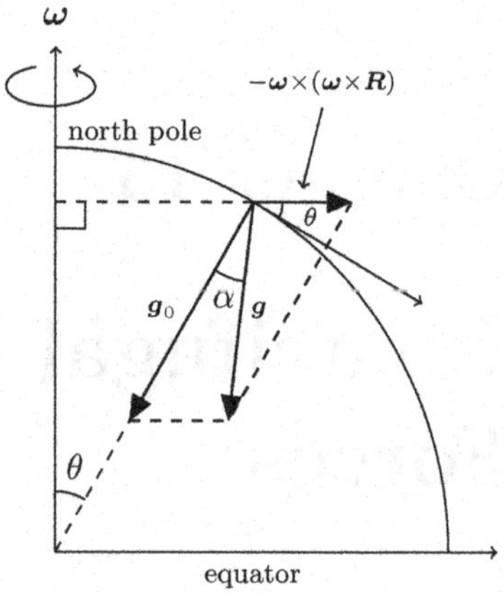

Fig. 15.2

Then

$$\dot{\phi}^2 R \sim 3.4\, cm/sec^2 \ll g \sim 980\, cm/sec^2 \,. \tag{15.6}$$

Also, from Fig. 15.2,

$$\alpha \sim \tan\alpha = \frac{g_h}{g_v} \sim \frac{\dot{\phi}^2 R \sin\theta\cos\theta}{g_0} \,. \tag{15.7}$$

Thus

$$\alpha_{max} = \alpha(\theta = 45^\circ) \sim \frac{\dot{\phi}^2 R}{2g_0} = 1.73 \times 10^{-3}\, rad \sim 0^\circ\, 6' \,. \tag{15.8}$$

According to (15.3) we have

$$g_{pole} = g_0 \qquad \text{at the poles}\,, \tag{15.9}$$

$$g_{eq} = g_0 - \dot{\phi}^2 R \qquad \text{at the equator}\,, \tag{15.10}$$

which implies that $\Delta g = g_{pole} - g_{eq}$ should be about $\dot{\phi}^2 R \sim 3.4\ cm/sec^2$. But we recall [from (14.51)] that Δg is actually about 5.2 cm/sec^2. The extra difference of 1.8 cm/sec^2 is precisely due to the effects that we considered in the last chapter, namely, the flattening of the earth at the poles and the non-uniform density of the earth.

Our second example regarding centrifugal forces is the shape of the surface of a rotating liquid. Consider a container filled with some liquid and made to rotate with an angular speed ω (Fig. 15.3). Our problem is to determine the

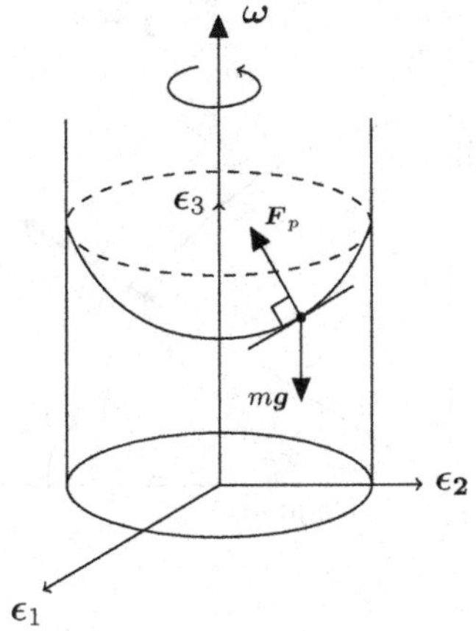

Fig. 15.3

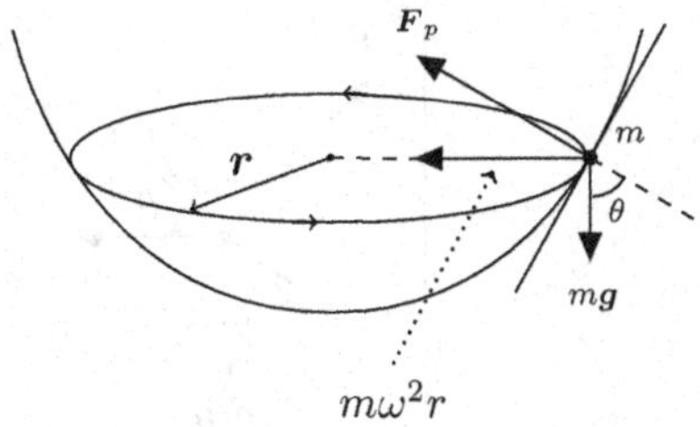

Fig. 15.4

shape of the surface of the fluid. We assume that all particles of the fluid are also rotating with the same angular speed, with respect to an inertial reference frame, say $\{\epsilon_i\}$. With respect to such a frame, a particle of the fluid of mass m on the surface is executing circular motion under the combined action of the gravitational force $m\boldsymbol{g}$ and the pressure force $\boldsymbol{F}_p$.

By considerations of hydrostatics, the pressure force $\boldsymbol{F}_p$ must be perpendicular to the surface of the liquid. From the viewpoint of the inertial frame $\{\epsilon_i\}$ (see Fig. 15.4), the horizontal component of the pressure force must provide for the centripetal force keeping m in circular orbit, while the vertical component balances the gravitational force. Thus

$$F_p \sin\theta = m\omega^2 r \, , \qquad F_p \cos\theta = mg \, , \tag{15.11}$$

where r is the radius of the circular orbit in question. The shape of the liquid surface is then determined by

$$\tan\theta = \frac{\omega^2 r}{g} \, . \tag{15.12}$$

The same conclusion can be reached if one analyzes the physics using a noninertial frame, say $\{\boldsymbol{e}_i\}$, that is fixed with respect to the fluid. From the viewpoint of such a frame, the surface of the fluid must be an equipotential surface under the combined action of the pressure, gravitational and centrifugal

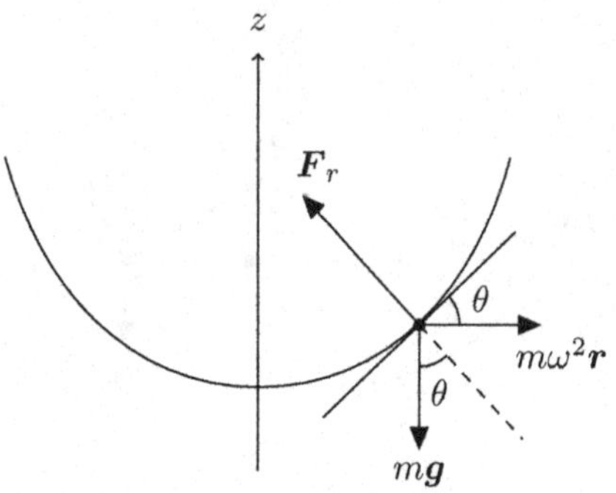

Fig. 15.5

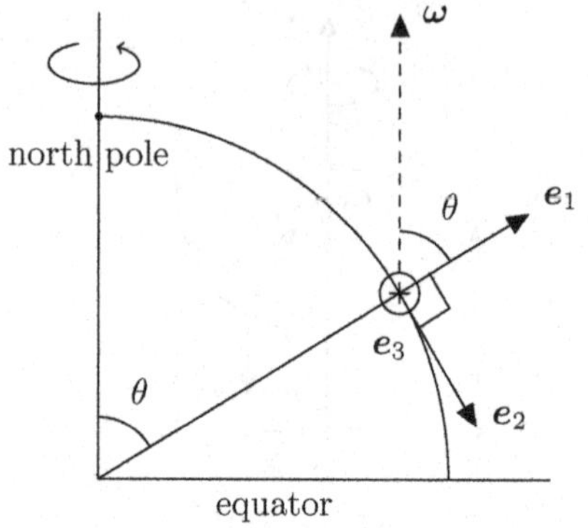

Fig. 15.6

forces. For a particle of mass m on the surface at a distance r from the z-axis, the centrifugal force has a magnitude $m\omega^2 r$, and is pointing away from the z-axis (Fig. 15.5). The pressure force $\boldsymbol{F}_p$ is still perpendicular to the surface of the fluid. The shape of this surface is again given by

$$\frac{dz}{dr} = \tan\theta = \frac{\omega^2 r}{g} . \tag{15.13}$$

The solution of this equation gives the shape of the fluid surface:

$$z = \frac{\omega^2 r^2}{2g} + z_0 , \tag{15.14}$$

which is a **surface of paraboloid**.

We will next turn our attention to an example of Coriolis forces: the effect of such forces on a point mass m falling freely under the action of gravity near the earth's surface, from the viewpoint of an orthonormal frame fixed to a particular point on the earth's surface [cf. Problem 12.1 and Fig. 12.2]. This frame is depicted again in Fig. 15.6.

In any noninertial frame $\{\boldsymbol{e}_i\}$, Coriolis forces are manifested when, with respect to that frame, an object has non-zero velocity $\boldsymbol{v} = \frac{dx^i}{dt}\boldsymbol{e}_i$. This force is given by (12.21). We adapt (12.19) to the present situation and write

$$m\frac{d^2x^i}{dt^2}\boldsymbol{e}_i = m\boldsymbol{g} - 2m\boldsymbol{\omega} \times \left(\frac{dx^i}{dt}\boldsymbol{e}_i\right) . \tag{15.15}$$

In the above equation the last term on the right-hand side (including the negative sign) is the Coriolis force $\boldsymbol{F}_C$. The acceleration due to gravity, $\boldsymbol{g}$, (including the centrifugal contribution discussed earlier in this chapter) is given by (15.1), while the angular velocity $\boldsymbol{\omega}$ [as given by (12.24)] describes the rotation of the earth about its axis. Let $\boldsymbol{r}$ be the position vector of the point mass with respect to the noninertial frame $\{\boldsymbol{e}_1, \boldsymbol{e}_2, \boldsymbol{e}_3\}$. We have

$$\boldsymbol{r} = x^1\, \boldsymbol{e}_1 + x^2\, \boldsymbol{e}_2 + x^3\, \boldsymbol{e}_3 \,, \tag{15.16}$$

$$\boldsymbol{v} = \frac{dx^i}{dt}\, \boldsymbol{e}_i = \dot{x}^1\, \boldsymbol{e}_1 + \dot{x}^2\, \boldsymbol{e}_2 + \dot{x}^3\, \boldsymbol{e}_3 \,, \tag{15.17}$$

$$\boldsymbol{\omega} = \omega \cos\theta\, \boldsymbol{e}_1 - \omega \sin\theta\, \boldsymbol{e}_2 \,, \tag{15.18}$$

where $\omega \equiv \dot{\phi}$ is the angular speed of the rotation of the earth. Then

$$\boldsymbol{F}_C = -2m\boldsymbol{\omega} \times \left(\frac{dx^i}{dt}\, \boldsymbol{e}_i \right) = -2m \begin{vmatrix} \boldsymbol{e}_1 & \boldsymbol{e}_2 & \boldsymbol{e}_3 \\ \omega\cos\theta & -\omega\sin\theta & 0 \\ \dot{x}^1 & \dot{x}^2 & \dot{x}^3 \end{vmatrix} \tag{15.19}$$

$$= -2m\omega\{-\dot{x}^3 sin\theta\, \boldsymbol{e}_1 - \dot{x}^3 \cos\theta\, \boldsymbol{e}_2 + (\dot{x}^2\cos\theta + \dot{x}^1 \sin\theta)\, \boldsymbol{e}_3\} \,.$$

For simplicity we will assume the following:

$$\boldsymbol{g} \sim \boldsymbol{g}_0 = -g_0\, \boldsymbol{e}_1 \,, \tag{15.20}$$

$$x^2 = x^3 \sim 0 \,, \qquad x^1 \sim h - \frac{1}{2} g_0 t^2 \,, \tag{15.21}$$

where h is the initial vertical height of the falling particle and $g_0 \sim 9.8\, m/sec^2$. Thus

$$\dot{x}^2 = \dot{x}^3 \sim 0 \,, \qquad \dot{x}^1 = -g_0 t \,, \tag{15.22}$$

and the Coriollis force, according to (15.19), is given by

$$\boldsymbol{F}_C \sim 2m\omega g_0 t \sin\theta\, \boldsymbol{e}_3 \qquad \text{(towards east)} \,. \tag{15.23}$$

It follows from the equation of motion (15.15) that

$$\ddot{x}^1 \sim -g_0 \,, \qquad \ddot{x}^2 \sim 0 \,, \qquad \ddot{x}^3 \sim 2\omega g_0 t \sin\theta \,. \tag{15.24}$$

This uncoupled system of equations has the solution [compare with (15.21)]

$$x^1 \sim h - \frac{1}{2}\, g_0 t^2 \,, \qquad x^2 \sim 0 \,, \qquad x^3 \sim \frac{1}{3}\, \omega g_0 t^3 \sin\theta \,. \tag{15.25}$$

At ground level, $x^1 = 0$, so that $t^2 = 2h/g_0$. It follows that the particle will hit ground at a distance

$$x^3 \sim \frac{1}{3}\, \omega g_0 \sin\theta \left(\frac{2h}{g_0} \right)^{3/2} = \frac{1}{3}\, \omega \sin\theta \left(\frac{8h^3}{g_0} \right)^{1/2} \tag{15.26}$$

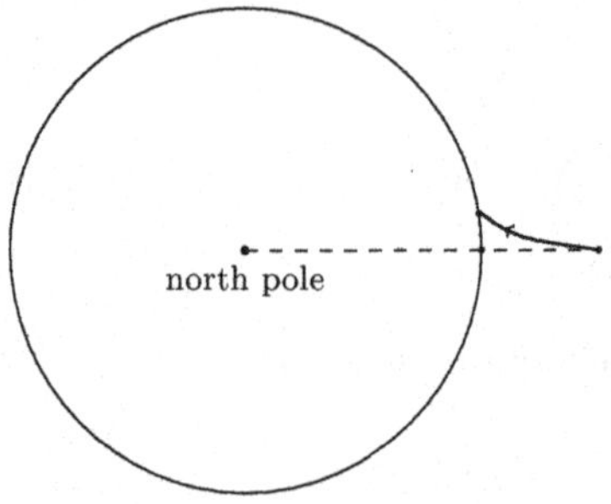

trajectory from the viewpoint
of the rotating frame

Fig. 15.7

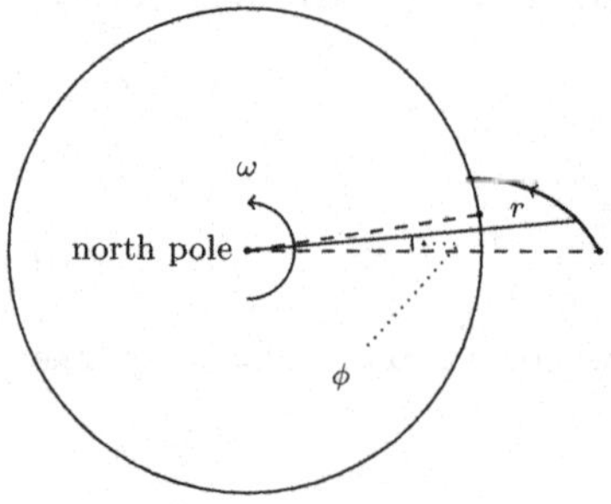

trajectory from the viewpoint
of the inertial frame

Fig. 15.8

due east of the position vertically below the point of release. For $h \sim 100\,m$ and $\theta = 45°$, and using $\omega \sim 7.29 \times 10^{-5}\,sec^{-1}$ [(15.5)], the above result yields a Coriolis deflection of $x^3 \sim 1.6\,cm$. For a given initial height h, the Coriolis deflection obviously assumes its maximum value at the equator and vanishes at the poles.

Problem 15.1 Verify the above quoted numerical result for Coriolis deflection.

It is instructive to understand the physical effect produced by the Coriolis force from the viewpoints of both the rotating (noninertial) frame $\{e_1, e_2, e_3\}$ and an inertial frame with respect to which the earth is rotating with an angular speed of ω. The trajectory of the falling particle with respect to the former (noninertial) frame can be obtained by eliminating t between the expressions for x^1 and x^3 in (15.25), and is illustrated qualitatively in Fig. 15.7. With respect to the latter (inertial) frame, the same trajectory is shown in Fig. 15.8. Initially, at height h, the angular speed of the particle is equal to that of the earth, namely, ω. As the particle falls, however, the angular speed $\dot{\phi}$ increases, due to the fact that the magnitude of the angular momentum $L = mr^2\dot{\phi}$ is conserved ($\dot{\phi}$ has to increase while r decreases). The conservation of angular momentum in this case follows from the fact that the gravitational force is a central force [recall (12.27) and (12.28)]. When the particle reaches the surface of the earth it gets ahead of the ground beneath it, which is rotating with constant angular speed ω, by an amount that is precisely equal to the Coriolis deflection as given by (15.26).

Problem 15.2 Obtain the trajectory of a falling particle from the viewpoint of the rotating (non-inertial) frame by eliminating t from the expressions for x^1 and x^3 in (15.25).

Chapter 16

Harmonic Oscillators: Fourier Transforms and Green's Functions

When the equation of motion of a physical system (mechanical or otherwise) is a *linear* differential equation, the solutions can be regarded as vectors in certain vector spaces and the powerful method of *integral transforms* (considered as a special of the theory of linear operators on vector spaces) will play an important role in the solution of the differential equation. In this and the next chapter we will consider the method of *Fourier transforms* (a special kind of integral transform) in the solution of such equations, and introduce the associated mathematical notion of *Green's functions*. This method gives an excellent illustration of the power of analytical tools (involving integrations) in solving linear algebra problems.

Consider the linear, ordinary, second-order differential equation (with constant coefficients)

$$\ddot{x} + 2\gamma\dot{x} + \omega_0^2 x = f(t) , \tag{16.1}$$

describing a damped, driven harmonic oscillator. In this equation

$$f(t) = \frac{F(t)}{m} , \qquad \gamma = \frac{\alpha}{2m} > 0 , \qquad \omega_0 = \sqrt{\frac{k}{m}} ,$$

where m is the mass of the oscillator, k is the force "spring") constant, $F(t)$ is the driving force, and the damping force is given by $-\alpha\dot{x}$, with $\alpha > 0$. Define the linear operator D by

$$D \equiv \frac{d^2}{dt^2} + 2\gamma\frac{d}{dt} + \omega_0^2 . \tag{16.2}$$

Then (16.1) can be written as

$$Dx(t) = f(t) . \tag{16.3}$$

The operator D is linear since, for any complex constants $\alpha, \beta \in \mathbb{C}$ and differentiable functions $x_1(t)$ and $x_2(t)$ [recall the definition of linearity given by (1.18)],

$$D(\alpha x_1 + \beta x_2) = \alpha D(x_1) + \beta D(x_2) \,. \tag{16.4}$$

The general solution of (16.1) can be written as

$$x(t) = x_c(t) + x_p(t) \,, \tag{16.5}$$

where

$$Dx_c(t) = 0 \,. \tag{16.6}$$

The functions x_c and x_p are called the **complementary solution** and **particular solution** of the differential equation (16.1), respectively. Equation (16.6) is often called the **homogeneous equation** corresponding to the **nonhomogeneous equation** (16.1).

We will find $x_c(t)$ and $x_p(t)$ as follows. Assume that

$$x_c(t) = \alpha e^{-i\omega t} \,, \tag{16.7}$$

where α and ω are as-yet undetermined constants. Substituting this expression for $x_c(t)$ into (16.6), we have

$$-\alpha e^{-i\omega t}(\omega^2 + 2i\gamma\omega - \omega_0^2) = 0 \,, \tag{16.8}$$

which implies

$$\omega^2 + 2i\gamma\omega - \omega_0^2 = 0 \,. \tag{16.9}$$

The above equation is called the **characteristic equation** for the linear operator D. Its solutions are given by

$$\omega_{\frac{1}{2}} = -i\gamma \pm \sqrt{\omega_0^2 - \gamma^2} \,. \tag{16.10}$$

Hence the general solution to the homogeneous equation (16.6) is given by

$$x_c(t) = \alpha e^{-i\omega_1 t} + \beta e^{-i\omega_2 t} \,, \tag{16.11}$$

where α and β are arbitrary constants which can be determined from the initial conditions $x(0)$ and $\dot{x}(0)$ of (16.6).

Imposing the physical requirement that $x_c(t)$ is real for all t, the set of all solutions to the homogeneous equation (16.6) forms a two-dimensional linear vector space isomorphic to $\mathbb{R}^2$ (the set of all ordered pairs of real numbers).

Problem 16.1 Prove the above statement. Hint: From (16.10), $i\omega_1 = \overline{i\omega_2}$. Thus, in order for $x_c(t)$ to be real, $\beta = \bar{\alpha}$ [in (16.11)].

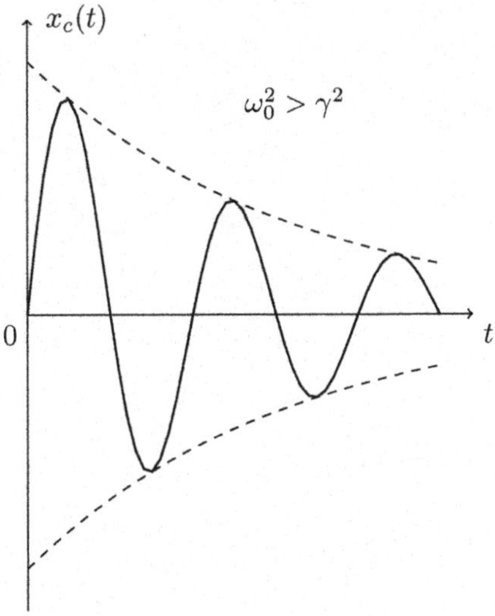

Fig. 16.1

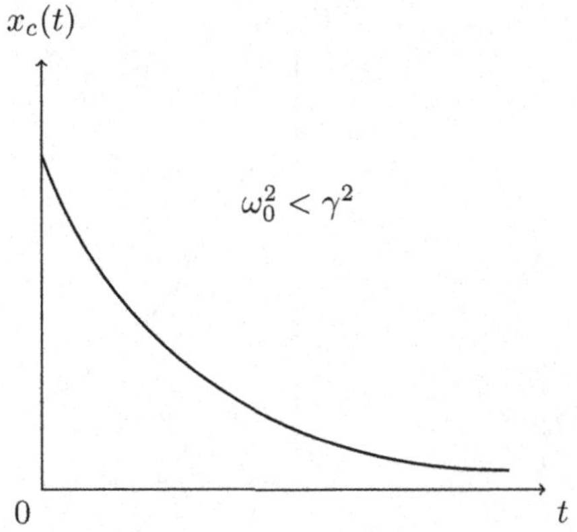

Fig. 16.2

Using the explicit expressions for ω_1 and ω_2 given by (16.10), a general complementary solution can be written as

$$x_c(t) = e^{-\gamma t}\left\{\alpha \exp\left(-it\sqrt{\omega_0^2 - \gamma^2}\right) + \beta \exp\left(it\sqrt{\omega_0^2 - \gamma^2}\right)\right\} . \tag{16.12}$$

For $\omega_0^2 > \gamma^2$ (the *underdamped* case), x_c is oscillatory but the amplitude decreases exponentially (see Fig. 16.1). If $\omega_0^2 < \gamma^2$ (the *overdamped* case), x_c decreases monotonically and exponentially (Fig. 16.2).
If $\omega_0 = \gamma$ (the *critically damped* case), the two *eigenvalues* of D are *degenerate*: $\omega_1 = \omega_2 = -i\gamma$. In this case, x_c is given by (Fig. 16.3)

$$x_c(t) = e^{-\gamma t}(\alpha + \beta t) . \tag{16.13}$$

Problem 16.2 Show that the critically damped solution given by (16.13) satisfies (16.1) when $\omega_0 = \gamma$.

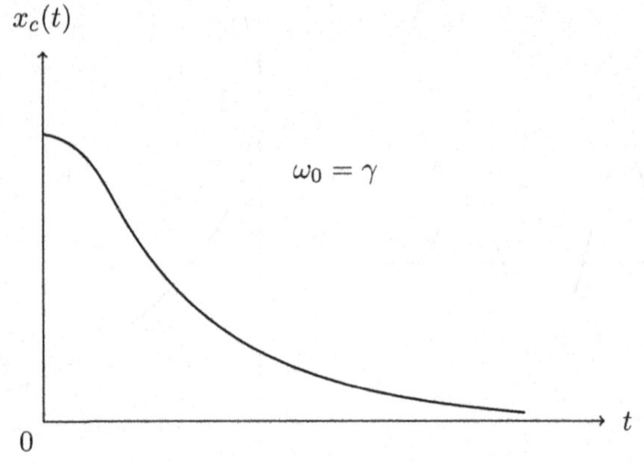

Fig. 16.3

Let us now study the solutions to the non-homogeneous equation (16.1). First consider a "monochromatic" driving force

$$f(t) = F(\omega)e^{-i\omega t} . \tag{16.14}$$

The *response* is the particular solution given by

$$x_p(t) = X(\omega)\, e^{-i\omega t} . \tag{16.15}$$

Substitution in (16.1) yields

$$-X(\omega)(\omega^2 + 2i\gamma\omega - \omega_0^2) = F(\omega) . \tag{16.16}$$

Thus

$$X(\omega) = -\frac{F(\omega)}{\omega^2 + 2i\gamma\omega - \omega_0^2} = -\frac{F(\omega)}{(\omega - \omega_1)(\omega - \omega_2)} \equiv G(\omega)F(\omega) . \tag{16.17}$$

The quantity

$$\boxed{G(\omega) = -\frac{1}{(\omega - \omega_1)(\omega - \omega_2)}} \tag{16.18}$$

is called the **Green's function** of the linear differential operator D. It is a *meromorphic function* of the complex variable ω, with first-order *poles* at ω_1 and ω_2, both in the lower-half ω-plane (see Fig. 16.4).

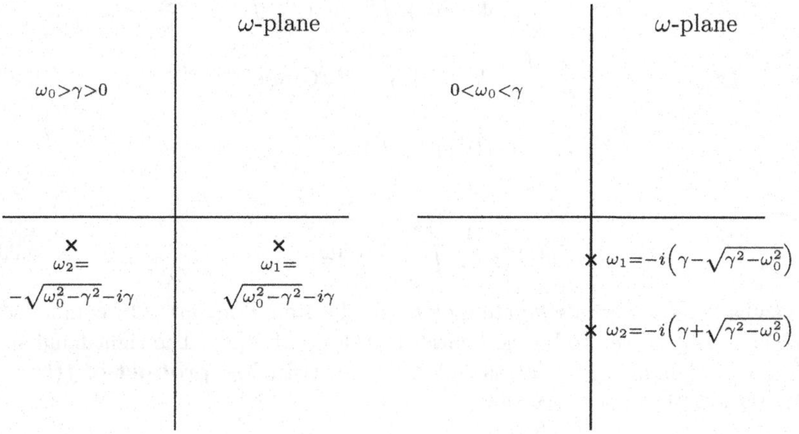

Fig. 16.4

Now let us consider a driving force composed of a linear superposition of monochromatic terms of the form (16.14):

$$f(t) = \frac{1}{2\pi}\int_{-\infty}^{\infty} d\omega\, F(\omega)e^{-i\omega t} \,. \tag{16.19}$$

The integral on the right-hand side of the above equation is called a **Fourier integral**. The pair $(f(t), F(\omega))$ is called a **Fourier transform pair**. One sometimes refers to this fact by writing

$$f(t) \overset{F.T.}{\longleftrightarrow} F(\omega) \,.$$

From (16.19) one can solve for $F(\omega)$ in terms of $f(t)$. The result (which follows from **Fourier's Theorem**) is

$$F(\omega) = \int_{-\infty}^{\infty} dt\, f(t)e^{i\omega t} \,. \tag{16.20}$$

Due to the linearity of D, the response to the driving force given by (16.19) is then

$$x(t) = \frac{1}{2\pi}\int_{-\infty}^{\infty} d\omega\, X(\omega)e^{-i\omega t} = \frac{1}{2\pi}\int_{-\infty}^{\infty} d\omega\, G(\omega)F(\omega)e^{-i\omega t} \,, \tag{16.21}$$

where the last equality follows from (16.17). To gain some physical insight into the above equation, we will manipulate it as follows. We have

$$\begin{aligned} x(t) &= \frac{1}{2\pi}\int_{-\infty}^{\infty} d\omega\, G(\omega) \int_{-\infty}^{\infty} dt'\, e^{i\omega t'} f(t')\, e^{-i\omega t} \\ &= \frac{1}{2\pi}\int_{-\infty}^{\infty} dt'\, f(t') \int_{-\infty}^{\infty} d\omega\, G(\omega) e^{-i\omega(t-t')} \\ &\equiv \int_{-\infty}^{\infty} dt'\, f(t') g(t-t')\,, \end{aligned} \tag{16.22}$$

where

$$g(t) \equiv \frac{1}{2\pi}\int_{-\infty}^{\infty} d\omega\, G(\omega) e^{-i\omega t} \tag{16.23}$$

is also called the *Green's function* of D (in the time domain). On comparison with (16.19) it is seen to be the Fourier transform of $G(\omega)$. The right-hand side of the last equality in (16.22) is called the **convolution product** of $f(t)$ and $g(t)$. By (16.21) we also see that

$$x(t) \overset{F.T.}{\longleftrightarrow} G(\omega)F(\omega)\,. \tag{16.24}$$

What is the physical meaning of $g(t)$? To answer this question, we need to introduce a very useful mathematical device, the *Dirac delta function*, as follows. We write

$$\begin{aligned} f(t) &= \frac{1}{2\pi}\int_{-\infty}^{\infty} d\omega\, e^{-i\omega t} F(\omega) = \frac{1}{2\pi}\int_{-\infty}^{\infty} d\omega\, e^{-i\omega t} \int_{-\infty}^{\infty} dt'\, f(t') e^{i\omega t'} \\ &= \int_{-\infty}^{\infty} dt'\, f(t') \left(\frac{1}{2\pi}\int_{-\infty}^{\infty} d\omega\, e^{i\omega(t'-t)} \right) \\ &\equiv \int_{-\infty}^{\infty} dt'\, f(t')\, \delta(t'-t)\,, \end{aligned} \tag{16.25}$$

where the **Dirac delta function** $\delta(t)$ is defined by

$$\delta(t) \equiv \frac{1}{2\pi}\int_{-\infty}^{\infty} d\omega\, e^{i\omega t}\,. \tag{16.26}$$

The above equation gives the so-called **integral representation** of the Dirac delta function. Putting $t = 0$ and $f(t') = 1$ in (16.25), we have the following important property of $\delta(t)$:

$$\int_{-\infty}^{\infty} \delta(t)\, dt = 1\,. \tag{16.27}$$

Equation (16.25) and the above equation suggests that $\delta(t)$ is an infinitely tall and infinitely narrow spiked function which is everywhere zero except at $t = 0$, such that the total area under the "graph" of the function is exactly 1. Mathematically speaking, however, $\delta(t)$ is not a function, in the proper mathematical

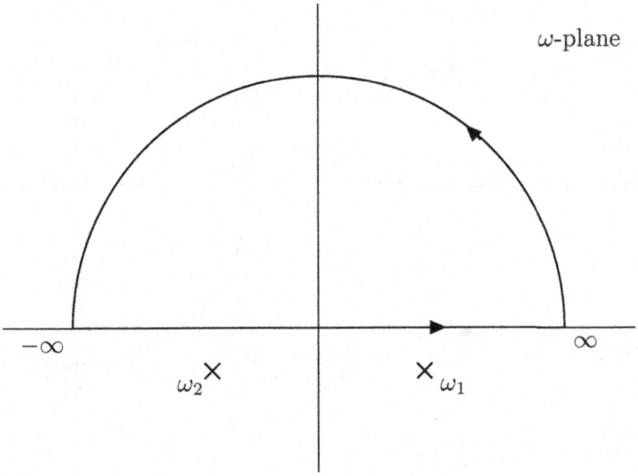

Fig. 16.5

sense of the word. It is what is known as a **distribution**, which is a map $\delta : S \to \mathbb{C}$, where S is the set of complex-valued "well-behaved" functions of a real variable whose domain includes the point $t = 0$. The map sends $f(t) \in S$ to $f(0) \in \mathbb{C}$. We represent this by the following equation [compare with (16.25)].

$$\int_{-\infty}^{\infty} dt\, f(t)\, \delta(t) = f(0)\ . \tag{16.28}$$

Suppose the driving force $f(t')$ in (16.22) is given by $\delta(t')$. Then (16.28) implies

$$x(t) = \int_{-\infty}^{\infty} dt'\, \delta(t')\, g(t - t') = g(t)\ . \tag{16.29}$$

Thus we have the following physical interpretation of the Green's function: *the Green's function $g(t)$ is the response to the δ-function "kick" applied to the oscillator at time $t = 0$.*

Let us now evaluate $g(t)$ using (16.23), with $G(\omega)$ given by (16.18). The required integral

$$g(t) = -\frac{1}{2\pi} \int_{-\infty}^{\infty} d\omega\, \frac{e^{-i\omega t}}{(\omega - \omega_1)(\omega - \omega_2)} \tag{16.30}$$

will be performed by *contour integration*. For $t < 0$, we close the contour by an infinite semicircle in the upper half ω-plane (Fig. 16.5). Since both poles of the

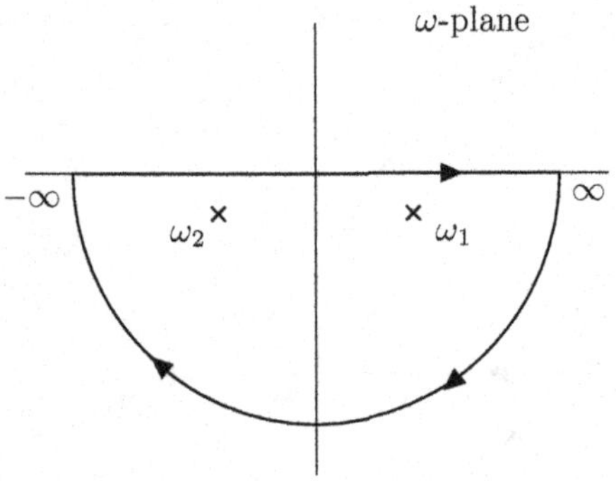

Fig. 16.6(a)

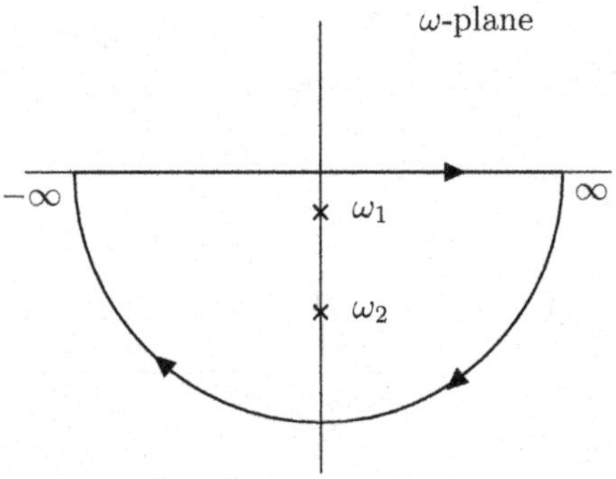

Fig. 16.6(b)

integrand (ω_1 and ω_2) are in the lower half plane, the *residue theorem* implies that

$$g(t) = 0 \,, \qquad t < 0 \,. \tag{16.31}$$

For $t > 0$, we close the contour by an infinite semicircle in the lower half plane. The positions of the two poles ω_1 and ω_2 are shown in Figs. 16.6(a) and 16.6(b) for the weak damping case ($\omega_0 > \gamma$) and the strong damping case ($\omega_0 < \gamma$), respectively.

We will evaluate the integral separately for these two cases. For weak damping ($\omega_0 > \gamma$), the two poles are given by

$$\omega_1 = \sqrt{\omega_0^2 - \gamma^2} - i\gamma \,, \quad \omega_2 = -\sqrt{\omega_0^2 - \gamma^2} - i\gamma \,. \tag{16.32}$$

Then, by the *residue theorem*,

$$\begin{aligned} g(t) &= \left(-\frac{1}{2\pi}\right)(-2\pi i)\left(\frac{e^{-i\omega_1 t}}{\omega_1 - \omega_2} + \frac{e^{-i\omega_2 t}}{\omega_2 - \omega_1}\right) \\ &= \frac{i}{\omega_1 - \omega_2}\left(e^{-i\omega_1 t} - e^{-i\omega_2 t}\right) \\ &= \frac{i}{2\sqrt{\omega_0^2 - \gamma^2}}\left\{e^{-it\left(\sqrt{\omega_0^2-\gamma^2}-i\gamma\right)} - e^{-it\left(-\sqrt{\omega_0^2-\gamma^2}-i\gamma\right)}\right\} \\ &= -\frac{ie^{-\gamma t}}{\sqrt{\omega_0^2 - \gamma^2}} \frac{\left(e^{i\sqrt{\omega_0^2-\gamma^2}\,t} - e^{-i\sqrt{\omega_0^2-\gamma^2}\,t}\right)}{2} \,. \end{aligned} \tag{16.33}$$

Thus, for weak damping, the Green's function in the time domain is given by

$$\boxed{g(t) = \frac{e^{-\gamma t}}{\sqrt{\omega_0^2 - \gamma^2}} \sin\left(\sqrt{\omega_0^2 - \gamma^2}\, t\right), \quad t \geq 0\,,\ \omega_0 > \gamma} \quad . \tag{16.34}$$

For strong damping ($\omega_0 < \gamma$), the two poles are given by

$$\omega_1 = -i\left(\gamma - \sqrt{\gamma^2 - \omega_0^2}\right), \quad \omega_2 = -i\left(\gamma + \sqrt{\gamma^2 - \omega_0^2}\right). \tag{16.35}$$

Similar to (16.33), we have

$$\begin{aligned} g(t) &= \frac{i}{\omega_1 - \omega_2}\left(e^{-i\omega_1 t} - e^{-i\omega_2 t}\right) \\ &= \frac{i}{2i\sqrt{\gamma^2 - \omega_0^2}}\left\{e^{-it\left(-i\gamma + i\sqrt{\gamma^2 - \omega_0^2}\right)} - e^{-it\left(-i\gamma - i\sqrt{\gamma^2 - \omega_0^2}\right)}\right\} \\ &= \frac{e^{-\gamma t}}{2\sqrt{\gamma^2 - \omega_0^2}}\left(e^{\sqrt{\gamma^2 - \omega_0^2}\, t} - e^{-\sqrt{\gamma^2 - \omega_0^2}\, t}\right). \end{aligned} \tag{16.36}$$

For strong damping, then, the Green's function in the time domain is

$$\boxed{g(t) = \frac{e^{-\gamma t}}{\sqrt{\gamma^2 - \omega_0^2}} \sinh\left(\sqrt{\gamma^2 - \omega_0^2}\, t\right), \quad t \geq 0\,,\ \omega_0 < \gamma} \quad . \tag{16.37}$$

Figures 16.7(a) and 16.7(b) show qualitative sketches for $g(t)$ for the weak and strong damping cases, respectively.

Problem 16.3 Calculate $g(t)$ for the critically damped case. Note that in this case the pole $\omega_1 = \omega_2$ is a *second order pole*.

The results (16.31), (16.34) and (16.37) can finally be put in (16.22) to yield the following expressions for the response $x_p(t)$ for both weak and strong damping cases:

$$x_p(t) = \int_{-\infty}^{t} dt'\, \frac{f(t')\, e^{-\gamma(t-t')} \sin\left(\sqrt{\omega_0^2 - \gamma^2}\,(t - t')\right)}{\sqrt{\omega_0^2 - \gamma^2}}\,, \quad \omega_0 > \gamma\,, \tag{16.38}$$

$$x_p(t) = \int_{-\infty}^{t} dt'\, \frac{f(t')\, e^{-\gamma(t-t')} \sinh\left(\sqrt{\gamma^2 - \omega_0^2}\,(t - t')\right)}{\sqrt{\gamma^2 - \omega_0^2}}\,, \quad \omega_0 < \gamma\,. \tag{16.39}$$

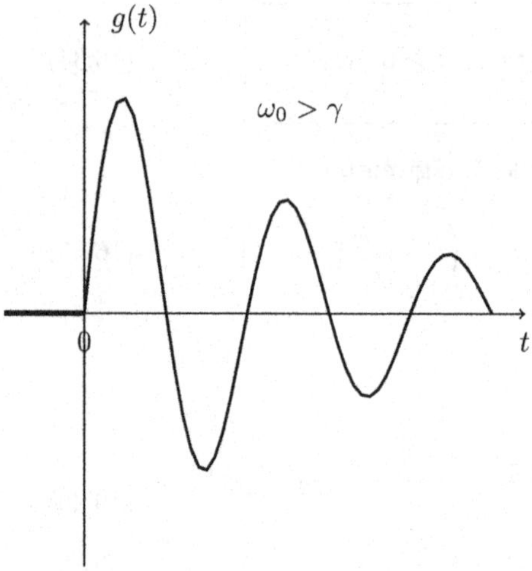

Fig. 16.7(a)

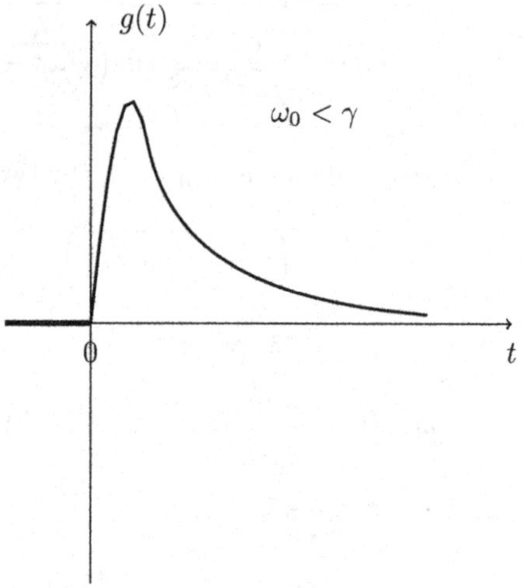

Fig. 16.7(b)

Note that the upper limit $t' = t$ in both of the above integrals is a consequence of the fact that $g(t-t') = 0$ for $t-t' < 0$, that is, $t' > t$ [cf. (16.31)]. Physically this means that the response at time t depends only on the driving force at earlier times. This is a manifestation of the **principle of causality**: the effect cannot precede the cause. The physical principle of causality has a deep mathematical counterpart, known as **Titchmarsh's Theorem**. This roughly states that *the Fourier transform of a* **causal function** *such as $g(t)$ (which is defined by the condition that $g(t) = 0$ for $t < 0$)* is analytic *in the upper half ω plane (all poles must be in the lower half plane).* We have seen that this is true for $G(\omega)$.

Fourier transform pairs satisfy an interesting property called *Parseval's theorem.* Let $(g(t), G(\omega))$ and $(f(t), F(\omega))$ be two Fourier transform pairs. We have, with overbars denoting complex conjugates,

$$\begin{aligned}\int_{-\infty}^{\infty} dt\, \overline{g(t)}\, f(t) &= \frac{1}{(2\pi)^2}\int_{-\infty}^{\infty} dt \int_{-\infty}^{\infty} d\omega\, F(\omega)\, e^{-i\omega t} \int_{-\infty}^{\infty} d\omega'\, \overline{G(\omega')}\, e^{i\omega' t} \\ &= \frac{1}{2\pi}\int_{-\infty}^{\infty} d\omega\, F(\omega) \int_{-\infty}^{\infty} d\omega'\, \overline{G(\omega')}\, \frac{1}{2\pi}\int_{-\infty}^{\infty} dt\, e^{i(\omega'-\omega)t} \\ &= \frac{1}{2\pi}\int_{-\infty}^{\infty} d\omega\, F(\omega) \int_{-\infty}^{\infty} d\omega'\, \overline{G(\omega')}\, \delta(\omega'-\omega)\,.\end{aligned} \tag{16.40}$$

We then have **Parseval's Theorem**:

$$\int_{-\infty}^{\infty} dt\, \overline{g(t)}\, f(t) = \frac{1}{2\pi} \int_{-\infty}^{\infty} d\omega\, \overline{G(\omega)}\, F(\omega)\,. \tag{16.41}$$

In particular, if $g(t) = f(t)$, the theorem gives

$$\int_{-\infty}^{\infty} |\, f(t)\,|^2\, dt = \frac{1}{2\pi} \int_{-\infty}^{\infty} |\, F(\omega)\,|^2\, d\omega\,. \tag{16.42}$$

This equation has an important application in *spectral analysis*. Suppose $f(t)$ is the wave function of a time-dependent signal. Then $|\, f(t)\,|^2$ is proportional to its intensity, and the left-hand side of (16.42) is the integrated intensity over time. Parseval's theorem in the form of (16.42) states that the intensity of the signal integrated over time is proportional to the *power spectrum* integrated over frequency.

Chapter 17

Classical Model of the Atom: Power Spectra

In this chapter we will present another illustration of the use of Fourier transform methods in the solution of a classical mechanical problem, more specifically, a classical electrodynamical problem.

Assume that an atom is modelled by a single electron of mass m and charge e embedded in a uniformly charged sphere (the nucleus) of radius a with total charge $-e$ (Fig. 17.1). This is known as *Thompson's Model* of the atom.
From the *Maxwell's equation*

$$\nabla \cdot \boldsymbol{E} = 4\pi\rho \,, \tag{17.1}$$

and Gauss' theorem of vector calculus [(8.5)] it can be shown that the electrostatic force on the electron due to the nucleus is given by

$$\boldsymbol{F} = -\frac{e^2 \boldsymbol{r}}{a^3} \,. \tag{17.2}$$

Problem 17.1 Prove the above statement.

Assume this is the only force acting on the electron, then the equation of motion is

$$m\ddot{\boldsymbol{r}} = -\frac{e^2 \boldsymbol{r}}{a^3} \,. \tag{17.3}$$

Equivalently,

$$\ddot{\boldsymbol{r}} + \omega_0^2 \boldsymbol{r} = 0 \,, \qquad \omega_0^2 \equiv \frac{e^2}{ma^3} \,. \tag{17.4}$$

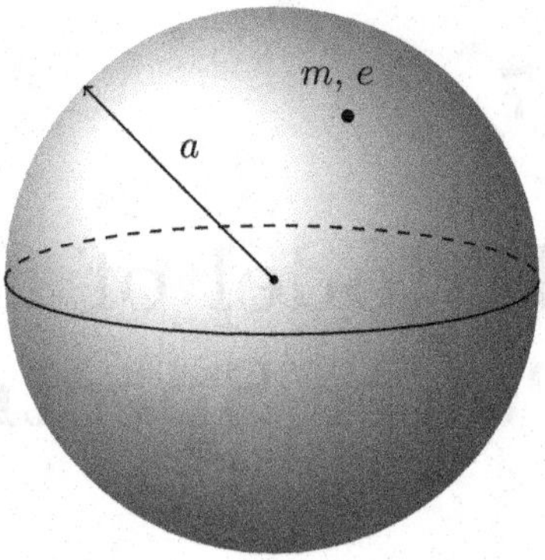

Fig. 17.1

This is the equation of motion of an undamped simple harmonic oscillator, with the general solution

$$\boldsymbol{r} = \boldsymbol{A}\cos\omega_0 t + \boldsymbol{B}\sin\omega_0 t\,, \tag{17.5}$$

where

$$\boldsymbol{A} = \boldsymbol{r}(0)\,, \qquad \boldsymbol{B} = \frac{\dot{\boldsymbol{r}}(0)}{\omega_0}\,. \tag{17.6}$$

The motion is obviously bound and periodic. In the absence of dissipative forces, the total energy

$$E = K + U = \frac{1}{2}m\dot{\boldsymbol{r}}^2 + \frac{1}{2}k\boldsymbol{r}^2\,, \quad k \equiv \omega_0^2 m\,, \tag{17.7}$$

is a constant of motion, and by (17.5), can be written as

$$E = \frac{1}{2}m\omega_0^2(A^2 + B^2) = \frac{1}{2}\left((\boldsymbol{r}(0))^2 + \frac{1}{\omega_0^2}(\dot{\boldsymbol{r}}(0))^2\right)\,. \tag{17.8}$$

Problem 17.2 Prove the above equation for the total energy.

From classical electrodynamics, however, an accelerating electron emits radiation and therefore will lose energy. We will use the following result for the time rate of energy loss of an accelerating charge:

$$\frac{dE}{dt} = -\frac{2e^2}{3c^3}\langle\, \ddot{\boldsymbol{r}}^2 \,\rangle\,, \tag{17.9}$$

where the angular brackets mean averaging over one cycle of the oscillatory motion of the electron. From (17.5),

$$\ddot{\boldsymbol{r}} = -\boldsymbol{A}\omega_0^2 \cos\omega_0 t - \boldsymbol{B}\omega_0^2 \sin\omega_0 t\,. \tag{17.10}$$

Hence

$$\ddot{\boldsymbol{r}}^2 = A^2\omega_0^4 \cos^2\omega_0 t + B^2\omega_0^4 \sin^2\omega_0 t - 2\boldsymbol{A}\cdot\boldsymbol{B}\,\omega_0^4 \sin\omega_0 t\cos\omega_0 t\,. \tag{17.11}$$

It follows that

$$\langle\, \ddot{\boldsymbol{r}}^2 \,\rangle \equiv \frac{1}{T}\int_0^T dt\, \ddot{\boldsymbol{r}}^2 = \frac{\omega_0}{2\pi}\int_0^{\frac{2\pi}{\omega_0}} dt\, \ddot{\boldsymbol{r}}^2 = \frac{\omega_0^4}{2}(A^2+B^2)\,. \tag{17.12}$$

Equation (17.9) then yields

$$\frac{dE}{dt} = -\frac{2e^2}{3c^3}\frac{\omega_0^4}{2}(A^2+B^2) = -\left(\frac{2e^2\omega_0^2}{3mc^3}\right)\left(\frac{1}{2}m\omega_0^2(A^2+B^2)\right) \equiv -\gamma E\,, \tag{17.13}$$

where

$$\gamma \equiv \frac{2e^2\omega_0^2}{3mc^3} \tag{17.14}$$

is the *decay constant* for E. We thus have

$$E(t) = E(0)e^{-\gamma t}\,. \tag{17.15}$$

We can model this energy loss by including a damping force $\boldsymbol{F}_d$ in addition to the electrostatic force $\boldsymbol{F}$ of (17.2). This can be done as follows. Starting from

$$m\ddot{\boldsymbol{r}} + k\boldsymbol{r} = \boldsymbol{F}_d\,, \tag{17.16}$$

we have

$$m\ddot{\boldsymbol{r}}\cdot\dot{\boldsymbol{r}} + k\boldsymbol{r}\cdot\dot{\boldsymbol{r}} = \boldsymbol{F}_d\cdot\dot{\boldsymbol{r}}\,, \tag{17.17}$$

which can be rewritten as

$$\frac{d}{dt}\left(\frac{1}{2}m\dot{\boldsymbol{r}}^2 + \frac{1}{2}k\boldsymbol{r}^2\right) = \frac{dE}{dt} = \boldsymbol{F}_d\cdot\dot{\boldsymbol{r}} = -\frac{2e^2}{3c^3}\langle\, \ddot{\boldsymbol{r}}^2 \,\rangle\,, \tag{17.18}$$

where the last equality follows from (17.9). Now

$$\ddot{\boldsymbol{r}}^2 = \frac{d}{dt}(\dot{\boldsymbol{r}}\cdot\ddot{\boldsymbol{r}}) - \dot{\boldsymbol{r}}\cdot\frac{d^3\boldsymbol{r}}{dt^3}\,. \tag{17.19}$$

Thus

$$\langle \boldsymbol{F}_d \cdot \dot{\boldsymbol{r}} \rangle = \frac{2e^2}{3c^3} \left\{ \left\langle \dot{\boldsymbol{r}} \cdot \frac{d^3\boldsymbol{r}}{dt^3} \right\rangle - \left\langle \frac{d}{dt}(\dot{\boldsymbol{r}} \cdot \ddot{\boldsymbol{r}}) \right\rangle \right\} . \tag{17.20}$$

In order to compare the average values in the right-hand side of the above equation, we assume that the damping is small and write

$$\boldsymbol{r} \sim \boldsymbol{A} \cos(\omega_0 t + \phi) , \tag{17.21}$$

$$\frac{d\boldsymbol{r}}{dt} \sim -\boldsymbol{A}\omega_0 \sin(\omega_0 t + \phi) , \tag{17.22}$$

$$\frac{d^2\boldsymbol{r}}{dt^2} \sim -\boldsymbol{A}\omega_0^2 \cos(\omega_0 t + \phi) , \tag{17.23}$$

$$\frac{d^3\boldsymbol{r}}{dt^3} \sim \boldsymbol{A}\omega_0^3 \sin(\omega_0 t + \phi) . \tag{17.24}$$

Hence

$$\frac{d}{dt}(\dot{\boldsymbol{r}} \cdot \ddot{\boldsymbol{r}}) \sim \frac{d}{dt}\left(\frac{A^2\omega_0^3}{2} \sin\{2(\omega_0 t + \phi)\} \right) = A^2\omega_0^4 \cos\{2(\omega_0 t + \phi)\} , \tag{17.25}$$

which implies

$$\left\langle \frac{d}{dt}(\dot{\boldsymbol{r}} \cdot \ddot{\boldsymbol{r}}) \right\rangle \sim 0 . \tag{17.26}$$

On the other hand,

$$\dot{\boldsymbol{r}} \cdot \frac{d^3\boldsymbol{r}}{dt^3} \sim -A^2\omega_0^4 \sin^2(\omega_0 t + \phi) . \tag{17.27}$$

Thus

$$\left\langle \dot{\boldsymbol{r}} \cdot \frac{d^3\boldsymbol{r}}{dt^3} \right\rangle \sim -\frac{A^2\omega_0^4}{2} . \tag{17.28}$$

It follows from (17.20) that

$$\boldsymbol{F}_d \cdot \dot{\boldsymbol{r}} \sim \frac{2e^2}{3c^3} \dot{\boldsymbol{r}} \cdot \frac{d^3\boldsymbol{r}}{dt^3} , \tag{17.29}$$

and thus

$$\boldsymbol{F}_d \sim \frac{2e^2}{3c^3} \frac{d^3\boldsymbol{r}}{dt^3} . \tag{17.30}$$

The equation of motion of an electron in the Thomson's model of the atom including *radiative effects* can finally be written as

$$m\ddot{\boldsymbol{r}} + \omega_0^2 \boldsymbol{r} - \frac{2e^2}{3c^3} \frac{d^3\boldsymbol{r}}{dt^3} = 0 . \tag{17.31}$$

We provide an approximate solution of this third-order equation as follows. Assume

$$\boldsymbol{r} = \boldsymbol{A} e^{i\omega t} . \tag{17.32}$$

Then

$$\dot{\boldsymbol{r}} = i\omega \boldsymbol{A}e^{i\omega t} , \quad \ddot{\boldsymbol{r}} = -\omega^2 \boldsymbol{A}e^{i\omega t} , \quad \frac{d^3\boldsymbol{r}}{dt^3} = -i\omega^3 \boldsymbol{A}e^{i\omega t} . \tag{17.33}$$

Substituting the above three quantities into the equation of motion (13.31) we obtain the following characteristic equation for the differential equation (17.31):

$$-\omega^2 + \omega_0^2 + \frac{2e^2}{3mc^3} i\omega^3 = 0 . \tag{17.34}$$

Problem 17.3 Verify the above characteristic equation by substituting (17.32) into (17.31).

In the third term on the left-hand side of (17.34) we will approximate $\dfrac{2e^2\omega^2}{3mc^3}$ by the constant γ [cf. (17.14)]. The third-degree characteristic equation (17.34) then appears as the quadratic equation

$$\omega^2 - i\gamma\omega - \omega_0^2 = 0 , \tag{17.35}$$

which has the solutions

$$\omega = \frac{i\gamma}{2} \pm \frac{\sqrt{4\omega_0^2 - \gamma^2}}{2} \sim \pm\omega_0 + i\frac{\gamma}{2} , \tag{17.36}$$

where in the last (approximate) equality we have assumed that $\gamma \ll \omega_0$. The approximate solution of (17.31) is then

$$\boldsymbol{r}(t) = e^{-\gamma t/2}(\boldsymbol{A}e^{i\omega_0 t} + \overline{\boldsymbol{A}}e^{-i\omega_0 t}) . \tag{17.37}$$

In the above equation, the complex conjugate of $\boldsymbol{A}$ appears in the second term on the right-hand side due to the requirement that $\boldsymbol{r}(t)$ is real.

We will now calculate the *power spectrum* $P(\omega)$ of the emitted radiation. Assume that the electron is released from rest at time $t = 0$, so that radiation begins to be emitted at this instant. Through Parseval's theorem [cf. (16.42)] and (17.9) we can define $P(\omega)$ by

$$\int_{-\infty}^{\infty} P(\omega)d\omega = \frac{2e^2}{3c^3}\int_0^{\infty} \ddot{\boldsymbol{r}}^2 dt . \tag{17.38}$$

Note that the lower time limit $t = 0$ is due to our assumption that the electron is released from rest at $t = 0$, that is, $\ddot{\boldsymbol{r}}(t) = 0$ for $t < 0$. A straightforward differentiation of (17.37) yields

$$\ddot{\boldsymbol{r}}(t) = \begin{cases} \left(\frac{\gamma}{2} - i\omega_0\right)^2 \boldsymbol{A} \exp\left\{-\left(\frac{\gamma}{2} - i\omega_0\right) t\right\} + \left(\frac{\gamma}{2} + i\omega_0\right)^2 \overline{\boldsymbol{A}} \exp\left\{-\left(\frac{\gamma}{2} + i\omega_0\right) t\right\}, \\ \qquad t > 0\,, \\ 0\,, \qquad t \leq 0\,. \end{cases} \tag{17.39}$$

Problem 17.4 Verify the above equation for the acceleration of the electron in the Thompson's model.

Let $\boldsymbol{g}(\omega)$ be the Fourier transform of $\ddot{\boldsymbol{r}}(t)$, that is,

$$\ddot{\boldsymbol{r}}(t) = \frac{1}{2\pi} \int_{-\infty}^{\infty} d\omega\, \boldsymbol{g}(\omega) e^{-i\omega t}\,, \qquad \boldsymbol{g}(\omega) = \int_{-\infty}^{\infty} dt\, \ddot{\boldsymbol{r}}(t) e^{i\omega t}\,. \tag{17.40}$$

Using (17.39) in the second equation of (17.40), we obtain

$$\begin{aligned} \boldsymbol{g}(\omega) &= \int_0^{\infty} dt\, \left(\frac{\gamma}{2} - i\omega_0\right)^2 \boldsymbol{A} \exp\left\{-\left(\frac{\gamma}{2} - i\omega_0 - i\omega\right) t\right\} \\ &\quad + \int_0^{\infty} dt\, \left(\frac{\gamma}{2} + i\omega_0\right)^2 \overline{\boldsymbol{A}} \exp\left\{-\left(\frac{\gamma}{2} + i\omega_0 - i\omega\right) t\right\} \\ &= \frac{\left(\frac{\gamma}{2} - i\omega_0\right)^2 \boldsymbol{A}}{\frac{\gamma}{2} - i(\omega_0 + \omega)} + \frac{\left(\frac{\gamma}{2} + i\omega_0\right)^2 \overline{\boldsymbol{A}}}{\frac{\gamma}{2} + i(\omega_0 - \omega)}\,. \end{aligned} \tag{17.41}$$

It is clear from the above equation that

$$\boldsymbol{g}(\omega) = \overline{\boldsymbol{g}(-\omega)}\,. \tag{17.42}$$

Problem 17.5 Show that (17.42) also follows from the requirement that $\ddot{\boldsymbol{r}}(t)$ is real.

From (17.40) we have

$$\int_0^\infty \ddot{\boldsymbol{r}}^2\,dt = \frac{1}{2\pi}\int_0^\infty dt \int_{-\infty}^\infty d\omega\, \boldsymbol{g}(\omega)e^{-i\omega t}\,\ddot{\boldsymbol{r}}(t)$$
$$= \frac{1}{2\pi}\int_{-\infty}^\infty d\omega\,\overline{\boldsymbol{g}(-\omega)}\cdot\int_0^\infty dt\,e^{-i\omega t}\,\ddot{\boldsymbol{r}}(t) = \frac{1}{2\pi}\int_\infty^{-\infty}(-d\omega')\overline{\boldsymbol{g}(\omega')}\int_0^\infty dt\,e^{i\omega' t}\,\ddot{\boldsymbol{r}}(t)$$
$$= \frac{1}{2\pi}\int_{-\infty}^\infty d\omega\,\overline{\boldsymbol{g}(\omega)}\int_0^\infty dt\,\ddot{\boldsymbol{r}}(t)e^{i\omega t}$$
$$= \frac{1}{2\pi}\int_{-\infty}^\infty d\omega\,\overline{\boldsymbol{g}(\omega)}\,\boldsymbol{g}(\omega) = \frac{1}{2\pi}\left(\int_{-\infty}^0 d\omega\,\overline{\boldsymbol{g}(\omega)}\,\boldsymbol{g}(\omega) + \int_0^\infty d\omega\overline{\boldsymbol{g}(\omega)}\,\boldsymbol{g}(\omega)\right)$$
$$= \frac{1}{2\pi}\left(\int_0^\infty d\omega\,\overline{\boldsymbol{g}(-\omega)}\,\boldsymbol{g}(-\omega) + \int_0^\infty d\omega\,\overline{\boldsymbol{g}(\omega)}\,\boldsymbol{g}(\omega)\right) = \frac{1}{\pi}\int_0^\infty d\omega\,\overline{\boldsymbol{g}(\omega)}\,\boldsymbol{g}(\omega)\,. \tag{17.43}$$

Since we are integrating only from 0 to ∞ (in ω), we can ignore the *anti-resonant term* in the explicit expression for $\boldsymbol{g}(\omega)$ given by (17.41), that is, the one proportional to $1/\left\{\frac{\gamma}{2} - i(\omega_0+\omega)\right\}$, and obtain

$$\frac{2e^2}{3c^3}\int_0^\infty \ddot{\boldsymbol{r}}^2\,dt = \frac{2e^2}{3\pi c^3}\int_0^\infty d\omega\,\frac{\left(\frac{\gamma}{2}+i\omega_0\right)^2\left(\frac{\gamma}{2}-i\omega_0\right)^2\,\boldsymbol{A}\cdot\overline{\boldsymbol{A}}}{\left(\frac{\gamma}{2}+i(\omega_0-\omega)\right)\left(\frac{\gamma}{2}-i(\omega_0-\omega)\right)}\,. \tag{17.44}$$

We assume that the radiative damping of the atom is weak, so that $\gamma/2 \ll \omega_0$. In this case the above integrand is sharply peaked near $\omega \sim \omega_0$ and the lower limit of the integral can be replaced by $-\infty$. On comparison with (17.38), we arrive at the following result for the power spectrum of a radiating (classical) atom:

$$\boxed{P(\omega) \sim \left(\frac{2e^2}{3\pi c^3}\right)\frac{\omega_0^4\,\boldsymbol{A}\cdot\overline{\boldsymbol{A}}}{(\omega-\omega_0)^2+\dfrac{\gamma^2}{4}}}\,. \tag{17.45}$$

The spectral line given by this equation is said to have a **Lorentzian lineshape**.

Chapter 18

Dynamical Systems and their Stabilities

In this chapter we will apply the ideas on *flows* and *tangent vector fields* introduced in Chapter 11 to an elementary study of some general features of *dynamical systems.* We will first use mainly two-dimensional linear systems to lay out some of the main ideas and techniques involved, because of their simplicity and the intuitive geometrical picture that they present. Towards the end of the chapter we will touch on various problems surrounding the important issue of the *stability of dynamical systems*, and present some central results that are applicable to non-linear systems in higher dimensions. In particular the so-called *Liapunov theory* on stability will be introduced. The concepts and results presented here will play an important role in the study of Hamiltonian dynamical systems towards the latter part of this text.

The second order homogeneous equation for the damped oscillator [cf. (16.1)]

$$\ddot{x} + 2\gamma\dot{x} + \omega_0^2 x = 0 \tag{18.1}$$

can be written as a system of two first-order equations:

$$\begin{aligned} \dot{x} &= y \,, \\ \dot{y} &= -\omega_0^2 x - 2\gamma y \,. \end{aligned} \tag{18.2}$$

We can also write this system in matrix form as

$$\begin{pmatrix} \dot{x} \\ \dot{y} \end{pmatrix} = \begin{pmatrix} 0 & 1 \\ -\omega_0^2 & -2\gamma \end{pmatrix} \begin{pmatrix} x \\ y \end{pmatrix} , \tag{18.3}$$

or

$$\dot{X} = AX \,, \tag{18.4}$$

where X is the column matrix and A is the 2×2 matrix appearing in (18.3).

In general, a **dynamical system** is a system of differential equations of the form [cf. (11.9)]

$$\frac{dX}{dt} = f(X,t) , \tag{18.5}$$

where $f : \mathbb{R}^n \times \mathbb{R} \to \mathbb{R}^n$ is a continuously differentiable function with domain and range as indicated. In applications in mechanics, the symbol t in the above equation usually stands for time. If $f(X,t)$ is independent of time, (18.5) becomes

$$\frac{dX}{dt} = f(X) . \tag{18.6}$$

A dynamical system of the form given by (18.6) is called an **autonomous system**. This kind of system can be written explicitly as a system of first-order differential equations as follows.

$$\begin{aligned} \frac{dx^1}{dt} &= f^1(x^1,\dots,x^n) , \\ \frac{dx^2}{dt} &= f^2(x^1,\dots,x^n) , \\ &\vdots \\ \frac{dx^n}{dt} &= f^n(x^1,\dots,x^n) . \end{aligned} \tag{18.7}$$

The system (18.7) [as well as the more general system (18.5)] is in general *non-linear.*

$f(X)$ is called the *vector field* of the dynamical system, whose components with respect to the *local coordinates* $(x^1,\dots,x^n)$ of a point $X \in \mathbb{R}^n$ are $f^1,\dots,f^n$. A solution $(x^1(t),\dots,x^n(t))$ of (18.7) is an *integral curve* of the vector field $f(X)$. The space in which X lives (in our case $\mathbb{R}^n$) is called the *phase space* of the dynamical system. The phase space of a dynamical system is in general a *differentiable manifold.*

The points X at which $f(X) = 0$ are of special importance in the theory of dynamical systems. They are called **equilibrium**, **fixed**, **stationary**, or **singular points** of the dynamical system (recall the discussion following Corollary 11.1). For example, $X = (x,y) = (0,0)$ is an equilibrium point of the system (18.2). The **stability** of solutions near equilibrium points will be of particular interest. To gain a qualitative understanding of the nature of solutions near equilibrium points, one often *linearizes* the dynamical system by Taylor expanding $f(X)$ around an equilibrium point, say, $\overline{X}$:

$$f(X) \sim f(\overline{X}) + Df(\overline{X})(X - \overline{X}) = Df(\overline{X})(X - \overline{X}) , \tag{18.8}$$

where $Df(\overline{X})$ denotes the derivative of f at the point $X = \overline{X}$. This derivative is actually a linear map from $\mathbb{R}^n$ to $L(\mathbb{R}^n,\mathbb{R}^n)$, the space of linear operators on

$\mathbb{R}^n$. $Df(\overline{X})$ has the following matrix representation:

$$Df(\overline{X}) = \begin{pmatrix} \frac{\partial f^1}{\partial x^1} & \frac{\partial f^1}{\partial x^2} & \cdots & \frac{\partial f^1}{\partial x^n} \\ \frac{\partial f^2}{\partial x^1} & \frac{\partial f^2}{\partial x^2} & \cdots & \frac{\partial f^2}{\partial x^n} \\ & \vdots & & \\ \frac{\partial f^n}{\partial x^1} & \frac{\partial f^n}{\partial x^2} & \cdots & \frac{\partial f^n}{\partial x^n} \end{pmatrix}_{X=\overline{X}} , \tag{18.9}$$

where all the derivatives are evaluated at the point $X = \overline{X}$. Near an equilibrium point $X = \overline{X}$, then, the behavior of the dynamical system (18.7) is described by the linearized system

$$\frac{dX}{dt} = Df(\overline{X})\, X , \tag{18.10}$$

if we choose a set of local coordinates so that $\overline{X} = 0$. The solutions of (18.10) depend on the *eigenvalues* of the linear operator $Df(\overline{X})$.

We will digress to state (without proofs) a few important facts on the theory of linear operators.

Theorem 18.1. *The solution to the system of equations $dX/dt = AX$, with A being constant, is*

$$X(t) = e^{At}\, X(0) , \tag{18.11}$$

where

$$e^{At} = \sum_{n=0}^{\infty} \frac{A^n t^n}{n!} . \tag{18.12}$$

Theorem 18.2. *(a) If $B = \Lambda^{-1} A \Lambda$ (a similarity transformation), then*

$$e^B = \Lambda^{-1} e^A \Lambda . \tag{18.13}$$

(b) If $[T, S] \equiv TS - ST = 0$, then

$$e^{S+T} = e^S\, e^T . \tag{18.14}$$

(c) For all linear operators S,

$$e^{-S} = (e^S)^{-1} . \tag{18.15}$$

Definition 18.1. *A matrix S is said to be* **diagonalizable** *(***semisimple***) if there exists an invertible matrix Λ such that $\Lambda^{-1} S \Lambda$ is diagonal.*

Definition 18.2. *A marix N is said to be* **nilpotent** *if there exists a positive integer m such that $N^m = 0$.*

Theorem 18.3. *(The* ***Primary Decomposition Theorem****) Any $A \in L(E, E)$ (the linear space of linear operators on E, where E is in general some complex vector space), has a unique decomposition $A = S + N$, such that $[S, N] = 0$, S is diagonalizable, and N is nilpotent.*

Theorem 18.4. *If*

$$B = \begin{pmatrix} a & -b \\ b & a \end{pmatrix}, \quad a, b \in \mathbb{R}, \tag{18.16}$$

then,

$$e^B = e^a \begin{pmatrix} \cos b & -\sin b \\ \sin b & \cos b \end{pmatrix}. \tag{18.17}$$

Theorem 18.5. *If A is the diagonal matrix given by*

$$A = \begin{pmatrix} \lambda_1 & \dots & 0 \\ \dots & \dots & \dots \\ 0 & \dots & \lambda_n \end{pmatrix}, \tag{18.18}$$

then

$$e^A = \begin{pmatrix} e^{\lambda_1} & \dots & 0 \\ \dots & \dots & \dots \\ 0 & \dots & e^{\lambda_n} \end{pmatrix}. \tag{18.19}$$

With the above facts we can now proceed to give a complete classification of two-dimensional linear dynamical systems. This is based on the following central result.

Theorem 18.6. *For any 2×2 real matrix A there exists an invertible matrix Λ such that $\Lambda^{-1} A \Lambda$ has one of the following forms:*

$$(1) \quad \begin{pmatrix} \lambda & 0 \\ 0 & \mu \end{pmatrix}, \qquad (2) \quad \begin{pmatrix} a & -b \\ b & a \end{pmatrix}, \qquad (3) \quad \begin{pmatrix} \lambda & 0 \\ 1 & \lambda \end{pmatrix},$$

where $\lambda, \mu, a, b \in \mathbb{R}$. Case (1) is obtained when the two eigenvalues λ and μ are real and the matrix A is diagonalizable (A is always diagonalizable when $\lambda \neq \mu$); (2) is obtained when the eigenvalues are $a \pm ib$; (3) is obtained when the two real eigenvalues are equal to each other and A is not diagonalizable.

The forms (1), (2) and (3) in the above theorem are called the **Jordan canonical forms** of the 2×2 matrix A.

According to Theorem 18.1, the solutions $X(t)$ to the equation $dX/dt = AX$, with A given by the cases (1), (2) in Theorem 18.6, are given by:

$$X(t) = \begin{pmatrix} x(t) \\ y(t) \end{pmatrix} = \begin{pmatrix} e^{\lambda t} & 0 \\ 0 & e^{\mu t} \end{pmatrix} \begin{pmatrix} x(0) \\ y(0) \end{pmatrix} , \qquad \text{for } A = \begin{pmatrix} \lambda & 0 \\ 0 & \mu \end{pmatrix} . \tag{18.20}$$

$$X(t) = \begin{pmatrix} x(t) \\ y(t) \end{pmatrix} = e^{at} \begin{pmatrix} \cos bt & -\sin bt \\ \sin bt & \cos bt \end{pmatrix} \begin{pmatrix} x(0) \\ y(0) \end{pmatrix} , \quad \text{for } A = \begin{pmatrix} a & -b \\ b & a \end{pmatrix} . \tag{18.21}$$

For case (3) we note that

$$A = \begin{pmatrix} \lambda & 0 \\ 1 & \lambda \end{pmatrix} = S + N , \tag{18.22}$$

where

$$S = \begin{pmatrix} \lambda & 0 \\ 0 & \lambda \end{pmatrix} , \quad N = \begin{pmatrix} 0 & 0 \\ 1 & 0 \end{pmatrix} , \tag{18.23}$$

with

$$[S, N] = 0 , \qquad N^2 = 0 . \tag{18.24}$$

Thus, by Theorem 18.2(b),

$$\begin{aligned} \exp\left(\begin{pmatrix} \lambda & 0 \\ 1 & \lambda \end{pmatrix} t \right) &= \exp\left(\begin{pmatrix} \lambda t & 0 \\ 0 & \lambda t \end{pmatrix} \right) \exp\left(\begin{pmatrix} 0 & 0 \\ t & 0 \end{pmatrix} \right) \\ &= e^{\lambda t} \left\{ 1 + \begin{pmatrix} 0 & 0 \\ t & 0 \end{pmatrix} \right\} = e^{\lambda t} \begin{pmatrix} 1 & 0 \\ t & 1 \end{pmatrix} . \end{aligned} \tag{18.25}$$

The solution with A given by case (3) is thus

$$\begin{pmatrix} x(t) \\ y(t) \end{pmatrix} = e^{\lambda t} \begin{pmatrix} 1 & 0 \\ t & 1 \end{pmatrix} \begin{pmatrix} x(0) \\ y(0) \end{pmatrix} , \quad \text{for } A = \begin{pmatrix} \lambda & 0 \\ 1 & \lambda \end{pmatrix} . \tag{18.26}$$

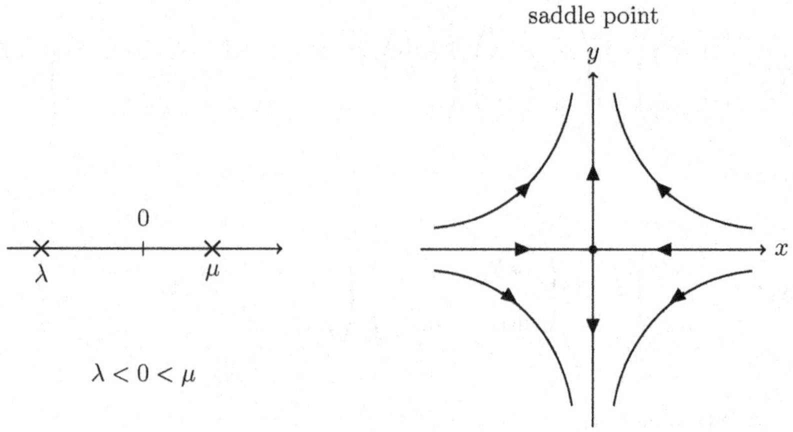

Fig. 18.1

The solutions (18.20), (18.21) and (18.26) can be understood more intuitively and geometrically by using **phase portraits**, which depict the integral curves of the vector field of the dynamical system in each case.

Case (1)a: The eigenvalues λ and μ of A satisfy $\lambda < 0 < \mu$. The point $(0,0)$ of phase space is a **saddle point**. The positions of the eigenvalues on the real axis and the phase portrait near a saddle point are shown in Fig. 18.1.
Case (1)b: If $\lambda < \mu < 0$, the solutions satisfy $\lim_{t\to\infty} X(t) = 0$, and $(0,0)$ is a **sink node**. If $0 < \lambda < \mu$, $\lim_{t\to-\infty} X(t) = 0$ and $(0,0)$ is a **source node**. Figures 18.2(a) (for the sink node) and 18.2(b) (for the source node) show the phase portraits near these nodes and the respective positions of the eigenvalues.
Cases (1)c: A is diagonalizable and the eigenvalues are degenerate. If $\lambda = \mu < 0$, then $\lim_{t\to\infty} X(t) = 0$ and $(0,0)$ is a **sink focus**. If $0 < \lambda = \mu$, then $\lim_{t\to-\infty} X(t) = 0$ and $(0,0)$ is a **source focus**. The positions of the eigenvalues and the phase portraits near the two different kinds of foci are shown in Fig. 18.3(a) (for the sink focus) and Fig. 18.3(b) (for the source focus).
Cases (2)a: $a = 0, b > 0$ or $b < 0$. The integral curves are closed curves representing *periodic solutions*, which satisfy $X(t + 2\pi/b) = X(t)$. These curves (circles) are given by the equation $x^2 + y^2 = (x(0))^2 + (y(0))^2$. The point $(0,0)$ is called a **center**. The positions of the eigenvalues and the phase portraits near a center are shown in Fig. 18.4. The circular integral curves run in the counterclockwise sense if $b > 0$ and in the clockwise sense if $b < 0$.
Cases (2)b: $a < 0, b > 0$ and $a < 0, b < 0$. For both cases we have $\lim_{t\to\infty} X(t) =$

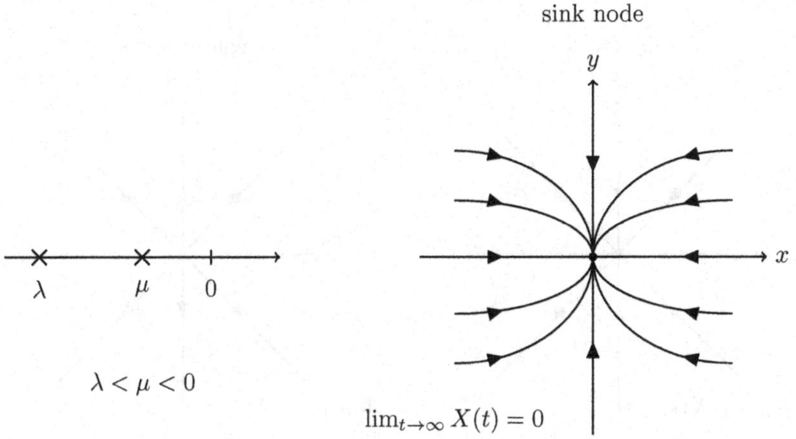

Fig. 18.2(a)

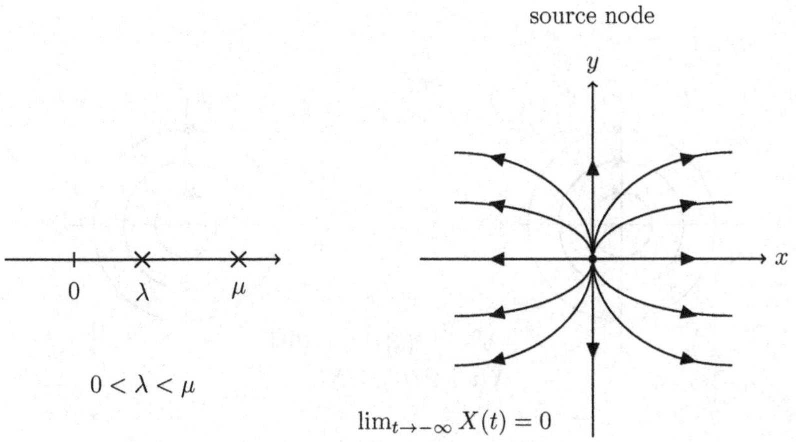

Fig. 18.2(b)

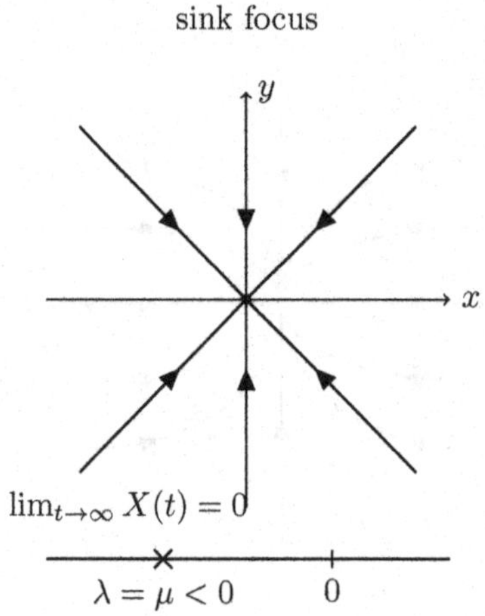

Fig. 18.3(a)

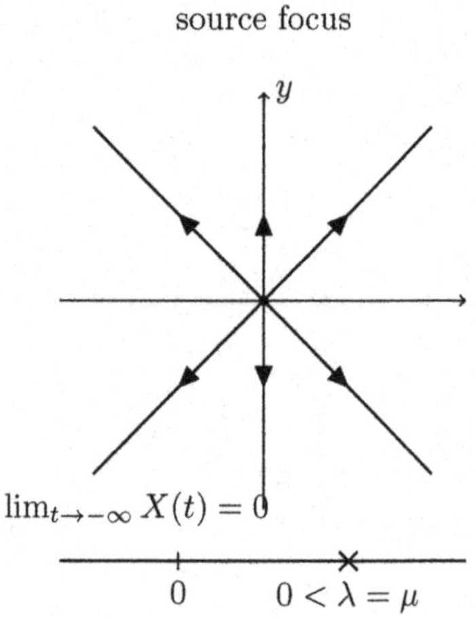

Fig. 18.3(b)

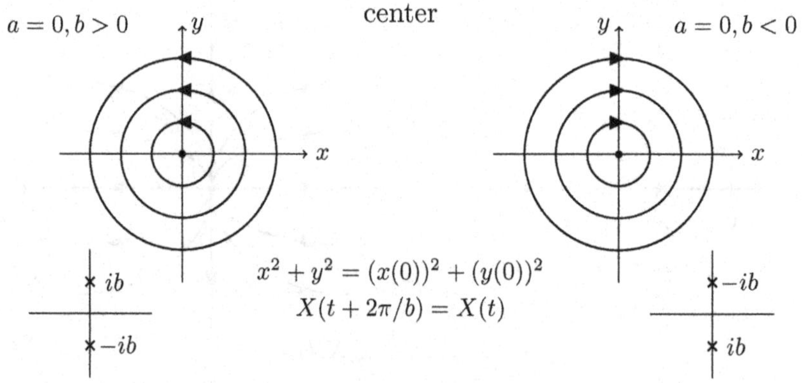

Fig. 18.4

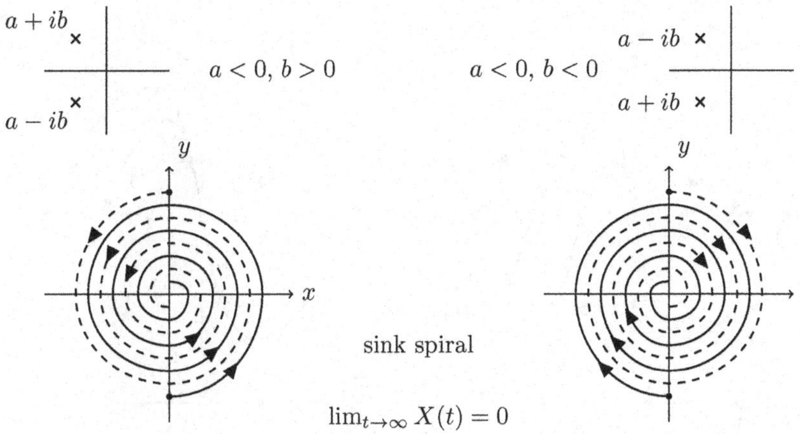

Fig. 18.5(a) Fig. 18.5(b)

0. The point $(0,0)$ is called a **sink spiral**. The positions of the eigenvalues and the corresponding phase portraits near the sinks are shown in Fig. 18.5(a) (for $b>0$) and Fig. 18.5(b) (for $b<0$).

Cases (2)c: $a>0, b>0$ and $a>0, b<0$. The solutions satisfy $\lim_{t\to-\infty} X(t) = 0$. The integral curves are the same as those of Case (2)b, but with the directions of the arrows reversed. The point $(0,0)$ is called a **source spiral**.

Case (3): $\lambda = \mu < 0$ or $0 < \lambda = \mu$ and A not diagonalizable. For negative degenerate eigenvalues, $\lim_{t\to\infty} X(t) = 0$ and the point $(0,0)$ is called a **sink with improper node**. For positive degenerate eigenvalues, $\lim_{t\to-\infty} X(t) = 0$ and the point $(0,0)$ is called a **source with improper node**. For both cases the phase (integral) curves are given by

$$x(t) = e^{\lambda t} x(0)\,, \qquad y(t) = e^{\lambda t}(tx(0) + y(0))\,. \tag{18.27}$$

The positions of the eigenvalues and the corresponding phase portraits are shown in Fig. 18.6.

When we transform the equation

$$\frac{dX'}{dt} = BX'\,, \tag{18.28}$$

with B being a canonical form back to the original coordinates to obtain

$$\frac{dX}{dt} = AX\,, \tag{18.29}$$

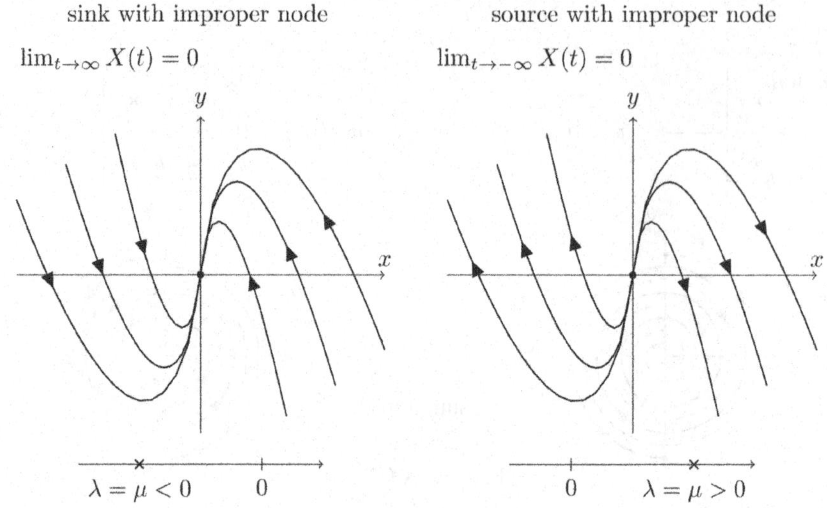

Figure 18.6:

where $B = \Lambda^{-1}A\Lambda$ for some invertible Λ, the above phase curves will in general be distorted but all the qualitative features will remain the same. This follows from the fact that *eigenvalues of a matrix are invariant under similarity transformations.*

Starting with (18.29) one can obtain (18.28) by multiplying on the left by Λ^{-1}. Thus

$$\Lambda^{-1}\frac{dX}{dt} = \frac{d}{dt}\Lambda^{-1}X = \Lambda^{-1}A\Lambda\Lambda^{-1}X \,. \tag{18.30}$$

On defining $X' \equiv \Lambda^{-1}X$ and setting $B \equiv \Lambda^{-1}A\Lambda$, we obtain (18.28).

For the remainder of the chapter we will provide a brief introduction to the theory of the *stability of dynamical systems.* (This is a vast subject. For a more in-depth study consult, for example, Hirsch and Smale 1974; Guckenheimer and Holmes, 1986; or the comprehensive treatise Katok and Hasselblatt, 1995.)

We will begin with a standard definition of the central concept of the *stability of equilibrium points* of a dynamical system (18.6).

Definition 18.3. *Suppose $\overline{X} \in W$ is an equilibrium point of the dynamical system (18.6), where $f : W \to \mathbb{R}^n$ is a C^1 (continuously differentiable) map from an open set $W \subset \mathbb{R}^n$ to the n-dimensional Euclidean space $\mathbb{R}^n$. $\overline{X}$ is said to be a* ***stable equilibrium point*** *if for every neighborhood U of $\overline{X}$ in W there is a neighborhood V of $\overline{X}$, with $V \subset U$, such that $X(0) \in V$ implies $X(t) \in U$ for all $t > 0$ (future stability). If, in addition, $\lim_{t\to\infty} X(t) = \overline{X}$, then $\overline{X}$ is said to be* ***asymptotically stable****. In a similar way, one can define stability for $t < 0$ (past stability).*

Note that *the concept of stability as defined above is invariant under coordinate transformations* $X = (x^1, \ldots, x^n) \to \zeta = (\zeta^1, \ldots, \zeta^n)$. Intuitively, an equilibrium point is (future) stable if nearby solutions remain nearby for all future times. It is also important to note that since in physical applications, an initial state can never be specified exactly, but only with a certain amount of uncertainty, an equilibrium point must be stable in order to be physically meaningful. An important immediate consequence of the above definition is the following theorem (stated without proof).

Theorem 18.7. ***(Liapunov)*** *A necessary condition for future stability is that no eigenvalue of the matrix $Df(\overline{X})$ has a positive real part. A sufficient condition is that all eigenvalues of this matrix have negative real parts.*

A special class of equilibrium points is the so-called *hyperbolic equilibrium points*, defined as follows.

Definition 18.4. *An equilibrium point $\overline{X}$ is said to be* **hyperbolic** *if the matrix $Df(\overline{X})$ has no eigenvalue with real part equal to zero. If the real parts are all negative, the hyperbolic point is called a* ***sink****. If the real parts are all positive, it is called a* ***source****. If both signs occur, the hyperbolic point is called a* ***saddle point****.*

Theorem 18.7 immediately implies the following result.

Corollary 18.1. *A hyperbolic equilibrium point is either unstable or asymptotically stable.*

Associated with an asymptotically stable equilibrium point is a *basin* of the point. By definition, if $\overline{X}$ is an asymptotically stable equilibrium point, there is a neighborhood U of $\overline{X}$ such that every solution curve of (18.6) that begins in U at $t = 0$ will tend towards $\overline{X}$ as $t \to \infty$.

Definition 18.5. *The union of all solution curves of a dynamical system that tend towards an asymptotically stable equilibrium point $\overline{X}$ as $t \to \infty$ is called the* **basin** *of the asymptotically stable equilibrium point, denoted by $B(\overline{X})$.*

In connection with hyperbolic points we also mention a basic theorem, known as **Hartman's Theorem**, which justifies the fact that one can study the qualitative behavior of a dynamical system near its hyperbolic equilibrium points by replacing the system by its linear approximation. This theorem is stated below without proof.

Theorem 18.8. ***(Hartman)*** *Suppose that near the hyperbolic equilibrium point $\overline{X} = 0$ a dynamical system can be written as*

$$\frac{dX}{dt} = f(X) = AX + g(X) , \tag{18.31}$$

where A is a constant matrix with no eigenvalue having real part zero (as guaranteed by the assumed hyperbolicity of the equilibrium point), $g(0) = 0$, and

$\partial g/\partial X = 0$ at $X = 0$. Then there exists a homeomorphism $X = u(\zeta)$ of a neighborhood of $\zeta = 0$ mapping the solution of (18.31) into those of the linear system

$$\frac{d\zeta}{dt} = A\zeta\,. \tag{18.32}$$

We will also mention a general theorem on stability for *complex* dynamical systems [which, of course, subsumes the real ones (18.6)] for all real values of the time t. This poses a very different mathematical problem from that for future stability. The theorem, known as the **Carathéodory-Cartan Theorem**, is stated below (without proof).

Theorem 18.9. ***(Carathéodory-Cartan)*** *Suppose a complex dynamical system is given by the system of differential equations*

$$\frac{dz}{dt} = f(z) = Az + \dots\,, \tag{18.33}$$

where $z = (z^1, \dots, z^n)$ and $f(z) = (f^1(z^i), \dots, f^n(z^i))$ are complex n-vectors, A is a constant matrix, $f(z)$ is holomorphic at $z = 0$, and $f(0) = 0$ ($z = 0$ is an equilibrium point). The following conditions are both necessary and sufficient conditions for $z = 0$ to be a stable equilibrium point for all real t:

(1) A is diagonalizable with purely imaginary eigenvalues.

(2) There exists a holomorphic mapping

$$z = u(\zeta) = \zeta + \dots\,, \tag{18.34}$$

where ζ is a complex n-vector, that maps (18.33) into the linear system [cf. (18.32)]

$$\frac{d\zeta}{dt} = A\zeta\,. \tag{18.35}$$

It is perhaps not surprising that the conditions are both sufficient. The remarkable fact is that they are also both necessary.

Apart from the information provided by the eigenvalues of $Df(\overline{X})$, the following result, called **Liapunov's Stability Theorem**, furnishes a powerful alternative method for detecting the stability properties of a dynamical system near equilibrium points without solving the differential equations corresponding to the system.

Theorem 18.10. ***(Liapunov's Stability Theorem)*** *Let $\overline{X} \in W \subset \mathbb{R}^n$ be an equilibrium point of a dynamical system (18.6). Suppose $V : U \to \mathbb{R}$ is a continuous function on a neighborhood $U \subset W$ of $\overline{X}$ and differentiable on $U/\{\overline{X}\}$ (U minus the point $\overline{X}$) such that the following properties are satisfied:*

(1) $V(\overline{X}) = 0$, and $V(X) > 0$ if $X \neq \overline{X}$;

(2) $\dot{V} \equiv dV(\varphi_t(X))/dt|_{t=0} \leq 0$ in $U/\{\overline{X}\}$, where $\varphi_t(X)$ is the flow of the dynamical system such that the solution to (18.6) passes through X at $t = 0$ (cf. Chapter 11).

Then the equilibrium point $\overline{X}$ is future stable (stable for all $t > 0$). If, in addition to the above two conditions, the function V also satisfies

(3) $\dot{V} < 0$ *in* $U/\{\overline{X}\}$,

then $\overline{X}$ is asymptotically stable. A function V satisfying properties (1) *and* (2) *is called a* ***Liapunov function****. If (3) is also satisfied, it is called a* ***strict Liapunov function****.*

Proof. (following Hirsch and Smale 1974) First assume that (1) and (2) hold. Let the closed ball

$$B_\delta(\overline{X}) \equiv \{X \in W \mid |X - \overline{X}| \leq \delta\} \subset U$$

of small enough radius δ and centered at $\overline{X}$ lie entirely in U. Let V_m be the minimum value of $V(X)$ on the boundary sphere $S_\delta(\overline{X})$ of $B_\delta(\overline{X})$. Then $V_m > 0$ by (1). Let U' be the set defined by

$$U' \equiv \{X \in B_\delta(\overline{X}) \mid V(X) < V_m\} \subset B_\delta(\overline{X}) \,.$$

Note that since $V(\overline{X}) = 0$, $\overline{X} \in U'$. Then no solution starting in U' can meet $S_\delta(\overline{X})$ at a later time, since by (2) $V(X)$ is non-increasing on solution curves. Thus every solution that starts in U' never leaves $B_\delta(\overline{X})$. This is sufficient to guarantee that $\overline{X}$ is a (future) stable equilibrium point. Now assume that (3) holds also, in addition to (1) and (2). $V(X)$ is then strictly decreasing [not only non-increasing, as (2) requires] on orbits starting in $U/\{\overline{X}\}$. Let one such orbit, $X(t)$, start in $U'/\{\overline{X}\}$, and suppose $X(t_n) \to X_0 \in B_\delta(\overline{X})$ for some infinite sequence $t_n \to \infty$. We wish to show that $X_0 = \overline{X}$. Assume the contrary, that is, $X_0 \neq \overline{X}$, and observe that

$$V(X(t)) > V(X_0) \qquad (\text{for all } t \geq 0) \,, \tag{18.36}$$

since $V(X(t))$ decreases monotonically in t and $V(X(t_n)) \to V(X_0)$ by continuity of V. Now let $Y(t)$ be a solution starting at X_0 (which is assumed to be not equal to $\overline{X}$), that is, $Y(0) = X_0$. Then by (3) we must have

$$V(Y(t)) < V(X_0) \qquad (\text{for all } t > 0) \,. \tag{18.37}$$

It follows by continuity of V again that, for every solution $Z(t')$ starting sufficiently near X_0, we have

$$V(Z(t')) < V(X_0) \qquad (\text{for all } t' > 0) \,. \tag{18.38}$$

Putting $Z(0) = X(t_N)$ for a sufficiently large integer N we arrive at the result

$$V(X(t_N + t')) < V(X_0) \qquad (t' > 0) \,, \tag{18.39}$$

which contradicts (18.36). Hence the premise $X_0 \neq \overline{X}$ cannot hold and we must have $X_0 = \overline{X}$. This shows that the equilibrium point $\overline{X}$ is the only possible limit point of the set $\{X(t) \mid t \geq 0\}$. The theorem is thus proved. □

Problem 18.1 Consider a unit mass moving under the influence of a conservative force field $\boldsymbol{F}(\boldsymbol{r}) = -\nabla\phi(\boldsymbol{r})$, where $\phi : W \subset \mathbb{R}^3 \to \mathbb{R}$ is the potential function. Newton's Second Law then results in the dynamical system

$$\frac{dx^i}{dt} = v^i \, , \qquad \frac{dv^i}{dt} = -\frac{\partial \phi}{\partial x^i} \qquad (i = 1, 2, 3) \, . \tag{18.40}$$

In this case $X = (x^1, x^2, x^3, v^1, v^2, v^3)$. Let $(\overline{x}^i, \overline{v}^i)$ be an equilibrium of (18.36). Then $\overline{v}^i = 0$ and $\nabla\phi(\overline{x}^i) = 0$. The total energy $E(x^i, v^i)$ is given by

$$E(x^i, v^i) = \frac{v^2}{2} + \phi(x^i) \, , \tag{18.41}$$

where $v^2 = (v^1)^2 + (v^2)^2 + (v^3)^2$. Show that

$$V(x^i, v^i) \equiv E(x^i, v^i) - E(\overline{x}^i, 0) \tag{18.42}$$

qualifies as a Liapunov function if $\phi(x^i)$ has a local minimum at $\overline{x}^i$, that is, if $\phi(x^i) > \phi(\overline{x}^i)$ for x^i near $\overline{x}^i$ $(x^i \neq \overline{x}^i)$. This result confirms the familiar fact, known as **Lagrange's Theorem**, that *an equilibrium point* $(\overline{x}^i, 0)$ *of a dynamical system for motion in a conservative force field is future stable if the potential energy has a local minimum at* $\overline{x}^i$.

Closely related to the concept of the stability of an equilibrium point is that of the stability of a *closed orbit*, defined as follows.

Definition 18.6. *Suppose $f : W \subset \mathbb{R}^n \to \mathbb{R}^n$ is a C^1 vector field for the dynamical system (18.6): $dX/dt = f(X)$, with the associated flow φ_t. Let $\gamma \subset W$ be a closed orbit of the flow (a periodic solution curve of the dynamical system). The orbit is said to be* ***asymptotically stable*** *if for every open set $U \subset W$ with $\gamma \subset U$, there exists an open set V with $\gamma \subset V \subset U$ such that $\varphi_t(V) \subset U$ for all $t > 0$ and*

$$\lim_{t\to\infty} d_{min}(\varphi_t(X), \gamma) = 0 \qquad \textit{for all } x \in V \, , \tag{18.43}$$

where $d_{min}(X, \gamma)$ is the minimum distance between X and a point in γ.

Asymptotically stable closed orbits have the intuitive property that after long enough times solution curves near one such orbit will evolve as if they had the same *period* as the orbit. This is expressed more precisely by means of the following definition and theorems (stated without proofs).

Definition 18.7. *A point X is said to have an* ***asymptotic period*** $\lambda \in \mathbb{R}$ *if it satisfies the asymptotic condition*

$$\lim_{t\to\infty} |\, \varphi_{t+\lambda}(X) - \varphi_t(X) \,| = 0 \, , \tag{18.44}$$

where φ_t is the flow of a dynamical system.

Theorem 18.11. *If γ is an asymptotically stable closed orbit with period λ, then there is a neighborhood U of γ ($\gamma \subset U$) such that every point of U has asymptotic period λ.*

Analogous to Liapunov's Theorem (Theorem 18.7), which gives conditions on the stability of equilibrium points $\overline{X}$ in terms of the eigenvalues of the linearized vector field $Df(\overline{X})$, we have the following theorem giving a sufficient condition on the stability of a closed orbit.

Theorem 18.12. *Let γ be a closed orbit of an n-dimensional dynamical system with period λ, and let $p \in \gamma$ be any point on the orbit. Let φ_t be the flow of the dynamical system. If $n-1$ of the eigenvalues of the linear map $D\varphi_\lambda(p)$ all have absolute values less than one, then γ is an asymptotically stable orbit.*

Note that 1 is always an eigenvalue of $D\varphi_\lambda(p)$ since, for a vector field $f(X)$,

$$D\varphi_\lambda(p)f(p) = f(p) \,. \tag{18.45}$$

The eigenvalue condition stated in the above theorem is, however, not a necessary condition for the asymptotic stability of a closed orbit. If it is satisfied, then the orbit is called a **periodic attractor**. Solution curves near a periodic attractor, in addition to having the same period as the attractor, can be described as being asymptotically "in phase" with the attractor. This is expressed more precisely in the following theorem.

Theorem 18.13. *Let γ be a periodic attractor whose flow is φ_t. Then there exists a neighborhood U of γ ($\gamma \subset U$) such that for every $X \in U$, there is a unique point $Y \in \gamma$ such that*

$$\lim_{t\to\infty} |\,\varphi_t(X) - \varphi_t(Y)\,| = 0 \,. \tag{18.46}$$

The proof of Theorem 18.12 and the general study of closed orbits are greatly facilitated by the concepts of *Poincaré sections* and *Poincaré maps*, which in turn lead to that of *discrete dynamical systems*. We will introduce these ideas now. Their uses will be illustrated more fully within the context of Hamiltonian dynamics in later parts of this book (see Chapter 42).

Definition 18.8. *Consider an n-dimensional dynamical system and a point $p \in W \subset \mathbb{R}^n$, where W is the phase space of the system. Let $U \subset W$ be an $(n-1)$-dimensional subspace of W containing p. An open subset $S \subset U$ with $p \in S$ is called a* ***Poincaré section*** *of the flow of the dynamical system at the point p if it is transverse to the flow at p, that is, if the vector field of the flow is not tangent to S at p.*

Definition 18.9. *Consider a flow φ_t of a dynamical system $dX/dt = f(X)$ with vector field $f : W \to \mathbb{R}^n$. Suppose γ is a closed orbit with period $\lambda > 0$ and containing the point $0 \in W$, and S is a Poincaré section at 0. Let $X \in S$ be a point sufficiently close to 0. Then there exists a neighborhood $U \subset W$ of 0 and*

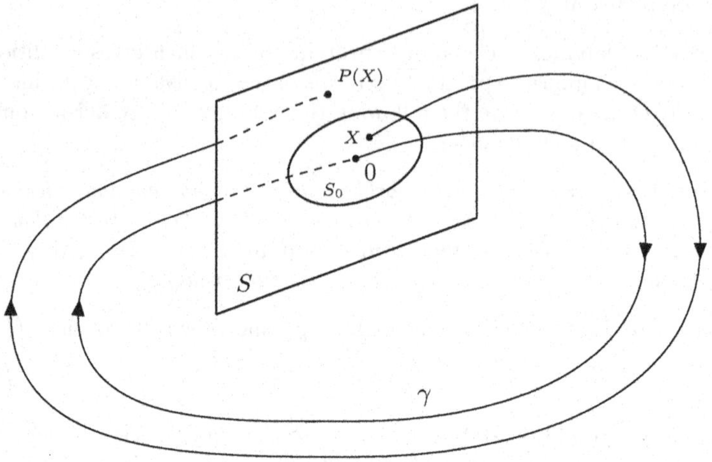

Fig. 18.7

a unique C^1 map $\tau : U \to \mathbb{R}$, such that $\varphi_{\tau(X)}(X) \in S$ for all $X \in U$, $\tau(0) = \lambda$, and $\tau(X) \approx \lambda$. The C^1 map $P : S_0 \to S$ (with $S_0 = S \cap U$) defined by

$$P(X) = \varphi_{\tau(X)}(X) \tag{18.47}$$

is called a ***Poincaré map*** *(see Fig. 18.7). Obviously the point $0 \in \gamma$ is a* ***fixed point*** *of the Poincaré map: $P(0) = 0$.*

Intuitively a Poincaré map P simply follows an orbit (in time) from a point of intersection of the orbit with a Poincaré section S until the next time the orbit intersects with S again. If P has a C^1 inverse also, then it is a *diffeomorphism.* Poincaré maps that are diffeomorphisms play a crucial role in the study of dynamical systems and their stabilities. This is in general achieved by studying the iterates P^n of the map ($n \in \mathbb{Z}$, $\mathbb{Z}$ being the set of integers). We state below (without proofs) two useful general theorems on Poincaré maps in the study of the stability of orbits.

Theorem 18.14. *Suppose $P : S_0 \to S$ is a Poincaré map for a closed orbit γ and $0 \in \gamma$. If a point $X \in S_0$ satisfies the condition $\lim_{n\to\infty} P^n(X) = 0$, then*

$$\lim_{t\to\infty} d_{min}(\varphi_t(X), \gamma) = 0\,, \tag{18.48}$$

where φ_t is the flow of the dynamical system and $d_{min}(X, \gamma)$ is the minimum distance between the point X and the orbit γ.

Similar to the concept of a *sink* for a continuous dynamical system (cf. Def. 18.4) we have the corresponding notion of a sink for discrete dynamical systems, defined in terms of the eigenvalues of the matrix of the differential $DP(X)$ of the Poincaré map.

Theorem 18.15. *Suppose γ is a closed orbit and $0 \in \gamma$. If 0 is a sink of the Poincaré map P, then γ is an asymptotically stable orbit (cf. Def. 18.6).*

We will end this chapter with some very brief introductory remarks on the notion of the *structural stability* of a flow (or its vector field) of a dynamical system under small perturbations of the vector field (for an in-depth and mathematically rigorous treatment, see Smale, 1967). This is a so-called *generic property* of the vector field. Its definition requires one to construct a *metric space*, or more generally, a topology, out of a set of vector fields on a certain phase space, so that the notion of neighborhoods of a particular vector field makes sense. Roughly speaking, we say that a vector field f is **structurally stable** in a certain region W of phase space if there exists a neighborhood of f such that if another vector field g is in that neighborhood, f and g are *topologically equivalent* on W. This means that there exists an orientation-preserving homeomorphism $H : W \to W$ such that for each $X \in W$,

$$H(\{\varphi_t(X) \,|\, t \geq 0\}) = \{\psi_t(H(X)) \,|\, t \geq 0\} , \tag{18.49}$$

where φ_t and ψ_t are the flows corresponding to the vector fields f and g, respectively, and the orientation-preserving property of H means that the orientations of the solution curves under φ_t and ψ_t are the same (if X is not an equilibrium point of either f or g).

Chapter 19

Many-Particle Systems and the Conservation Principles

In this chapter we will return to Newton's Laws and apply them formally to systems of many particles, with a view mainly to subsequent applications in rigid-body dynamics. The useful notion of the *center of mass* will be introduced and its importance discussed. In essence this concept highlights many subtleties and introduces significant simplifications involved in the formulation of (1) the laws of motion for a system of particles using Newton's Laws (originally stated explicitly for single particles), and (2) the conservation principles of linear momentum, angular momentum, and energy.

Consider a system of N point-particles, of masses m_α, $\alpha = 1, \dots, N$ in arbitrary motion. Let O be an inertial frame and O' be an arbitrary moving frame (Fig. 19.1); $\boldsymbol{r}_\alpha$ and $\boldsymbol{r}'_\alpha$ be the position vectors of the α-th particle with respect to O and O' respectively; and $\boldsymbol{R}$ be the position of the origin of O' with respect to O. Then

$$\boldsymbol{r}_\alpha = \boldsymbol{r}'_\alpha + \boldsymbol{R}\,, \tag{19.1}$$

and, on differentiation of this equation with respect to the time t, the velocities of the particles are given by

$$\boldsymbol{v}_\alpha = \boldsymbol{v}'_\alpha + \boldsymbol{V}\,, \tag{19.2}$$

where $\boldsymbol{V} \equiv d\boldsymbol{R}/dt$.

The total linear momentum $\boldsymbol{P}$ of the system with respect to O is given by

$$\boldsymbol{P} = \sum_\alpha m_\alpha \boldsymbol{v}_\alpha = \sum_\alpha m_\alpha(\boldsymbol{v}'_\alpha + \boldsymbol{V}) = \sum_\alpha m_\alpha \boldsymbol{v}'_\alpha + M\boldsymbol{V} = \boldsymbol{P}' + M\boldsymbol{V}\,, \tag{19.3}$$

where $M \equiv \sum_\alpha m_\alpha$ is the total mass of the system and

$$\boldsymbol{P}' \equiv \sum_\alpha m_\alpha \boldsymbol{v}'_\alpha \tag{19.4}$$

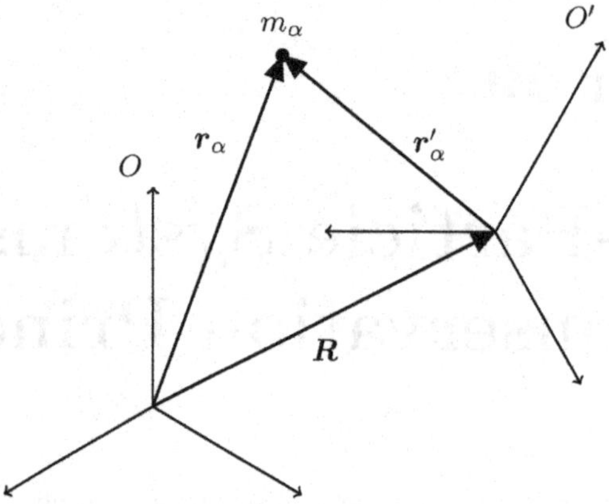

Fig. 19.1

is the total linear momentum of the system with respect to O'. Suppose $\boldsymbol{R}$ is chosen so that

$$M\boldsymbol{R} = \sum_\alpha m_\alpha \boldsymbol{r}_\alpha \,. \tag{19.5}$$

Then

$$\boldsymbol{P} = \frac{d}{dt}\left(\sum_\alpha m_\alpha \boldsymbol{r}_\alpha\right) = M\frac{d\boldsymbol{R}}{dt} = M\boldsymbol{V} \,. \tag{19.6}$$

It follows from (19.3) that, with respect to this choice of the origin of O',

$$\boldsymbol{P}' = 0 \,. \tag{19.7}$$

A reference frame O' with respect to which the total linear momentum $\boldsymbol{P}'$ vanishes is called a **center-of-mass frame**. The vector $\boldsymbol{R}$ satisfying (19.5) specifies a point called the **center-of-mass** of the system of particles, and $\boldsymbol{R}$ itself is called the center-of-mass position vector.

Each mass in the system is in general subjected to *external forces* (forces arising from sources external to the system); and we assume that all *internal forces* act in a pairwise fashion and obey Newton's Third Law: $\boldsymbol{F}_{\alpha\beta} = -\boldsymbol{F}_{\beta\alpha}$, where $\boldsymbol{F}_{\alpha\beta}$ is the force on the α-th particle due to the β-th particle. *The net force on all the particles is then just the sum of the external forces* since the sum of all the internal forces cancel out pairwise. We write

$$\boldsymbol{F}_{net} = \boldsymbol{F}_{ext} = \sum_{\alpha=1}^{N} \boldsymbol{F}_{\alpha\,(ext)} \,, \tag{19.8}$$

where $\boldsymbol{F}_{\alpha\,(ext)}$ is the net external force acting on the α-th particle. Now, applying Newton's Second Law to each of the particles and let $\boldsymbol{F}_\alpha$ be the net force acting on the α-th particle (including both external and internal forces), we have

$$\boldsymbol{F}_{net} = \sum_{\alpha=1}^{N} \boldsymbol{F}_\alpha = \sum_{\alpha=1}^{N} \frac{d\boldsymbol{p}_\alpha}{dt} = \frac{d}{dt}\sum_{\alpha=1}^{N} \boldsymbol{p}_\alpha = \frac{d\boldsymbol{P}}{dt} \,. \tag{19.9}$$

Combining the above two equations and using (19.6) (which applies to $\boldsymbol{R}_{CM}$, the position vector of the origin of a center-of-mass frame), we have

$$\frac{d\boldsymbol{P}}{dt} = M\frac{d^2\boldsymbol{R}_{CM}}{dt^2} = \boldsymbol{F}_{ext} \,. \tag{19.10}$$

The second equality of the above equation can be interpreted as follows: *The motion of the CM of a multiparticle system of total mass M is the same as that of a point mass of the same mass subjected to a force equal to* $\boldsymbol{F}_{ext}$ *(the net external force on the system).* Thus *a CM frame is not an inertial frame unless the net external force vanishes.* Equation (19.10) also implies the following very important and useful conservation principle:

Principle of Conservation of Linear Momentum (for a system of particles): *The total linear momentum of a system of particles is conserved if the net external force acting on it is zero.*

Next we consider the total angular momentum

$$\boldsymbol{L} = \sum_\alpha \boldsymbol{r}_\alpha \times \boldsymbol{p}_\alpha = \sum_\alpha \boldsymbol{r}_\alpha \times m_\alpha \boldsymbol{v}_\alpha \,. \tag{19.11}$$

From (19.2) we have

$$\begin{aligned} \boldsymbol{L} &= \sum_\alpha m_\alpha \boldsymbol{r}_\alpha \times (\boldsymbol{v}'_\alpha + \boldsymbol{V}) \\ &= \sum_\alpha m_\alpha \boldsymbol{r}_\alpha \times \boldsymbol{v}'_\alpha + \Big(\sum_\alpha m_\alpha \boldsymbol{r}_\alpha\Big) \times \boldsymbol{V} \\ &= \sum_\alpha m_\alpha (\boldsymbol{r}'_\alpha + \boldsymbol{R}) \times \boldsymbol{v}'_\alpha + M\boldsymbol{R} \times \boldsymbol{V} \\ &= \sum_\alpha m_\alpha \boldsymbol{r}'_\alpha \times \boldsymbol{v}'_\alpha + \boldsymbol{R} \times \sum_\alpha m_\alpha \boldsymbol{v}'_\alpha + M\boldsymbol{R} \times \boldsymbol{V} \,. \end{aligned} \tag{19.12}$$

If O' is a center-of-mass frame, then $\sum_\alpha m_\alpha \boldsymbol{v}'_\alpha = \boldsymbol{P}' = 0$, and

$$\boldsymbol{L} = \boldsymbol{L}' + \boldsymbol{R} \times \boldsymbol{P} \,, \tag{19.13}$$

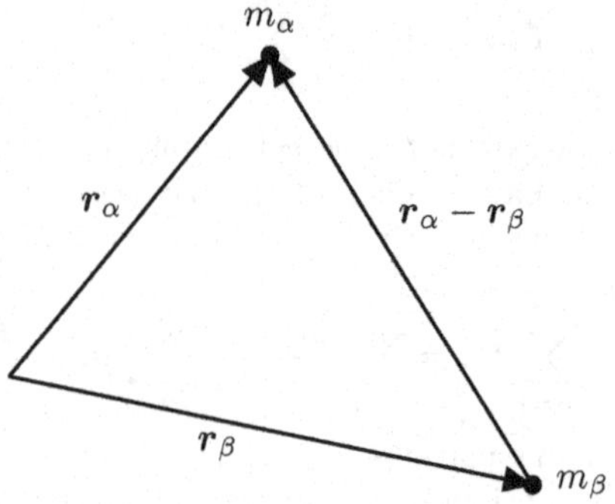

Fig. 19.2

where

$$\boldsymbol{L}' \equiv \sum_\alpha m_\alpha \boldsymbol{r}'_\alpha \times \boldsymbol{v}'_\alpha \tag{19.14}$$

is the total angular momentum in the center-of-mass (CM) frame O'.

We will now consider the net torque $\boldsymbol{\tau}$ in relation to the total angular momentum. Let $\boldsymbol{F}^{(e)}_\alpha$ and $\boldsymbol{F}^{(i)}_\alpha$ be the external and internal forces on the α-th particle, respectively. Then

$$\boldsymbol{\tau} = \sum_\alpha \boldsymbol{r}_\alpha \times (\, \boldsymbol{F}^{(e)}_\alpha + \boldsymbol{F}^{(i)}_\alpha \,) \, . \tag{19.15}$$

We will show that the internal forces do not contribute to the net torque, provided $\boldsymbol{F}_{\alpha\beta}$ (the force on m_α by m_β), for arbitrary α and β, is along the direction of $\boldsymbol{r}_\alpha - \boldsymbol{r}_\beta$, in other words, along the straight line joining m_α and m_β (see Fig. 19.2). We have

$$\begin{aligned} \sum_\alpha \boldsymbol{r}_\alpha \times \boldsymbol{F}^{(i)}_\alpha &= \sum_\alpha \boldsymbol{r}_\alpha \times \sum_{\beta \neq \alpha} \boldsymbol{F}_{\alpha\beta} = \sum_{\alpha \neq \beta} \boldsymbol{r}_\alpha \times \boldsymbol{F}_{\alpha\beta} \\ &= \sum_{\alpha < \beta} (\, \boldsymbol{r}_\alpha \times \boldsymbol{F}_{\alpha\beta} + \boldsymbol{r}_\beta \times \boldsymbol{F}_{\beta\alpha} \,) = \sum_{\alpha < \beta} (\, \boldsymbol{r}_\alpha - \boldsymbol{r}_\beta \,) \times \boldsymbol{F}_{\alpha\beta} \, , \end{aligned} \tag{19.16}$$

where the last equality follows from Newton's third law, which implies that $\boldsymbol{F}_{\beta\alpha} = -\boldsymbol{F}_{\alpha\beta}$. This quantity obviously vanishes if $\boldsymbol{F}_{\alpha\beta}$ is along the direction of $\boldsymbol{r}_\alpha - \boldsymbol{r}_\beta$. Thus

$$\boldsymbol{\tau} = \sum_\alpha \boldsymbol{r}_\alpha \times \boldsymbol{F}^{(e)}_\alpha \, . \tag{19.17}$$

Now

$$\boldsymbol{L} = \sum_\alpha \boldsymbol{r}_\alpha \times m_\alpha \boldsymbol{v}_\alpha \,, \tag{19.18}$$

which implies that

$$\frac{d\boldsymbol{L}}{dt} = \sum_\alpha \boldsymbol{r}_\alpha \times m_\alpha \frac{d\boldsymbol{v}_\alpha}{dt} = \sum_\alpha \boldsymbol{r}_\alpha \times (\, \boldsymbol{F}^{(e)}_\alpha + \boldsymbol{F}^{(i)}_\alpha \,) = \sum_\alpha \boldsymbol{r}_\alpha \times \boldsymbol{F}^{(e)}_\alpha \,. \tag{19.19}$$

It follows from (19.18) that

$$\boldsymbol{\tau} = \frac{d\boldsymbol{L}}{dt} \,. \tag{19.20}$$

Note that this equation, which applies to a system of particles, looks exactly the same as (12.28), which applies to a single particle. It also implies, together with (19.17), the following important conservation principle:

Principle of Conservation of Angular Momentum (for a system of particles): *The total angular momentum of a system of particles is conserved if the net torque due to external forces vanishes.*

The net torque can also be written in terms of the position vectors $\boldsymbol{r}'_\alpha$ as follows:

$$\boldsymbol{\tau} = \sum_\alpha (\, \boldsymbol{R} + \boldsymbol{r}'_\alpha \,) \times \boldsymbol{F}^{(e)}_\alpha = \boldsymbol{R} \times \sum_\alpha \boldsymbol{F}^{(e)}_\alpha + \sum_\alpha \boldsymbol{r}'_\alpha \times \boldsymbol{F}^{(e)}_\alpha \,. \tag{19.21}$$

By (19.9),

$$\boldsymbol{R} \times \sum_\alpha \boldsymbol{F}^{(e)}_\alpha = \boldsymbol{R} \times \frac{d\boldsymbol{P}}{dt} = \frac{d}{dt}(\boldsymbol{R} \times \boldsymbol{P}) \,, \tag{19.22}$$

where the last equality follows from the fact that

$$\frac{d\boldsymbol{R}}{dt} = \boldsymbol{V} = \frac{\boldsymbol{P}}{M} \,. \tag{19.23}$$

We can then write (19.21) as

$$\boldsymbol{\tau} = \frac{d}{dt}(\boldsymbol{R} \times \boldsymbol{P}) + \sum_\alpha \boldsymbol{r}'_\alpha \times \boldsymbol{F}^{(e)}_\alpha \,. \tag{19.24}$$

On the other hand, (19.13) implies

$$\frac{d\boldsymbol{L}}{dt} = \frac{d}{dt}(\boldsymbol{R} \times \boldsymbol{P}) + \frac{d\boldsymbol{L}'}{dt} \,. \tag{19.25}$$

Comparing with (19.24) and recalling (19.20) we obtain

$$\frac{d\boldsymbol{L}'}{dt} = \sum_\alpha \boldsymbol{r}'_\alpha \times \boldsymbol{F}^{(e)}_\alpha \,, \tag{19.26}$$

which is similar in form to (19.19). This is interesting because the CM (center-of-mass) frame is in general not an inertial frame, and yet the equation of motion (19.26) retains the same form as that with respect to an inertial frame [Eq. (19.19)].

Finally we consider the total energy of the system. The kinetic energy K of the system with respect to the inertial frame O is:

$$K = \frac{1}{2}\sum_\alpha m_\alpha \boldsymbol{v}_\alpha^2 \,. \tag{19.27}$$

Using (19.2) we have

$$\begin{aligned} K &= \frac{1}{2}\sum_\alpha m_\alpha(\boldsymbol{v}'_\alpha \cdot \boldsymbol{v}'_\alpha + 2\boldsymbol{V}\cdot\boldsymbol{v}'_\alpha + \boldsymbol{V}^2) \\ &= \frac{1}{2}M\boldsymbol{V}^2 + \boldsymbol{V}\cdot\boldsymbol{P}' + K' \,, \end{aligned} \tag{19.28}$$

where

$$K' \equiv \frac{1}{2}\sum_\alpha m_\alpha \boldsymbol{v}'_\alpha \cdot \boldsymbol{v}'_\alpha \tag{19.29}$$

is the total kinetic energy of the system with respect to O'. If we choose O' to be a center-of-mass frame, then $\boldsymbol{P}' = 0$ [recall (19.7) and the development immediately preceding the equation], and we have the following transformation rule for the kinetic energy between an inertial frame and a center-of-mass frame for a system of particles:

$$K = K' + \frac{1}{2}M\boldsymbol{V}^2 \,. \tag{19.30}$$

We will assume that all forces (external and internal) are *conservative* so that [recall (12.34) and (12.35)] there exists a potential energy function $U(\boldsymbol{r}_1, \ldots, \boldsymbol{r}_N)$ such that

$$dU = -\sum_{\alpha=1}^{N}\left(\boldsymbol{F}_\alpha^{(e)} + \boldsymbol{F}_\alpha^{(i)}\right)\cdot d\boldsymbol{r}_\alpha \,. \tag{19.31}$$

Hence, in a manner entirely similar to the derivation of the *work-energy theorem* for a single particle system leading to (12.32), we have

$$\begin{aligned} dU &= -\sum_\alpha \frac{d\boldsymbol{p}_\alpha}{dt}\cdot d\boldsymbol{r}_\alpha = -\sum_\alpha d\boldsymbol{p}_\alpha\cdot\boldsymbol{v}_\alpha = -\sum_\alpha m_\alpha d\boldsymbol{v}_\alpha\cdot\boldsymbol{v}_\alpha \\ &= -d\left(\sum_\alpha \frac{m_\alpha \boldsymbol{v}_\alpha\cdot\boldsymbol{v}_\alpha}{2}\right) = -dK \,. \end{aligned} \tag{19.32}$$

Setting the total mechanical energy of the system

$$E = K + U \,, \tag{19.33}$$

we have the following conservation theorem:

Principle of Conservation of Energy (for a system of particles): *The total mechanical energy of a system of particles is conserved if all external and internal forces acting on the particles are conservative.*

We will conclude this chapter by looking at the potential energy in more detail. We separate this energy into external and internal parts by writing

$$U = U^{(e)} + U^{(i)} , \tag{19.34}$$

where $U^{(e)}$ and $U^{(i)}$ are given by

$$dU^{(e)} = -\sum_\alpha \boldsymbol{F}_\alpha^{(e)} \cdot d\boldsymbol{r}_\alpha , \qquad dU^{(i)} = -\sum_\alpha \boldsymbol{F}_\alpha^{(i)} \cdot d\boldsymbol{r}_\alpha . \tag{19.35}$$

We have

$$dU^{(e)} = -\sum_\alpha \boldsymbol{F}_\alpha^{(e)} \cdot d(\boldsymbol{r}'_\alpha + \boldsymbol{R}) = -\sum_\alpha \boldsymbol{F}_\alpha^{(e)} \cdot d\boldsymbol{r}'_\alpha - \sum_\alpha \boldsymbol{F}_\alpha^{(e)} \cdot d\boldsymbol{R} . \tag{19.36}$$

The second term on the right-hand side of the last equality yields, since all the internal forces sum to zero,

$$-\left(\sum_\alpha \boldsymbol{F}_\alpha^{(e)}\right) \cdot d\boldsymbol{R} = -\left(\sum_\alpha \boldsymbol{F}_\alpha\right) \cdot d\boldsymbol{R} = -\frac{d\boldsymbol{P}}{dt} \cdot d\boldsymbol{R} = -d\boldsymbol{P} \cdot \frac{d\boldsymbol{R}}{dt} = -d\boldsymbol{P} \cdot \boldsymbol{V} . \tag{19.37}$$

Recalling that the total linear momentum $\boldsymbol{P}$ is given by $\boldsymbol{P} = M\boldsymbol{V} + \boldsymbol{P}'$ [cf. (19.3)], and assuming that O' is a CM frame (so that $\boldsymbol{P}' = 0$), we have

$$-\sum_\alpha \boldsymbol{F}_\alpha^{(e)} \cdot d\boldsymbol{R} = -d(M\boldsymbol{V}) \cdot \boldsymbol{V} = -d\left(\frac{M\boldsymbol{V} \cdot \boldsymbol{V}}{2}\right) . \tag{19.38}$$

This result can be interpreted as the conservation of energy for the motion of a point mass of mass M and velocity $\boldsymbol{V}$ located at the center of mass of the system of particles, and subjected to a force equal to the net external force on the system. By virtue of this result, Eq. (19.36) implies that

$$dU^{(e)} = -\sum_\alpha \boldsymbol{F}_\alpha^{(e)} \cdot d\boldsymbol{r}'_\alpha - d\left(\frac{MV^2}{2}\right) . \tag{19.39}$$

The principle of conservation of energy, together with (19.30), (19.33) and (19.34), then give

$$dE = dK + dU^{(e)} + dU^{(i)} = dK' - \sum_\alpha \boldsymbol{F}_\alpha^{(e)} \cdot d\boldsymbol{r}'_\alpha + dU^{(i)} = 0 . \tag{19.40}$$

If we define the potential energy $U'^{(e)}$ due to external conservative forces with respect to the center-of-mass (CM) frame O' (up to an arbitrary constant) by

$$dU'^{(e)} \equiv -\sum_\alpha \boldsymbol{F}_\alpha^{(e)} \cdot d\boldsymbol{r}'_\alpha , \tag{19.41}$$

Eq. (19.40) then indicates that the total energy with respect to O', defined by

$$E' \equiv K' + U'^{(e)} + U^{(i)} \tag{19.42}$$

is also conserved. Analogous to the manipulations in (19.16) for the calculation of torques, the internal potential energy $U^{(i)}$ can be expressed as

$$\begin{aligned} -\,dU^{(i)} &= \sum_{\alpha} \boldsymbol{F}_{\alpha}^{(i)} \cdot d\boldsymbol{r}_{\alpha} = \sum_{\alpha\neq\beta} \boldsymbol{F}_{\alpha\beta} \cdot d\boldsymbol{r}_{\alpha} = \sum_{\alpha<\beta} (\boldsymbol{F}_{\alpha\beta} \cdot d\boldsymbol{r}_{\alpha} + \boldsymbol{F}_{\beta\alpha} \cdot d\boldsymbol{r}_{\beta}) \\ &= \sum_{\alpha<\beta} (\boldsymbol{F}_{\alpha\beta} \cdot d\boldsymbol{r}_{\alpha} - \boldsymbol{F}_{\alpha\beta} \cdot d\boldsymbol{r}_{\beta}) = \sum_{\alpha<\beta} \boldsymbol{F}_{\alpha\beta} \cdot d(\boldsymbol{r}_{\alpha} - \boldsymbol{r}_{\beta}) = \sum_{\alpha<\beta} \boldsymbol{F}_{\alpha\beta} \cdot d(\boldsymbol{r}'_{\alpha} - \boldsymbol{r}'_{\beta})\,. \end{aligned} \tag{19.43}$$

The last equality holds because $\boldsymbol{r}_{\alpha\beta} \equiv \boldsymbol{r}_{\alpha} - \boldsymbol{r}_{\beta} = \boldsymbol{r}'_{\alpha} - \boldsymbol{r}'_{\beta} \equiv \boldsymbol{r}'_{\alpha\beta}$. On assuming again that $\boldsymbol{F}_{\alpha\beta}$ is a function of $\boldsymbol{r}_{\alpha\beta}$ only, we have

$$dU^{(i)} = -\sum_{\alpha<\beta} \boldsymbol{F}_{\alpha\beta}(\boldsymbol{r}'_{\alpha\beta}) \cdot d\boldsymbol{r}'_{\alpha\beta}\,. \tag{19.44}$$

Thus *the internal potential energy can be expressed in terms of only coordinates with respect to the center-of-mass frame.*

Chapter 20

Rigid Body Dynamics: The Euler-Poisson Equations of Motion

Consider an orthonormal moving frame $\{\boldsymbol{e}_1, \boldsymbol{e}_2, \boldsymbol{e}_3\}$ rigidly attached to a rigid body in motion with respect to an inertial frame $\{\boldsymbol{\epsilon}_1, \boldsymbol{\epsilon}_2, \boldsymbol{\epsilon}_3\}$, with the origin of the moving frame located at the center of mass (CM) of the rigid body (Fig. 20.1). The moving frame under consideration is thus a center-of-mass (CM) frame.

Assume for the moment that the rigid body is composed of discrete point masses, with the position vector of the α-th mass (of mass m_α) (with respect to the moving frame) denoted by $\boldsymbol{r}_\alpha$. (Note that this quantity was denoted by $\boldsymbol{r}'_\alpha$ in the last chapter.) Let $\boldsymbol{\omega}$ be the instantaneous angular velocity of the rigid body as defined by (4.25). Using (4.13) in (19.29), the total kinetic energy K of the rigid body (with respect to the $\{\boldsymbol{\epsilon}_i\}$ frame) is given by

$$K = \frac{1}{2} M \boldsymbol{V}^2 + \frac{1}{2} \sum_\alpha m_\alpha (\boldsymbol{\omega} \times \boldsymbol{r}_\alpha) \cdot (\boldsymbol{\omega} \times \boldsymbol{r}_\alpha) \,, \tag{20.1}$$

where M is the total mass of the rigid body and $\boldsymbol{V}$ is the velocity of the center of mass of the rigid body.

Problem 20.1 Use the vector identity (1.61) and the "bac minus cab" identity (1.59) to show that

$$(\boldsymbol{\omega} \times \boldsymbol{r}_\alpha) \cdot (\boldsymbol{\omega} \times \boldsymbol{r}_\alpha) = \omega^2 r_\alpha^2 - (\boldsymbol{\omega} \cdot \boldsymbol{r}_\alpha)^2 \,. \tag{20.2}$$

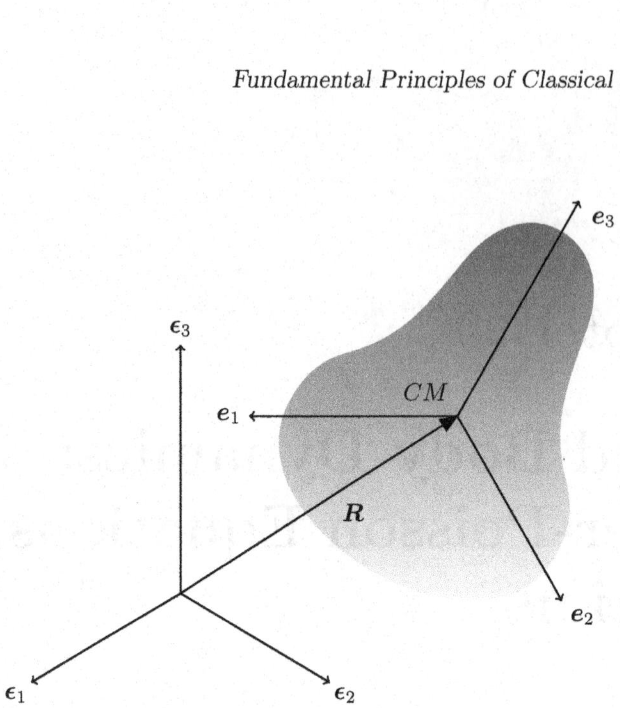

Fig. 20.1

Using the above result, (20.1) can be rewritten as

$$K = \frac{1}{2}M\boldsymbol{V}^2 + \frac{1}{2}\sum_\alpha m_\alpha\{\,\omega^2 r_\alpha^2 - (\boldsymbol{\omega}\cdot\boldsymbol{r}_\alpha)^2\}\,. \tag{20.3}$$

The second term on the right-hand side of the above equation can be interpreted as the kinetic energy of rotation of the rigid body with angular velocity $\boldsymbol{\omega}$ about an axis passing through the center of mass of the rigid body. This term, which we will now denote by K_{rot}, can be further manipulated using the tensorial index notation as follows:

$$\begin{aligned} K_{rot} &= \frac{1}{2}\sum_\alpha m_\alpha\left(\omega_i\omega_j\,\delta^{ij}(x_\alpha)_l^2 - \omega_i\omega_j(x_\alpha)^i(x_\alpha)^j\right) \\ &= \frac{1}{2}\,\omega_i\omega_j\sum_\alpha m_\alpha\left(\delta^{ij}(x_\alpha)_l^2 - (x_\alpha)^i(x_\alpha)^j\right)\,, \end{aligned} \tag{20.4}$$

where $\omega_i = \omega^i$ are the components of the angular velocity vector $\boldsymbol{\omega}$ defined in (4.25), and $(x_\alpha)_l = (x_\alpha)^l$ are the components in $\boldsymbol{r}_\alpha = (x_\alpha)^l\boldsymbol{e}_l$. Defining the **moment of inertia tensor**, which is a symmetric (2,0)-type tensor [cf. discussion following (1.32)] by

$$I^{ij} \equiv \sum_\alpha m_\alpha\left(\,\delta^{ij}\,(x_\alpha)^l(x_\alpha)_l - (x_\alpha)^i(x_\alpha)^j\right)\,, \tag{20.5}$$

we can write the total kinetic energy as

$$K = \frac{1}{2} M\mathbf{V}^2 + \frac{1}{2} I^{ij} \omega_i \omega_j . \qquad (20.6)$$

Note that to describe the motion of a rigid body, we need six degrees of freedom: the three coordinates X, Y, Z for the position of the center of mass, and three angles specifying the orientation of the orthonormal moving frame.

We will now make a transition from a discrete-mass system to a continuous mass distribution and write the inertia tensor in terms of an integral over the volume occupied by the distribution:

$$I^{ij} = \int d^3r\, \rho \left(\delta^{ij} x^l x_l - x^i x^j \right) , \qquad (20.7)$$

where ρ is the local mass density of the rigid body. In matrix form we can write, on setting $x^1 = x, x^2 = y, x^3 = z$,

$$I^{ij} = \int d^3r\, \rho \begin{pmatrix} y^2 + z^2 & -xy & -xz \\ -yx & x^2 + z^2 & -yz \\ -zx & -zy & x^2 + y^2 \end{pmatrix} . \qquad (20.8)$$

Since I^{ij} is symmetric, it is diagonalizable. The axes of the CM frame $\{\mathbf{e}'_1, \mathbf{e}'_2, \mathbf{e}'_3\}$ with respect to which I^{ij} is diagonal are called **principal axes of inertia**, and the diagonal elements of the diagonalized matrix are called the **principal moments of inertia**, usually denoted by I^1, I^2, I^3. Expressed with respect to the principal axes, the rotational kinetic energy of a rigid body is given by

$$K_{rot} = \frac{1}{2} \left(I^1 (\omega_1)^2 + I^2 (\omega_2)^2 + I^3 (\omega_3)^2 \right) , \qquad (20.9)$$

where $\omega_1, \omega_2, \omega_3$ are components of the angular velocity $\boldsymbol{\omega}$ with respect to the principal axes $\mathbf{e}'_1, \mathbf{e}'_2, \mathbf{e}'_3$. From (20.8) we see that

$$I^1 + I^2 = \int d^3r\, \rho (x^2 + y^2 + 2z^2) \geq \int d^3r\, \rho (x^2 + y^2) = I^3 . \qquad (20.10)$$

Hence *none of the principal moments of inertia can exceed the sum of the other two.*

Rigid bodies can be classified according to their principal moments of inertia into three different types: 1) if I^1, I^2, I^3 are all different, then the rigid body is called an **asymmetric top** (Fig. 20.2); 2) if $I^1 = I^2 \neq I^3$, then the rigid body is called a **symmetric top** (Fig. 20.3); 3) if $I^1 = I^2 = I^3$, then the rigid body is called a **spherical top** (Fig. 20.4).

Consider two orthonormal frames $\{\mathbf{e}_i\}$ and $\{\mathbf{e}'_i\}$ that are parallelly displaced from each other (Fig. 20.5), with the origin of the former located at the center of mass of a rigid body, and the origin of the latter away from the center of

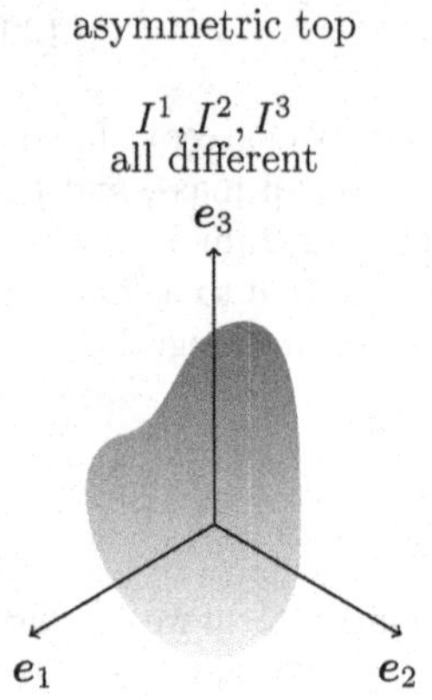

Fig. 20.2

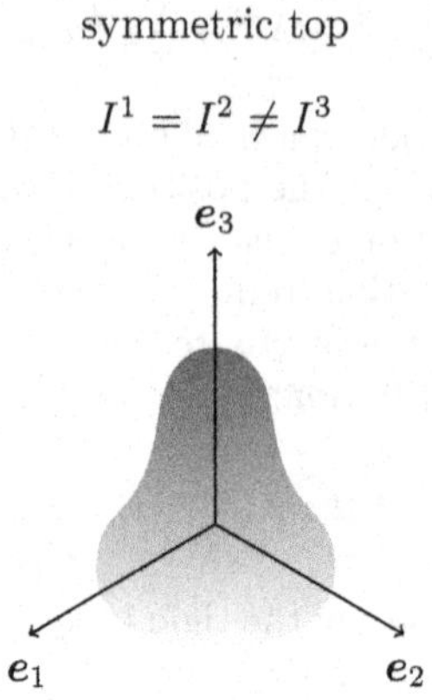

Fig. 20.3

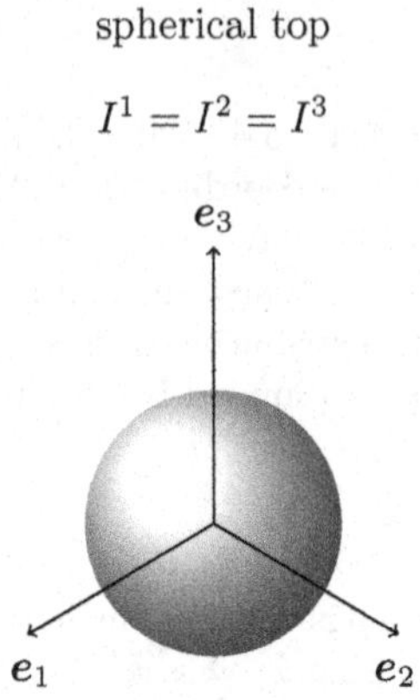

Fig. 20.4

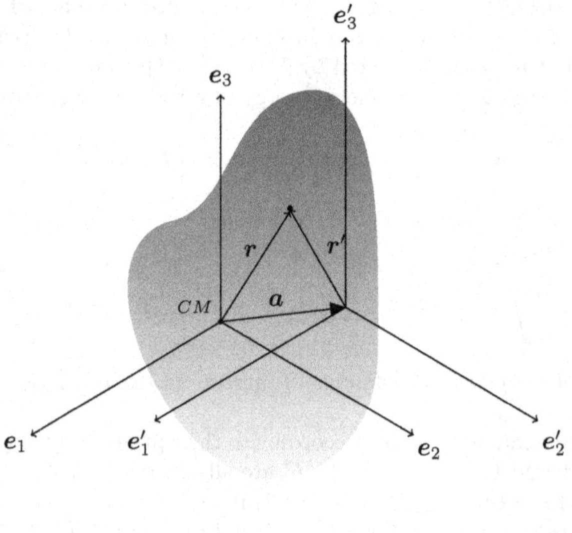

Fig. 20.5

mass and having a position vector $\boldsymbol{a}$. Since $\boldsymbol{r} = \boldsymbol{r}' + \boldsymbol{a}$, we can write the inertia tensor $(I')^{ij}$ with respect to the $\{\mathbf{e}'_i\}$ frame as follows.

$$\begin{aligned}(I')^{ij} &= \int d^3r\,\rho\,\left(\delta^{ij}(x')^l(x')_l - (x')^i(x')^j\right)\\ &= \int d^3r\,\rho\,\left\{\delta^{ij}(x^l - a^l)(x_l - a_l) - (x^i - a^i)(x^j - a^j)\right\}\,.\end{aligned} \tag{20.11}$$

Simplifying the right-hand side of the second equality, and remembering that $\int d^3r\,\rho\,x^i = 0$ (since the origin O is at the center of mass), we have

$$(I')^{ij} = I^{ij} + M\left(a^2\delta^{ij} - a^i a^j\right)\,, \tag{20.12}$$

where M is the total mass of the rigid body. The above result implies the **parallel axis theorem** for moments of inertia of a rigid body when the two frames are parallelly displaced from each other.

Problem 20.2 Show that (20.12) follows from (20.11) and justify that it does in fact imply the parallel axes theorem for the calculation of moments of inertia.

Problem 20.3 In this problem we will investigate the effect of the angular momentum of a bicycle's wheels on the stability of the bicycle and the rider. When upright, the center-of-mass of a bike and rider is at a height $2L$ above the ground. Each of the two wheels has mass m, radius L, and moment of inertia mL^2 about an axis through the center of the wheel. The bicycle travels with velocity V in a circular path of radius R. Assume that $\omega \gg \Omega$, where ω is the angular speed of the wheels and Ω the angular speed of the bike around the circular path. Show that the bicycle leans from the vertical at an angle ϕ given by

$$\tan\phi \approx \frac{V^2}{Rg}\left(1 + \frac{m}{M}\right)\,,$$

where M is the total mass of the bike and the rider.
Hint: Calculate the total torque about the center of mass and set that equal to the time rate of change of the angular momentum with respect to the center of mass.

Let us now calculate the angular momentum $\boldsymbol{L}$ of a rigid body with respect to a body-fixed frame $\{\mathbf{e}_1, \mathbf{e}_2, \mathbf{e}_3\}$ whose origin is located at a fixed-point of the body, not necessarily the center of mass. We have

$$\boldsymbol{L} = \int d^3r\,\rho\,\boldsymbol{r}\times\boldsymbol{v} = \int d^3r\,\rho\,\boldsymbol{r}\times(\boldsymbol{\omega}\times\boldsymbol{r}) = \int d^3r\,\rho\,\left(r^2\boldsymbol{\omega} - \boldsymbol{r}(\boldsymbol{r}\cdot\boldsymbol{\omega})\right)\,. \tag{20.13}$$

With

$$\boldsymbol{\omega} = \omega_1\,\mathbf{e}_1 + \omega_2\,\mathbf{e}_2 + \omega_3\,\mathbf{e}_3\,, \tag{20.14}$$

we can write in component form (using tensorial notation),

$$L^i = \int d^3r\, \rho\, (x^l x_l \omega^i - x^i x^k \omega_k) = \int d^3r\, \rho\, (\delta^{ik} x^l x_l - x^i x^k)\, \omega_k\,. \tag{20.15}$$

Equation (20.7) then implies that

$$L^i = I^{ik} \omega_k\,, \tag{20.16}$$

where I^{ik} are the components of the inertia tensor with respect to the body-fixed frame $\{\boldsymbol{e}_1, \boldsymbol{e}_2, \boldsymbol{e}_3\}$, which, again, is not necessarily a CM frame.

If $\{\boldsymbol{e}_1, \boldsymbol{e}_2, \boldsymbol{e}3\}$ is a principal frame,

$$L^1 = I^1 \omega_1\,, \quad L^2 = I^2 \omega_2\,, \quad L^3 = I^3 \omega_3\,, \tag{20.17}$$

or

$$\boldsymbol{L} = I^1 \omega_1 \boldsymbol{e}_1 + I^2 \omega_2 \boldsymbol{e}_2 + I^3 \omega_3 \boldsymbol{e}_3\,. \tag{20.18}$$

For a spherical top ($I^1 = I^2 = I^3 = I$), the above equation assumes its simplest form:

$$\boldsymbol{L} = I \boldsymbol{\omega}\,. \tag{20.19}$$

The dynamics of a rigid body is determined by the equation of motion (19.20). Writing $\boldsymbol{L} = L^i \boldsymbol{e}_i$, the equation of motion can be written as

$$\frac{d\boldsymbol{L}}{dt} = \dot{\boldsymbol{L}} + \boldsymbol{\omega} \times \boldsymbol{L} = \boldsymbol{\tau}\,, \tag{20.20}$$

where $\boldsymbol{\tau}$ is the net torque due to external forces [cf. (19.17)], and

$$\dot{\boldsymbol{L}} \equiv \frac{dL^1}{dt} \boldsymbol{e}_1 + \frac{dL^2}{dt} \boldsymbol{e}_2 + \frac{dL^3}{dt} \boldsymbol{e}_3\,. \tag{20.21}$$

Evaluating the vector cross product in (20.20) and using (20.17) for $\boldsymbol{L}$, (20.19) can be written in terms of components (with respect to principal axes) as

$$\begin{aligned} I^1 \dot{\omega}_1 + \omega_2 L^3 - \omega_3 L^2 &= \tau^1\,, \\ I^2 \dot{\omega}_2 + \omega_3 L^1 - \omega_1 L^3 &= \tau^2\,, \\ I^3 \dot{\omega}_3 + \omega_1 L^2 - \omega_2 L^1 &= \tau^3\,. \end{aligned} \tag{20.22}$$

Using (20.17) on the left-hand sides of the above equations for the L^i, the equations can be rewritten in the standard form:

$$\begin{aligned} I^1 \dot{\omega}_1 + (I^3 - I^2) \omega_2 \omega_3 &= \tau^1\,, \\ I^2 \dot{\omega}_2 + (I^1 - I^3) \omega_3 \omega_1 &= \tau^2\,, \\ I^3 \dot{\omega}_3 + (I^2 - I^1) \omega_1 \omega_2 &= \tau^3\,. \end{aligned} \tag{20.23}$$

The above equations are known as the **Euler's equations for rigid-body dynamics**. These constitute a set of three coupled non-linear differential equations, and are in general difficult to solve, even if $\boldsymbol{\tau}$ is constant. The difficulty arises in part from the fact that the components of the net torque τ^1, τ^2, τ^3 are with respect to the body-fixed frame and depend on the unknown orientation of the rigid body. To see the mathematical consequences of this fact more explicitly, we consider the general case of an asymmetric top with a space-fixed point O under the action of the gravitational force $M\boldsymbol{g}$, which acts at the CM of the body. Let $\boldsymbol{z}$ be the unit vector pointing from O along the vertically upward direction, and $\boldsymbol{R}$ be the position vector of the CM with respect to the point O. We express the components of these vectors with respect to the body-fixed principal frame $\{\boldsymbol{e}_1, \boldsymbol{e}_2, \boldsymbol{e}_3\}$ with origin at O as follows:

$$\boldsymbol{z} = z_1\,\boldsymbol{e}_1 + z_2\,\boldsymbol{e}_2 + z_3\,\boldsymbol{e}_3\,, \qquad (\boldsymbol{z}\cdot\boldsymbol{z} = 1)\,, \tag{20.24}$$

$$\boldsymbol{R} = R_1\,\boldsymbol{e}_1 + R_2\,\boldsymbol{e}_2 + R_3\,\boldsymbol{e}_3\,. \tag{20.25}$$

Note that R_1, R_2, R_3 are constants while z_1, z_2, z_3 are functions of the time t as the rigid body (with fixed point O) moves about. We then have

$$\boldsymbol{\tau} = \boldsymbol{R}\times M\boldsymbol{g} = -Mg\,\boldsymbol{R}\times\boldsymbol{z} = Mg\,\boldsymbol{z}\times\boldsymbol{R}\,. \tag{20.26}$$

So

$$\tau^1 = Mg\,(z_2R_3 - z_3R_2)\,,\quad \tau^2 = Mg\,(z_3R_1 - z_1R_3)\,,\quad \tau^3 = Mg\,(z_1R_2 - z_2R_1)\,. \tag{20.27}$$

Thus the three Euler equations (20.23) actually involve six unknowns: $\omega_1, \omega_2, \omega_3$, z_1, z_2, z_3, to be solved as functions of time t . To be able to solve for them we need three more independent equations relating the ω's to the z's. These can be obtained from (20.24) as follows. We have, since $\boldsymbol{z}$ is a space-fixed vector, and on recalling (4.8), (4.10) and (4.11),

$$\begin{aligned} \frac{d\boldsymbol{z}}{dt} = 0 &= \dot{z}_i\,\boldsymbol{e}_i + z_i\frac{D\boldsymbol{e}_i}{dt} = \dot{z}_i\,\boldsymbol{e}_i + z_i\frac{\omega_i^j}{dt}\,\boldsymbol{e}_j = \dot{z}_i\,\boldsymbol{e}_i + z_i\varphi_i^j\,\boldsymbol{e}_j \\ &= \dot{z}_i\,\boldsymbol{e}_i + \epsilon_i{}^j{}_k z_i\omega_k\,\boldsymbol{e}_j = (\dot{z}_i + \epsilon^i{}_{kj}\omega_k z_j)\,\boldsymbol{e}_i\,. \end{aligned} \tag{20.28}$$

In the above equation all repeated indices are summed over. Thus

$$\dot{z}_i = \sum_{kj}\epsilon_{ijk}\,\omega_k z_j \qquad (i = 1,2,3)\,. \tag{20.29}$$

These are the three additional equations that we sought. The six unknowns ω_i and z_i are not independent. In addition to the constraint

$$\boldsymbol{z}\cdot\boldsymbol{z} = (z_1)^2 + (z_2)^2 + (z_3)^2 = 1 \tag{20.30}$$

(imposed by the fact that $\boldsymbol{z}$ is a unit vector), there is an additional constraint:

$$\boldsymbol{z}\cdot\boldsymbol{L} = I^1 z_1\omega_1 + I^2 z_2\omega_2 + I^3 z_3\omega_3 = \text{constant}\,. \tag{20.31}$$

This expresses the fact that the projection of the angular momentum $\boldsymbol{L}$ along the direction of the force $M\boldsymbol{g} = -Mg\,\boldsymbol{z}$ must be a constant of motion, as follows from

$$\frac{d}{dt}(\boldsymbol{z}\cdot\boldsymbol{L}) = \boldsymbol{z}\cdot\frac{d\boldsymbol{L}}{dt} = \boldsymbol{z}\cdot\boldsymbol{\tau} = \boldsymbol{z}\cdot(\boldsymbol{R}\times M\boldsymbol{g}) = 0\,. \tag{20.32}$$

The three Euler equations of motion (20.23) for ω_i with the right-hand sides given by (20.27), together with the three additional equations (20.29) for z_i, form a system of six coupled, non-linear, first-order differential equations to be solved for the six unknown functions $\omega_i(t)$ and $z_i(t)$. They are known as the **Euler-Poisson equations of motion** for a spinning top with one fixed point in space, and are of the form (18.7). On defining $\mathcal{I}$ to be the 3×3 diagonal matrix given by

$$\mathcal{I}_{ij} = \delta_{ij} I^i / Mg \,, \tag{20.33}$$

and further defining

$$\dot{\boldsymbol{\omega}} \equiv \dot{\omega}_1\, \boldsymbol{e}_1 + \dot{\omega}_2\, \boldsymbol{e}_2 + \dot{\omega}_3\, \boldsymbol{e}_3 \,, \tag{20.34}$$

$$\dot{\boldsymbol{z}} \equiv \dot{z}_1\, \boldsymbol{e}_1 + \dot{z}_2\, \boldsymbol{e}_2 + \dot{z}_3\, \boldsymbol{e}_3 \,, \tag{20.35}$$

the Euler-Poisson equations can be written compactly as the following two coupled vector equations:

$$\mathcal{I}\dot{\boldsymbol{\omega}} = (\mathcal{I}\boldsymbol{\omega}) \times \boldsymbol{\omega} + \boldsymbol{z} \times \boldsymbol{R} \,, \qquad \dot{\boldsymbol{z}} = \boldsymbol{z} \times \boldsymbol{\omega} \,, \tag{20.36}$$

where $\mathcal{I}\boldsymbol{\omega}$ is understood to be the vector given by $\left(\sum_j \mathcal{I}_{ij}\omega_j\right) \boldsymbol{e}_i$. The solutions of the Euler-Poisson equations (20.36) depend on six adjustable parameters: the three principal moments I^1, I^2, I^3 and the three center-of-mass coordinates R_1, R_2, R_3.

Problem 20.4 Show explicitly that the Euler-Poisson equations in the vector equation form (20.36) follow from the Euler equations (20.23) and the subsidiary equations (20.29) for $\dot{z}_i$.

Problem 20.5 Show that the form of the Euler-Poisson equations (20.36) are also valid for a more general situation than that discussed in the text: the case when the rigid body is acted upon by an external force field that has *axial symmetry*, where the axis of symmetry is space-fixed. Let $\boldsymbol{z}$ be the unit vector along the symmetry axis of the force field, which is assumed to be derivable from a potential function $V(z_1, z_2, z_3)$. Show that in this case the Euler-Poisson equations take the form

$$\mathcal{I}\dot{\boldsymbol{\omega}} = (\mathcal{I}\boldsymbol{\omega}) \times \boldsymbol{\omega} + \boldsymbol{z} \times \partial_{\boldsymbol{z}} V \,, \qquad \dot{\boldsymbol{z}} = \boldsymbol{z} \times \boldsymbol{\omega} \,, \tag{20.37}$$

where

$$\partial_{\boldsymbol{z}} V \equiv \frac{\partial V}{\partial z_1}\, \boldsymbol{e}_1 + \frac{\partial V}{\partial z_2}\, \boldsymbol{e}_2 + \frac{\partial V}{\partial z_3}\, \boldsymbol{e}_3 \,. \tag{20.38}$$

We will conclude this chapter with the study of some approximate solutions of the Euler equations of motion: we will first use (20.20) directly to study the

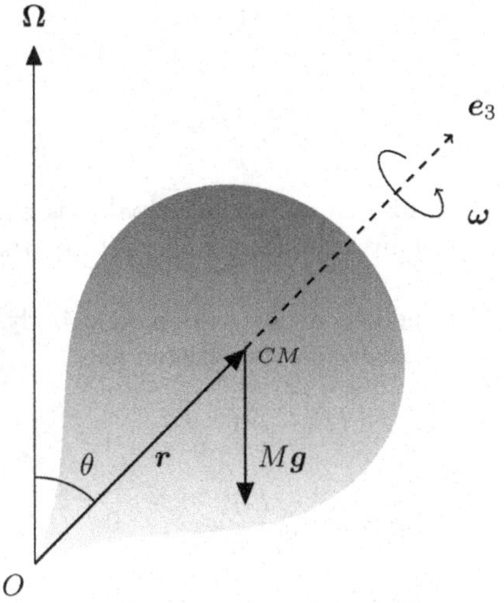

Fig. 20.6

special case of the dynamics of a fast spinning symmetric top, and then use the Euler equations (20.23) to analyze the free rotations of rigid bodies (external torque vanishes). The solutions for more general cases involving external forces will be discussed in Chapter 35 using the Hamiltonian formulation.

Consider a spinning symmetric top ($I^1 = I^2$) pivoted at an apex O (which is not the center of mass) and under the action of gravity (Fig. 20.6). Suppose the spinning axis (axis of rotation) is initially along the principal axis $\boldsymbol{e}_3$, and the spinning is fast enough so that

$$Mgr \ll I^3(\omega_3)^2 \,. \tag{20.39}$$

In other words, the torque exerted by the gravitational force is negligible in comparison with the rotational kinetic energy of the top. Then ω_1 and ω_2 remain small at all times and by (20.17),

$$\boldsymbol{L} \sim I^3\omega_3\,\boldsymbol{e}_3 \,, \tag{20.40}$$

with $\omega_3 \sim$ constant. Equation (20.19) then implies that

$$\begin{aligned} \frac{d\boldsymbol{L}}{dt} &\sim I^3\omega_3\,\frac{d\boldsymbol{e}_3}{dt} = \boldsymbol{\tau}_{gravitational} \\ &= \boldsymbol{r}\times M\boldsymbol{g} = Mr\,\boldsymbol{e}_3\times\boldsymbol{g} = -Mr\,\boldsymbol{g}\times\boldsymbol{e}_3 \,. \end{aligned} \tag{20.41}$$

From the above two equations we obtain

$$\frac{de_3}{dt} \sim \boldsymbol{\Omega} \times \boldsymbol{e}_3 \,, \tag{20.42}$$

where

$$\boldsymbol{\Omega} \equiv -\frac{Mr\boldsymbol{g}}{I^3\omega_3} \,. \tag{20.43}$$

Thus while the top is spinning about its principal axis $\boldsymbol{e}_3$, that axis precesses around the upward vertical direction (along $\boldsymbol{\Omega}$) with precessional angular speed Ω.

We will now consider the free rotations of a rigid body. In this case $\tau^i = 0$ and the Euler's equations (20.22) can be written as

$$\begin{aligned} \dot{\omega}_1 + \frac{(I^3 - I^2)}{I^1}\omega_2\omega_3 = 0 \,, \\ \dot{\omega}_2 + \frac{(I^1 - I^3)}{I^2}\omega_3\omega_1 = 0 \,, \\ \dot{\omega}_3 + \frac{(I^2 - I^1)}{I^3}\omega_1\omega_2 = 0 \,. \end{aligned} \tag{20.44}$$

For the case of a symmetric top ($I^1 = I^2$), the third of the above equations immediately implies that $\omega_3 =$ constant. The first two equations can then be recast as

$$\frac{d\omega_1}{dt} = -\Omega\omega_2 \qquad \frac{d\omega_2}{dt} = \Omega\omega_1 \,, \tag{20.45}$$

where

$$\Omega \equiv \left(\frac{I^3 - I^1}{I^1}\right)\omega_3 = \text{constant} \,. \tag{20.46}$$

To solve the coupled equations (20.29), we use the same trick employed in the solutions of (13.27). In other words, let $\omega \equiv \omega_1 + i\omega_2$. Then the coupled equations (20.29) imply that

$$\frac{d\omega}{dt} = i\Omega\omega \,, \tag{20.47}$$

whose solution is simply $\omega(t) = A\,\exp(i\Omega t)$, or

$$\omega_1(t) = A\cos(\Omega t) \,, \qquad \omega_2(t) = A\sin(\Omega t) \,, \tag{20.48}$$

where $A = \sqrt{(\omega_1)^2 + (\omega_2)^2}$ = a real constant. Thus the projection of $\boldsymbol{\omega}$ on the $\boldsymbol{e}_1 - \boldsymbol{e}_2$ plane rotates about the $\boldsymbol{e}_3$ axis with the constant angular speed Ω (Fig. 20.7). Taking into consideration that $\boldsymbol{\omega}$ also has a constant component ω_3 along the $\boldsymbol{e}_3$ axis, we have the picture that the angular velocity vector $\boldsymbol{\omega}$ precesses around $\boldsymbol{e}_3$ with the constant angular speed Ω (see Fig. 20.8). Since the angular momentum vector $\boldsymbol{L}$ is given by

$$\boldsymbol{L} = I^1(\omega_1\boldsymbol{e}_1 + \omega_2\boldsymbol{e}_2) + I^3\omega_3\boldsymbol{e}_3 \,, \tag{20.49}$$

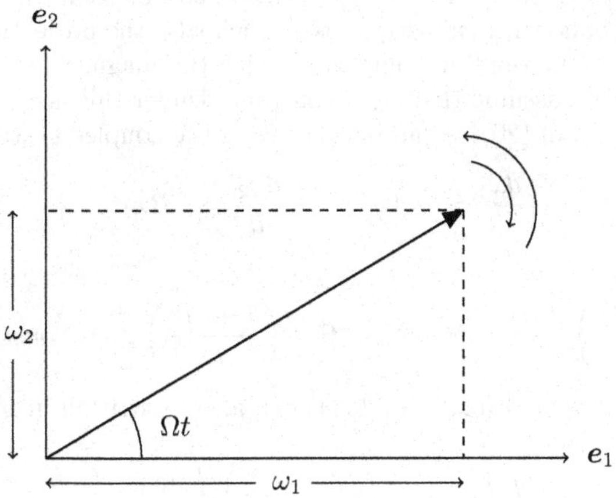

Fig. 20.7

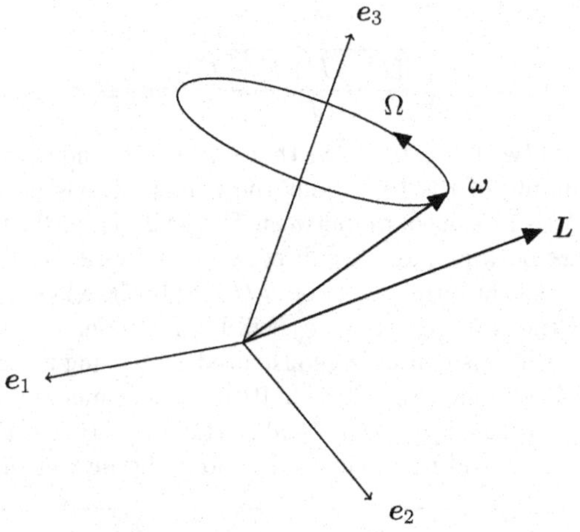

Fig. 20.8

(remember that we are considering a symmetrical top with $I^1 = I^2$), $\boldsymbol{L}$ also precesses around the symmetry axis $\boldsymbol{e}_3$ with the same angular speed Ω (Fig. 20.8).

Let us finally study the free rotations of an asymmetric top ($I^1 \neq I^2 \neq I^3$) using Euler's equations in the form (20.44). We assume that initially the top is made to spin around the symmetry axis $\boldsymbol{e}_3$ such that the projection of $\boldsymbol{\omega}$ on the $\boldsymbol{e}_1 - \boldsymbol{e}_2$ plane is very small in comparison with the magnitude of $\boldsymbol{\omega}$. For small time t we can then assume that $\omega_3 \sim$ constant. Under this last assumption the first two equations of (20.44) can be written as the coupled system

$$\frac{d\omega_1}{dt} = -A\omega_2 \,, \qquad \frac{d\omega_2}{dt} = -B\omega_1 \,, \tag{20.50}$$

where

$$A \equiv \left(\frac{I^3 - I^2}{I^1}\right)\omega_3 \sim \text{constant} \,, \quad B \equiv \left(\frac{I^1 - I^3}{I^2}\right)\omega_3 \sim \text{constant} \,, \tag{20.51}$$

and in general $A \neq B$. Equation (20.34) can also be written in matrix form:

$$\frac{d}{dt}\begin{pmatrix}\omega_1 \\ \omega_2\end{pmatrix} = \begin{pmatrix} 0 & -A \\ -B & 0\end{pmatrix}\begin{pmatrix}\omega_1 \\ \omega_2\end{pmatrix} \,. \tag{20.52}$$

Regardless of the values of A and B, the eigenvalues of the 2×2 matrix on the right-hand side of the above equation are distinct. According to Theorem 18.6, the dynamical system (20.36) then belongs to either case (1) or case (2) of that Theorem.

If I^3 is either the largest or the smallest of the principal moments (Fig. 20.9), then $\lambda^2 = AB < 0$ and

$$\lambda_{\pm} = \pm i\sqrt{\frac{|(I^3 - I^2)(I^1 - I^3)|}{I^1 I^2}}\,\omega_3 \equiv \pm ib \,. \tag{20.53}$$

In this case, (18.21) (with $a = 0$) shows that both $\omega_1(t)$ and $\omega_2(t)$ are oscillatory, and thus the spinning of the top about the principal axis $\boldsymbol{e}_3$ is stable. If, on the other hand, I^3 is intermediate between I^1 and I^2 (Fig. 20.10), $AB > 0$ and the eigenvalues are both real and are directly given by (20.54) below. Equation (18.20) then implies that either $\omega_1(t)$ or $\omega_2(t)$, or both, will exhibit exponential growth, and thus the spinning about $\boldsymbol{e}_3$ will be unstable.

The above conclusions can be demonstrated by tossing a rectangular object (such as a book) in the air (Fig. 20.11). If the object has uniform mass distribution, it is not hard to get it to spin stably (for short times) about its longest or shortest axis; but it will not spin stably about the intermediate one.

Problem 20.6 Show that the eigenvalues λ of the 2×2 matrix in (20.52) are given by $\lambda^2 = AB$, or

$$\lambda_{\pm} = \pm\sqrt{\frac{(I^3 - I^2)(I^1 - I^3)}{I^1 I^2}}\,\omega_3 \,. \tag{20.54}$$

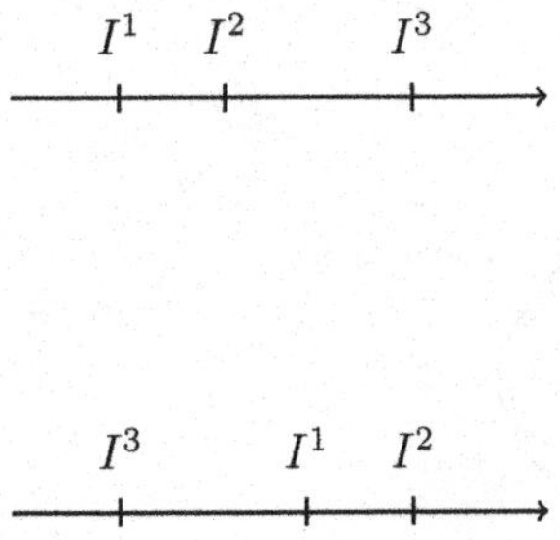

Fig. 20.9

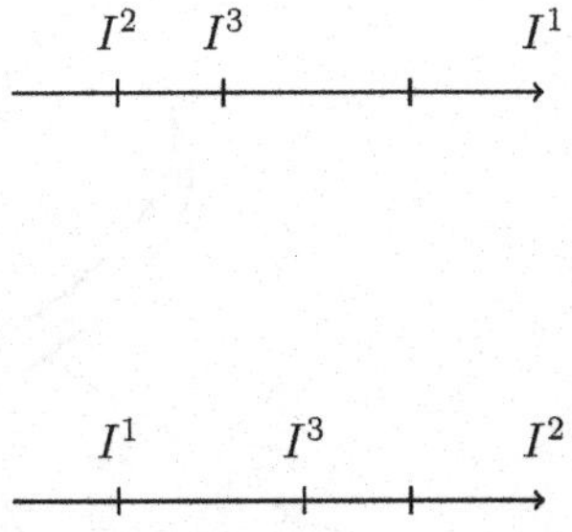

Fig. 20.10

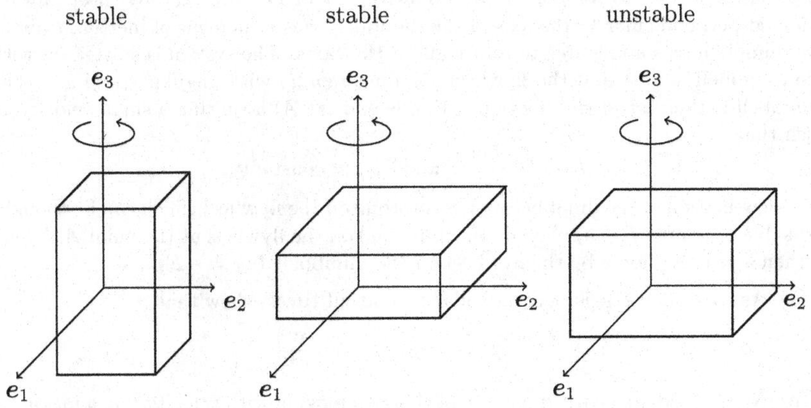

Fig. 20.11

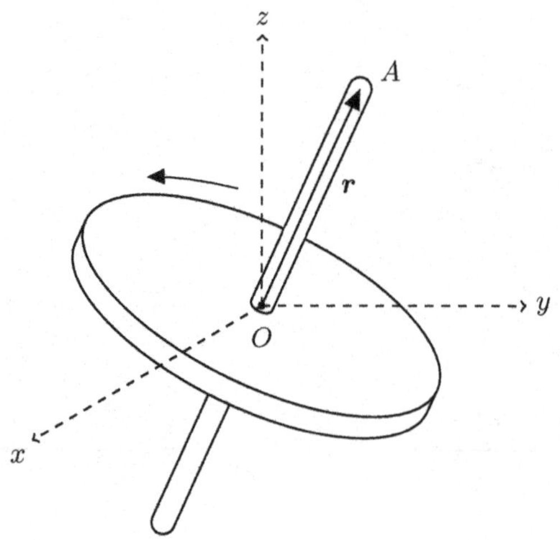

Fig. 20.12

Problem 20.7 Consider a dumbbell-flywheel system (see Fig. 20.12) with its CM fixed at the origin O of some inertial frame. The dumbbell, with a moment of inertia I_d about a normal to its axis, is free to swivel about O; while the flywheel (always oriented perpendicular to the axis of the dumbbell) has a moment of inertia I_f about the dumbbell axis and is free to rotate about that axis. The system is started off with the dumbbell at rest and the flywheel spinning *rapidly* with angular speed ω_f. The dumbbell is then set in *slow* motion with one end (at A) acquiring a small velocity $\boldsymbol{v}$, such that

$$I_f\omega_f \gg \frac{I_d v}{r} \quad \text{and} \quad \omega_f \approx \text{constant} .$$

The slow motion of the dumbbell causes a torque on the flywheel which can be thought of as due to a force $\boldsymbol{F}_{fd}$ applied by the dumbbell on the flywheel at the point A. There is then a reaction force by the flywheel on the dumbbell $\boldsymbol{F}_{df} = -\boldsymbol{F}_{fd}$.

(a) Assume that $\boldsymbol{F}_{df}$ is perpendicular to $\boldsymbol{r}$ at all times, show that

$$\boldsymbol{F}_{df} = \boldsymbol{v} \times \left(-I_f\omega_f \frac{\boldsymbol{r}}{r^3}\right) .$$

(b) Now consider a charged particle of charge e moving with velocity $\boldsymbol{v}$ in a magnetic field $\boldsymbol{B}$ due to a **magnetic monopole** of magnetic charge M (so that $\boldsymbol{B}$ is given by Coulomb's law). Write an expression for the Lorentz force $\boldsymbol{F}_L$ on the charged particle. Compare this expression with that for $\boldsymbol{F}_{df}$ given in part (a). What role does $I_f\omega_f$ play in the electromagnetic scenario?

Chapter 21

Topology and Systems with Holonomic Constraints: Homology and de Rham Cohomology

The fundamental importance of topology in many areas of physics has been increasingly recognized, by both physicists and mathematicians, in recent decades. Examples are geometrical (Berry) phases in classical and quantum physics, the quantum Hall effect, and many other problems in quantum field theory and condensed matter physics. In this chapter we will introduce some basic topological notions relevant to many of these physics problems, specifically the twin concepts of *homology* and *de-Rham cohomology*, within a classical mechanical context. We will attempt to do so in a way that largely bypasses formal mathematical developments, using the physically intuitive and familiar example of the mechanical work done by force fields in systems with holonomic constraints. This example will also demonstrate, yet again, how differential forms, which are needed for the definition of the *de Rham groups*, can be used with great advantage in physics calculations.

Let us first consider particles moving in 3-dimensional Euclidean space ($\mathbb{R}^3$) under the action of a force field $\boldsymbol{F}(\boldsymbol{r})$. The work done by $\boldsymbol{F}(\boldsymbol{r})$ along a path C in $\mathbb{R}^3$ is given by [cf. (12.31)]

$$W = \int_C \boldsymbol{F}(\boldsymbol{r}) \cdot d\boldsymbol{r} \,. \tag{21.1}$$

In terms of Cartesian coordinates the above equation can be written as

$$W = \int_C F_x dx + F_y dy + F_z dz \,. \tag{21.2}$$

The 1-form

$$w = F_x(x,y,z)\,dx + F_y(x,y,z)\,dy + F_z(x,y,z)\,dz \tag{21.3}$$

is actually independent of the choice of (local) coordinates (Cartesian, spherical, or other kinds), and has been referred to as the *work 1-form* [cf. Chapter 8, following (8.4)]. The work done along a path C then appears as

$$W = \int_C w\,. \tag{21.4}$$

Motions in other than the 3-dimensional Euclidean space $\mathbb{R}^3$ are readily realized (and visualized) as constrained motions in submanifolds of $\mathbb{R}^3$ [cf. Chapter 10]. If the constraints are holonomic, the equations of motion can be cast entirely by means of *D'Alambert's Principle* in terms of the local coordinates q^i of the submanifold, which can serve as *generalized coordinates* of the mechanical system. [We will not use this principle explicitly here; the reader may refer to standard texts, such as Goldstein 1965, for details, and also the discussion following (25.57). The notion of generalized coordinates will be discussed more fully in Chapter 31, where the Lagrangian formalism of classical mechanics will be introduced.] For example, motion confined on the surface of a 2-dimensional sphere (S^2) can be completely described by the two generalized coordinates θ and ϕ (the polar and azimuth angles, respectively). One can then imagine that the effects of the forces of constraint are partially encoded in the topology of the submanifold. In this context, the work 1-form can be expressed as

$$w = \sum_{i=1}^{n} F_i(q^1,\ldots,q^n)\,dq^i\,, \tag{21.5}$$

where n is the dimension of the submanifold, and F_i denotes the i-th component of the *generalized force*. In this chapter we will be particularly concerned with the cases where $n = 1$ or 2; but the rules for the calculation of exterior derivatives introduced in Chapter 6 will allow us to treat the general case for arbitrary n. We recognize that with the work 1-form w given by (21.5), its exterior derivative is, according to (7.8),

$$dw = (\nabla \times \boldsymbol{F}) \cdot d\boldsymbol{a}\,, \tag{21.6}$$

where $d\boldsymbol{a}$ is an oriented area element in $\mathbb{R}^3$. Furthermore, (7.12) and the fact that $d^2 = 0$ (Theorem 6.2) imply the familiar fact in vector calculus that the divergence of a curl always vanishes.

At this point we introduce two important notions concerning differential forms – *closed* and *exact forms.*

Definition 21.1. *If a differential form ω (of arbitrary order) is such that $d\omega = 0$, then it is said to be a* ***closed form.*** *If an n-form ω is such that $\omega = d\lambda$, where λ is an $(n-1)$-form, then ω is said to be an* ***exact form.***

Thus the work one-form corresponding to a curl-free force field must be a closed form; and a *conservative force field* must give rise to an exact work one-form w, such that

$$w = d\lambda = -dU \,, \tag{21.7}$$

where $U = -\lambda$ is the *potential energy function* corresponding to the conservative force field. Equivalently, the condition (21.7) can be given as

$$\oint_C w = 0 \tag{21.8}$$

for *any* closed path C.

Theorem 6.2 ($d^2 = 0$) and (21.7) imply that an exact form is always closed. In other words, a conservative force field is always curl-free. But is the converse true? Can we find curl-free force fields and closed loops in a manifold such that the loop integral, or work done by these fields, over the loops is non-zero? We shall see that the answer to the last question is not automatically in the negative, as casually stated in many textbooks, and depends critically on the topology of the manifold on which the force fields are defined.

First consider some familiar force fields in Euclidean 3-space ($\mathbb{R}^3$) such as the constant field $\boldsymbol{F}(\boldsymbol{r}) = m\boldsymbol{g}$ (the gravitational field near the surface of the earth) and the central field $\boldsymbol{F}(\boldsymbol{r}) = \alpha\boldsymbol{r}/r^3$ (the Coulomb or Newtonian gravitational field). In the first case the potential energy function $U(\boldsymbol{r})$ as defined by (21.7) is given by $U(\boldsymbol{r}) = -m\boldsymbol{g}\cdot\boldsymbol{r}$. Thus $w = -dU = m\boldsymbol{g}\cdot d\boldsymbol{r}$ and

$$\oint_C w = m\boldsymbol{g}\cdot\oint_C d\boldsymbol{r} = 0 \,. \tag{21.9}$$

In the second case, because of the singularity at $\boldsymbol{r} = 0$, the underlying manifold is considered to be $M = \mathbb{R}^3 - \{0\}$ (Euclidean 3-space with a hole at the origin). For this field we have

$$w = \boldsymbol{F}\cdot d\boldsymbol{r} = \frac{\alpha}{r^3}\,\boldsymbol{r}\cdot d\boldsymbol{r} = d\left(-\frac{\alpha}{r}\right) = -dU \,, \tag{21.10}$$

where the potential function is given by

$$U(\boldsymbol{r}) = \frac{\alpha}{r} \,, \qquad (r \neq 0) \,. \tag{21.11}$$

Thus,

$$\oint_C w = -\oint_C dU = 0 \,, \tag{21.12}$$

where C is any closed curve in $\mathbb{R}^3 - \{0\}$ (that is, not crossing the origin). Equations (21.9) and (21.12) indicate that the above two examples both represent conservative force fields. It is also clear that, in both cases, $dw = (\nabla\times\boldsymbol{F})\cdot d\boldsymbol{a} = 0$, that is, the force fields are curl-free.

Next consider a general curl-free force field $\boldsymbol{F}(\boldsymbol{r})$ defined on either $\mathbb{R}^3$, $\mathbb{R}^3 - \{0\}$, the two-dimensional plane $\mathbb{R}^2$, or the two-dimensional sphere S^2.

As discussed above, the curl-free condition implies that the corresponding work one-form $w = \boldsymbol{F}(\boldsymbol{r}) \cdot d\boldsymbol{r}$ satisfies $dw = 0$ (w is closed). Since in all these spaces any closed path C is the boundary of a two-dimensional region A (a *topological* property that we express by writing $C = \partial A$, where ∂ is the boundary operator), we have

$$\oint_C \boldsymbol{F} \cdot d\boldsymbol{r} = \int_{\partial A} \boldsymbol{F} \cdot d\boldsymbol{r} = \int_A \nabla \times \boldsymbol{F} \cdot d\boldsymbol{a} = 0 \,, \tag{21.13}$$

where in the second equality we have used the Stokes theorem in vector calculus, and the last follows from the assumption that $\boldsymbol{F}$ is curl-free. Thus the work done along an arbitrary closed path vanishes, so that there must exist some potential function $U(\boldsymbol{r})$ such that $w = -dU$ (w is exact). The above argument shows that *in a space where any closed path is the boundary of a surface, the curl of a vector field vanishes if and only if the field is expressible as the gradient of a scalar potential, that is, if and only if the field is conservative.* Stated in terms of differential forms we have the following equivalent mathematical fact:

Theorem 21.1. *Suppose w is a differential one-form defined on a manifold M and every closed path in M is the boundary of some region in M, then w is closed if and only if it is exact.*

The important point to note in the above theorem is that a topological (global) property of the manifold on which a differential form is defined determines the properties of the differential form (an object defined in terms of local coordinates).

We will now consider examples of mechanical systems defined on submanifolds of $\mathbb{R}^3$ with curl-free force fields that are not conservative. Let us first consider the one-dimensional case of a force field on S^1 (the 1-sphere), which is topologically just the unit circle in $\mathbb{R}^2$. Choose the local coordinate to be the polar angle θ, with $0 \le \theta \le 2\pi$. Since the top form on this manifold is a 1-form (the manifold being 1-dimensional), all 1-forms are necessarily closed. But there exist many non-exact 1-forms on S^1. For example, consider the work 1-form

$$w = \sin^2\theta \, d\theta \,. \tag{21.14}$$

Clearly $dw = 0$, and, choosing the closed loop C to be S^1 itself,

$$\int_C w = \int_0^{2\pi} \sin^2\theta \, d\theta = \pi \neq 0 \,. \tag{21.15}$$

Thus $w = \sin^2\theta\, d\theta$ is not an exact form. It may be thought that since we can write $\sin^2\theta \, d\theta = dU$, where $U(\theta) = \theta/2 - \sin 2\theta/4$, w should be exact. But this expression for U is only valid *locally*, and fails to be globally valid because $U(0) \neq U(2\pi)$. Thus U is not a 0-form over all of S^1. On the other hand, the work 1-form $w = \cos\theta \, d\theta$ is both closed and exact, as can be verified easily. In this case $w = dU(\theta)$, where $U(\theta) = \sin\theta$. The fact that there are closed forms that are not exact on S^1 is dictated by the topological property that there are closed paths in S^1 which are not boundaries. Indeed, the closed path $C = S^1$ is not the boundary of anything in S^1.

As our next example we consider the work 1-form on the manifold $\mathbb{R}^2 - \{0\}$ (the two-dimensional Euclidean plane with the origin removed) given by

$$w = -\frac{\alpha y}{2\pi r^2}dx + \frac{\alpha x}{2\pi r^2}dy \,, \tag{21.16}$$

where $r^2 = x^2 + y^2$ and α is a positive constant. The corresponding force field (more simply expressed in polar coordinates) is

$$\boldsymbol{F}(\boldsymbol{r}) = \frac{\alpha}{2\pi r}\,\boldsymbol{e}_\theta \,, \tag{21.17}$$

where $\boldsymbol{e}_\theta$ is the unit vector along the positive (anticlockwise) θ-direction. It can be verified straightforwardly by direct calculation, using the rules of exterior differentiation introduced in Chapter 6, that

$$dw = -\frac{\alpha}{2\pi}\frac{\partial}{\partial y}\left(\frac{y}{r^2}\right) dy \wedge dx + \frac{\alpha}{2\pi}\frac{\partial}{\partial x}\left(\frac{x}{r^2}\right) dx \wedge dy = 0 \,. \tag{21.18}$$

Thus w is closed. Choosing the closed path C on our manifold to be the circle $x^2 + y^2 = R^2$, $R \neq 0$, and the orientation of C to be anticlockwise, we can also establish by direct calculation [most easily using (21.17)] that

$$\int_C w = \int_C \boldsymbol{F}(\boldsymbol{r}) \cdot d\boldsymbol{r} = \int_0^{2\pi} \frac{\alpha}{2\pi R} R d\theta = \alpha \neq 0 \,. \tag{21.19}$$

Thus w is not exact. As in the previous example, the fact that there exists a closed form that is not exact is dictated by the topological property that, on the manifold $M = \mathbb{R}^2 - \{0\}$, a circle with its center at the origin is not the boundary of any two-dimensional region in M, since the origin is not part of M.

Our last example for curl-free 1-forms that are not exact will consist of a work 1-form defined on a 2-torus (T^2), topologically an "inner tube" with one hole, or a 2-dimensional sphere with one "handle". The local coordinates can be chosen to be θ_1 and θ_2, with $0 \leq \theta_1, \theta_2 \leq 2\pi$. We can imagine θ_1 to be an angle parametrizing the position of a loop around the inner tube, and θ_2 to be an angle serving the same purpose for a loop around the cross section of the tube. Let a 1-form w be given by

$$w = A(\theta_1)d\theta_1 + B(\theta_2)d\theta_2 \,, \tag{21.20}$$

where $A(\theta_1)$ is a function of only θ_1 and $B(\theta_2)$ is a function of only θ_2, and both are periodic functions with period 2π. Clearly, $dw = 0$ on applying the rules of exterior differentiation, so w is closed. Consider the closed loops C_1 and C_2 characterized above, defined by the conditions $\theta_2 =$ constant and $\theta_1 =$ constant, respectively. We see that

$$\oint_{C_1} w = \int_0^{2\pi} A(\theta_1)\, d\theta_1 \,, \qquad \oint_{C_2} w = \int_0^{2\pi} B(\theta_2)\, d\theta_2 \,. \tag{21.21}$$

Since there are infinitely many periodic functions such that the above integrals do not vanish, we conclude that the work 1-form w is not exact. Once again, this is accounted for by the topology of the 2-torus: neither closed loop C_1 or C_2 is the boundary of any two-dimensional region on T^2.

Having obtained through the above examples a glimpse of the relationship between differential forms and the topologies of the manifolds on which they are defined, we will now engage in a somewhat more detailed (but informal) look at topology through the use of algebraic methods – a subject known as *algebraic topology.* It is outside the scope of this text to give a complete presentation of the relevant introductory notions in algebra. In any case, only some elementary concepts in group theory are required here. The reader may consult any number of standard texts for a quick reference (see, for example, Lam 2006, Bamberg and Sternberg, 1991).

Given a manifold M consider an arbitrary closed 1-form w defined on it. In general, the integrals of w over two different *oriented closed loops* C_1 and C_2 will not be equal to each other. (An oriented loop C, also called a **one-cycle**, is one with a given sense of traversal, so that $-C$ means C traversed in the reverse sense.) If, however, $C_1 - C_2$ is the boundary of some 2-dimensional region $A \subset M$, that is, if $C_1 - C_2 = \partial A$, then

$$\oint_{C_1} w - \oint_{C_2} w = \int_{C_1 - C_2} w = \int_{\partial A} w = \int_A dw = 0\,, \tag{21.22}$$

where the third equality follows from Stokes Theorem (Theorem 8.1) and the fourth from the assumption that w is closed (that is, $dw = 0$). Two closed oriented loops C_1 and C_2 in M are said to be **homologous** if there exists a 2-dimensional region $A \subset M$ such that $C_1 - C_2 = \partial A$. Equation (21.22) shows that an arbitrary closed 1-form yields the same result (a real number) when integrated over homologous loops. Thus an *equivalence class* of homologous loops (denoted by $[C]$, where C is any *representative* in the class) can be considered as a real function on closed 1-forms. The set of equivalence classes of homologous loops in a manifold M has a linear as well as an abelian group structure, and is called the **first homology group** of M with real coefficients, denoted $H_1(M, \mathbb{R})$, when the loops are allowed to be multiplied by real numbers. For a given loop C and a real number a we define the loop aC by the requirement $(aC, w) = a(C, w)$ for all 1-forms w [recall (8.36)]. Thus $H_1(M, \mathbb{R})$ is both a vector space over the reals and an abelian group, with the identity of the group being the equivalence class of all closed loops that are also boundaries. It can be expressed as a *quotient group* as follows:

$$H_1(M, \mathbb{R}) = \frac{Z_1}{B_1}\,, \tag{21.23}$$

where Z_1 is the group of all 1-cycles in M and B_1 is the subgroup of Z_1 consisting of all 1-cycles in M that are also boundaries. For manifolds in which closed loops are always boundaries, our earlier development shows that the result of integrating any closed 1-form over closed loops is always zero. Thus, for these manifolds, $H_1(M, \mathbb{R})$ is the trivial group consisting of only the identity element, in this case the constant real function which maps all closed 1-forms to the real number 0. In particular, we have the following results:

$$H_1(\mathbb{R}^3, \mathbb{R}) = H_1(\mathbb{R}^2, \mathbb{R}) = H_1(S^2, \mathbb{R}) = \{0\} \tag{21.24}$$

for the 3-dimensional Euclidean space $\mathbb{R}^3$, the 2-dimensional Euclidean space $\mathbb{R}^2$, and the 2-dimensional sphere S^2. These results also follow from the intuitive fact that every closed loop in any one of the three manifolds is the boundary of a 2-dimensional region.

For $M = S^1$ (the circle), we recognize intuitively that there is only one kind of closed loops: an integral multiple of complete windings around S^1 itself. Let us represent one such oriented winding by nC, where n is a non-zero integer and C is an oriented winding once around S^1. The only closed loop which is also a boundary is a point in S^1, which can be represented by the integer 0 ($0.C$). We can generalize the coefficients of C to real numbers and represent the group of closed loops in S^1 by aC, where a is an arbitrary real number. It is clear that any two closed loops of the form aC and bC with $a \neq b$ are not homologous since $aC - bC$ cannot be a boundary. We thus have the result that

$$H_1(S^1, \mathbb{R}) = \mathbb{R} , \tag{21.25}$$

that is, *the first homology group of the* 1*-dimensional sphere is isomorphic to the group of real numbers*, where the group multiplication is just ordinary addition.

For $M = \mathbb{R}^2 - \{0\}$ (the 2-dimensional plane with the origin removed), all closed loops can be generated by two kinds of loops: those that encircle the origin and those that do not. A closed loop of the latter kind is obviously a boundary of some region in M, and, as before, can be represented by the number 0. A representative of the former category can again be represented by aC, where a is a real number and C is an oriented closed loop encircling the origin once. Consider two of these loops aC_1 and bC_2, with $a \neq b$. We have

$$C \equiv aC_1 - bC_2 = (a - b)C_1 + b(C_1 - C_2) = (a - b)C_1 + b\partial A , \tag{21.26}$$

where A is a region bounded by $C_1 - C_2$ (in this case an annulus). Since C_1 is not a boundary, neither is C. So aC_1 is not homologous to bC_2. We thus have

$$H_1(\mathbb{R}^2 - \{0\}, \mathbb{R}) = \mathbb{R} . \tag{21.27}$$

For $M = T^2$ (the 2-torus) all representatives of closed loops can be generated by three kinds of loops: a loop on the surface which can be shrunk continuously to a point (one that is the boundary of a region in M), denoted by C_0, and two loops characterized each by constant θ_2 and constant θ_1, denoted by C_1 and C_2, respectively [cf. discussion immediately preceding (21.20)]. The equivalence class $[C_0]$ is represented by the number 0, while a general closed loop nonhomologous to C_0 can be represented as $aC_1 + bC_2$, where a and b are real numbers. Thus a general element in $H_1(T^2, \mathbb{R})$ can be written as $a[C_1] + b[C_2]$; and we have

$$H_1(T^2, \mathbb{R}) = \mathbb{R} \oplus \mathbb{R} , \tag{21.28}$$

which is an abelian group as well as a 2-dimensional real vector space.

We note that while the first homology group $H_1(M, \mathbb{R})$ provides important information on the topology of M, this information is far from complete. Indeed, as seen from (21.24), (21.25) and (21.27), many topological manifolds with distinct topologies share the same homology group.

Now consider a given element $[C] \in H_1(M, \mathbb{R})$, that is, an equivalence class of homologous loops in M to which the loop C belongs. Two arbitrary closed 1-forms w_1 and w_2 defined on M will in general yield different results when integrated over C. However, if $w_1 - w_2$ is exact, that is, if there exists a function U on M such that $w_1 - w_2 = dU$, then,

$$\oint_C w_1 - \oint_C w_2 = \oint_C (w_1 - w_2) = \oint_C dU = 0 \,. \tag{21.29}$$

Two closed 1-forms w_1 and w_2 defined on a manifold M are said to be **cohomologous** if $w_1 - w_2$ is an exact form, that is, if there exists a function U, defined *globally* on M, such that $w_1 - w_2 = dU$. Thus cohomologous 1-forms yield the same result (a real number) when integrated over an arbitrary loop. Analogous to the set of equivalence classes of homologous loops in M, the set of equivalences classes of cohomologous 1-forms on M also has a linear and an abelian group structure. It is called the **first de-Rham cohomology group** of M with real coefficients, denoted by $H^1(M, \mathbb{R})$, with the identity element being the equivalence class of all exact 1-forms on M. In view of our previous discussion on homologous loops, an element in $H_1(M, \mathbb{R})$, that is, an equivalence class of homologous loops in M, can be viewed as a real-valued function on $H^1(M, \mathbb{R})$, that is, a function acting on equivalence classes of cohomologous 1-forms on M. Analogous to Eq. (21.23), $H^1(M, \mathbb{R})$ can also be expressed as a quotient group:

$$H^1(M, \mathbb{R}) = \frac{Z^1}{B^1} \,, \tag{21.30}$$

where Z^1 is the group of all closed 1-forms (1-**cocycles**) defined on M, and B^1 is the subgroup of Z^1 consisting of all closed 1-forms that are also exact. An element of $H^1(M, \mathbb{R})$ is usually written $[w]$, which denotes the equivalence class of cohomologous 1-forms to which w belongs.

We are now ready to state a special case of a deep and powerful mathematical result known as the **de-Rham theorem**:

Theorem 21.2. *The first homology group and the first cohomology group of an arbitrary manifold are dual to each other as vector spaces and are hence isomorphic.*

Thus the analogs of Eqs. (13.24), (13.25), (13.27) and (13.28) are

$$H^1(\mathbb{R}^3, \mathbb{R}) = H^1(\mathbb{R}^2, \mathbb{R}) = H^1(S^2, \mathbb{R}) = \{0\} \,, \tag{21.31}$$

$$H^1(S^1, \mathbb{R}) = H^1(\mathbb{R}^2 - \{0\}, \mathbb{R}) = \mathbb{R} \,, \tag{21.32}$$

$$H^1(T^2, \mathbb{R}) = \mathbb{R} \oplus \mathbb{R} \,. \tag{21.33}$$

The above stated results for H^1 and H_1 provide a glimpse into the deep connection between differential (local) and topological (global) properties of a manifold. The de Rham Theorem in fact goes much further. It turns out that one can define homology and cohomology groups of arbitrary order, and the duality relationship stated above for the first-order groups holds for the corresponding groups of arbitrary order. (For more details, the reader may consult, for example, Singer and Thorpe 1976.)

Chapter 22

Connections on Vector Bundles: Affine Connections on Tangent Bundles

In Chapter 4 we introduced the so-called *affine connection* on the tangent bundle of a differential manifold, and saw how the kinematical notion of the angular velocity was based on it. In Chapter 5 the tangent bundle T_0^1 was generalized to the tensor bundle T_s^r, which was pointed out as a special example of a vector bundle. In this chapter we will formulate the general notion of a *vector bundle* and introduce the notion of the *connection* (covariant derivative) on such a bundle. In the next several chapters we will illustrate by means of specific examples how these mathematical structures can be applied in classical mechanics.

In the abstract formulation of a vector bundle one starts with two spaces E and M, usually taken to be differentiable manifolds, and a real or complex vector space V (although in classical mechanics one works mainly with real vector spaces). Mappings satisfying certain rules are then introduced between these spaces or Cartesian products of these spaces to identify the structure of the bundle, in a manner similar to the construction of the tensor bundle T_s^r [cf. (5.14) to (5.16) and Fig. 5.4]. The details are given in the following definition.

Definition 22.1. *Suppose E and M are differentiable manifolds, $\pi : E \to M$ is a smooth surjective (onto) map, and V is a q-dimensional vector space (either real or complex). If there exist an open covering $\{U, W, Z, \dots\}$ of M and a corresponding set of maps $\{\varphi_U, \varphi_W, \varphi_Z, \dots\}$ that satisfy the following three conditions, then the triplet (E, M, π) (sometimes also denoted by $\pi : E \to M$) is called a (real or complex) q-dimensional* ***vector bundle*** *on M (depending on whether V is real or complex):*

1) Every map $\varphi_U : U \times V \to \pi^{-1}(U)$ is a diffeomorphism; and for any

$x \in U, \boldsymbol{v} \in V$,

$$\pi \circ \varphi_U(x, \boldsymbol{v}) = x \,. \tag{22.1}$$

2) For any $x \in U$ and $\boldsymbol{v} \in V$, the map $\varphi_{U,x} : V \to \pi^{-1}(x)$ defined by

$$\varphi_{U,x}(\boldsymbol{v}) \equiv \varphi_U(x, \boldsymbol{v}) \tag{22.2}$$

is a homeomorphism (bijective and bi-continuous).

3) For any $x \in U \cap W \neq \emptyset$, the map $g_{UW} : U \cap W \to GL(V)$ [$GL(V)$ being the general linear group of the vector space V] defined by

$$g_{UW}(x) \equiv \varphi_{W,x}^{-1} \circ \varphi_{U,x} \in GL(V) \tag{22.3}$$

is smooth.

*E is called the **total space**, M the **base space**, π the **bundle projection map**, V the **typical fiber**, and each φ_U a **local trivialization** of the vector bundle. The functions in the set $\{g_{UW}\}$ are called the **transition functions** of the vector bundle. For any $x \in M$, $E_x \equiv \pi^{-1}(x)$ is called the **fiber** of the vector bundle E over the point x.*

Essentially, a vector bundle is a "bundling" or "glueing" together, by means of the transition functions, of isomorphic vector spaces at different points of the base space in such a way that the "glueing" operation preserves the linear structure of the vector spaces [recall the discussion after (5.16)]. The family of transition functions then determines the *structure* of the vector bundle. The simplest structure is given by that of a *trivial bundle*, defined as follows.

Definition 22.2. *If a vector bundle $\pi : E \to M$ with typical fiber V is globally a product, that is, if $E = M \times V$, then the bundle is said to be **trivial**.*

The projection map $\pi : M \times V \to M$ of a trivial bundle is given by $\pi(x, \boldsymbol{v}) = x$, for any $x \in M$ and $\boldsymbol{v} \in V$. Thus, for any $x \in M$, $\pi^{-1}(x) = \{x\} \times V$ [or $\pi^{-1}(M) = M \times V$], and a trivial bundle is said to possess a **global trivialization**. The existence of the invertible functions φ_U amounts to the fact that *locally, a vector bundle is always a product.*

Problem 22.1 Show that the family of transition functions of a vector bundle satisfies the following properties:

a) For any $x \in U$, $g_{UU}(x) : V \to V$ is the identity map $id : V \to V$.

b) If $x \in U \cap W \cap Z \neq \emptyset$, than

$$g_{UW}(x) \circ g_{WZ}(x) \circ g_{ZU}(x) = id : V \to V \,. \tag{22.4}$$

The commonplace physics notion of a *vector field* [such as a velocity field or an electric field defined on the physical (Euclidean) space $M = \mathbb{R}^3$, for example], is more appropriately defined mathematically within the setting of a vector bundle.

Definition 22.3. *Let $\pi : E \to M$ be a vector bundle, and $s : M \to E$ be a smooth map. If*

$$\pi \circ s = id : M \to M \ , \tag{22.5}$$

then s is called a smooth **section** *(smooth* **vector field***) of the vector bundle.*

Suppose that, for a given smooth section s of the vector bundle $\pi : E \to M$, there exists a point $x \in M$, with $x \in U \cap W \neq \emptyset$, such that $\varphi_U^{-1}(s(x)) = (x, \mathbf{0})$, where $\mathbf{0}$ is the zero-vector in V. Then $s(x) = \varphi_U(x, \mathbf{0}) = \varphi_{U,x}(\mathbf{0})$. It follows that $\varphi_{W,x}^{-1}(s(x)) = (\varphi_{W,x}^{-1} \circ \varphi_{U,x}^{-1})(\mathbf{0}) = [g_{UW}(x)](\mathbf{0}) = \mathbf{0}$, where the last equality is implied by the fact that $g_{UW}(x) \in GL(V)$. Thus $s(x)$ is mapped by every local trivialization to the zero-vector in V. Such a point $x \in M$ is said to be a **singular point** of the smooth section s. We have encountered this notion in the case of a smooth tangent vector field already (see the discussion following Corollary 11.1). A smooth section of a vector bundle which is non-singular everywhere does not always exist for a particular base space M. As in the case of the tangent bundle, these kinds of sections (vector fields) reflect specific topological properties of M.

As discussed at the end of Chapter 5, the set of all smooth sections on the vector bundle $\pi : E \to M$, denoted by $\Gamma(E)$, is a vector space as well as a $C^\infty(M)$-module (sections can be multiplied by smooth functions on M and remain sections). We can now introduce the general notion of the *connection on a vector bundle.* This is the basic mathematical concept behind the physics of gauge fields.

Definition 22.4. *A* **connection** *on a vector bundle $\pi : E \to M$ is a map $D : \Gamma(E) \to \Gamma(T^*(M) \otimes E)$ [where $T^*(M)$ is the cotangent bundle of M] which satisfies the following conditions:*

(1) For any $s_1, s_2 \in \Gamma(E)$,

$$D(s_1 + s_2) = D(s_1) + D(s_2) \ . \tag{22.6}$$

(2) For any $s \in \Gamma(E)$ and $\alpha \in C^\infty(M)$,

$$D(\alpha s) = d\alpha \otimes s + \alpha D s \ . \tag{22.7}$$

The connection D as defined above generalizes the notion of the *covariant derivative* introduced in (4.8), and in fact is called the **covariant derivative** of the section s. Note that if we let $\alpha = -1$, then by condition (2) above, $D(-s) = -Ds$. So D maps a zero section to a zero section and hence must be a linear map.

A fundamental fact on connections is stated in the following theorem (without proof).

Theorem 22.1. *A connection always exists on a vector bundle.*

Now suppose $\boldsymbol{X}$ is a smooth tangent field on M [a smooth section on the tangent bundle $T(M)$]. One can then introduce the notion of the *directional covariant derivative* of a section s of a vector bundle $\pi : E \to M$ along $\boldsymbol{X}$.

Definition 22.5. *Suppose* $\pi : E \to M$ *is a vector bundle,* $\boldsymbol{X}$ *is a smooth tangent field on* M*, and* s *is a smooth section of* E*, then the* **directional covariant derivative** *of the section* s *along* $\boldsymbol{X}$*,* $D_{\boldsymbol{X}}s$*, is given by*

$$D_{\boldsymbol{X}} s \equiv \langle \boldsymbol{X}, Ds \rangle \in \Gamma(E) , \tag{22.8}$$

where $\langle \, , \rangle$ *is the pairing between a vector field in the tangent bundle* $T(M)$ *and one in the cotangent bundle* $T^*(M)$ *[cf. (1.36)].*

It is not difficult to check that the directional covariant derivative satisfies the following properties: If $\boldsymbol{X}, \boldsymbol{Y} \in T(M)$ (smooth tangent fields on M), $s, s_1, s_2 \in \Gamma(E)$ (smooth sections of E), and $\alpha \in C^\infty(M)$ (smooth function on M), then

$$D_{\boldsymbol{X}+\boldsymbol{Y}} s = D_{\boldsymbol{X}} s + D_{\boldsymbol{Y}} s , \tag{22.9a}$$
$$D_{\alpha \boldsymbol{X}} = \alpha D_{\boldsymbol{X}} \, s , \tag{22.9b}$$
$$D_{\boldsymbol{X}}(s_1 + s_2) = D_{\boldsymbol{X}} s_1 + D_{\boldsymbol{X}} s_2 , \tag{22.9c}$$
$$D_{\boldsymbol{X}}(\alpha s) = \langle \boldsymbol{X} , d\alpha \rangle \, s + \alpha \, D_{\boldsymbol{X}} s . \tag{22.9d}$$

Problem 22.2 Verify the above four properties [Eqs. (22.9a) to (22.9d)] of the directional covariant derivative.

Given a connection D on a vector bundle, the directional covariant derivatives $D_{\boldsymbol{X}}$ and $D_{\boldsymbol{Y}}$ for arbitrary tangent vector fields $\boldsymbol{X}$ and $\boldsymbol{Y}$ on the base manifold in general fail to commute. In the next chapter we will introduce the *curvature matrix of two-forms* to measure the degree of non-commutativity of covariant derivatives.

We will now see that the equation defining affine connections on a tangent bundle [Eq. (4.8)] remains essentially the same for the more general connections on vector bundles. Suppose the typical fiber of the vector bundle $\pi : E \to M$ is an q-dimensional vector space and M is an n-dimensional manifold. Consider a coordinate neighborhood $(U; x^i)$ with local coordinates $x^1, \dots, x^n$. Choose q smooth sections $s_1, \dots, s_q$ which are linearly independent everywhere in U. These then constitute a **local frame field** of E on U, and the set $\{dx^i \otimes s_\alpha; 1 \le i \le n, 1 \le \alpha \le q\}$ forms a basis of $T_p^* \otimes E_p$ for all points $p \in U$, E_p being the fiber above the point $p \in U$. We can thus write, locally in U,

$$Ds_\alpha = A^\beta_{\alpha i} \, dx^i \otimes s_\beta \qquad (1 \le \alpha \le q) , \tag{22.10}$$

where $A^\beta_{\alpha i} \in C^\infty(U)$ are smooth functions on U. The one-forms

$$\omega^\beta_\alpha \equiv A^\beta_{\alpha i} \, dx^i \qquad (1 \le \alpha, \beta \le q) , \tag{22.11}$$

constitute the **connection matrix** (ω_α^β) with respect to the local frame field $\{s_\alpha\}$. Written in terms of the connection matrix, Eq. (22.10) appears as

$$Ds_\alpha = \omega_\alpha^\beta \otimes s_\beta \, , \tag{22.12}$$

which is the generalization of (4.8) to vector bundles. Note that in the above three equations we have used latin indices to denote base-msanifold coordinate indices and Greek indices to denote the vector components of the fiber space. Equation (22.12) can also be written as the matrix equation

$$DS = \omega \otimes S \, , \tag{22.13}$$

where $\otimes$ indicates matrix multiplication as well as the Cartesian product, and

$$S \equiv \begin{pmatrix} s_1 \\ \vdots \\ s_q \end{pmatrix} , \qquad \omega \equiv \begin{pmatrix} \omega_1^1 & \omega_1^2 & \dots & \omega_1^q \\ & & \vdots & \\ \omega_q^1 & \omega_q^2 & \dots & \omega_q^q \end{pmatrix} . \tag{22.14}$$

This matrix equation is obviously the generalization of (4.16). As in the discussion leading up to the gauge transformation rule (4.19), the connection matrix (ω_α^β) depends on the choice of the local frame field, and under a change of local frame fields – a **gauge transformation** – given by

$$S' = gS \, , \quad \text{or} \quad (s')_\alpha = g_\alpha^\beta \, s_\beta \, , \tag{22.15}$$

where (g_α^β) is an invertible matrix whose elements are smooth functions on U $[\, g_\alpha^\beta(x^i) \in C^\infty(U)$ and $det(g) \neq 0\,]$, the same steps giving the derivation of (4.19) lead to an identical result for the gauge transformation rule for connections on vector bundles:

$$\omega' = (dg)g^{-1} + g\omega g^{-1} \, . \tag{22.16}$$

As pointed out in Chapter 4 [following (4.19)], this transformation equation distinguishes a connection matrix from a tensorial object, since it implies that the vanishing of a connection matrix is not an invariant property under a local transformation of frame fields. This is confirmed more rigorously by the following theorem.

Theorem 22.2. *Suppose D is a connection on a vector bundle $\pi : E \to M$, and p is a point in M. Then there exists a local frame field S in a coordinate neighborhood of p such that the connection matrix ω with respect to S is zero at p.*

Proof. Choose a coordinate neighborhood $(U; x^i)$ of p such that $x^i(p) = 0$, $1 \leq i \leq n$, where n is the dimension of the base manifold M. Let S' be a local frame field on U with corresponding connection matrix $\omega' = (\omega'^\beta_\alpha)$, where

$$\omega'^\beta_\alpha = A'^\beta_{\alpha i} \, dx^i \, , \tag{22.17}$$

and $A'^{\beta}_{\alpha i}$ are smooth functions on U. Now let

$$g^{\beta}_{\alpha} \equiv \delta^{\beta}_{\alpha} - A'^{\beta}_{\alpha i}(p) \cdot x^i \ . \tag{22.18}$$

Then $g = (g^{\beta}_{\alpha})$ is the identity matrix at p; and there must exist a neighborhood $V \subset U$ of p in which g is non-degenerate. Thus $S = gS'$ is a local frame field in V. Since $dg(p) = -\omega'(p)$, Eq. (22.16) implies that the connection matrix ω with respect to S at the point p is given by

$$\omega(p) = (dg\, g^{-1} + g\omega' g^{-1})(p) = -\omega'(p) + \omega'(p) = 0 \ . \tag{22.19}$$

We conclude that S is the desired local frame field of the theorem. □

The notion of a connection on a vector bundle is equivalent to the notion of *parallel displacement* or *parallel translation* of a section (or a vector field). For a given connection D on a vector bundle, if a section s of the bundle satisfied the condition

$$Ds = 0 \ , \tag{22.20}$$

then s is called a **parallel section** (with respect to the given connection). The zero section is obviously a parallel section (with respect to any connection). But in general, for an arbitrary connection on a vector bundle, non-zero parallel sections may not exist. We will see in a subsequent chapter (Chapter 27) that this is the case for connections with non-zero *curvature*. Rather than parallel sections in general, we are more often interested in the notion of the parallel translation of a section along a given curve on the base manifold of a vector bundle.

Definition 22.6. *Suppose C is a parametrized curve in the base manifold M of a vector bundle, and $\boldsymbol{X}$ is a tangent vector field along C. If a section s of the vector bundle satisfies the following equation for the directional covariant derivative of a connection D:*

$$D_{\boldsymbol{X}}\, s = 0 \ , \tag{22.21}$$

then the section s is said to be **parallel** *along the curve C.*

Locally, parallel sections emerge from the solution of a system of first order differential equations. Let C be parametrized in a local coordinate neighborhood by a set of parametrized equations for the local coordinates: $x^i = x^i(t)$, $1 \le i \le n$, where n is the dimension of the base manifold M and t is the parameter. By (5.1) and (5.5) we can write

$$\boldsymbol{X} = \sum_{i=1}^{n} \frac{dx^i}{dt} \frac{\partial}{\partial x^i} \ . \tag{22.22}$$

Let $S = \{s_1, \dots, s_q\}$ be a local frame field in U. Then a section s can be expressed with respect to S locally as

$$s = \sum_{\alpha=1}^{q} \lambda^{\alpha}(x^i)\, s_{\alpha} \ , \tag{22.23}$$

where the λ^α's are local functions on U. By using (22.8) and (22.9), it is not difficult to see that a section s is a parallel section along C if and only if the coefficients λ^α satisfy the following system of equations:

$$\langle \boldsymbol{X}\,,\, Ds\,\rangle = \sum_{\alpha=1}^{q}\left(\frac{d\lambda^\alpha}{dt} + A^\alpha_{\beta i}\,\frac{dx^i}{dt}\,\lambda^\beta\right)\, s_\alpha = 0\,, \tag{22.24}$$

where

$$\frac{d\lambda^\alpha}{dt} = \frac{\partial\lambda^\alpha}{\partial x^i}\,\frac{dx^i}{dt}\,, \tag{22.25}$$

and $A^\alpha_{\beta i}$ are the connections defined in (22.11). Since $\{s_\alpha\}$ is linearly independent, (22.24) implies the following system of first-order ordinary differential equations:

$$\boxed{\frac{d\lambda^\alpha}{dt} + A^\alpha_{\beta i}\,\frac{dx^i}{dt}\,\lambda^\beta = 0\,, \qquad 1 \leq \alpha \leq q}\;. \tag{22.26}$$

Problem 22.3 Use (22.8) and (22.9) to verify the system of equations (22.24).

By a fundamental result of the theory of ordinary differential equations, a unique solution for the system of equations (22.26) exists for any initial conditions. Thus we see that any vector $\boldsymbol{v} \in E_p$ at a given point p on the curve C uniquely determines a vector field that is parallel along the curve C. This vector field is called the **parallel displacement** or **parallel translation** of the vector $\boldsymbol{v}$ along the curve C. The parallel displacement along C obviously introduces isomorphisms among fibers of the vector bundle at different points on the curve C in the base manifold, which leads to the notion of *holonomy.*

Definition 22.7. *Let C be a smooth parametrized curve in a manifold M, and D be a connection on the vector bundle $\pi : E \to M$. Suppose $C(t_0) = p$ and $C(t_1) = q$ ($t_0 < t_1$) are points on the curve C. Then the linear map $H(C;D) : \pi^{-1}(p) \to \pi^{-1}(q)$ defined by*

$$H(C;D)(\boldsymbol{v}) = \boldsymbol{w}\,, \tag{22.27}$$

where $\boldsymbol{v}$ is an arbitrary vector in $\pi^{-1}(p)$, and $\boldsymbol{w}$ is the result of parallelly translating $\boldsymbol{v}$ along the curve C from p to q, is called the ***holonomy map*** *from p to q along the curve C.*

We state without proofs the following two facts concerning the holonomy map and parallel displacements: (1) The holonomy map $H(C;D)$ is independent of the parametrization of the curve C; and (2) The parallel displacement, and hence the holonomy map, is invariant under a change of local frame fields (a gauge transformation).

An important situation arises when the curve C is a closed loop: $C(t_0) = C(t_1) = p$. In this case $H(C;D)$ is an invertible linear operator on the vector space $E_q = \pi^{-1}(p)$. The set of operators

$$H_p \equiv \{ H(C;D) \,|\, C(t_0) = C(t_1) = p \} \tag{22.28}$$

corresponding to all loops at $p \in M$ then forms a subgroup of $GL(m)$, where m is the dimension of the fiber space E_p. This subgroup is called the **holonomy group** of the connection D with reference to the point $p \in M$.

We will now revisit the *affine connection*, that is, the connection on the tangent bundle TM of a differentiable manifold M, first introduced in Chapter 4. Suppose $dim(M) = n$ and $(U; x^i)$ is a coordinate neighborhood of M. The **natural frame field** on U is defined by $\{s_i = \partial/\partial x^i\}$, $i = 1, \dots, n$. Then an **affine connection** D on U is given by [recall (22.12)]

$$Ds_i = \omega_i^j \otimes s_j = \Gamma^j_{ik} dx^k \otimes s_j \;, \tag{22.29}$$

where the Γ^j_{ik} are smooth functions on U. These connection coefficients of the affine connection D with respect to the local coordinates x^i [which play the same role that the connection coefficients $A^\beta_{\alpha i}$ in (22.10) do for a general vector bundle] are often called **Christoffel symbols**, a term that is of common usage in the physics literature on general relativity.

Problem 22.4 Let $(U; x^i)$ and $(W; y^i)$ be two coordinate neighborhoods of a differentiable manifold M such that $U \bigcap W \neq \emptyset$. The Christoffel symbols of an affine connection D on TM with respect to the coordinate neighborhoods are defined by

$$D\left(\frac{\partial}{\partial x^i}\right) = \Gamma^j_{ik} dx^k \otimes \frac{\partial}{\partial x^j} \;, \qquad D\left(\frac{\partial}{\partial y^i}\right) = (\Gamma')^j_{ik} dy^k \otimes \frac{\partial}{\partial y^j} \;. \tag{22.30}$$

Use (22.17), the gauge transformation rule for an arbitrary vector bundle, to show that the gauge transformation rule for Christoffel symbols are given by

$$(\Gamma')^j_{ik} = \frac{\partial x^l}{\partial y^i}\frac{\partial x^p}{\partial y^k}\frac{\partial y^j}{\partial x^m}\Gamma^m_{lp} + \frac{\partial^2 x^l}{\partial y^i \partial y^k}\frac{\partial y^j}{\partial x^l} \;. \tag{22.31}$$

Note that the presence of the second term on the right-hand side of the above equation shows that Γ^k_{ij} is not a tensor field.

Problem 22.5 Show that the **affine covariant derivative** of a vector field $\boldsymbol{X} = X^i \dfrac{\partial}{\partial x^i}$ corresponding to an affine connection is given by

$$D\left(X^i \frac{\partial}{\partial x^i}\right) = X^i{}_{,j} dx^j \otimes \frac{\partial}{\partial x^i} \;, \tag{22.32}$$

where the affine directional covariant derivative coefficients ${X^i}_{,j}$ (the affine covariant derivative of X^i along the direction x^j) are given by

$$ {X^i}_{,j} \equiv \frac{\partial X^i}{\partial x^j} + \Gamma^i_{kj} X^k \,. \tag{22.33} $$

Given an affine connection D on M (a connection on the tangent bundle TM), there is an induced connection (also denoted by D) on the (dual) cotangent bundle T^*M defined in terms of $D(dx^i)$ by

$$ d\left\langle \frac{\partial}{\partial x^i}, dx^j \right\rangle = \left\langle D\left(\frac{\partial}{\partial x^i}\right), dx^j \right\rangle + \left\langle \frac{\partial}{\partial x^i}, D(dx^j) \right\rangle , \tag{22.34} $$

where, in analogy to (22.29), we write

$$ D(dx^i) = (\omega^*)^i_j \otimes dx^j \,. \tag{22.35} $$

Note that both sides of (22.34) are one-forms. The left-hand side vanishes by virtue of the duality relationship (5.6). Thus, on using the first of Eqs. (22.30) together with (5.6), we have

$$ 0 = \Gamma^j_{ik} dx^k + (\omega^*)^j_i \,. \tag{22.36} $$

This implies

$$ (\omega^*)^j_i = -\Gamma^j_{ik} dx^k \,. \tag{22.37} $$

Thus the affine covariant derivative of a one-form (cotangent vector field) $X_i\, dx^i$ is given by

$$ D(X_i\, dx^i) = X_{i,j}\, dx^j \otimes dx^i \,, \tag{22.38} $$

where, in analogy to (22.33),

$$ X_{i,j} \equiv \frac{\partial X_i}{\partial x^j} - \Gamma^k_{ij} X_k \,. \tag{22.39} $$

The above equation, together with (22.33), are basic formulas in classical *tensor analysis*. The equations (22.32), (22.33), (22.38) and (22.39), together with the "Leibniz rule", allow us to calculate the affine covariant derivative of a mixed-type tensor field of arbitrary rank. If T is an (r, s)-type tensor field, then DT is an $(r, s+1)$-type tensor field. For example, for the $(2,1)$-type tensor field expressed locally by $T = T^{ij}_k\, dx^k \otimes \frac{\partial}{\partial x^i} \otimes \frac{\partial}{\partial x^j}$, we have

$$ \begin{aligned} DT &= dT^{ij}_k \otimes dx^k \otimes \frac{\partial}{\partial x^i} \otimes \frac{\partial}{\partial x^j} + T^{ij}_k\, D(dx^k) \otimes \frac{\partial}{\partial x^i} \otimes \frac{\partial}{\partial x^j} \\ &= T^{ij}_k\, dx^k \otimes D\left(\frac{\partial}{\partial x^i}\right) \otimes \frac{\partial}{\partial x^j} + T^{ij}_k\, dx^k \otimes \frac{\partial}{\partial x^i} \otimes D\left(\frac{\partial}{\partial x^j}\right) . \end{aligned} \tag{22.40} $$

Problem 22.6 Show that for the (2, 1)-type tensor field T in (22.40), the "Leibniz rule" given by that equation leads to the result

$$DT = T^{ij}_{k,l}\, dx^l \otimes dx^k \otimes \frac{\partial}{\partial x^i} \otimes \frac{\partial}{\partial x^j}\,, \tag{22.41}$$

where

$$T^{ij}_{k,l} \equiv \frac{\partial T^{ij}_k}{\partial x^l} + \Gamma^i_{hl} T^{hj}_k + \Gamma^j_{hl} T^{ih}_k - \Gamma^h_{kl} T^{ij}_h\,. \tag{22.42}$$

The notion of the parallel displacement of a vector along a curve in the general setting of a vector bundle can be adapted to the special case of the parallel displacement of a tangent vector along a curve in the setting of a tangent bundle. Historically this special case was first studied by Levi-Civita, and is known as **Levi-Civita parallelism**. It gives rise to the important geometric notion of *self-parallel curves*, or *geodesics* in Riemannian geometry. Let $C(t)$: $x^i(t)$ be a parametrized curve in an n-dimensional Riemannian manifold M (see Chapter 27), where t (usually the time in the context of classical mechanics) is the parameter. By (22.26), the tangent vector field

$$\boldsymbol{X}(t) = X^i(t)\left(\frac{\partial}{\partial x^i}\right)_{C(t)}$$

on the curve C is parallel along C if the functions $X^i(t)$ satisfy the following set of first-order differential equations:

$$\boxed{\frac{dX^i}{dt} + X^j \Gamma^i_{jk} \frac{dx^k}{dt} = 0\,, \qquad i = 1, \dots, n}\ . \tag{22.43}$$

Given the initial conditions $X^i(0)$, that is, some vector $\boldsymbol{X}(0) \in T_{C(0)}M$, $\boldsymbol{X}(t)$ at any point in the curve $C(t)$ is determined uniquely by the process of parallel translation. This property of parallel translation depends crucially on the fact that the system of equations (22.43) is linear, so that a global solution in a finite interval $0 < t < T$ within the range of definition of the parametrized curve is guaranteed. If the equations were non-linear, only a solution within an infinitesimal neighborhood $0 < t < \epsilon$ is guaranteed.

Of particular interest is the case of **self parallelism**, when the tangent vector field $\boldsymbol{X}_C$ *of* the curve $C(t)$ is parallel along C. We say that the curve C is *self-parallel*, or a **geodesic**. In this case, $X^i = dx^i/dt$, and (22.43) gives the following condition for a curve $C(t) : x^i(t)$ to be a geodesic:

$$\boxed{\frac{d^2x^i}{dt^2} + \Gamma^i_{jk} \frac{dx^j}{dt}\frac{dx^k}{dt} = 0\,, \qquad i = 1, \dots, n}\ . \tag{22.44}$$

This is a system of second-order ordinary differential equations. The theory of such equations guarantees the following important result, which will be stated without proof.

Theorem 22.3. *There exists a unique geodesic through a given point of a Riemannian manifold which is tangent to a given tangent vector at that point.*

However, due to the non-linear nature of the system of equations (22.44) only the existence of local solutions is guaranteed. In general, global geodesics may not exist.

The geometric notion of geodesics plays a fundamental role in the *general theory of relativity*. For a fuller mathematical discussion see Chapter 29.

Chapter 23

The Parallel Translation of Vectors: The Foucault Pendulum

Starting with this chapter we will use a few simple examples in classical mechanics to illustrate the usefulness of the mathematical notions introduced in the last chapter.

The Foucault pendulum is simply a very long pendulum so that the rotation of the plane of oscillation of the pendulum during the course of a day is easily observable (when one is away from the equator). This rotation can be understood very intuitively, and also as a formal application of the method of fictitious forces that we introduced earlier when we use a frame of reference (noninertial) attached to the surface of the earth (cf. Problem 12.1). In this chapter, however, we will use it as a tool to explore the geometrical relationship between forces in general and the *curvature* of space, in the spirit of the *general theory of relativity*, where the gravitational force is understood as a manifestation of the curvature of spacetime. The mathematical notion of curvature, as a derivative notion from that of connections on vector bundles, will be introduced more carefully in Chapters 25 and 27.

Let us imagine a pendulum at the north pole, and a two-dimensional noninertial frame $\{e_1, e_2\}$ (with e_1 due south and e_2 due east) attached to the rotating earth with origin at the north pole (Fig. 23.1). With respect to an observer O at rest in an inertial frame (who in fact observes that the earth spins around its axis with angular speed ω), it is clear that the plane of oscillation of the pendulum is fixed in time as the earth rotates. From the viewpoint of an observer O' at rest with respect to the $\{e_1, e_2\}$ frame, however, the earth is not moving. Instead, the plane of oscillation of the pendulum rotates clockwise with the angular speed ω (Fig. 23.2).

At a location away from the poles, one can similarly choose a noninertial frame $\{e_1, e_2\}$ attached to the earth, with e_1 along the direction of increasing

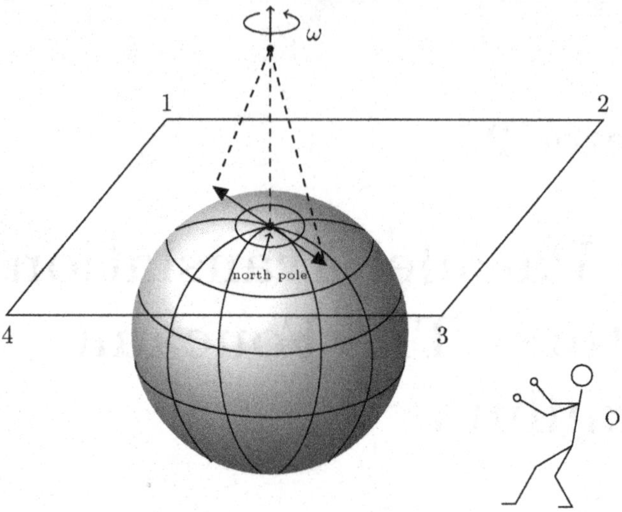

Fig. 23.1

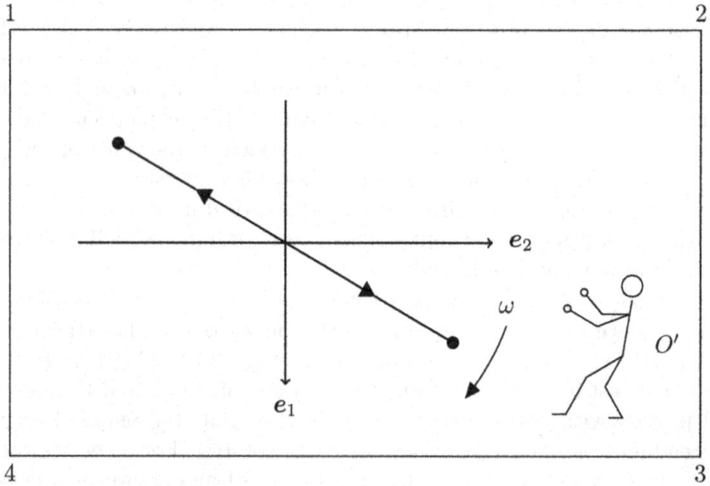

Fig. 23.2

colatitude θ (due south) and e_2 along the direction of increasing longitude ϕ (due east) (Fig. 23.3). With respect to this local frame, the plane of oscillation of the pendulum rotates with an angular speed $\omega\cos\theta$, clockwise in the northern hemisphere and counterclockwise in the southern hemisphere (Fig. 23.4).

At the equator (where $\theta = \pi/2$), the plane of oscillation is fixed ($\omega\cos\theta = 0$). The oscillation of the pendulum (in the $e_1 e_2$ plane) is shown in Fig. 23.5. Writing the position vector of the pendulum bob as

$$\boldsymbol{x} = x^1\,\boldsymbol{e}_1 + x^2\,\boldsymbol{e}_2 \tag{23.1}$$

(where we have ignored the vertical component), the equations of motion at the equator are obviously given by those of a free harmonic oscillator:

$$\frac{d^2x^1}{dt^2} + \omega_0^2 x^1 = 0\,, \qquad \frac{d^2x^2}{dt^2} + \omega_0^2 x^2 = 0\,, \qquad (\theta = \pi/2)\,, \tag{23.2}$$

where ω_0 is the natural frequency of oscillation of the pendulum. Away from the equator, then, the rotation of the plane of oscillation can be attributed to a fictitious force, which we will call the ω-force. This fictitious force arises as a result of the choice of the noninertial frame $\{e_1, e_2\}$ for the description of the motion of the pendulum. We are not particularly concerned with the explicit form of the ω-force at this point, and will simply for the time being write the coupled equations of motion including the effects of this force as

$$\frac{d^2x^1}{dt^2} = -\omega_0^2 x^1 + \dots\,, \qquad \frac{d^2x^2}{dt^2} = -\omega_0^2 x^2 + \dots\,, \qquad (\theta \neq \pi/2)\,, \tag{23.3}$$

where the dots in both equations represent the effects of the ω-force.

Now introduce a moving frame $\{e_1', e_2'\}$ which rotates with the plane of oscillation of the pendulum (Fig. 23.6). It is clear that, from the viewpoint of an observer O'' at rest with respect to this frame, the ω-force vanishes. This is similar to the situation of *Einstein's elevator*: a person inside a freely falling elevator and at rest with respect to it would observe no acceleration for a freely falling ball, and therefore would infer that there is no gravitational force acting on the ball. In other words, the effect of the gravitational force can be eliminated by the introduction of a suitably chosen noninertial (accelerating) frame. Equivalently, the gravitational force can be mimicked by a suitably chosen accelerating frame. (Imagine an elevator accelerating upwards with an acceleration $g = 9.8\,m/sec^2$.) This is called the **Principle of Equivalence**. When Einstein realized this he called it the "happiest thought of my life".

Going back to the $\{e_1, e_2\}$ frame we realize that the appearance of the fictitious ω-force is due to the fact that $De_i/dt \neq 0$. But we recall from (4.8) that

$$\frac{De_i}{dt} = \frac{\omega_i^j}{dt}\,e_j = \varphi_i^j\,e_j\,, \tag{23.4}$$

where (ω_i^j) is an affine connection matrix of one-forms. Hence, in order to eliminate the ω-force, we need to choose a frame in which (φ_i^j) vanishes. We

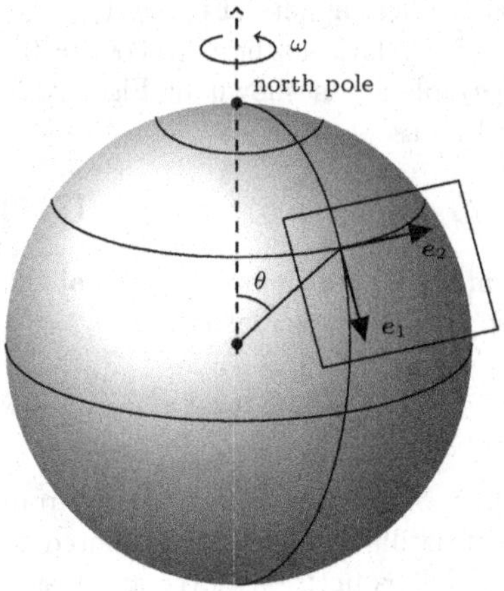

Fig. 23.3

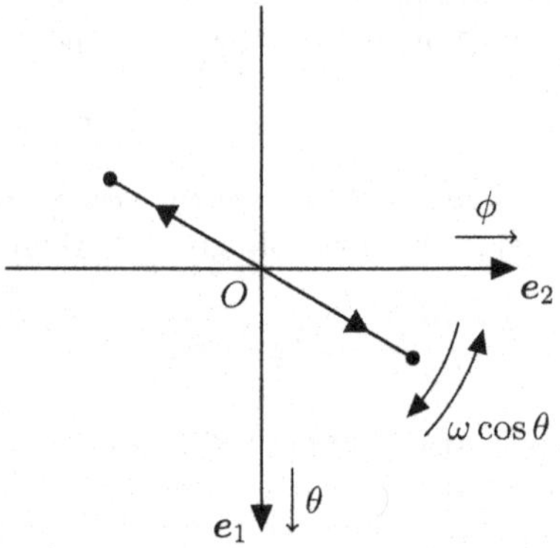

Fig. 23.4

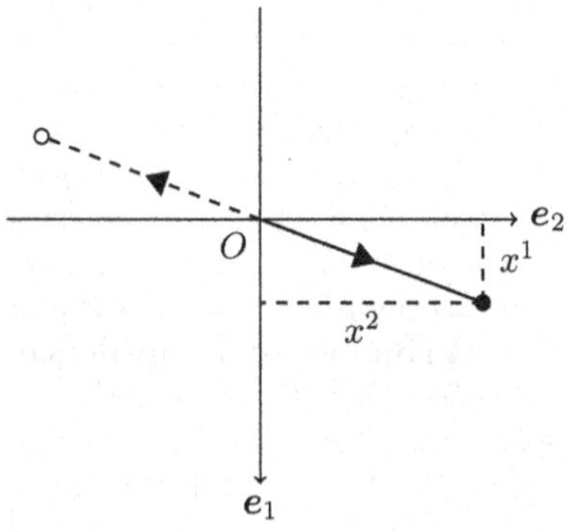

Fig. 23.5

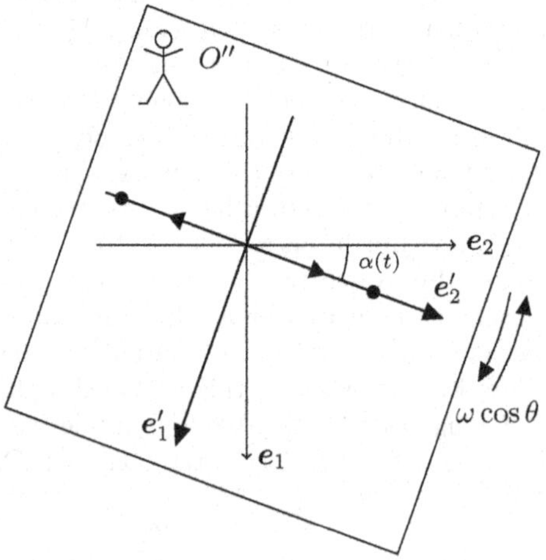

Fig. 23.6

will show that this is precisely the frame $\{\mathbf{e}'_1, \mathbf{e}'_2\}$, which rotates together with the plane of oscillation of the pendulum.

From (12.14), the equations of motion with respect to the $\{\mathbf{e}_1, \mathbf{e}_2\}$ frame are given by

$$\frac{d^2x^i}{dt^2}\,\mathbf{e}_i = -\omega_0^2\, x^i\, \mathbf{e}_i - 2\frac{dx^i}{dt}\,\varphi_i^j\, \mathbf{e}_j - x^j \left(\frac{d\varphi_j^i}{dt} + \varphi_j^k \varphi_k^i \right) \mathbf{e}_i \,. \tag{23.5}$$

The 2×2 matrix of 1-forms (ω_i^j) (with respect to $\{\mathbf{e}_1, \mathbf{e}_2\}$ is given by a restriction of the 3×3 matrix in (4.32) (by eliminating the first row and first column):

$$(\omega_i^j) = \begin{pmatrix} 0 & \cos\theta\, d\phi \\ -\cos\theta\, d\phi & 0 \end{pmatrix} . \tag{23.6}$$

Thus $(\varphi_i^j) = (\omega_i^j/dt) \neq 0$ unless $\theta = \pi/2$. Suppose $\{\mathbf{e}'_1, \mathbf{e}'_2\}$ is related to $\{\mathbf{e}_1, \mathbf{e}_2\}$ by a rotation (through an angle α in the clockwise sense) as shown in Fig. 23.7. Then we have [compare with (1.7)]

$$\begin{pmatrix} \mathbf{e}'_1 \\ \mathbf{e}'_2 \end{pmatrix} = \begin{pmatrix} \cos\alpha & -\sin\alpha \\ \sin\alpha & \cos\alpha \end{pmatrix} \begin{pmatrix} \mathbf{e}_1 \\ \mathbf{e}_2 \end{pmatrix} , \tag{23.7}$$

or

$$\mathbf{e}'_i = g_i^j\, \mathbf{e}_j \,, \tag{23.8}$$

where g is the 2×2 matrix on the right-hand side of (23.7). By the gauge transformation rule (22.16) the connection matrix of one-forms ω' with respect to $\{\mathbf{e}'_1, \mathbf{e}'_2\}$ is then

$$\omega' = (dg)g^{-1} + g\omega g^{-1} \,. \tag{23.9}$$

Using (23.6) for ω, it follows that

$$((\omega')_i^j) = \begin{pmatrix} 0 & \cos\theta\, d\phi - d\alpha \\ -\cos\theta\, d\phi + d\alpha & 0 \end{pmatrix} . \tag{23.10}$$

Problem 23.1 Verify (23.10) by direct calculation, using (23.9).

Thus $\varphi' = \omega'/dt = 0$ if

$$\frac{d\alpha}{dt} = \cos\theta\, \frac{d\phi}{dt} = \omega\, \cos\theta \,, \tag{23.11}$$

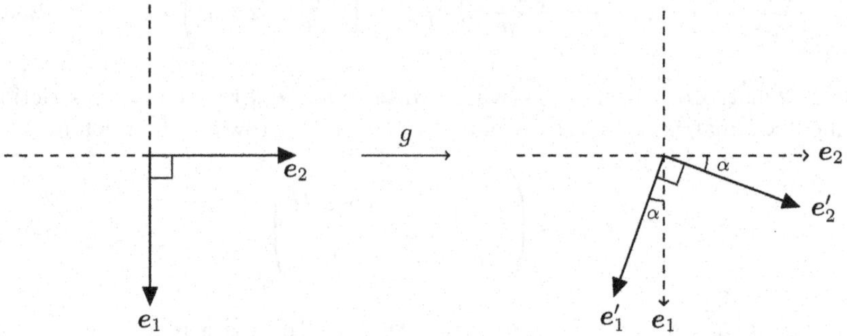

Fig. 23.7

where $\dot{\phi} = \omega$ is the angular speed of the earth's rotation about its axis. Referring to Figs. 23.6 and 23.7, if we take the e'_2 axis to be along the pendulum's plane of oscillation, then indeed the $\{e'_1, e'_2\}$ frame is the moving frame attached to the earth's surface but rotating in unison with the Foucault pendulum.

Geometrically, an affine connection matrix (ω_i^j) defined on a Riemannian manifold, which is a structure imposed on the manifold so as to enable *covariant differentiation* of tangent vector fields, gives rise to the *Riemann curvature* of the manifold. We will see how this occurs in more detail in Chapter 27. In the present context, the Riemannian manifold under consideration is the 2-dimensional spherical surface (surface of the globe). The local frames $\{e_1, e_2\}$ and $\{e'_1, e'_2\}$ each constitutes a basis set of the tangent space at the point of the spherical surface to which the (common) origin of the frames is attached. As the earth rotates, we can imagine the origin of these frames moving along a latitude circle. The two frames also move, according to the equations

$$\frac{De_i}{dt} = \varphi_i^j \, e_j \; , \qquad \frac{De'_i}{dt} = (\varphi')_i^j \, e'_j \; . \tag{23.12}$$

But since $\varphi' = 0$,

$$\frac{De'_i}{dt} = 0 \; . \tag{23.13}$$

As we saw in the last chapter, a tangent vector which satisfies the above equation as it is transported along a curve on the underlying manifold is said to be *parallelly translated* (or *parallelly displaced*) along the curve.

Problem 23.2 Assuming that the earth's angular speed of rotation $\dot{\phi} \equiv \omega$ is constant, use (23.6) in (23.5) to show that

$$\begin{aligned} \frac{d^2x^1}{dt^2} &= -(\omega_0^2 - \omega^2\cos^2\theta)x^1 + 2\frac{dx^2}{dt}\omega\cos\theta\,, \\ \frac{d^2x^2}{dt^2} &= -(\omega_0^2 - \omega^2\cos^2\theta)x^2 - 2\frac{dx^1}{dt}\omega\cos\theta\,. \end{aligned} \tag{23.14}$$

Problem 23.3 Solve the coupled system of equations (23.14) by writing $x = x^1 + ix^2$ and convince yourself that the solution describes the motion of a Foucault pendulum.

Problem 23.4 Consider a vector initially tangent to a latitude line ($\theta =$ constant) on a spherical surface. With θ (colatitude) and ϕ (longitude) as local coordinates ($x^1 = \theta, x^2 = \phi$) of the two-dimensional spherical surface, calculate all the Christoffel symbols Γ^i_{jk} using (23.6) for the connection matrix (ω_i^j) and (22.29) for the definition of the Christoffel symbols. Then solve the system of two first-order coupled differential equations (22.43) and describe how the vector behaves as it is parallelly transported along the latitude line. Show that upon parallel translation of the vector over one complete turn around the latitude circle, the vector does *not* return to its original position, but will have been rotated by an angle of $2\pi\cos\theta$. This is an example of the phenomenon of *holonomy* discussed in Chapter 22 [following (22.28)].

Chapter 24

Geometric Phases, Gauge Fields, and the Mechanics of Deformable Bodies: The "Falling Cat" Problem

We will first introduce a very simplified model of a "falling cat", which will provide a glimpse into the mechanics of *deformable* (non-rigid) bodies. We will see that the physical concept of *gauge fields*, as it is related to the differential geometric notion of *geometric phases*, which is in turn based on that of connections of vector bundles (discussed in Chapter 22), will emerge naturally. The developments in this example are adapted from an analysis presented by R. Littlejohn and M. Reinsch (Littlejohn and Reinsch 1997). A more general differential geometric formulation of this problem [based on work by Shapere and Wilczek (see Shapere and Wilczek 1989)] will be presented at the end of the chapter. For more in-depth discussions on the general topic of geometric phases in classical mechanics, the reader may refer to, for example, the monograph by Chruściński and Jamiołkowski (Chruściński and Jamiołkowski 2004).

Consider a cat with most of its mass concentrated in its head, the rear portion of its torso, and the extremities of its limbs, which are tied together (see Fig. 24.1). As it falls, suppose the only external force on it is its weight, which acts through its *center of mass* (CM). We will model the dynamics of this falling cat by a system of three massless rigid rods of equal length R with equal masses m attached to their free ends. Rods 1 and 2 are joined at the CM position O, while rod 3 is attached to rod 2. The frame Oxy then represents the (non-rotating) CM frame of the falling cat. The system of rods and masses are free to move on the xy plane with only the joint O fixed (with respect to the CM frame), so that the angles θ, α and β are all variables (see Fig. 24.2). The angle θ is an *orientational coordinate* of the system, while α and β are the

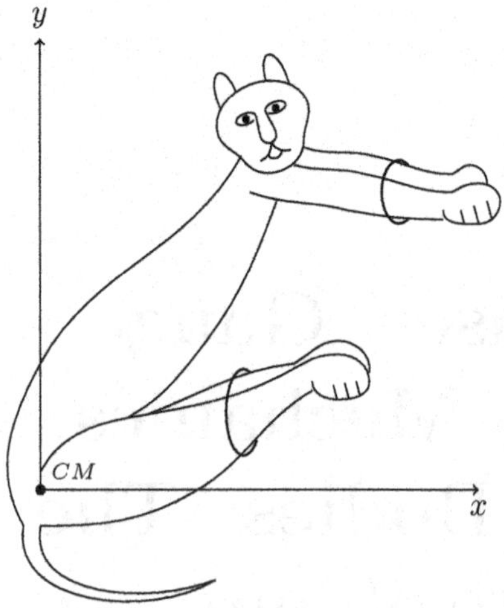

Fig. 24.1

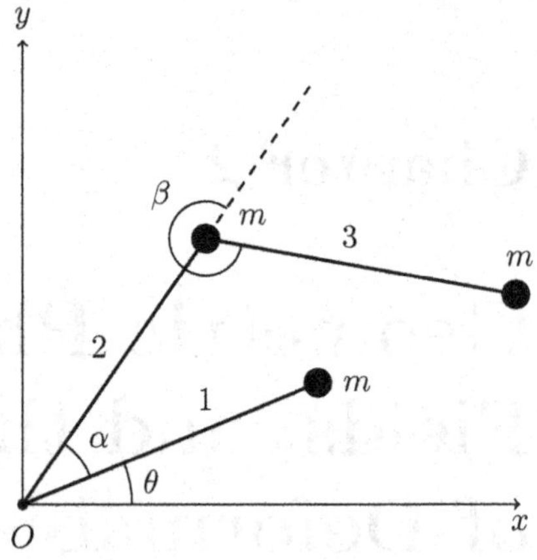

Fig. 24.2

shape coordinates. There is obviously no external torque with respect to the CM frame, so that the total angular momentum with respect to this frame (Oxy) is conserved. Only internal forces act at the joint and elsewhere to cause changes in shape and orientation.

Assume initially the system has zero angular momentum with respect to the CM frame. Let the Cartesian coordinates of the three masses be denoted by $(x^i, y^i, 0)$, where $i = 1, 2, 3$. Since, for a single mass, the angular momentum is given by

$$\boldsymbol{L} = \boldsymbol{r} \times \boldsymbol{p} = m(x\dot{y} - y\dot{x})\,\boldsymbol{k}\,, \tag{24.1}$$

where $\boldsymbol{k}$ is the unit vector perpendicular to the xy plane along the third axis of the right-handed xyz frame, the condition of the conservation of angular momentum (with respect to the $Oxyz$ CM frame) can be expressed as

$$\boldsymbol{L} = m\sum_{i=1}^{3}\left(x^i\frac{dy^i}{dt} - y^i\frac{dx^i}{dt}\right) = 0\,. \tag{24.2}$$

On multiplying the above expression by dt we obtain the differential expression

$$\sum_{i=1}^{3}(x^i dy^i - y^i dx^i) = 0\,. \tag{24.3}$$

The geometry of Fig. 24.2 implies that

$$\begin{aligned} x^1 &= R\cos\theta\,, & y^1 &= R\sin\theta\,,\\ x^2 &= R\cos(\alpha+\theta)\,, & y^2 &= R\sin(\alpha+\theta)\,,\\ x^3 &= R\cos(\alpha+\theta)+R\cos(\alpha+\beta+\theta)\,, & y^3 &= R\sin(\alpha+\theta)+R\sin(\alpha+\beta+\theta)\,. \end{aligned}$$

Calculating the differentials of the left-hand sides of the above group of equations and substituting into (24.3) we obtain, after simplification,

$$(4+2\cos\beta)\,d\theta + (3+2\cos\beta)\,d\alpha + (1+\cos\beta)\,d\beta = 0\,. \tag{24.4}$$

Problem 24.1 Verify by direct calculation that (24.4) follows from (24.3).

It follows directly from (24.4) that we can write

$$d\theta = -A_\alpha(\alpha,\beta)\,d\alpha - A_\beta(\alpha,\beta)\,d\beta\,, \tag{24.5}$$

where

$$A_\alpha = \frac{3+2\cos\beta}{4+2\cos\beta}\,, \qquad A_\beta = \frac{1+\cos\beta}{4+2\cos\beta}\,. \tag{24.6}$$

As we will explain, the quantities A_α and A_β can be regarded as the two components of a **vector potential**, or **gauge potential** (similar to the vector potential responsible for a magnetic field in Maxwell's electrodynamics).

The following geometrical picture emerges from our treatment thus far. The two-dimensional space with local coordinates α and β can be regarded as the base manifold of the fiber bundle describing our system. We will call it the *shape manifold.* It is just a square (with length of each side equal to 2π) in the 2-dimensional $\alpha\beta$ plane, with opposite sides identified. The manifold is thus topologically a **2-torus** (as can be seen in Fig. 24.3). Each point on the torus represents a particular shape of the (model) falling cat, which is specified by a pair of values for the angles α and β. Now, with each shape (each point point in the base manifold), we can associate a particular orientation, that is, a point in the *fiber space* that we will call the *orientation manifold*, coordinatized by the single local variable θ ($0 \le \theta \le 2\pi$). Apparently this fiber space is just the 1-dimensional manifold S_1, the so-called 1-sphere, or a circle. The actual value of the orientation coordinate for a point on the circle depends on an arbitrary choice of the point (on the circle) corresponding to $\theta = 0$. This is called a choice of **gauge**. So long as the joint O is fixed, the orientations of the x and y axes for each pair of shape values (α,β) can be arbitrary: the physics of the system of rods and masses (the falling cat as described by our model) does not depend on how we locally choose to measure the fiber space (orientation) variable θ. This

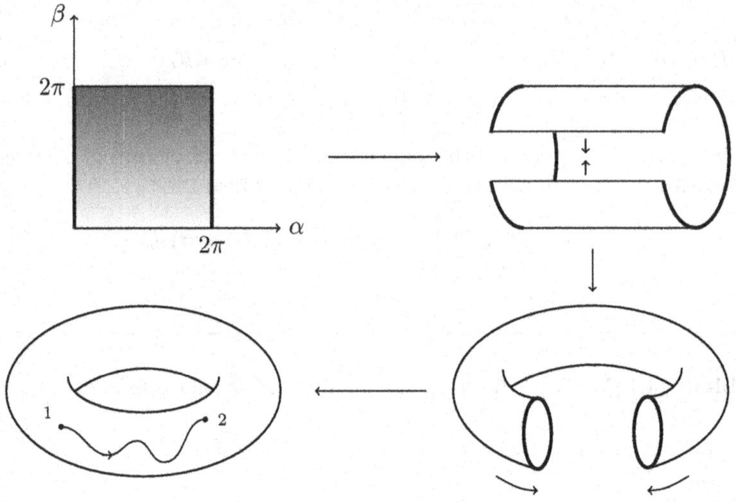

Fig. 24.3

is called the **principle of gauge invariance**, a deep and far-reaching principle in physics.

What does the gauge potential $\boldsymbol{A}$ do? According to (24.5), it governs how the orientation coordinate θ changes as the shape variables α and β are varied. As one moves from point 1 to point 2 in the base (shape) manifold along some arbitrary path, one can integrate the equation to obtain the net resulting change in θ. The vector potential thus "connects" points on different fibers (sitting over different points in the base manifold) in a specific way, *depending on the specific path traversed in the base manifold.* It thus plays the same role as the mathematical notion of the *connection* introduced in Chapter 22; and indeed is mathematically equivalent to it. So *the physical concept of the gauge field is equivalent to the mathematical concepts of the connection and the covariant derivative of vector fields.* This equivalence, as evidenced through diverse physical phenomena (such as different gauge field theories in quantum field theory), underlies the importance of differential geometry in physics.

In the present context, the gauge potential components A_i (where $i = \alpha, \beta$) are special cases of the connection matrix components $A^{\beta}_{\alpha i}$ in (22.11). Here the fiber space is 1-dimensional so the connection matrix of 1-forms is a 1 by 1 matrix. The accumulated change of the fiber coordinate θ along a certain path in the base manifold as determined by the gauge potential (connection) A_i is called the **geometric phase**. This is a special case of the *holonomy map* introduced in Def. 22.7. As we will see below for the present example, *the net change in the fiber coordinate, or the geometric phase, over a closed loop in the base manifold does not vanish in general.*

Writing the 1 by 1 connection "matrix" of one forms as

$$\omega = A_\alpha \, d\alpha + A_\beta \, d\beta \,, \tag{24.7}$$

we have, on exterior differentiation [cf. (6.13)],

$$d\omega = \left(\frac{\partial A_\beta}{\partial \alpha} - \frac{\partial A_\alpha}{\partial \beta} \right) d\alpha \wedge d\beta \,. \tag{24.8}$$

Suppose D is an arbitrary surface with boundary on the shape manifold (the base manifold 2-torus) bounded by the closed loop ∂D, then Eqs. (24.5) and (24.7) imply, by virtue of Stokes' Theorem (Theorem 8.1) [Eq. (8.33)], that the geometric phase accumulated over the closed loop ∂D is given by

$$(\Delta\theta)_{\partial D} = -\oint_{\partial D} (A_\alpha d\alpha + A_\beta d\beta) = -\oint_D \left(\frac{\partial A_\beta}{\partial \alpha} - \frac{\partial A_\alpha}{\partial \beta} \right) d\alpha \wedge d\beta \,. \tag{24.9}$$

The area integral on the right-hand side of the second equality in the above equation does not vanish in general, and gives the *flux* of the antisymmetric **field-strength** tensor

$$F_{\alpha\beta} \equiv \left(\frac{\partial A_\beta}{\partial \alpha} \right) - \left(\frac{\partial A_\alpha}{\partial \beta} \right) \tag{24.10}$$

derived from the gauge potential through the bounded surface D. Geometrically, $F_{\alpha\beta}$ is the *curvature tensor* arising from the connection $\boldsymbol{A}$. The mathematical notion of the curvature arising from a connection will be formally introduced in Chapter 27.

The cat changes its shape as it falls; but when it resumes its original shape, its orientation has changed also! The falling cat has acquired a geometric phase, depending on how he stretches his limbs and body as he falls, in other words, depending on the path that he has traversed in the shape manifold as he falls. There is an analogous phenomenon in quantum mechanics known as the **Aharonov-Bohm Effect**, in which an electron looping around a magnetic flux tube acquires a geometric phase in its wave function calculated in a similar way as by (24.9), but with the gauge potential $\boldsymbol{A}$ being the vector potential giving rise to the magnetic field in the flux tube. For a more in-depth discussion of this effect and its geometrical setting, the reader may consult, for example, Lam 2009.

Problem 24.2 Consider the closed loop $\partial D : a \to b \to c \to d \to a$ in the shape manifold which consists of a square of side $\pi/2$ (see Fig. 24.4). Compute the change in

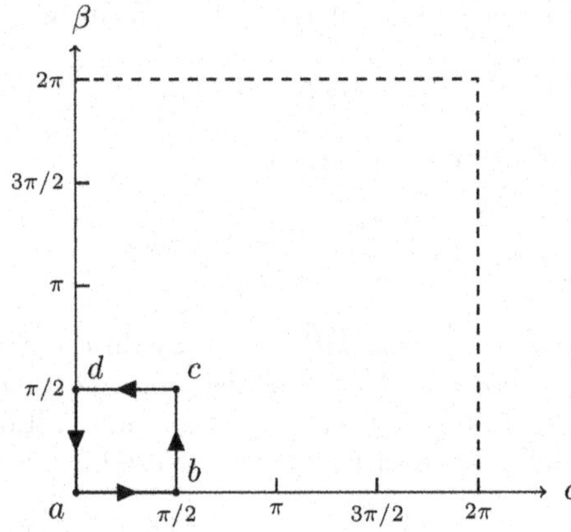

Fig. 24.4

the orientation coordinate θ over each leg of the square loop by integrating Eq. (24.5). Show that

$$(\Delta\theta)_{a\to b} = -75^\circ\ , \quad (\Delta\theta)_{b\to c} = -(\Delta\theta)_{d\to a} = -27.68^\circ\ , \quad (\Delta\theta)_{c\to d} = 67.5^\circ\ .$$

Thus the total change in the geometric phase over the entire loop is $(\Delta\theta)_{\partial D} = -7.5^\circ$. Verify this last result directly by evaluating the flux integral

$$-\oint_D F_{\alpha\beta}\, d\alpha \wedge d\beta\ .$$

Start with the configuration at the point a ($\alpha = \beta = 0$) with $\theta = 0$. Draw a figure for this configuration. Draw four more figures showing how the system of rods (or the model cat with his limbs tied up) look at the points b, c and d in the shape manifold, and when the system finally returns to point a.

We should note that the fiber bundle introduced in this chapter is not a vector bundle as defined in Chapter 22, but rather what is known as a *principal bundle*, in which the typical fiber does not have the structure of a vector space, but is a *Lie group*. In the present case, the principal bundle can be written $\pi : E \to T^2$, where the base manifold T^2 is the 2-torus and the total space is locally the product $T^2 \times U(1)$. In other words the typical fiber is the unitary group $U(1)$ of dimension 1 (topologically a 1-sphere S_1, or the circle) whose local coordinate is θ. The fundamental notion of the connection on a vector bundle introduced in Chapter 22 can be extended to that on a principal bundle. In the case of principal bundles with Lie groups as structure groups, as is the case in this chapter or others of importance in physics, the connections are basically the same, where the connection matrices are so-called *Lie algebra-valued one forms* on the base manifold, with the Lie algebra being a matrix representation of the Lie algebra of the structure (Lie) group. We will discuss principal bundles and Lie algebras in more detail in Chapter 28.

Let us consider the motion of deformable bodies more generally with changing orientations in three-dimensional Euclidean space, such as that of a more realistic falling cat or a springboard diver. Assume that each time-varying (unoriented) **shape** is described by a point $\theta = \{\theta^1, \dots, \theta^n\}$ in an n-dimensional manifold $\mathcal{S}$, designating the *shape space*, and each *orientation* by an orthonormal frame $\{\boldsymbol{e}_1, \boldsymbol{e}_2, \boldsymbol{e}_3\}$ in $\mathbb{R}^3$. The mathematical setup is then the *frame bundle* $\pi : E \to \mathcal{S}$, where the total space E is locally $\mathbb{R}^n \times SO(3)$ [with the *fiber* being $SO(3)$], and the *base manifold* is the shape space $\mathcal{S}$.

We assume that the only external force on the body acts through the *center of mass* at all times (such as the gravitational force on a falling cat or a diver) so that with respect to a center-of-mass frame the net torgue is zero and consequently *the total angular momentum is conserved.* Without loss of generality we can set the total momentum to be zero. *The change in orientation is caused by only internal forces which serve to deform the shape of the body as it moves through space.* The sequence of unoriented shape changes in time is described by a *parametrized curve* $\gamma(t) = \{\theta^i(t)\}$ in $\mathcal{S}$.

With this mathematical setup, a *gauge field structure* naturally emerges by applying the laws of classical mechanics to this problem.

We begin by picking a **frame field** $\{\epsilon_1(t), \epsilon_2(t), \epsilon_3(t)\}$ on $\gamma(t)$. In other words, we associate an arbitrary *orientation* with a shape of the body as its shape changes along the path $\gamma(t)$. For example, one may use the *principal axes* corresponding to a particular shape. We will refer to this as the *unoriented* frame field. Let the *actual* (physical) orientation corresponding to a particular shape $\{\theta^1(t), \dots, \theta^N(t)\} \in \gamma(t)$ be specified by the frame $\{\varepsilon_1(t), \varepsilon_2(t), \varepsilon_3(t)\}$, referred to as the *oriented* frame field (see Fig. 24.5).

Suppose

$$\epsilon_i(t) = R_i^j(t)\, \varepsilon_j(t) \,, \tag{24.11}$$

where (R_i^j) is a rotation matrix in $SO(3)$ to be determined (as a function of the time t). In matrix notation, with both $\boldsymbol{\epsilon}$ and $\boldsymbol{\varepsilon}$ being column matrices,

$$\boldsymbol{\epsilon} = R\boldsymbol{\varepsilon} \,. \tag{24.12}$$

Since $\{\varepsilon_i\}$ describes the actual physical orientations, we can write

$$\frac{D\varepsilon_i}{dt} = \omega_i^j \varepsilon_j \,, \tag{24.13}$$

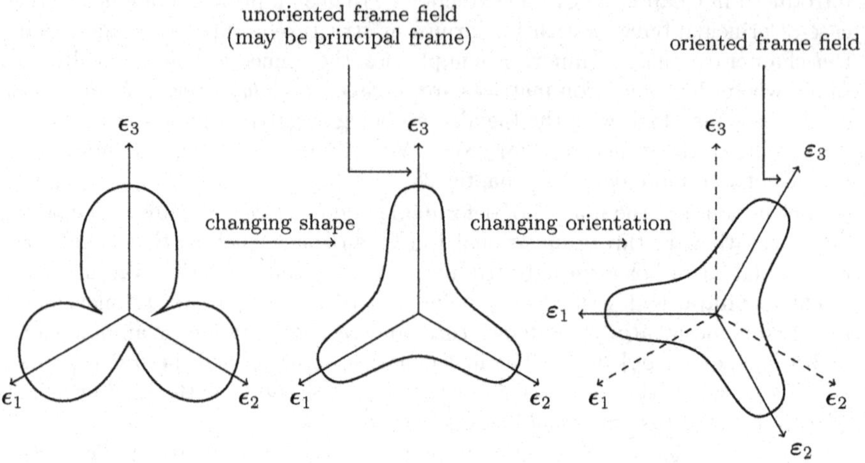

Fig. 24.5

where D is a *covariant derivative*, and ω is a connection matrix of one-forms describing the physical rotations from the unoriented frame field $\{\epsilon_i\}$. For convenience we will pick the unoriented frame field to be space-fixed, so that

$$D\epsilon_i = 0\,. \tag{24.14}$$

We then have, from (24.12),

$$\begin{aligned} 0 = D\epsilon = D(R\varepsilon) &= DR\,\varepsilon + R\,D\varepsilon = dR\,R^{-1}\,\epsilon + R\,\omega\varepsilon \\ &= dR\,R^{-1}\,\epsilon + R\omega R^{-1}\,\epsilon = (dRR^{-1} + R\omega R^{-1})\,\epsilon\,. \end{aligned} \tag{24.15}$$

Thus

$$dR\,R^{-1} + R\omega R^{-1} = 0\,, \tag{24.16}$$

which implies

$$\omega = -R^{-1}\,dR\,. \tag{24.17}$$

Under a *gauge transformation* $\varepsilon \to \varepsilon' = g\varepsilon$, we have

$$\epsilon = R\varepsilon = Rg^{-1}\,\varepsilon'\,, \tag{24.18}$$

that is

$$\epsilon = R'\,\varepsilon'\,, \tag{24.19}$$

where

$$R' = R\,g^{-1}\,. \tag{24.20}$$

Hence the connection matrix ω' with respect to the frame field $\boldsymbol{\varepsilon}'$ is given by

$$\begin{aligned}\omega' &= -(R')^{-1}\,dR' = -(Rg^{-1})^{-1}\,d(Rg^{-1})\\ &= -gR^{-1}\,(dR\,g^{-1} + R\,dg^{-1}) = g\,(-R^{-1}\,dR)\,g^{-1} - g\,dg^{-1}\,.\end{aligned} \tag{24.21}$$

Since $g\,dg^{-1} = -(dg)\,g^{-1}$ (which follows from the fact that $gg^{-1} = 1$), we have

$$\omega' = g\omega g^{-1} + (dg)\,g^{-1}\,. \tag{24.22}$$

This is exactly the gauge transformation rule for a connection matrix [cf. (4.19)]. So (24.17) is indeed a valid expression for a connection.

Now we recall that with respect to the $\{\varepsilon_i\}$ frame field, the connection matrix ω_i^j is related to the angular velocity $\boldsymbol{\omega} = \omega^1\,\varepsilon_1 + \omega^2\,\varepsilon_2 + \omega^3\,\varepsilon_3$ by the following:

$$\frac{\omega_i^j}{dt} = \varphi_i^j = \varepsilon_i{}^j{}_k\,\omega^k\,. \tag{24.23}$$

The components of the angular velocity ω^i can be calculated by expressing the total conserved angular momentum of the shape-changing object with respect to the frame field $\{\epsilon_i\}$. We have, on formally assuming that the object is composed of a number of discrete mass points $m^{(n)}$, each with position vector $\boldsymbol{r}^{(n)} = \tilde{x}^{(n)i}\,\boldsymbol{\varepsilon}_i = x^i\,\boldsymbol{\epsilon}_i$,

$$\begin{aligned}\boldsymbol{L} &= \sum_n m^{(n)}\boldsymbol{r}^{(n)} \times \frac{d}{dt}\left(\tilde{x}^{(n)i}\,\boldsymbol{\varepsilon}_i\right)\\ &= \sum_n m^{(n)}\boldsymbol{r}^{(n)} \times \left(\dot{\tilde{x}}^{(n)i}\,\boldsymbol{\varepsilon}_i + \tilde{x}^{(n)i}\,\frac{D\boldsymbol{\varepsilon}_i}{dt}\right)\\ &= \varepsilon^i_{jk}\sum_n m^{(n)}\,\tilde{x}^{(n)j}\dot{\tilde{x}}^{(n)k}\,\boldsymbol{\varepsilon}_i + \sum_n m^{(n)}\,\boldsymbol{r}^{(n)} \times \left(\tilde{x}^{(n)i}\,\frac{\omega_i^j}{dt}\,\boldsymbol{\varepsilon}_j\right)\,.\end{aligned} \tag{24.24}$$

On defining

$$\tilde{L}^i \equiv \varepsilon^i_{jk}\sum_n m^{(n)}\,\tilde{x}^{(n)j}\dot{\tilde{x}}^{(n)k}\,, \tag{24.25}$$

the first term on the right-hand side of (24.24) appears as $\tilde{L}^i\,\boldsymbol{\varepsilon}_i$, while the second term can be manipulated as follows:

$$\begin{aligned}
&\sum_n m^{(n)}\, \boldsymbol{r}^{(n)} \times \left(\tilde{x}^{(n)i}\, \frac{\omega_i^j}{dt}\, \boldsymbol{\varepsilon}_j \right) = \sum_n m^{(n)}\, \varepsilon^i_{jk} \tilde{x}^{(n)j} \tilde{x}^{(n)l}\, \frac{\omega_l^k}{dt}\, \boldsymbol{\varepsilon}_i \\
&= \sum_n m^{(n)}\, \varepsilon^i_{jk} \tilde{x}^{(n)j} \tilde{x}^{(n)l} \varepsilon_l{}^k{}_m \omega^m\, \boldsymbol{\varepsilon}_i = \sum_n m^{(n)}\, \varepsilon^{ki}{}_j \varepsilon_{kml}\, \tilde{x}^{(n)j} \tilde{x}^{(n)l} \omega^m\, \boldsymbol{\varepsilon}_i \\
&= \sum_n m^{(n)} (\delta^i_m \delta_{jl} - \delta^i_l \delta_{jm})\, \tilde{x}^{(n)j} \tilde{x}^{(n)l} \omega^m\, \boldsymbol{\varepsilon}_i \\
&= \sum_n m^{(n)}\, (\tilde{x}^{(n)j} \tilde{x}^{(n)}_j \omega^i - \tilde{x}^{(n)}_j \tilde{x}^{(n)i} \omega^j)\, \boldsymbol{\varepsilon}_i \\
&= \sum_n m^{(n)} \left((\tilde{x}^{(n)})^2 \delta^i_j - \tilde{x}^{(n)i} \tilde{x}^{(n)}_j \right) \omega^j\, \boldsymbol{\varepsilon}_i \equiv \tilde{I}^i_j \omega^j\, \boldsymbol{\varepsilon}_i \;,
\end{aligned} \tag{24.26}$$

where

$$\tilde{I}^i_j \equiv \delta^i_j (\tilde{x}^{(n)})^2 - \tilde{x}^{(n)i} \tilde{x}^{(n)}_j = \tilde{I}^j_i \tag{24.27}$$

is the *symmetric* inertia tensor for a particular shape with respect to the frame $\{\boldsymbol{\varepsilon}_i\}$. Setting $\boldsymbol{L} = 0$ and using (24.24), we then have

$$\tilde{I}^i_j \omega^j = -\tilde{L}^i \tag{24.28}$$

(note the negative sign), which implies

$$\omega^k = -(\tilde{I}^{-1})^k_l \tilde{L}^l \,. \tag{24.29}$$

By (24.23) the connection matrix (24.13) (with respect to the frame field $\{\boldsymbol{\varepsilon}_i\}$) is then given by

$$\varphi^j_i \equiv \frac{\omega_i^j}{dt} = \Gamma^j_{ik}\, \frac{d\theta^k}{dt} = -\varepsilon_i{}^j{}_k\, (\tilde{I}^{-1})^k_l \tilde{L}^l \,. \tag{24.30}$$

Equation (24.17) also implies

$$\left(R^{-1} \frac{dR}{dt} \right)^j_i = \varepsilon_i{}^j{}_k\, (\tilde{I}^{-1})^k_l \tilde{L}^l \,. \tag{24.31}$$

Note that even though the angular momentum $\boldsymbol{L}$ vanishes, the "apparent" components $\tilde{L}^i$ of the angular momentum defined in (24.25) (with respect to the frame field $\boldsymbol{\varepsilon}_i$) for the calculation of the connection matrix ω do not necessarily vanish. This somewhat confusing situation is ultimately traceable to the fact that the angular velocity (derived from the connection matrix) and hence the angular momentum does not transform tensorially, but like a connection. So if the initial angular momentum of a deformable body vanishes, the rotation matrix R determining the change of orientation of the body while its shape changes is determined completely by the conservation of angular momentum.

Given a parametrized curve $\theta(t)$ in shape space, our objective is to calculate the time dependence of the rotation matrix $R_i^j(t)$. The differential equations governing this time evolution can be obtained from the condition $D\epsilon_i/dt = 0$ [cf. (24.14)]. By (24.11) again

$$\frac{dR_i^j}{dt}\,\varepsilon_j + R_i^j\,\frac{D\varepsilon_j}{dt} = \frac{dR_i^l}{dt}\,\varepsilon_l + R_i^j\varphi_j^l\,\varepsilon_l = 0\,. \tag{24.32}$$

This yields the following set of coupled first-order differential equations for the rotation matrix elements $R_i^j(t)$:

$$\frac{dR_i^j}{dt} + R_i^l\,\varphi_l^j = 0\,, \tag{24.33}$$

or

$$\frac{dR_i^j}{dt} + R_i^l\,\Gamma_{lk}^j(\theta^1,\ldots,\theta^N)\,\frac{d\theta^k}{dt} = 0\,, \tag{24.34}$$

where the Christoffel symbols Γ_{ik}^j of the connection matrix ω_i^j are obtained from the last equality of (24.30). Note that only the matrix elements in each row are coupled together.

Problem 24.3 Verify (24.31) by calculating $R^{-1}\dot{R}$ directly. Start by writing the total angular momentum in terms of the frame field $\{\epsilon_i\}$:

$$\boldsymbol{L} = \varepsilon^i{}_{jk}\sum_n m^{(n)}\,x^{(n)j}\,\dot{x}^{(n)k}\,\epsilon_i\,. \tag{24.35}$$

Then impose the conservation requirement $\boldsymbol{L} = 0$ and, since $D\epsilon_i/dt = 0$, set

$$L^i = \varepsilon^{ijk}\sum_n m^{(n)}\,x_j^{(n)}\,\dot{x}_k^{(n)} = 0\,. \tag{24.36}$$

Proceed by using

$$x_i^{(n)}(t) = R_i^j(t)\,\tilde{x}_j^{(n)}(t) \tag{24.37}$$

in (24.35).

Chapter 25

Force and Curvature

In this chapter we will illustrate the intimate relationship between the physical concept of force and the geometrical concept of curvature, within the classical mechanical setting of constrained motion on a curved surface. In so doing we will also develop some elementary, but quite powerful differential geometric techniques, based on Cartan's method of moving frames [cf. Chapter 9], for the study of two-dimensional curved surfaces. The well-known classical results presented here were first discovered by the eminent 18th century mathematician F. Gauss, who was in fact universally recognized as the founder of modern differential geometry. As a historical note, at least among most physicists, differential geometry (in the form of *Riemannian geometry*) was not recognized as being a vital mathematical tool in physics until the advent of Einstein's general theory of relativity in the second decade of the 20th century – a classical theory of gravitation in which the gravitational force was in fact explained in terms of the curvature of four-dimensional spacetime.

We begin by casting Newton's Second Law in a more geometrical light. Consider a classical trajectory $\boldsymbol{r}(t)$ of a particle of mass m and set $d\boldsymbol{r} = ds\, \boldsymbol{e}_1$, where ds is the element of arclength of the trajectory and $\boldsymbol{e}_1$ the unit *tangent vector* to the trajectory in the forward direction. The instantaneous speed v is given by $v = ds/dt$. Then $d\boldsymbol{r}/dt = v\boldsymbol{e}_1$ and the force is given by $\boldsymbol{F} = md^2\boldsymbol{r}/dt^2 = md(v\boldsymbol{e}_1)/dt$. We define the local **geodesic curvature** $\kappa_g(s)$ of a **space curve** by

$$d\boldsymbol{e}_1/ds = \kappa_g\, \boldsymbol{e}_2 \;, \tag{25.1}$$

where $\boldsymbol{e}_2$ is the unit *normal vector* to the space curve as determined by (25.1), such that $\{\boldsymbol{e}_1, \boldsymbol{e}_2, \boldsymbol{e}_3\}$ forms an oriented orthonormal frame along the curve. This formula is one of the so-called **Frenet formulas** in classical differential geometry (the other two being $d\boldsymbol{e}_2/ds = -\kappa_g \boldsymbol{e}_1 + \tau \boldsymbol{e}_3$ and $d\boldsymbol{e}_3/ds = -\tau\boldsymbol{e}_2$, where $\tau(s)$ is called the *torsion* of the curve). The Frenet formulas, which together can be written in the matrix form

$$\frac{d}{ds}\begin{pmatrix} \boldsymbol{e}_1 \\ \boldsymbol{e}_2 \\ \boldsymbol{e}_3 \end{pmatrix} = \begin{pmatrix} 0 & \kappa_g & 0 \\ -\kappa_g & 0 & \tau \\ 0 & -\tau & 0 \end{pmatrix}\begin{pmatrix} \boldsymbol{e}_1 \\ \boldsymbol{e}_2 \\ \boldsymbol{e}_3 \end{pmatrix},$$

in fact trace back (though not historically) to the notion of the covariant derivative of vector fields restricted (pulled back) to the curve [cf. (4.8)]. So we can write

$$D\boldsymbol{e}_1 = \omega_1^2\, \boldsymbol{e}_2 , \qquad D\boldsymbol{e}_2 = \omega_2^1\, \boldsymbol{e}_1 + \omega_2^3\, \boldsymbol{e}_3 ,$$

where $\omega_1^2 = \kappa_g(s)ds$ and $\omega_2^3 = \tau(s)ds$ are 1-forms on $\mathbb{R}^3$ pulled back to the trajectory curve.

Since $d\boldsymbol{e}_1/dt = (d\boldsymbol{e}_1/ds)(ds/dt) = v\, d\boldsymbol{e}_1/ds$, we have the following somewhat unconventional expression for the force:

$$\boldsymbol{F} = m\left(v^2\kappa_g\, \boldsymbol{e}_2 + \frac{dv}{dt}\, \boldsymbol{e}_1\right) . \tag{25.2}$$

This expression exhibits the direct relationship between force and curvature in a transparent and simple way. It is interesting to note immediately that in the case of uniform circular motion, where $dv/dt = 0$ and $\kappa_g = 1/r$ (r being the radius of the circle), the above expression reduces to the familiar one [$\boldsymbol{F} = (mv^2/r)\, \boldsymbol{e}_2$] for the *centripetal force.*

To further illustrate the fundamental relationship between force and curvature we consider the kinematics of a point particle constrained to move on a smooth, oriented two-dimensional surface S, viewed as being embedded in the three-dimensional Euclidean space $\mathbb{E}^3$. For this purpose we adopt the formalism presented in the first part of Chapter 9 for *Cartan's method of moving frames* and develop a framework for a differential geometric description of the surface, eventually leading to an *intrinsic* definition of the local (Gaussian) curvature of the surface. As we shall see, the possibility of defining and determining the curvature of a two-dimensional manifold intrinsically, without referring to its embedding in a higher-dimensioanl manifold, was a momentous discovery by Gauss that initiated the modern viewpoint in the study of differential geometry. This intrinsic viewpoint was later developed fully by B. Riemann, as Riemannian geometry, for spaces of arbitrary dimensions (see Chapter 27).

Introduce an orthonormal frame field $\{\boldsymbol{e}_i\}$ ($i = 1, 2, 3$) everywhere on S such that $\boldsymbol{e}_1$ and $\boldsymbol{e}_2$ are always tangent to the surface and $\boldsymbol{e}_3$ is normal to the surface, and the orientation of $\{\boldsymbol{e}_i\}$ is consistent with a chosen orientation of $\mathbb{E}^3$ (right- or left-handed). Obviously, $\{\boldsymbol{e}_1, \boldsymbol{e}_2, \boldsymbol{e}_3\}$ induces an oriented frame field $\{\boldsymbol{e}_1, \boldsymbol{e}_2\}$ on S, considered as a submanifold of $\mathbb{E}^3$. Such a frame field is usually called a **Darboux frame field** on a submanifold. Let $\boldsymbol{R}$ be the position vector of a point $p \in S$ with respect to a fixed frame, (9.10) then specializes to

$$d\boldsymbol{R} = \omega^1\boldsymbol{e}_1 + \omega^2\boldsymbol{e}_2 , \quad \omega^3 = 0 , \tag{25.3}$$

where the one-forms ω^i are given by (9.8). The first of the above equations implies that the square of the length of the line element in S, or the **first fundamental form** of S, is given by

$$I = d\boldsymbol{R} \cdot d\boldsymbol{R} = (\omega^1)^2 + (\omega^2)^2 \,, \tag{25.4}$$

and that the *area element* on S is

$$dA = \omega^1 \wedge \omega^2 \,. \tag{25.5}$$

The structure equations of (9.13), written out explicitly, read

$$d\omega^1 = \omega^2 \wedge \omega_2^1 \,, \quad d\omega^2 = \omega^1 \wedge \omega_1^2 \,, \tag{25.6}$$

$$0 = \omega^1 \wedge \omega_1^3 + \omega^2 \wedge \omega_2^3 \,, \tag{25.7}$$

while those of (9.14) give

$$d\omega_1^2 = \omega_1^3 \wedge \omega_3^2 \,, \tag{25.8}$$

$$d\omega_1^3 = \omega_1^2 \wedge \omega_2^3 \,, \quad d\omega_2^3 = \omega_2^1 \wedge \omega_1^3 \,. \tag{25.9}$$

In the classical differential geometry of surfaces, (25.8) is known as **Gauss' equation**, and the equations in (25.9) are known as **Codazzi's equations.**

Since ω^1 and ω^2 are linearly independent, it follows from (25.7) and *Cartan's Lemma* [cf. discussion following (2.35)] that

$$\omega_1^3 = h_{11}\omega^1 + h_{12}\omega^2 \,, \quad \omega_2^3 = h_{21}\omega^1 + h_{22}\omega^2 \,, \quad h_{12} = h_{21} \,, \tag{25.10}$$

where h_{ij} $(i, j = 1, 2)$ are functions on S. These functions turn out to be coefficients of the **second fundamental form**, defined as

$$II \equiv -d\boldsymbol{R} \cdot d\boldsymbol{e}_3 = \omega^1\omega_1^3 + \omega^2\omega_2^3 \,, \tag{25.11}$$

where the second equality follows from the equation for $d\boldsymbol{e}_3$ in (9.10) and the antisymmetry condition (9.16). Indeed, applying (25.10) to (25.11) we have

$$II = h_{11}(\omega^1)^2 + 2h_{12}\omega^1\omega^2 + h_{22}(\omega^2)^2 \,. \tag{25.12}$$

Note that the First and Second Fundamental Forms as given by (25.4) and (25.11), respectively, are *not* exterior differential forms.

The second fundamental form gives rise to a self-adjoint transformation $W : T_x(S) \to T_x(S)$ on the tangent space at each point $x \in S$, called the **Weingarten transformation**, defined by:

$$W(\boldsymbol{X}) \equiv -\langle \boldsymbol{X}, d\boldsymbol{e}_3 \rangle = \langle \boldsymbol{X}, \omega_1^3 \rangle \, \boldsymbol{e}_1 + \langle \boldsymbol{X}, \omega_2^3 \rangle \, \boldsymbol{e}_2 \,, \quad \boldsymbol{X} \in T_x(S) \,. \tag{25.13}$$

With respect to the frame field $\{\boldsymbol{e}_1, \boldsymbol{e}_2\}$ on S, the matrix representation of the Weingarten transformation is then given by

$$W = \begin{pmatrix} h_{11} & h_{21} \\ h_{12} & h_{22} \end{pmatrix} . \tag{25.14}$$

The eigenvalues of this matrix are

$$\kappa_1, \kappa_2 = H \pm \sqrt{H^2 - K}\ , \tag{25.15}$$

where

$$H \equiv \frac{1}{2}(h_{11} + h_{22})\ , \qquad K \equiv h_{11}h_{22} - h_{12}^2 = \kappa_1\kappa_2\ . \tag{25.16}$$

κ_1, κ_2, H and K are all independent of the choice of Darboux frames. κ_1 and κ_2 are called the **principal curvatures** of the surface S at a point $x \in S$. If $\kappa_1 \neq \kappa_2$, they are the extremal (maximum and minimum) values of the normal curvature. The corresponding eigenvectors are called the **principal axes** of S at x. Geometrically $h_{ii}(x)$ is the **normal curvature** of the curve on S (at the point x along the direction $\boldsymbol{e}_i$) formed by the intersection of the surface S and the plane determined by the normal vector $\boldsymbol{e}_3$ and $\boldsymbol{e}_i$. H and K are called the **mean curvature** and the **Gaussian (total) curvature** of S at x, respectively.

From the second equation of (9.10) we obtain by projection the induced *Levi-Civita connection* (specified by a covariant derivative D) on the tangent bundle $\pi : T(S) \to S$ [cf. discussion following (9.16)]:

$$D\boldsymbol{e}_1 = \omega_1^2 \boldsymbol{e}_2\ , \quad D\boldsymbol{e}_2 = \omega_2^1 \boldsymbol{e}_1\ . \tag{25.17}$$

This connection is determined uniquely by (25.6) and the antisymmetry property $\omega_1^2 + \omega_2^1 = 0$ [cf. (9.16)]. On the other hand, using (25.10) in the Gauss equation (25.8), we have

$$d\omega_1^2 = -\omega_1^3 \wedge \omega_2^3 = -(h_{11}h_{22} - h_{12}^2)\,\omega^1 \wedge \omega^2 = -K\,\omega^1 \wedge \omega^2\ , \tag{25.18}$$

where the last equality follows from the second equation of (25.16). This equation is a most remarkable result. It shows that *the Gaussian curvature K, being determined by $d\omega_1^2, \omega^1$ and ω^2, is an intrinsic quantity of the surface S* (**Gauss' Theorema Egregium**). On the other hand, (25.1) implies that the (geodesic) curvature of a space curve cannot be defined intrinsically (since it is given in terms of a normal direction to the tangent direction of the curve).

The geometrical meaning of the Gaussian curvature at some point on a 2-dimensional surface S can be appreciated more intuitively by means of the so-called **Gauss map** $G : S \longrightarrow S^2 \subset \mathbb{R}^3$, where S^2 is a unit 2-dimensional sphere, defined as follows. For any point $p \in S$, the normal vector $\boldsymbol{e}_3$ to the surface S at the point is specified. Parallelly transport $\boldsymbol{e}_3$ in $\mathbb{R}^3$ from p to the center of the sphere. G then maps $p \in S$ to a point on the surface of the sphere marked by the intersection of the the tip of the unit vector $\boldsymbol{e}_3$ and the sphere (see Fig. 25.1). We write

$$G(p) = \boldsymbol{e}_3(p)\ . \tag{25.19}$$

Similar to (25.4), the **third fundamental form** of the surface S is defined by

$$III \equiv d\boldsymbol{e}_3 \cdot d\boldsymbol{e}_3 = (\omega_3^1)^2 + (\omega_3^2)^2\ , \tag{25.20}$$

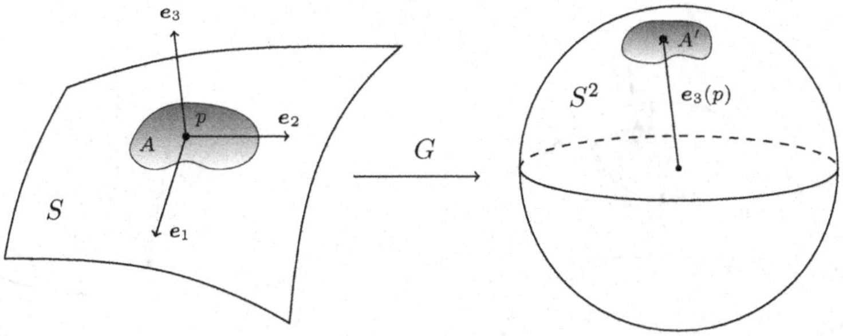

Fig. 25.1

where the second equality follows from $de_3 = \omega_3^1 e_1 + \omega_3^2 e_2$. This can be considered as the pullback of the first fundamental form (length element squared) on the unit sphere S^2 by the Gauss map back to the surface S:

$$G^*((ds_{S^2})^2) = (\omega_3^1)^2 + (\omega_3^2)^2 \, . \tag{25.21}$$

Thus, if $d\sigma$ is an area element on S^2, then

$$G^*(d\sigma) = \omega_3^1 \wedge \omega_3^2 \, . \tag{25.22}$$

If we denote the area element $\omega^1 \wedge \omega^2$ on S by dA, then (25.18) implies that the Gaussian curvature K is given by

$$K = \frac{G^*(d\sigma)}{dA} \, . \tag{25.23}$$

This result yields the following intuitive interpretation of K. Suppose the image of a domain D on the surface S under the Gauss map is a domain D' on the spherical surface S^2; and the area of D is A while that of D' is A'. Then the magnitude of the Gaussian curvature of S at the point p is given by

$$|K(p)| = \lim_{D \to p} \frac{A'}{A} \, . \tag{25.24}$$

It is immediately clear that if S is a planar surface, then the Gauss map would send all points in S to a single point on S^2. So for any region D in S with area $A \neq 0$, $A' = 0$, which implies $K = 0$.

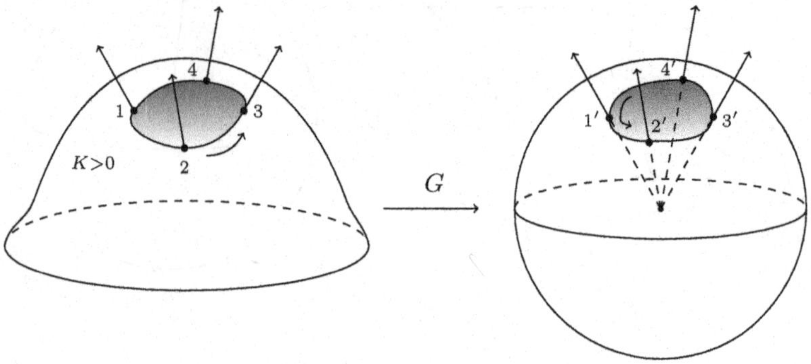

Fig. 25.2

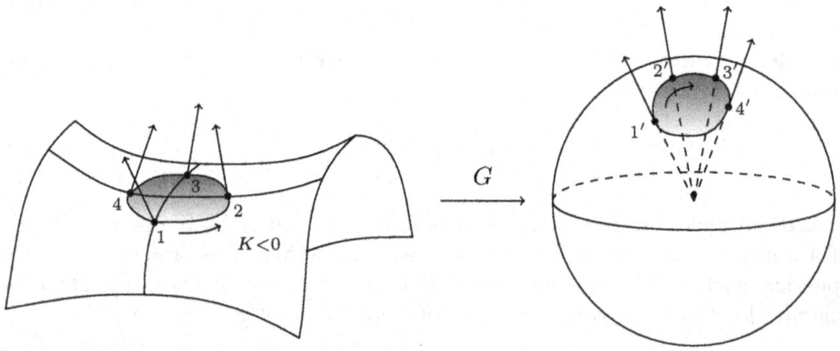

Fig. 25.3

The sign of the Gaussian curvature K also carries geometric significance. Points $p \in S$ at which $K(p)$ is positive, negative, and zero are called **elliptic points**, **hyperbolic points**, and **parabolic points**, respectively. Regions on a surface consisting entirely of elliptic or hyperbolic points can be characterized with the help of the Gauss map in a simple way. If we traverse a small closed curve without *double points* around an elliptic point in an elliptic region, the image points of the curve under the Gauss map on S^2 will also be a closed curve without double points, traversed in the *same sense* as the original curve (see Fig. 25.2). A small closed curve without double points around a hyperbolic point in a hyperbolic region will also be mapped by the Gauss map onto a closed curve without double points on S^2, but this time the sense of traversal is reversed (see Fig. 25.3). We see that an elliptic region is "cupped shaped", while a hyperbolic region is "saddle-shaped". A typical surface consisting of an elliptic region and a hyperbolic region separated by a curve consisting of parabolic points is a bell-shaped surface (see Fig. 25.4). Such a curve (at which $K = 0$) must be present since K must vary continuously on the surface. Another surface of this kind is a torus. The image under the Gauss map of the boundary curve (without double points) of a region on the torus consisting of elliptic, hyperbolic, and parabolic points is shown in Fig. 25.5. Note that all parabolic points map to a single point on S^2, which is a double point on the image curve – a figure eight. For an introductory but in-depth discussion of the geometry and topology of two-dimensional surfaces, the reader may consult the classic work by Hilbert and Cohn-Vossen (Hilbert and Cohn-Vossen 1952).

Problem 25.1 Consider a spherical surface (two-dimensional) of radius r, with local coordinates θ (polar angle) and ϕ (azimuthal angle). If we imagine that the surface is embedded in three-dimensional Euclidean space, the arclength element ds in Euclidean space is given by

$$ds^2 = dr^2 + r^2\,d\theta^2 + r^2\sin^2\theta\,d\phi^2\ .$$

The *First Fundamental Form* (25.4) of the spherical surface is then (on setting $dr = 0$)

$$I = (\omega^1)^2 + (\omega^2)^2 = r^2\,d\theta^2 + r^2\sin^2\theta\,d\phi^2\ .$$

This implies that

$$\omega^1 = r\,d\theta\ , \qquad \omega^2 = r\sin\theta\,d\phi\ ,$$

and so the *area element* on the spherical surface is

$$\omega^1 \wedge \omega^2 = r^2\sin\theta\,d\theta\wedge d\phi\ .$$

Taking into account the radial as the third dimension, we can write $\omega^3 = dr$. The connection matrix of one-forms given by (4.32) can be rearranged (with $\boldsymbol{e}_1 = \boldsymbol{e}_\theta, \boldsymbol{e}_2 = \boldsymbol{e}_\phi, \boldsymbol{e}_3 = \boldsymbol{e}_r$) as:

$$(\omega_i^j) = \begin{pmatrix} 0 & \cos\theta\,d\phi & -d\theta \\ -\cos\theta\,d\phi & 0 & -\sin\theta\,d\phi \\ d\theta & \sin\theta\,d\phi & 0 \end{pmatrix}\ .$$

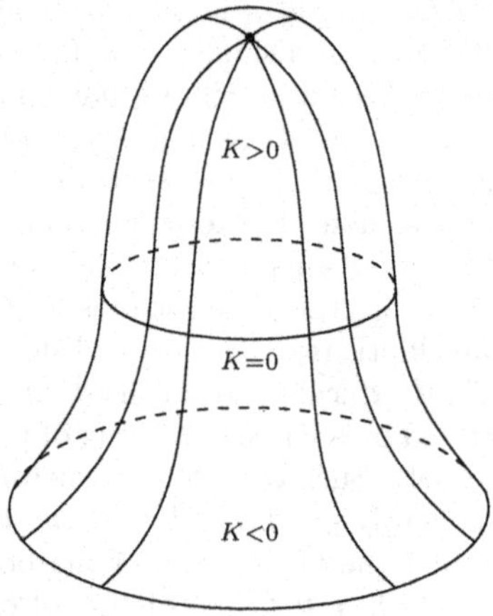

Fig. 25.4

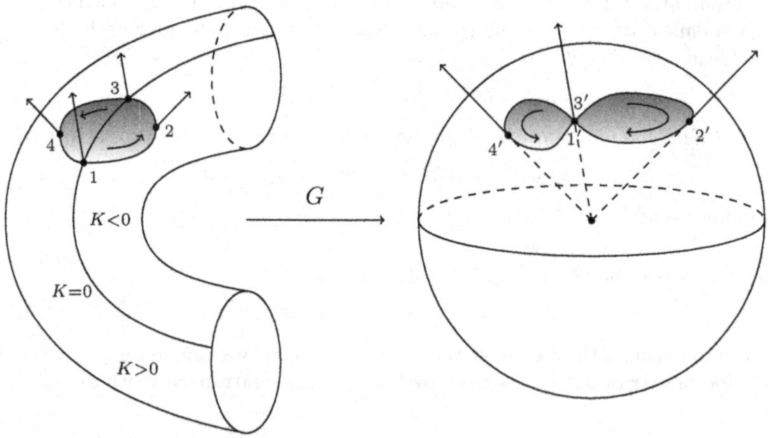

Fig. 25.5

(a) With the above explicit expressions for the *coframe field* ω^i ($i = 1, 2, 3$) and the connection matrix elements ω_i^j ($i, j = 1, 2, 3$) verify all the *integrability conditions* given by (25.6) to (25.9).

(b) Use (25.10) to check that, with respect to the frame field $\{e_1, e_2\}$, the *Weingarten transformation* matrix is given by

$$W = \begin{pmatrix} h_{11} & h_{12} \\ h_{21} & h_{22} \end{pmatrix} = \begin{pmatrix} -1/r & 0 \\ 0 & -1/r \end{pmatrix} .$$

Thus for a spherical surface of radius r, the *normal curvatures* of arcs tangent to e_1 and e_2 are both given by $h_{11} = h_{22} - -1/r$; the *principal curvatures* κ_1 and κ_2 are both given by $\kappa_1 = \kappa_2 = -1/r$, and the *Gaussian curvature* is given by $K = 1/r^2$.

(c) Use the explicit expressions for ω^1, ω^2 and ω_1^2 in (25.18), that is, $d\omega_1^2 = -K\,\omega^1 \wedge \omega^2$, to verify the result for the Gaussian curvature K obtained in (b).

(d) Use the explicit expressions for ω^i, ω_i^j and h_{ij} to verify that the right-hand sides of (25.11) and (25.12) both lead to the same expression for the *Second Fundamental Form* for the spherical surface of radius r:

$$II = -r\,d\theta^2 - r\sin^2\theta\,d\phi^2 .$$

Problem 25.2 Consider a two-dimensional torus generated by the surface of revolution about the z-axis of a circle perpendicular to the xy-plane and of radius R_0. The center of the circle lies on the xy-plane at a distance of $R\,(> R_0)$ from the origin. Let α be the angular coordinate around the circle and ϕ be the azimuthal angular coordinate (see Fig. 25.6).

(a) Show that the First Fundamental Form (arclength element squared) of the toroidal surface is given by

$$ds^2 = I = (\omega^1)^2 + (\omega^2)^2 = R_0^2\,d^2\alpha + (R - R_0\cos\alpha)^2\,d^2\phi .$$

Hence we can make the following identifications:

$$\omega^1 = R_0\,d\alpha , \qquad \omega^2 = (R - R_0\cos\alpha)\,d\phi .$$

The area element on the toroidal surface is then

$$\omega^1 \wedge \omega^2 = R_0(R - R_0\cos\alpha)\,d\alpha \wedge d\phi .$$

Let the unit-vector frame field $\{e_1, e_2\}$ be as shown in Fig. 25.6, satisfying the duality condition $\langle\, e_i\,,\,\omega^j\,\rangle = \delta_i^j$. Thus

$$e_1 = e_\alpha = \frac{1}{R_0}\frac{\partial}{\partial\alpha} , \qquad e_2 = e_\phi = \frac{1}{(R - R_0\cos\alpha)}\frac{\partial}{\partial\phi} , \qquad e_3 = e_{R_0} = \frac{\partial}{\partial R_0} ,$$

where e_3 is the unit normal vector to the toroidal surface.

(b) Verify that the *connection matrix* of one-forms given by

$$(\omega_i^j) = \begin{pmatrix} 0 & \sin\alpha\, d\phi & d\alpha \\ -\sin\alpha\, d\phi & 0 & -\cos\alpha\, d\phi \\ -d\alpha & \cos\alpha\, d\phi & 0 \end{pmatrix}$$

satisfies the structure equations (25.6) and (25.7). (This result for the connection matrix will be derived from first principles in Chapter 27.)

(c) Use the above result and (25.18) to show that the Gaussian curvature K for the toroidal surface is given by

$$K = -\frac{\cos\alpha}{R_0(R - R_0\cos\alpha)} \,.$$

Note that for $-\pi/2 < \alpha < \pi/2$ (the inner surface), $K < 0$ and the torus consists of *hyperbolic points*; while for $\pi/2 < \alpha < 3\pi/2$ (the outer surface) $K > 0$ and the torus consists of *elliptic points*. The two circles ($\alpha = \pi/2$ and $\alpha = 3\pi/2$) consist of *parabolic points*, at which $K = 0$.

(d) Use (25.10) to show that, with respect to the frame field $\{\mathbf{e}_1, \mathbf{e}_2\}$, the *Weingarten transformation matrix* is given by

$$W = \begin{pmatrix} h_{11} & h_{12} \\ h_{21} & h_{22} \end{pmatrix} = \begin{pmatrix} \dfrac{1}{R_0} & 0 \\ 0 & -\dfrac{\cos\alpha}{R - R_0\cos\alpha} \end{pmatrix} .$$

Since this matrix is diagonal, the diagonal elements give the *principal curvatures* of the torus.

(e) Use (25.16) to verify that the Gaussian curvature is given by the result in (c) and that the *mean curvature* H is given by

$$H = \frac{R - 2R_0\cos\alpha}{R_0(R - R_0\cos\alpha)} \,.$$

Equation (25.18) provides an important example of the process of *transgression*, which we will now explain. The orthonormal *frame bundle* on the surface S can be identified with the *circle bundle* of unit tangent vectors $\pi : E \to S$ on S, where the local fiber coordinate of the latter can be taken to be the angle α between a tangent vector $\mathbf{e}'_1$ and $\mathbf{e}_1$, with $\{\mathbf{e}_1, \mathbf{e}_2\}$ being a local frame field on S (or a local section of E). The pullback forms (recall Def. 6.2) $(\omega')^i \equiv \pi^*(\omega^i)$ $(i = 1, 2)$, $(\omega')_1^2 \equiv \pi^*(\omega_1^2)$ are globally well-defined linearly independent 1-forms on E. Thus they constitute a coframe field on E, which implies that the total space E is **parallelizable**, even though the base manifold S may not be. In fact, setting $\mathbf{e}'_i = g_i^j\, \mathbf{e}_j$ with

$$(g_i^j) = \begin{pmatrix} \cos\alpha & \sin\alpha \\ -\sin\alpha & \cos\alpha \end{pmatrix} , \tag{25.25}$$

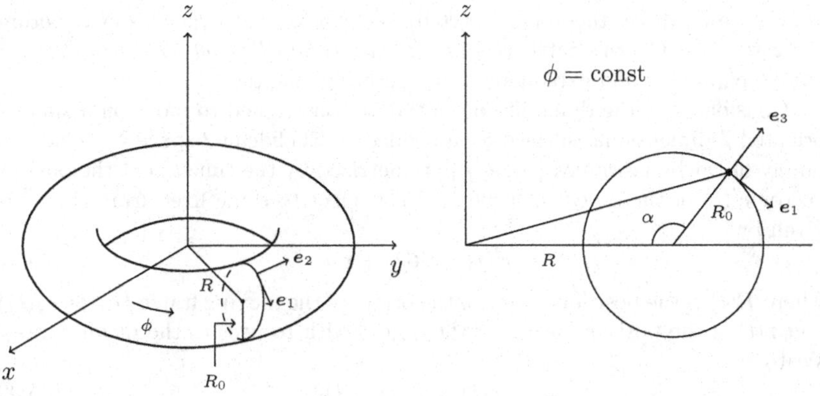

Fig. 25.6

we have

$$(\omega')^1 = (\cos\alpha)\,\omega^1 + (\sin\alpha)\,\omega^2\,, \quad (\omega')^2 = -(\sin\alpha)\,\omega^1 + (\cos\alpha)\,\omega^2\,, \tag{25.26}$$

so that

$$(\omega')^1 \wedge (\omega')^2 = \omega^1 \wedge \omega^2\,. \tag{25.27}$$

Also, the gauge transformation rule (4.19) implies

$$(\omega')^2_1 = -(\omega')^1_2 = \omega^2_1 + d\alpha\,, \quad (\omega')^1_1 = (\omega')^2_2 = 0\,. \tag{25.28}$$

Since α is a fiber coordinate, the first of the above equations confirms that $(\omega')^2_1$ is a differential one-form on the total space E. Thus, on exterior differentiation,

$$d(\omega')^2_1 = d\omega^2_1 = -K\,\omega^1 \wedge \omega^2\,, \tag{25.29}$$

where the second equality follows from (25.18). This crucial equation has to be understood very carefully. The second equality ($d\omega^2_1 = -K\,\omega^1 \wedge \omega^2$) implies that the curvature 2-form $K\,\omega^1 \wedge \omega^2$, as a closed 2-form on S, is only locally exact (as follows from **Poincaré's lemma**). The first equality,

$$d(\omega')^2_1 = -K\,\omega^1 \wedge \omega^2\,, \tag{25.30}$$

however, says that the pullback of the curvature form to the total bundle space E is globally exact. It is the exterior derivative of $(\omega')^2_1$, which is regarded as a global connection 1-form on E. Equation (25.22) states that the restriction of this connection to a fiber is $d\alpha$. Stated in another way, $(\omega')_1^2$ is the extension of $d\alpha$ to a form on E, whose exterior derivative is a form on S. Such a construction is called **transgression**. Furthermore, (25.24) implies that the pullback of the curvature form on S, being an exact form on E, can be integrated on S via pullback and Stokes theorem [see (26.6) in the next chapter]. This fact contains the essence of Chern's intrinsic proof of the *Gauss-Bonnet-Chern theorem.* It will be explained and exploited in the following chapter.

Consider now the dynamics of an object constrained to move on a smooth, oriented 2-dimensional surface S embedded in Euclidean 3-space $\mathbb{E}^3$. Choose a family of Darboux frames $\{\boldsymbol{e}_1, \boldsymbol{e}_2\}$ parametrized by the time t. Let the position vector $\boldsymbol{r}'(t)$ of the object at time t with respect to some fixed frame in $\mathbb{E}^3$ be given by

$$\boldsymbol{r}'(t) = \boldsymbol{R}(t) + \boldsymbol{r}(t) \,, \tag{25.31}$$

where $\boldsymbol{R}(t)$ is the position vector of the origin of the moving frame $\{\boldsymbol{e}_1(t), \boldsymbol{e}_2(t)\}$, and $\boldsymbol{r}(t)$ is the position vector of the object with respect to the moving frame. Write

$$\boldsymbol{r}(t) = x^i(t)\, \boldsymbol{e}_i(t) \,. \tag{25.32}$$

Then, from (9.1) and (9.2) [recall the derivations of (12.10) and (12.14)],

$$\frac{d\boldsymbol{r}'}{dt} = \varphi^i\, \boldsymbol{e}_i + \frac{dx^i}{dt}\, \boldsymbol{e}_i + x^i \varphi_i^j\, \boldsymbol{e}_j \,, \tag{25.33}$$

$$\frac{d^2\boldsymbol{r}'}{dt^2} = \frac{d\varphi^i}{dt}\, \boldsymbol{e}_i + \varphi^i \varphi_i^j\, \boldsymbol{e}_j + \frac{d^2x^i}{dt^2}\, \boldsymbol{e}_i + 2\frac{dx^i}{dt}\, \varphi_i^j\, \boldsymbol{e}_j + x^i \frac{d\varphi_i^j}{dt}\, \boldsymbol{e}_j + x^i \varphi_i^j \varphi_j^k\, \boldsymbol{e}_k \,, \tag{25.34}$$

where $\varphi^i(t) \equiv \omega^i/dt$ and $\varphi_i^j(t) \equiv \omega_i^j/dt$ [cf. (4.10)]. Suppose the net external force on the object is zero. Then the components of the acceleration with respect to the moving frame $\{\boldsymbol{e}_1, \boldsymbol{e}_2\}$ are given by

$$\frac{d^2x^i}{dt^2} = -\frac{d\varphi^i}{dt} - \varphi^j \varphi_j^i - 2\frac{dx^j}{dt}\, \varphi_j^i - x^j\, \frac{d\varphi_j^i}{dt} - x^j \varphi_j^k \varphi_k^i \,, \quad i = 1, 2 \,, \tag{25.35}$$

with the repeated indices summed from 1 to 3 and $X^3 = 0, \varphi^3 = 0$. The right-hand side of this equation (multiplied by the mass of the object) represents the non-inertial forces. As we saw in Chapter 12, the third term, depending on the velocity of the object with respect to the moving frame, is the Coriolis force; and the last term is the centrifugal force. Written out explicitly, (25.29) gives the following (in general) non-linear coupled equations for x^1 and x^2:

$$\frac{d^2x^1}{dt^2} = -\frac{d\varphi^1}{dt} + \varphi^2 \varphi_1^2 + 2\varphi_1^2\, \frac{dx^2}{dt} + \frac{d\varphi_1^2}{dt}\, x^2 + \left[(\varphi_1^2)^2 + (\varphi_1^3)^2\right] x^1 + \varphi_1^3 \varphi_2^3\, x^2 \,, \tag{25.36}$$

$$\frac{d^2x^2}{dt^2} = -\frac{d\varphi^2}{dt} - \varphi^1 \varphi_1^2 - 2\varphi_1^2\, \frac{dx^1}{dt} - \frac{d\varphi_1^2}{dt}\, x^1 + \left[(\varphi_1^2)^2 + (\varphi_2^3)^2\right] x^2 + \varphi_1^3 \varphi_2^3\, x^1 \,. \tag{25.37}$$

According to (25.10), we have

$$\varphi_1^3 = h_{11}\varphi^1 + h_{12}\varphi^2\,, \quad \varphi_2^3 = h_{21}\varphi^1 + h_{22}\varphi^2\,; \quad h_{12} = h_{21}\,. \tag{25.38}$$

If we choose e_1 and e_2 to be along the *principal axes* [see discussion following (25.16)], so that $h_{11} = \kappa_1$, $h_{22} = \kappa_2$, $h_{12} = h_{21} = 0$, the above conditions reduce to $\varphi_1^3 = \kappa_1\varphi^1$, $\varphi_2^3 = \kappa_2\varphi^2$, which imply $\varphi_1^3\varphi_2^3 = K\,\varphi^1\varphi^2$, where K is the Gaussian curvature of the surface S at a point $x \in S$ [cf. (25.16)]. The coupled equations (25.33) and (25.34), written with respect to the principal axes, then appear as

$$\frac{d^2x^1}{dt^2} = -\frac{d\varphi^1}{dt} + \varphi^2\varphi_1^2 + 2\varphi_1^2\frac{dx^2}{dt} + \left(\frac{d\varphi_1^2}{dt} + K(t)\,\varphi^1\varphi^2\right)x^2 + \left[(\varphi_1^2)^2 + (\kappa_1(t))^2(\varphi^1)^2\right]x^1\,, \tag{25.39}$$

$$\frac{d^2x^2}{dt^2} = -\frac{d\varphi^2}{dt} - \varphi^1\varphi_1^2 - 2\varphi_1^2\frac{dx^1}{dt} - \left(\frac{d\varphi_1^2}{dt} - K(t)\,\varphi^1\varphi^2\right)x^1 + \left[(\varphi_1^2)^2 + (\kappa_2(t))^2(\varphi^2)^2\right]x^2\,. \tag{25.40}$$

The right-hand sides of these equations contain three classes of terms. First, the terms involving the second time derivatives $d\varphi^1/dt$ and $d\varphi^2/dt$ reflect the contribution of the translational accelerations of the origin of the moving frame to non-inertial forces. Secondly, the terms proportional to $K, (\kappa_1)^2$ and $(\kappa_2)^2$, which will be called the curvature terms, display clearly the explicit dependence of non-inertial forces on the local values of the *Gaussian curvature* [cf. (25.16)] and the *principal curvatures* [cf. (25.15)]. These two classes of terms cannot in general be transformed away by a local change of Darboux frames (gauge transformation). Due to the general gauge transformation rule for connection one-forms (4.19) and more specifically (25.22) for the present situation, however, the third class of terms, those involving φ_1^2 and its time derivative, can be gauge-transformed away locally. Indeed, by choosing a local Darboux frame field $\{e_1', e_2'\}$ so that $(\omega')_1^2 = 0$ [and thus $(\varphi')_1^2 = (\omega')_1^2/dt = 0$] locally, this can be achieved. With respect to such a local frame field, then, (25.17) implies that

$$\frac{De_i'}{dt} = 0\,, \quad i = 1, 2\,. \tag{25.41}$$

This is recognized to be the condition for parallel displacement [corresponding to the induced Levi-Civita connection of (25.17)] of the tangent vectors e_i' along a curve C in the two-dimensional surface S traversed by the origin of the moving frame $\{e_1', e_2'\}$. Given an arbitrary choice of Darboux frame field $\{e_1, e_2\}$ along the curve C, with respect to which the induced Levi-Civita connection is ω_1^2, (25.22) implies that the condition for $\{e_1', e_2'\}$ being a parallelly displaced Darboux frame (or a parallel frame) along C is

$$\frac{d\alpha}{dt} + \varphi_1^2 = 0\,, \tag{25.42}$$

where α is the angle between $\mathbf{e}_1$ and $\mathbf{e}'_1$, measured in the positive sense given by the orientation of $\{\mathbf{e}_1, \mathbf{e}_2\}$ [see discussion before (25.19)].

With respect to a parallel frame, the equations of motion (25.33) and (25.34) reduce to

$$\frac{d^2x^1}{dt^2} = -\frac{d\varphi^1}{dt} + (\varphi_1^3)^2\, x^1 + \varphi_1^3\varphi_2^3\, x^2 \,, \tag{25.43}$$

$$\frac{d^2x^2}{dt^2} = -\frac{d\varphi^2}{dt} + (\varphi_2^3)^2\, x^2 + \varphi_1^3\varphi_2^3\, x^1 \,. \tag{25.44}$$

In these equations φ_1^3 and φ_2^3 are given by (25.32), and φ^1, φ^2 are the components of the velocity of the origin of the moving parallel frame along the directions of its axes. We give the matrix elements of the Weingarten transformation with respect to the parallel frame explicitly as follows

$$h_{11} = \kappa_1 \cos^2\alpha + \kappa_2 \sin^2\alpha \,, \quad h_{22} = \kappa_1 \sin^2\alpha + \kappa_2 \cos^2\alpha \,, \tag{25.45}$$

$$h_{12} = h_{21} = (\kappa_2 - \kappa_1)\sin\alpha\cos\alpha \,, \tag{25.46}$$

where the angle α is defined in the statement immediately following (25.36) (with $\{\mathbf{e}_1, \mathbf{e}_2\}$ being the principal axes), and is required to satisfy (25.36) (with φ_1^2 given with respect to the principal axes). The results of the above two equations are computed from the similarity transformation

$$g \begin{pmatrix} \kappa_1 & 0 \\ 0 & \kappa_2 \end{pmatrix} g^{-1} \,,$$

where g is given by (25.19).

The curvature terms, which are the last two terms in each of the right-hand sides of (25.37) and (25.38), also depend directly on the velocity components φ^1 and φ^2 of the origin of the moving parallel frame. Hence, either when the curvatures can be locally neglected, or in the limit of infinitely slow (*adiabatic*) translational motion of the moving frame (φ^1, $\varphi^2 \to 0$), the curvature terms vanish. In addition, in the adiabatic limit, the translational acceleration terms $-d\varphi^i/dt$ also vanish. When both of these sets of conditions apply, the non-inertial forces entirely vanish, but only locally. The idea that, at least locally, forces that depend on the geometrical structure of the manifold on which motion takes place (the non-inertial forces) can be transformed away is the basis of *the principe of equivalence.*

Adopting a somewhat different viewpoint, one can think of an observer living on the two-dimensional surface S who is oblivious to the existence of the third dimension along $\mathbf{e}_3$, but who is well aware of the geometry of the surface as determined by the first fundamental form (25.4) (the Riemannian metric) and the associated Levi-Civita connection given by (25.17). Such an observer would formulate the equations of motion analogous to (25.30) and (25.31) without the curvature terms (which are due to the existence of the third dimension). Assuming also that the translational acceleration of the moving frame vanishes, our observer would conclude that, at least locally, the non-inertial forces can be eliminated completely by parallel displacement of the axes of the moving frame. Suppose the curvature terms were not in fact negligible, and experimental results indeed do not support the theoretical conclusion mentioned above. Theorists among our two-dimensional beings might then be motivated to explain the discrepancy between theoretical prediction based on a two-dimensional world and experimental observation by postulating the existence of an extra dimension. This kind of reasoning is prototypical of that behind the so-called *Kaluza-Klein theories*, a class of theories designed to unify the fundamental forces of nature by the inclusion of extra dimensions, in addition to those of 4-dimensional spacetime.

Problem 25.3 Find the equations of motion on a torus.

To close this chapter on the intimate connection between force and curvature we will bring to light a historical formulation of Newton's Laws of Motion that is not commonly presented in most textbooks of mechanics, but is highly relevant to our present concern (for a more complete discussion see the classic monograph: Whittaker 1965). This is the so-called *Gauss' Principle of Least Constraint* and its corollary *Hertz' Principle of Least Curvature.*

For simplicity we will just consider the case of a single particle constrained to move on a two-dimensional surface S embedded in 3-dimensional Euclidean space, specified by a single (in general *non-holonomic*) *constraint* equation

$$\sum_{i=1}^{3} f_i(x^1, x^2, x^3)\, dx^i = 0 \,. \tag{25.47}$$

Gauss' Principle of Least Constraint states that the quantity

$$K(\boldsymbol{r}(t)) \equiv \left(\frac{d^2\boldsymbol{r}}{dt^2} - \frac{\boldsymbol{F}(\boldsymbol{r})}{m} \right)^2 , \tag{25.48}$$

where $\boldsymbol{F}(\boldsymbol{r})$ is the net force on the particle other than the constraint forces, must be a minimum for the actual trajectory $\boldsymbol{r}_0(t)$, among all the *kinematically possible trajectories*, defined to be all those curves on S which, at a particular time t, share the same position and velocity as the actual trajectory.

To prove this principle we will follow the method presented (for more general cases of multiple-particle dynamics) by Whittaker (see Whittaker 1965). First it follows from (25.47) that, for any kinematically possible trajectory $x^i(t)$ (as defined above),

$$f_i \frac{dx^i}{dt} = 0 \,. \tag{25.49}$$

Differentiating once more with respect to the time t we have

$$f_i \, \ddot{x}^i + \frac{\partial f_i}{\partial x^j} \, \dot{x}^i \dot{x}^j = 0 \, . \tag{25.50}$$

The actual trajectory $x_0^i(t)$ must necessarily satisfy this equation also. So we have

$$f_i \, \ddot{x}_0^i + \frac{\partial f_i}{\partial x_0^j} \, \dot{x}_0^i \dot{x}_0^j = 0 \, . \tag{25.51}$$

Since $\dot{x}^i = \dot{x}_0^i$, by virtue of the definition of kinematically possible paths, we have, on subtracting (25.51) from (25.50),

$$f_i \, (\ddot{x}^i - \ddot{x}_0^i) = 0 \, . \tag{25.52}$$

Since the functions f_i are in general arbitrary, (25.49) and (25.52) imply

$$\ddot{x}^i - \ddot{x}_0^i = c \, \dot{x}^i \qquad (c = \text{constant for all } i) \, . \tag{25.53}$$

The i-th component of the constraint force is given by $m\ddot{x}^i - F^i$. Since these forces do no work when the particle moves along a kinematically possible trajectory (they are perpendicular to the surface S), we have

$$\sum_i (m\ddot{x}^i - F^i) \, \dot{x}^i = 0 \, . \tag{25.54}$$

It then follows from (25.53) that

$$\sum_i (m\ddot{x}^i - F^i) \, (\ddot{x}^i - \ddot{x}_0^i) = 0 \, . \tag{25.55}$$

This equation can be straightforwardly shown to be equivalent to the following.

$$\sum_i \left(\ddot{x}^i - \frac{F^i}{m} \right)^2 = \sum_i \left(\ddot{x}_0^i - \frac{F^i}{m} \right)^2 + \sum_i (\ddot{x}^i - \ddot{x}_0^i)^2 \, . \tag{25.56}$$

Since the second term on the right-hand side is larger than or equal to zero, we must have

$$\sum_i \left(\ddot{x}^i - \frac{F^i}{m} \right)^2 \geq \sum_i \left(\ddot{x}_0^i - \frac{F^i}{m} \right)^2 , \tag{25.57}$$

which is what is asserted by Gauss' Principle of Least Constraint. Note that because of (25.54), this principle is equivalent to *D'Lambert's Principle of Virtual Work.*

If the net force $\boldsymbol{F}$ responsible for motion on the surface S vanishes, then Gauss' Principle of Least Constraint asserts that

$$K_{(\boldsymbol{F}=0)}(\boldsymbol{r}(t)) = \sum_i \left(\frac{d^2 x^i(t)}{dt^2} \right)^2 = \left| \frac{d^2 \boldsymbol{r}}{dt^2} \right|^2 \tag{25.58}$$

is minimum when $x^i(t) = x_0^i(t)$. Now consider the tangent plane $T_{\boldsymbol{r}}S$ at the point $\boldsymbol{r} \in S$. This will be a two-dimensional real vector space. Choose $\boldsymbol{e}_1$ to be a unit vector along the direction of the velocity of a kinematically possible trajectory at $\boldsymbol{r}$, and $\boldsymbol{e}_2$ to be an orthogonal unit vector in $T_{\boldsymbol{r}}S$, so that the *orientation* of the frame $\{\boldsymbol{e}_1, \boldsymbol{e}_2\}$ is consistent with that of S. Let s be the arclength along the trajectory. Then we have $d\boldsymbol{r} = ds\, \boldsymbol{e}_1$. If we are just interested in the geometry of the trajectory and not in the speed with which the particle moves along it, we can set the speed to be constant, or just set the time parameter t to be the arclength s along the trajectory: $t = s$. In either case $v =$ constant, and $dv/dt = 0$. Following the development at the beginning of this chapter, we see that [cf. (25.2)]

$$\frac{d^2\boldsymbol{r}}{dt^2} = v^2 \kappa_g\, \boldsymbol{e}_2\,, \tag{25.59}$$

where κ_g is the *geodesic curvature* of the trajectory at the point $\boldsymbol{r}$. In view of (25.58) we see that *if the externally applied force* $\boldsymbol{F}$ *vanishes, then the constrained motion of a particle on a surface is such that* $\mathbf{r}(t)$ *must have minimum geodesic curvature at each time* t*, among all kinematically possible paths.* This statement, which is a special case of Gauss' Principle of Lease Constraint, is called **Hertz' Principle of Least Curvature**. Effectively it is the same as Newton's First Law. We will explore the geometrical properties of geodesic curves in the context of classical mechanics in more detail in Chapters 29 and 30.

Problem 25.4 Show that the equations (25.55) and (25.56) are algebraically equivalent.

Chapter 26

The Gauss-Bonnet-Chern Theorem and Holonomy

As we have seen in Chapters 23 and 24, many problems in classical mechanics involve the study of *holonomy*, which is concerned with the behavior of a tangent vector when it is parallelly translated over a closed loop in a Riemannian manifold with curvature. Specifically the tangent vector does not necessarily return to itself on traversing a closed loop. Recalling the necessary condition (25.35) for a vector field $\{e'_i\}$ on a two-dimensional surface to be parallelly translated along a curve $C(t)$ on the surface and the first of Eqs. (25.22), we can calculate the angular mismatch α_C (holonomy) over a closed loop $C(t)$ by integrating the equation

$$d\alpha = -\omega_1^2 \,, \tag{26.1}$$

where the connection one form ω_1^2 giving rise to the curvature is expressed with respect to an arbitrary frame field $\{e_i\}$. In this chapter we will show that the result of the integration is equal to the integral of the Gaussian curvature K over the area enclosed by the loop modulo 2π, namely,

$$\alpha_C = \int_M K\,\omega^1 \wedge \omega^2 + 2m\pi \,, \tag{26.2}$$

where m is an integer and $C = \partial M$ with the proper orientation. This result will be derived with the help of some elegant and powerful techniques introduced in a legendary *intrinsic* proof of the classical *Gauss-Bonnet theorem* for arbitrary dimensions by S. S. Chern (intrinsic in the sense that the manifold in question is not assumed to be embedded in a higher-dimensional manifold). As a result of this ground-breaking proof, which opened the door to many developments in modern differential geometry, this central theorem was often known as *the Gauss-Bonnet-Chern Theorem*.

We first present the main ideas of Chern's proof in the simple case of a two-dimensional surface without boundary. (The main ideas and techniques of this elegant proof for manifolds of arbitrary dimension are already present in the two-dimensional case.) To begin we introduce two key concepts in the geometry and topology of manifolds: (1) the **index** I_p at a *singular point* $p \in S$ of a vector field on S is roughly (and intuitively) defined as the number of turns that the tangent field vector makes on traversing the oriented boundary curve of an infinitesimal neighborhood containing the singular point once around the singular point (always an integer); and (2) the **Euler characteristic** χ (always an integer) of a two-dimensional manifold S is a *topological invariant* of the manifold given by $\chi = f - e + v$, where f, e and v are the number of faces, number of edges, and the number of vertices of any *triangulation* of S. (For more detailed discussions of these concepts, the reader may consult any standard text on differential geometry, for example, Chern, Chen and Lam 1999.) The following theorem (stated withut proof) gives a fundamental relationship between these two concepts.

Theorem 26.1. ***(Poincaré-Hopf)*** *The sum of the indices at the various singular points of a smooth tangent vector field on a compact oriented 2-dimensional Riemannian manifold with finitely many singular points is equal to the Euler characteristic of the manifold.*

As an illustration of the Poincaré-Hopf Theorem, consider a particular triangulation of a compact, two-dimensional manifold without boundary. Construct a smooth tangent vector field on this manifold as follows. The geometrical center of each triangular face, the mid-point of each boundary line of each triangular face, and each vertex of each triangular face, are all singular points, with indices equal to $+1, -1$ and $+1$, respectively, for each of these three types of singular points (see Fig. 26.1). Thus the sum of the singular points is given by

$$\sum_p I_p = f - e + v = \chi \, ,$$

by definition of the Euler characteristic, which is independent of the triangulation.

Suppose S is a *compact* two-dimensional Riemannian manifold *without boundary*. Let θ be a tangent vector field (*local section*) on the *circle bundle* of unit tangent vectors $\pi : E \to S$ [see discussion preceding (25.19)] with only one *singular point* $p \in S$. (The proof is not essentially different for a field with multiple singular points.) According to the Poincaré-Hopf Index Theorem, such a section always exists, by choosing $I_p = \chi$, where I_p is the index of the vector field θ at p and χ is the Euler characteristic of S. Let $S' = S - \{p\}$. Then θ is globally defined on S' (θ does not have any singular point in S') and $\pi \circ \theta : S' \to S'$ is the identity map. Equation (25.24), written out more carefully in terms of the pullback map π^*, becomes

$$d(\omega')_1^2 = -\pi^*(K\, dA) \, , \tag{26.3}$$

where $(\omega')_1^2$ is a *global* one-form on the total bundle space E. A local one-form ω_1^2 defined only in S' is then given by the pullback form

$$\omega_1^2 \equiv \theta^*((\omega')_1^2) \, . \tag{26.4}$$

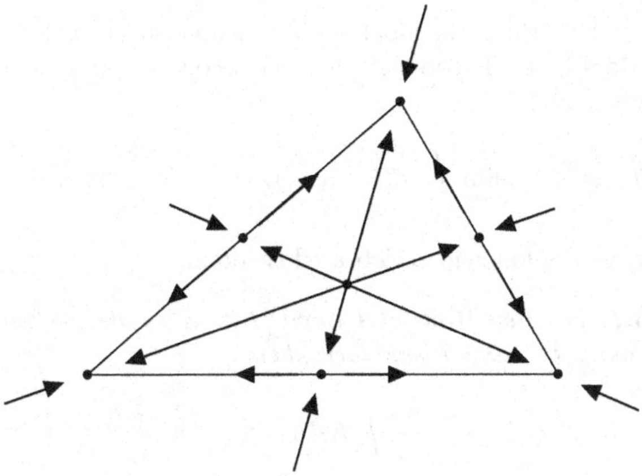

Fig. 26.1

Then we have, on S',

$$d\omega_1^2 = d(\theta^*((\omega')_1^2)) = \theta^*(d(\omega')_1^2) = -\theta^*\pi^*(KdA) = -KdA\,, \tag{26.5}$$

where the second equality follows from (6.35) and the last from the fact that $\pi \circ \theta$ is the identity on S'. This result is just the second equality of (25.23). Now, since K is continuous at p,

$$\int_S KdA = \int_{S'} KdA = -\int_{S'} d(\theta^*((\omega')_1^2)) = -\lim_{\epsilon\to 0}\int_{S_\epsilon} d(\theta^*((\omega')_1^2))\,, \tag{26.6}$$

where $S_\epsilon \equiv S - \Delta_\epsilon$ with Δ_ϵ being an ϵ-ball neighborhood of p ($\epsilon > 0$). By Stoke's theorem we then have

$$\int_S KdA = \lim_{\epsilon\to 0}\int_{\partial\Delta_\epsilon} \theta^*((\omega')_1^2)\,, \tag{26.7}$$

since S is without boundary (so $\partial S_\varepsilon = -\partial\Delta_\varepsilon$). Next choose a local section $\hat{\theta}$ defined everywhere in Δ_ϵ, including the point p, that is, $\hat{\theta}$ is everywhere non-singular in Δ_ϵ. By a gauge transformation [cf. the first of Eqs. (25.22)] we have, on $\partial\Delta_\epsilon$,

$$\theta^*((\omega')_1^2) = d\alpha + \hat{\theta}^*((\omega')_1^2)\,. \tag{26.8}$$

It follows from (26.7) that

$$\int_S KdA = \lim_{\epsilon \to 0} \int_{\partial \Delta_\epsilon} d\alpha + \lim_{\epsilon \to 0} \int_{\partial \Delta_\epsilon} \hat{\theta}^*((\omega')_1^2) \,. \tag{26.9}$$

The first integral on the right-hand side by definition yields $2\pi I_p = 2\pi\chi$ (by the Poincaré-Hopf Index Theorem), while the second integral vanishes by using Stoke's theorem again:

$$\lim_{\epsilon \to 0} \int_{\partial \Delta_\epsilon} \hat{\theta}^*((\omega')_1^2) = \lim_{\epsilon \to 0} \int_{\Delta_\epsilon} d(\hat{\theta}^*((\omega')_1^2)) = -\lim_{\epsilon \to 0} \int_{\Delta_\epsilon} KdA = 0 \,. \tag{26.10}$$

We thus arrive at the following celebrated theorem.

Theorem 26.2. ***(Gauss-Bonnet-Chern)*** *If S is a compact two-dimensional Riemannian manifold without boundary, then*

$$\frac{1}{2\pi} \int_S KdA = \chi \,, \tag{26.11}$$

where K is the Gaussian curvature, dA is an area element, and χ is the Euler characteristic of S.

Although we have only sketched the proof of this theorem for the special case of a two-dimensional surface without boundary, the result as stated in the above equation (when suitably generalized) is valid for a manifold without boundary of arbitrary dimsnsion. This remarkable theorem was the first major result in modern differential geometry that demonstrated the deep relationship between local properties, such as the curvature, and global (topological) properties, such as the Euler characteristic, of a differential manifold. The crucial element of the proof sketched above is the use that is made of the global one-form $(\omega')_1^2$ on E obtained by the process of transgression (cf. discussion in the last chapter). As Chern insightfully remarked: this one-form was available neither to Gauss nor to Bonnet (Chern 1990).

Equation (26.11) for the Gauss-Bonnet-Chern Theorem can be generalized to the following form for a 2-dimensional surface M with boundary ∂M (see Chern 1967):

$$\int_{\partial M} \kappa_g ds + \int_M KdA + \sum_i (\pi - \alpha_1) = 2\pi\chi \,, \tag{26.12}$$

where κ_g is the geodesic curvature and ds the element of arclength of the boundary curve, respectively, $\pi - \alpha_i$ are the exterior angles of the vertices of ∂S, and χ is the Euler characteristic of M (see Fig. 26.2). It is illuminating to see that for a plane triangle ($\kappa_g = 0, K = 0, \chi = 1$), the Gauss-Bonnet-Chern Theorem reads $3\pi - (\alpha_1 + \alpha_2 + \alpha_3) = 2\pi$, and thus reduces to the familiar fact in Euclidean geometry that the sum of the three (interior) angles of a plane triangle is π.

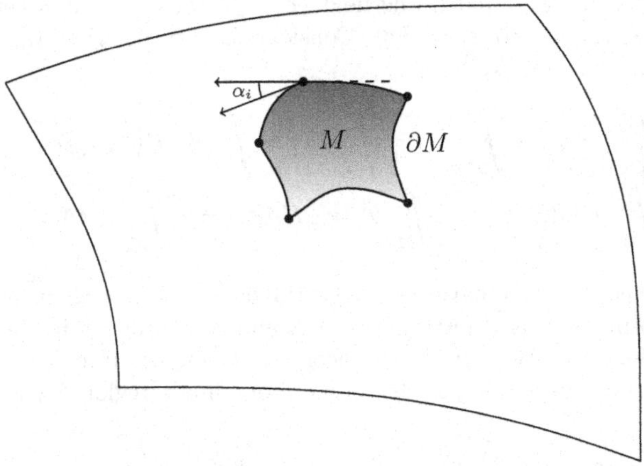

Fig. 26.2

Problem 26.1 For two-dimensional compact surfaces, the Euler characteristic is given by $\chi = 2 - 2g$, where g is the *genus* (number of handles) of the surface. Thus, for a spherical surface $g = 0$ and $\chi = 2$. Verify the Gauss-Bonnet-Chern Theorem for a spherical surface of radius r. Recall that the Gaussian curvature K for a spherical surface of radius r is given by $K = 1/r^2$ [cf. Problem 25.1(b)].

Problem 26.2 Verify the Gauss-Bonnet-Chern Theorem for the two-dimensional toroidal surface ($g = 1$ and hence $\chi = 0$) specified in Problem 25.2. Use the results for the area element $\omega^1 \wedge \omega^2$ and the Gaussian curvature K given in that problem.

Problem 26.3 For a spherical triangle on a spherical surface of radius R whose three sides are made up of segments of great circles, the area A is given by

$$A = R^2(\alpha_1 + \alpha_2 + \alpha_3 - \pi) ,$$

where $\alpha_1, \alpha_2, \alpha_3$ are the three angles of the triangle, each one understood to be the angle between the two tangent lines to two adjacent sides of the triangle. Use the version of the Gauss-Bonnet-Chern Theorem given by (26.12) to verify this result.

Returning to the calculation of holonomy we consider an arbitrary oriented closed curve C in S. Choose a vector field θ on S with a lone singular point p

such that p lies in the region M_1 bounded by C, that is, $C = \partial M_1$. Assuming S to be without boundary, the complement of M_1, denoted by M_2, is then bounded by $-C$, that is, $-C = \partial M_2$. In (26.1) we suppose ω_1^2 to be given with respect to an oriented frame field $\{\mathbf{e}_1, \mathbf{e}_2\}$, where $\mathbf{e}_1$ at a certain point $x \in S$ $(x \neq p)$ is given by $\mathbf{e}_1 = \theta(x)$. Consider the closed region $M_\epsilon \equiv M_1 - \Delta_\epsilon$, then $C - \partial\Delta_\epsilon = \partial M_\epsilon$. We have

$$\begin{aligned}\int_{C-\partial\Delta_\epsilon} \omega_1^2 &= \int_{\partial M_\epsilon} \omega_1^2 = \int_{M_\epsilon} d\omega_1^2 = \int_{M_\epsilon} d\left(\theta^*((\omega')_1^2)\right) \\ &= \int_{M_\epsilon} \theta^*(d(\omega')_1^2) = -\int_{M_\epsilon} \theta^*\left(\pi^*(KdA)\right) = -\int_{M_\epsilon} KdA\,.\end{aligned} \tag{26.13}$$

In the above equation we have made essential use of the transgression technique of pulling forms back and forth between E and S introduced in the discussion above on the Gauss-Bonnet-Chern theorem. It follows from (26.1) that the holonomy α_C corresponding to the loop C bounding a region M_1 is given by

$$\begin{aligned}\alpha_C &= -\int_C \omega_1^2 = -\lim_{\epsilon\to 0}\left(-\int_{M_\epsilon} KdA + \int_{\partial\Delta_\epsilon} \omega_1^2\right) \\ &= \int_{M_1} KdA - 2\pi I_p\,,\end{aligned} \tag{26.14}$$

where the second term on the right-hand side of the last equality is obtained by using (26.8) and (26.10). Had we chosen to calculate α_C by considering $-C$ as the boundary of M_2, we would have

$$\begin{aligned}\alpha_C &= -\int_C \omega_1^2 = \int_{-C} \omega_1^2 = \int_{\partial M_2} \omega_1^2 = \int_{M_2} d\omega_1^2 \\ &= \int_{M_2} d(\theta^*(\omega')_1^2) = \int_{M_2} \theta^* d(\omega')_1^2 = -\int_{M_2} \theta^*\pi^*(KdA) = -\int_{M_2} KdA \\ &= -\left(2\pi\chi - \int_{M_1} KdA\right) = \int_{M_1} KdA - 2\pi I_p\,,\end{aligned} \tag{26.15}$$

which is the same result obtained in (26.13). In the next to last equality we have used the Gauss-Bonnet-Chern theorem (26.11).

Problem 26.4 Consider a tangent vector being parallelly translated once around a latitude circle C (with colatitude θ) on a spherical surface of radius r. Use the holonomy result (26.2), and the fact that the Gaussian curvature at any point on the spherical surface is given by $K = 1/r^2$, to verify that the angular mismatch α_C is given by

$$\alpha_C = 2\pi(1 - \cos\theta)\,.$$

Check that one obtains the same result whether the area integral is performed over the top cap or the bottom cap of the sphere. (*Hint*: Recall from Problem 25.1 that we can set $\omega^1 = r\, d\theta$, $\omega^2 = r \sin\theta\, d\phi$.)

Chapter 27

The Curvature Tensor in Riemannian Geometry

Spurred by the success of Einstein's *general theory of relativity*, in which the curvature of a 4-dimensional pseudo-Riemannian manifold describes the gravitational force, Cartan proposed in 1925 a generalization of a main result of his moving frames approach to the study of affine spaces [given by (9.14)] which would in one clean stroke broaden the notion of the *Riemannian curvature tensor* to that of the *curvature* associated with an arbitrary *connection* on a vector bundle (see Cartan 1937). Instead of (9.14), Cartan introduced the following generalization, which in effect defines the **curvature matrix** of two-forms Ω_i^j:

$$d\omega_i^j = \omega_i^k \wedge \omega_k^j + \Omega_i^j \,, \tag{27.1}$$

where ω_i^j is an arbitrary connection matrix of one-forms. The above equation is frequently written in the (index-free) abbreviated form

$$\boxed{\Omega = d\omega - \omega \wedge \omega} \,, \tag{27.2}$$

where $\wedge$ means the exterior product of one-forms as well as matrix multiplication. Notice that in the case of a two-dimensional surface (where $i, j = 1, 2$), $\omega \wedge \omega = 0$, hence it follows from (27.1) and the result (25.18) for the Gaussian curvature K that $\Omega_1^2 = -K\,\omega^1 \wedge \omega^2$. From now on, Eq. (27.2) will be taken to be the definition of the curvature matrix of 2-forms associated with a connection matrix of 1-forms ω on a general vector bundle.

The fact that the matrix of two-forms Ω as defined above is indeed a tensorial quantity can be demonstrated straightforwardly. Under a local change of frame fields (gauge change) given by (4.14), the connection matrix transforms as [cf. (4.19)]

$$\omega' g = dg + g\omega \,. \tag{27.3}$$

Exteriorly differentiating this equation [using the rule (6.11) and Theorem 6.2] we have

$$d\omega'\, g - \omega' \wedge dg = dg \wedge \omega + g\, d\omega \; . \tag{27.4}$$

Equation (27.3) also gives

$$dg = \omega'\, g - g\omega \; . \tag{27.5}$$

Using this result in (27.4) we then obtain

$$d\omega'\, g - \omega' \wedge (\omega'\, g - g\omega) = (\omega'\, g - g\omega) \wedge \omega + g\, d\omega \; , \tag{27.6}$$

which simplifies to

$$(d\omega' - \omega' \wedge \omega')g - g(d\omega - \omega \wedge \omega) \; . \tag{27.7}$$

Recognizing that the curvature 2-form with respect to the new frame is given by $\Omega' = d\omega' - \omega' \wedge \omega'$, we finally have the following transformation rule for the curvature 2-form under a local gauge change:

$$\Omega' = g\,\Omega\, g^{-1} \; . \tag{27.8}$$

This is precisely the transformation rule for a $(1,1)$ tensor [recall (1.31), in which S plays the role of our g^{-1} here]. We stress again the difference between the transformation rule for Ω and that for ω [as given by (4.19)], and remind the reader that the latter is a non-tensorial quantity.

Any curvature matrix of two-forms satisfies the so-called **Bianchi identity**:

$$\boxed{d\Omega = \omega \wedge \Omega - \Omega \wedge \omega} \; . \tag{27.9}$$

This can be obtained easily by exteriorly differentiating both sides of (27.2):

$$\begin{aligned} d\Omega &= d^2\omega - (d\omega \wedge \omega - \omega \wedge d\omega) \\ &= -(\Omega + \omega \wedge \omega) \wedge \omega + \omega \wedge (\Omega + \omega \wedge \omega) = \omega \wedge \Omega - \Omega \wedge \omega \; . \end{aligned} \tag{27.10}$$

In the second equality above we have used (27.2) to substitute $\Omega + \omega \wedge \omega$ for $d\omega$.

Problem 27.1 There is a subtlety and potential source of confusion in the notation used in the definition of the curvature two-form in (27.2) that has to do with antisymmetric nature of exterior products. Throughout this text the convention used for denoting matrix elements such as ω_i^j is that the subscript i and the superscript j are the row and column indices, respectively. Show that if this convention is reversed, namely, if the superscript j and the subscript i are respectively interpreted to be the row and column indices instead, then the gauge transformation rule (27.3) would appear as

$$\omega' = g^{-1}\, dg + g^{-1}\, \omega\, g \; ,$$

the definition for the curvature (27.2) would appear as

$$\Omega = d\omega + \omega \wedge \omega \; ,$$

and the tensorial transformation rule (27.8) for the curvature would appear as

$$\Omega' = g^{-1}\,\Omega\, g\,.$$

In the above equations, the connection matrix ω and the local gauge change matrix g are still defined by [cf. (4.8) and (4.14)]

$$De_i = \omega_i^j\, e_j\,, \qquad e_i' = g_i^j(x)\, e_j\,,$$

where x denotes the local coordinates on the base manifold. The reader is cautioned that the convention of this problem is also commonly used in the literature.

We recall the Christoffel symbols Γ^j_{ik} introduced in (22.29) and write

$$\omega_i^j = \Gamma^j_{ik}(x^1,\dots,x^n)\, dx^k\,, \tag{27.11}$$

where $x^1,\dots,x^n$ are the local coordinates of an n-dimensional base manifold, and i, j are indices referring to a basis in the fiber (vector) space, whose dimension is in general different from n. Equation (27.2) can then be used to write the curvature matrix elements (which are 2-forms) in terms of the Christoffel symbols as follows:

$$\begin{aligned}\Omega_i^j &= d\omega_i^j - \omega_i^k \wedge \omega_k^j \\ &= \frac{\partial \Gamma^j_{ik}}{\partial x^l}\, dx^l \wedge dx^k - \Gamma^h_{il}\Gamma^j_{hk}\, dx^l \wedge dx^k \\ &= \frac{1}{2}\left(\frac{\partial \Gamma^j_{il}}{\partial x^k} - \frac{\partial \Gamma^j_{ik}}{\partial x^l} + \Gamma^h_{il}\Gamma^j_{hk} - \Gamma^h_{ik}\Gamma^j_{hl}\right) dx^k \wedge dx^l \\ &\equiv \frac{1}{2} R^j_{ikl}\, dx^k \wedge dx^l \quad .\end{aligned} \tag{27.12}$$

The quantity within the parentheses on the right-hand side of the last equality in the above equation is known as the **curvature tensor**:

$$\boxed{R^j_{ikl} = \frac{\partial \Gamma^j_{il}}{\partial x^k} - \frac{\partial \Gamma^j_{ik}}{\partial x^l} + \Gamma^h_{il}\Gamma^j_{hk} - \Gamma^h_{ik}\Gamma^j_{hl}} \quad . \tag{27.13}$$

Note that in the above equation only i and j are the tensorial indices [cf. (27.8)].

We will now study the curvature tensor in the context of Riemannian geometry, which is the geometry of a differentiable manifold determined by a so-called *Riemannian metric*, defined as follows [recall the discussion of the general notion of a metric (a prescription for measuring distances and angles) given in Chapter 1]:

Definition 27.1. *A* ***Riemannian metric*** *G on an n-dimensional differentiable manifold M is a symmetric, positive-definite, covariant $(0,2)$-tensor field on M. A manifold endowed with such a metric is called a* ***Riemannian manifold***.

G can be written locally in terms of its components g_{ij} as

$$G = g_{ij}(x^1, \ldots, x^n)\, dx^i \otimes dx^j \;, \tag{27.14}$$

where the dx^i are regarded as basis vectors in the *natural coframe field* on M. This tensor field then acts on two arbitrary tangent vector fields $\boldsymbol{X} = X^i \dfrac{\partial}{\partial x^i}$ and $\boldsymbol{Y} = Y^i \dfrac{\partial}{\partial x^i}$ according to

$$G(\boldsymbol{X}, \boldsymbol{Y}) = g_{ij}(x)\, X^i Y^j \;. \tag{27.15}$$

The components g_{ij} constitute the **metric tensor** of the Riemannian manifold. Symmetry of G means $g_{ij} = g_{ji}$. Positive-definiteness of G means that for all tangent vector fields $\boldsymbol{X}$, $G(\boldsymbol{X}, \boldsymbol{X}) \geq 0$, with the equality holding only for $\boldsymbol{X} = 0$. Note that a positive-definite metric is necessarily *non-degenerate*, which means that (again recall the discussion in Chapter 1), if $G(\boldsymbol{X}, \boldsymbol{Y}) = 0$ for all tangent vector fields $\boldsymbol{X}$, then $\boldsymbol{Y} = 0$. From now on we will often drop the tensor product sign $\otimes$ in (27.14) for brevity (conforming to the custom usually adopted in the physics literature). A smooth manifold endowed with a symmetric, non-positive-definite, non-degenerate metric is called a *pseudo-Riemannian manifold.* An important physical example is the 4-dimensional *spacetime* of special and general relativity. Our development below will apply mainly to Riemannian manifolds.

The Riemannian metric G gives prescriptions for the determinations of lengths of tangent vectors, angles between tangent vectors, and arclengths of curves on a Riemannian manifold M as follows. The length of a tangent vector $\boldsymbol{X} \in T_x(M)$ is given by

$$|\,\boldsymbol{X}\,| \equiv \sqrt{G(\boldsymbol{X}, \boldsymbol{X})}\;. \tag{27.16}$$

The angle θ between two tangent vectors $\boldsymbol{X}, \boldsymbol{Y} \in T_x M$ is determined by

$$\cos\theta \equiv \frac{G(\boldsymbol{X}, \boldsymbol{Y})}{|\,\boldsymbol{X}\,|\,|\,\boldsymbol{Y}\,|}\;. \tag{27.17}$$

This, of course, is reminiscent of the ordinary scalar product of two vectors given by (1.43). The above equations give meanings to the notions of *orthogonality* and *orthonormality* of tangent vectors in a Riemannian manifold. The **arclength** of a *parametrized curve* $C : [\,t_0, t_1\,] \to M$ [specified by the functions $x^i(t)$] on M is given by

$$\int_{t_0}^{t_1} ds = \int_{t_0}^{t_1} \sqrt{G\left(\frac{dx^i}{dt}\frac{\partial}{\partial x^i}\,,\;\frac{dx^j}{dt}\frac{\partial}{\partial x^j}\right)}\, dt = \int_{t_0}^{t_1} \sqrt{g_{ij}(x(t))\,\frac{dx^i}{dt}\frac{dx^j}{dt}}\, dt\;. \tag{27.18}$$

This equation allows us to write the *arclength element* squared – the First Fundamental form [cf. (25.4)] – symbolically as

$$ds^2 = g_{ij}(x)\, dx^i dx^j \;. \tag{27.19}$$

Since ds^2 is invariant under a local change of coordinates $x^i \to x'^i$ we can write

$$g_{kl}(x)\, dx^k dx^l = g'_{ij}(x')\, dx'^i dx'^j \,. \tag{27.20}$$

On the other hand (cf. Problem 6.7),

$$g_{kl}\, dx^k dx^l = g_{kl}\, \frac{\partial x^k}{\partial x'^i} \frac{\partial x^l}{\partial x'^j}\, dx'^i dx'^j \,. \tag{27.21}$$

Thus the transformation rule for the metric tensor under a local change of coordinates is

$$g'_{ij}(x') = g_{kl}(x)\, \frac{\partial x^k}{\partial x'^i} \frac{\partial x^l}{\partial x'^j} \,. \tag{27.22}$$

Non-degeneracy of the metric G implies that the matrix (g_{ij}) is invertible. Its inverse will be denoted by (g^{ij}), so we have

$$g^{ik} g_{kj} = g_{jk} g^{ki} = \delta^i_j \,. \tag{27.23}$$

The metric gives rise to an inner (scalar) product on the tangent space $T_x(M)$ according to (27.15). As discussed in Chapter 1, it then provides an isomorphism between $T_x(M)$ and its dual, the cotangent space $T^*_x(M)$. If $\boldsymbol{X} \in T_x(M)$ is a tangent vector, then there exists a *unique* cotangent vector $\boldsymbol{X}^* \in T^*_x(M)$ such that

$$\langle \boldsymbol{Y} , \boldsymbol{X}^* \rangle = G(\boldsymbol{Y}, \boldsymbol{X}) \,, \qquad \text{for all } \boldsymbol{Y} \in T_x(M) \,. \tag{27.24}$$

Writing $\boldsymbol{X}$ and $\boldsymbol{X}^*$ in component form with respect to the natural bases $\{ \partial/\partial x^i \}$ and $\{dx^i\}$, respectively:

$$\boldsymbol{X} = X^i \frac{\partial}{\partial x^i} \,, \qquad \boldsymbol{X}^* = X_i\, dx^i \,, \tag{27.25}$$

(27.24) implies

$$X_i = g_{ij} X^j \,. \tag{27.26}$$

It then follows from (27.23) that

$$X^i = g^{ij} X_j \,. \tag{27.27}$$

In the physics literature one frequently speaks of using the metric tensor and its inverse to lower and raise indices, respectively. Using the covariant vector X_i and the contravariant vector Y^i, and recalling the symmetry property of the metric G, one can then write

$$G(\boldsymbol{X}, \boldsymbol{Y}) = g_{ij}\, X^i Y^j = X^i Y_i = g_{ji}\, X^i Y^j = X_j Y^j \,. \tag{27.28}$$

When a curvature two-form (27.2) arises from an **affine connection**, that is, a connection on the tangent bundle TM, the curvature tensor R^j_{ikl} defined by (27.12) is called the **Riemann curvature tensor**, a tensorial object where all four indices are tensorial and assume values from 1 to n, with $n = dim(M)$.

The Riemann curvature tensor is then a (1, 3)-type tensor and can be written with respect to natural frame fields as

$$R = R^j_{ikl} \frac{\partial}{\partial x^j} \otimes dx^i \otimes dx^k \otimes dx^l \,. \tag{27.29}$$

Given tangent vector fields $\boldsymbol{X} = X^i \partial_i$ and $\boldsymbol{Y} = Y^j \partial_j$ (where $\partial_i \equiv \partial/\partial x^i$), it follows from (27.12), (2.30) and (5.4) that

$$\langle \boldsymbol{X} \wedge \boldsymbol{Y} \,,\, \Omega^j_i \rangle = R^j_{ikl}\, X^k Y^l \,. \tag{27.30}$$

The Riemann curvature then gives rise to a map $R(\boldsymbol{X}, \boldsymbol{Y}) : \Gamma(TM) \longrightarrow \Gamma(TM)$ (mapping a tangent vector field to another tangent vector field). For a tangent vector field $\boldsymbol{Z} = Z^i \partial_i$ we define the action of the so-called **Riemann curvature operator** $R(\boldsymbol{X}, \boldsymbol{Y})$ on $\boldsymbol{Z}$ as follows:

$$R(\boldsymbol{X}, \boldsymbol{Y})\, \boldsymbol{Z} \equiv Z^i \langle \boldsymbol{X} \wedge \boldsymbol{Y} \,,\, \Omega^j_i \rangle\, \partial_j = R^j_{ikl}\, Z^i X^k Y^l\, \partial_j \,. \tag{27.31}$$

Writing $\boldsymbol{X} = \partial_k, \boldsymbol{Y} = \partial_l$ and $\boldsymbol{Z} = \partial_i$ in (27.31) we see that

$$R^j_{ikl} = \langle\, R(\partial_k, \partial_l)\, \partial_i \,,\, dx^j \,\rangle \,. \tag{27.32}$$

The Riemann curvature operator can also be expressed in coordinate-free form in terms of directional covariant derivatives as

$$R(\boldsymbol{X}, \boldsymbol{Y}) = D_{\boldsymbol{X}} D_{\boldsymbol{Y}} - D_{\boldsymbol{Y}} D_{\boldsymbol{X}} - D_{[\boldsymbol{X}\,,\,\boldsymbol{Y}]} \,, \tag{27.33}$$

where the directional covariant derivatives have been defined in (22.8) and the Lie bracket in (6.22). This result gives a geometrical interpretation of the curvature operator: *it measures the non-commutativity of directional covariant derivatives.*

Problem 27.2 Verify (27.30).

Problem 27.3 Use (27.8) and (27.12) to show that, under a local change of coordinates $x^i \to x'^i(x^j)$, the tensorial transformation rule for the Riemann curvature tensor is

$$(R')^j_{ikl} = R^q_{prs} \frac{\partial x'^j}{\partial x^q} \frac{\partial x^p}{\partial x'^i} \frac{\partial x^r}{\partial x'^k} \frac{\partial x^s}{\partial x'^l} \,. \tag{27.34}$$

Verify that this result conforms with the general tensor transformation rule given by (1.32).

Problem 27.4 Recall the definitions of the directional covariant derivative [(22.8)] and the Lie bracket [(6.22)], and use (27.31) and the result in Theorem 6.3 [(6.14)] to verify (27.33) for the Riemann curvature operator.

Definition 27.2. *Suppose an affine connection matrix with respect to a frame-field $\{\mathbf{e}_i\}$ (not necessarily orthonormal) on a Riemannian manifold is given by $D\mathbf{e}_i = \omega_i^j\, \mathbf{e}_j$, and $\{\omega^i\}$ is the coframe field dual to $\{\mathbf{e}_i\}$. The affine connection is said to be* **torsion free** *if the following integrability condition is satisfied [cf. (9.13)]*

$$d\omega^j = \omega^i \wedge \omega_i^j \,. \tag{27.35}$$

Let $\{\mathbf{e}_i\}$ be the natural frame field, that is $\mathbf{e}_i = \partial/\partial x^i = \partial_i$, where x^i are the local coordinates. Then $\omega^i = dx^i$. Writing the connection matrix elements in terms of the Christoffel symbols Γ^j_{ik} [cf. (27.11)] the above torsion-free condition implies $\Gamma^j_{ik}\, dx^i \wedge dx^k = 0$, which in turn implies $\Gamma^j_{ik} = \Gamma^j_{ki}$. We define the **torsion tensor** T^j_{ik} with respect to the natural frame field $\{\partial_i\}$ to be the $(1,2)$-type tensor

$$T^j_{ik} \equiv \Gamma^j_{ki} - \Gamma^j_{ik} \,. \tag{27.36}$$

It is obvious that the torsion tensor satisfies the following symmetry property:

$$T^j_{ik} = -T^j_{ki} \,. \tag{27.37}$$

An affine connection D is then torsion free if the torsion tensor T^j_{ik} of D vanishes.

Problem 27.5 Use (22.31) (the gauge transformation rule for the Christoffel symbols) to verify that the torsion tensor as defined by (27.36) indeed satisfies the tensorial transformation rule

$$(T')^j_{ik} = \frac{\partial y^j}{\partial x^m} \frac{\partial x^p}{\partial y^i} \frac{\partial x^l}{\partial y^k}\, T^m_{pl} \tag{27.38}$$

under the local change of coordinates $x^i \to y^i(x^j)$.

In terms of local coordinates (x^i) the torsion tensor can be expressed as

$$T = T^j_{ik}\, \frac{\partial}{\partial x^j} \otimes dx^i \otimes dx^k \,.$$

Thus it can be viewed as a map $T : \Gamma(T(M)) \times \Gamma(T(M)) \longrightarrow \Gamma(T(M))$: for any two tangent vector fields $\boldsymbol{X} = X^i\partial_i, \boldsymbol{Y} = Y^i\partial_i \in \Gamma(T(M))$, we have

$$T(\boldsymbol{X},\boldsymbol{Y}) = T^j_{ik} X^i Y^k\, \frac{\partial}{\partial x^j} \,. \tag{27.39}$$

The above result can also be expressed in the coordinate-free form

$$T(\boldsymbol{X},\boldsymbol{Y}) = D_{\boldsymbol{X}}\boldsymbol{Y} - D_{\boldsymbol{Y}}\boldsymbol{X} - [\,\boldsymbol{X}\,,\,\boldsymbol{Y}\,] \,. \tag{27.40}$$

Problem 27.6 Verify the result (27.40) for the torsion tensor by using (27.36), (27.39), and the analytical expression for directional covariant differentiation given by (22.33). Use (27.40), (27.33), and the *Jacobi identity* [(1.46)] applied to the tangent vector fields $\boldsymbol{X}, \boldsymbol{Y}, \boldsymbol{Z}$ to show that, for a Riemann curvature operator $R(\boldsymbol{X}, \boldsymbol{Y})$ derived from a torsion-free connection,

$$R(\boldsymbol{X}, \boldsymbol{Y})\,\boldsymbol{Z} + R(\boldsymbol{Y}, \boldsymbol{Z})\,\boldsymbol{X} + R(\boldsymbol{Z}, \boldsymbol{X})\,\boldsymbol{Y} = 0\,. \tag{27.41}$$

This result (valid only for the curvature operator of a torsion-free affine connection) is sometimes also referred to as the **first Bianchi identity**. From the Bianchi identity (27.9) for a general affine connection, show that, the curvature operator of a torsion-free affine connection also satisfies the following property: For any tangent vector fields $\boldsymbol{X}, \boldsymbol{Y}, \boldsymbol{Z}, \boldsymbol{W}$,

$$D_{\boldsymbol{X}}\left(R(\boldsymbol{Y}, \boldsymbol{Z})\,\boldsymbol{W}\right) + D_{\boldsymbol{Y}}\left(R(\boldsymbol{Z}, \boldsymbol{X})\,\boldsymbol{W}\right) + D_{\boldsymbol{Z}}\left(R(\boldsymbol{X}, \boldsymbol{Y})\,\boldsymbol{W}\right) = 0\,, \tag{27.42}$$

a result that is sometimes known as the **second Bianchi identity**. Show that in component form this result can be expressed as

$$D_h R^j_{ikl} + D_k R^j_{ilh} + D_l R^j_{ihk} = 0\,, \tag{27.43}$$

where $D_h \equiv D_{\partial/\partial x^h}$ and

$$D_h R^j_{ikl} = R^j_{ikl,h}\,,$$

with the right-hand side written in the notation of (22.42). Note the cyclic order of the vector fields $(\boldsymbol{X}, \boldsymbol{Y}, \boldsymbol{Z})$ in (27.41) and (27.42) and the indices (h, k, l) in (27.43).

Problem 27.7 On the 2-dimensional Euclidean plane one can introduce the local (plane-polar) coordinates $x^1 = r, x^2 = \theta$. With respect to the orthonormal frame field $e_1 = e_r = \partial_r$, $e_2 = e_\theta = (1/r)\,\partial_\theta$ (with the dual coframe field $\omega^1 = dr$, $\omega^2 = r\,d\theta$), an affine connection D is given by

$$\begin{pmatrix} De_1 \\ De_2 \end{pmatrix} = \begin{pmatrix} 0 & d\theta \\ -d\theta & 0 \end{pmatrix} \begin{pmatrix} e_1 \\ e_2 \end{pmatrix}\,.$$

Show that this connection is torsion free by verifying that (27.35), namely, the integrability condition $d\omega^j = \omega^i \wedge \omega_i^j$, is satisfied. Show that $T(e_1, e_2) = 0$ by showing directly that

$$D_{e_2}\, e_1 = \frac{1}{r}\, e_2\,, \qquad D_{e_1}\, e_2 = 0\,, \qquad [\, e_1\,, e_2\,] = -\frac{1}{r}\, e_2\,. \tag{27.44}$$

Problem 27.8 Show that the affine connection given by (27.42) can equivalently be specified by

$$D\begin{pmatrix} \partial_r \\ \partial_\theta \end{pmatrix} = \begin{pmatrix} 0 & d\theta/r \\ -r\,d\theta & dr/r \end{pmatrix} \begin{pmatrix} \partial_r \\ \partial_\theta \end{pmatrix}\,.$$

Hence show that

$$D_{\partial_r}\,\partial_\theta = D_{\partial_\theta}\,\partial_r = \frac{1}{r}\,\partial_\theta\,. \tag{27.45}$$

This result, together with the obvious fact that $[\partial_r, \partial_\theta] = 0$, clearly corroborates the torsion free criterion (27.41).

From the definition of the torsion tensor given in (27.36) it is apparent that from any affine connection D with Christoffel symbols Γ^j_{ik}, one can always construct a torsion-free connection D' with Christoffel symbols

$$\tilde{\Gamma}^j_{ik} = \frac{1}{2}(\Gamma^j_{ik} + \Gamma^j_{ki}) . \tag{27.46}$$

Motivated by Def. 27.2 we can define the **torsion form** $\mathcal{T}$ (a vector-valued two-form) by

$$\mathcal{T}^i \equiv d\omega^i - \omega^j \wedge \omega^i_j . \tag{27.47}$$

Expressed with respect to the natural frame field, this 2-form is related to the torsion tensor defined above by [compare with the expression relating the curvature two-form to the curvature tensor given by (27.12)]

$$\mathcal{T}^i = \frac{1}{2} T^i_{jk}\, dx^j \wedge dx^k . \tag{27.48}$$

Indeed, with $\omega^i = dx^i$ and $\omega^i_j = \Gamma^i_{jk}\, dx^k$, we have

$$\begin{aligned} d\omega^i - \omega^j \wedge \omega^i_j &= 0 - \Gamma^i_{jk}\, dx^j \wedge dx^k = \frac{1}{2}(\Gamma^i_{kj} - \Gamma^i_{jk})\, dx^j \wedge dx^k \\ &= \frac{1}{2} T^i_{jk}\, dx^j \wedge dx^k . \end{aligned} \tag{27.49}$$

Denoting the row matrix of 1-forms $(\omega^1, \dots \omega^n)$ by θ [recall that this notation was introduced in (9.8)] we can write

$$\boxed{\mathcal{T} = d\theta - \theta \wedge \omega} . \tag{27.50}$$

This equation together with (27.2) (defining the curvature two-form) are called **Cartan's structure equations** of the connection ω. They are valid independent of the choice of frame fields.

To accompany the property of torsion freeness of an affine connection, we will introduce the important notion of *metric compatibility* of an affine connection.

Definition 27.3. *An affine connection D on a Riemannian (or pseudo-Riemannian) manifold with metric $G = (g_{ij})$ is said to be* ***metric compatible*** *with G if*

$$DG = 0 . \tag{27.51}$$

In (27.51) the action of the covariant derivative D on cotangent vector fields is given by (22.38) and (22.39): If $D\partial_i = \omega^j_i\, \partial_j$, then

$$D(dx^i) = -\omega^i_j\, dx^j . \tag{27.52}$$

Writing $G = g_{ij}\, dx^i \otimes dx^j$ we have

$$\begin{aligned} DG &= dg_{ij}\, dx^i \otimes dx^j - g_{ij}\omega_l^i\, dx^l \otimes dx^j - g_{ij}\omega_l^j\, dx^i \otimes dx^l \\ &= (dg_{ij} - \omega_i^k\, g_{kj} - \omega_j^k g_{ik})\, dx^i \otimes dx^j \,. \end{aligned} \tag{27.53}$$

Thus the condition for metric compatibility can also be stated as

$$dg_{ij} = g_{ik}\omega_j^k + g_{kj}\omega_i^k \,. \tag{27.54}$$

This equation actually applies to any frame field with respect to which G and ω are expressed. With respect to an orthonormal frame field $\{\boldsymbol{e}_i\}$, $g_{ij} = G(\boldsymbol{e}_i, \boldsymbol{e}_j) = \delta_{ij}$, and so (27.54) simplifies to [cf. (4.24)]

$$\omega_i^j + \omega_j^i = 0 \,. \tag{27.55}$$

With respect to the natural frame field the metric compatibility condition can be written

$$dg_{ij} - g_{ik}\Gamma^k_{jl}\, dx^l - g_{kj}\Gamma^k_{il}\, dx^l = 0 \,. \tag{27.56}$$

This condition guarantees that the inner product $g_{ij}X^iY^j$ between two vectors $\boldsymbol{X} = X^i\,\partial_i$ and $\boldsymbol{Y} = Y^i\,\partial_i$ remain constant when the two vectors are parallelly displaced along a curve $C : x^i(t)$ if the parallel displacement is done with respect to a metric compatible connection. Indeed, by (22.43), X^i and Y^i on parallel translation change according to

$$\frac{dX^i}{dt} + X^j\,\Gamma^i_{jk}\,\frac{dx^k}{dt} = 0 \,, \qquad \frac{dY^i}{dt} + Y^j\,\Gamma^i_{jk}\,\frac{dx^k}{dt} = 0 \,. \tag{27.57}$$

Hence, along the curve C,

$$\begin{aligned} \frac{d}{dt}\left(g_{ij}\, X^iY^j\right) &= \frac{dg_{ij}}{dt}\, X^iY^j + g_{ij}\,\frac{dX^i}{dt}\, Y^j + g_{ij}\, X^i\,\frac{dY^j}{dt} \\ &= \left(\frac{dg_{ij}}{dt} - g_{kj}\,\Gamma^k_{ih}\,\frac{dx^h}{dt} - g_{ik}\,\Gamma^k_{jh}\,\frac{dx^h}{dt}\right) X^iY^j \,. \end{aligned}$$

Thus the right-hand side vanishes if the metric compatibility condition (27.56) is satisfied. In particular, under a metric compatible connection, the length of vectors and the angle between vectors remain unchanged under parallel translation. This is the geometrical meaning of the term metric compatibility.

Problem 27.9 Suppose D is a metric-compatible affine connection on a Riemannian manifold M. Write $\boldsymbol{A}\cdot\boldsymbol{B} \equiv G(\boldsymbol{A},\boldsymbol{B})$, where G is the metric tensor and $\boldsymbol{A},\boldsymbol{B}$ are any two tangent vector fields on M. Show that, for any tangent vector fields $\boldsymbol{X},\boldsymbol{Y},\boldsymbol{Z}$ on M,

$$\boldsymbol{X}\,(\boldsymbol{Y}\cdot\boldsymbol{Z}) = (D_{\boldsymbol{X}}\boldsymbol{Y})\cdot\boldsymbol{Z} + \boldsymbol{Y}\cdot(D_{\boldsymbol{X}}\boldsymbol{Z}) \,, \tag{27.58}$$

where the left-hand side is the action of $\boldsymbol{X}$ on the scalar function $\boldsymbol{Y}\cdot\boldsymbol{Z}$ defined in (5.2) and (5.19). *Hint*: Use the metric compatibility condition (27.56).

We will now state and prove the following important theorem.

Theorem 27.1. ***(Fundamental Theorem of Riemanian Geometry).*** *On every Riemannian or pseudo-Riemannian manifold, there exists a unique torsion-free and metric-compatible affine connection.*

Proof. Let M be a Riemannian or pseudo-Riemannian manifold, and g_{ij} is a Riemannian metric on M with respect to a natural frame field. Suppose $\omega_i^j = \Gamma^j_{ik}\, dx^k$ is an affine connection on M that is both torsion-free and metric compatible. Then by (27.36) and (27.54),

$$\Gamma^j_{ik} = \Gamma^j_{ki} \,, \tag{27.59}$$

$$dg_{ij} = \left(g_{il}\,\Gamma^l_{jk} + g_{lj}\,\Gamma^l_{ik}\right) dx^k \,. \tag{27.60}$$

Define the $(0,3)$-tensor Γ_{ijk} by

$$\Gamma_{ijk} \equiv g_{lj}\,\Gamma^l_{ik} \,. \tag{27.61}$$

Then (27.59) and (27.60) imply

$$\Gamma_{ijk} = \Gamma_{kji} \,, \tag{27.62}$$

$$\frac{\partial g_{ij}}{\partial x^k} = \Gamma_{ijk} + \Gamma_{jik} \,. \tag{27.63}$$

We can obtain two more independent equations analogous to (27.63) by performing two even permutations of the indices (ijk) in that equation:

$$(ijk) \to (jki) \Longrightarrow \quad \frac{\partial g_{jk}}{\partial x^i} = \Gamma_{jki} + \Gamma_{kji} \,, \tag{27.64}$$

$$(ijk) \to (kij) \Longrightarrow \quad \frac{\partial g_{ki}}{\partial x^j} = \Gamma_{kij} + \Gamma_{ikj} \,. \tag{27.65}$$

Adding (27.64) to (27.65) and then subtracting (27.63) from the result, and using the symmetry property (27.62) for Γ_{ijk}, we have

$$\boxed{\Gamma_{ikj} = \frac{1}{2}\left(\frac{\partial g_{ik}}{\partial x^j} + \frac{\partial g_{jk}}{\partial x^i} - \frac{\partial g_{ij}}{\partial x^k}\right)} \,. \tag{27.66}$$

Multiplying both sides by g^{kl} and summing over k, and then interchanging the indices k and l, we obtain the equivalent equation

$$\boxed{\Gamma^k_{ij} = \frac{1}{2} g^{kl}\left(\frac{\partial g_{il}}{\partial x^j} + \frac{\partial g_{jl}}{\partial x^i} - \frac{\partial g_{ij}}{\partial x^l}\right)} \,. \tag{27.67}$$

The equation (27.66) [or (27.67)] shows that the torsion-free and metric-compatible connection (specified by the Christoffel symbols Γ^j_{ik}), if it exists, is determined *uniquely* by the metric g_{ij}. Conversely, it is quite obvious that the $(0,3)$-type tensor defined by (27.66) satisfies (27.59) and (27.60), the conditions of torsion-freeness and metric-compatibility, respectively, for a connection. Finally, the

(1, 2)-type tensor as defined by (27.67) can be shown to satisfy the transformation rule (22.31) for the Christoffel symbols specifying an affine connection on M. This demonstrates the existence of a torsion-free and metric-compatible affine connection on a Riemannian (or pseudo-Riemannian) M, and the theorem is proved. □

The unique affine connection on a Riemannian (or pseudo-Riemannian) manifold that is both torsion-free and metric-compatible is called the **Levi-Civita connection**. The tensors Γ_{ijk} and Γ^j_{ik} are called **Christoffel symbols of the first and second kind**, respectively, in classical tensor analysis.

Recall that for two-dimensional Riemannian manifolds (most intuitively recognized as two-dimensional surfaces embedded in three-dimensional Euclidean space), the *Gaussian curvature* [cf. (25.16) and (25.24)] plays a central role in the study of their geometries. We will now introduce a generalization of the Gaussian curvature, called the *sectional curvature*, which will play a similar role in n-dimensional Riemannian manifolds, for $n > 2$.

First define the (0, 4)-tensor R_{ijkl}, also called the *Riemann curvature tensor*, as follows:

$$R_{ijkl} \equiv g_{hj}\, R^h_{ikl} \,, \tag{27.68}$$

where g_{hj} is the metric tensor. From (27.13), one obtains the analogous formula

$$R_{ijkl} = \frac{\partial \Gamma_{ijl}}{\partial x^k} - \frac{\partial \Gamma_{ijk}}{\partial x^l} + \Gamma^h_{ik}\,\Gamma_{jhl} - \Gamma^h_{il}\,\Gamma_{jhk} \,. \tag{27.69}$$

Equation (27.12) then implies that we have the covariant curvature two-form

$$\Omega_{ij} = \frac{1}{2}\, R_{ijkl}\, dx^k \wedge dx^l \,. \tag{27.70}$$

Problem 27.10 Show that the covariant curvature tensor R_{ijkl} satisfies the following symmetry properties:

$$(1) \qquad R_{ijkl} = -R_{jikl} = -R_{ijlk} \,, \tag{27.71}$$

$$(2) \qquad R_{ijkl} + R_{iklj} + R_{iljk} = 0 \,, \tag{27.72}$$

$$(3) \qquad R_{ijkl} = R_{klij} \,. \tag{27.73}$$

The quantities R_{ijkl} are actually the components of a (0, 4)-tensor R with respect to the natural frame field $\partial_i = \partial/\partial x^i$, so that we can write

$$R = R_{ijkl}\, dx^i \otimes dx^j \otimes dx^k \otimes dx^l \,. \tag{27.74}$$

R can also be viewed as a multilinear function $R : T_x(M) \times T_x(M) \times T_x(M) \times T_x(M) \longrightarrow \mathbb{R}$, defined, for $\boldsymbol{X}, \boldsymbol{Y}, \boldsymbol{Z}, \boldsymbol{W} \in T_x(M)$, by

$$R(\boldsymbol{X}, \boldsymbol{Y}, \boldsymbol{Z}, \boldsymbol{W}) = \langle\, \boldsymbol{X} \otimes \boldsymbol{Y} \otimes \boldsymbol{Z} \otimes \boldsymbol{W} \,,\, R \,\rangle \,, \tag{27.75}$$

where the pairing $\langle\, ,\, \rangle$ is defined by (2.2). If we express $\boldsymbol{X}, \boldsymbol{Y}, \boldsymbol{Z}, \boldsymbol{W}$ in terms of the natural frame field also: $\boldsymbol{X} = X^i\,\partial_i$, $\boldsymbol{Y} = Y^i\,\partial_i$, $\boldsymbol{Z} = Z^i\,\partial_i$, $\boldsymbol{W} = W^i\,\partial_i$, we have

$$R(\boldsymbol{X}, \boldsymbol{Y}, \boldsymbol{Z}, \boldsymbol{W}) = R_{ijkl}\, X^i Y^j Z^k W^l\,, \tag{27.76}$$

with

$$R_{ijkl} = R(\partial_i\,, \partial_j\,, \partial_k\,, \partial_l)\,. \tag{27.77}$$

In view of (27.31) giving the action of the map $R(\boldsymbol{X}, \boldsymbol{Y}) : \Gamma(TM) \longrightarrow \Gamma(TM)$, we also have

$$R(\boldsymbol{X}, \boldsymbol{Y}, \boldsymbol{Z}, \boldsymbol{W}) = G(R(\boldsymbol{Z}, \boldsymbol{W})\boldsymbol{X}\,, \boldsymbol{Y})\,, \tag{27.78}$$

where G is the Riemann metric tensor.

Problem 27.11 Show that the symmetry properties of the covariant curvature tensor R given in Problem 27.10 can be restated as follows:

$$(1)\quad R(\boldsymbol{X}, \boldsymbol{Y}, \boldsymbol{Z}, \boldsymbol{W}) = -R(\boldsymbol{Y}, \boldsymbol{X}, \boldsymbol{Z}, \boldsymbol{W}) = -R(\boldsymbol{X}, \boldsymbol{Y}, \boldsymbol{W}, \boldsymbol{Z})\,, \tag{27.79}$$

$$(2)\quad R(\boldsymbol{X}, \boldsymbol{Y}, \boldsymbol{Z}, \boldsymbol{W}) + R(\boldsymbol{X}, \boldsymbol{Z}, \boldsymbol{W}, \boldsymbol{Y}) + R(\boldsymbol{X}, \boldsymbol{W}, \boldsymbol{Y}, \boldsymbol{Z}) = 0\,, \tag{27.80}$$

$$(3)\quad R(\boldsymbol{X}, \boldsymbol{Y}, \boldsymbol{Z}, \boldsymbol{W}) = R(\boldsymbol{Z}, \boldsymbol{W}, \boldsymbol{X}, \boldsymbol{Y})\,. \tag{27.81}$$

The $(0, 2)$ metric tensor G gives rise to the $(0, 4)$ tensor $\mathcal{G}$, defined as follows:

$$\mathcal{G}(\boldsymbol{X}, \boldsymbol{Y}, \boldsymbol{Z}, \boldsymbol{W}) \equiv G(\boldsymbol{X}, \boldsymbol{Z})\, G(\boldsymbol{Y}, \boldsymbol{W}) - G(\boldsymbol{X}, \boldsymbol{W})\, G(\boldsymbol{Y}, \boldsymbol{Z})\,, \tag{27.82}$$

for arbitrary vector fields $\boldsymbol{X}, \boldsymbol{Y}, \boldsymbol{Z}, \boldsymbol{W}$ on M. It is easy to verify that $\mathcal{G}$ satisfies the same three properties (for R) given in Problem 27.11. In particular, we see that if $\boldsymbol{X}$ and $\boldsymbol{Y}$ are linearly independent tangent vectors in $T_x(M)$, then $\mathcal{G}(\boldsymbol{X}, \boldsymbol{Y}, \boldsymbol{X}, \boldsymbol{Y})$ is precisely the square of the area of the parallelogram determined by $\boldsymbol{X}$ and $\boldsymbol{Y}$, since

$$\begin{aligned} \mathcal{G}(\boldsymbol{X}, \boldsymbol{Y}, \boldsymbol{X}, \boldsymbol{Y}) &= G(\boldsymbol{X}, \boldsymbol{X})\, G(\boldsymbol{Y}, \boldsymbol{Y}) - (G(\boldsymbol{X}, \boldsymbol{Y}))^2 \\ &= |\boldsymbol{X}|^2 \cdot |\boldsymbol{Y}|^2 - (\boldsymbol{X} \cdot \boldsymbol{Y})^2 = |\boldsymbol{X}|^2\, |\boldsymbol{Y}|^2\, \sin^2\theta\,, \end{aligned} \tag{27.83}$$

where θ is the angle between $\boldsymbol{X}$ and $\boldsymbol{Y}$.

Suppose $\boldsymbol{X}$ and $\boldsymbol{Y}$ span a two-dimensional subspace $E \subset T_x(M)$, and $\boldsymbol{X}'$ and $\boldsymbol{Y}'$ are another two linearly independent vectors spanning the same subspace. Then we have the linear relationship

$$\begin{pmatrix} \boldsymbol{X}' \\ \boldsymbol{Y}' \end{pmatrix} = \begin{pmatrix} a & b \\ c & d \end{pmatrix} \begin{pmatrix} \boldsymbol{X} \\ \boldsymbol{Y} \end{pmatrix}, \tag{27.84}$$

where the 2×2 matrix on the right-hand-side is an invertible one, so that $ad - bc \neq 0$. By making use of the multi-linearity property of the tensors R

and $\mathcal{G}$ [as functions on $(\boldsymbol{X}, \boldsymbol{Y}, \boldsymbol{Z}, \boldsymbol{W})$] and their symmetry properties given in Problem 27.11, we have

$$R(\boldsymbol{X}', \boldsymbol{Y}', \boldsymbol{X}', \boldsymbol{Y}') = (ad - bc)^2 R(\boldsymbol{X}, \boldsymbol{Y}, \boldsymbol{X}, \boldsymbol{Y}) , \tag{27.85}$$

$$\mathcal{G}(\boldsymbol{X}', \boldsymbol{Y}', \boldsymbol{X}', \boldsymbol{Y}') = (ad - bc)^2 \mathcal{G}(\boldsymbol{X}, \boldsymbol{Y}, \boldsymbol{X}, \boldsymbol{Y}) . \tag{27.86}$$

Thus

$$\frac{R(\boldsymbol{X}', \boldsymbol{Y}', \boldsymbol{X}', \boldsymbol{Y}')}{\mathcal{G}(\boldsymbol{X}', \boldsymbol{Y}', \boldsymbol{X}', \boldsymbol{Y}')} = \frac{R(\boldsymbol{X}, \boldsymbol{Y}, \boldsymbol{X}, \boldsymbol{Y})}{\mathcal{G}(\boldsymbol{X}, \boldsymbol{Y}, \boldsymbol{X}, \boldsymbol{Y})} . \tag{27.87}$$

This implies that the quantity on the right-hand side of the above equation is independent of the choice of linearly independent vectors spanning the subspace $E \subset T_x(M)$ and only depends on E. We are then led to the following definition.

Definition 27.4. *Let $E \subset T_x(M)$ be a two-dimensional subspace of the tangent space at a point $x \in M$ of a Riemannian manifold M. Then the quantity*

$$K(x, E) \equiv -\frac{R(\boldsymbol{X}, \boldsymbol{Y}, \boldsymbol{X}, \boldsymbol{Y})}{\mathcal{G}(\boldsymbol{X}, \boldsymbol{Y}, \boldsymbol{X}, \boldsymbol{Y})} , \tag{27.88}$$

which is a function of E (independent of the choice of the linearly independent vectors $\boldsymbol{X}, \boldsymbol{Y} \in E$ spanning E), is called the **sectional curvature** *of the Riemannian manifold M at (x, E).*

For a two-dimensional Riemannian manifold, the *Gaussian curvature* K is given [cf. (25.18)] by

$$d\omega_1^2 = -K\, \omega^1 \wedge \omega^2 , \tag{27.89}$$

where $\{\omega^1, \omega^2\}$ is co-frame field dual to an orthonormal frame field $\{\boldsymbol{e}_1, \boldsymbol{e}_2\}$, and the torsion-free and metric compatible connection form ω_1^2 given by

$$D\boldsymbol{e}_1 = \omega_1^2\, \boldsymbol{e}_2 . \tag{27.90}$$

Since ω_i^j is antisymmetric due to the condition of metric-compatibility [cf. (27.55)], $(\omega \wedge \omega)_1^2 = 0$, and (27.2) implies that

$$d\omega_1^2 = \Omega_1^2 . \tag{27.91}$$

With respect to an orthonormal frame field the $(0, 2)$ covariant curvature matrix of 2-forms Ω_{ij} is given by

$$\Omega_{ij} = g_{jk}\, \Omega_i^k = \delta_{jk}\, \Omega_i^k = \Omega_i^j . \tag{27.92}$$

So $\Omega_1^2 = \Omega_{12}$, and it follows from (27.89) that the Gaussian curvature is given by

$$K = -\frac{\Omega_{12}}{\omega^1 \wedge \omega^2} . \tag{27.93}$$

By (27.70),

$$\Omega_{12} = \frac{1}{2} R_{12kl}\, \omega^k \wedge \omega^l = R_{1212}\, \omega^1 \wedge \omega^2 , \tag{27.94}$$

where R_{1212} is with respect to the coframe field $\{\omega^1, \omega^2\}$. We then have

$$K = -R_{1212} = -R(\boldsymbol{e}_1, \boldsymbol{e}_2, \boldsymbol{e}_1, \boldsymbol{e}_2) . \tag{27.95}$$

Since $\{\boldsymbol{e}_1, \boldsymbol{e}_2\}$ is an orthonormal frame field, $G(\boldsymbol{e}_1, \boldsymbol{e}_1) = G(\boldsymbol{e}_2, \boldsymbol{e}_2) = 1$ and $G(\boldsymbol{e}_1, \boldsymbol{e}_2) = G(\boldsymbol{e}_2, \boldsymbol{e}_1) = 0$. Hence (27.82) implies that $\mathcal{G}(\boldsymbol{e}_1, \boldsymbol{e}_2, \boldsymbol{e}_1, \boldsymbol{e}_2) = 1$. It follows from (27.95) and Definition 27.4 that, *for a two-dimensional Riemannian manifold, the Gaussian (total) curvature is the same as the sectional curvature.*

We will state two theorems (without proofs) which illustrate the importance of the sectional curvature of a Riemannian manifold. (For the proofs, the reader may consult, for example, Chern, Chen and Lam 1999.)

Theorem 27.2. *The curvature tensor R_{ijkl} of a Riemannian manifold M at any point $x \in M$ is uniquely determined by the sectional curvatures of all the two-dimensional tangent subspaces at x.*

Theorem 27.3. *Let M be a connected n-dimensional Riemannian manifold with $n > 2$. If the sectional curvatures $K(x, E)$ at every point $x \in M$ only depend on x and not on the tangent subspace $E \subset T_x(M)$, that is, $K(x, E) = K(x)$ for all $x \in M$, then M is in fact a constant curvature manifold: $K(x)$ is a constant over the entire manifold.*

Problem 27.12 If the sectional curvature $K(x, E)$ at a point $x \in M$ of a Riemannian manifold M is independent of the tangent subspace $E \subset T_x(M)$, that is, $K(x, E) = K(x)$, show that, for any $\boldsymbol{X}, \boldsymbol{Y}, \boldsymbol{Z}, \boldsymbol{W} \in T_x(M)$,

$$R(\boldsymbol{X}, \boldsymbol{Y}, \boldsymbol{Z}, \boldsymbol{W}) = -K(x)\, G(\boldsymbol{X}, \boldsymbol{Y}, \boldsymbol{Z}, \boldsymbol{W}) . \tag{27.96}$$

This fact turns out to be essential in the proof of Theorem 27.3 above. Follow the steps below.

(a) Define the covariant $(0, 4)$-tensor S by

$$S(\boldsymbol{X}, \boldsymbol{Y}, \boldsymbol{Z}, \boldsymbol{W}) \equiv R(\boldsymbol{X}, \boldsymbol{Y}, \boldsymbol{Z}, \boldsymbol{W}) + K(x)\, G(\boldsymbol{X}, \boldsymbol{Y}, \boldsymbol{Z}, \boldsymbol{W}) . \tag{27.97}$$

Then S obviously satisfies the symmetry properties in Problem 27.10, and we have, from the definition of the sectional curvature,

$$S(\boldsymbol{X}, \boldsymbol{Y}, \boldsymbol{X}, \boldsymbol{Y}) = 0 . \tag{27.98}$$

It follows that

$$S(\boldsymbol{X} + \boldsymbol{Z}, \boldsymbol{Y}, \boldsymbol{X} + \boldsymbol{Z}, \boldsymbol{Y}) = 0 . \tag{27.99}$$

(b) Expand (27.99) using the multi-linearity property of S to deduce that

$$S(\boldsymbol{X}, \boldsymbol{Y}, \boldsymbol{Z}, \boldsymbol{Y}) = 0 . \tag{27.100}$$

So we have

$$S(\boldsymbol{X}, \boldsymbol{Y} + \boldsymbol{W}, \boldsymbol{Z}, \boldsymbol{Y} + \boldsymbol{W}) = 0 . \tag{27.101}$$

(c) Expand (27.101) and use (27.98) to deduce that

$$S(\boldsymbol{X},\boldsymbol{Y},\boldsymbol{Z},\boldsymbol{W}) = -S(\boldsymbol{X},\boldsymbol{W},\boldsymbol{Z},\boldsymbol{Y}) = -S(\boldsymbol{X},\boldsymbol{Z},\boldsymbol{Y},\boldsymbol{W})\,. \tag{27.102}$$

(d) Use property (3) in Problem 27.10 to finally show that

$$S(\boldsymbol{X},\boldsymbol{Y},\boldsymbol{Z},\boldsymbol{W}) = 0\,. \tag{27.103}$$

This is the same as (27.96), the result to be proved.

Note that (27.96) can be expressed in component form as

$$R_{ijkl}(x) = -K(x)\left(g_{ik}(x)g_{jl}(x) - g_{il}(x)g_{jk}(x)\right)\,. \tag{27.104}$$

The $(0,4)$ covariant Riemann curvature tensor R gives rise to the very useful $(0,2)$ covariant **Ricci tensor** by *contraction*, denoted *Ric*, and defined as follows. For any two tangent vectors $\boldsymbol{X},\boldsymbol{Y} \in T_x(M)$,

$$Ric\,(\boldsymbol{X},\boldsymbol{Y}) \equiv \sum_{i=1}^{n} R(\boldsymbol{e}_i,\boldsymbol{X},\boldsymbol{e}_i,\boldsymbol{Y})\,, \tag{27.105}$$

where $\{\boldsymbol{e}_1,\dots,\boldsymbol{e}_n\}$ is an orthonormal basis in $T_x(M)$. The above definition is independent of the choice of the orthonormal basis. Suppose $\{\boldsymbol{e}'_i\}$ is another orthonormal basis, and the two bases are related by the orthogonal transformation (a_i^j), that is, $\boldsymbol{e}'_i = a_i^j\,\boldsymbol{e}_j$, and, as matrices, $(a)\,(a)^T = 1$ [where $(a)^T$ is the transpose of (a)]. Then, by multi-linearity of R,

$$\sum_i^n R(\boldsymbol{e}'_i,\boldsymbol{X},\boldsymbol{e}'_i,\boldsymbol{Y}) = \sum_{i,j,l} a_i^j a_i^l\, R(\boldsymbol{e}_j,\boldsymbol{X},\boldsymbol{e}_l,\boldsymbol{Y}) = \sum_{jl} \delta_{jl}\, R(\boldsymbol{e}_j,\boldsymbol{X},\boldsymbol{e}_l,\boldsymbol{Y})$$
$$= \sum_j R(\boldsymbol{e}_j,\boldsymbol{X},\boldsymbol{e}_j,\boldsymbol{Y})\,,$$

where the second equality follows from the orthogonality of the matrix (a_i^j). By the symmetry property (3) of R given in Problem 27.11 [cf. (27.81)] it is clear that *Ric* is a symmetric tensor:

$$Ric\,(\boldsymbol{X},\boldsymbol{Y}) = Ric\,(\boldsymbol{Y},\boldsymbol{X})\,. \tag{27.106}$$

Suppose $\boldsymbol{e}$ is a unit vector in $T_x(M)$ [$G(\boldsymbol{e},\boldsymbol{e}) = 1$]. Then we can define a function $\boldsymbol{e} \mapsto Ric\,(\boldsymbol{e},\boldsymbol{e})$ on the *tangent unit sphere* S_x at the point $x \in M$, where $S_x = \{\boldsymbol{X} \in T_x(M)\,|\,G(\boldsymbol{X},\boldsymbol{X}) = 1\}$. This function is called the **Ricci curvature** of the Riemannian manifold M at the point x along the direction $\boldsymbol{e}$. On choosing an orthonormal basis $\{\boldsymbol{e}_1,\dots,\boldsymbol{e}_n\}$ so that $\boldsymbol{e}_1 = \boldsymbol{e}$, we have

$$Ric\,(\boldsymbol{e},\boldsymbol{e}) = R(\boldsymbol{e},\boldsymbol{e},\boldsymbol{e},\boldsymbol{e}) + \sum_{i=2}^{n} R(\boldsymbol{e}_i,\boldsymbol{e},\boldsymbol{e}_i,\boldsymbol{e}) = \sum_{i=2}^{n} R(\boldsymbol{e}_i,\boldsymbol{e},\boldsymbol{e}_i,\boldsymbol{e})\,. \tag{27.107}$$

Thus, by (27.95), the Ricci curvature is the negative of the sum of $n-1$ sectional curvatures. By further contraction of the Ricci tensor one can obtain a C^∞ (smooth) function $\mathcal{S}(x)$ on M, called the **scalar curvature**, equal to the negative of the sum of all sectional curvatures corresponding to all the 2-dimensional tangent subspaces spanned by vector pairs among the basis vectors $e_1, \ldots, e_n$ of $T_x(M)$:

$$\mathcal{S}(x) \equiv \sum_{i,j}^{n} R(e_i, e_j, e_i, e_j) . \tag{27.108}$$

Problem 27.13 Show that in component form the Ricci tensor Ric and the scalar curvature $\mathcal{S}$ are given respectively, with respect to any basis in $T_x(M)$, by

$$Ric_{ij} = R^k_{ijk} = g^{hk} R_{ihjk} , \tag{27.109}$$

$$\mathcal{S} = g^{ij} Ric_{ij} . \tag{27.110}$$

In formulating the *general theory of relativity* (or a general theory of gravitational forces) Einstein sought an equation which describes how a 4-dimensional Pseudo-Riemannian manifold (4-dimensional spacetime) is curved by the presence of energy and momentum. In classical mechanics, the flow of energy and momentum is completely described by a symmetric $(0,2)$-type covariant tensor $T_{\mu\nu}$, called the **energy-momentum tensor**. The laws of local energy and momentum conservation are expressed by the following *divergenceless condition*:

$$D^\mu T_{\mu\nu} = 0 , \tag{27.111}$$

where $D^\mu \equiv g^{\mu\nu} D_\nu$ [for the definition of the covariant derivative D_μ recall (27.43)]. After much trial and error, Einstein settled on equating (up to a constant to maintain dimensional consistency) the so-called *Einstein tensor* $G_{\mu\nu}$ to the energy-momentum tensor $T_{\mu\nu}$. The **Einstein tensor**, which by requirement has to be a divergenceless tensor [in the sense of (27.111)], is given in terms of the Ricci tensor Ric_{ij} and the scalar curvature $\mathcal{S}$ as follows:

$$G_{\mu\nu} \equiv Ric_{\mu\nu} - \frac{1}{2} g_{\mu\nu} \mathcal{S} , \tag{27.112}$$

where $g_{\mu\nu}$ is the metric tensor. The resulting so-called **Einstein's equation** is

$$G_{\mu\nu} = 8\pi G \, T_{\mu\nu} , \tag{27.113}$$

where $G = 6.67 \times 10^{-11} \, N \cdot m^2/kg^2$ is Newton's gravitational constant [cf. (14.4)], and (part of) the natural units c (speed of light) $= 1$ is used. [Note that, of course, $G_{\mu\nu}$ is not to be confused with $g_{\mu\nu}$, and Newton's constant G is not to be confused with the metric tensor G (defined in Def. 27.1).]

Problem 27.14 Use the second Bianchi identity in the form of (27.43) and the fact that the metric tensor is divergenceless:

$$D^{\mu} g_{\mu\nu} = 0 , \tag{27.114}$$

which is actually equivalent to the metric compatibility condition of D [cf. (27.56)], to show that the Einstein tensor is divergenceless:

$$D^{\mu} G_{\mu\nu} = 0 . \tag{27.115}$$

Problem 27.15 The **Laplace-Beltrami operator** (or *Laplace operator*) Δ on a scalar function f on a Riemannian manifold M is a second-order differential operator defined as follows:

$$\Delta f \equiv tr\, D^2 f , \tag{27.116}$$

where D is an affine connection on M, so that $D^2 f = D(Df) = D(df)$ is a symmetric covariant rank $(0,2)$ tensor on M, and tr, called the **trace operator**, is defined for any rank $(0,2)$ covariant tensor $S(x)$ on M by

$$tr\, S(x) \equiv \sum_i^n S(\boldsymbol{e}_i, \boldsymbol{e}_i) , \tag{27.117}$$

where $n = dim\,(M)$ and $\{\boldsymbol{e}_1, \dots, \boldsymbol{e}_n\}$ is an orthonormal basis in $T_x(M)$. Show that:

(a) The definition of the trace operator as given above is independent of the choice of the orthonormal basis in $T_x(M)$.

(b) With $g_{ij} = G(\partial/\partial x^i \,, \partial/\partial x^j)$, we have

$$\Delta f = \sum_{i,j} g^{ij}\, D^2 f \left(\frac{\partial}{\partial x^i} , \frac{\partial}{\partial x^j} \right) . \tag{27.118}$$

(c) With $|g| \equiv$ determinant of g_{ij}, we have

$$\Delta f = \frac{1}{\sqrt{|g|}} \sum_j \frac{\partial}{\partial x^j} \left(\sum_i g^{ij} \sqrt{|g|}\, \frac{\partial f}{\partial x^i} \right) . \tag{27.119}$$

This obviously reduces to the classical Laplace operator on Euclidean $\mathbb{R}^n$ (where $g_{ij} = \delta_{ij}$):

$$\Delta f = \nabla^2 f = \sum_i \frac{\partial^2 f}{\partial (x^i)^2} .$$

The formula in (c) above indicates that the Laplace-Beltrami operator Δ is a second order linear **elliptic** partial differential operator.

Chapter 28

Frame Bundles and Principal Bundles, Connections on Principal Bundles

Much of our mathematical development up to this point centering around *Cartan's method of moving frames* and its applications in classical mechanics is based on the notion of *frame fields* on a Riemannian manifold. This construct leads to the notion of a *frame bundle*, and then to that of a *connection on a frame bundle*, which in turn can be abstracted and generalized immediately to the notions of a *principal bundle* and a *connection on a principal bundle*. Although these latter more abstract mathematical constructs are not usually applied directly to physical problems, an understanding and appreciation of them, even at an introductory level, will lead to deeper insights and a certain mathematical facility in the use of their *associated vector bundles*, with which we have been mainly concerned in our quest for the applications of differential geometry in classical mechanics. In this chapter we will present an elementary introduction to these highly abstract and intricate concepts. As usual, we will attempt to base our discussion on geometrical intuition and concrete calculational procedures, and will not be overly concerned with mathematical rigor.

Recall that a frame at a point $x \in M$, where M is an n-dimensional Riemannian manifold, is a set of n linearly independent vectors $\boldsymbol{e}_1, \dots, \boldsymbol{e}_n$ in the tangent space $T_x(M)$ at the point x. We denote one such frame by $(x;\, \boldsymbol{e}_1, \dots, \boldsymbol{e}_n)$. One can then take the set P of all frames at all points of M, define a natural (surjective) projection $\pi : P \to M$ by

$$\pi(x;\, \boldsymbol{e}_1, \dots, \boldsymbol{e}_n) = x \; , \tag{28.1}$$

and give P a differentiable structure (inherited from that of M) to construct a so-called **frame bundle** $\pi : P \to M$ on M.

The differentiable structure on the frame bundle P is given as follows. In a local coordinate neighborhood $(U;\, x^i)$ of M, there is the *natural frame field* $(\partial_1, \dots, \partial_n)$, where $\partial_i = \partial/\partial x^i$. Then an arbitrary frame $(x;\, \mathbf{e}_1, \dots, \mathbf{e}_n)$ on U can be expressed as

$$\mathbf{e}_i = X_i^k\, \partial_k \,, \qquad 1 \le i \le n \,, \tag{28.2}$$

where the $n \times n$ real matrix (X_i^k) is an element of $GL(n, \mathbb{R})$, the group of non-singular $n \times n$ real matrices. One can then define a map

$$\varphi_U : U \times GL(n, \mathbb{R}) \longrightarrow \pi^{-1}(U) \,,$$

called a **local trivialization** (or local product structure) of the frame bundle P, such that for any $x \in U$ and $(X_i^k) \in GL(n, \mathbb{R})$,

$$\varphi_U(x,\, (X_i^k)) = (x;\, \mathbf{e}_1, \dots, \mathbf{e}_n) \,. \tag{28.3}$$

The set of $n + n^2$ real numbers $(x^i,\, X_j^k)$ can then be regarded as local coordinates for a point $p = (x;\, \mathbf{e}_1, \dots, \mathbf{e}_n) \in P$. Hence the frame bundle P is an $(n + n^2)$-dimensional manifold. Now let $(V;\, y^i)$ be another local coordinate neighborhood of M such that $U \bigcap V \neq \emptyset$. The local transformation of coordinates in $U \bigcap V$ is given by a set of smooth functions $f^i(x^1, \dots, x^n)$:

$$y^i = f^i(x^1, \dots, x^n) \,, \qquad i = 1, \dots, n \,, \tag{28.4}$$

with the corresponding transformation between the natural bases $(\partial/\partial x^i)$ and $(\partial/\partial y^i)$:

$$\frac{\partial}{\partial x^i} = \frac{\partial y^j}{\partial x^i} \frac{\partial}{\partial y^j} \,. \tag{28.5}$$

Let $(y^i,\, Y_j^k)$ be the local coordinates of $p = (x;\, \mathbf{e}_1, \dots, \mathbf{e}_n) \in P$, with $x \in U \cap V$, corresponding to another local trivialization $\varphi_V : V \times GL(n, \mathbb{R}) \to \pi^{-1}(V)$. We require that

$$\mathbf{e}_i = X_i^k \frac{\partial}{\partial x^k} = Y_i^k \frac{\partial}{\partial y^k} \,, \tag{28.6}$$

which implies, as follows from (28.5),

$$Y_i^k = X_i^j \frac{\partial y^k}{\partial x^j} \,. \tag{28.7}$$

These equations, together with (28.4), constitute the coordinate transformations for any two coordinate neighborhoods $\pi^{-1}(U)$ and $\pi^{-1}(V)$ in P, where $U \cap V \neq \emptyset$ [which necessarily implies that $\pi^{-1}(U) \cap \pi^{-1}(V) \neq \emptyset$], and thus determine the differentiable structure on P.

As in vector bundles, one can construct a family of **transition functions** on frame bundles, $g_{UV} : U \cap V \to GL(n, \mathbb{R})$ $(U \cap V \neq \emptyset)$, as follows. For $X \in GL(n, \mathbb{R})$ and $x \in U$, define

$$\varphi_{U,x}(X) \equiv \varphi_U(x, X) \,. \tag{28.8}$$

Then $\varphi_{U,x} : GL(n,\mathbb{R}) \to \pi^{-1}(x)$ is a homomorphism, and so the *typical fiber* of the frame bundle P is $GL(n,\mathbb{R})$. The transition functions are given by

$$g_{UV}(x) = \varphi_{V,x}^{-1} \circ \varphi_{U,x} \,. \qquad (28.9)$$

The quantity on the right-hand side acts on an $X \in GL(n,\mathbb{R})$ to produce a $Y \in GL(n,\mathbb{R})$ according to (28.7), and is equivalent to the *right multiplication* of X by the *Jacobian matrix* $(\partial y^k/\partial x^j)_x \in GL(n,\mathbb{R})$. Since these Jacobian matrices (one at each $x \in U \cap V$) also determine the transition functions of the *tangent bundle* $T(M)$, we say that the frame bundle $\pi : P \to M$ is the *principal bundle* associated with the tangent bundle $T(M)$. The typical fiber as well as the structure group of the frame bundle are both $GL(n,\mathbb{R})$. Thus frame bundles, and principal bundles (which we will define below) in general, are not vector bundles.

The structure group $GL(n,\mathbb{R})$ acts naturally on the frame bundle P by a *left action* L_g, $g \in GL(n,\mathbb{R})$, which preserves the fibers (mapping a fiber onto itself). This left action is defined by

$$L_g\,(x;\, \boldsymbol{e}_1,\dots,\boldsymbol{e}_n) = (x;\, (\boldsymbol{e}')_1,\dots,(\boldsymbol{e}')_n)\,, \quad \text{with} \quad (\boldsymbol{e}')_i = g_i^j\, \boldsymbol{e}_j \,. \qquad (28.10)$$

The fiber-preserving property of L_g can be expressed by

$$\pi \circ L_g = \pi \,, \qquad (28.11)$$

for all $g \in GL(n,\mathbb{R})$. The left action also satisfies

$$L_{g_1 g_2} = L_{g_1} \circ L_{g_2} \qquad \text{for all} \qquad g_1,\, g_2 \in GL(n,\mathbb{R}) \,. \qquad (28.12)$$

We will now discuss how the notion of an affine connection on a tangent bundle as defined in previous chapters (cf. Def. 22.4), which is essentially a local definition, can be formulated in a global manner in the context of frame bundles. Suppose an affine connection D is given on a Riemannian manifold M. For a parametrized curve $\gamma : [0,1] \to M$ with $\gamma(0) = x \in M$ and a given point $p = (x;\, \{\boldsymbol{e}_i\})$ in the frame bundle P, there exists a **natural lift** of γ to P, denoted by $\tilde{\gamma}_p : [0,1] \to P$, such that $\pi(\tilde{\gamma}_p(t)) = \gamma(t)$, and where $\tilde{\gamma}_p(t) = (\gamma(t);\, \boldsymbol{e}_1(t),\dots,\boldsymbol{e}_n(t))$, with each $\boldsymbol{e}_i(t)$ being the *parallel translation* of $\boldsymbol{e}_i$ along the curve $\gamma(t)$ as determined by the given affine connection D [according to (22.43)]. This natural lift satisfies the property

$$L_g\left(\tilde{\gamma}_{(x;\,\{\boldsymbol{e}_i\})}\right) = \tilde{\gamma}_{(x;\,\{g\boldsymbol{e}_i\})} \quad \text{for all} \;\; g \in GL(n,\mathbb{R}) \,. \qquad (28.13)$$

Now fix $x \in M$ and $p = (x;\, \{\boldsymbol{e}_i\}) \in P$. Consider all curves $\gamma : [0,1] \to M$ with starting point $\gamma(0) = x$ and their corresponding natural lifts $\tilde{\gamma}_p : [0,1] \to P$, with starting points $\tilde{\gamma}_p(0) = p$. When $\dot{\gamma}(0) \equiv d\gamma/dt|_{t=0}$ ranges over all of the tangent space $T_x(M)$, the totality of the corresponding $d\tilde{\gamma}/dt|_{t=0}$ constitutes a subspace $\mathcal{H}_p$, called the **horizontal subspace**, of the tangent space $T_p(P)$ on the frame bundle. We assert without proof that $\mathcal{H}_p$ is n-dimensional, where n

is the dimension of M, and that they project, through the bundle projection map π, to tangent spaces on M:

$$\pi_*(\mathcal{H}_p) = T_x(M) , \qquad \text{with} \qquad \pi(p) = x . \tag{28.14}$$

This ensures that, at each point $p \in P$, the tangent space $T_p(P)$ can be written as a direct sum of vector subspaces:

$$T_p(P) = \mathcal{H}_p \oplus \mathcal{V}_p , \tag{28.15}$$

where the **vertical subspaces** $\mathcal{V}_p$ are tangent spaces to the fiber $\pi^{-1}(x)$, or equivalently, $\pi_*(\mathcal{V}_p) = 0$. Equation (28.13) can be re-expressed in terms of the horizontal subspaces as follows:

$$(L_g)_* \mathcal{H}_p = \mathcal{H}_{L_g(p)} . \tag{28.16}$$

Having constructed a *horizontal subspace field* on a frame bundle $\pi : P \to M$ corresponding to a particular affine connection D on the base manifold M, one can reverse the procedure and show that (the proof will not be presented here) any horizontal subspace field $\mathcal{H}_p$ satisfying the three conditions (28.14), (28.15) and (28.16) above determines uniquely an affine connection D on the base manifold M such that $\mathcal{H}_p$ is precisely the horizontal subspace field constructed from D according to the parallel translation procedure described above. Thus *the notion of an affine connection on M and that of a horizontal subspace field on the frame bundle of M satisfying the aforementioned three conditions are one and the same.* We can now define the concept of the *connection on a frame bundle*:

Definition 28.1. *Suppose $\pi : P \to M$ is a frame bundle on a Riemannian manifold M. A choice of a horizontal subspace field $\mathcal{H}_p \subset T_p(P)$ satisfying the three conditions (28.14), (28.15), and (28.16) is called a* ***connection on a frame bundle****.*

Before proceeding to the general notions of a principal bundle and a connection on a principal bundle, we will present a very useful and instructive method of the analytical construction of horizontal subspaces of a frame bundle. This procedure will also establish the fact that, topologically, the frame bundle of a Riemannian manifold M is simpler than that of M, namely, a frame bundle of M is always *parallellizable* while M is not so in general (see below).

Let $(U; x^i)$ and $(V; y^i)$ be two coordinate neighborhoods in M, with $U \cap V \neq \emptyset$. Then we have two corresponding coordinate neighborhoods $\pi^{-1}(U)$ and $\pi^{-1}(V)$ in P with non-vanishing intersections, whose local coordinates are (x^i, X^k_j) and (y^i, Y^k_j), respectively. On $U \cap V$,

$$dy^i = \frac{\partial y^i}{\partial x^j}\, dx^j , \tag{28.17}$$

while it follows from (28.7) that

$$(X^{-1})^j_i = \frac{\partial y^k}{\partial x^i}\, (Y^{-1})^j_k , \tag{28.18}$$

where X^{-1} and Y^{-1} are the inverse matrices of X and Y, respectively. The above two equations imply that

$$(X^{-1})^j_i \, dx^i = (Y^{-1})^j_i \, dy^i \,. \tag{28.19}$$

This shows that the n one-forms θ^i defined by $[n = dim\,(M)]$

$$\theta^i \equiv (X^{-1})^i_j \, dx^j \tag{28.20}$$

are independent of local coordinates on P and are thus *global* and linearly independent one-forms on P.

Equation (28.20) can be inverted to give

$$dx^i = X^i_j \, \theta^j \,. \tag{28.21}$$

Hence the *Pfaffian system* of equations

$$\theta^i = 0 \,, \qquad i = 1, \dots, n \,, \tag{28.22}$$

is equivalent to

$$dx^i = 0 \,, \qquad i = 1, \dots, n \,, \tag{28.23}$$

in every coordinate neighborhood $\pi^{-1}(U)$ of P. Clearly, the above system of Pfaffian equations is *completely integrable*, with an *integral submanifold* in P given by

$$x^i = \text{constant} \,, \qquad i = 1, \dots, n \,, \tag{28.24}$$

which is precisely the fiber $\pi^{-1}(x)$ above $x \in M$. The tangent subspace field tangent to the fiber is then the *vertical subspace field* $\mathcal{V}_p$ introduced above [in (28.15)]. Thus the Pfaffian system (28.22) determines the n^2-dimensional vertical subspace field $\mathcal{V}_p$.

Problem 28.1 Show by computing $d\theta^i$ [with θ^i as defined by (28.20)] to obtain a differential 2-form on P, and on the basis of the *Frobenius Theorem* (Theorem 10.2), that the Pfaffain system of equations (28.22) is integrable.

Now suppose that an affine connection D is given on M by means of an $n \times n$ matrix of one-forms ω^j_i satisfying $D\dfrac{\partial}{\partial x^i} = \omega^j_i \dfrac{\partial}{\partial x^j}$, where $\left\{\dfrac{\partial}{\partial x^i}\right\}$ is a local (natural) frame field on a coordinate neighborhood $(x^i;\, U)$ in M. Equation (28.2) implies that

$$De_i = \left(dX^k_i + X^j_i \omega^k_j\right) \otimes \frac{\partial}{\partial x^k} \,. \tag{28.25}$$

Since $(x^i,\, X^k_j)$ are local coordinates on P, the quantities

$$DX^k_i \equiv dX^k_i + X^j_i \omega^k_j = dX^k_i + X^j_i \Gamma^k_{jl} dx^l \tag{28.26}$$

are local differential one-forms on $\pi^{-1}(U) \subset P$. We will show that these are in fact global one-forms on P as well. Under the local gauge transformation of natural frame fields on $U \cap V$ given by

$$\frac{\partial}{\partial y^i} = g_i^j \, \frac{\partial}{\partial x^j} \,, \qquad \text{with} \qquad g_i^j = \frac{\partial x^j}{\partial y^i} \,, \tag{28.27}$$

the gauge-transformed affine connection matrix $(\omega')_i^j$ is given, according to (22.16), by

$$(\omega')_i^j = d\left(\frac{\partial x^k}{\partial y^i}\right)\frac{\partial y^j}{\partial x^k} + \frac{\partial x^k}{\partial y^i}\,\omega_k^l\,\frac{\partial y^j}{\partial x^l} \,. \tag{28.28}$$

Thus, by (28.26),

$$\begin{aligned} DY_i^k &= dY_i^k + Y_i^j\,(\omega')_j^k \\ &= dX_i^j\,\frac{\partial y^k}{\partial x^j} + X_i^j\,d\left(\frac{\partial y^k}{\partial x^j}\right) + Y_i^j\left(d\left(\frac{\partial x^l}{\partial y^j}\right)\frac{\partial y^k}{\partial x^l} + \frac{\partial x^l}{\partial y^j}\,\omega_l^m\,\frac{\partial y^k}{\partial x^m}\right) . \end{aligned} \tag{28.29}$$

Using (28.7) for Y_i^k and recognizing that $\dfrac{\partial x^l}{\partial y^j}\dfrac{\partial y^k}{\partial x^l} = \delta_j^k$, the third term on the right-hand side of the second equality in the above equation can be written as

$$\begin{aligned} &-X_i^h\,\frac{\partial y^j}{\partial x^h}\,\frac{\partial x^l}{\partial y^j}\,d\left(\frac{\partial y^k}{\partial x^l}\right) + X_i^h\,\frac{\partial y^j}{\partial x^h}\,\frac{\partial x^l}{\partial y^j}\,\omega_l^m\,\frac{\partial y^k}{\partial x^m} \\ &= -X_i^h\,\delta_h^l\,d\left(\frac{\partial y^k}{\partial x^l}\right) + X_i^h\,\delta_h^l\,\omega_l^m\,\frac{\partial y^k}{\partial x^m} = -X_i^l\,d\left(\frac{\partial y^k}{\partial x^l}\right) + X_i^l\,\omega_l^m\,\frac{\partial y^k}{\partial x^m} \,. \end{aligned} \tag{28.30}$$

Equation (28.29) then yields

$$dY_i^k + Y_i^j\,(\omega')_j^k = (dX_i^j + X_i^l\,\omega_l^j)\,\frac{\partial y^k}{\partial x^j} \,, \tag{28.31}$$

or, by the definition given in (28.26),

$$DY_i^k = DX_i^j\,\frac{\partial y^k}{\partial x^j} \,. \tag{28.32}$$

To facilitate our development, we will write (28.18) and (28.32) in matrix form, with the matrix notations $X \equiv (X_i^j)$, $Y \equiv (Y_i^j)$ and $\partial y/\partial x \equiv (\partial y^j/\partial x^i)$:

$$X^{-1} = \frac{\partial y}{\partial x}\cdot Y^{-1} \,, \tag{28.33}$$

$$DY = DX\cdot\frac{\partial y}{\partial x} \,. \tag{28.34}$$

These equations then imply

$$DX\cdot X^{-1} = DY\cdot\left(\frac{\partial y}{\partial x}\right)^{-1}\cdot\left(\frac{\partial y}{\partial x}\right)\cdot Y^{-1} = DY\cdot Y^{-1} \,, \tag{28.35}$$

or

$$(X^{-1})^j_k\, DX^k_i = (Y^{-1})^j_k\, DY^k_i\,. \tag{28.36}$$

Hence the differential 1-forms

$$\theta^j_i \equiv (X^{-1})^j_k\, DX^k_i = (X^{-1})^j_k\,(dX^k_i + X^l_i\omega^k_l) = (X^{-1})^j_k\,(dX^k_i + X^l_i\Gamma^k_{lm}dx^m)\,, \tag{28.37}$$

where (ω^k_l) is the affine connection matrix of 1-forms with respect to the local natural frame field on M, are independent of the choice of local coordinates on P, and are therefore global 1-forms on P. This result, together with that stated immediately after (28.20), allows us to conclude that the $n + n^2$ one forms θ^i [(28.20)] and θ^j_i [(28.37)] are linearly independent everywhere on the frame bundle P, and therefore constitute a global *coframe field* on P. Thus there is a global *frame field* on P also, and we say that *the frame bundle P is* ***parallelizable***. In this sense the frame bundle is simpler in topological structure than the base manifold, which is not always parallelizable.

It is instructive to rewrite (28.37) as follows. Remembering that (X^j_i) as a matrix is a group element in $GL(n,\mathbb{R})$, we rewrite it as $(X^j_i) = g$. In matrix notation, then, on denoting the matrix of one-forms (θ^j_i) by θ, Equation (28.37) reappears as

$$\theta = dg\, g^{-1} + g\,\omega\, g^{-1}\,. \tag{28.38}$$

The right-hand side looks identical to that in (22.16) [or (4.19)], which is the equation for the gauge transformation rule for the connection ω. But it has to be interpreted differently. Whereas in (22.16), the various quantities are one-forms on the base manifold M [$g(x)$ is a function of the local coordinates of $x \in M$], the quantities in (28.38) are one-forms on the total space P (the g^j_i are considered as fiber coordinates). So the one-forms θ^j_i are in fact pullback forms of ω by the bundle projection map π, and we can write

$$\theta = \pi^*(\omega)\,. \tag{28.39}$$

This is an example of a *transgression* phenomenon on the frame bundle [see discussion following (25.30)].

In the same way that the Pfaffian system $\theta^i = 0$ determines the n^2-dimensional vertical subspace field $\mathcal{V}_p$ on P, the Pfaffian system

$$\theta^j_i = 0\,, \qquad 1 \le i, j \le n\,, \tag{28.40}$$

determines an n-dimensional tangent subspace field $\mathcal{H}_p$ on P: If $\mathbb{H} \in \mathcal{H}_p$, then $\langle\, \mathbb{H}\,,\, \theta^j_i\,\rangle_p = 0$. We claim that this is the *horizontal subspace field* introduced above. Furthermore, the vertical and horizontal subspace fields so determined satisfy the three conditions (28.14), (28.15) and (28.16). We state this important result (without proof) in the form of the following theorem.

Theorem 28.1. *Suppose $\pi : P \to M$ is the frame bundle of a Riemannian (or pseudo-Riemannian) manifold M equipped with an affine connection D on the tangent bundle $T(M)$. The global one forms θ^i [(28.20)] and θ^j_i [(28.37)] on P*

determine, by means of the Pfaffian systems of equations (28.22) and (28.38), the vertical and horizontal tangent subspace fields on P, $\mathcal{V}_p$ and $\mathcal{H}_p$ respectively, which satisfy the following properties:

(1) *At each point $p \in P$, the tangent space $T_p(P)$ can be decomposed as the direct sum of the vertical and horizontal subspaces:*

$$T_p(P) = \mathcal{V}_p \oplus \mathcal{H}_p \,. \tag{28.41}$$

(2) *The horizontal subspaces of $T_p(P)$ project onto the tangent spaces of the base manifold M under the bundle projection π, that is, if $\pi(p) = x \in M$, we have the differential projection*

$$\pi_*(\mathcal{H}_p) = T_x(M) \,. \tag{28.42}$$

(3) *The horizontal subspace field $\mathcal{H}_p$ is invariant under the left action L_g [$g \in GL(n, \mathbb{R})$, defined by (28.10)] of the group $GL(n, \mathbb{R})$ on P:*

$$(L_g)_* \mathcal{H}_p = \mathcal{H}_{L_g p} \,. \tag{28.43}$$

Problem 28.2 For the coframe field θ^j_i on a frame bundle P defined by (28.37) and a group element $g \in GL(n, \mathbb{R})$, show that

$$(L_g)^* \, (\theta^j_i) = g^k_i \, \theta^l_k \, (g^{-1})^j_l \,, \tag{28.44}$$

where L_g is the left action on P defined by (28.10). The operation on the right-hand side on the coframe field matrix (θ^j_i) by g is called the action of the **adjoint representation** of g, $Ad(g)$, on (θ^j_i):

$$Ad(g) \, (\theta^j_i) \equiv g^k_i \, \theta^l_k \, (g^{-1})^j_l \,. \tag{28.45}$$

Based on the relatively concrete notion of the frame bundle as developed above, one can generalize to the more abstract notion of the *principal bundle*, defined as follows.

Definition 28.2. *A* ***principal fiber bundle*** *consists of a manifold P (the* **total space***), a Lie group G, a* **base manifold** *M, and a projection map $\pi : P \to M$ such that the following conditions hold:*

(1) *For each $g \in G$ there is a left-action diffeomorphism $L_g : P \to P$ [written $L_g(p) = gp$] such that*

 (a) *$(g_1 g_2)p = g_1(g_2 p)$, for all $g_1, g_2 \in G$ and $p \in P$;*

(b) $ep = p$ *for all* $p \in P$*, where* e *is the identity in* G*;*

(c) G ***acts freely*** *on* P *to the left, that is, there are no fixed points of* L_g *if* $g \neq e$*. Equivalently, if* $gp = p$ *for some* $p \in P$*, then* $g = e$*.*

(2) The base manifold M *is the quotient space of the total space* P *with respect to the left action of* G *on* P*, and the projection* $\pi : P \to P$ *is smooth and surjective:*

$$\pi^{-1}(\pi(p)) = \{gp \,|\, g \in G\} . \tag{28.46}$$

For $x \in M$*,* $\pi^{-1}(x)$ *is called the* ***fiber*** *above* x*. All fibers are diffeomorphic to* G*, but there is no natural group structure on any fiber.*

(3) P *is locally* ***trivial:*** *for each* $x \in M$ *there is a neighborhood* U *of* x *such that* $\pi^{-1}(U)$ *is diffeomorphic to* $U \times G$*. In other words, there is a diffeomorphism* $T_U : \pi^{-1}(U) \to U \times G$ *given by*

$$T_U(p) = (\pi(p)\, , \varphi_U(p))\, , \tag{28.47}$$

where the map $\varphi_U : \pi^{-1}(U) \to G$ *has the property*

$$\varphi_U(gp) = g\, \varphi_U(p)\, . \tag{28.48}$$

The diffeomorphism T_U *is called a* ***local trivialization****, or a* ***local choice of gauge*** *(in the physics literature).*

The structure of a principal bundle is determined by the **transition functions**, defined similarly to those of vector bundles (cf. Def. 22.1).

Definition 28.3. *Suppose* T_U *and* T_V *are two local trivializations of a principal bundle* $\pi : P \to M$ *with Lie group* G*, where* $U \cap V \neq \emptyset$*. The* ***transition function*** $G_{UV} : U \cap V \to G$ *from* T_U *to* T_V *is defined, for* $x \in U \cap V$*, by*

$$g_{UV}(x) = \varphi_V^{-1}(p) \cdot \varphi_U(p)\, , \tag{28.49}$$

where p *is any point in* $\pi^{-1}(x)$*.*

Problem 28.3 Show that the transitions $g_{UV}(x)$, $(x \in U \cap V \neq \emptyset)$ as defined above does not depend on the choice of the point p in the total space P.

Problem 28.4 Show that the transition functions as defined above satisfy the following properties:

$$g_{UU}(x) = e\, , \quad \text{for all } x \in U\, ; \tag{28.50}$$

$$g_{VU}(x) = (g_{UV}(x))^{-1}\, , \quad \text{for all } x \in U \cap V\, ; \tag{28.51}$$

$$g_{UV}(x)\, g_{VW}(x)\, g_{WU}(x) = e\, , \quad \text{for all } x \in U \cap V \cap W\, . \tag{28.52}$$

In the above equations e is the identity element in the Lie group G of the principal bundle.

Principal bundles are classified according to their degrees of non-triviality. We will not delve into the fascinating and important problem of the classification of principal bundles (which actually has significant applications in physics). Instead, we will simply define the notion of a *trivial bundle*, and state (without proof) a necessary and sufficient condition for a principal bundle to be trivial (Theorem 28.2 below). (For more details on this topic the reader may consult, for example, Lam 2006.)

Definition 28.4. *A principal bundle $\pi : P \to M$ is said to be* ***trivial*** *if there exists a* ***global trivialization****, that is, a diffeomorphism $T_M : P \to M \times G$. In other words, the principal bundle is trivial if P is globally a product.*

To state the theorem referred to above we need one more definition.

Definition 28.5. *A* ***local section*** *of a principal bundle $\pi : P \to M$ is a map $\sigma : U \to P$, where U is a neighborhood in M, such that $\pi \circ \sigma(x) = x$ for any $x \in U$. A* ***global section****, if it exists, is a map $\sigma : M \to P$ such that $\pi \circ \sigma(x) = x$ for any $x \in M$.*

We then have the following theorem.

Theorem 28.2. *A principal bundle is trivial if and only if it has a global section.*

We will conclude this chapter with three equivalent definitions of the notion of a *connection on a principal bundle.* Although it derives from the notion of a connection on a vector bundle, the definitions look superficially very different from that of the latter, and from each other. They represent the end products of a highly distilled process of generalizations, and in fact were not cast in their final forms until the middle of the 20th century. In this process, the elucidation of the properties of a horizontal subspace field on a frame bundle played a vital role. We will, however, not demonstrate the equivalence of the definitions here. The interested reader can consult, for example, Lam 2006.

Definition 28.6. *A* ***connection on a principal bundle*** *$\pi : P \to M$ with Lie group G is a smooth assignment to each point $p \in P$ of an n-dimensional subspace $\mathcal{H}_p$ of the tangent space $T_p(P)$ (n being the dimension of the base manifold M) such that the following properties are satisfied:*

(1)

$$T_p(P) = \mathcal{H}_p \oplus \mathcal{V}_p \, , \tag{28.53}$$

where $\mathcal{V}_p$ is the subspace of $T_p(P)$ defined by

$$\mathcal{V}_p \equiv \{ \mathbb{V} \in T_p(P) \,|\, \pi_*(\mathbb{V}) = 0 \} \, . \tag{28.54}$$

(2) The subspace field $\mathcal{H}_p$ is invariant *under the* left action L_g *of the Lie group* G*:*

$$(L_g)_* \mathcal{H}_p = \mathcal{H}_{gp} \tag{28.55}$$

for all $g \in G$*.*

The linear subspaces $\mathcal{V}_p$ and $\mathcal{H}_p$ are called the ***vertical subspace*** *and a* ***horizontal subspace*** *of $T_p(P)$, respectively.*

The reader should compare the above definition with Theorem 28.1.

To better understand the next definition the reader may need a few fundamental geometrical facts concerning the relationship between a Lie group and its Lie algebra (the relevant information may be found, for example, in Chern, Chen and Lam 1999, or Lam 2006). In particular, we note that a Lie group G is simultaneously a group and a differentiable manifold, and the **Lie algebra** $\mathcal{G}$ of the Lie group G, in addition to being an algebra, can also be identified as the tangent space at the identity of G (considered as a manifold). Equation (28.44) and the definition of the action of the *adjoint representation* of G given by (28.45) should also be noted in preparation.

Definition 28.7. *Let $\mathcal{G}$ be the Lie algebra of a Lie group G, which is the structure group of a principal bundle $\pi : P \to M$. For $A \in \mathcal{G}$, define a vector field $A^\sharp$ on P, called the* ***fundamental field****, by*

$$A^\sharp(p) \equiv \frac{d}{dt}(\exp(At)p)_{t=0} \,. \tag{28.56}$$

A ***connection*** *on the principal bundle is a Lie algebra-valued one-form ω on P satisfying the following two properties:*

$$(1) \quad \omega\left(A^\sharp(p)\right) = A \,, \tag{28.57}$$

$$(2) \quad (L_g)^*(\omega) = \mathcal{A}d(g)\,(\omega)\,, \qquad \textit{for all } g \in G\,, \tag{28.58}$$

where $\mathcal{A}d(g) : \mathcal{G} \to \mathcal{G}$ is the adjoint representation of G [whose action is defined by (28.45)].

We characterize property (2) above by saying that the connection one-form ω is *ad-invariant* under left-actions by the Lie group G.

Problem 28.5 Show that property (2) in Def. 28.7 for the ad-invariance of a connection ω on a principal bundle can be stated in more detail as

$$\omega_{gp}((L_g)_*\mathbb{X}_p) = (\mathcal{A}d(g))\,(\omega_p(\mathbb{X}_p))\,, \tag{28.59}$$

for all $g \in G$, $p \in P$ and $\mathbb{X}_p \in T_p(P)$.

The last definition of the connection on a principal bundle is cast in the form of Lie algebra-valued one-forms on the base manifold, rather than one-forms on the total space, as in Def. 28.7. This definition is more in line with our more familiar understanding of connections on vector bundles [when compared with (22.16) or (4.19)], and is the one most suitable for physics applications (and applications in this book).

Definition 28.8. *A* ***connection*** *on a principal bundle* $\pi : P \to M$ *with structure (Lie) group* G *is an assignment to each local trivialization* $T_U : \pi^{-1}(U) \to U \times G$ *(each choice of gauge) a Lie algebra-valued (local) one-form* ω_U *on* U *which satisfies the following gauge transformation rule (between the local trivializations* T_U *and* T_V*, with* $U \cap V \neq \emptyset$*):*

$$\omega_V = (dg_{UV})\, g_{UV}^{-1} + g_{UV}\, \omega_U\, g_{UV}^{-1} \,, \tag{28.60}$$

where g_{UV} *are the transition functions defined in Def. 28.3.*

With this definition we have come full circle back to (4.19). At this juncture the reader is urged to review our earlier development leading up to that equation, and should appreciate much better why we chose to highlight it with so much fanfare at various parts of the text. Indeed, there is a direct line leading from the basic classical mechanical concept of the angular velocity of rigid bodies to the relatively sophisticated differential geometric concept of connections on fiber bundles!

Chapter 29

Calculus of Variations, the Euler-Lagrange Equations, the First Variation of Arclength and Geodesics

Beginning with this chapter we will enter into an in-depth study of the so-called *Lagrangian and Hamiltonian formulations* of the principles of classical mechanics. These are complimentary formulations that have provided very potent and powerful mathematical tools for the solutions of classical mechanical problems, and indeed, for the development of physical theories beyond classical mechanics, such as quantum theory and even quantum field theory. As we will see, both of these formulations rest on a bedrock of geometrical principles, for which much of our mathematical development thus far should provide useful preparative groundwork.

We will start with an introduction to the mathematical problem of the *calculus of variations*, and discuss its relevance in classical mechanics.

In ordinary calculus we determine the **extrema** (maxima or minima) of functions by taking derivatives. For functions of one variable $y = f(x)$ we set $dy/dx = 0$ and solve for values of x that satisfy this equation. For functions $f(x^1, \dots, x^n)$ of n variables ($n > 1$), the extrema are obtained by setting

$$\frac{\partial f}{\partial x^i} \equiv \partial_i f = 0\,, \quad i = 1, \dots, n\,. \tag{29.1}$$

In the **calculus of variations** we are concerned with the extrema of functions whose domains are infinite dimensional vector spaces, for example, the space of curves, $\mathcal{C}$, in three-dimensional Euclidean space. We write these functions as

$$\Phi : \mathcal{C} \to \mathbb{R}\,, \quad C \mapsto a\,, \tag{29.2}$$

where $C \in \mathcal{C}$ is a curve in the space $\mathcal{C}$ and $a \in \mathbb{R}$ ($\mathbb{R}$ being the set of real numbers). Such a function is sometimes called a **functional**. More generally, the domain of a functional is a vector space of functions of a certain type (a function space). A basic example of a functional is the *length functional*, which maps a *parametrized curve* in an n-dimensional Riemannian manifold to its length. For a parametrized curve C (with parameter t) defined by $x^1(t), \dots, x^n(t)$, the length functional is given by [cf. (27.18)]

$$\Phi(C) = \int ds = \int_{t_1}^{t_2} dt \sqrt{g_{ij} \frac{dx^i}{dt} \frac{dx^j}{dt}} , \tag{29.3}$$

where g_{ij} is the *Riemannian metric*, $i, j = 1, \dots, n$, and t_1, t_2 are real numbers such that $t_1 \leq t_2$.

Problem 29.1 Show that the length functional for curves on a Euclidean plane with rectangular coordinates (x, y) is given by

$$\Phi(y(x)) = \int_{t_1}^{t_2} dt \sqrt{\dot{x}^2 + \dot{y}^2} = \int_{x_1}^{x_2} dx \sqrt{1 + \left(\frac{dy}{dx}\right)^2} . \tag{29.4}$$

Analogous to the fundamental concept of the differentiability of ordinary functions, we have the more general concept of the *differentiability of functionals*, given as follows.

Definition 29.1. *A functional $\Phi : \mathcal{C} \to \mathbb{R}$ (with $\mathcal{C}$ being a space of parametrized curves) is said to be* ***differentiable*** *if*

$$\Phi(C + h) - \Phi(C) = F(h) + R(h) , \tag{29.5}$$

for all curves $C \in \mathcal{C}$ and all ***variation curves*** *$h \in \mathcal{C}$ satisfying $|h^i(t)| < \epsilon$ and $|\partial h^i(t)/\partial t| < \epsilon$ for some $\epsilon > 0$, such that the functional R is of order h^2 (which means $|R| <$ constant $\times\, \epsilon^2$ and written $R(h) = O(h^2)$), and F is a linear functional. Linearity of F means that*

$$F(ah_1 + bh_2) = aF(h_1) + bF(h_2) , \quad h_1, h_2 \in \mathcal{C} , \; a, b \in \mathbb{R} . \tag{29.6}$$

The linear part F, also denoted $\Phi'(C)$, is called the ***differential****, or the* ***first variation****, of the functional Φ (Fig. 29.1).*

In classical mechanics we deal with functionals of the form

$$\Phi(C) = \int_{t_1}^{t_2} dt\, L(x^i(t), \dot{x}^i(t), t) , \quad i = 1, \dots, n , \tag{29.7}$$

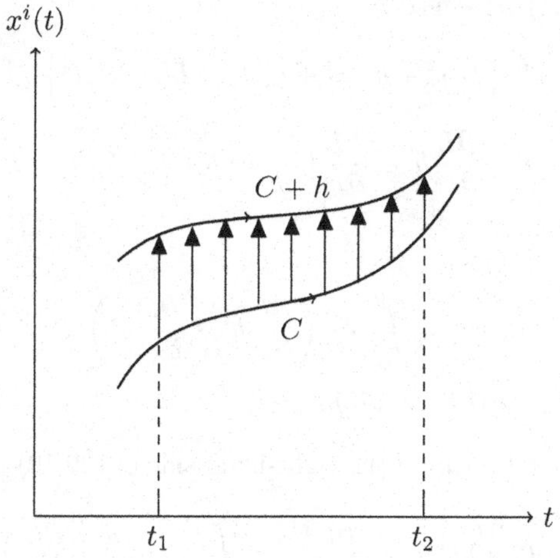

Fig. 29.1

where the $x^i(t)$ define a parametrized curve in the n-dimensional **configuration space** M of a classical mechanical system, $\dot{x}^i(t) \equiv dx^i/dt$, and L, called the **Lagrangian function**, is a differentiable function of the $2n+1$ variables x^i, $\dot{x}^i$ and t. Recalling our discussion of the tangent bundle in Chapter 5, we see that L is a function on $T(M)$, the tangent bundle of the configuration manifold M, if it does not depend explicitly on the time t. The $\{\dot{x}^i\}$ are precisely the components of a tangent vector at the point in the base manifold M whose local coordinates are $\{x^i\}$.

We immediately have the following basic theorem in the calculus of variations.

Theorem 29.1. *The functional given by (29.7) is differentiable, and its first variation (differential) $\Phi'(C)$ is given by*

$$\Phi'(C) = F(h) = \int_{t_1}^{t_2} \left\{ \frac{\partial L}{\partial x^i} - \frac{d}{dt}\left(\frac{\partial L}{\partial \dot{x}^i}\right) \right\} h^i(t)\, dt + \left. \left(\frac{\partial L}{\partial \dot{x}^i} h^i\right) \right|_{t_1}^{t_2} . \qquad (29.8)$$

Proof. We have

$$
\begin{aligned}
&\Phi(C+h) - \Phi(C) \\
&= \int_{t_1}^{t_2} dt \left\{ L(x^i + h^i, \dot{x}^i + \dot{h}^i, t) - L(x^i, \dot{x}^i, t) \right\} \\
&= \int_{t_1}^{t_2} dt \left(\frac{\partial L}{\partial x^i} h^i + \frac{\partial L}{\partial \dot{x}^i} \dot{h}^i \right) + O(h^2) \equiv F(h) + R(h) \,,
\end{aligned}
\tag{29.9}
$$

where

$$F(h) = \int_{t_1}^{t_2} dt \left(\frac{\partial L}{\partial x^i} h^i + \frac{\partial L}{\partial \dot{x}^i} \dot{h}^i \right) , \tag{29.10}$$

$$R(h) = O(h^2) \,. \tag{29.11}$$

Integrating the second term on the right-hand side of (29.10) by parts, we have

$$\int_{t_1}^{t_2} dt \, \frac{\partial L}{\partial \dot{x}^i} \dot{h}^i = h^i \left. \frac{\partial L}{\partial \dot{x}^i} \right|_{t_1}^{t_2} - \int_{t_1}^{t_2} dt \, h^i \frac{d}{dt} \left(\frac{\partial L}{\partial \dot{x}^i} \right) . \tag{29.12}$$

Putting the above result back into (29.10) for $F(h)$ we obtain the theorem. □

Given a differentiable functional, its extremal, if one exists, is defined as follows (again in a manner analogous to ordinary calculus).

Definition 29.2. *An* ***extremal****, or* ***critical curve****, of a differentiable functional $\Phi(C)$ is a curve C such that the differential $F(h; C) = 0$ for all variation curves $h(t)$.*

To proceed further we will need the following Lemma, which will be stated without proof.

Lemma 29.1. *If a continuous function $f(t)$ satisfies the condition*

$$\int_{t_1}^{t_2} f(t) h(t) \, dt = 0 \,, \tag{29.13}$$

for all continuous functions $h(t)$ with $h(t_1) = h(t_2) = 0$, then $f(t) = 0$.

The above Lemma and Theorem 22.1 then lead to the following theorem, which is the main result of this chapter.

Theorem 29.2. *A parametrized curve C in an n-dimensional differentiable manifold (characterized by n coordinate functions $x^i(t)$, $i = 1, \dots, n$) is a critical curve (an extremal) of the functional*

$$\Phi(C) = \int_{t_1}^{t_2} dt \, L(x^i, \dot{x}^i, t) \tag{29.14}$$

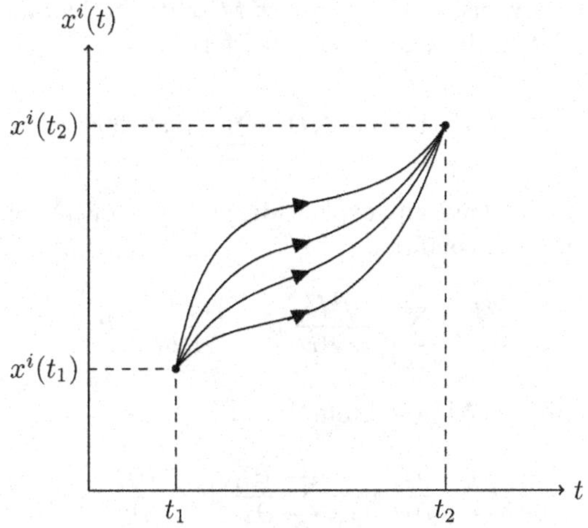

Fig. 29.2

on the space of curves passing through the points $x(t_1) = x_1$ and $x(t_2) = x_2$ (Fig. 29.2) if and only if

$$\boxed{\frac{d}{dt}\left(\frac{\partial L}{\partial \dot{x}^i}\right) - \frac{\partial L}{\partial x^i} = 0\,, \quad i = 1, \ldots, n}\quad , \tag{29.15}$$

along the curve $x^i(t)$.

Equations (29.15) are called the **Euler-Lagrange equations.**

The Euler-Lagrange equations (29.15) motivate one to define the so-called **Lagrangian derivatives**, $\mathcal{L}_{x^i} f$, of a function $f(x^i, \dot{x}^i, t)$:

$$\mathcal{L}_{x^i} f \equiv \frac{\partial f}{\partial x^i} - \frac{d}{dt}\left(\frac{\partial f}{\partial \dot{x}^i}\right) . \tag{29.16}$$

It will be useful to determine to what extent a function f is determined by its n Lagrangian derivatives. Suppose two functions $g(x^i, \dot{x}^i, t)$ and $h(x^i, \dot{x}^i, t)$ share the same Lagrangian derivatives, so that $\mathcal{L}_{x^i} g = \mathcal{L}_{x^i} h \;\; (i = 1, \ldots, n)$. For an arbitrary function $f(x^i, \dot{x}^i, t)$ we can write

$$\mathcal{L}_{x^i} f = \frac{\partial f}{\partial x^i} - \sum_{j=1}^{n}\left(\frac{\partial^2 f}{\partial \dot{x}^i \partial x^j}\,\dot{x}^j + \frac{\partial^2 f}{\partial \dot{x}^i \partial \dot{x}^j}\,\ddot{x}^j\right) - \frac{\partial^2 f}{\partial \dot{x}^i \partial t}\,, \tag{29.17}$$

where the right-hand side is to be regarded as a function of the $3n+1$ variables $x^i, \dot{x}^i, \ddot{x}^i$ and t. Let $f = g - h$, then $\mathcal{L}_{x^i} f = 0$. The coefficient of $\ddot{x}^j$ in the above expression must then vanish, and we have $\partial^2 f / \partial \dot{x}^i \partial \dot{x}^j = 0$ for all $i, j = 1, \ldots, n$. This forces $f(x^i, \dot{x}^i, t)$ to have the functional form

$$f(x^i, \dot{x}^i, t) = f_0(x^i, t) + \sum_{i=1}^{n} f_i(x^j, t)\, \dot{x}^i \,, \tag{29.18}$$

where f_0 and f_i do not have any explicit dependences on $\dot{x}^i$. Substituting this expression for f into the condition

$$\frac{\partial f}{\partial x^i} - \sum_{j=1}^{n} \frac{\partial^2 f}{\partial \dot{x}^i \partial x^j}\, \dot{x}^j - \frac{\partial^2 f}{\partial \dot{x}^i \partial t} = 0 \tag{29.19}$$

[which follows from (29.17)], we obtain

$$\frac{\partial f_0}{\partial x^i} + \sum_{j=1}^{n} \frac{\partial f_j}{\partial x^i}\, \dot{x}^j - \sum_{j=1}^{n} \frac{\partial f_i}{\partial x^j}\, \dot{x}^j - \frac{\partial f_i}{\partial t} = 0 \,. \tag{29.20}$$

Since all the x^i and $\dot{x}^i$ are independent, it follows from this equation that

$$\frac{\partial f_0}{\partial x^i} = \frac{\partial f_i}{\partial t} \,, \qquad \frac{\partial f_j}{\partial x^i} = \frac{\partial f_i}{\partial x^j} \,. \tag{29.21}$$

These conditions imply that there exists a local function $v(x^i, t)$ (explicitly independent of the $\dot{x}^i$) whose total differential dv is given by

$$dv = f_0\, dt + f_i\, dx^i \,. \tag{29.22}$$

Indeed, on exteriorly differentiating the right-hand side of the above equation, we find that

$$\begin{aligned}
d(f_0 dt + f_i dx^i) &= df_0 \wedge dt + df_i \wedge dx^i \\
&= \sum_{i=1}^{n} \frac{\partial f_0}{\partial x^i}\, dx^i \wedge dt + \sum_{i=1}^{n} \left(\frac{\partial f_i}{\partial t}\, dt \wedge dx^i + \sum_{j=1}^{n} \frac{\partial f_i}{\partial x^j}\, dx^j \wedge dx^i \right) \\
&= \sum_{i=1}^{n} \left(\frac{\partial f_0}{\partial x^i} - \frac{\partial f_i}{\partial t} \right) dx^i \wedge dt + \sum_{i,j} \frac{\partial f_i}{\partial x^j}\, dx^j \wedge dx^i \\
&= \sum_{i,j} \frac{\partial f_i}{\partial x^j}\, dx^j \wedge dx^i = \sum_{i<j} \left(\frac{\partial f_i}{\partial x^j} - \frac{\partial f_j}{\partial x^i} \right) dx^j \wedge dx^i = 0 \,,
\end{aligned} \tag{29.23}$$

where we have made use of (29.21). The assertion (29.22), that the right-hand side of that equation is locally exact, then follows from Poincaré's lemma. Thus, by (29.18),

$$g(x^i, \dot{x}^i, t) - h(x^i, \dot{x}^i, t) = f(x^i, \dot{x}^i, t) = f_0 + f_i\, dx^i = \frac{dv(x^i, t)}{dt} \,. \tag{29.24}$$

This states that the Lagrangian derivatives of a function $f(x^i, \dot{x}^i, t)$ [given by (29.16)], as functions of $x^i, \dot{x}^i, \ddot{x}^i, t$, determine the function up to an additive function which is the total derivative respect to t of a function $v(x^i, t)$ not explicitly dependent on any $\dot{x}^i$. This result will be very useful when we discuss the notion of *canonical transformations* in Hamiltonian dynamics in Chapter 36.

Problem 29.2 Show algebraically that the Euler-Lagrange equations (29.15) are invariant under a general invertible coordinate transformation

$$(x^1, \dots, x^n) \longrightarrow (\xi^1, \dots, \xi^n)$$

given by twice continuously differentiable functions of $n+1$ variables:

$$x^i = x^i(\xi^1, \dots, \xi^n\,;\, t) \qquad (i = 1, \dots, n)\,.$$

Note that invertibility requires that the Jacobian determinant satisfies

$$\left| \frac{\partial x^i}{\partial \xi^j} \right| \neq 0\,.$$

Proceed as follows.

(a) Embed the solution $x^i(t)$ of the extremum problem into a family of admissable functions $x^i(\alpha; t)$ $(-1 < \alpha < 1)$ which satisfy $x^i(t) = x^i(0, t)$, $x^i(\alpha; t_1) = x^i(t_1)$ and $x^i(\alpha; t_2) = x^i(t_2)$. It is also required that both $x^i(\alpha; t)$ and $\partial x^i(\alpha; t)/\partial t$ are continuously differentiable with respect to α. The above coordinate transformation applied to the family $x^i(\alpha; t)$ yields the family of functions $\xi^i(\alpha; t)$. We will consider the partial derivatives of the two families of functions $x^i(\alpha; t)$ and $\xi^i(\alpha; t)$. It is apparent that

$$\frac{\partial x^i}{\partial \dot{\xi}^j} = 0\,, \qquad \frac{\partial x^i}{\partial \alpha} = \frac{\partial x^i}{\partial \xi^j}\frac{\partial \xi^j}{\partial \alpha} \qquad \dot{x}^i = \frac{\partial x^i}{\partial t} + \frac{\partial x^i}{\partial \xi^j}\dot{\xi}^j\,.$$

It follows from the last equation that

$$\frac{\partial \dot{x}^i}{\partial \dot{\xi}^j} = \frac{\partial x^i}{\partial \xi^j}\,.$$

For a Lagrangian function $L(x^i, \dot{x}^i; t)$ define the quantity

$$s \equiv \sum_{i=1}^{n} \frac{\partial L}{\partial \dot{x}^i}\frac{\partial x^i}{\partial \alpha}\,.$$

Show that s is invariant under the coordinate transformation $\{x^i\} \to \{\xi^i\}$, that is,

$$\sum_{i=1}^{n} \frac{\partial L}{\partial \dot{x}^i}\frac{\partial x^i}{\partial \alpha} = \sum_{i=1}^{n} \frac{\partial L}{\partial \dot{\xi}^i}\frac{\partial \xi^i}{\partial \alpha}\,.$$

This implies that

$$\frac{ds}{dt} = \sum_{i=1}^{n} \left(\frac{d}{dt}\left(\frac{\partial L}{\partial \dot{x}^i}\right)\frac{\partial x^i}{\partial \alpha} + \frac{\partial L}{\partial \dot{x}^i}\frac{\partial \dot{x}^i}{\partial \alpha} \right)$$

is also invariant under the same transformation.

(b) Show that

$$\frac{\partial L}{\partial \alpha} = \sum_{i=1}^{n} \left(\frac{\partial L}{\partial x^i} \frac{\partial x^i}{\partial \alpha} + \frac{\partial L}{\partial \dot{x}^i} \frac{\partial \dot{x}^i}{\partial \alpha} \right)$$

is invariant under the coordinate transformation $\{x^i\} \to \{\xi^i\}$, that is,

$$\sum_{i=1}^{n} \left(\frac{\partial L}{\partial x^i} \frac{\partial x^i}{\partial \alpha} + \frac{\partial L}{\partial \dot{x}^i} \frac{\partial \dot{x}^i}{\partial \alpha} \right) = \sum_{i=1}^{n} \left(\frac{\partial L}{\partial \xi^i} \frac{\partial \xi^i}{\partial \alpha} + \frac{\partial L}{\partial \dot{\xi}^i} \frac{\partial \dot{\xi}^i}{\partial \alpha} \right) .$$

This result, together with that in (a), implies that

$$A \equiv \frac{\partial L}{\partial \alpha} - \frac{ds}{dt}$$

is invariant under the coordinate transformation.

(c) We can write the Euler-Lagrange equations in terms of Lagrangian derivatives as $\mathcal{L}_{x^i} L = 0 \;\; (i = 1, \dots, n)$. Show that the transformation rule for the Lagrangian derivatives under the coordinate transformation $\{x^i\} \to \{\xi^i\}$ is given by

$$\mathcal{L}_{\xi^i} L = \sum_{j=1}^{n} \frac{\partial x^j}{\partial \xi^i} \mathcal{L}_{x^j} L \qquad (i = 1, \dots n) .$$

This result is valid for any α in the interval $-1 < \alpha < 1$. Hence it must be valid in particular for $\alpha = 0$ and thus for the solution of the extremum problem $x^i = x^i(t)$. It follows that the Euler-Lagrange equations are invariant under general coordinate transformations.

As examples on the use of the Euler-Lagrange equations we will first determine the extremum path between two points on a Euclidean plane and two points on a cylindrical surface. Consider first the Euclidean plane. From (29.4) for the length functional we set

$$L(x, y, \dot{x}, \dot{y}, t) = \sqrt{\dot{x}^2 + \dot{y}^2} . \tag{29.25}$$

Thus

$$\frac{\partial L}{\partial \dot{x}} = \frac{\dot{x}}{\sqrt{\dot{x}^2 + \dot{y}^2}} , \qquad \frac{\partial L}{\partial \dot{y}} = \frac{\dot{y}}{\sqrt{\dot{x}^2 + \dot{y}^2}} , \qquad \frac{\partial L}{\partial x} = \frac{\partial L}{\partial y} = 0 , \tag{29.26}$$

and the Euler-Lagrange equations (29.15) imply

$$\frac{d}{dt}\left(\frac{\dot{x}}{\sqrt{\dot{x}^2 + \dot{y}^2}} \right) = \frac{d}{dt}\left(\frac{\dot{y}}{\sqrt{\dot{x}^2 + \dot{y}^2}} \right) = 0 . \tag{29.27}$$

Let

$$\frac{\dot{x}}{\sqrt{\dot{x}^2 + \dot{y}^2}} = a = \text{constant} , \qquad \frac{\dot{y}}{\sqrt{\dot{x}^2 + \dot{y}^2}} = b = \text{constant} . \tag{29.28}$$

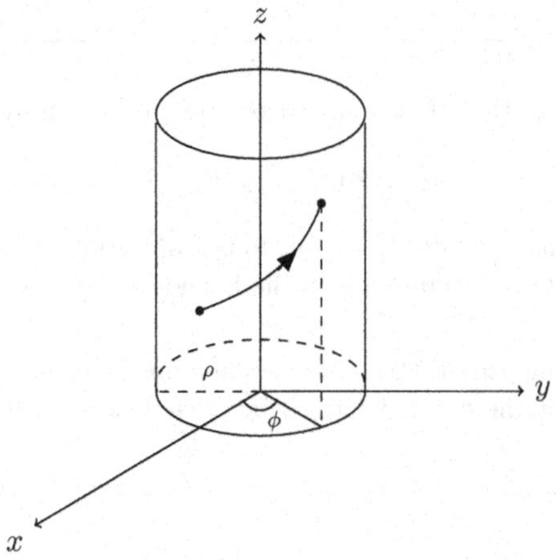

Fig. 29.3

Then

$$\frac{\dot{x}}{a} - \frac{\dot{y}}{b} = \frac{d}{dt}\left(\frac{x}{a} - \frac{y}{b}\right) = 0\,, \tag{29.29}$$

which implies that

$$y = \alpha x + \beta\,, \tag{29.30}$$

where α and β are constants. Thus the extremum (in fact the shortest) path joining two points on a Euclidean plane is a straight line.

On a cylindrical surface of radius ρ we can use cylindrical coordinates ϕ and z, which are related to the rectangular coordinates x, y, z by (Fig. 29.3)

$$x = \rho\cos\phi\,, \quad y = \rho\sin\phi\,. \tag{29.31}$$

The arclength element ds is then given by

$$\begin{aligned} ds^2 &= (dx)^2 + (dy)^2 + (dz)^2 \\ &= \rho^2\sin^2\phi\,(d\phi)^2 + \rho^2\cos^2\phi\,(d\phi)^2 + (dz)^2 = \rho^2(d\phi)^2 + (dz)^2\,. \end{aligned} \tag{29.32}$$

For the length functional we thus set the Lagrangian to be

$$L = \sqrt{\rho^2\dot{\phi}^2 + \dot{z}^2}\,. \tag{29.33}$$

An application of the Euler-Lagrange equations (29.15) with $x^1 = z$ and $x^2 = \phi$ then gives

$$z = \alpha\phi + \beta\,, \tag{29.34}$$

with α and β being constants. This equation says that the path of shortest length between two points on a cylindrical surface is a spiral with constant pitch $dz/d\phi$ (which may be zero).

Problem 29.3 Show that the parametrized straight line on the Euclidean plane

$$x^1(t) = at + x^1(0)\,, \qquad x^2(t) = bt + x^2(0)\,,$$

satisfy the *geodesic equation* (22.44) [or (29.48) below]. (Hint: What are the *Christoffel symbols* for the *Levi-Civita connection* on the Euclidean plane?)

Problem 29.4 Compute the Christoffel symbols for the two-dimensional cylindrical surface and show that the curve described by (29.34) satisfies the geodesic equation (22.44).

As our next example for the use of the Euler-Lagrange equations we recall (29.3) [or (27.18)] for the arclength in a Riemannian manifold with a metric $g_{\mu\nu}$. We will rewrite the equation as

$$\Phi(C) = \int ds = \int d\tau\, L(x^\mu, \dot{x}^\mu)\,, \tag{29.35}$$

where the Lagrangian function $L(x^\mu, \dot{x}^\mu)$ is given by

$$L(x^\mu, \dot{x}^\mu) = \sqrt{g_{\mu\nu}\,\dot{x}^\mu \dot{x}^\nu}\,, \tag{29.36}$$

with $\dot{x}^\mu = dx^\mu/d\tau$, and τ being a parameter of the parametrized curve C (that is not necessarily the time t). In a relativistic setting, for example, τ is usually taken to be the *proper time* – the time with respect to a *Lorentz frame* that is at rest with respect to a moving object.

Problem 29.5 Show that the arclength functional (29.35) with the Lagrangian function given by (29.36) is **reparametrization invariant** under an arbitrary reparametrization $\tau \to \tau'(\tau)$.

In the present context, the Euler-Lagrange equations (29.15) can be written as

$$\frac{d}{d\tau}\frac{\partial L}{\partial\left(\dfrac{dx^\mu}{d\tau}\right)} = \frac{\partial L}{\partial x^\mu}\,. \tag{29.37}$$

We have, from (29.36),

$$\frac{\partial L}{\partial x^\mu} = \frac{(\partial_\mu g_{\rho\lambda})\,\dot{x}^\rho \dot{x}^\lambda}{2L(\tau)}\,, \qquad \frac{\partial L}{\partial \dot{x}^\mu} = \frac{g_{\mu\nu}\,\dot{x}^\nu}{L(\tau)}\,. \tag{29.38}$$

So, from the second equation above,

$$\frac{d}{d\tau}\left(\frac{\partial L}{\partial \dot{x}^\mu}\right) = \frac{1}{L(\tau)}\frac{d}{d\tau}(g_{\mu\nu}\,\dot{x}^\nu) - \frac{dL(\tau)/d\tau}{L^2(\tau)}\,g_{\mu\nu}\,\dot{x}^\nu\,. \tag{29.39}$$

Substituting into the Euler-Lagrange equations (29.37) and canceling out one factor of $1/L(\tau)$, we have

$$\frac{d}{d\tau}(g_{\mu\nu}\,\dot{x}^\nu) - \frac{1}{2}(\partial_\mu g_{\rho\lambda})\,\dot{x}^\rho\dot{x}^\lambda = \frac{g_{\mu\nu}}{L(\tau)}\frac{dL(\tau)}{d\tau}\,\dot{x}^\nu\,, \tag{29.40}$$

which implies

$$g_{\mu\nu}\,\ddot{x}^\nu + (\partial_\rho g_{\mu\nu})\,\dot{x}^\rho\dot{x}^\nu - \frac{1}{2}(\partial_\mu g_{\rho\lambda})\,\dot{x}^\rho\dot{x}^\lambda = \frac{g_{\mu\nu}}{L(\tau)}\frac{dL(\tau)}{d\tau}\,\dot{x}^\nu\,. \tag{29.41}$$

Making the replacements of indices $\mu \to \nu$, $\nu \to \lambda$ in this equation we obtain

$$g_{\nu\lambda}\,\ddot{x}^\lambda - \frac{1}{2}(\partial_\nu g_{\rho\lambda})\,\dot{x}^\rho\dot{x}^\lambda + (\partial_\rho g_{\nu\lambda})\,\dot{x}^\rho\dot{x}^\lambda = \frac{g_{\nu\lambda}}{L(\tau)}\frac{dL(\tau)}{d\tau}\,\dot{x}^\lambda\,. \tag{29.42}$$

Multiplying this equation by $g^{\mu\nu}$ and summing over ν, and remembering that $g^{\mu\nu}g_{\nu\lambda} = \delta^\mu_\lambda$, we further obtain

$$\ddot{x}^\mu - \frac{1}{2}g^{\mu\nu}(\partial_\nu g_{\rho\lambda})\,\dot{x}^\rho\dot{x}^\lambda + g^{\mu\nu}(\partial_\rho g_{\nu\lambda})\,\dot{x}^\rho\dot{x}^\lambda = \frac{1}{L(\tau)}\frac{dL(\tau)}{d\tau}\,\dot{x}^\mu\,. \tag{29.43}$$

We split up the third term on the left-hand side as follows

$$g^{\mu\nu}(\partial_\rho g_{\nu\lambda})\,\dot{x}^\rho\dot{x}^\lambda = \frac{1}{2}g^{\mu\nu}(\partial_\rho g_{\nu\lambda})\,\dot{x}^\rho\dot{x}^\lambda + \frac{1}{2}g^{\mu\nu}(\partial_\lambda g_{\nu\rho})\,\dot{x}^\lambda\dot{x}^\rho\,, \tag{29.44}$$

where the second term, which is equal to the first term, has been obtained from the first term by making the interchange of dummy indices $\rho \leftrightarrow \lambda$. Equation (29.43) can then be written as

$$\ddot{x}^\mu + \frac{1}{2}g^{\mu\nu}(\partial_\rho g_{\lambda\nu} + \partial_\lambda g_{\rho\nu} - \partial_\nu g_{\rho\lambda})\,\dot{x}^\rho\dot{x}^\lambda = \frac{1}{L(\tau)}\frac{dL(\tau)}{d\tau}\,\dot{x}^\mu\,. \tag{29.45}$$

Recalling (27.67) for the Christoffel symbols the above equation becomes

$$\frac{d^2x^\mu}{d\tau^2} + \Gamma^\mu_{\rho\lambda}\frac{dx^\rho}{d\tau}\frac{dx^\lambda}{d\tau} = \frac{1}{L(\tau)}\frac{dL(\tau)}{d\tau}\frac{dx^\mu}{d\tau}\,. \tag{29.46}$$

This is the equation for an extremum path specified by the parametrized curve $x^\mu(\tau)$. By (29.35), $L(\tau) = ds/d\tau$. If the parameter τ is related to the arclength s by a linear relationship:

$$s = \alpha\tau + \beta \qquad (\alpha,\, \beta = \text{constants})\,, \tag{29.47}$$

then $ds/d\tau$ = constant and $dL(\tau)/d\tau = d^2s/d\tau^2 = 0$. The right-hand side of (29.46) then vanishes, and that equation reduces precisely to the *geodesic equation* given by (22.44), now written as

$$\frac{d^2x^\mu}{d\tau^2} + \Gamma^\mu_{\rho\lambda}\frac{dx^\rho}{d\tau}\frac{dx^\lambda}{d\tau} = 0 . \qquad (29.48)$$

A parameter that satisfies (29.47) is called an **affine parameter**. Conversely, a geodesic curve associated with a *metric-compatible* affine connection *must* be parametrized by an affine parameter. Indeed, under the metric-compatibility requirement that the length of a tangent vector $dx^\mu/d\tau$ to the geodesic curve $x^\mu(\tau)$ (as determined by the metric tensor $g_{\mu\nu}$) must remain constant under parallel translation along the geodesic curve, it follows from (27.15) and (27.19) that

$$\sqrt{g_{\mu\nu}\frac{dx^\mu}{d\tau}\frac{dx^\nu}{d\tau}} = \frac{ds}{d\tau} = \text{constant} . \qquad (29.49)$$

The geodesic equation (29.48) is essentially the content of *Newton's First Law*, viewed in retrospect with the hindsight of Einstein's *general theory of relativity*, which posits that a massive particle moves along a geodesic curve in 4-dimensional curved *spacetime*, whose curvature produces the gravitational field.

The Euler-Lagrange equations for the variation of arclengths in a Riemannian manifold, which result in (29.46), can be cast in a more geometric light in terms of the affine connection (covariant directional derivative) $D_{\boldsymbol{T}}$, where $\boldsymbol{T} = (dx^\mu/dt)\,\partial/\partial x^\mu$ is a tangent vector to a parametrized curve $C = (x^\mu(t))$, in fact, the velocity vector. The derivation given in Chapter 22 of the geodesic equation (29.48), that is, (29.46) with the right-hand side set equal to zero, shows that that equation follows from the *auto-parallel* requirement

$$D_{\boldsymbol{T}}\boldsymbol{T} = 0 . \qquad (29.50)$$

It is not difficult to see, on recalling (29.36), that the Euler-Lagrange equations in the present context, (29.46), is equivalent to

$$D_{\boldsymbol{T}}\left(\frac{\boldsymbol{T}}{\sqrt{G(\boldsymbol{T},\boldsymbol{T})}}\right) = 0 , \qquad (29.51)$$

where $G(\boldsymbol{T},\boldsymbol{T}) = g_{\mu\nu}\,\dot{x}^\mu\dot{x}^\nu = L^2$. As noted in our discussion immediately above (29.49), this equation implies the geodesic equation (29.48) if the connection D is metric-compatible, that is, if $G(\boldsymbol{T},\boldsymbol{T})$ is constant upon parallel translation of $\boldsymbol{T}$.

The first variation of the arclength determines an extremum curve or a geodesic, but it is not sufficient to determine whether the curve has minimum length. (In other words, *a geodesic curve joining two points in a Riemannian manifold is not necessarily the shortest curve between the two points.*) To do so requires the so-called *second variation of arclength*. In order to prepare for a

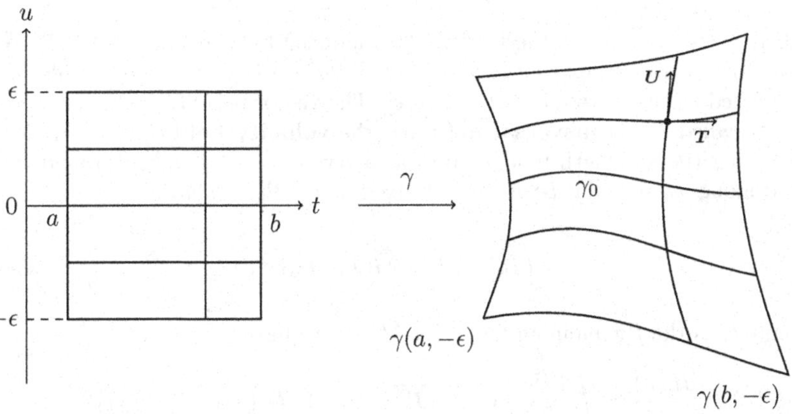

Fig. 29.4

discussion of the second variation (in the next chapter), we will revisit the first variation more formally using the tools of Riemannian geometry, and in doing so lay the groundwork for the exploration of some key geometrical properties of geodesic curves which will play vital roles in the study of the structure of a Riemannian manifold. We first define formally the notion of the *variation* of a smooth curve (see Fig. 29.4).

Definition 29.3. *Let $C : [a,b] \to M$ be a smooth parametrized curve in a Riemannian manifold M. A* **variation** *of C is a smooth map $\gamma : [a,b] \times [-\epsilon, \epsilon] \to M$ (from the rectangle $[a,b] \times [\epsilon, \epsilon] \subset \mathbb{R}^2$ to M) such that $\gamma(t,0) = C(t)$, $t \in [a,b]$. The curve $C(t)$ is called the* **base curve** *of the variation. For each $u \in [-\epsilon, \epsilon]$, the parametrized curve $\gamma_u : [a,b] \to M$ defined by $\gamma_u(t) = \gamma(t,u)$ is called a curve in the variation (a t-curve); while the parametrized curve $\gamma_t(u) \equiv \gamma(t,u)$ (with t fixed) is called a* **transversal curve** *(a u-curve) of the variation.*

Note that in this definition it has *not* been assumed that the curves in the variation $\gamma_u(t)$ start and end at the same points as the base curve (as it was in Theorem 29.2), that is, it is not necessarily true that $\gamma_u(a) = \gamma_0(a)$ and $\gamma_u(b) = \gamma_0(b)$, for all $u \in [-\epsilon, \epsilon]$.

The variation map γ in the above definition gives rise to two tangent vector fields $\boldsymbol{T}$ and $\boldsymbol{U}$, along the t-curves and u-curves in M, respectively (see Fig.

29.4):

$$\boldsymbol{T}(\gamma(t,u)) \equiv \gamma_* \left(\frac{\partial}{\partial t}(t,u) \right) , \qquad \boldsymbol{U}(\gamma(t,u)) \equiv \gamma_* \left(\frac{\partial}{\partial u}(t,u) \right) , \tag{29.52}$$

where γ_* is the tangent (derivative) map induced by γ, defined in (6.27). Note that in the above equations, $\partial/\partial t$ and $\partial/\partial u$ are tangent vector fields in $\mathbb{R}^2$ restricted to the rectangle $[a,b] \times [-\epsilon, \epsilon]$. The vector fields $\boldsymbol{U}$ and $\boldsymbol{T}$ are respectively called the **transversal field** and the **velocity field** of the variation γ. Denoting the arclength functional of a t-curve by $L(u) \equiv L(\gamma_u(t))$, and the $\boldsymbol{T}$ field along a particular t-curve (with fixed u) by $\boldsymbol{T}_u$, we have

$$L(u) = \int_a^b \sqrt{G(\boldsymbol{T}_u, \boldsymbol{T}_u)}\, dt , \tag{29.53}$$

where G is the Riemannian metric of M. So we have

$$\begin{aligned} L'(u) \equiv \frac{dL(u)}{du} &= \int_a^b \frac{\partial}{\partial u} \sqrt{G(\boldsymbol{T}_u, \boldsymbol{T}_u)}\, dt = \int_a^b \boldsymbol{U}\left(\sqrt{G(\boldsymbol{T}_u, \boldsymbol{T}_u)}\right) dt \\ &= \frac{1}{2} \int_a^b \frac{\boldsymbol{U}\left(G(\boldsymbol{T}_u, \boldsymbol{T}_u)\right)}{\sqrt{G(\boldsymbol{T}_u, \boldsymbol{T}_u)}}\, dt = \frac{1}{2} \int_a^b \frac{\boldsymbol{U}(\boldsymbol{T}_u \cdot \boldsymbol{T}_u)}{|\boldsymbol{T}_u|}\, dt , \end{aligned} \tag{29.54}$$

where on the right-hand side of the last equality we have used the "scalar product" notation $\boldsymbol{X} \cdot \boldsymbol{Y}$ for $G(\boldsymbol{X}, \boldsymbol{Y})$ [cf. (27.58)], and have written $|\boldsymbol{T}_u|$ for the length $\sqrt{G(\boldsymbol{T}_u, \boldsymbol{T}_u)}$ of the vector $\boldsymbol{T}_u$. Using (27.58) and the fact that the metric tensor G is symmetric (so that $\boldsymbol{X} \cdot \boldsymbol{Y} = \boldsymbol{Y} \cdot \boldsymbol{X}$), we then have

$$L'(u) = \int_a^b \frac{\boldsymbol{T}_u \cdot (D_{\boldsymbol{U}} \boldsymbol{T}_u)}{|\boldsymbol{T}_u|}\, dt , \tag{29.55}$$

where D is the torsion-free and metric-compatible Levi-Civita connection. From (29.52) and the fact that $[\partial/\partial t\, , \partial/\partial u] = 0$, we have

$$[\boldsymbol{T}\, , \boldsymbol{U}] = 0 . \tag{29.56}$$

Hence (27.40) for the torsion tensor and the fact that D is torsion-free imply that

$$D_{\boldsymbol{T}} \boldsymbol{U} = D_{\boldsymbol{U}} \boldsymbol{T} . \tag{29.57}$$

It follows from (29.55) that

$$L'(u) = \int_a^b \frac{\boldsymbol{T}_u \cdot (D_{\boldsymbol{T}_u} \boldsymbol{U})}{|\boldsymbol{T}_u|}\, dt . \tag{29.58}$$

Applying (27.58) again, we have

$$\frac{\boldsymbol{T}_u}{|\boldsymbol{T}_u|} \cdot (D_{\boldsymbol{T}_u} \boldsymbol{U}) = \boldsymbol{T}_u \left(\boldsymbol{U} \cdot \frac{\boldsymbol{T}_u}{|\boldsymbol{T}_u|} \right) - \boldsymbol{U} \cdot \left(D_{\boldsymbol{T}_u} \left(\frac{\boldsymbol{T}_u}{|\boldsymbol{T}_u|} \right) \right) . \tag{29.59}$$

Hence (29.58) can be rewritten as

$$L'(u) = \int_a^b dt \left[\frac{\partial}{\partial t} \left(\boldsymbol{U} \cdot \frac{\boldsymbol{T}_u}{|\boldsymbol{T}_u|} \right) - \boldsymbol{U} \cdot \left(D_{\boldsymbol{T}_u} \left(\frac{\boldsymbol{T}_u}{|\boldsymbol{T}_u|} \right) \right) \right] . \tag{29.60}$$

This is called the formula for the **first variation of arclength** in Riemannian geometry. For variations with fixed end-points, the transversal field $\boldsymbol{U}$ satisfies $\boldsymbol{U}(\gamma(a,u)) = \boldsymbol{U}(\gamma(b,u)) = 0$. The *boundary term* (first term) of the above integral vanishes, and the extremum condition for the base curve (for which $u = 0$) can be expressed as

$$L'(0) = \int_a^b \boldsymbol{U} \cdot \left(D_{\boldsymbol{T}} \left(\frac{\boldsymbol{T}}{|\boldsymbol{T}|} \right) \bigg|_{u=0} \right) dt = 0 . \tag{29.61}$$

For arbitrary transversal fields $\boldsymbol{U}$, this is equivalent to (29.51). Imposing the condition that D is metric compatible so that $|\boldsymbol{T}|$ remains constant (constant *speed*) under parallel translation along the base curve, the geodesic equation (29.50) results.

Chapter 30

The Second Variation of Arclength, Index Forms, and Jacobi Fields

To determine whether a geodesic joining two points in a Riemannian manifold has the shortest distance among all neighboring paths between the points one needs to derive an equation for the *second variation of arclength*, in analogy to calculating the second derivative in ordinary calculus to determine whether a function has a local maximum or minimum. In this process, one will need to explore the geometrical properties of geodesics in some detail, and introduce new analytic tools, such as *exponential maps*, *index forms*, and *Jacobi fields*. We will see that in the equation for the second variation, the Riemann curvature tensor plays a vital role. Furthermore, properties of geodesic curves together with the possibility of rendering the Riemannian manifold into a *metric space* through a *distance function* induced by the Riemannian metric allow the understanding of the relationship between curvature and topology at a deep level. Most theorems in this regard mentioned in the present chapter will be stated without proof. (Readers interested in the proofs may consult, for example, Spivak, Vol. III 1975, do Carmo 1993, and Chern, Chen and Lam 1999.)

We will first introduce a special coordinate system near an arbitrary point $x \in M$ of a Riemannian manifold M such that the coordinates of any point on a geodesic emanating from x are linear functions of the arclength. These coordinates facilitate greatly the analysis of the local properties of geodesics in the neighborhood of a point, and exist by virtue of the theory of ordinary differential equations applied to the second order geodesic equation (22.44). In essence, this theory asserts that there exists, for any point x in the coordinate neighborhood $(U; x^i)$ of M, a neighborhood $W \subset U$ of x in which a unique solution to (22.44) is given by

$$x^i(t) = f^i(t, (x_0)^k, X^k) . \tag{30.1}$$

This solution satisfies the initial conditions

$$f^i(0,(x_0)^k,X^k) = (x_0)^i\,, \qquad \left.\frac{dx^i}{dt}\right|_{t=0} = \left.\frac{\partial f^i}{\partial t}\right|_{t=0} = X^i\,, \tag{30.2}$$

for $|t| \le 1$ and $||\boldsymbol{X}|| \equiv \sqrt{\sum_{i=1}^n (X^i)^2}$ (**norm** of the vector $\boldsymbol{X}$) less than some positive number r [with $n = dim\,(M)$]. Since

$$f^i(ct,(x_0)^k,X^k)|_{t=0} = (x_0)^i \quad \text{and} \quad \left.\frac{\partial f^i(ct,(x_0)^k,X^k)}{\partial t}\right|_{t=0} = cX^i \quad (0 \le c \le 1)\,, \tag{30.3}$$

the uniqueness property of the solutions of second order differential equations also implies that the functions f^i satisfy

$$f^i(ct,(x_0)^k,X^k) = f^i(t,(x_0)^k,cX^k)\,, \tag{30.4}$$

for all non-negative numbers $c \le 1$. By the second equation in (30.2) the vector $\boldsymbol{X} = (X^1,\dots,X^n)$ can be considered as a vector in the tangent space $T_x(M)$. One can then define a *diffeomorphism* $\exp_x : T_x(M) \to M$, called the **exponential map**, which maps rays (shorter than r) through the zero vector $\boldsymbol{0} \in T_x(M)$ to unique geodesics in M passing through x. Specifically, for $\boldsymbol{X} \in T_x(M)$ such that $||\boldsymbol{X}|| < r$,

$$(\exp_x \boldsymbol{X})^i \equiv f^i(1,(x_0)^k,X^k)\,. \tag{30.5}$$

Setting $c = 0$ and $t = 1$ in (30.4), and using the first equation in (30.2), we have

$$f^i(1,(x_0)^k,0) = f^i(0,(x_0)^k,X^k) = (x_0)^i\,. \tag{30.6}$$

Thus, for fixed x, the exponential map $\exp_x$ provides a smooth map from a neighborhood of the origin in the tangent space $T_x(M) = \mathbb{R}^n$ to a neighborhood of x in M. As the parameter t varies between 0 and 1, tX^i describes a straight line in $T_x(M)$ along the direction of $\boldsymbol{X}$ and starting from the origin, and $\exp_x(t\boldsymbol{X})$ traces out a geodesic curve starting from $x \in M$ and tangent to $\boldsymbol{X}$ at x. One has the geometric picture that, corresponding to each tangent vector $\boldsymbol{X}$ in the *ball* $B_x(r) \subset T_x(M)$ given by

$$B_x(r) \equiv \{\boldsymbol{X} \in T_x(M)\,|\,||\boldsymbol{X}|| < r\}\,, \tag{30.7}$$

there is a unique geodesic curve, called a **radial geodesic**, emanating from x (see Fig. 30.1). One is then led to the following basic theorem concerning geodesics in Riemannian manifolds, which will be stated without proof.

Theorem 30.1. *Let M be a Riemannian manifold. For every $x \in M$, there exists a neighborhood W of x such that any two points in W can be connected by a geodesic curve which sits entirely in W.*

The neighborhood W specified in this theorem is called a **geodesic convex neighborhood** (associated with x). It can be identified with the image of

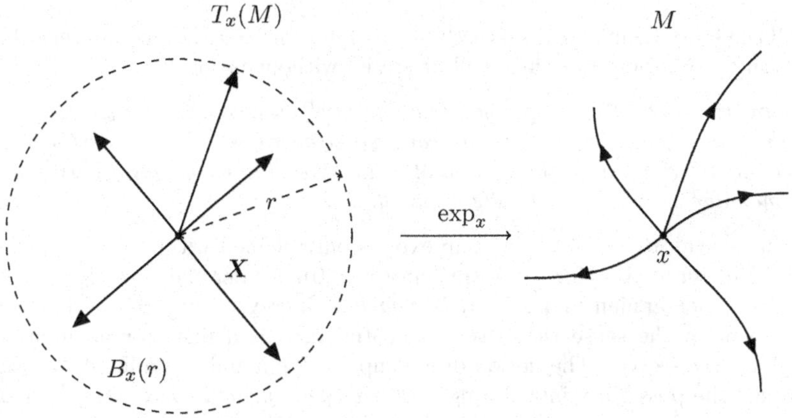

Fig. 30.1

$B_x(r)$ under the exponential map $\exp_x$, which will be denoted by $\mathcal{B}_x(r)$. Since $\exp_x$ is a diffeomorphism, the local coordinates y^i of any point $y \in \mathcal{B}_x(r)$ $(y \neq x)$ can be identified with the components of the vector $(\exp_x)^{-1}(y) \in T_x(M)$. Thus

$$y^i = tX^i \;, \qquad (\text{for some } \; t \text{ in } 0 < t \leq 1) \;, \tag{30.8}$$

with $t = 0$ corresponding to the point x, that is $x^i = 0$. These local coordinates are called **geodesic normal coordinates**.

Now let $S_x(\delta)$ be a *sphere* in $T_x(M)$ given by

$$S_x(\delta) \equiv \{\boldsymbol{X} \in T_x(M) \,|\, ||\boldsymbol{X}|| = \delta < r\} \;. \tag{30.9}$$

Then one can define the *hypersphere* $\Sigma_x(\delta)$ in M, called a **geodesic sphere** of radius δ centered at x, to be the image of $S_x(\delta)$ under $\exp_x$:

$$\Sigma_x(\delta) \equiv \exp_x\left(S_x(\delta)\right) \;. \tag{30.10}$$

As seen above, each $\boldsymbol{X} \in S_x(\delta)$ yields a radial geodesic

$$\gamma_x(t, \boldsymbol{X}) = \exp_x(t\boldsymbol{X}) \;, \qquad 0 \leq t \leq 1 \;, \tag{30.11}$$

which intersects all geodesic spheres centered at x with radii $\leq r$. We then have the following important Lemma (stated without proof).

Lemma 30.1. ***(The Gauss Lemma)*** *A radial geodesic curve $\gamma_x(t, \boldsymbol{X})$ intersects all geodesic spheres centered at x orthogonally with respect to the Riemannian metric.*

The Gauss Lemma leads directly to the following central theorem regarding the length of geodesics (which will be stated without proof).

Theorem 30.2. *The unique geodesic curve which connects any point $x \in M$ to any other point in the geodesic convex neighborhood W of x is the shortest curve in W connecting the two points. In other words, geodesic curves in a Riemannian manifold are locally minimal.*

In general, the exponential map $\exp_x$ is only defined within a ball $B_x(r) \subset T_x(M)$ of finite size (for a certain finite $r > 0$). Under the special condition that the Riemannian manifold M is *complete*, however, $\exp_x$ can be infinitely extended, in the sense that the exponential map is defined for an infinitely large ball ($r \to \infty$). The notion of a complete Riemannian manifold M arises through the possibility introducing a *metric space structure* into M, so that M is endowed with the usual metric-space topology. This is done by introducing a **distance function** $d : M \times M \to [0, \infty)$ in M, defined by

$$d(x, y) \equiv inf\, L(\gamma) \,, \qquad (30.12)$$

where the *infimum* (greatest lower bound) is that of the lengths of all piecewise smooth curves γ joining the points x and y in M, as determined by the Riemannian metric. Note that the definition makes use of the infimum of the lengths of curves rather than the shortest curve, because *in general, there may not exist a shortest curve joining two arbitrary points in a Riemannian manifold.* It can be straightforwardly verified that this definition satisfies the following standard requirements of a distance function:

$$(1) \quad d(x, y) \geq 0 \,, \quad \text{(the equality holds if and only if } x = y) \,, \qquad (30.13a)$$

$$(2) \quad d(x, y) = d(y, x) \,, \qquad (30.13b)$$

$$(3) \quad d(x, y) \leq d(x, z) + d(z, y) \quad \text{(triangle inequality)} \,. \qquad (30.13c)$$

Note that condition (1) above does not apply to pseudo-Riemannian manifolds. In a metric space one can define the notion of a *Cauchy sequence of points.*

Definition 30.1. *Let $\{x_n\}$ be an infinite sequence of points in a metric space M with a distance function d. If for any given positive number ϵ there exists a positive integer $N(\epsilon)$ such that $d(x_n, x_m) < \epsilon$ for any $n, m > N(\epsilon)$, then the sequence is said to be a* ***Cauchy sequence*** *in M.*

The notion of a *complete Riemannian manifold* is then made precise by the following definition.

Definition 30.2. *Suppose a Riemannian manifold M is rendered into a metric space by means of a distance function induced by the Riemannian metric. If every Cauchy sequence in M converges, then M is said to be a* ***complete Riemannian manifold.***

The concept of completeness is the most important one in the study of global properties of a Riemannian manifold. As an illustration, we will give (without proof) the following basic theorem concerning this concept.

Theorem 30.3. ***(Hopf-Rinow)***. *Suppose M is a connected Riemannian manifold. Then the following statements are equivalent:*

(1) M is complete.

(2) For all $x \in M$, the exponential map $\exp_x$ is defined on the entire tangent space $T_x(M)$. (Equivalently, any geodesic curve in M can be infinitely extended.)

(3) Every closed and bounded subset of M is compact.

This theorem leads to the following useful and important facts.

Corollary 30.1. *A connected, compact Riemannian manifold is always complete.*

Corollary 30.2. *A complete Riemannian manifold cannot be a proper, open submanifold of another connected Riemannian manifold. (We say that a complete Riemannian manifold is* ***non-extendable****.)*

Corollary 30.3. *In a complete Riemannian manifold, any two points can be connected by a minimal geodesic curve.*

We will now calculate a formula for the second variation of arclength in Riemannian geometry. From the result for the first variation given by (29.49) we have

$$L''(u) = \int_a^b \frac{\partial}{\partial u}\left(\frac{\boldsymbol{T}\cdot D_{\boldsymbol{T}}\boldsymbol{U}}{|\boldsymbol{T}|}\right) dt \,, \tag{30.14}$$

where we have suppressed the subscript u for the tangent vector field $\boldsymbol{T}_u$, in order to simplify the writing of the following equations. The integrand in the above equation can be expressed, on recalling (29.43), as,

$$\begin{aligned} &\boldsymbol{U}\left(\frac{\boldsymbol{T}\cdot D_{\boldsymbol{T}}\boldsymbol{U}}{|\boldsymbol{T}|}\right) \\ &= \boldsymbol{U}\left(\frac{1}{\sqrt{\boldsymbol{T}\cdot\boldsymbol{T}}}\right)(\boldsymbol{T}\cdot D_{\boldsymbol{T}}\boldsymbol{U}) + \frac{1}{|\boldsymbol{T}|}\left(D_{\boldsymbol{U}}\boldsymbol{T}\cdot D_{\boldsymbol{T}}\boldsymbol{U} + \boldsymbol{T}\cdot D_{\boldsymbol{U}}D_{\boldsymbol{T}}\boldsymbol{U}\right) . \end{aligned} \tag{30.15}$$

Recalling (29.47), (29.48) (since D is assumed to be torsion-free) and the derivation of (29.45), we have

$$\begin{aligned} &\boldsymbol{U}\left(\frac{\boldsymbol{T}\cdot D_{\boldsymbol{T}}\boldsymbol{U}}{|\boldsymbol{T}|}\right) \\ &= -\frac{(\boldsymbol{T}\cdot D_{\boldsymbol{T}}\boldsymbol{U})^2}{|\boldsymbol{T}|^3} + \frac{1}{|\boldsymbol{T}|}(D_{\boldsymbol{T}}\boldsymbol{U}\cdot D_{\boldsymbol{T}}\boldsymbol{U}) + \frac{1}{|\boldsymbol{T}|}\boldsymbol{T}\cdot(D_{\boldsymbol{U}}D_{\boldsymbol{T}}\boldsymbol{U}) \\ &= -\frac{1}{|\boldsymbol{T}|^3}\left\{\boldsymbol{T}(\boldsymbol{T}\cdot\boldsymbol{U}) - (D_{\boldsymbol{T}}\boldsymbol{T})\cdot\boldsymbol{U}\right\}^2 + \frac{1}{|\boldsymbol{T}|}|D_{\boldsymbol{T}}\boldsymbol{U}|^2 \\ &\qquad + \frac{1}{|\boldsymbol{T}|}\left[\boldsymbol{T}\cdot\{-R(\boldsymbol{T},\boldsymbol{U})\boldsymbol{U} + D_{\boldsymbol{T}}D_{\boldsymbol{U}}\boldsymbol{U}\}\right] , \end{aligned} \tag{30.16}$$

where in the last equality we have made use of (27.58), and (27.33) for the appearance of the curvature operator $R(\boldsymbol{T},\boldsymbol{U})$ [again recalling (29.47)]. Assuming that the second derivative is evaluated at the *critical base curve*, and recalling that the length $|\boldsymbol{T}|$ remains constant under parallel translation (because D is metric-compatible), we have

$$D_{\boldsymbol{T}}\left(\frac{\boldsymbol{T}}{|\boldsymbol{T}|}\right) = D_{\boldsymbol{T}}\boldsymbol{T} = \frac{d|\boldsymbol{T}|}{dt} = 0\,. \tag{30.17}$$

Also, by (27.58),

$$\boldsymbol{T}\cdot(D_{\boldsymbol{T}}D_{\boldsymbol{U}}\boldsymbol{U}) = \boldsymbol{T}(\boldsymbol{T}\cdot D_{\boldsymbol{U}}\boldsymbol{U}) - D_{\boldsymbol{U}}\boldsymbol{U}\cdot D_{\boldsymbol{T}}\boldsymbol{T} = \boldsymbol{T}(\boldsymbol{T}\cdot D_{\boldsymbol{U}}\boldsymbol{U}) = \frac{\partial}{\partial t}(\boldsymbol{T}\cdot D_{\boldsymbol{U}}\boldsymbol{U})\,. \tag{30.18}$$

It then follows from (30.14) that

$$L''(0) = \left.\frac{(\boldsymbol{T}\cdot D_{\boldsymbol{U}}\boldsymbol{U})}{|\boldsymbol{T}|}\right|_a^b + \int_a^b dt\left[\frac{|D_{\boldsymbol{T}}\boldsymbol{U}|^2}{|\boldsymbol{T}|} - \frac{\boldsymbol{T}\cdot\{R(\boldsymbol{T},\boldsymbol{U})\boldsymbol{U}\}}{|\boldsymbol{T}|} - \frac{1}{|\boldsymbol{T}|^3}\left(\frac{\partial}{\partial t}(\boldsymbol{T}\cdot\boldsymbol{U})\right)^2\right]. \tag{30.19}$$

Looking at the term in the integrand involving the curvature operator $R(\boldsymbol{T},\boldsymbol{U})$, Eqs. (27.78) and (27.81) imply that

$$\begin{aligned}&\boldsymbol{T}\cdot R(\boldsymbol{T},\boldsymbol{U})\boldsymbol{U} = R(\boldsymbol{T},\boldsymbol{U})\boldsymbol{U}\cdot\boldsymbol{T} = G(R(\boldsymbol{T},\boldsymbol{U})\boldsymbol{U},\boldsymbol{T}) = R(\boldsymbol{U},\boldsymbol{T},\boldsymbol{T},\boldsymbol{U})\\ &= R(\boldsymbol{T},\boldsymbol{U},\boldsymbol{U},\boldsymbol{T}) = G(R(\boldsymbol{U},\boldsymbol{T})\boldsymbol{T},\boldsymbol{U}) = \boldsymbol{U}\cdot R(\boldsymbol{U},\boldsymbol{T})\boldsymbol{T}\,.\end{aligned} \tag{30.20}$$

Equation (30.19) can then be written as

$$L''(0) = \left.\frac{\boldsymbol{T}\cdot(D_{\boldsymbol{U}}\boldsymbol{U})}{|\boldsymbol{T}|}\right|_a^b + \int_a^b dt\left[\frac{|D_{\boldsymbol{T}}\boldsymbol{U}|^2}{|\boldsymbol{T}|} + \frac{\boldsymbol{U}\cdot R(\boldsymbol{T},\boldsymbol{U})\boldsymbol{T}}{|\boldsymbol{T}|} - \frac{1}{|\boldsymbol{T}|^3}\left(\frac{\partial}{\partial t}(\boldsymbol{T}\cdot\boldsymbol{U})\right)^2\right]. \tag{30.21}$$

This is the formula for **the second variation of arclength** evaluated at a critical curve. As in (29.51) for the first variation, the *boundary term* (the first term on the right-hand side) vanishes for variations with fixed end-points [$\boldsymbol{U}(a) = \boldsymbol{U}(b) = 0$]. Significantly, the second variation formula involves the Riemann curvature operator whereas the first variations formula [(29.51)] does not.

We will now define the **index form** $I(\boldsymbol{V},\boldsymbol{W})$ on smooth vector fields $\boldsymbol{V}$ and $\boldsymbol{W}$ along a geodesic $\gamma(t)$ with end-points $\gamma(a)$ and $\gamma(b)$, and tangent vector field $\boldsymbol{T}$:

$$I(\boldsymbol{V},\boldsymbol{W}) \equiv \int_a^b dt\,\frac{1}{|\boldsymbol{T}|}\left[D_{\boldsymbol{T}}\boldsymbol{V}\cdot D_{\boldsymbol{T}}\boldsymbol{W} + R(\boldsymbol{T},\boldsymbol{V})\boldsymbol{T}\cdot\boldsymbol{W}\right]. \tag{30.22}$$

The second variation formula (30.21) then assumes the form

$$L''(0) = I(\boldsymbol{U},\boldsymbol{U}) + \left.\frac{\boldsymbol{T}\cdot(D_{\boldsymbol{U}}\boldsymbol{U})}{|\boldsymbol{T}|}\right|_a^b - \int_a^b dt\,\frac{1}{|\boldsymbol{T}|^3}\left(\frac{\partial}{\partial t}(\boldsymbol{T}\cdot\boldsymbol{U})\right)^2. \tag{30.23}$$

This can be simplified further by introducing the component of $\boldsymbol{U}$ that is orthogonal to $\boldsymbol{T}$:

$$\boldsymbol{U}_{\perp} \equiv \boldsymbol{U} - \left(\boldsymbol{U} \cdot \frac{\boldsymbol{T}}{|\boldsymbol{T}|}\right) \frac{\boldsymbol{T}}{|\boldsymbol{T}|} . \tag{30.24}$$

It is not difficult to show that, under the conditions (30.17),

$$D_{\boldsymbol{T}}\boldsymbol{U}_{\perp} \cdot D_{\boldsymbol{T}}\boldsymbol{U}_{\perp} = D_{\boldsymbol{T}}\boldsymbol{U} \cdot D_{\boldsymbol{T}}\boldsymbol{U} - \frac{1}{|\boldsymbol{T}|^2}\left(\frac{\partial}{\partial t}(\boldsymbol{T} \cdot \boldsymbol{U})\right)^2 . \tag{30.25}$$

In addition, (27.78) and the symmetry properties of the Riemannian curvature tensor (27.79) imply that

$$\boldsymbol{U} \cdot R(\boldsymbol{T}, \boldsymbol{U})\boldsymbol{T} = \boldsymbol{U}_{\perp} \cdot R(\boldsymbol{T}, \boldsymbol{U}_{\perp})\boldsymbol{T} . \tag{30.26}$$

We then have the following more compact form for the second variation equation (30.23):

$$L''(0) = I(\boldsymbol{U}_{\perp}, \boldsymbol{U}_{\perp}) + \left. \frac{\boldsymbol{T} \cdot (D_{\boldsymbol{U}}\boldsymbol{U})}{|\boldsymbol{T}|} \right|_a^b . \tag{30.27}$$

It is clear from (30.23) that a useful criterion to determine whether a geodesic is a minimal geodesic is the following Lemma.

Lemma 30.2. *If a transversal variation field $\boldsymbol{U}(t)$ of a geodesic $\gamma(t)$ keeps the end-points fixed, then $I(\boldsymbol{U}, \boldsymbol{U}) < 0$ implies that the geodesic cannot be a minimal geodesic.*

Problem 30.1 Verify the properties (30.25) and (30.26) for the transversal field along a geodesic curve.

Let us return to the index form as defined by (30.22). Since $d|\boldsymbol{T}|/dt = 0$ [cf. (30.17)], the index form $I(\boldsymbol{J}, \boldsymbol{V})$ can be rewritten as, after using (27.58) and integrating out a boundary term,

$$I(\boldsymbol{J}, \boldsymbol{V}) = \frac{1}{\boldsymbol{T}} \boldsymbol{V} \cdot D_{\boldsymbol{T}}\boldsymbol{J} \Big|_a^b - \frac{1}{|\boldsymbol{T}|} \int_a^b dt\, \boldsymbol{V} \cdot [D_{\boldsymbol{T}}D_{\boldsymbol{T}}\boldsymbol{J} + R(\boldsymbol{J}, \boldsymbol{T})\boldsymbol{T}] . \tag{30.28}$$

A vector field $\boldsymbol{J}$ along a geodesic curve $\gamma(t)$ with velocity vector field $\boldsymbol{T}$ is said to be a **Jacobi field** if it satisfies the equation

$$D_{\boldsymbol{T}}D_{\boldsymbol{T}}\boldsymbol{J} + R(\boldsymbol{J}, \boldsymbol{T})\boldsymbol{T} = 0 . \tag{30.29}$$

This equation, called the **Jacobi equation**, is equivalent to a set of coupled second-order ordinary differential equations. To see this, we choose an orthonormal frame field $\{\boldsymbol{e}_1(t), \dots, \boldsymbol{e}_n(t)\}$ *parallelly translated* along the geodesic and write

$$\boldsymbol{J}(t) = J^i(t)\, \boldsymbol{e}_i(t) , \qquad D_{\boldsymbol{T}}\boldsymbol{e}_i = 0 . \tag{30.30}$$

Then

$$D_{\boldsymbol{T}} = \langle \boldsymbol{T}, dJ^i \rangle \, e_i = \frac{dJ^i}{dt} \, e_i \,, \tag{30.31}$$

and

$$D_{\boldsymbol{T}} D_{\boldsymbol{T}} \boldsymbol{J} = \frac{d^2 J^i}{dt^2} \, e_i \,. \tag{30.32}$$

Choose $\boldsymbol{e}_1$ to be the unit tangent vector field along the geodesic, that is, $\boldsymbol{e}_1 = \boldsymbol{T}/|\boldsymbol{T}|$. Using (27.31) for the Riemann curvature operator, we have

$$\begin{aligned} R(\boldsymbol{J}, \boldsymbol{T})\boldsymbol{T} &= R(\boldsymbol{J}, |\boldsymbol{T}|\boldsymbol{e}_1)\,|\boldsymbol{T}|\boldsymbol{e}_1 = |\boldsymbol{T}|^2 \, R^j_{ikl}\,(\boldsymbol{e}_1)^i J^k (\boldsymbol{e}_1)^l \, \boldsymbol{e}_j \\ &= |\boldsymbol{T}|^2 \, R^j_{ikl} \, \delta^i_1 J^k \delta^l_1 \, \boldsymbol{e}_j = |\boldsymbol{T}|^2 \, R^j_{1k1} J^k \, \boldsymbol{e}_j \,, \end{aligned} \tag{30.33}$$

where R^j_{ikl} are the components of the Riemann curvature tensor with respect to $\{\boldsymbol{e}_i\}$. The Jacobi equation (30.29) thus becomes

$$\frac{d^2 J^i}{dt^2} \, \boldsymbol{e}_i + |\boldsymbol{T}|^2 \, R^j_{1k1} J^k \, \boldsymbol{e}_j = 0 \,. \tag{30.34}$$

This is equivalent to the following set of coupled second order differential equations:

$$\frac{d^2 J^i}{dt^2} + |\boldsymbol{T}|^2 \, R^i_{1k1} J^k = 0; , \qquad (i = 1, \ldots, n) \,. \tag{30.35}$$

Note that $|\boldsymbol{T}|$ is constant along a geodesic. The theory of second order differential equations then imply that *if* $\gamma : [a, b] \to M$ *is a given geodesic curve, and* $\boldsymbol{V}, \boldsymbol{W} \in T_{\gamma(a)}(M)$, *then there exists a unique Jacobi field* $\boldsymbol{J}(t)$ *along* γ *such that* $\boldsymbol{J}(a) = \boldsymbol{V}$ *and* $D_{\boldsymbol{T}}\boldsymbol{J}|_{t=a} = \boldsymbol{W}$. This fact also implies that *the zeroes of a Jacobi field along a geodesic are discrete, unless the Jacobi field is the zero-field along the geodesic.*

We have the following basic theorem on Jacobi fields (stated without proof):

Theorem 30.4. *Suppose* $\gamma(t)$ *is a geodesic and* $\boldsymbol{U}$ *is a vector field along* $\gamma(t)$. *Then* $\boldsymbol{U}$ *is a Jacobi field if and only if* $\boldsymbol{U}$ *is the transversal field (see Def. 29.3) of a variation* $\gamma_u(t)$ *of the critical base curve* $\gamma_0(t) = \gamma(t)$, *restricted to* $\gamma_0(t)$.

Problem 30.2 Suppose $\boldsymbol{U}$ is a Jacobi field along a geodesic curve $\gamma(t)$. Then there exist two real numbers a and b such that

$$\boldsymbol{U} = \boldsymbol{U}_\perp + (at + b)\,\boldsymbol{T} \,, \tag{30.36}$$

where $\boldsymbol{T}$ is a tangent vector field along γ, $\boldsymbol{U}_\perp$ is also a Jacobi field, and $\boldsymbol{U}_\perp \cdot \boldsymbol{T} = 0$.

Jacobi fields constitute an indispensable tool in the study of the properties of geodesics in Riemannian geometry. In particular the relationship between the sectional curvatures and geodesics can be studied through the notion of *conjugate points* on a geodesic, which can be characterized by the behavior of Jacobi fields at such points. In the remainder of this chapter, we will identify some noteworthy properties of the index form in relation to conjugate points and Jacobi fields, and finally state an important theorem, the *Cartan-Hadamard Theorem*, that illustrates this relationship.

The notion of conjugate points on a geodesic is defined as follows.

Definition 30.3. *Suppose at a point $x \in M$ of a Riemannian manifold, the derivative map of the exponential map, $(\exp_x)_*$, is* ***degenerate*** *at a tangent vector $\boldsymbol{V} \in T_x(M)$, that is, there exists some non-zero vector $\boldsymbol{W} \in T_{\boldsymbol{V}}(T_x(M))$ such that $(\exp_x)_*\boldsymbol{W} = 0 \in T_{\exp_x \boldsymbol{V}}(M)$, then the point $\gamma(1) = \exp_x \boldsymbol{V}$ on the geodesic $\gamma(t) = \exp_x t\boldsymbol{V}$, $0 \leq t \leq 1$, is said to be* ***conjugate*** *to the point $\gamma(0) = x$.*

Given a geodesic $\gamma(t) = \exp_x(t\boldsymbol{V})$ starting at a point $x \in M$, we can construct a variation according to

$$\gamma(t,u) = \exp_x[t(\boldsymbol{V} + u\boldsymbol{W})] \,. \tag{30.37}$$

Each t-curve in this variation is a geodesic, and by Theorem 30.4, the transversal field $\boldsymbol{U}$ of this variation is a Jacobi field. It is clear from the explicit form of this variation that

$$\boldsymbol{U}(t=0) = 0 \qquad \text{and} \qquad \boldsymbol{U}(t=1) = (\exp_x)_*(\boldsymbol{W}) \,. \tag{30.38}$$

We then have the following Lemma.

Lemma 30.3. *The point $\gamma(1)$ is conjugate to the point $\gamma(0)$ on a geodesic $\gamma(t) = \exp_x t\boldsymbol{V}$ $(0 \leq t \leq 1,\ \boldsymbol{V} \in T_x(M))$, or equivalently, $(\exp_x)_*$ is degenerate at $\boldsymbol{V}$, if and only if there exists a non-zero Jacobi field $\boldsymbol{J}(t)$ along $\gamma(t)$ such that $\boldsymbol{J}(0) = \boldsymbol{J}(1) = 0$.*

The following group of lemmas (Lemmas 30.4 to 30.7) are important results on the relationship between conjugate points on a geodesic and index forms along a geodesic. In these lemmas, $\mathcal{V}$ will denote the vector space of all smooth vector fields $\boldsymbol{W}(t)$ on a geodesic $\gamma(t)$ $(a \leq t \leq b)$ joining $\gamma(a)$ and $\gamma(b)$ such that $\boldsymbol{W}(a) = \boldsymbol{W}(b) = 0$.

Lemma 30.4. *(1) If no point along a geodesic $\gamma(t)$ $(a \leq t \leq b)$ is conjugate to $\gamma(a)$, then $I(\boldsymbol{W},\boldsymbol{W}) \geq 0$ for all $\boldsymbol{W}(t) \in \mathcal{V}$, and the equality holds only if $\boldsymbol{W} = 0$.*
(2) If $\gamma(b)$ is conjugate to $\gamma(a)$ on a geodesic $\gamma(t)$ $(a \leq t \leq b)$, but for all $a < \tau < b$, $\gamma(\tau)$ is not conjugate to $\gamma(a)$, then $I(\boldsymbol{W},\boldsymbol{W}) \geq 0$ for all $\boldsymbol{W} \in \mathcal{V}$, and the equality holds only if $\boldsymbol{W}$ is a Jacobi field.

Lemma 30.5. *For a geodesic $\gamma(t)$ $(a \leq t \leq b)$ a necessary and sufficient condition for a point $\gamma(c)$ $(a < c < b)$ to be conjugate to $\gamma(a)$ is that there exists a vector field $\boldsymbol{W} \in \mathcal{V}$ such that $\boldsymbol{W} \cdot \boldsymbol{T} = 0$ (orthogonal to the velocity field of the geodesic) and $I(\boldsymbol{W},\boldsymbol{W}) < 0$.*

Lemma 30.6. *Suppose $\gamma(t)\,(a \leq t \leq b)$ is a geodesic containing no conjugate points. Let $\boldsymbol{W}(t)$ and $\boldsymbol{J}(t)$ be smooth vector fields on $\gamma(t)$ such that $\boldsymbol{W}(a) = \boldsymbol{J}(a)$, $\boldsymbol{W}(b) = \boldsymbol{J}(b)$, and $\boldsymbol{J}(t)$ is a Jacobi field. Then $I(\boldsymbol{W},\boldsymbol{W}) \geq I(\boldsymbol{J},\boldsymbol{J})$, and the equality holds if and only if $\boldsymbol{W} = \boldsymbol{J}$.*

Lemma 30.7. *$I(\boldsymbol{V},\boldsymbol{W}) = 0$ for all $\boldsymbol{W} \in \mathcal{V}$ if and only if $\boldsymbol{V}$ is a Jacobi field.*

We finally give the *Cartan-Hadamard Theorem*, a basic theorem in Riemannian geometry relating the *global* properties of a Riemannian manifold to its curvature and the properties of conjugate points on its geodesics.

Theorem 30.5. ***(Cartan-Hadamard)*** *(1) Suppose M is a complete Riemannian manifold with negative sectional curvatures everywhere. Then for every point $x \in M$, the exponential map $\exp_x : T_x(M) \to M$ does not have conjugate points.*
(2) Suppose M is a simply-connected, complete Riemannian manifold, and for some point $x \in M$, the exponential map $\exp_x : T_x(M) \to M$ does not have conjugate points. Then $\exp_x$ is a diffeomorphism.

Proof. We will only provide a proof of (1). Since M is complete, the exponential map is infinitely exdendable by the Hopf-Rinow Theorem (Theorem 30.3). So one can define the function $f : [0,\infty) \to \mathbb{R}$ by $f(t) = \boldsymbol{U}(t)\cdot\boldsymbol{U}(t)$, where $\boldsymbol{U}(t)$ is the transversal field along a geodesic $\gamma(t)$ starting at $x \in M$, and $\boldsymbol{U}(0) = 0$. We claim that under the premises of (1), the second derivative $\ddot{f} \geq 0$. To show this we calculate $\ddot{f}$ as follows. First

$$\dot{f} = \frac{d}{dt}(\boldsymbol{U}\cdot\boldsymbol{U}) = \boldsymbol{T}(\boldsymbol{U}\cdot\boldsymbol{U}) = 2\boldsymbol{U}\cdot(D_{\boldsymbol{T}}\boldsymbol{U})\,. \tag{30.39}$$

Hence

$$\ddot{f} = 2\boldsymbol{T}[\boldsymbol{U}\cdot(D_{\boldsymbol{T}}\boldsymbol{U})] = 2\left[(D_{\boldsymbol{T}}D_{\boldsymbol{T}}\boldsymbol{U})\cdot\boldsymbol{U} + (D_{\boldsymbol{T}}\boldsymbol{U})\cdot(D_{\boldsymbol{T}}\boldsymbol{U}))\right]. \tag{30.40}$$

Since $\boldsymbol{U}$ is a Jacobi field (by Theorem 30.4), it satisfies the Jacobi equation (30.29). So

$$D_{\boldsymbol{T}}D_{\boldsymbol{T}}\boldsymbol{U} = -R(\boldsymbol{U},\boldsymbol{T})\boldsymbol{T}\,. \tag{30.41}$$

Thus

$$\ddot{f} = 2\left[(D_{\boldsymbol{T}}\boldsymbol{U})\cdot(D_{\boldsymbol{T}}\boldsymbol{U}) - \boldsymbol{U}\cdot(R(\boldsymbol{U},\boldsymbol{T})\boldsymbol{T})\right]. \tag{30.42}$$

On the other hand, by (27.78) and the definition of the *sectional curvature* [(27.88)], if we let E be the tangent subspace spanned by $\boldsymbol{T}$ and $\boldsymbol{U}$ at each point of the geodesic, we have

$$\begin{aligned}(R(\boldsymbol{U},\boldsymbol{T})\boldsymbol{T})\cdot\boldsymbol{U} &= R(\boldsymbol{T},\boldsymbol{U},\boldsymbol{U},\boldsymbol{T}) = -R(\boldsymbol{T},\boldsymbol{U},\boldsymbol{T},\boldsymbol{U})\\ &= K(x(t),E)\,\mathcal{G}(\boldsymbol{T},\boldsymbol{U},\boldsymbol{T},\boldsymbol{U}) = K(x(t),E)\left[\,|\boldsymbol{T}|^2|\boldsymbol{U}|^2 - (\boldsymbol{T}\cdot\boldsymbol{U})^2\,\right],\end{aligned} \tag{30.43}$$

where $K(x(t),E)$ is the sectional curvature corresponding to the tangent subspace E at the point $x(t)$ along the geodesic, and the $(0,4)$-tensor $\mathcal{G}$ is defined by (27.82). By the *Schwarz inequality* of vector spaces equipped with a non-negative scalar product, the term within the square brackets on the right-hand side of the last equality in the above equation is always ≥ 0. It follows from (30.42) that $\ddot{f} \geq 0$ if the sectional curvatures are always negative. Suppose statement (1) of the theorem does not hold. Then there is a conjugate point along $\gamma(t)$ for some $t_0 > 0$. By Lemma 30.3 there exists a Jacobi field $\boldsymbol{J}(t)$ along $\gamma(t)$ such that $\boldsymbol{J}(0) = \boldsymbol{J}(t_0) = 0$. By Theorem 30.4, this Jacobi field must be the transversal field $\boldsymbol{U}(t)$. So the function $f(t)$ has the property that $f(0) = f(t_0) = 0$, and for $0 < t < t_0$, $\ddot{f}(t) \geq 0$. This cannot be the case unless, for $0 \leq t \leq t_0$, $f(t) = \boldsymbol{U}(t) \cdot \boldsymbol{U}(t) = 0$. This in turn implies that, for $0 \leq t \leq t_0$, $\boldsymbol{U}(t)$ is identically zero. We have thus arrived at a contradiction, and so (1) must hold. □

Chapter 31

The Lagrangian Formulation of Classical Mechanics: Hamilton's Principle of Least Action, Lagrange Multipliers in Constrained Motion

Conside the **action functional** [cf. (29.7)] of a classical mechanical system:

$$\Phi(C) = \int_{t_1}^{t_2} dt\, L \,, \tag{31.1}$$

where L, called the **Lagrangian**, is a function on the tangent bundle of the *configuration space (manifold)* of the mechanical system, in a sense to be specified below. It is given by

$$L = T - U \,, \tag{31.2}$$

where, for an N-particle system, the kinetic energy T and the potential energy U are given respectively by

$$T = \frac{1}{2}\sum_{\alpha=1}^{N} m_\alpha \dot{\boldsymbol{r}}_\alpha^2 \,, \tag{31.3}$$

$$U = U(\boldsymbol{r}_1, \dots, \boldsymbol{r}_N) \,. \tag{31.4}$$

The components of $\boldsymbol{r}_\alpha$ can be labelled $x^1, x^2, x^3, \ldots, x^{3N-1}, x^{3N}$. The Euler-Lagrange equations [cf. (29.15)] are then

$$\frac{d}{dt}\left(\frac{\partial L}{\partial \dot{x}^i}\right) - \frac{\partial L}{\partial x^i} = 0\,, \quad i = 1, \ldots, 3N\,. \tag{31.5}$$

Using Cartesian coordinates we can write

$$T = \frac{1}{2}\left\{ m_1\left(\frac{dx^1}{dt}\right)^2 + m_1\left(\frac{dx^2}{dt}\right)^2 + m_1\left(\frac{dx^3}{dt}\right)^2 + \cdots + m_N\left(\frac{dx^{3N-2}}{dt}\right)^2 + m_N\left(\frac{dx^{3N-1}}{dt}\right)^2 + m_N\left(\frac{dx^{3N}}{dt}\right)^2 \right\}. \tag{31.6}$$

Hence

$$\frac{\partial L}{\partial \dot{x}^i} = \frac{\partial T}{\partial \dot{x}^i} = m_\alpha \dot{x}^i\,, \qquad \frac{\partial L}{\partial x^i} = -\frac{\partial U}{\partial x^i}\,. \tag{31.7}$$

The Euler-Lagrange equations then imply

$$\frac{d}{dt}\left(m_\alpha \dot{x}^i\right) = -\frac{\partial U}{\partial x^i}\,, \quad i = 1, \ldots, 3N\,, \tag{31.8}$$

which are precisely Newton's equations of motion.

The equivalence of the Euler-Lagrange equations (31.5) and Newton's equations of motion (31.8) implies that *the extremals of the action functional are precisely the classical trajectories obtained from Newton's second law.* This is the statement of **Hamilton's Principle of Least Action**.

The Euler-Lagrange equations are invariant with respect to changes in local coordinates of the configuration manifold Q. It is customary in classical mechanics to denote a general set of local coordinates of an N-particle system by $(q^i, \ldots q^{3N})$, where each q^i is a function of the $3N$ variables $(x^1, \ldots, x^{3N})$. The $\{q^i\}$ are called **generalized (canonical) coordinates** of the mechanical system, and will be considered as the local coordinates of Q (considered as a differentiable manifold); while the $\dot{q}^i$ are then the local coordinates of a point in the tangent space $T_q(Q)$ [recall (5.1)]. This is the reason why *the Lagrangain $L(q^i, \dot{q}^i)$ can be considered as a function on the tangent bundle $T_*(Q)$ of the configuration manifold Q, with $(q^i, \dot{q}^i)$ being the local coordinates of the bundle.* The time evolution of any set of generalized coordinates $q^i(t)$ is then determined by the Euler-Lagrange equations:

$$\boxed{\frac{d}{dt}\left(\frac{\partial L}{\partial \dot{q}^i}\right) - \frac{\partial L}{\partial q^i} = 0\,, \quad i = 1, \ldots, 3N}\,. \tag{31.9}$$

The quantities

$$p_i \equiv \frac{\partial L}{\partial \dot{q}^i} \tag{31.10}$$

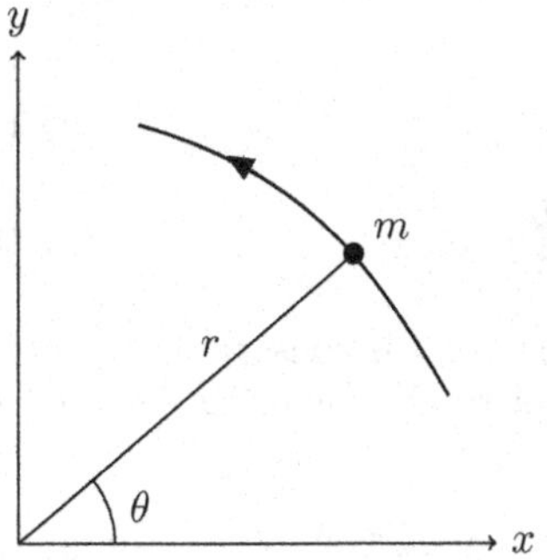

Fig. 31.1

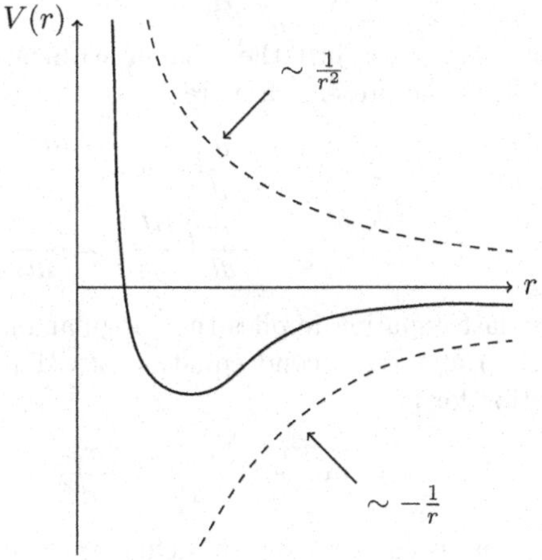

Fig. 31.2

are called the **generalized (canonical) momenta**, and the quantities $\partial L/\partial q^i$ are called the **generalized forces**. On recalling our discussion of *cotangent vectors* in Chapter 5, especially that surrounding Eqs. (5.2) to (5.5), we see that the p_i can be considered as components of the dual vector to the tangent vector whose components are $\dot{q}^i$. This fact will be explained in more detail in Chapter 33, on Hamiltonian dynamics.

A generalized coordinate q^i is said to be a **cyclic coordinate** if the Lagrangian L does not depend *explicitly* on it, that is, if $\partial L/\partial q^i = 0$. It is clear from the Euler-Lagrange equations that if a coordinate q^i is cyclic, then

$$\frac{d}{dt}\left(\frac{\partial L}{\partial \dot{q}^i}\right) = \frac{dp_i}{dt} = 0\,, \tag{31.11}$$

which implies that $p_i =$ constant. We thus have the important result that *the generalized momentum corresponding to a cyclic coordinate is conserved.*

As an example of generalized coordinates and momenta let us consider the planar motion of a particle of mass m in a central potential field [$U(\boldsymbol{r}) = U(r)$: the potential energy only depends on the magnitude of the position vector of the particle]. Instead of the Cartesian coordinates x and y we will use the plane polar coordinates r and θ as generalized coordinates (Fig. 31.1). The Lagrangian is given by

$$L = T - U = \frac{m}{2}(\dot{r}^2 + r^2\dot{\theta}^2) - U(r)\,. \tag{31.12}$$

Equation (31.10) then gives the following generalized momenta:

$$p_r = \frac{\partial L}{\partial \dot{r}} = m\dot{r} \,, \qquad p_\theta = \frac{\partial L}{\partial \dot{\theta}} = mr^2\dot{\theta} \,. \tag{31.13}$$

Note that p_θ is just the angular momentum. The Euler-Lagrange equations (31.9) in the present case are

$$\frac{d}{dt}(m\dot{r}) = -\frac{\partial U}{\partial r} + mr\dot{\theta}^2 \,, \tag{31.14}$$

$$\frac{d}{dt}\left(\frac{\partial L}{\partial \dot{\theta}}\right) = \frac{dp_\theta}{dt} = \frac{\partial L}{\partial \theta} = 0 \,. \tag{31.15}$$

The last equation implies that angular momentum p_θ is conserved. Equation (31.14) and the second equation of (31.13) lead to the following equation of motion for r:

$$m\frac{d^2r}{dt^2} = -\frac{\partial U}{\partial r} + \frac{p_\theta^2}{mr^3} = -\frac{\partial}{\partial r}\left(U + \frac{p_\theta^2}{2mr^2}\right) \,. \tag{31.16}$$

Since p_θ is constant we can define an effective potential energy V:

$$V \equiv U + \frac{p_\theta^2}{2mr^2} \,, \tag{31.17}$$

and our two-dimensional problem is reduced to a one-dimensional one in the coordinate r. This is specified by

$$m\frac{d^2r}{dt^2} = -\frac{dV}{dr} \,. \tag{31.18}$$

For the Kepler problem studied in Chapter 13, in which $U = -GMm/r$, the effective potential

$$V(r) = -\frac{GMm}{r} + \frac{p_\theta^2}{2mr^2} \tag{31.19}$$

is sketched in Fig. 31.2.

Frequently we encounter situations where motion on a configuration manifold [with local (canonical) coordinates $(q^1, \dots, q^n)$] is constrained. According to the discussion in Chapter 10, whether the constraints are holonomic or not, the quantities dq^i are no longer linearly independent and thus the Euler-Lagrange equations (31.9) (as a set of n independent equations) can no longer be deduced from the extremalization condition of Chapter 29. The Lagrangian formulation of classical mechanics introduced earlier will then require a revision that follows from the so-called **Lagrange method of undetermined multipliers**, which we will now introduce.

Consider a Lagrangian $L(q^1, \dots, q^n; \dot{q}^1, \dots, \dot{q}^n)$, where the n generalized coordinates q^i are not independent, but are related by m (with $m < n$) constraint Pffaffian equations of the form [cf. (10.2)]

$$\sum_{k=1}^{n} a_{lk}(q^i)\, dq^k = 0 \,, \quad l = 1, \dots, m \,. \tag{31.20}$$

Recall that the extremalization of the classical action

$$\int_{t_1}^{t_2} L(q^i, \dot{q}^i)\, dt$$

leads to the variational condition

$$\int_{t_1}^{t_2} dt \sum_{k=1}^{n} \left(\frac{\partial L}{\partial q^k} - \frac{d}{dt}\frac{\partial L}{\partial \dot{q}^k} \right) \delta q^k = 0\,. \tag{31.21}$$

But now we *cannot* claim the usual form of the Euler-Lagrange equations:

$$\frac{\partial L}{\partial q^k} - \frac{d}{dt}\frac{\partial L}{\partial \dot{q}^k} = 0\,, \quad k = 1, \ldots, n\,,$$

since the n δq^k's are not independent.

To incorporate the effects of the constraints we will introduce m arbitrary constants $\lambda_1, \ldots, \lambda_m$, and write

$$\lambda_l \sum_{k=1}^{n} a_{lk}\, \delta q^k = 0\,, \quad l = 1, \ldots, m\,, \tag{31.22}$$

which follows trivially from (31.20). These constants will be called **Lagrange multipliers**. Summing the above expression over l and integrating over t we obtain

$$\int_{t_1}^{t_2} dt \sum_{l=1}^{m} \lambda_l a_{lk}\, \delta q^k = 0\,. \tag{31.23}$$

Adding (31.21) and (31.23) we arrive at

$$\int_{t_1}^{t_2} dt \sum_{k=1}^{n} \left(\frac{\partial L}{\partial q^k} - \frac{d}{dt}\left(\frac{\partial L}{\partial \dot{q}^k}\right) + \sum_{l=1}^{m} \lambda_l a_{lk} \right) \delta q^k = 0\,. \tag{31.24}$$

So far the Lagrange multipliers are still arbitrary; and of the n q^k's in the above expression, only $n - m$ can be picked to be independent. We will let these be the first $n - m$ of the q^k's. Then choose the m Lagrange multipliers $\lambda_1, \ldots, \lambda_m$ such that the following m equations are satisfied:

$$\frac{\partial L}{\partial q^k} - \frac{d}{dt}\left(\frac{\partial L}{\partial \dot{q}^k}\right) + \sum_{l=1}^{m} \lambda_l a_{lk} = 0\,, \quad k = n - m + 1, \ldots, n\,. \tag{31.25}$$

With the λ's determined by the above equations, (31.24) then implies

$$\int_{t_1}^{t_2} dt \sum_{k=1}^{n-m} \left(\frac{\partial L}{\partial q^k} - \frac{d}{dt}\left(\frac{\partial L}{\partial \dot{q}^k}\right) + \sum_{l=1}^{m} \lambda_l a_{lk} \right) \delta q^k = 0\,, \tag{31.26}$$

where the $n - m$ δq^k's are independent. Thus

$$\frac{\partial L}{\partial q^k} - \frac{d}{dt}\left(\frac{\partial L}{\partial \dot{q}^k}\right) + \sum_{l=1}^{m} \lambda_l a_{lk} = 0\,, \quad k = 1, \ldots, n - m\,. \tag{31.27}$$

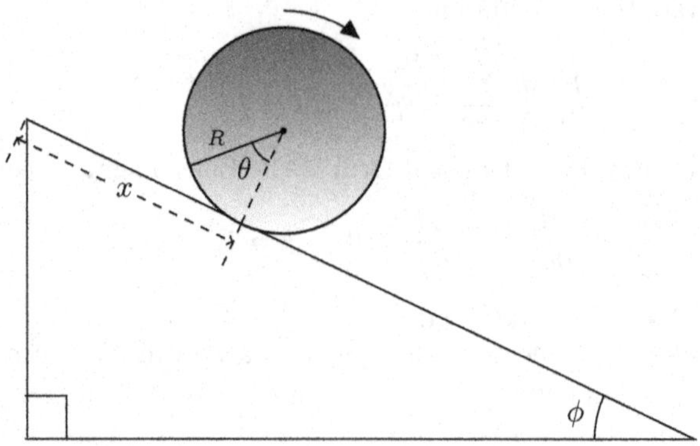

Fig. 31.3

Combining (31.25) and (31.27) we then have the complete set of *modified Euler-Lagrange equations*:

$$\frac{\partial L}{\partial q^k} - \frac{d}{dt}\left(\frac{\partial L}{\partial \dot{q}^k}\right) + \sum_{l=1}^{m} \lambda_l a_{lk} = 0 \,, \quad k = 1, \dots, n \,, \tag{31.28}$$

which, together with the set [cf. (31.20)]

$$\sum_{k=1}^{n} a_{lk}\, \dot{q}^k = 0 \,, \quad l = 1, \dots, m \,, \tag{31.29}$$

constitute a system of $n+m$ equations in the $n+m$ unknowns $q^1, \dots, q^n, \lambda_1, \dots, \lambda_m$. The quantities

$$Q_k \equiv \sum_{l=1}^{m} \lambda_l a_{lk} \,, \quad k = 1, \dots, n \,, \tag{31.30}$$

can be interpreted as generalized *forces of constraint*. Thus the method of undetermined Lagrange multipliers not only provides solutions for the $q^i(t)$ but also yields explicit expressions for the forces of constraint, which may be otherwise difficult to determine.

Consider the example of an object with a circular cross section rolling down an incline plane without slipping, as shown in Fig. 31.3. The generalized coordinates can be chosen to be x and θ. These are not independent, but are related by

$$x = R\theta \ , \tag{31.31}$$

where R is the radius of the circular cross section. This equation implies

$$dx - Rd\theta = 0 \ , \tag{31.32}$$

which is of the form of (31.20). In this case $n = 2$ and $m = 1$, so there is only one Lagrange multiplier λ. The Lagrangian is given by

$$L(x, \theta, \dot{x}, \dot{\theta}) = \frac{1}{2}m\dot{x}^2 + \frac{1}{2}I\dot{\theta}^2 - mg(l - x)\sin\phi \ , \tag{31.33}$$

where I is the moment of inertia of the object about an axis through its center of mass and m is the mass of the object. The modified Lagrange equations (31.28) appear as

$$mg\sin\phi - m\ddot{x} + \lambda = 0 \ , \tag{31.34}$$

$$-I\ddot{\theta} - \lambda R = 0 \ , \tag{31.35}$$

while the constraint equation (31.29) appears as

$$\dot{x} - R\dot{\theta} = 0 \ . \tag{31.36}$$

Thus

$$\ddot{\theta} = \frac{\ddot{x}}{R} \tag{31.37}$$

and (31.35) then implies

$$\lambda = -\frac{I}{R^2}\ddot{x} \ . \tag{31.38}$$

Substituting this in (31.34) and solving for $\ddot{x}$, we obtain

$$\ddot{x} = \frac{g\sin\phi}{1 + \dfrac{I}{mR^2}} \ . \tag{31.39}$$

This is the acceleration of the center of mass of the object moving down the incline. The solution for $\theta(t)$ can then be obtained by integrating (31.37). Finally (31.38) and (31.39) give

$$\lambda = -\frac{I}{R^2}\left(\frac{g\sin\phi}{1 + \dfrac{I}{mR^2}}\right) \ . \tag{31.40}$$

This force of constraint can be interpreted as the frictional force giving rise to the rolling motion of the object. The negative sign indicates that it acts along the negative x-direction, that is, up the incline.

Problem 31.1 A wedge (inclination θ) is pushed along a table top with constant horizontal acceleration $A\mathbf{e}_1$. A block of mass m slides without friction on the wedge. Model the motion of the block by means of a Lagrangian

$$L(x, \dot{x}) = \frac{1}{2}m\dot{x}^2 - U(x) \,,$$

where x is the horizontal displacement of the block from the vertical side of the wedge, and $U(x)$ is an effective potential energy.

(a) Find $U(x)$ (express in terms of m, g, θ and A, besides x).

(b) Calculate the value of A so that the block does not slide either up or down the wedge.

Problem 31.2 A pendulum consists of a mass m_1 capable of sliding without friction on a horizontal bar and another mass m_2 tied to m_1 by a massless, non-stretchable string of length L. Choose x (the position of m_1 from some fixed point on the horizontal bar) and ϕ (the angular displacement from the vertical of the pendulum) to be the generalized coordinates of the system.

(a) Construct the Lagrangian $L(x, \phi, \dot{x}, \dot{\phi})$.

(b) Use the result in (a) to deduce the Euler-Lagrange equations for this system.

(c) Deduce the second-order ordinary differential equation for only the variable $\phi(t)$. Simplify the equation on assuming $m_1 = m_2$ and small oscillations ($\phi \ll 1$).

Problem 31.3 A stick of length L and mass M initially stands upright on a frictionless horizontal tabletop and starts falling. Assume the stick has a uniform mass distribution. Use the method of undetermined Lagrange multipliers to find (a) the speed of the center of mass (CM) of the stick as a function of the position of the CM (specified by the variables y and θ, where y is the distance of the CM from its initial position and θ measures the rotation angle of the stick from its initial vertical position about an axis through the CM), and (b) the reaction force on the stick as a function of the position of the CM.

Problem 31.4 A spherical ball (with uniform mass distribution) of radius ρ rolls without slipping inside a hemispherical bowl of radius $R > 2\rho$. Choose θ and ϕ to be the canonical coordinates, where θ is the angular displacement from the vertical of a line joining the center of the bowl to the center of the ball, and ϕ is the angle describing the rotational motion of the ball about an axis through its center. For small angles θ show that the frequency of oscillation of the ball about the stable position is given by

$$\omega \approx \sqrt{\frac{g(R-\rho)}{(R-\rho)^2 + \dfrac{2R^2}{5}}} \,.$$

Hints: Obtain the constraint relation between θ and ϕ and use the method of undetermined Lagrange multipliers.

Problem 31.5 Consider a balanced skate moving on a rough inclined plane, in such a way that the velocity of the center of mass of the skate is always along the direction of the length of the skate. Let the mass of the skate be m and the angle of inclination of the plane be θ. Choose a coordinate system so that the xy-plane is the inclined plane, the y-axis is along the horizontal direction, and the positive x-axis points downward. Let ϕ be the angle between the length of the skate and the x-axis. The constraint relation between the generalized coordinates, chosen to be x, y and ϕ, is [cf. Problem 10.3]

$$\sin\phi\,\dot{x} = \cos\phi\,\dot{y}\ . \tag{31.41}$$

(a) Show that the Lagrangian is given by

$$L(x, y, \phi;\, \dot{x}, \dot{y}, \dot{\phi}) = \frac{m}{2}\,(\dot{x}^2 + \dot{y}^2) + \frac{1}{2}\,I\dot{\phi}^2 + mg\sin\theta\,x\ , \tag{31.42}$$

where I is the moment of inertia of skate about a perpendicular axis through its center of mass and θ is a constant quantity.

(b) Show that the modified Euler-Lagrange equations are

$$\frac{\partial L}{\partial x} - \frac{d}{dt}\left(\frac{\partial L}{\partial \dot{x}}\right) + \lambda\sin\phi = 0\ , \tag{31.43a}$$

$$\frac{\partial L}{\partial y} - \frac{d}{dt}\left(\frac{\partial L}{\partial \dot{y}}\right) - \lambda\cos\phi = 0\ , \tag{31.43b}$$

$$\frac{\partial L}{\partial \phi} - \frac{d}{dt}\left(\frac{\partial L}{\partial \dot{\phi}}\right) = 0\ , \tag{31.43c}$$

where λ is the single Lagrange multiplier of this problem.

(c) Show that the above three equations reduce to the following set:

$$m\ddot{x} - \lambda\sin\phi = mg\sin\theta\ , \tag{31.44a}$$

$$m\ddot{y} + \lambda\cos\phi = 0\ , \tag{31.44b}$$

$$\ddot{\phi} = 0\ . \tag{31.44c}$$

These equations, together with the constraint relation (31.41), will determine the four unknowns $x(t), y(t), \phi(t)$ and λ of this problem.

(d) Show that a solution to the problem is given by

$$\phi(t) = \omega t \qquad (\phi(0) = 0)\ , \tag{31.45a}$$

$$x(t) = \frac{g\sin\theta}{2\omega^2}\,\sin^2\omega t\ , \tag{31.45b}$$

$$y(t) = \frac{g\sin\theta}{2\omega^2}\left(\omega t - \frac{1}{2}\,\sin(2\omega t)\right)\ . \tag{31.45c}$$

It is interesting to note that the motion along the x-direction is bounded so that the skate will not fall off the inclined plane.

Chapter 32

Small Oscillations and Normal Modes

We consider a mechanical system, described by the generalized coordinates $\{Q^i\}$, which possesses local minima of the potential energy function $U(Q^i)$. In the neighborhood of each local minimum, the system exhibits stable behavior; whereas near local maxima, unstable behavior results instead (as will be demonstrated below) (Fig. 32.1). The positions $Q_0 = \{Q_0^i\}$ (in configuration space) of the minima (and maxima) can be determined by the equations

$$\left(\frac{\partial U}{\partial Q^i}\right)_{Q_0} = 0\,, \qquad i = 1, \dots, n\,. \tag{32.1}$$

Near a local minimum, let

$$Q^i = Q_0^i + q^i\,. \tag{32.2}$$

We have, on Taylor expanding $U(Q^i)$,

$$U(Q^1, \dots, Q^n) = U(Q_0^1, \dots, Q_0^n) + \left(\frac{\partial U}{\partial Q^i}\right)_{Q_0} q^i + \frac{1}{2}\left(\frac{\partial^2 U}{\partial Q^i \partial Q^j}\right)_{Q_0} q^i q^j + \dots\,. \tag{32.3}$$

Setting $U(Q_0^1, \dots, Q_0^n) = 0$ and using (32.1) we can write

$$U(q^1, \dots, q^n) = \frac{1}{2} U_{ij}\, q^i q^j\,, \tag{32.4}$$

where

$$U_{ij} \equiv \left(\frac{\partial^2 U}{\partial q^i \partial q^j}\right)_{Q_0}\,. \tag{32.5}$$

The U matrix with elements U_{ij} is clearly symmetrical:

$$U_{ij} = U_{ji}\,. \tag{32.6}$$

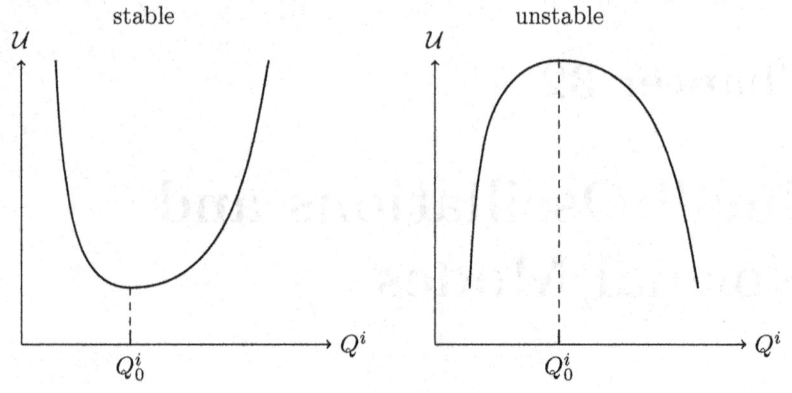

Fig. 32.1

Thus, near a local minimum, the small displacements q^i can be used as generalized coordinates of the system.

The kinetic energy $\mathcal{T}$ can in general be written as a quadratic form of the generalized velocities $\dot{Q}^i$:

$$\mathcal{T} = \frac{1}{2} m_{ij}(Q^1, \ldots, Q^n)\, \dot{Q}^i \dot{Q}^j \,, \tag{32.7}$$

where again, near Q_0^i, the m_{ij} can be expanded as

$$m_{ij}(Q^1, \ldots, Q^n) = m_{ij}(Q_0^i, \ldots, Q_0^n) + \left(\frac{\partial m_{ij}}{\partial Q^k}\right)_{Q_0} q^k + \ldots \,. \tag{32.8}$$

Since, in general, the $m_{ij}(Q_0^i)$ do not vanish, we will only keep the leading term in the above expansion and write

$$\mathcal{T} = \frac{1}{2} T_{ij}\, \dot{q}^i \dot{q}^j \,, \tag{32.9}$$

where

$$T_{ij} \equiv m_{ij}(Q_0^1, \ldots, Q_0^n) \,. \tag{32.10}$$

Just as for U_{ij}, T_{ij} is also symmetric, since in the sum over i and j in (32.7) each term is symmetric in an interchange between i and j. Thus

$$T_{ij} = T_{ji} \,. \tag{32.11}$$

Using the generalized coordinates $(q^1, \dots, q^n)$ the Lagrangian L can be expressed as

$$L(q^1, \dots, q^n) = \frac{1}{2}\left(T_{ij}\,\dot{q}^i\dot{q}^j - U_{ij}\,q^iq^j\right). \tag{32.12}$$

Using the symmetry conditions (32.6) and (32.11), it is easily seen that

$$\frac{\partial L}{\partial q^i} = -U_{ij}\,q^j\,, \qquad \frac{\partial L}{\partial \dot{q}^i} = T_{ij}\,\dot{q}^j\,. \tag{32.13}$$

Problem 32.1 Verify the equations in (32.13) by using the explicit expression for the Lagrangian given by (32.12).

The Euler-Lagrange equations (31.9) then lead to the following coupled equations of motion for the generalized coordinates q^i:

$$T_{ij}\,\ddot{q}^j + U_{ij}\,q^j = 0\,, \quad i = 1, \dots, n\,. \tag{32.14}$$

To solve these equations let

$$q^i(t) = a^i e^{-i\omega t}\,. \tag{32.15}$$

Substitution in (32.14) then yields

$$(U_{ij} - \omega^2 T_{ij})\,a^j = 0\,, \quad i = 1, \dots, n\,, \tag{32.16}$$

which is a set of n homogeneous algebraic equations in the n unknowns $a^1, \dots, a^n$. It is a well-known fact in linear algebra that for non-trivial solutions to exist, one must have

$$\det(U_{ij} - \omega^2 T_{ij}) = 0\,, \tag{32.17}$$

which is an n-th degree algebraic equation in ω^2. The *fundamental theorem of algebra* then implies that there are in general n solutions for ω^2, some of which may be repeated.

Let us now digress to look at the case when T_{ij} is proportional to the identity matrix; or, without loss of generality, when T is the identity matrix: $T_{ij} = \delta_{ij}$. (The latter condition can be achieved from the former by normalizing the generalized coordinates q^i.) Then

$$\mathcal{T} = \frac{1}{2}\delta_{ij}\,\dot{q}^i\dot{q}^j = \frac{1}{2}\sum_i (\dot{q}^i)^2\,, \tag{32.18}$$

and, writing $\omega^2 = \lambda$, (32.17) reduces to

$$U_{ij}\,a^j = \lambda a^i\,. \tag{32.19}$$

In matrix (operator) notation, the above equation can be written as

$$Ua = \lambda a \,, \tag{32.20}$$

where U represents either the matrix U_{ij} or the operator whose matrix representation is U_{ij} with respect to some specific choice of basis vectors in the n-dimensional linear vector space on which U acts; and a represents a vector in the same space, or a column vector with the n components $a^1, \ldots, a^n$ (with respect to the same basis set). Note that the left-hand side of (32.19) represents a matrix multiplication.

An equation of the form (32.20) in general defines what is called an **eigenvalue problem**, in which one is supposed to solve for all possible scalars λ and all possible vectors a satisfying the equation. The possible λ's and the possible a's are called the **eigenvalues** and **eigenvectors**, respectively, of the operator U. Analogous to (32.17), the eigenvalues are determined by the **characteristic equation**

$$\det(U - \lambda) = 0 \,, \tag{32.21}$$

where, on the left-hand side of the above equation, λ is understood to mean λI, with I being the identity matrix. Similar to (32.17), Eq. (32.21) is an n-th degree algebraic equation in the unknown λ, with n roots in general (some possibly repeated). Corresponding to each eigenvalue $\lambda^{(k)}$, we have an eigenvector $a^{(k)}$, so that (32.20) can be written

$$Ua^{(k)} = \lambda^{(k)} a^{(k)} \,. \tag{32.22}$$

Note that $a^{(k)}$ is not unique if $\lambda^{(k)}$ is a repeated root of the characteristic equation. If it occurs as a root m_k times then we say that the eigenvalue $\lambda^{(k)}$ has **multiplicity** m_k, or that it is m_k-fold degenerate.

Since U is a real symmetric matrix, a well-known theorem of linear algebra asserts that it is diagonalizable: there exists an orthonormal basis set with respect to which the matrix representation of U is diagonal. Furthermore, the basis vectors in this set are all eigenvectors of U. If the n eigenvalues are all distinct, then there is a unique normalized eigenvector corresponding to each eigenvalue. If a particular eigenvalue $\lambda^{(k)}$ has multiplicity m_k, then the **eigenspace** of $\lambda^{(k)}$ (that is, the *invariant subspace* under U consisting of all eigenvectors of U corresponding to the eigenvalue $\lambda^{(k)}$) is m_k-dimensional, and one can certainly choose m_k of these eigenvectors that are orthonormalized to be the basis of the eigenspace. This process is called **Schmidt orthogonalization**. Assume that we have carried out this process, and obtained an orthonormal basis set of n eigenvectors of U. Call these $a^{(j)}, j = 1, \ldots, n$. Define an $n \times n$ matrix A whose columns are made up of the column vectors $a^{(j)}$, that is,

$$A_{ij} \equiv (a^{(j)})^i \,, \tag{32.23}$$

where i and j are the row index and column index, respectively. For example, we may have a situation where $n = 8$, but there are only three distinct eigenvalues $\lambda^{(1)}, \lambda^{(2)}$ and $\lambda^{(3)}$, with $\lambda^{(1)}$ being nondegenerate (multiplicity $m_1 = 1$), $\lambda^{(2)}$

being 3-fold degenerate, and $\lambda^{(3)}$ being 4-fold degenerate. We may then arrange the Schmidt orthonormalized set $\{a^{(j)}\}$ so that $a^{(1)}$ is an eigenvector corresponding to the eigenvalue $\lambda^{(1)}$; $a^{(2)}, a^{(3)}$ and $a^{(4)}$ are eigenectors corresponding to the eigenvalue $\lambda^{(2)}$; and $a^{(5)}, a^{(6)}, a^{(7)}$ and $a^{(8)}$ are eigenvectors corresponding to the eigenvalue $\lambda^{(3)}$. Rename the n eigenvalues as λ^j, $j = 1, \dots, n$, so that if there are only p distinct eigenvalues $\lambda^{(1)}, \lambda^{(2)}, \dots, \lambda^{(p)}$ (with $p \leq n$) then

$$\begin{aligned} &\lambda^1 = \cdots = \lambda^{m_1} = \lambda^{(1)} , \\ &\lambda^{m_1+1} = \cdots = \lambda^{m_1+m_2} = \lambda^{(2)} , \\ &\vdots \\ &\lambda^{n-m_p+1} = \cdots = \lambda^n = \lambda^{(p)} , \end{aligned} \tag{32.24}$$

with

$$m_1 + m_2 + \cdots + m_p = n . \tag{32.25}$$

In the above example with $n = 8$, we would have $\lambda^1 = \lambda^{(1)}, \lambda^2 = \lambda^3 = \lambda^4 = \lambda^{(2)}$ and $\lambda^5 = \lambda^6 = \lambda^7 = \lambda^8 = \lambda^{(3)}$.

With this convention (32.22) can be rewritten as

$$\sum_l U_{il} A_{lj} = A_{ij}\lambda^j = \sum_l A_{il}\delta_{lj}\lambda^j , \tag{32.26}$$

where in the last two expressions j is not summed over. On further defining a diagonal matrix Λ whose diagonal elements are the eigenvalues λ^j, that is,

$$\Lambda_{ij} = \delta_{ij}\lambda^j , \tag{32.27}$$

where again, the i and j are row and column indices, respectively, (32.24) can be expressed as

$$\sum_l U_{il} A_{lj} = \sum_l A_{il} \Lambda_{lj} . \tag{32.28}$$

In matrix form this is

$$UA = A\Lambda . \tag{32.29}$$

Now since the columns of the matrix A represent orthonormalized vectors, $\det(A) \neq 0$ and A^{-1} must exist. It follows that the above equation is equivalent to

$$A^{-1}UA = \Lambda . \tag{32.30}$$

Thus A is the matrix which diagonalizes U.

We will now go back to the problem specified by (32.16) and write it in operator notation as

$$Ua = \lambda T a , \tag{32.31}$$

where $\lambda \equiv \omega^2$. Using the same notation introduced above (A_{ij} for the matrix of eigenvectors and λ^i for the possibly degenerate eigenvalues) we have

$$\begin{aligned}(UA)_{ij} &= \sum_l U_{il}A_{lj} = \sum_l \lambda^j T_{il}A_{lj} \quad \text{(no sum over } j\text{)} \\ &= \sum_l \lambda^j T_{il} \sum_m \delta_{mj}A_{lm} \quad \text{(no sum over } j\text{)} \\ &= \sum_{lm} T_{il}A_{lm}\Lambda_{mj} = (TA\Lambda)_{ij} \, .\end{aligned} \tag{32.32}$$

Equivalently we have the matrix equation [compare with (32.29)]

$$UA = TA\Lambda \, . \tag{32.33}$$

It will first be shown that the eigenvalues λ^i are all real and positive. From (32.32) we have

$$\sum_l U_{il}A_{lj} = \lambda^j \sum_l T_{il}A_{lj} \, . \tag{32.34}$$

Complex-conjugation (denoted below by overbars) of this equation yields, since U and T are both real,

$$\sum_l U_{il}\overline{A_{lj}} = \overline{\lambda^j} \sum_l T_{il}\overline{A_{lj}} \, . \tag{32.35}$$

Since U and T are also both symmetric, the above equation can be rewritten as

$$\sum_l U_{li}\overline{A_{lj}} = \overline{\lambda^j} \sum_l T_{li}\overline{A_{lj}} \, . \tag{32.36}$$

Now in (32.36), make the interchange of indices $i \leftrightarrow l$ and let $j \to k$. This leads to

$$\sum_i U_{il}\overline{A_{ik}} = \overline{\lambda^k} \sum_i T_{il}\overline{A_{ik}} \, . \tag{32.37}$$

Next multiply (32.34) by $\overline{A_{ik}}$ and sum over i; and then multiply (32.37) by A_{lj} and sum over l. The following equations result.

$$\sum_{il} U_{il}A_{lj}\overline{A_{ik}} = \lambda^j \sum_{il} T_{il}A_{lj}\overline{A_{ik}} \, , \tag{32.38}$$

$$\sum_{il} U_{il}A_{lj}\overline{A_{ik}} = \overline{\lambda^k} \sum_{il} T_{il}A_{lj}\overline{A_{ik}} \, . \tag{32.39}$$

Subtracting the latter from the former, we have

$$(\lambda^j - \overline{\lambda^k}) \sum_{il} T_{il}A_{lj}\overline{A_{ik}} = 0 \, . \tag{32.40}$$

Let $j = k$. Then

$$(\lambda^j - \overline{\lambda^j}) \sum_{il} T_{il}A_{lj}\overline{A_{ij}} = 0 \, . \tag{32.41}$$

The sum over i and l is real, since it can be seen that its complex conjugate is equal to itself:

$$\sum_{il} T_{il}\overline{A_{lj}}A_{ij} = \sum_{il} T_{li}A_{ij}\overline{A_{lj}} = \sum_{il} T_{il}A_{lj}\overline{A_{ij}} \,, \tag{32.42}$$

where we have used the facts that T is both real and symmetric, and the right-hand side of the last equality is obtained from the left-hand side by making the interchange $i \leftrightarrow l$. In fact, the sum in (32.41) is *positive definite.* This means that it is always real and positive, except when $A_{ij} = 0$ for all i. Hence (27.41) implies that $\lambda^j - \overline{\lambda^j} = 0$, and thus λ^j must be real for all j. Note that the sum in (32.41) can be interpreted as the square of the length of the vector $a^{(j)}$ [cf. (32.23)], with T_{il} playing the role of the *metric tensor.*

Problem 32.2 Show that the sum in (32.41) is positive definite.
Hint: Start by writing $A_{lj} = \alpha_{lj} + i\beta_{lj}$ and require the sum to be real. Then show that the sum in question reduces to the sum of two kinetic energy-like terms with non-zero velocities, which, on physical grounds, must be real and positive.

Now multiply (32.34) by A_{ij} and sum over i:

$$\sum_{il} U_{il}A_{lj}A_{ij} = \lambda^j \sum_{il} T_{il}A_{lj}A_{ij} \,, \tag{32.43}$$

which implies

$$\lambda^j = \frac{\sum_{il} U_{il}A_{ij}A_{lj}}{\sum_{il} T_{il}A_{ij}A_{lj}} \,. \tag{32.44}$$

According to (32.9) and (32.4), the denominator is a kinetic energy-like term (always positive definite) while the numerator is a potential energy-like term (positive definite near local minima). Hence (32.44) implies that the eigenvalues λ^j are always positive. We have thus shown that *the eigenvalues λ^j are all both real and positive.* Since $\lambda = \omega^2$, Eq. (32.15) guarantees the stability of small oscillations near local minima of the potential function.

The reality of all eigenvalues allows us to choose all the $A_{ik} = (a^{(k)})^i$, that is, all components of eigenvectors, to be real also. Thus (32.40) can be rewritten as

$$(\lambda^j - \lambda^k)\sum_{il} T_{il}A_{lj}A_{ik} = 0 \,. \tag{32.45}$$

For j and k corresponding to distinct eigenvalues ($\lambda^j \neq \lambda^k$), it is clear from the above equation that

$$\sum_{il} T_{il}A_{lj}A_{ik} = \sum_{il} (A^T)_{ki}T_{il}A_{lj} = (A^T T A)_{kj} = 0 \,, \tag{32.46}$$

where A^T is the transpose matrix of A. Note that this is a statement that, under the metric T_{il}, the scalar product of the eigenvectors a^j and a^k vanishes (the eigenvectors are orthogonal).

If $\lambda^j = \lambda^k = \lambda^{(p)}$ (the p-th distinct eigenvalue) is m_p-fold degenerate (where $m_p > 1$), then, as in the eigenvalue problem (32.19) [equivalently (32.20)], the eigenspace consisting of all eigenvectors corresponding to the eigenvalue $\lambda^{(p)}$ is m_p-dimensional, and one can use the Schmidt orthogonalization process to select m_p of these eigenvectors that are orthonormalized to be the basis of the eigenspace. Assume that this process has been carried out for all eigenspaces corresponding to degenerate eigenvalues. Then, with $A_{ij} = (a^{(j)})^i$, where all the $a^{(j)}$, $j = 1, \ldots, n$ are orthonormalized, we have

$$(A^T T A)_{kj} = \delta_{kj} \; , \quad k, j = 1, \ldots, n \; , \tag{32.47}$$

or, in matrix notation,

$$A^T T A = 1 \; , \tag{32.48}$$

where the 1 on the right-hand side represents the identity matrix. Equation (32.48) expresses the orthonormality condition of the basis set $a^{(j)}$, $j = 1, \ldots, n$, under the metric T_{ij}.

Problem 32.3 Show that all eigenvectors corresponding to a degenerate eigenvalue for the problem specified by (32.31) constitute a subspace of the vector space on which U and T act as linear operators.

Problem 32.4 Show explicitly how to Schmidt orthonormalize a basis in an eigenspace of dimension 3.

With the orthonormality condition (32.48) for A, Eq. (32.33) is equivalent to, on multiplying by A^T on the left,

$$A^T U A = \Lambda \; , \tag{32.49}$$

where Λ is the diagonal matrix whose diagonal elements are precisely the eigenvalues (occurring with multiplicities) [recall (32.27)]. Thus U is diagonalized by the matrix A whose columns are the orthonormalized eigenvectors of the eigenvalue problem given by (32.31).

We are now ready to return to the solutions of (32.14). From (32.15) and our development up to this point we can write a general solution as

$$q^i(t) = \sum_k a^{(k)i} e^{-i\omega_k t} \; , \tag{32.50}$$

where the sum is over all the eigenvalues ω_k determined by (32.16) (with degeneracies) and $a^{(k)}$ are the corresponding orthonormalized eigenvectors. The vibrational motion of the system with all the q^i having the same frequency equal to a particular eigenvalue ω_k is called a **normal mode** of vibration. Let us now define a new set of generalized coordinates, ζ^i, obtained from the q^i by the following linear transformation:

$$q^i \equiv A_{ij}\zeta^j \,, \tag{32.51}$$

where the A_{ij} has been defined by (32.23). In matrix notation we can write

$$q = A\zeta \,, \tag{32.52}$$

where q and ζ are both column matrices. We will see that using the new coordinates ζ^i, both the kinetic and potential energies can be expressed as linear superpositions of decoupled terms each corresponding to a normal mode of vibration.

Recalling (32.4) we can write

$$\mathcal{U} = \frac{1}{2}U_{ij}q^i q^j = \frac{1}{2}\begin{pmatrix} q^1 & \dots & \dots & q^n \end{pmatrix} \begin{pmatrix} U_{11} & U_{12} & \dots & U_{1n} \\ U_{21} & U_{22} & \dots & \dots \\ & \vdots & & \\ U_{n1} & \dots & \dots & U_{nn} \end{pmatrix} \begin{pmatrix} q^1 \\ \vdots \\ \vdots \\ q^n \end{pmatrix} , \tag{32.53}$$

or, in compact matrix notation,

$$\mathcal{U} = q^T U q \,, \tag{32.54}$$

where q^T is the transpose of the column matrix q, in other words, the row matrix appearing on the right-hand side of (32.53). From (32.52), on the other hand,

$$q^T = \zeta^T A^T \,. \tag{32.55}$$

It follows that

$$\mathcal{U} = \frac{1}{2}\zeta^T A^T U A \zeta = \frac{1}{2}\zeta^T \Lambda \zeta \,, \tag{32.56}$$

where we have used (32.49) and the diagonal matrix Λ is defined by (32.27). In component form, (32.56) appears as

$$\mathcal{U} = \frac{1}{2}\sum_k \omega_k^2\, \zeta_k^2 \,. \tag{32.57}$$

Similarly, from (32.9), the kinetic energy can be written

$$\mathcal{T} = \dot{q}^T T \dot{q} = \frac{1}{2}\,\dot{\zeta}^T A^T T A \dot{\zeta} = \frac{1}{2}\,\dot{\zeta}^T \dot{\zeta} \,, \tag{32.58}$$

where we have used the normalization condition (32.48). Equivalently, in component form,

$$\mathcal{T} = \frac{1}{2}\sum_k (\dot{\zeta}_k)^2 . \tag{32.59}$$

The coordinates $\zeta_k(t)$ are called the **normal coordinates** of the vibrating system. Equations (32.57) and (32.59) demonstrate that, in terms of the normal coordinates, both the kinetic energy and potential energy can be expressed as a sum of decoupled terms.

In place of (32.12) the Lagrangian can be written in terms of the normal coordinates as

$$L = \mathcal{T} - \mathcal{U} = \frac{1}{2}\sum_k \{(\dot{\zeta}_k)^2 - \omega_k^2 \zeta_k^2\} . \tag{32.60}$$

The Euler-Lagrange equations

$$\frac{d}{dt}\left(\frac{\partial L}{\partial \dot{\zeta}_k}\right) = \frac{\partial L}{\partial \zeta_k} , \quad k = 1, \ldots, n , \tag{32.61}$$

become

$$\ddot{\zeta}_k + \omega_k^2 \zeta_k = 0 , \quad k = 1, \ldots, n . \tag{32.62}$$

Thus, *in terms of the normal coordinates, the vibrating system behaves like a set of decoupled oscillators with different frequencies.* The solutions of the above equations are trivially given by

$$\zeta_k(t) = c_k \, e^{-i\omega_k t} , \quad k = 1, \ldots, n . \tag{32.63}$$

As an example of an application of the theory developed in this chapter we consider the linear triatomic molecule (Fig. 32.2). The interatomic forces are modeled by springs with spring constant k, and the equilibrium distances between the atoms are all equal to b. From left to right in Fig. 32.2, the atomic masses are m, M and m. The position coordinates Q^1, Q^2, Q^3 of the atoms are measured from some arbitrary origin (on the line of the molecule).

Suppose the minimum of the potential energy $\mathcal{U}$ occurs at $\{Q_0^1, Q_0^2, Q_0^3\}$. Then, obviously,

$$Q_0^2 - Q_0^1 = Q_0^3 - Q_0^2 = b . \tag{32.64}$$

Let q^1, q^2 and q^3 be defined as in (32.2). Then

$$\begin{aligned}
\mathcal{U} &= \frac{k}{2}(Q^2 - Q^1 - b)^2 + \frac{k}{2}(Q^3 - Q^2 - b)^2 \\
&= \frac{k}{2}\{(q^2 - q^1)^2 + (q^3 - q^2)^2\} \\
&= \frac{k}{2}\{(q^1)^2 + 2(q^2)^2 + (q^3)^2 - 2q^1q^2 - 2q^2q^3\} \\
&\equiv \frac{1}{2} U_{ij} q^i q^j .
\end{aligned} \tag{32.65}$$

Thus the matrix U with matrix elements U_{ij} is given explicitly by

$$U = k \begin{pmatrix} 1 & -1 & 0 \\ -1 & 2 & -1 \\ 0 & -1 & 1 \end{pmatrix} . \tag{32.66}$$

The kinetic energy is

$$\mathcal{T} = \frac{m}{2}\{(\dot{q}^1)^2 + (\dot{q}^3)^2\} + \frac{M}{2}(\dot{q}^2)^2 \equiv \frac{1}{2} T_{ij}\dot{q}^i\dot{q}^j . \tag{32.67}$$

Thus the matrix T is given explicitly by

$$T = \begin{pmatrix} m & 0 & 0 \\ 0 & M & 0 \\ 0 & 0 & m \end{pmatrix} . \tag{32.68}$$

The characteristic equation (32.17) is then

$$\det(U - \omega^2 T) = \begin{vmatrix} k - \omega^2 m & -k & 0 \\ -k & 2k - \omega^2 M & -k \\ 0 & -k & k - \omega^2 m \end{vmatrix} = 0 , \tag{32.69}$$

which is equivalent to the following third-degree algebraic equation in ω^2:

$$\omega^2 (k - \omega^2 m)\{\omega^2 Mm - k(M + 2m)\} = 0 . \tag{32.70}$$

Problem 32.5 Check that the algebraic equation (32.70) follows from the characteristic equation (32.69).

We then have the following distinct values for the eigenfrequencies:

$$\omega_1 = 0 , \qquad \omega_2 = \sqrt{\frac{k}{m}} , \qquad \omega_3 = \sqrt{\frac{k}{m}\left(1 + \frac{2m}{M}\right)} . \tag{32.71}$$

The eigenvector $a^{(j)}$, with components $(a^{(j)})^i \equiv A_{ij}$, corresponding to the j-th eigenvalue ω_j is determined up to a normalization constant by the following system of equations [c.f. (32.16)]:

$$(k - (\omega_j)^2 m)A_{1j} - kA_{2j} = 0 \,, \tag{32.72a}$$

$$-kA_{1j} + (2k - (\omega_j)^2 M)A_{2j} - kA_{3j} = 0 \,, \tag{32.72b}$$

$$-kA_{2j} + (k - (\omega_j)^2 m)A_{3j} = 0 \,. \tag{32.72c}$$

The normalization constants are fixed by the normalization condition (32.48), which, in the present context, gives

$$\begin{pmatrix} A_{11} & A_{21} & A_{31} \\ A_{12} & A_{22} & A_{32} \\ A_{13} & A_{23} & A_{33} \end{pmatrix} \begin{pmatrix} m & 0 & 0 \\ 0 & M & 0 \\ 0 & 0 & m \end{pmatrix} \begin{pmatrix} A_{11} & A_{12} & A_{13} \\ A_{21} & A_{22} & A_{23} \\ A_{31} & A_{32} & A_{33} \end{pmatrix} = \begin{pmatrix} 1 & 0 & 0 \\ 0 & 1 & 0 \\ 0 & 0 & 1 \end{pmatrix} \,. \tag{32.73}$$

Equating the diagonal elements on both sides we have

$$m(A_{11}^2 + A_{31}^2) + MA_{21}^2 = 1 \,, \tag{32.74a}$$

$$m(A_{12}^2 + A_{32}^2) + MA_{22}^2 = 1 \,, \tag{32.74b}$$

$$m(A_{13}^2 + A_{33}^2) + MA_{23}^2 = 1 \,. \tag{32.74c}$$

Problem 32.6 Check that all three equations in (32.74) follow from (32.73).

The normal mode $\omega_1 = 0$ clearly corresponds to rigid translation of the molecule along its axis (Fig. 32.3a). For this mode, the solution for (32.72) is

$$A_{11} = A_{21} = A_{31} \,. \tag{32.75}$$

The normalization condition (32.74a) then fixes these components of the eigenvector to be

$$A_{11} = A_{21} = A_{31} = \frac{1}{\sqrt{2m + M}} \,. \tag{32.76}$$

For the normal mode $\omega_2 = \sqrt{k/m}$, Eq.(32.72a) [or (32.72c)] implies

$$A_{22} = 0 \,. \tag{32.77}$$

It then follows from (32.72b) that

$$A_{12} = -A_{32} \,. \tag{32.78}$$

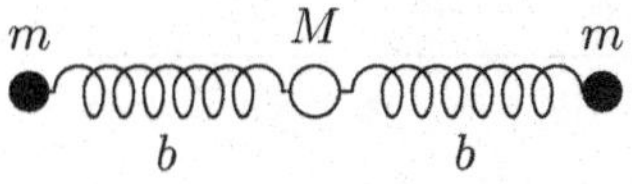

Fig. 32.2

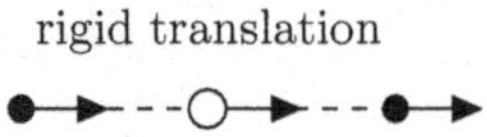

Fig. 32.3a

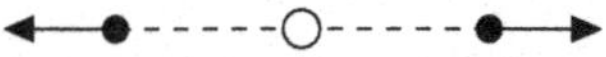

Fig. 32.3b

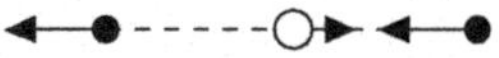

Fig. 32.3c

Equation (32.74b) then normalizes the components of the eigenvector to be

$$A_{12} = -A_{32} = \frac{1}{\sqrt{2m}} \,, \qquad A_{22} = 0 \,. \tag{32.79}$$

The normal mode ω_2 is illustrated in Fig. 32.3b.

For the normal mode ω_3 [cf. (32.71)], Eqs. (32.72a) and (32.72c) imply that

$$A_{13} = A_{33} \,. \tag{32.80}$$

Equation (32.72b) then shows that

$$\frac{A_{23}}{A_{13}} = -\frac{2m}{M} \,. \tag{32.81}$$

The normalization condition (32.74c) then yields

$$A_{13} = A_{33} = \frac{1}{\sqrt{2m\left(1 + \dfrac{2m}{M}\right)}} \,, \qquad A_{23} = \frac{-2}{\sqrt{2M\left(2 + \dfrac{M}{m}\right)}} \,. \tag{32.82}$$

The normal mode ω_3 is illustrated in Fig. 32.3c.

Note that in principle each of the solutions for the eigenvectors given by (32.76), (32.79) and (32.82) is only determined up to a complex constant of absolute value 1. If we require all the solutions to be real, as we have done, then each of the solutions can be multiplied by a factor of -1 without changing the physics.

Problem 32.7 Verify (32.76), (32.79) and (32.82) for the normalization constants.

Chapter 33

The Hamiltonian Formulation of Classical Mechanics: Hamilton's Equations of Motion

The Lagrangian formulation of classical mechanics discussed in Chapter 31 is based on the Lagrangian function $L(q^i, \dot{q}^i)$, which is a function on the *tangent bundle* of the configuration manifold. In this chapter we introduce the *Hamiltonian formulation of classical mechanics*, which is based on the *Hamiltonian function*, a function on the *cotangent bundle* of the configuration manifold. We recall (cf. Chapter 5) that a cotangent space at a point in the configuration manifold is the dual vector space to the tangent space at that point; and that the cotangent bundle is the union of cotangent spaces at all points of the configuration manifold, with an induced differentiable structure from the tangent bundle.

To construct a function on the cotangent bundle from a given function on the tangent bundle, we first introduce the *Legendre Transform* of a function of a single variable as follows. Let $y = f(x)$ be a *convex function*: $f''(x) > 0$. The **Legendre transform** of $f(x)$ is a function $g(p)$ (of a new variable p) constructed as follows. Choose a number p. Draw the straight line $y = px$. Then find the point $x(p)$ on the x-axis such that, at that point, the vertical distance between the straight line $y = px$ and the curve $y = f(x)$ is a maximum (Fig. 33.1).

Let

$$F(p, x) = px - f(x) \,. \tag{33.1}$$

Then $x(p)$ is determined by the condition $\partial F/\partial x = 0$, that is

$$p - f'(x) = 0 \,. \tag{33.2}$$

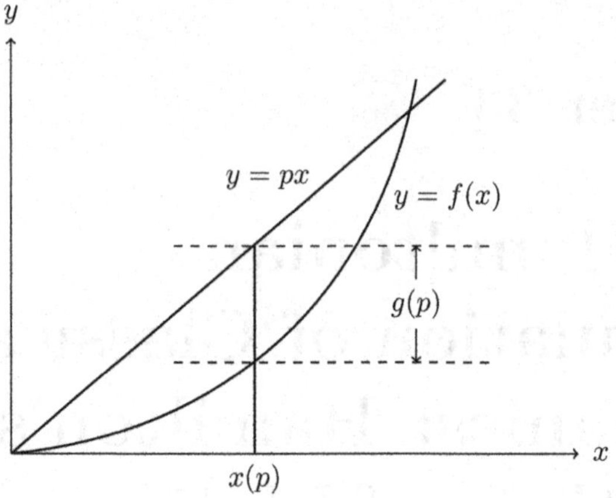

Fig. 33.1

This equation can be solved for x in terms of p. Since $f(x)$ is convex, the solution is unique. Then the Legendre transform $g(p)$ of $f(x)$ is defined by

$$g(p) \equiv F(p, x(p)) \,. \tag{33.3}$$

As a simple example consider $f(x) = mx^2/2$. Let $F(p, x) = px - mx^2/2$. Then $\partial F/\partial x = 0$ implies $p - mx = 0$, or $x = p/m$. Thus

$$g(p) = F(p, x(p)) = p\left(\frac{p}{m}\right) - \frac{m}{2}\left(\frac{p}{m}\right)^2 = \frac{p^2}{2m} \,. \tag{33.4}$$

We state the following useful theorem without proof.

Theorem 33.1. *The Legendre transform is* **involutive**, *that is, if $g(p)$ is the Legendre transform of $f(x)$, then $f(x)$ is also the Legendre transform of $g(p)$.*

We can generalize the above definition of the Legendre transform to the case of a function of several variables. Let $f(x^1, \dots, x^n)$ be convex, that is, the quadratic form

$$\frac{\partial^2 f}{\partial x^i \partial x^j}\, dx^i dx^j$$

is positive definite; and let

$$F(p_1, \dots, p_n\,;\, x^1, \dots, x^n) \equiv p_i x^i - f(x^1, \dots, x^n) \,. \tag{33.5}$$

Determine the n functions

$$x^1(p_1, \dots, p_n) \,, \quad \dots \quad , x^n(p_1, \dots, p_n)$$

by the conditions

$$\frac{\partial F}{\partial x^1} = 0 \,, \quad \dots \quad , \frac{\partial F}{\partial x^n} = 0 \,, \tag{33.6}$$

which imply

$$p_1 = \frac{\partial f}{\partial x^1} \,, \quad \dots \quad , p_n = \frac{\partial f}{\partial x^n} \,. \tag{33.7}$$

Then the Legendre transform $g(p_1, \dots, p_n)$ of $f(x^1, \dots, x^n)$ is given by

$$g(p_1, \dots, p_n) = F(p_1, \dots, p_n \,;\, x^1(p_1, \dots, p_n), \dots, x^n(p_1, \dots, p_n)) \,. \tag{33.8}$$

If $(x^1, \dots, x^n)$ are considered as components of a vector in a given vector space, the $(p_1, \dots, p_n)$ are the components of a *covector* in the dual space of the given vector space (Recall Def. 1.2). *Thus Legendre transforms turn functions on a given vector space to functions on the dual space.*

Let us now return to the Euler-Lagrange equations of classical mechanics [cf. (31.9)]:

$$\frac{d}{dt}\left(\frac{\partial L}{\partial \dot{q}^i}\right) - \frac{\partial L}{\partial q^i} = 0 \,, \quad i = 1, \dots, n \,, \tag{33.9}$$

where n is the total number of degrees of freedom. Assume that the Lagrangian $L(q, \dot{q})$ is a convex function with respect to each $\dot{q}^i$, namely, the quadratic form

$$\frac{\partial L}{\partial \dot{q}^i \partial \dot{q}^j}\, d\dot{q}^i d\dot{q}^j$$

is positive definite. Define the canonical momenta p_i by [cf. (31.10)]

$$p_i \equiv \frac{\partial L}{\partial \dot{q}^i} \,, \quad i = 1, \dots, n \,. \tag{33.10}$$

As indicated in Chapter 31, the p_i are local coordinates on the cotangent bundle of the configuration manifold, whose local coordinates are q^i. We can arrive at this conclusion in a more mathematically precise way as follows. At a given point $q \in Q$ of the configuration manifold Q, one considers the map $\lambda_q : T_q Q \to T_q^* Q$ from the tangent space $T_q Q$ to the cotangent space $T_q^* Q$, defined by differentiating the restriction of the Langrangian function $L(q, \dot{q})$ to $T_q Q$ (restricted from the whole tangent bundle $T_* Q$ to the tangent space at q):

$$\lambda_q(\dot{q}) \equiv \partial_{\dot{q}} L(q, \dot{q}) = (\partial_{\dot{q}^1} L, \dots, \partial_{\dot{q}^n} L) = (p_1, \dots, p_n) \,.$$

It is clear that

$$dL|_{T_q Q} = \frac{\partial L}{\partial \dot{q}^i}\, d\dot{q}^i = p_i(q; \dot{q}^1, \dots, \dot{q}^n)\, d\dot{q}^i$$

is a one-form on the tangent space $T_q Q$, and so must be a smooth rank-one covariant [or rank $(0,1)$] tansor field on $T_q Q$. In other words, λ_q maps a vector in the tangent space T_qQ to one in the cotangent space T_q^*Q. Equation (33.10) can be inverted to solve for the $\dot{q}^i$ uniquely in terms of the p_i and q^i [λ_q is invertible under the condition that $L(q,\dot{q})$ is convex with respect to the $\dot{q}$]. We then obtain the **Hamiltonian function**

$$H(p,q) = p\,(\lambda_q)^{-1}(p) - L(q,(\lambda_q)^{-1}(p))\,, \quad \text{for all } p \in T_q^*Q\,,$$

that is,

$$\boxed{H(p_i,q^i;t) = p_i\dot{q}^i(p_j,q^j) - L(q^i,\dot{q}^i(p_j,q^j);t)}\,, \tag{33.11}$$

which is the Legendre transform of the Lagrangian $L(q^i,\dot{q}^i;t)$ with respect to the $\dot{q}^i$. In the above equation all occurrences of $\dot{q}^i$ on the right-hand side are expressed in terms of $\{p_i,q^i\}$ through (33.10). Thus, as discussed above, *The Hamiltonian H is a function on the cotangent bundle of the configuration manifold.* This cotangent bundle, whose local coordinates are (q^i,p_i), is sometimes called **phase space** in physics (especially in *statistical mechanics*).

Now

$$dH = \frac{\partial H}{\partial p_i}\,dp_i + \frac{\partial H}{\partial q^i}\,dq^i + \frac{\partial H}{\partial t}\,dt\,. \tag{33.12}$$

On the other hand, by (33.11),

$$\begin{aligned} d(p_i\dot{q}^i - L) &= p_i d\dot{q}^i + \dot{q}^i dp_i - \frac{\partial L}{\partial q^i}\,dq^i - \frac{\partial L}{\partial \dot{q}^i}\,d\dot{q}^i - \frac{\partial L}{\partial t}\,dt \\ &= \dot{q}^i dp_i - \frac{\partial L}{\partial q^i}\,dq^i - \frac{\partial L}{\partial t}\,dt\,, \end{aligned} \tag{33.13}$$

where on the right-hand side of the first equality we have set $\partial L/\partial\dot{q}^i = p_i$. Comparing the coefficients of dp_i, dq^i and dt in the above two equations, we have

$$\dot{q}^i = \frac{\partial H}{\partial p_i}\,, \quad -\frac{\partial L}{\partial q^i} = \frac{\partial H}{\partial q^i}\,, \quad \frac{\partial H}{\partial t} = -\frac{\partial L}{\partial t}\,. \tag{33.14}$$

By the Euler-Lagrange equations (33.9) and the definition of the canonical momenta p_i given by (33.10), we have

$$\frac{\partial L}{\partial q^i} = \frac{d}{dt}\left(\frac{\partial L}{\partial \dot{q}^i}\right) = \dot{p}_i\,. \tag{33.15}$$

The first two equations of (33.14) then become

$$\boxed{\frac{dq^i}{dt} = \frac{\partial H}{\partial p_i}\,, \quad \frac{dp_i}{dt} = -\frac{\partial H}{\partial q^i}\,, \qquad i = 1,\dots,n}\,. \tag{33.16}$$

These are called **Hamilton's equations of motion**. They apply not only to mechanical systems, in which $L = T - U$, but to general Lagrangians. We see that the Euler-Lagrange equations (33.9), which is a system of n second order differential equations, are equivalent to the Hamilton's equations, a system of $2n$ first order differential equations.

For further development, we will need the following lemma.

Lemma 33.1. *Suppose $f(x^1, \dots, x^n)$ is the quadratic form*

$$f(x^1, \dots, x^n) = a_{ij}\, x^i x^j \,. \tag{33.17}$$

Then

$$f(x^1, \dots, x^n) = g(p_1, \dots, p_n) \,, \tag{33.18}$$

where $g(p_1, \dots, p_n) = p_i x^i - f$ is the Legendre transform of f, and the $\{x^i\}$ and $\{p_i\}$ are related by (33.7).

Proof. We use **Euler's theorem** on homogeneous functions: Suppose $f(x^1, \dots, x^n)$ is a *homogeneous function* of degree k, that is, f is of the form

$$f(x^1, \dots, x^n) = a_{i_1 \dots i_k}\, x^{i_1} \dots x^{i_k} \,. \tag{33.19}$$

Then

$$\frac{\partial f}{\partial x^i}\, x^i = kf \,. \tag{33.20}$$

We thus have

$$\begin{aligned} g(p_1, \dots, p_n) &= p_i x^i - f(x^1, \dots, x^n) = \frac{\partial f}{\partial x^i} x^i - f(x^1, \dots, x^n) \\ &= 2f - f = f(x^1, \dots, x^n) \,, \end{aligned} \tag{33.21}$$

since by assumption, f is a quadratic form and hence of degree 2. □

Theorem 33.2. *For mechanical systems, in which the Lagrangian is given by $L = T - U$, where the kinetic energy T is a quadratic form:*

$$T = \frac{1}{2}\, a_{ij}(q^1, \dots, q^n)\, \dot{q}^i \dot{q}^j \,, \tag{33.22}$$

and the potential energy U is a function only of the canonical coordinates:

$$U = U(q^1, \dots, q^n) \,, \tag{33.23}$$

the Hamiltonian H is the total energy:

$$H = T + U \,. \tag{33.24}$$

Proof. We have

$$\begin{aligned} H &= p_i \dot{q}^i - L = \frac{\partial L}{\partial \dot{q}^i}\, \dot{q}^i - L = \frac{\partial T}{\partial \dot{q}^i}\, \dot{q}^i - L \\ &= 2T - L = 2T - (T - U) = T + U \,, \end{aligned} \tag{33.25}$$

where in the fourth equality we have used Euler's theorem (33.20). □

As a simple example of the above theorem, consider the one-dimensional problem specified by

$$T = \frac{1}{2} m\dot{q}^2 , \quad U = U(q) . \tag{33.26}$$

Then

$$p = \frac{\partial L}{\partial \dot{q}} = \frac{\partial T}{\partial \dot{q}} = m\dot{q} \tag{33.27}$$

and

$$H = p\dot{q} - L = \frac{p^2}{m} - (T - U) = \frac{p^2}{m} - \left(\frac{1}{2} m \left(\frac{p}{m} \right)^2 - U \right) , \tag{33.28}$$

that is,

$$H = \frac{p^2}{2m} + U(q) . \tag{33.29}$$

Hamiltonian's equation of motion then take the form

$$\dot{q} = \frac{\partial H}{\partial p} = \frac{p}{m} , \qquad \dot{p} = -\frac{\partial H}{\partial q} = -\frac{\partial U}{\partial q} , \tag{33.30}$$

which together imply

$$m\ddot{q} = -\frac{\partial U}{\partial q} . \tag{33.31}$$

This is precisely Newton's equation of motion.

Theorem 33.3. *Suppose the Hamiltonian H is equal to the total energy. If H does not depend explicitly on the time t: $\partial H/\partial t = 0$, then the total energy is conserved.*

Proof. We write

$$dH = \sum_i \left(\frac{\partial H}{\partial q^i} dq^i + \frac{\partial H}{\partial p_i} dp_i \right) + \frac{\partial H}{\partial t} dt . \tag{33.32}$$

Then, from Hamilton's equations (33.16), we have

$$\begin{aligned} \frac{dH}{dt} &= \sum_i \left(\frac{\partial H}{\partial q^i} \dot{q}^i + \frac{\partial H}{\partial p_i} \dot{p}_i \right) + \frac{\partial H}{\partial t} \\ &= \sum_i \left\{ \left(\frac{\partial H}{\partial q^i} \right) \left(\frac{\partial H}{\partial p_i} \right) + \left(\frac{\partial H}{\partial p_i} \right) \left(-\frac{\partial H}{\partial q^i} \right) \right\} + \frac{\partial H}{\partial t} = \frac{\partial H}{\partial t} . \end{aligned} \tag{33.33}$$

The theorem follows. □

Recall that q^i is cyclic if $\partial L/\partial q^i = 0$. By (33.11) it is easily seen that q^i is cyclic if and only if $\partial H/\partial q^i = 0$. The second of the Hamilton's equations (33.16) then implies that *if q^i is cyclic then p_i is conserved.* A conserved quantity is called a **constant of motion** or a **first integral** of a classical mechanical system. By Theorem 33.3, if $\partial H/\partial t = 0$ (the Hamiltonian does not depend explicitly on time) then the total energy H is a constant of motion.

If, for a classical mechanical system with n degrees of freedom, one can find n functionally independent constants of motion satisfying certain additional mathematical conditions (which will be introduced in Chapter 36), the system is said to be **integrable**, in the sense that Hamilton's equations can be solved completely by algebraic operations and quadratures (evaluations of integrals). It is a fact of life, however, that most systems are not integrable. The general strategy to solve a classical mechanical problem within the framework of the Hamiltonian formulation is to make use of some symmetry properties of the system (for example, spherical symmetry of the potential in a central field problem) to choose a system of generalized coordinates q^i such that, when expressed in terms of these coordinates, H is independent of as many of them as possible. In other words, we try to find as many cyclic coordinates as possible. The corresponding generalized momenta are then first integrals. The problem is then reduced to one with a smaller number of coordinates (degrees of freedom). Integrable systems and the conditions for integrability will be considered in more detail from a mathematical viewpoint in Chapter 36.

Let us consider again the problem of the motion of a particle of mass m moving in a central potential field. Using the polar coordinates r and θ as the generalized coordinates, the Lagrangian L, and the canonical momenta p_r and p_θ, are given by (31.12) and (31.13) respectively. The latter imply

$$\dot{r} = \frac{p_r}{m}, \qquad \dot{\theta} = \frac{p_\theta}{mr^2}. \tag{33.34}$$

It follows from (33.11) that the Hamiltonian is given by

$$H(p_r, p_\theta; r, \theta) = \frac{p_r^2}{2m} + \frac{p_\theta^2}{2mr^2} + U(r). \tag{33.35}$$

Problem 33.1 Derive the above equation from (33.11).

Note that θ is a cyclic coordinate. Hamilton's equations are then

$$\dot{r} = \frac{\partial H}{\partial p_r} = \frac{p_r}{m}, \tag{33.36}$$

$$\dot{\theta} = \frac{\partial H}{\partial p_\theta} = \frac{p_\theta}{mr^2}, \tag{33.37}$$

$$\dot{p}_r = -\frac{\partial H}{\partial r} = \frac{p_\theta^2}{mr^3} - \frac{dU}{dr}, \tag{33.38}$$

$$\dot{p}_\theta = -\frac{\partial H}{\partial \theta} = 0. \tag{33.39}$$

The last equation implies that the angular momentum p_θ is a constant of motion. We thus have two constants of motion:

$$H = T + U = E \quad \text{(total energy)} , \tag{33.40}$$

$$p_\theta = mr^2\dot{\theta} = L \quad \text{(angular momentum)} . \tag{33.41}$$

Since the system is two-dimensional, it is integrable (see Theorem 36.3). By (33.36) we can write

$$\frac{p_r^2}{2m} = \frac{m\dot{r}^2}{2} . \tag{33.42}$$

Then (33.40) implies

$$\frac{m\dot{r}^2}{2} = E - \left(\frac{L^2}{2mr^2} + U(r) \right) , \tag{33.43}$$

or equivalently,

$$\frac{dr}{dt} = \sqrt{\frac{2}{m}\left\{ E - \left(U(r) + \frac{L^2}{2mr^2} \right) \right\}} . \tag{33.44}$$

The classical trajectory $r(t)$ can then be found by direct integration. Once $r(t)$ is found, $\dot{\theta} = L/(mr^2)$ can be integrated also. This example illustrates the fact that *every two-dimensional system which has a cyclic coordinate is integrable.*

As another example consider the one-dimensional simple harmonic oscillator with generalized coordinate q such that the Lagrangian is given by

$$L = \frac{1}{2}(\dot{q}^2 - q^2) . \tag{33.45}$$

Then the generalized momentum p is given by

$$p = \frac{\partial L}{\partial \dot{q}} = \dot{q} , \tag{33.46}$$

and the Hamiltonian H is

$$H = p\dot{q} - L = p^2 - \frac{1}{2}(p^2 - q^2) , \tag{33.47}$$

or

$$H(p, q) = \frac{1}{2}(p^2 + q^2) . \tag{33.48}$$

Thus Hamilton's equations

$$\dot{q} = \frac{\partial H}{\partial p} , \quad \dot{p} = -\frac{\partial H}{\partial q} , \tag{33.49}$$

lead to

$$\dot{q} = p , \quad \dot{p} = -q . \tag{33.50}$$

In matrix notation the above two equations can be written as the single one

$$\frac{d}{dt}\begin{pmatrix} q \\ p \end{pmatrix} = \begin{pmatrix} 0 & 1 \\ -1 & 0 \end{pmatrix}\begin{pmatrix} q \\ p \end{pmatrix} . \tag{33.51}$$

This is recognized as a dynamical system of type (2) in Theorem 18.6 (with $a = 0$ and $b = -1$). From (18.21) the solution is given by

$$\begin{pmatrix} q(t) \\ p(t) \end{pmatrix} = \begin{pmatrix} \cos t & \sin t \\ -\sin t & \cos t \end{pmatrix}\begin{pmatrix} q(0) \\ p(0) \end{pmatrix} . \tag{33.52}$$

It follows that

$$q(t) = \cos t\, q(0) + \sin t\, p(0) , \tag{33.53}$$
$$p(t) = -\sin t\, q(0) + \cos t\, p(0) . \tag{33.54}$$

Clearly,

$$p^2 + q^2 = (p(0))^2 + (q(0))^2 . \tag{33.55}$$

Equation (33.48) then implies that the total energy is conserved.

Problem 33.2 If

$$A = \begin{pmatrix} 0 & 1 \\ -1 & 0 \end{pmatrix} , \tag{33.56}$$

show that

$$e^{At} = \cos t + A \sin t , \tag{33.57}$$

by computing explicitly

$$e^{At} = 1 + At + \frac{A^2t^2}{2!} + \frac{A^3t^3}{3!} + \dots , \tag{33.58}$$

and observing that

$$A^2 = -1 , \quad A^4 = 1 , \quad A^6 = -1 , \quad A^8 = 1 , \dots \text{ etc.} \tag{33.59}$$

Problem 33.3 Consider a system with a 2-dimensional configuration manifold with generalized coordinates x, y. The Lagrangian for the system is given by

$$L(x, y, \dot{x}, \dot{y}) = \dot{x}\dot{y} - \omega^2 xy + \gamma(x\dot{y} - \dot{x}y) ,$$

where ω and γ are positive constants satisfying $\gamma < \omega$.

(a) Write down the Euler-Lagrange equations for x and y.
(b) Identify a physical system that these equations would describe.
(c) Construct the Hamiltonian function $H(x, y, p_x, p_y)$ from the Legendre transform of the Lagrangian with respect to the tangent coordinates $\dot{x}$ and $\dot{y}$.
(d) Show that Hamilton's equations of motion for this Hamiltonian yield the same differential equations for $x(t)$ and $y(t)$ as given in (a).

Problem 33.4 This is a "toy" *three-body problem*. Consider the motion of an object subjected to the gravitational pulls of the earth and the moon. The mass of the object is negligible in comparison to either the earth's mass or the moon's mass, so that the motion of the earth-moon system is not affected by the object. Consider a rotating frame $\{\boldsymbol{e}_1, \boldsymbol{e}_2\}$ rotating with *constant* angular speed ω relative to a space-fixed frame $\{\boldsymbol{\delta}_1, \boldsymbol{\delta}_2\}$, where the center-of-mass of the earth-moon system is at the origin of both frames, and the line joining the earth and the moon is always along the $\boldsymbol{e}_1$ axis. Write

$$\boldsymbol{r} = q^i \boldsymbol{e}_i \,, \qquad \boldsymbol{v} = \dot{q}^i \boldsymbol{e}_i \,.$$

Then [cf. (12.19)]

$$\ddot{q}^i \boldsymbol{e}_i = -\nabla U - 2\boldsymbol{\omega} \times \boldsymbol{v} - \boldsymbol{\omega} \times (\boldsymbol{\omega} \times \boldsymbol{r}) \,,$$

where the last two terms on the right-hand side are non-inertial forces per unit mass (Coriolis and centrifugal respectively), and U is the potential energy (per unit mass) of the object due to the gravitational pulls of the earth and the moon.

(a) Show that the above equation of motion gives rise to the following set of coupled equations:

$$\ddot{q}^1 = -\frac{\partial U}{\partial q^1} + 2\omega \dot{q}^2 + \omega^2 q^1 \,,$$
$$\ddot{q}^2 = -\frac{\partial U}{\partial q^2} - 2\omega \dot{q}^1 + \omega^2 q^2 \,.$$

(b) Show that the two coupled equations in (a) are the same as the Euler-Lagrange equations with the Lagrangian given by

$$L = \frac{1}{2}[(\dot{q}^1)^2 + (\dot{q}^2)^2] - \omega(q^2 \dot{q}^1 - q^1 \dot{q}^2) - U(q^1, q^2) + \frac{\omega^2}{2}((q^1)^2 + (q^2)^2) \,.$$

(c) Give expressions for the conjugate momenta $p_i = \partial L / \partial \dot{q}^i$, and show that the Hamiltonian is given by

$$H(q^1, q^2, p_1, p_2) = \frac{1}{2}(p_1 + \omega q^2)^2 + \frac{1}{2}(p_2 - \omega q^1)^2 + U(q^1, q^2) - \frac{\omega^2}{2}[(q^1)^2 + (q^2)^2] \,.$$
$$= \frac{1}{2}[(p_1)^2 + (p_2)^2] + \omega(p_1 q^2 - p_2 q^1) + U(q^1, q^2) \,.$$

(d) Show that Hamiltonian's equations of motion are also equivalent to the equations of motion given in (a).

Problem 33.5 Consider the damped undriven harmonic oscillator problem described by the equation [cf. (16.1)]

$$m\ddot{q} + b\dot{q} + kq = 0 \,.$$

(a) Show that the Euler-Lagrange equation for the Lagrangian

$$L(q,\dot{q},t) = e^{bt/m}\left(\frac{m}{2}\dot{q}^2 - \frac{k}{2}q^2\right)$$

yields the above equation of motion for $q(t)$.

(b) Do a Legendre transformation and show that the Hamiltonian corresponding to this Lagrangian is given by

$$H(p,q,t) = e^{-bt/m}\frac{p^2}{2m} + e^{bt/m}\frac{k}{2}q^2\,.$$

The fact that this Hamiltonian depends on the time t explicitly indicates that the energy is not conserved, as expected.

(c) Write Hamilton's equations of motion for $q(t)$ and $p(t)$. Show that these together yield the second-order equation for q given at the beginning of this problem.

We will close this chapter by showing how to construct the Lagrangian and Hamiltonian functions for the fundamental problem of *classical (Maxwell) electrodynamics*: the classical motion of a charged particle in an electromagnetic field. The equation of motion, according to the **Lorentz force law**, is given by

$$\frac{d}{dt}(m\boldsymbol{v}) = e\boldsymbol{E} + \frac{e}{c}\,\boldsymbol{v}\times\boldsymbol{B}\,, \tag{33.60}$$

where m is the mass of a charged particle of charge e moving with velocity $\boldsymbol{v}$, c is the speed of light, and $\boldsymbol{E}$ and $\boldsymbol{B}$ are the electric and magnetic fields, respectively. The latter are given in terms of the vector potential $\boldsymbol{A}$ and the electrostatic potential ϕ by

$$\boldsymbol{E} = -\nabla\phi - \frac{1}{c}\frac{\partial \boldsymbol{A}}{\partial t}\,,\qquad \boldsymbol{B} = \nabla\times\boldsymbol{A}\,. \tag{33.61}$$

Writing (33.60) in tensor notation we have

$$\begin{aligned} m\frac{d\dot{x}^i}{dt} &= eE^i + \frac{e}{c}\,\varepsilon^i{}_{jk}\,\dot{x}^j B^k = eE^i + \frac{e}{c}\,\varepsilon^i{}_{jk}\varepsilon^{kl}{}_m\,\dot{x}^j\partial_l A^m \\ &= eE^i + \frac{e}{c}\,\left(\delta^{il}\delta_{jm} - \delta^i_m\delta^l_j\right)\,\dot{x}^j\partial_l A^m \\ &= eE^i + \frac{e}{c}\left(\sum_j \dot{x}^j\partial_i A^j - \dot{x}^j\partial_j A^i\right) \\ &= e\left(-\partial_i\phi - \frac{1}{c}\frac{\partial A^i}{\partial t}\right) + \frac{e}{c}\left(\sum_j \dot{x}^j\partial_i A^j - \dot{x}^j\partial_j A^i\right)\,. \end{aligned} \tag{33.62}$$

Noting that

$$\frac{dA^i}{dt} = \dot{x}^j \partial_j A^i + \frac{\partial A^i}{\partial t} \,, \tag{33.63}$$

Eq. (33.62) reduces to

$$\frac{d}{dt}\left(m\dot{x}^i + \frac{e}{c}A^i\right) = -\partial_i \left(e\phi - \frac{e}{c}\sum_j \dot{x}^j A^j\right) \quad (i = 1,2,3) \,. \tag{33.64}$$

This set of equations can be identified with the set of *Euler-Lagrange equations* [cf. (33.9)]

$$\frac{d}{dt}\left(\frac{\partial L}{\partial \dot{x}^i}\right) = \frac{\partial L}{\partial x^i} \qquad (i = 1,2,3) \tag{33.65}$$

if the *Lagrangian* is given by

$$L(x^i, \dot{x}^i) = \frac{m}{2}\sum_j (\dot{x}^j)^2 - e\phi(x^i) + \frac{e}{c}\sum_j \dot{x}^j A^j(x^i, t) \,. \tag{33.66}$$

From this Lagrangian we can obtain the *canonical momenta* p_i:

$$p_i = \frac{\partial L}{\partial \dot{x}^i} = m\dot{x}^i + \frac{e}{c}A^i \,. \tag{33.67}$$

The *Hamiltonian* $H(x^i, p_i)$ (as a function of x^i and p_i) is then obtained from the equaton

$$H(x^i, p_i) = \dot{x}^i p_i - L(x^i, \dot{x}^i) \,, \tag{33.68}$$

in which the $\dot{x}^i$'s are expressed in terms of p_i and x^i through (33.67). Instead of doing this directly it will be a bit easier to first calculate the right-hand side of (33.68) in terms of x^i and $\dot{x}^i$ by using (33.66) and (33.67). Thus

$$\begin{aligned} &\dot{x}^i p_i - L \\ &= \sum_i \dot{x}^i \left(m\dot{x}^i + \frac{e}{c}A^i\right) - \frac{m}{2}\sum_i (\dot{x}^i)^2 + e\phi - \frac{e}{c}\sum_i \dot{x}^i A^i \\ &= \frac{m}{2}\sum_i (\dot{x}^i)^2 + e\phi \,. \end{aligned} \tag{33.69}$$

Finally (33.67) and (33.68) imply

$$\boxed{H(\boldsymbol{x}, \boldsymbol{p}) = \frac{\left(\boldsymbol{p} - \frac{e}{c}\boldsymbol{A}\right)^2}{2m} + e\phi(\boldsymbol{x})} \quad . \tag{33.70}$$

The replacement of the momentum $\boldsymbol{p}$ in the above expression by $\boldsymbol{p} - (e/c)\boldsymbol{A}$ specifies the so-called **minimal coupling** between a charged particle and an electromagnetic field.

Let us now consider the relativistic situation. For a spacetime point (event) $x^\mu = (x^0, x^1, x^2, x^3) = (ct, x, y, z)$ in **Minkowski space** with **Lorentz metric**

$$\eta_{\mu\nu} = \begin{pmatrix} -1 & 0 & 0 & 0 \\ 0 & 1 & 0 & 0 \\ 0 & 0 & 1 & 0 \\ 0 & 0 & 0 & 1 \end{pmatrix} , \tag{33.71}$$

the square of the invariant length (expressed in terms of the **proper time** τ) is given by

$$-c^2 d\tau^2 = -c^2 dt^2 + d\boldsymbol{x}^2 , \tag{33.72}$$

where

$$d\boldsymbol{x}^2 = dx^2 + dy^2 + dz^2 . \tag{33.73}$$

Noting that in a given Lorentz frame the 3-velocity $\boldsymbol{v}$ is given by $\boldsymbol{v} = d\boldsymbol{x}/dt$, Eq. (33.72) gives

$$d\tau = dt\sqrt{1 - \frac{v^2}{c^2}} = \frac{dt}{\gamma} , \tag{33.74}$$

where $v^2 = \boldsymbol{v} \cdot \boldsymbol{v}$ and

$$\gamma \equiv \frac{1}{\sqrt{1 - \dfrac{v^2}{c^2}}} . \tag{33.75}$$

Now define the relativistic 4-velocity u^ν:

$$u^\nu \equiv \frac{dx^\nu}{d\tau} . \tag{33.76}$$

Thus

$$u^0 = \frac{dx^0}{d\tau} = c\frac{dt}{d\tau} = \gamma c \tag{33.77}$$

and

$$(u^1, u^2, u^3) = \left(\frac{dx^1}{d\tau}, \frac{dx^2}{d\tau}, \frac{dx^3}{d\tau}\right) = \gamma\left(\frac{dx^1}{dt}, \frac{dx^2}{dt}, \frac{dx^3}{dt}\right) . \tag{33.78}$$

So we can write

$$u^\nu = \gamma\,(c, \boldsymbol{v}) . \tag{33.79}$$

The invariant length squared of the 4-velocity is given by

$$\eta_{\mu\nu} u^\nu u^\mu = u_\mu u^\mu = \gamma^2\,(-c^2 + v^2) = -c^2 . \tag{33.80}$$

With the covariant 4-vector potential A_μ given by

$$A_\mu = (A_0, A_1, A_2, A_3) = (-\phi, \boldsymbol{A}) , \tag{33.81}$$

the **electromagnetic field tensor** $F_{\mu\nu}$ is an antisymmetric (rank two) covariant tensor defined by

$$F_{\mu\nu} = \partial_\mu A_\nu - \partial_\nu A_\mu \,. \tag{33.82}$$

Using (33.61) it can be verified easily that

$$F_{\mu\nu} = \begin{pmatrix} 0 & -E^1 & -E^2 & -E^3 \\ E^1 & 0 & B^3 & -B^2 \\ E^2 & -B^3 & 0 & B^1 \\ E^3 & B^2 & -B^1 & 0 \end{pmatrix} . \tag{33.83}$$

Now define the 4-force

$$f_\mu \equiv \frac{e}{c} F_{\mu\nu} u^\nu \,. \tag{33.84}$$

It can be straightforwardly verified using (33.80) and (33.84) that

$$(f_1, f_2, f_3) = (f^1, f^2, f^3) = \gamma(e\boldsymbol{E} + \frac{e}{c}\boldsymbol{v} \times \boldsymbol{B}) \tag{33.85}$$

and

$$f_0 = -f^0 = -\frac{e}{c}\gamma \boldsymbol{E} \cdot \boldsymbol{v} \,. \tag{33.86}$$

We assert that the relativistic equation of motion is

$$\frac{d}{d\tau}(mu^\nu) = f^\nu \,. \tag{33.87}$$

Indeed, recalling (33.80) and taking the spatial components of the above equation we have

$$\frac{d}{d\tau}(m\gamma\boldsymbol{v}) = m\frac{d}{dt}(\gamma\boldsymbol{v})\,\frac{dt}{d\tau} = \gamma\frac{d}{dt}\left(\frac{m\boldsymbol{v}}{\sqrt{1-\frac{v^2}{c^2}}}\right) = \gamma(e\boldsymbol{E} + \frac{e}{c}\boldsymbol{v} \times \boldsymbol{B}) \,. \tag{33.88}$$

Hence

$$\frac{d}{dt}\left(\frac{m\boldsymbol{v}}{\sqrt{1-\frac{v^2}{c^2}}}\right) = e\boldsymbol{E} + \frac{e}{c}\boldsymbol{v} \times \boldsymbol{B} \,. \tag{33.89}$$

This yields the expected non-relativistic limit given by (33.60).

Taking the time component of (33.88) we obtain

$$\frac{d}{dt}(m\gamma c)\,\frac{dt}{d\tau} = \gamma\frac{d}{dt}(m\gamma c) = f^0 = \gamma\frac{e}{c}\boldsymbol{E} \cdot \boldsymbol{v} \,. \tag{33.90}$$

Thus

$$\frac{d}{dt}\left(\frac{mc^2}{\sqrt{1-\frac{v^2}{c^2}}}\right) = e\boldsymbol{E}\cdot\boldsymbol{v}\,. \tag{33.91}$$

In the non-relativistic limit ($v \ll c$)

$$\frac{mc^2}{\sqrt{1-\frac{v^2}{c^2}}} \approx mc^2\left(1+\frac{v^2}{2c^2}\right) = mc^2 + \frac{mv^2}{2}\,, \tag{33.92}$$

and (33.92) reduces to

$$\frac{d}{dt}\left(\frac{m\boldsymbol{v}^2}{2}\right) = e\boldsymbol{E}\cdot\boldsymbol{v}\,. \tag{33.93}$$

It is seen that this equation follows from (33.60) when we take the scalar product of that equation with $\boldsymbol{v}$.

Based on (33.93) we will identify the total energy E of a free particle by

$$E = \frac{mc^2}{\sqrt{1-\frac{v^2}{c^2}}} = \gamma mc^2\,. \tag{33.94}$$

We can then define the 4-momentum p^μ as follows:

$$p^\mu = (E/c, m\frac{d\boldsymbol{x}}{d\tau}) = (\gamma mc, \gamma m\frac{d\boldsymbol{x}}{dt}) = m(u^0, \boldsymbol{u})\,. \tag{33.95}$$

It follows from (33.81) that

$$p^\mu p_\mu = -m^2c^2\,. \tag{33.96}$$

Writing the 3-momentum $\boldsymbol{p}$ as

$$\boldsymbol{p} = \gamma m\frac{d\boldsymbol{x}}{dt} = \gamma m\boldsymbol{v}\,, \tag{33.97}$$

we have, from (33.97),

$$-m^2c^2 = -\frac{E^2}{c^2} + \boldsymbol{p}^2\,, \tag{33.98}$$

or

$$E^2 = \boldsymbol{p}^2c^2 + m^2c^4\,. \tag{33.99}$$

We will now try to obtain the relativistic Lagrangian and Hamiltonian. The *classical action Ldt* must be Lorentz invariant, so we define a Lorentz invariant Lagrangian L' by

$$Ldt = L'd\tau = L'dt\sqrt{1-\frac{v^2}{c^2}}\,. \tag{33.100}$$

Hence

$$L = L'\sqrt{1 - \frac{v^2}{c^2}} = \frac{L'}{\gamma} . \tag{33.101}$$

Based on the expression for the non-relativistic Lagrangian given by (33.66) we conjecture that

$$L' = \frac{e}{c} A_\mu u^\mu + \alpha , \tag{33.102}$$

where α is a Lorentz invariant constant. From (33.66) we also know that the non-relativistic field-free Lagrangian is given by

$$L_{\text{non-relativistic, field-free}} = \frac{mv^2}{2} + constant . \tag{33.103}$$

On the other hand, (33.102) and (33.103) imply that

$$L_{\text{relativistic, field-free}} = \alpha\sqrt{1 - \frac{v^2}{c^2}} \approx \alpha\,(1 - \frac{v^2}{2c^2}) . \tag{33.104}$$

Comparison of the above two equations shows that

$$\alpha = -mc^2 . \tag{33.105}$$

Thus (33.103) gives

$$L' = \frac{e}{c} A_\mu u^\mu - mc^2 . \tag{33.106}$$

Using (33.80) for u^μ and (33.82) for A_μ we have

$$L' = \frac{e}{c}\gamma\,(-c\phi + \boldsymbol{A}\cdot\boldsymbol{v}) . \tag{33.107}$$

Equation (33.102) then gives the following expression for L:

$$L = -mc^2\sqrt{1 - \frac{v^2}{c^2}} + \frac{e}{c}\,\boldsymbol{v}\cdot\boldsymbol{A} - e\phi . \tag{33.108}$$

The components of the canonical momentum $\boldsymbol{p}$ are then given by

$$p_i = \frac{\partial L}{\partial \dot{x}^i} \qquad (i = 1, 2, 3) . \tag{33.109}$$

Thus

$$\begin{aligned} p_i &= \frac{\partial}{\partial \dot{x}^i}\left(-mc^2\sqrt{1 - \sum_j \dot{x}^j/c^2}\right) + \frac{e}{c}A_i \\ &= -mc^2\,\frac{1}{2}\,\frac{-2\dot{x}^i/c^2}{\sqrt{1-\frac{v^2}{c^2}}} + \frac{e}{c}A_i = \frac{m\dot{x}^i}{\sqrt{1-\frac{v^2}{c^2}}} + \frac{e}{c}A_i , \end{aligned} \tag{33.110}$$

or

$$\boldsymbol{p} = \gamma m\boldsymbol{v} + \frac{e}{c}\boldsymbol{A}\,. \tag{33.111}$$

It follows that the Hamiltonian is given by

$$\begin{aligned} H &= p_i\dot{x}^i - L \\ &= \gamma m\boldsymbol{v}\cdot\boldsymbol{v} + \frac{e}{c}\boldsymbol{v}\cdot\boldsymbol{A} - \left(-mc^2\sqrt{1-\frac{v^2}{c^2}} + \frac{e}{c}\boldsymbol{v}\cdot\boldsymbol{A} - e\phi\right) \\ &= \frac{mv^2}{\sqrt{1-\frac{v^2}{c^2}}} + mc^2\sqrt{1-\frac{v^2}{c^2}} + e\phi = \frac{mc^2}{\sqrt{1-\frac{v^2}{c^2}}} + e\phi\,. \end{aligned} \tag{33.112}$$

Now, from (33.112),

$$\left(\boldsymbol{p} - \frac{e}{c}\boldsymbol{A}\right)^2 = \gamma^2 m^2 v^2 = -m^2c^2 + \frac{m^2c^2}{1-\frac{v^2}{c^2}}\,. \tag{33.113}$$

Hence

$$\frac{mc}{\sqrt{1-\frac{v^2}{c^2}}} = \sqrt{\left(\boldsymbol{p} - \frac{e}{c}\boldsymbol{A}\right)^2 + m^2c^2}\,. \tag{33.114}$$

Equation (33.113) then gives the following expression for the relativistic Hamiltonian

$$\boxed{H(\boldsymbol{x},\boldsymbol{p},t) = c\sqrt{\left(\boldsymbol{p} - \frac{e}{c}\boldsymbol{A}\right)^2 + m^2c^2} + e\phi(\boldsymbol{x})}\,. \tag{33.115}$$

In the non-relativistic limit ($v \ll c$),

$$H \approx mc^2\left(1 + \frac{\left(\boldsymbol{p} - \frac{e}{c}\boldsymbol{A}\right)^2}{2m^2c^2}\right) + e\phi = mc^2 + \frac{\left(\boldsymbol{p} - \frac{e}{c}\boldsymbol{A}\right)^2}{2m} + e\phi\,. \tag{33.116}$$

Subtracting the rest energy term mc^2 we obtain again the non-relativistic result (33.70).

Chapter 34

Symmetry and Conservation

In this chapter we will demonstrate, using the Lagrangian and Hamiltonian formulations, that the three fundamental conservation principles of classical mechanics: the conservation of energy, linear momentum, and angular momentum, are all consequences of certain *spatial-temporal symmetries* of classical mechanical systems. The relationship between conservation and symmetry in fact applies to a much broader context than classical mechanics; indeed, it is the cornerstone of modern theoretical physics. The general mathematical theorem stating that *corresponding to every symmetry there is a conserved quantity* is called **Noether's theorem**.

We recall from the last chapter that a *constant of motion* (or *first integral*) is a function $f(q^i, \dot{q}^i)$ which remains constant in time, where $\{q^i\}$ are the generalized coordinates. Such a function is also called a *conserved quantity*.

Suppose a system is *time-translation variant*, then the Lagrangian cannot depend explicitly on the time t. Thus it follows from the Euler-Lagrange equations (33.9) that

$$\begin{aligned} \frac{dL}{dt} &= \frac{\partial L}{\partial q^i}\dot{q}^i + \frac{\partial L}{\partial \dot{q}^i}\ddot{q}^i = \dot{q}^i \frac{d}{dt}\left(\frac{\partial L}{\partial \dot{q}^i}\right) + \frac{\partial L}{\partial \dot{q}^i}\ddot{q}^i \\ &= \frac{d}{dt}\left(\dot{q}^i \frac{\partial L}{\partial \dot{q}^i}\right) . \end{aligned} \tag{34.1}$$

Since $p_i = \partial L / \partial \dot{q}^i$, the above equation immediately implies

$$\frac{d}{dt}\left(p_i \dot{q}^i - L\right) = 0\,, \tag{34.2}$$

or, from (33.11),

$$\frac{dH}{dt} = 0\,. \tag{34.3}$$

where H is the Hamiltonian function. Equation (34.3) is of course a statement of the conservation of energy. We thus have the following principle: *conservation of energy is a consequence of time-translation symmetry.*

Next suppose a system is *space-translation invariant.* This means that under a spatial translation (Fig. 25.1)

$$\boldsymbol{r}_\alpha \to \boldsymbol{r}_\alpha + \boldsymbol{\epsilon} \tag{34.4}$$

for all α, where $\boldsymbol{r}_\alpha$ is the position of the α-th particle in a many-particle system, $\delta L = 0$. Now

$$\delta L = \frac{\partial L}{\partial q^i}\delta q^i = \sum_\alpha \frac{\partial L}{\partial \boldsymbol{r}_\alpha} \cdot \delta \boldsymbol{r}_\alpha = \boldsymbol{\epsilon} \cdot \sum_\alpha \frac{\partial L}{\partial \boldsymbol{r}_\alpha} . \tag{34.5}$$

Hence the requirement that $\delta L = 0$ for any $\boldsymbol{\epsilon}$ implies that

$$\sum_\alpha \frac{\partial L}{\partial \boldsymbol{r}_\alpha} = 0 . \tag{34.6}$$

Letting $\boldsymbol{v}_\alpha = \dot{\boldsymbol{r}}_\alpha$ and recalling the Euler-Lagrange equations, the above equation can be written

$$\frac{d}{dt}\sum_\alpha \frac{\partial L}{\partial \boldsymbol{v}_\alpha} = 0 . \tag{34.7}$$

But from (31.10), $\boldsymbol{p}_\alpha = \partial L/\partial \boldsymbol{v}_\alpha$. Hence (34.7) implies that

$$\sum_\alpha \boldsymbol{p}_\alpha \equiv \boldsymbol{P} = \text{constant} . \tag{34.8}$$

This of course is a statement of the conservation of the total linear momentum. Thus, *conservation of linear momentum is a consequence of space-translation symmetry.*

Equation (34.6) has a simple physical meaning. We have

$$\frac{\partial L}{\partial \boldsymbol{r}_\alpha} = -\frac{\partial U}{\partial \boldsymbol{r}_\alpha} = \boldsymbol{F}_\alpha , \tag{34.9}$$

where $\boldsymbol{F}_\alpha$ is the force on the α-th particle. Thus (34.6) implies that

$$\sum_\alpha \boldsymbol{F}_\alpha = 0 . \tag{34.10}$$

In other words, the net force on the system vanishes. In particular, for a two-body system in the absence of external forces, we have

$$\boldsymbol{F}_1 + \boldsymbol{F}_2 = 0 , \tag{34.11}$$

which is *Newton's Third Law* (see Chapter 12).

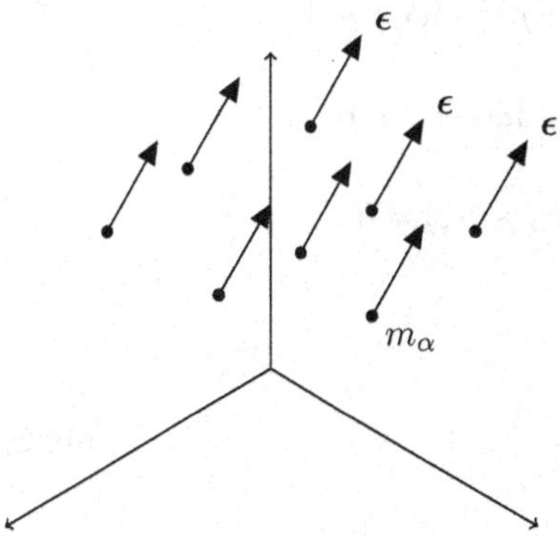

Fig. 34.1

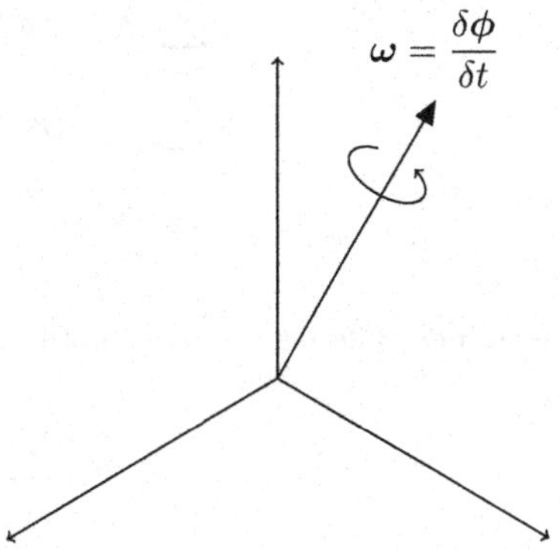

Fig. 34.2

Finally we suppose the system is *rotationally invariant.* Suppose an infinitesimal rotation is described by the angular velocity vector (Fig. 34.2)

$$\boldsymbol{\omega} = \frac{\delta \boldsymbol{\phi}}{\delta t} \,. \tag{34.12}$$

Then we have [cf. (4.13)]

$$\boldsymbol{v} = \boldsymbol{\omega} \times \boldsymbol{r} \,, \tag{34.13}$$

from which it follows that

$$\delta \boldsymbol{r} = \delta \boldsymbol{\phi} \times \boldsymbol{r} \,, \tag{34.14}$$

$$\delta \boldsymbol{v} = \delta \boldsymbol{\phi} \times \boldsymbol{v} \,. \tag{34.15}$$

Under the rotation with $\delta \boldsymbol{r}$ and $\delta \boldsymbol{v}$ given by the above two equations, rotational invariance requires that

$$\delta L = \sum_\alpha \left(\frac{\partial L}{\partial \boldsymbol{r}_\alpha} \cdot \delta \boldsymbol{r}_\alpha + \frac{\partial L}{\partial \boldsymbol{v}_\alpha} \cdot \delta \boldsymbol{v}_\alpha \right) = 0 \,. \tag{34.16}$$

Using

$$\frac{\partial L}{\partial \boldsymbol{v}_\alpha} = \boldsymbol{p}_\alpha \,, \tag{34.17}$$

$$\frac{\partial L}{\partial \boldsymbol{r}_\alpha} = -\frac{\partial U}{\partial \boldsymbol{r}_\alpha} = -\frac{\partial}{\partial \boldsymbol{r}_\alpha}(T+U) = -\frac{\partial H}{\partial \boldsymbol{r}_\alpha} = \dot{\boldsymbol{p}}_\alpha \,, \tag{34.18}$$

where the last equality of (34.18) follows from Hamilton's equations, we have

$$\begin{aligned}
&\sum_\alpha \left(\frac{\partial L}{\partial \boldsymbol{r}_\alpha} \cdot \delta \boldsymbol{r}_\alpha + \frac{\partial L}{\partial \boldsymbol{v}_\alpha} \cdot \delta \boldsymbol{v}_\alpha \right) \\
&= \sum_\alpha \left(\dot{\boldsymbol{p}}_\alpha \cdot (\delta \boldsymbol{\phi} \times \boldsymbol{r}_\alpha) + \boldsymbol{p}_\alpha \cdot (\delta \boldsymbol{\phi} \times \boldsymbol{v}_\alpha) \right) \\
&= \sum_\alpha \left(\delta \boldsymbol{\phi} \cdot (\boldsymbol{r}_\alpha \times \dot{\boldsymbol{p}}_\alpha) + \delta \boldsymbol{\phi} \cdot (\boldsymbol{v}_\alpha \times \boldsymbol{p}_\alpha) \right) \\
&= \delta \boldsymbol{\phi} \cdot \sum_\alpha (\boldsymbol{r}_\alpha \times \dot{\boldsymbol{p}}_\alpha + \dot{\boldsymbol{r}}_\alpha \times \boldsymbol{p}_\alpha) = 0 \, ,
\end{aligned} \tag{34.19}$$

for all $\delta\boldsymbol{\phi}$. The last equality implies

$$\frac{d}{dt} \left(\sum_\alpha \boldsymbol{r}_\alpha \times \boldsymbol{p}_\alpha \right) = 0 \, , \tag{34.20}$$

which is the statement of the conservation of the total angular momentum of the system. Thus *conservation of angular momentum is a consequence of rotational invariance.*

Chapter 35

Symmetric Tops

In this chapter we will use the problem of symmetric tops to further illustrate the use of Hamilton's equations of motion. We will first introduce a useful set of generalized coordinates for the description of the rotational motion of rigid bodies, the so-called *Euler angles.*

Imagine a rigid body with its three principal axes of inertia originally aligned with a space-fixed orthonormal frame $\{\epsilon_1, \epsilon_2, \epsilon_3\}$. Call this orientation the standard orientation. An arbitrary orientation of the rigid body can be reached from the standard orientation by making three successive rotations as follows (Fig. 35.1):

1. Rotate about the ϵ_3-axis by an angle ϕ; call the new axes $\{e_1'', e_2', e_3'\}$.

2. Rotate about the e_2' axis by an angle θ; call the new axes $\{e_1', e_2', e_3\}$.

3. Rotate about the e_3 axis by an angle ψ; call the final axes $\{e_1, e_2, e_3\}$.

(In Fig. 35.1, the frame drawn with thick lines in each diagram represents the end result of the rotations up to each of the above three steps.) The three angles ϕ, θ and ψ $(0 \le \phi < 2\pi,\ 0 \le \theta \le \pi,\ 0 \le \psi < 2\pi)$ are called the **Euler angles** specifying the orientation of a rigid body. Using the Euler angles, the angular velocity $\boldsymbol{\omega}$ can be written as

$$\boldsymbol{\omega} = \dot{\phi}\, \epsilon_3 + \dot{\theta}\, e_2' + \dot{\psi}\, e_3 \,. \tag{35.1}$$

Let us consider a symmetric top with e_3 as the symmetry axis ($I^1 = I^2$). Because of the axial symmetry any two orthogonal axes on the e_1-e_2 plane together with e_3 will constitute a set of principal axes. We will choose $\{e_1', e_2', e_3\}$ as a set of principal axes with respect to which components of $\boldsymbol{\omega}$ will be calculated. From Fig. 35.1 it is easily seen that

$$\epsilon_3 = (\epsilon_3 \cdot e_1')\, e_1' + (\epsilon_3 \cdot e_3)\, e_3 = -\sin\theta\, e_1' + \cos\theta\, e_3 \,. \tag{35.2}$$

Equation (35.1) then implies

$$\boldsymbol{\omega} = (-\dot{\phi}\sin\theta)\, e_1' + \dot{\theta}\, e_2' + (\dot{\psi} + \dot{\phi}\cos\theta)\, e_3 \,. \tag{35.3}$$

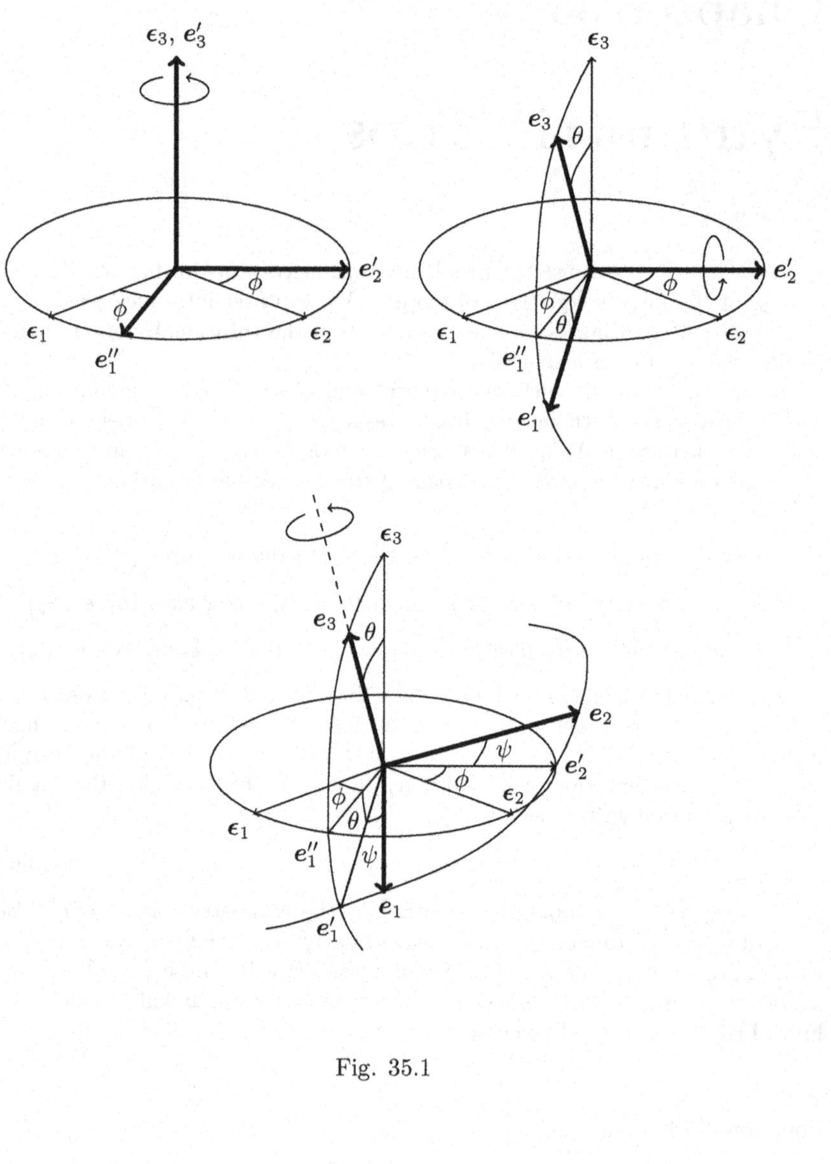

ϵ_3, e'_3
e'_2
ϕ
ϕ
ϵ_1
ϵ_2
e''_1
ϵ_3
e_3
θ
e'_2
ϕ
ϕ
θ
ϵ_1
ϵ_2
e''_1
e'_1
ϵ_3
e_3
θ
e_2
ψ
e'_2
ϕ
ϵ_2
ϕ
θ
ϵ_1
e''_1
ψ
e_1
e'_1

Fig. 35.1

It follows from (20.9) that

$$T_{rot} = \frac{I^1}{2}(\dot{\phi}^2 \sin^2\theta + \dot{\theta}^2) + \frac{I^3}{2}(\dot{\psi} + \dot{\phi}\cos\theta)^2 \,. \tag{35.4}$$

Suppose the symmetric top is fixed at a point on the symmetry axis lower than the center of mass (CM), and the only external force is the gravitational force acting at the center of mass. (Refer back to Fig. 20.6, where the vertical axis shown is along the direction of ϵ_3.) From (19.29) and (19.30) the total kinetic energy is given by

$$T = \frac{1}{2}MV^2 + T_{rot} \,, \tag{35.5}$$

where M is the mass of the top and $\boldsymbol{V}$ is the velocity of the center of mass. As the top moves the CM moves on the surface of a sphere of radius r. Hence

$$\frac{1}{2}M\boldsymbol{V}^2 = \frac{1}{2}Mr^2(\dot{\theta}^2 + \sin^2\theta\,\dot{\phi}^2) \,. \tag{35.6}$$

Combining the last three equations we have

$$T = \frac{1}{2}(I^1 + Mr^2)(\dot{\theta}^2 + \dot{\phi}^2 \sin^2\theta) + \frac{I^3}{2}(\dot{\psi} + \dot{\phi}\cos\theta)^2 \,. \tag{35.7}$$

Realizing that the potential energy is given by

$$U = Mgr\cos\theta \,, \tag{35.8}$$

and defining

$$I' \equiv I^1 + Mr^2 \,, \tag{35.9}$$

the Lagrangian can be written as

$$L = \frac{I'}{2}(\dot{\theta}^2 + \dot{\phi}^2 \sin^2\theta) + \frac{I^3}{2}(\dot{\psi} + \dot{\phi}\cos\theta)^2 - Mgr\cos\theta \,. \tag{35.10}$$

The canonical momenta are then given by

$$p_\phi = \frac{\partial L}{\partial \dot{\phi}} = I'\dot{\phi}\sin^2\theta + I^3\cos\theta\,(\dot{\psi} + \dot{\phi}\cos\theta) \,, \tag{35.11}$$

$$p_\theta = \frac{\partial L}{\partial \dot{\theta}} = I'\dot{\theta} \,, \tag{35.12}$$

$$p_\psi = \frac{\partial L}{\partial \dot{\psi}} = I^3(\dot{\psi} + \dot{\phi}\cos\theta) \,. \tag{35.13}$$

From these equations one can solve for $\dot{\phi}$, $\dot{\theta}$ and $\dot{\psi}$ to yield

$$\dot{\phi} = \frac{p_\phi - p_\psi\cos\theta}{I'\sin^2\theta} \,, \tag{35.14}$$

$$\dot{\theta} = \frac{p_\theta}{I'} \,, \tag{35.15}$$

$$\dot{\psi} = \frac{p_\psi}{I^3} - \left(\frac{p_\phi - p_\psi\cos\theta}{I'\sin^2\theta}\right)\cos\theta \,. \tag{35.16}$$

Substituting these expressions in (35.7) and (35.8) for T and U, respectively, we then have the following expression for the Hamiltonian:

$$H = T + U = \frac{(p_\phi - p_\psi \cos\theta)^2}{2I' \sin^2\theta} + \frac{p_\theta^2}{2I'} + \frac{p_\psi^2}{2I^3} + Mgr\cos\theta\,. \tag{35.17}$$

Problem 35.1 Verify the expression in (35.17) for the Hamiltonian of a symmetric top.

A glance at the Hamiltonian reveals that ϕ and ψ are cyclic coordinates ($\partial H/\partial\phi = \partial H/\partial\psi = 0$). Thus, by Hamilton's equations,

$$\dot{p}_\phi = -\frac{\partial H}{\partial \phi} = 0\,, \quad \dot{p}_\psi = -\frac{\partial H}{\partial \psi} = 0\,, \tag{35.18}$$

which imply

$$p_\phi = \text{constant}\,, \qquad p_\psi = \text{constant}\,. \tag{35.19}$$

Since $\partial H/\partial t = 0$, the total energy E is also conserved (cf. Theorem 33.3). We have thus found three constants of motion and the problem is *integrable* (cf. discussion in Chapter 33), since it is reduced to a one-dimensional one (in θ) with

$$H = \frac{p_\theta^2}{2I'} + U_{eff}(\theta)\,, \tag{35.20}$$

where U_{eff} is an effective potential defined by

$$U_{eff}(\theta) \equiv \frac{(p_\phi - p_\psi\cos\theta)^2}{2I'\sin^2\theta} + \frac{p_\psi^2}{2I^3} + Mgr\cos\theta\,. \tag{35.21}$$

The equation of motion for θ is given by the Hamilton equation

$$\dot{p}_\theta = -\frac{\partial H}{\partial \theta}\,, \tag{35.22}$$

which, by (35.12) and (35.20), is equivalent to,

$$I'\frac{d^2\theta}{dt^2} = -\frac{dU_{eff}(\theta)}{d\theta}\,. \tag{35.23}$$

This equation can in principle be integrated directly. We will, however, not solve it explicitly, but will instead attempt to gain a qualitative understanding of the solutions by examining the shape of the effective potential $U_{eff}(\theta)$, a sketch of which is given in Fig. 35.2 for the case $p_\phi \neq \pm p_\psi$. (For this case the vertical lines $\theta = 0$ and $\theta = \pi$ are asymptotes to the potential energy curve.)

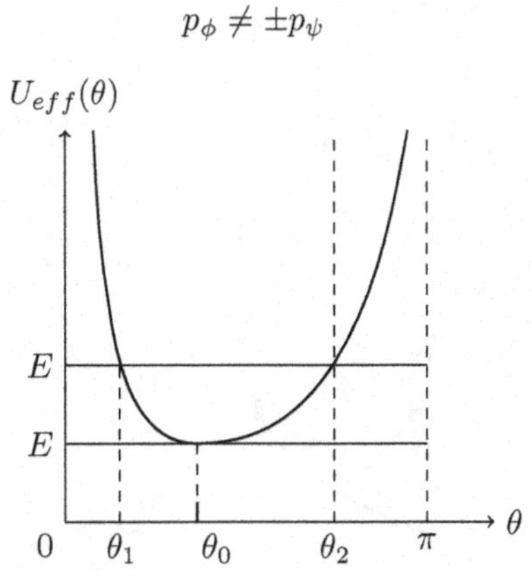

Fig. 35.2

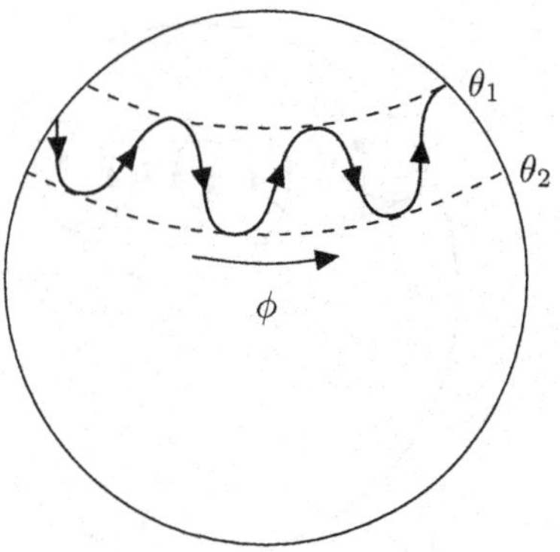

Fig. 35.3

Energy conservation is expressed by the following equation:

$$\frac{p_\theta^2}{2I'} + U_{eff}(\theta) = E = \text{constant} . \tag{35.24}$$

Obviously the classically allowed values of θ must be in the region where $E \geq U_{eff}(\theta)$. If $E = U_{eff}(\theta_0)$, then $\theta = \theta_0 =$ constant. Equations (35.4) and (35.19) then imply that $\dot{\phi} =$ constant, and the motion of the top is a steady precession of the symmetry axis $\mathbf{e}_3$ around the vertical direction with constant angular speed $\dot{\phi}$ and inclined at constant angle $\theta = \theta_0$. This case was already treated by an approximate method in Chapter 20 [recall the discussion surrounding (20.25)].

If $E > U_{eff}(\theta_0)$, the motion is periodic with θ assuming values in the range $\theta_1 \leq \theta \leq \theta_2$. The top "nods" at the same time as ϕ changes in value. The time evolution of θ and ϕ can then be represented by tracing the path of the CM (or any point on the symmetry axis except the fixed point) on the surface of a sphere (since any point on the surface of a sphere can be coordinatized by θ and ϕ). [For the steady precession case discussed above, this path is just a latitude circle (with constant θ).]

From (35.14) we see that $\dot{\phi}$ vanishes only when $\cos\theta = p_\phi / p_\psi$. We thus consider the following three cases according to the value of $\cos^{-1}(p_\phi/p_\psi)$.

(1) $\cos^{-1}\left(\dfrac{p_\phi}{p_\psi}\right) \notin [\theta_1, \theta_2]$. Since $\dot{\phi}$ is never zero, it cannot change in sign. Thus ϕ increases (or decreases) monotonically as θ oscillates between θ_1

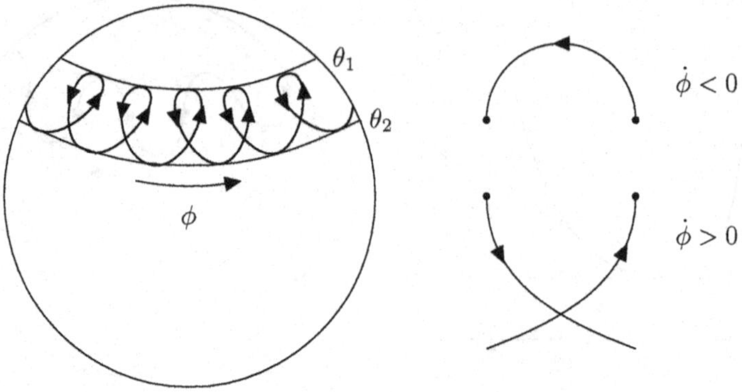

Fig. 35.4

$$\theta_1 = \cos^{-1}\left(\frac{p_\phi}{p_\psi}\right)$$

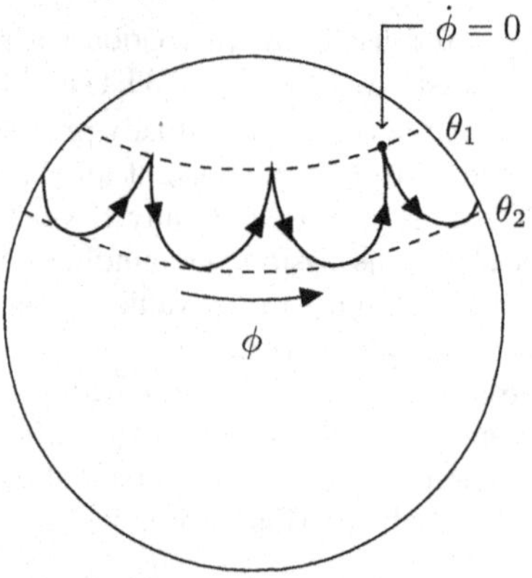

Fig. 35.5

$$\theta_2 = \cos^{-1}\left(\frac{p_\phi}{p_\psi}\right)$$

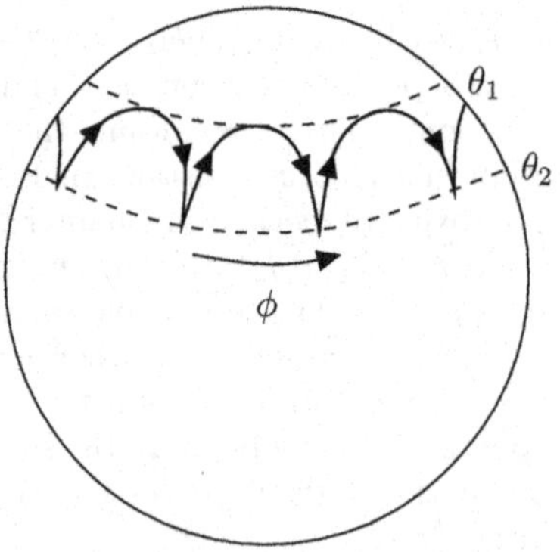

Fig. 35.6

and θ_2 (see Fig. 35.3).

(2) $\theta_1 < \cos^{-1}\left(\frac{p_\phi}{p_\psi}\right) < \theta_2$. In this case $\dot{\phi}$ vanishes at a value of θ intermediate between θ_1 and θ_2. Hence ϕ both increases and decreases depending on the value of θ. A possible motion is shown in Fig. 35.4.

(3) (a) $\theta_1 = \cos^{-1}\left(\frac{p_\phi}{p_\psi}\right)$, (b) $\theta_2 = \cos^{-1}\left(\frac{p_\phi}{p_\psi}\right)$. A possible motion for (a) is shown in Fig. 35.5 and one for (b) is shown in Fig. 35.6.

Problem 35.2 A symmetric top can spin with its symmetry axis aligned vertically ($\theta = 0$) only if $p_\phi = p_\psi = I_3\omega_3$. Assume this condition is satisfied.

(a) Show that the effective potential $U(\theta)$ can be written as

$$U(\theta) = \left(\frac{I_3^2\omega_3^2}{2I_1}\right)\tan^2\frac{\theta}{2} + \frac{I_3\omega_3^2}{2} + Mgh\cos\theta\,.$$

(b) Expand $U(\theta)$ as a power series in θ, retaining only the lowest order terms.

(c) Find the critical frequency ω_0 such that the vertical top will be stable if $\omega_3 > \omega_0$ and unstable if $\omega_3 < \omega_0$. Sketch the graphs of $U(\theta)$ for these cases.

Hint: The vertical top is stable if $U(0)$ is a minimum and unstable if $U(0)$ is a maximum.

Chapter 36

Canonical Transformations and the Symplectic Group

The notion of canonical transformations is at the heart of the mathematical structure underlying Hamilton's equations of motion, both from an analytical and a geometrical viewpoint. It also provides for an essential tool towards the solution of these equations. In this chapter we will focus on the analytical and algebraic developments, leaving the geometrical one for Chapter 38.

We recall our preliminary discussion of the concept of integrability in Chapter 33, where it was mentioned that a basic strategy for the solution of Hamilton's equations of motion was to choose an appropriate set of canonical coordinates so that as many canonical momenta as possible become constants of motion. If this can be achieved, then the dimensionality of the problem will be effectively reduced. The transformation from an initial set of canonical variables (which are usually selected on the basis of the physical context) to a new set such that the strategy becomes effective, is thus not unrestricted, but must satisfy the requirement that the set of Hamilton's equations of motion is invariant under the transformation. Such transformations are usually referred to as *canonical transformations* in the classical mechanics literature (see, for example, Goldstein 1965). We will, however, introduce a slightly more restrictive requirement to serve as the basic definition of a canonical transformation based on the notion of the Lagrangian derivative introduced in Chapter 29 (following the treatment by Siegel and Moser 1971), such that the more intuitive requirement of the invariance of Hamilton's equations is a necessary condition of the definition. We will see that the important algebraic notion of the *symplectic group* emerges naturally from this definition. As we will come to appreciate, the symplectic group offers a very natural organizational framework for the study of canonical transformations.

Consider the cotangent bundle T^*Q (of an n-dimensional configuration manifold Q) with local coordinates (q^i, p_i). We will investigate invertible transformations $(q^i, p_i) \to (Q^i, P_i)$ of the form

$$q^i = q^i(Q^j, P_j)\,, \quad p_i = p_i(Q^j, P_j) \qquad (i = 1, \dots, n)\,. \tag{36.1}$$

If the Lagrangian

$$L(q^i, \dot{q}^i, p_i, \dot{p}_i, t) \equiv p_i \dot{q}^i - H(q^i, p_i, t) \tag{36.2}$$

is viewed as a function of the indicated variables in the above expression, then, by (29.16), we have the following expressions for the Lagrangian derivatives:

$$\mathcal{L}_{q^i} L = -\frac{\partial H(q^j, p_j)}{\partial q^i} - \dot{p}_i\,, \qquad \mathcal{L}_{p_i} L = \dot{q}^i - \frac{\partial H(q^j, p_j)}{\partial p_i}\,. \tag{36.3}$$

Note that as defined by (36.2) L is explicitly independent of $\dot{p}_i$, hence the second equation in (36.3). Hamilton's equations of motion (33.16) then amount to

$$\mathcal{L}_{q^i} L = \mathcal{L}_{p_i} L = 0\,. \tag{36.4}$$

Given a Lagrangian function $L(q^i, \dot{q}^i, t)$ and a coordinate transformation (36.1), we can define a function $L(Q^i, \dot{Q}^i, P_i, \dot{P}_i, t)$ by

$$L(Q^i, \dot{Q}^i, P_i, \dot{P}_i, t) \equiv L(q^i(Q^j, P_j), \dot{q}^i(Q^j, P_j, \dot{Q}^j, \dot{P}_j), t)\,. \tag{36.5}$$

Guided by (36.3) we can now define the notion of canonical transformations.

Definition 36.1. *Given a Lagrangian function $L(q^i, \dot{q}^i, t)$ a transformation of local canonical coordinates of the form (36.1) is said to be a* ***canonical transformation*** *if the Lagrangian derivatives of the function $L(Q^i, \dot{Q}^i, P_i, \dot{P}_i, t)$ [as defined by (36.5)] retain their invariant (Hamiltonian) forms as given by (36.3), that is,*

$$\mathcal{L}_{Q^i} L = -\frac{\partial H'(Q^i, P_i, t)}{\partial Q^i} - \dot{P}_i\,, \qquad \mathcal{L}_{P_i} L = \dot{Q}^i - \frac{\partial H'(Q^i, P_i, t)}{\partial P_i}\,, \tag{36.6}$$

for some function $H'(Q^i, P_i, t)$ (which can be considered as the transformed Hamiltonian function).

We note immediately that the function $L'(Q^i, \dot{Q}^i, P_i, \dot{P}_i, t)$ defined by

$$L'(Q^i, \dot{Q}^i, P_i, \dot{P}_i, t) \equiv P_i \dot{Q}^i - H'(Q^i, P_i, t) \tag{36.7}$$

for an arbitrary Hamiltonian function $H'(Q^i, P_i, t)$ automatically satisfies (36.6).

Suppose $\{q^i, p_i\} \to \{Q^i, P_i\}$ is a canonical transformation. Since the functions L [given by (36.5)] and L' [given by (36.7)] share the same Lagrangian derivatives, they must be related by

$$L = L' + \frac{dv(Q^i, P_i, t)}{dt}\,, \tag{36.8}$$

where $v(Q^i, P_i, t)$ is a function explicitly independent of $\dot{Q}^i$ and $\dot{P}_i$, according to (29.24). Thus it follows from (36.2) and (36.7) that

$$\frac{dv(Q^i, P_i, t)}{dt} = H' - H + \dot{q}^i p_i - \dot{Q}^i P_i \, . \tag{36.9}$$

Since

$$\dot{q}^i = \frac{\partial q^i}{\partial Q^j}\, \dot{Q}^i + \frac{\partial q^i}{\partial P_j}\, \dot{P}_i \, , \tag{36.10}$$

Eq. (36.9) can be written as

$$\frac{dv(Q^i, P_i, t)}{dt} = H' - H + \left(\frac{\partial q^i}{\partial Q^j} p_i - P_j \right) \dot{Q}^j + \left(\frac{\partial q^i}{\partial P_j} p_i \right) \dot{P}_j \, . \tag{36.11}$$

Therefore

$$\frac{\partial v(Q^i, P_i, t)}{\partial Q^i} = \frac{\partial q^j}{\partial Q^i}\, p_j - P_i \, , \tag{36.12}$$

$$\frac{\partial v(Q^i, P_i, t)}{\partial P_i} = \frac{\partial q^j}{\partial P_i}\, p_j \, , \tag{36.13}$$

$$H'(Q^i, P_i, t) = H(q^i(Q^j, P_j), p_i(Q^j, P_j)) + \frac{\partial v(Q^i, P_i, t)}{\partial t} \, . \tag{36.14}$$

The function $v(Q^i, P_i, t)$ is called a **generating function** of the canonical transformation $\{q^i, p_i\} \to \{Q^i, P_i\}$, and (36.14) determines the new Hamiltonian in terms of the old one and the generating function.

We now wish to show that the set of canonical transformations possesses a group structure. Consider the $2n \times 2n$ matrix

$$\mathcal{M} = \begin{pmatrix} \mathcal{A} & \mathcal{B} \\ \mathcal{C} & \mathcal{D} \end{pmatrix} \tag{36.15}$$

corresponding to a canonical transformation $\{q^i, p_i\} \to \{Q^i, P_i\}$, where each of the entries is an $n \times n$ *Jacobian matrix* defined as follows:

$$\mathcal{A} \equiv \left(\frac{\partial q^i}{\partial Q^j} \right) , \quad \mathcal{B} \equiv \left(\frac{\partial q^i}{\partial P_j} \right) , \quad \mathcal{C} \equiv \left(\frac{\partial p_i}{\partial Q^j} \right) , \quad \mathcal{D} \equiv \left(\frac{\partial p_i}{\partial P_j} \right) . \tag{36.16}$$

Since the canonical transformation is supposed to be invertible, we assume that the determinant $|\mathcal{M}| \neq 0$. Now the partial derivatives of the function $v(Q^i, P_i, t)$ introduced in (36.8) clearly satisfies the properties

$$\frac{\partial^2 v}{\partial Q^i \partial Q^j} = \frac{\partial^2 v}{\partial Q^j \partial Q^i} \, , \qquad \frac{\partial^2 v}{\partial Q^i \partial P_j} = \frac{\partial^2 v}{\partial P_j \partial Q^i} \, , \qquad \frac{\partial^2 v}{\partial P_i \partial P_j} = \frac{\partial^2 v}{\partial P_j \partial P_i} \, . \tag{36.17}$$

It follows from (36.12) and (36.13) that the above three equations are equivalent, respectively, to the following three:

$$\frac{\partial^2 q^k}{\partial Q^j \partial Q^i}\,p_k + \frac{\partial q^k}{\partial Q^i}\frac{\partial p_k}{\partial Q^j} = \frac{\partial^2 q^k}{\partial Q^i \partial Q^j}\,p_k + \frac{\partial q^k}{\partial Q^j}\frac{\partial p_k}{\partial Q^i}\,, \tag{36.18}$$

$$\frac{\partial^2 q^k}{\partial P_j \partial Q^i}\,p_k + \frac{\partial q^k}{\partial Q^i}\frac{\partial p_k}{\partial P_j} - \delta_{ij} = \frac{\partial^2 q^k}{\partial Q^i \partial P_j}\,p_k + \frac{\partial q^k}{\partial P_j}\frac{\partial p_k}{\partial Q^i}\,, \tag{36.19}$$

$$\frac{\partial^2 q^k}{\partial P_j \partial P_i}\,p_k + \frac{\partial q^k}{\partial P_i}\frac{\partial p_k}{\partial P_j} = \frac{\partial^2 q^k}{\partial P_i \partial P_j}\,p_k + \frac{\partial q^k}{\partial P_j}\frac{\partial p_k}{\partial P_i}\,. \tag{36.20}$$

These immediately reduce to

$$\frac{\partial q^k}{\partial Q^i}\frac{\partial p_k}{\partial Q^j} = \frac{\partial q^k}{\partial Q^j}\frac{\partial p_k}{\partial Q^i}\,,\quad \frac{\partial q^k}{\partial Q^i}\frac{\partial p_k}{\partial P_j} - \delta_{ij} = \frac{\partial q^k}{\partial P_j}\frac{\partial p_k}{\partial Q^i}\,,\quad \frac{\partial q^k}{\partial P_i}\frac{\partial p_k}{\partial P_j} = \frac{\partial q^k}{\partial P_j}\frac{\partial p_k}{\partial P_i}\,, \tag{36.21}$$

which can be written compactly in terms of the matrices defined in (36.16) as follows:

$$\mathcal{C}^T\mathcal{A} = \mathcal{A}^T\mathcal{C}\,,\qquad \mathcal{D}^T\mathcal{A} - \mathcal{B}^T\mathcal{C} = I\,,\qquad \mathcal{D}^T\mathcal{B} = \mathcal{B}^T\mathcal{D}\,, \tag{36.22}$$

where T means transpose and I denotes the $n \times n$ identity matrix. Introducing the $2n \times 2n$ matrix

$$\mathcal{J} \equiv \begin{pmatrix} 0 & I \\ -I & 0 \end{pmatrix}\,, \tag{36.23}$$

and using the definition for the matrix $\mathcal{M}$ given by (36.15), we see that the matrix equations (36.22) can be combined to yield the single matrix equation

$$\boxed{\mathcal{M}^T\mathcal{J}\mathcal{M} = \mathcal{J}}\quad . \tag{36.24}$$

Problem 36.1 Verify the assertion that the matrix equations (36.22) are equivalent to the partial derivative equations (36.21).

Problem 36.2 Verify that the single matrix equation (36.24) is equivalent to the three matrix equations (36.22).

We have thus shown that when the matrix $\mathcal{M}$ corresponds to a canonical transformation it necessarily satisfies (36.24). Conversely, it is not difficult to show that (36.24) is also a sufficient condition for the transformation to be canonical.

A matrix $\mathcal{M}$ that satisfies (36.24) is said to be a **symplectic matrix**. Since $\det(\mathcal{J}) = |\mathcal{J}| = 1$, it follows from (36.24) that $|\mathcal{M}|^2 = 1$, so $(\mathcal{M})^{-1}$ exists. Using (36.24) again we have

$$(\mathcal{M}^{-1})^T \mathcal{J} \mathcal{M}^{-1} = (\mathcal{M}^T)^{-1} \mathcal{M}^T \mathcal{J} \mathcal{M} \mathcal{M}^{-1} = \mathcal{J} , \tag{36.25}$$

where in the first equality we have also used the fact that for any invertible matrix A, $(A^{-1})^T = (A^T)^{-1}$. Thus if $\mathcal{M}$ is symplectic, so is $\mathcal{M}^{-1}$. Furthermore, if $\mathcal{M}_1$ and $\mathcal{M}_2$ are both symplectic, so is the product matrix $\mathcal{M}_1\mathcal{M}_2$, since

$$(\mathcal{M}_1\mathcal{M}_2)^T \mathcal{J} \mathcal{M}_1\mathcal{M}_2 = \mathcal{M}_2^T \mathcal{M}_1^T \mathcal{J} \mathcal{M}_1 \mathcal{M}_2 = \mathcal{M}_2^T \mathcal{J} \mathcal{M}_2 = \mathcal{J} . \tag{36.26}$$

Thus the set of $2n \times 2n$ symplectic matrices forms a group, called the **symplectic group**. In conclusion we have proved the following important theorem.

Theorem 36.1. *The set of local transformations $\{q^i, p_i\} \to \{Q^i, P_i\}$ on the cotangent bundle T^*M of a configuration manifold M given by the functional relationships $q^i = q^i(Q^j, P_j)$ and $p_i = p_i(Q^j, P_j)$ is canonical if and only if the Jacobian transformation matrix $\mathcal{M}$ [defined by (36.15) and (36.16)] is a symplectic matrix identically in (Q^i, P_i) in a suitable domain of these variables. In other words the set of local canonical transformations forms a symplectic group.*

Problem 36.3 It is clear that canonical transformations (as defined in Def. 36.1) leave the form of Hamilton's equations of motion invariant. However, the set of transformations that leave Hamilton's equations of motion invariant is slightly more general than the set of canonical transformations. Show that this latter set satisfies the following slightly more general condition than the symplectic condition given by (36.24):

$$\mathcal{M}^T \mathcal{J} \mathcal{M} = \lambda \mathcal{J} , \tag{36.27}$$

where $\lambda \neq 0$ is a constant scalar. Proceed according to the following steps.

(a) Denote collectively the canonical coordinates $(q^1, \dots, q^n, p_1, \dots, p_n)$ by $(z_1, \dots, z_{2n})$. Show that Hamilton's equations of motion $\dot{q}^i = \partial H/\partial p_i$, $\dot{p}_i = -\partial H/\partial q^i$ can be written as the single matrix equation

$$\dot{z} - \mathcal{J} \partial_z H = 0 , \tag{36.28}$$

where z is the column matrix (z_i), $\partial_z H$ the column matrix $(\partial H/\partial z_i)$, and $\mathcal{J}$ is the $2n \times 2n$ matrix defined in (36.23).

(b) Consider a transformation of canonical coordinates $\{q^i, p_i\} \to \{Q^i, P_i\}$ with $(Q^1, \dots, Q^n, P_1, \dots, P_n)$ denoted by $(Z_1, \dots, Z_{2n})$. It will be useful to consider transformations which may be explicitly time-dependent: $z_i = z_i(Z_j, t)$. Let the Jacobian matrix of the transformation be denoted by $\mathcal{M}$ [compare with (36.15) and (36.16)]:

$$\mathcal{M}_{ij} = \frac{\partial z_i}{\partial Z_j} . \tag{36.29}$$

Show that

$$\mathcal{M}^{-1}(\dot{z} - \mathcal{J}\partial_z H) = \dot{Z} + \mathcal{M}^{-1}\partial_t z - \mathcal{M}^{-1}\mathcal{J}(\mathcal{M}^T)^{-1}\partial_Z H \,, \tag{36.30}$$

where $\partial_t z$ and $\partial_Z H$ are the column matrices $(\partial_t z_i(Z_j, t))$ and $(\partial_{Z_i} H(z_j(Z_k, t)))$, respectively, with the indicated explicit dependence on the variables Z_i and t.

(c) In order for the transformation $\{z_i\} \to \{Z_i\}$ to leave Hamilton's equations of motion (36.28) invariant, the right-hand side of (36.30) must have the form $\dot{Z} - \mathcal{J}\partial_Z H'$, for some new Hamiltonian function $H'(Z_i, t)$. Show that this new Hamiltonian function must satisfy the matrix equation

$$\partial_Z H' = \mathcal{J}^{-1}\mathcal{M}^{-1}\mathcal{J}(\mathcal{M}^T)^{-1}\partial_Z H - \mathcal{J}^{-1}\mathcal{M}^{-1}\partial_t z \,. \tag{36.31}$$

(d) Define the following $2n \times 2n$ matrices:

$$\mathcal{P} = (\mathcal{P}_{ij}) \equiv \mathcal{J}^{-1}\mathcal{M}^{-1}\mathcal{J}(\mathcal{M}^T)^{-1} \,, \quad \mathcal{Q} = (\mathcal{Q}_{ij}) \equiv -\mathcal{J}^{-1}\mathcal{M}^{-1} \,. \tag{36.32}$$

Equation (36.31) can then be rewritten as

$$\partial_Z H' = \mathcal{P}\,\partial_Z H + \mathcal{Q}\,\partial_t z \,. \tag{36.33}$$

Integrability of this equation requires

$$\frac{\partial^2 H'}{\partial Z_l \partial Z_k} = \frac{\partial^2 H'}{\partial Z_k \partial Z_l} \,. \tag{36.34}$$

Show from (36.33) that this condition is equivalent to

$$\sum_{r=1}^{2n} \partial_{Z_l} \left(\mathcal{P}_{kr}\, \partial_{Z_r} H + \mathcal{Q}_{kr}\, \partial_t z_r\right) = \sum_{r=1}^{2n} \partial_{Z_k} \left(\mathcal{P}_{lr}\, \partial_{Z_r} H + \mathcal{Q}_{lr}\, \partial_t z_r\right) \,. \tag{36.35}$$

(e) Equation (36.35) must be valid for an arbitrary Hamiltonian function $H(Z_i, t) \equiv H(z_i(Z_j, t))$. It then implies

$$\sum_{r=1}^{2n} \mathcal{P}_{kr}\, \partial_{Z_l}(\partial_{Z_r} H) = \sum_{r=1}^{2n} \mathcal{P}_{lr}\, \partial_{Z_k}(\partial_{Z_r} H) \,, \tag{36.36}$$

$$\sum_{r=1}^{2n} \partial_{Z_l}(\mathcal{P}_{kr})\, \partial_{Z_r} H = \sum_{r=1}^{2n} \partial_{Z_k}(\mathcal{P}_{lr})\, \partial_{Z_r} H \,. \tag{36.37}$$

$$\sum_{r=1}^{2n} \partial_{Z_l}(\mathcal{Q}_{kr} \partial_t z_r) = \sum_{r=1}^{2n} \partial_{Z_k}(\mathcal{Q}_{lr} \partial_t z_r) \,. \tag{36.38}$$

(f) Equation (36.36) implies that the matrix $\mathcal{P}\partial^2 H$ is symmetric, where $\partial^2 H = (\partial_{Z_i}(\partial_{Z_j} H))$ is also symmetric. From this condition show that $\mathcal{P}$ is proportional to the identity matrix: $\mathcal{P} = \theta(Z_i, t) I$, where $\theta(Z_i, t)$ is some function of the indicated variables. This can be achieved with the following steps. Consider a square matrix A satisfying the condition that AB is symmetric for any arbitrary symmetric matrix B of the same size. (i) show that $\sum_k A_{ik} B_{jk} = \sum_k A_{jk} B_{ik}$. (ii) Show that A must also be symmetric by choosing B to be the identity matrix. (iii) Show that A must additionally be diagonal by choosing $B_{ij} = b_i\, \delta_{ij}$, with $b_i \neq b_j$ for $i \neq j$. So we can write $A_{ij} = A_i\, \delta_{ij}$. (iv) Finally, by choosing B to be an arbitrary symmetric matrix, show that $A_i = A_j$.

(g) Show that the result in (f), together with (36.37), imply that $\mathcal{P}$ differs from the identity matrix only by a scalar factor that does not depend explicitly on Z_i:

$$\mathcal{P} = \theta(t)\, I\ . \tag{36.39}$$

(h) It is easily verified from (36.23) that $\mathcal{J}^{-1} = -\mathcal{J}$. Use this fact and the definitions of the matrices $\mathcal{P}$ and $\mathcal{Q}$ [(36.32)] to show that

$$\mathcal{P}^{-1}\mathcal{Q} = \mathcal{M}^{-1}\mathcal{J}\ . \tag{36.40}$$

Hence show that this result, together with (36.39), imply that

$$\mathcal{Q} = \theta(t)\, \mathcal{M}^T\mathcal{J}\ , \tag{36.41}$$

where $\theta(t)$ is the scalar function introduced in (36.39).

(i) Multiply (36.41) on the right by the column matrix $\partial_t z$ and define the column matrix

$$u \equiv \mathcal{J}\, \partial_t z\ . \tag{36.42}$$

Equation (36.38) then implies that

$$\partial_{Z_l}(\mathcal{M}^T u)_k = \partial_{Z_k}(\mathcal{M}^T u)_l\ , \tag{36.43}$$

where $(\mathcal{M}^T u)_k$ is the k-th entry in the column matrix $\mathcal{M}^T u$. Recalling the definition of $\mathcal{M}$ given by (36.29), show that this result implies that

$$\sum_{r=1}^{2n}(\partial_{Z_k} z_r)(\partial_{Z_l} u_r) = \sum_{r=1}^{2n}(\partial_{Z_l} z_r)(\partial_{Z_k} u_r)\ . \tag{36.44}$$

(j) Show that the left-hand side of (36.44) is equal to $(\mathcal{M}^T\mathcal{J}(\partial_t\mathcal{M}))_{kl}$. Hence the right-hand side must be equal to $(\mathcal{M}^T\mathcal{J}(\partial_t\mathcal{M}))^T_{kl}$.

(k) Equation (36.44) is thus equivalent to the matrix equation

$$\mathcal{M}^T\mathcal{J}(\partial_t\mathcal{M}) = (\mathcal{M}^T\mathcal{J}(\partial_t\mathcal{M}))^T\ . \tag{36.45}$$

Show that this is equivalent to

$$\mathcal{M}^T\mathcal{J}(\partial_t\mathcal{M}) + \partial_t\mathcal{M}^T\mathcal{J}\mathcal{M} = 0\ . \tag{36.46}$$

(l) Convince yourself that (36.46) is equivalent to

$$\partial_t(\mathcal{M}^T\mathcal{J}\mathcal{M}) = 0\ , \tag{36.47}$$

that is, $\mathcal{M}^T\mathcal{J}\mathcal{M}$ is explicitly time-independent. From the definition of $\mathcal{P}$ given by the first equation of (36.32) show that

$$\mathcal{P} = -\mathcal{J}^{-1}(\mathcal{M}^T\mathcal{J}\mathcal{M})^{-1}\ . \tag{36.48}$$

Thus (36.47) implies that $\mathcal{P}$ is explicitly time-independent also. Equation (36.39) then implies that $\theta(t)$ in fact does not depend on the time t. So

$$\mathcal{P} = \theta\, I\ , \tag{36.49}$$

where θ is a constant scalar.

(m) Finally, deduce from the definition of $\mathcal{P}$ [(36.32)] that

$$\mathcal{M}^T \mathcal{J} \mathcal{M} = \left(\frac{1}{\theta}\right) \mathcal{J} . \tag{36.50}$$

This is the sought-after result (36.27), if we identify $1/\theta$ with λ.

Problem 36.4 Show that the trivial transformation

$$q^i = Q^i , \qquad p_i = \lambda P_i \tag{36.51}$$

satisfies (36.27). Show that all transformations satisfying (36.27) can be obtained by combining the one given by (36.51) with a canonical transformation [one that satisfies (36.24)].

We will now identify the algebraic origin of symplectic matrices. As defined by (36.24) these matrices can be understood as representations of linear transformations on an even-dimensional real vector space under which a certain *antisymmetric, non-degenerate* and *bilinear* real two-form on the vector space remains invariant. Suppose ω is an antisymmetric, non-degenerate, bilinear two-form on a $2n$-dimensional real vector space $\mathbb{V}$, whose matrix representation with respect to a certain basis of $\mathbb{V}$:

$$\{e_1, \dots, e_n, e_{n+1}, \dots, e_{2n}\} = \{\boldsymbol{\xi}_1, \dots, \boldsymbol{\xi}_n, \boldsymbol{\eta}_1, \dots, \boldsymbol{\eta}_n\} ,$$

called the **canonical basis**, is given by [recall (36.23)]

$$\omega = \mathcal{J} = \begin{pmatrix} 0 & I \\ -I & 0 \end{pmatrix} , \tag{36.52}$$

where 0 denotes the $n \times n$ zero matrix, I denotes the $n \times n$ identity matrix, and $\omega_{ij} = \omega(e_i, e_j)$. In other words,

$$\omega(\boldsymbol{\xi}_i, \boldsymbol{\xi}_j) = \omega(\boldsymbol{\eta}_i, \boldsymbol{\eta}_j) = 0 \quad (i, j = 1, \dots, n) , \tag{36.53}$$

$$\omega(\boldsymbol{\xi}_i, \boldsymbol{\eta}_j) = -\omega(\boldsymbol{\eta}_j, \boldsymbol{\xi}_i) = \delta_{ij} \quad (i, j = 1, \dots, n) . \tag{36.54}$$

This two-form is clearly antisymmetric and non-degenerate. Suppose f is a linear transformation on $\mathbb{V}$. Invariance of ω under the transformation f means that

$$f^* \omega = \omega , \tag{36.55}$$

or, for any two vectors $\boldsymbol{e}, \boldsymbol{e}' \in \mathbb{V}$,

$$(f^*\omega)(\boldsymbol{e}, \boldsymbol{e}') \equiv \omega(f(\boldsymbol{e}), f(\boldsymbol{e}')) = \omega(\boldsymbol{e}, \boldsymbol{e}') . \tag{36.56}$$

Now let the matrix representation of f with respect to the basis $\{e_i\}$ be F_i^j, that is [cf. (1.24)], $f(e_i) = F_i^j\, e_j$. Let $\theta_{ij} = \theta(e_i, e_j)$ be the matrix representation of an arbitrary bilinear two-form θ with respect to the basis $\{\boldsymbol{e}_i\}$. Then, with respect to the basis $\boldsymbol{e}'_i \equiv f(\boldsymbol{e}_i)$, the matrix representation of θ is given by

$$\theta' = F^T \theta F \,, \tag{36.57}$$

where the matrix F is regarded as the matrix with elements F_i^j (j being the row index and i being the column index), and θ is regarded as the matrix with elements θ_{ij} (i being the row index and j being the column index). Indeed,

$$\begin{aligned} \theta'_{ij} &= \theta(\boldsymbol{e}'_i, \boldsymbol{e}'_j) = \theta(F_i^k \boldsymbol{e}_k, F_j^l \boldsymbol{e}_l) = F_i^k F_j^l \, \theta(\boldsymbol{e}_k, \boldsymbol{e}_l) = F_i^k F_j^l \, \theta_{kl} \\ &= \sum_{k,l} (F^T)_k^i \theta_{kl} F_j^l \,, \end{aligned} \tag{36.58}$$

where the third equality follows from the bilinearity property of θ. For the antisymmetric bilinear form ω defined by (36.52), Eq. (36.56) then shows that a linear transformation f keeps ω invariant if $F^T \mathcal{J} F = \mathcal{J}$. Recalling (36.24) as the definition for symplectic matrices, we have the following definition.

Definition 36.2. *A linear transformation f on an even-dimensional vector space $\mathbb{V}$ is said to be a* ***symplectic transformation*** *if its matrix representation $F = (F_i^j)$ ($f(\boldsymbol{e}_i) = F_i^j \, \boldsymbol{e}_j$) with respect to the canonical basis $\{\boldsymbol{e}_i\}$ for a given antisymmetric bilinear form ω satisfies*

$$F^T \mathcal{J} F = \mathcal{J} \,, \tag{36.59}$$

where $\mathcal{J}$ is the matrix representation of ω with respect to its canonical basis [cf. (36.53) and (36.54)]. In the above matrix equation the matrix element F_i^j is the element in the j-th row and the i-th column of the matrix F.

If $dim(\mathbb{V}) = 2n$ and F is written in the form [cf. (36.15)]

$$F = \begin{pmatrix} A & B \\ C & D \end{pmatrix} , \tag{36.60}$$

where each entry is an $n \times n$ block matrix, it is easily verified that the condition (36.59) requires that

$$C^T A = A^T C \,, \qquad D^T A - B^T C = I \,, \qquad D^T B = B^T D \,, \tag{36.61}$$

which is exactly analogous to (36.15).

Write a symplectic matrix F as $F = I + u + \dots$, where u is a infinitesimal matrix: $|u_{ij}| \ll 1$. By (36.59) we have

$$F^T \mathcal{J} F = (I + u^T + \dots) \mathcal{J} (I + u + \dots) = \mathcal{J} + u^T \mathcal{J} + \mathcal{J} u + \dots = \mathcal{J} \,. \tag{36.62}$$

This implies that

$$u^T \mathcal{J} + \mathcal{J} u = 0 \,. \tag{36.63}$$

We will take this equation to be the defining condition for a matrix u to be *infinitesimally symplectic.* This immediately leads to the notion of **infinitesimal symplectic transformations**. We have the following theorems for infinitesimal symplectic transformations.

Theorem 36.2. *The following statements concerning an infinitesimal symplectic transformation u on an even-dimensional vector space $\mathbb{V}$ are equivalent to each other [in (i) and (ii) u denotes the matrix representation of the symplectic transformation with respect to the canonical basis of $\mathbb{V}$]:*

(i) $u^T\mathcal{J} + \mathcal{J}u = 0$.

(ii) If one writes in block matrix form

$$u = \begin{pmatrix} a & b \\ c & d \end{pmatrix} ,$$

then $d = -a^T$, $c = c^T$, $b = b^T$.

(iii) $\omega(u\mathbf{e}, \mathbf{e}') + \omega(\mathbf{e}, u\mathbf{e}') = 0$, *for any* $\mathbf{e}, \mathbf{e}' \in \mathbb{V}$.

Proof. We will adopt the following strategy: (i) ⇒ (ii) ⇒ (iii) ⇒ (i). Assuming (i) we have

$$\begin{aligned} & u^T\mathcal{J} + \mathcal{J}u \\ &= \begin{pmatrix} a^T & c^T \\ b^T & d^T \end{pmatrix}\begin{pmatrix} 0 & I \\ -I & 0 \end{pmatrix} + \begin{pmatrix} 0 & I \\ -I & 0 \end{pmatrix}\begin{pmatrix} a & b \\ c & d \end{pmatrix} \\ &= \begin{pmatrix} -c^T & a^T \\ -d^T & b^T \end{pmatrix} + \begin{pmatrix} c & d \\ -a & -b \end{pmatrix} = \begin{pmatrix} c - c^T & a^T + d \\ -(a+d^T) & b^T - b \end{pmatrix} = 0 . \end{aligned} \tag{36.64}$$

Statement (ii) then follows.

Let $\{\boldsymbol{\xi}_1, \dots, \boldsymbol{\xi}_n, \boldsymbol{\eta}_1, \dots, \boldsymbol{\eta}_n\}$ be the canonical basis of ω and write

$$u\,\boldsymbol{\xi}_i = a_i^j\,\boldsymbol{\xi}_j + b_i^j\,\boldsymbol{\eta}_j , \qquad u\,\boldsymbol{\eta}_i = c_i^j\,\boldsymbol{\xi}_j + d_i^j\,\boldsymbol{\eta}_j . \tag{36.65}$$

Let two arbitrary vectors $\mathbf{e}$ and $\mathbf{e}'$ in $\mathbb{V}$ be given with respect to the canonical basis by

$$\mathbf{e} = x^i\,\boldsymbol{\xi}_i + y^i\,\boldsymbol{\eta}_i , \qquad \mathbf{e}' = x'^k\,\boldsymbol{\xi}_k + y'^k\,\boldsymbol{\eta}_k . \tag{36.66}$$

We then have

$$\begin{aligned} u\,\mathbf{e} &= x^i\,(u\,\boldsymbol{\xi}_i) + y^i\,(u\,\boldsymbol{\eta}_i) = x^i a_i^j\,\boldsymbol{\xi}_j + x^i b_i^j\,\boldsymbol{\eta}_j + y^i c_i^j\,\boldsymbol{\xi}_j + y^i d_i^j\,\boldsymbol{\eta}_j \\ &= (x^i a_i^j + y^i c_i^j)\,\boldsymbol{\xi}_j + (x^i b_i^j + y^i d_i^j)\,\boldsymbol{\eta}_j . \end{aligned} \tag{36.67}$$

Similarly,

$$u\,\mathbf{e}' = (x'^i a_i^j + y'^i c_i^j)\,\boldsymbol{\xi}_j + (x'^i b_i^j + y'^i d_i^j)\,\boldsymbol{\eta}_j\,. \tag{36.68}$$

Hence

$$\begin{aligned}
&\omega(u\,\mathbf{e}, \mathbf{e}') \\
&= (x^i a_i^j + y^i c_i^j)\,x'^k\,\omega(\boldsymbol{\xi}_j, \boldsymbol{\xi}_k) + (x^i a_i^j + y^i c_i^j)\,y'^k\,\omega(\boldsymbol{\xi}_j, \boldsymbol{\eta}_k) \\
&\quad + (x^i b_i^j + y^i d_i^j)\,x'^k\,\omega(\boldsymbol{\eta}_j, \boldsymbol{\xi}_k) + (x^i b_i^j + y^i d_i^j)\,y'^k\,\omega(\boldsymbol{\eta}_j, \boldsymbol{\eta}_k) \\
&= (x^i a_i^j + y^i c_i^j)\,y'^k\,\delta_{jk} + (x^i b_i^j + y^i d_i^j)\,x'^k\,(-\delta_{jk}) \\
&= \sum_{i,j} \left(x^i y'^j a_i^j + y^i y'^j c_i^j - x^i x'^j b_i^j - y^i x'^j d_i^j \right),
\end{aligned} \tag{36.69}$$

where the second equality follows from (36.53) and (36.54). Similarly

$$\begin{aligned}
\omega(\mathbf{e}, u\mathbf{e}') &= x^k\,(x'^i b_i^j + y'^i d_i^j)\,\delta_{kj} - y^k\,(x'^i a_i^j + y'^i c_i^j)\,\delta_{kj} \\
&= \sum_{i,j} \left(x^j x'^i b_i^j + x^j y'^i d_i^j - y^j x'^i a_i^j - y^j y'^i c_i^j \right) \\
&= \sum_{i,j} \left(x^i x'^j b_j^i + x^i y'^j d_j^i - y^i x'^j a_j^i - y^i y'^j c_j^i \right),
\end{aligned} \tag{36.70}$$

where in the last equality we have interchanged the indices i and j. From the above two equations it follows that

$$\begin{aligned}
&\omega(u\mathbf{e}, \mathbf{e}') + \omega(\mathbf{e}, u\mathbf{e}') \\
&= \sum_{i,j} \left(x^i y'^j\,(a_i^j + d_j^i) - x'^j y^i\,(a_j^i + d_i^j) + y^i y'^j\,(c_i^j - c_j^i) + x^i x'^j\,(b_j^i - b_i^j) \right).
\end{aligned} \tag{36.71}$$

Thus statement (ii) implies statement (iii).

We also see from (36.71) that, since x^i, y^i, x'^i and y'^i are completely arbitrary constants, statement (iii) implies statement (ii). But from the third equality of (36.64), (ii) $\Rightarrow$ (i). Thus, finally, (iii) $\Rightarrow$ (i), and the theorem is proved. □

Theorem 36.3. *If λ is an eigenvalue of an infinitesimal symplectic transformation, so is $-\lambda$. Complex eigenvalues of an infinitesimal symplectic transformation must occur in complex conjugate pairs.*

Proof. The characteristic polynomial $P(\lambda)$ of an infinitesimal symplectic transformation u is given by

$$P(\lambda) \equiv det(u - \lambda I)\,, \tag{36.72}$$

and the eigenvalues of u are obtained from the solutions of $P(\lambda) = 0$. Suppose the dimension of the vector space on which u acts is $2n$. We have

$$\begin{aligned} P(\lambda) &= det(u - \lambda I) = det[\mathcal{J}(u - \lambda I)] = det(\mathcal{J}u - \lambda\mathcal{J}) \\ &= det(-u^T\mathcal{J} - \lambda\mathcal{J}) = det[-(u^T + \lambda)\mathcal{J}] = (-1)^{2n}\, det[(u^T + \lambda)\mathcal{J}] \\ &= det(u^T + \lambda I) = det[(u + \lambda I)^T] = det(u + \lambda I) = det[u - (-\lambda)I] \\ &= P(-\lambda)\,, \end{aligned} \tag{36.73}$$

where the second equality follows from the fact that $det(\mathcal{J}) = 1$ and the fourth from (36.63) (the condition that u being infinitesimal symplectic). Thus $P(\lambda) = 0$ if and only if $P(-\lambda) = 0$, and the first statement of the theorem is proved. The second statement is obvious since, u being a real matrix, the coefficients of its characteristic polynomial are all real. □

Chapter 37

Generating Functions and the Hamilton-Jacobi Equation

In this chapter we will consider various methods to generate canonical transformations $\{q^i, p_i\} \to \{Q^i, P_i\}$, with the ultimate goal of simplifying a given set of Hamilton's equations of motion.

The starting point is the *generating function* $v(Q^i, P_i, t)$ as introduced in (36.8). We will assume the general case where the Lagrangian function $L(q^i, \dot{q}^i, t)$ may be explicitly time-dependent. Introduce a function $F_1(q^i, Q^i, t)$ (note the variables) defined by

$$F_1(q^i, Q^i, t) \equiv v(Q^i, P_i(q^j, Q^j), t) , \tag{37.1}$$

where $P_i(q^j, Q^j)$ as functions of the old q's and the new Q's have been obtained from solutions of the equations $q^i = q^i(Q^j, P_j)$ [first set of equations in (36.1)]. In order for the solutions to exist we require, by the *Inverse Function Theorem* of multivariable calculus, that [cf. (36.16)]

$$\left| \frac{\partial q^i}{\partial P_j} \right| \equiv |\mathcal{B}| \neq 0 . \tag{37.2}$$

(Note that there are canonical transformations for which this condition does not hold, as, for example, the trivial identity transformation $Q^i = q^i, P_i = p_i$.) Canonical transformations satisfying (37.2) will be called *canonical transformations of the first kind.* Then we have, on applying (36.12) and (36.13),

$$\frac{\partial v}{\partial P_i} = \frac{\partial F_1}{\partial q^j} \frac{\partial q^j}{\partial P_i} = p_j \frac{\partial q^j}{\partial P_i} , \tag{37.3}$$

$$\frac{\partial v}{\partial Q^i} = \frac{\partial F_1}{\partial q^j} \frac{\partial q^j}{\partial Q^i} + \frac{\partial F_1}{\partial Q^i} = p_j \frac{\partial q^j}{\partial Q^i} - P_i . \tag{37.4}$$

The second equalities of each of the above two equations imply

$$\boxed{p_i = \frac{\partial F_1(q^j, Q^j, t)}{\partial q^i}\,, \qquad P_i = -\frac{\partial F_1(q^j, Q^j, t)}{\partial Q^i}}\,. \tag{37.5}$$

It also follows from (36.14) that the new Hamiltonian function is related to the old one by

$$H'(Q^i, P_i, t) = H(q^i, p_i, t) + \frac{\partial F_1}{\partial t}\,, \tag{37.6}$$

since from (37.1), $\partial F_1/\partial t = \partial v/\partial t$. Note that (37.5) allows us to write

$$dF_1 = p_i\, dq^i - P_i\, dQ^i + \frac{\partial F_1}{\partial t}\, dt\,. \tag{37.7}$$

If $F_1(q^i, Q^i, t)$ is known, the $2n$ new canonical coordinates Q^i and P_i can be solved in terms of (q^j, p_j, t) from the $2n$ equations (37.5). The solutions are guaranteed to exist since (37.2) implies $|\partial P_i/\partial q^j| \neq 0$. By the second equation of (37.5) this implies

$$\left|\frac{\partial^2 F_1}{\partial q^i \partial Q^j}\right| \neq 0\,. \tag{37.8}$$

This condition guarantees that the first set of n equations in (37.5) can be solved for Q^i in terms of (q^j, p_j, t). The second set of n equations in (37.5) can then be solved for P_i in terms of (q^j, p_j, t) also. The function $F_1(q^i, Q^i, t)$ satisfying (37.8) and (37.5) for a canonical transformation $\{q^i, p_i\} \to \{Q^i, P_i\}$ is known as a *generating function of the first kind.*

Problem 37.1 Show that the generating function of the first kind

$$F_1(q^i, Q^i) = \sum_{i=1}^{n} q^i Q^i \tag{37.9}$$

results in the canonical transformation $Q^i = p_i, P_i = -q^i$.

In a similar manner we can treat canonical transformations for which [cf. (36.16)]

$$\left|\frac{\partial q^i}{\partial Q^j}\right| \equiv |\mathcal{A}| \neq 0\,. \tag{37.10}$$

These will be referred to as *canonical transformations of the second kind.* Rather than proceeding directly as above, we will convert a canonical transformation that satisfies (37.10) to one that satisfies (37.2). Suppose $\{q^i, p_i\} \to \{Q^i, P_i\}$ is a canonical transformation that satisfies (37.10), and let it be followed by the canonical transformation $\{Q^i, P_i\} \to \{Q'^i, P'_i\}$:

$$Q^i = -P'_i\,, \qquad P_i = Q'^i\,. \tag{37.11}$$

The latter is clearly a canonical transformation that satisfies (37.2). The resulting composite transformation $\{q^i, p_i\} \to \{Q'^i, P'_i\}$, by virtue of the fact that canonical transformations are elements of the symplectic group (as established in the last chapter), is also a canonical transformation, and satisfies (37.2). Let $v'(Q^i, P_i)$ be the generating function for the first canonical transformation that satisfies

$$L = L' + \frac{dv'(Q^i, P_i, t)}{dt}\,, \tag{37.12}$$

[cf. (36.8)], where L' is defined by (36.7); and $v''(Q'^i, P'_i)$ be the generating function for the second that satisfies

$$L' = L'' + \frac{dv''(Q'^i, P'_i)}{dt}\,, \tag{37.13}$$

with $L'' = P'_i\dot{Q}'^i - H''(Q'^i, P'_i, t)$ for some function H'' with the indicated variables [cf. (36.7)]. We thus have $d(v'+v'')/dt = L - L''$, and $v \equiv v' + v''$ is the generating function for the composite canonical transformation $\{q^i, p_i\} \to \{Q'^i, P'_i\}$. By the result of Problem (37.1), v'' is a generating function of the first kind and thus can be written

$$v''(Q'^i, P'_i) = v''(Q^i, Q'^i) = \sum_i Q^i Q'^i = Q^i P_i\,, \tag{37.14}$$

where the second equality is a result of the specific canonical transformation (37.11). Since the composite canonical transformation satisfies (37.2), v can be written as a generating function of the first kind (for the composite transformation) whose variables are the old and new canonical positions:

$$v(q^i, Q'^i, t) = v(q^i, P_i, t)\,, \tag{37.15}$$

where the change from Q'^i to P_i in the arguments follows from (37.11). We thus have

$$v'(Q^i(q^j, P_j), P_i, t) = v(q^i, P_i, t) - P_i Q^i(q^j, P_j)\,, \tag{37.16}$$

with, as follows from the second equation of (37.5),

$$Q^i = -P'_i = \frac{\partial v(q^j, Q'^j, t)}{\partial Q'^i} = \frac{\partial v(q^j, P_j, t)}{\partial P_i}\,. \tag{37.17}$$

With this equation $v(q^i, P_i)$ is seen as the *Legendre transform* of a function $V(q^i, Q^i)$ with respect to Q^i:

$$v(q^i, P_i, t) = V(q^i, Q^i(q^j, P_j), t) + P_i Q^i(q^j, P_j) \tag{37.18}$$

such that

$$P_i = -\frac{\partial V(q^j, Q^j, t)}{\partial Q^i} . \tag{37.19}$$

Comparing (37.18) with (37.16), we see that $V(q^i, Q^i, t)$ has to satisfy

$$V(q^i, Q^i(q^j, P_j), t) = v'(Q^i(q^j, P_j), P_i, t) . \tag{37.20}$$

We will rename $v(q^i, P_i, t)$ by $F_2(q^i, P_i, t)$ and write

$$F_2(q^i, P_i, t) \equiv V(q^i, Q^i(q^j, P_j), t) + P_i\, Q^i(q^j, P_j) . \tag{37.21}$$

In the above equation [as well as in (37.16) and (37.18)] Q^i can be expressed in the functional form $Q^i = Q^i(q^j, P_j)$ by virtue of the fact that the canonical transformation under consideration $\{q^i, p_i\} \to \{Q^i, P_i\}$ satisfies (37.10). We also have, from (37.11) and the first equation in (37.5),

$$p_i = \frac{\partial v(q^j, Q'^j, t)}{\partial q^i} = \frac{\partial v(q^j, P_j, t)}{\partial q^i} . \tag{37.22}$$

Incorporating this equation and (37.17) we can write

$$\boxed{p_i = \frac{\partial F_2(q^j, P_j, t)}{\partial q^i}, \qquad Q^i = \frac{\partial F_2(q^j, P_j, t)}{\partial P_i}} , \tag{37.23}$$

in analogy to (37.5). These equations can be used to solve for the new canonical coordinates (Q^i, P_i) in terms of the old ones (q^i, p_i). The solutions are guaranteed to exist due to (37.10) and the second equation of (37.22), which together imply, in analogy to (37.8),

$$\left| \frac{\partial^2 F_2}{\partial q^j \partial P_i} \right| \neq 0 . \tag{37.24}$$

The function $F_2(q^i, P_i, t)$ can then be used as a generating function for a canonical transformation $\{q^i, p_i\} \to \{Q^i, P_i\}$ that satisfies (37.10), and is called a *generating function of the second kind.* Since $\partial v''/\partial t = 0$ [from (37.14)], it follows from (37.6) that $H''(Q'^i, P'_i, t) = H'(Q^i, P_i, t)$. On the other hand, considering the composite canonical transformation $\{q^i, p_i\} \to \{Q'^i, P'_i\}$, we have $H''(Q'^i, P'_i, t) = H(q^i, p_i, t) + \partial v/\partial t$. Thus, for a canonical transformation generated by $F_2(q^i, P_i, t)$, the old and new Hamiltonian functions are related by

$$H'(Q^i, P_i, t) = H(q^i, p_i, t) + \frac{\partial F_2}{\partial t} . \tag{37.25}$$

In analogy to (37.7) we have

$$dF_2 = p_i\, dq^i + Q^i\, dP_i + \frac{\partial F_2}{\partial t}\, dt . \tag{37.26}$$

In the literature, $F_2(q^i, P_i, t)$ is often denoted by $S(q^i, P_i, t)$, the standard symbol used for the *classical action.* This quantity is also often called **Hamilton's principal function.**

At this point it will be useful to introduce the following theorem.

Theorem 37.1. *Every canonical transformation can be expressed as the composition of two canonical transformations each of which is either of the first kind or the second kind.*

Proof. Referring to (36.15) we see that every canonical transformation is one for which either $|\mathcal{B}| \neq 0$ (the first kind), $|\mathcal{A}| \neq 0$ (the second kind), $|\mathcal{D}| \neq 0$ or $|\mathcal{C}| \neq 0$. We will refer to the last two kinds as canonical transformations of the third and fourth kinds, respectively. Of course, a particular canonical transformation can belong to more than one of these categories. The theorem is obvious for canonical transformations of the first two kinds.

Consider a canonical transformation $\mathcal{M}$ written in the matrix form (36.15). Let $\mathcal{M}' = \mathcal{J}\mathcal{M}$, with $\mathcal{J}$ given by (36.23). So $\mathcal{M} = \mathcal{J}^{-1}\mathcal{M}'$, with

$$\mathcal{M}' = \begin{pmatrix} 0 & I \\ -I & 0 \end{pmatrix} \begin{pmatrix} \mathcal{A} & \mathcal{B} \\ \mathcal{C} & \mathcal{D} \end{pmatrix} = \begin{pmatrix} \mathcal{C} & \mathcal{D} \\ -\mathcal{A} & -\mathcal{B} \end{pmatrix} . \tag{37.27}$$

$\mathcal{J}$ is clearly a canonical transformation of the first kind, as is $\mathcal{J}^{-1} = -\mathcal{J}$. The former represents the canonical transformation $q^i = P_i, p_i = -Q^i$, while the latter the canonical transformation $q^i = -P_i, p_i = Q^i$. If $\mathcal{M}$ is a canonical transformation of the third kind (with $|\mathcal{A}| = |\mathcal{B}| = |\mathcal{C}| = 0, |\mathcal{D}| \neq 0$), then $\mathcal{M}'$ is obviously a canonical transformation of the first kind. If, on the other hand, $\mathcal{M}$ is a canonical transformation of the fourth kind ($|\mathcal{A}| = |\mathcal{B}| = |\mathcal{D}| = 0, |\mathcal{C}| \neq 0$), then $\mathcal{M}'$ is a canonical transformation of the second kind. So in both cases the theorem is also proved. □

Problem 37.2 Show by explicit matrix multiplication, in a manner similar to (37.27), how a canonical transformation of the first kind can be converted into a canonical transformation of the second kind.

In order to arrive at generating functions [as Legendre transforms in analogy to (37.21)] for canonical transforms of the third and fourth kinds, we proceed as follows. First note that, with a symplectic matrix $\mathcal{M}$ representing a canonical transform $\{q^i, p_i\} \to \{Q^i, P_i\}$ as given by (36.15) and (36.16), we have

$$\mathcal{M}^{-1} = \begin{pmatrix} \dfrac{\partial Q^i}{\partial q^j} & \dfrac{\partial Q^i}{\partial p_j} \\ \dfrac{\partial P_i}{\partial q^j} & \dfrac{\partial P_i}{\partial p_j} \end{pmatrix} , \tag{37.28}$$

where both Q^i and P_i are regarded as functions of (q^j, p_j). On the other hand, $\mathcal{M}^{-1}$ can also be calculated quite simply from the requirement (36.27). This equation implies

$$\mathcal{J}\mathcal{M} = (\mathcal{M}^T)^{-1}\mathcal{J} = (\mathcal{M}^{-1})^T\mathcal{J} . \tag{37.29}$$

Hence

$$(\mathcal{M}^{-1})^T = \mathcal{J}\mathcal{M}\mathcal{J}^{-1} = -\mathcal{J}\mathcal{M}\mathcal{J} \,. \tag{37.30}$$

Using the matrix forms (36.15) for $\mathcal{M}$ and (36.23) for $\mathcal{J}$, it is straightforwardly seen that

$$(\mathcal{M}^{-1})^T = \begin{pmatrix} \mathcal{D} & -\mathcal{C} \\ -\mathcal{B} & \mathcal{A} \end{pmatrix} , \tag{37.31}$$

and hence

$$\mathcal{M}^{-1} = \begin{pmatrix} \mathcal{D}^T & -\mathcal{B}^T \\ -\mathcal{C}^T & \mathcal{A}^T \end{pmatrix} . \tag{37.32}$$

Comparing with (37.28) we see that $|\mathcal{D}| \neq 0$ implies $|\partial Q^i/\partial q^j| \neq 0$, and $|\mathcal{C}| \neq 0$ implies $|\partial P_i/\partial q^j| \neq 0$. It follows that for canonical transformations of the third kind ($|\mathcal{D}| \neq 0$), q^i can be expressed as functions $q^i(p_j, Q^j)$, and for canonical transformations of the fourth kind ($|\mathcal{C}| \neq 0$), q^i can be expressed as functions $q^i(p_j, P_j)$.

Consider the following sequence of canonical transformations

$$\{Q'^i P'_i\} \xrightarrow{v''} \{q^i, p_i\} \xrightarrow{v'} \{Q^i, P_i\} \,,$$

where v'' is the generating function for the canonical transformation $Q'^i = -p_i, P'_i = q^i$, and v' is a generating function for a canonical transformation of either the third or fourth kind. Then we have the composite canonical transformation

$$\{Q'^i, P'_i\} \xrightarrow{v} \{Q^i, P_i\} \,,$$

where $v = v'' + v'$ is the generating function for this transformation. It is quite obvious that if v' generates a canonical transformation of the third kind, then v generates one of the first kind; while if v' generates a canonical transformation of the fourth kind, then v generates one of the second kind. Analogous to (37.14) we have

$$v''(q^i, p_i) = \sum_i q^i Q'^i = -p_i q^i \,. \tag{37.33}$$

(Note that a generating function can either be written as a function of the old canonical coordinates or the new ones.) We then have

$$v(Q'^i, P'_i, t) = v'(q^i, p_i, t) - p_i q^i \,, \tag{37.34}$$

where the arguments of the generating functions are written in terms of the respective old coordinates. Since $Q'^i = -p_i$ and $P'_i = q^i$ we can write the above equation as

$$v(p_i, q^i, t) = v'(q^i, p_i, t) - p_i q^i \,. \tag{37.35}$$

Consider first the case where v' generates a canonical transformation of the third kind, in which case, as discussed above, it is possible to write $q^i = q^i(p_j, Q^j)$. We will define a function $F_3(p_i, Q^i, t)$ by

$$F_3(p_i, Q^i, t) \equiv v(p_i, q^i(p_j, Q^j), t) \ . \tag{37.36}$$

Since $v(Q'^i, P'_i, t)$ generates a canonical transformation of the first kind, we have

$$q^i = P'_i = \frac{\partial v}{\partial Q'^i} = -\frac{\partial v}{\partial p_i} = -\frac{\partial F_3(p_i, Q^i, t)}{\partial p_i} \ , \tag{37.37}$$

where the second equality follows from the first equation of (37.5). At the same time, the second equation of (37.5) implies

$$P_i = -\frac{\partial v}{\partial Q^i} = -\frac{\partial F_3(p_i, Q^i, t)}{\partial Q^i} \ . \tag{37.38}$$

We will incorporate the above two equations to write

$$\boxed{q^i = -\frac{\partial F_3(p_i, Q^i, t)}{\partial p_i} \ , \qquad P_i = -\frac{\partial F_3(p_i, Q^i, t)}{\partial Q^i}} \ . \tag{37.39}$$

These are analogous to (37.5) and (37.23) for the generating functions of the first and second kinds, respectively. As before, the condition $|\partial P_i/\partial p_j| \neq 0$ (for canonical transformations of the third kind), together with the second of the above equations, imply that

$$\left| \frac{\partial^2 F_3}{\partial p_j \partial Q^i} \right| \neq 0 \ , \tag{37.40}$$

which is sufficient for (37.39) to yield the solutions $Q^i(q^j, p_j)$ and $P_i(q^j, p_j)$. The function $F_3(p_i, Q^i, t)$ is called a *generating function of the third kind*. The three Hamiltonian functions (written as functions of their natural variables) are related by $H'(P_i, Q^i, t) = H_0(P'_i, Q'^i, t) + \partial v/\partial t$ and $H(p_i, q^i, t) = H_0(P'_i, Q'^i, t)$ (since $\partial v''/\partial t = 0$). It follows that, for a canonical transformation of the third kind,

$$H'(Q^i, P_i, t) = H(q^i, p_i, t) + \frac{\partial F_3}{\partial t} \ . \tag{37.41}$$

By (37.39) we have, analogous to (37.26),

$$dF_3 = -q^i \, dp_i - P_i \, dQ^i + \frac{\partial F_3}{\partial t} \, dt \ . \tag{37.42}$$

Writing (37.35) as

$$F_3(p_i, Q^i, t) = v(p_i, q^i(p_j, Q^j), t) = v'(q^i(p_j, Q^j), p_i, t) - p_i q^i(p_j, Q^j) \ , \tag{37.43}$$

we see that $F_3(p_i, Q^i, t)$ is the Legendre transform of a function $V(q^i, Q^i, t)$ with respect to q^i:

$$F_3(p_i, Q^i, t) = V(q^i(p_j, Q^j), Q^i, t) - p_i q^i(p_j, Q^j) , \tag{37.44}$$

such that

$$p_i = \frac{\partial V(q^j, Q^j, t)}{\partial q^i} , \tag{37.45}$$

and

$$V(q^i(p_j, Q^j), Q^i, t) = v'(q^i(p_j, Q^j), p_i, t) . \tag{37.46}$$

With a canonical transformation of the fourth kind, the q^i can be expressed as functions $q^i(p_j, P_j)$ and we can rewrite (37.43) as

$$F_4(p_i, P_i, t) \equiv v(p_i, q^i(p_j, P_j), t) = v'(q^i(p_j, P_j), p_i, t) - p_i q^i(p_j, P_j) . \tag{37.47}$$

As in the case for $F_3(p_i, Q^i, t)$ the function $F_4(p_i, P_i, t)$ is seen as the Legendre transform of a function $V(q^i, P_i, t)$ with respect to q^i:

$$F_4(p_i, P_i, t) = V(q^i(p_j, P_j), P_i, t) - p_i q^i(p_j, P_j) , \tag{37.48}$$

such that

$$p_i = \frac{\partial V(q^j, P_j, t)}{\partial q^i} , \tag{37.49}$$

and

$$V(q^i(p_j, P_j), P_i, t) = v'(q^i(p_j, P_j), p_i, t) . \tag{37.50}$$

Since now $v(Q'^i, P'_i, t)$ generates a canonical transformation of the second kind, we have, by the first equation of (37.23),

$$q^i = P'_i = \frac{\partial v}{\partial Q'^i} = -\frac{\partial v}{\partial p_i} = -\frac{\partial F_4}{\partial p_i} , \tag{37.51}$$

and also, by the second equation of (37.23),

$$Q^i = \frac{\partial v}{\partial P_i} = \frac{\partial F_4}{\partial P_i} . \tag{37.52}$$

Putting the above two equations together we have

$$\boxed{q^i = -\frac{\partial F_4(p_j, P_j, t)}{\partial p_i} , \qquad Q^i = \frac{\partial F_4(p_j, P_j, t)}{\partial P_i}} . \tag{37.53}$$

The condition $|\partial Q^i / \partial p_j| \neq 0$ (for canonical transformations of the fourth kind) ensures that

$$\left| \frac{\partial^2 F_4}{\partial p_i \partial P_j} \right| \neq 0 , \tag{37.54}$$

which is sufficient for (37.52) to yield the solutions $Q^i(q^j, p_j)$ and $P_i(q^j, p_j)$. The function $F_4(p_i, P_i, t)$ is called a *generating function of the fourth kind.* By exactly the same reason which leads to (37.41), we have, for canonical transformations of the fourth kind,

$$H'(Q^i, P_i, t) = H(q^i, p_i, t) + \frac{\partial F_4}{\partial t} \; . \tag{37.55}$$

Finally, (37.39) implies

$$dF_4 = -q^i \, dp_i + Q^i \, dP_i + \frac{\partial F_4}{\partial t} \, dt \; . \tag{37.56}$$

The generating functions $F_1(q^i, Q^i, t)$, $F_2(q^i, P_i, t)$, $F_3(p_i, Q^i, t)$ and $F_4(p_i, P_i, t)$ furnish another useful criterion for a transformation being canonical. Consider the differential (or the exterior derivative) of F_1 as given by (37.7). Exteriorly differentiating dF_1 we have

$$0 = d^2 F_1 = dp_i \wedge dq^i - dP_i \wedge dQ^i + d\left(\frac{\partial F_1}{\partial t} \, dt \right) \; . \tag{37.57}$$

But

$$\begin{aligned} & d\left(\frac{\partial F_1(q^i, Q^i, t)}{\partial t} \, dt \right) \\ &= \frac{\partial^2 F_1}{\partial t \, \partial q^i} \, dq^i \wedge dt + \frac{\partial^2 F_1}{\partial t \, \partial Q^i} \, dQ^i \wedge dt + \frac{\partial^2 F_1}{\partial t^2} \, dt \wedge dt \\ &= \frac{\partial p_i}{\partial t} \, dq^i \wedge dt - \frac{\partial P_i}{\partial t} \, dQ^i \wedge dt = 0 \; , \end{aligned} \tag{37.58}$$

where the second equality follows from (37.5) and the fact that $dt \wedge dt = 0$. It then follows from (37.57) that, under a canonical transformation $\{q^i, p_i\} \to \{Q^i, P_i\}$ of the first kind,

$$\boxed{dp_i \wedge dq^i = dP_i \wedge dQ^i} \; . \tag{37.59}$$

It is easy to check that the same result is obtained on exteriorly differentiating dF_2, dF_3 and dF_4 [by using the expressions for these differentials given by (37.26), (37.42) and (37.56)]. Hence the important result (37.59) is true for any canonical transformation in general. In other words, *the two-form $dp_i \wedge dq^i$ is invariant under canonical transformations.* It is called the *symplectic form* in the cotangent bundle T^*Q. We will discuss the geometrical significance of this result in Chapter 39.

Conversely, if we assume the validity of (37.59), we can infer the existence of four different kinds of generating functions that would yield canonical transformations. Indeed, making use of the rules of exterior differentiation [cf. (6.13)], the vanishing of the two-form $dp_i \wedge dq^i - dP_i \wedge dQ^i$ can be expressed in four different ways:

$$d(p_i\, dq^i - P_i\, dQ^i) = 0\,, \tag{37.60a}$$
$$d(p_i\, dq^i + Q^i\, dP_i) = 0\,, \tag{37.60b}$$
$$d(-q^i\, dp_i - P_i\, dQ^i) = 0\,, \tag{37.60c}$$
$$d(-q^i\, dp_i + Q^i\, dP_i) = 0\,. \tag{37.60d}$$

Since, as indicated, all the expressions above within parentheses are closed one-forms, by Poincaré's Lemma, each is also locally exact:

$$p_i\, dq^i - P_i\, dQ^i = d\mathcal{F}_1(q^i, Q^i)\,, \tag{37.61a}$$
$$p_i\, dq^i + Q^i\, dP_i = d\mathcal{F}_2(q^i, P_i)\,, \tag{37.61b}$$
$$-q^i\, dp_i - P_i\, dQ^i = d\mathcal{F}_3(p_i, Q^i)\,, \tag{37.61c}$$
$$-q^i\, dp_i + Q^i\, dP_i = d\mathcal{F}_4(p_i, P_i)\,, \tag{37.61d}$$

where the four functions $\mathcal{F}_1, \mathcal{F}_2, \mathcal{F}_3$ and $\mathcal{F}_4$ satisfy

$$\frac{\partial \mathcal{F}_1(q^j, Q^j)}{\partial q^i} = p_i\,, \qquad \frac{\partial \mathcal{F}_1(q^j, Q^j)}{\partial Q^i} = -P_i\,, \tag{37.62a}$$
$$\frac{\partial \mathcal{F}_2(q^j, P^j)}{\partial q^i} = p_i\,, \qquad \frac{\partial \mathcal{F}_2(q^j, P^j)}{\partial P^i} = Q_i\,, \tag{37.62b}$$
$$\frac{\partial \mathcal{F}_3(p_j, Q^j)}{\partial p_i} = -q^i\,, \qquad \frac{\partial \mathcal{F}_3(p_j, Q^j)}{\partial Q^i} = -P_i\,, \tag{37.62c}$$
$$\frac{\partial \mathcal{F}_4(p_j, P^j)}{\partial p_i} = -q^i\,, \qquad \frac{\partial \mathcal{F}_4(p_j, P_j)}{\partial P_i} = Q^i\,. \tag{37.62d}$$

We can then construct functions $F_1(q^i, Q^i, t), F_2(q^i, P_i, t), F_3(p_i, Q^i, t)$ and $F_4(p_i, P_i, t)$ such that

$$d(F_\alpha - \mathcal{F}_\alpha) = \frac{\partial F_\alpha}{\partial t}\, dt \qquad (\alpha = 1, 2, 3, 4)\,. \tag{37.63}$$

It follows that the functions F_α satisfy (37.5), (37.23), (37.39) and (37.53) for $\alpha = 1, 2, 3, 4$, respectively. These functions will then be the generating functions for general canonical transformations, as discussed earlier. We have thus proved the following useful theorem.

Theorem 37.2. *A transformation $\{q^i, p_i\} \to \{Q^i, P_i\}$ on the cotangent bundle T^*Q of a configuration manifold Q is canonical if and only if the two-form $dp_i \wedge dq^i$ on T^*Q is invariant under the transformation, that is, if and only if $dp_i \wedge dq^i = dP_i \wedge dQ^i$.*

Problem 37.3 By computing $dP \wedge dQ$ use Theorem 37.2 to show that the following transformation of canonical coordinates for a 1-dimensional system:

$$Q = q^\alpha \cos\beta p \,, \qquad P = q^\alpha \sin\beta p$$

is canonical only if $\alpha = 1/2$ and $\beta = 2$. Assuming these values for α and β construct a generating function of the first kind $F_1(q, Q)$ for the canonical transformation by requiring that $dF_1 = pdq - PdQ$. Show by integrating this expression that

$$F_1(q, Q(p,q)) = q(p - \frac{1}{4}\sin 4p) \,.$$

This is most straightforwardly done by expressing P and Q in terms of p and q in the expression for dF_1.

Problem 37.4 Consider the following transformation of canonical coordinates:

$$P_i = \frac{p_i}{\sum_j (p_j)^2} \,, \qquad Q^i = q^i \sum_j (p_j)^2 - 2p_i(q^j p_j) \,.$$

(a) Compute $dP_i \wedge dQ^i$ and use Theorem 37.2 to show that the transformation is canonical.

(b) Define

$$a \equiv \sum_i (q^i)^2 \,, \qquad b \equiv q^i p_i \,, \qquad c \equiv \sum_i (p_i)^2 \,,$$

$$A \equiv \sum_i (Q^i)^2 \,, \qquad B \equiv Q^i P_i \,, \qquad C \equiv \sum_i (P_i)^2 \,.$$

Show that

$$A = ac^2 \,, \qquad B = -b \,, \qquad C = \frac{1}{c} \,,$$

and hence deduce that

$$AC - B^2 = ac - b^2 \,.$$

(c) Compute Hamiltonian's principal function S (generating function of the second kind) for this canonical transformation [apart from a term which depends linearly on the time t and does not depend on (q^i, p_i) {see (37.75) below}] by requiring that $dS = p_i dq^i + Q^i dP_i$. Proceed by writing $Q^i dP_i$ in terms of the old coordinates q^i and p_i, and integrating the differential expression. Show that, by using the results in (b),

$$S(q^i, P_i) = \frac{q^i P_i}{\sum_j (P_j)^2} \,.$$

(d) Verify that $S(q^i, P_i)$ as given explicitly above yields the correct canonical transformation, by showing that $\partial S/\partial q^i = p_i$ and $\partial S/\partial P_i = Q^i$.

This canonical transformation is known as the **Levi-Civita transformation**.

Problem 37.5 Consider the problem of the two-dimensional harmonic oscillator with the Hamiltonian

$$H(q^i, p_i) = \frac{1}{2m}\left(p_1^2 + p_2^2\right) + \frac{1}{2}m\omega^2\left((q^1)^2 + (q^2)^2\right) \,,$$

and the transformation of canonical coordinates

$$Q^1 = \tan^{-1}\left(m\omega\,\frac{q^1}{p_1}\right), \qquad Q^2 = \tan^{-1}\left(m\omega\,\frac{q^2}{p_2}\right),$$

$$P_1 = \frac{1}{2m\omega}\left(p_1^2 + m^2\omega^2(q^1)^2\right), \qquad P_2 = \frac{1}{2m\omega}\left(p_2^2 + m^2\omega^2(q^2)^2\right).$$

(a) Show that the inverse transformation is given by

$$q^1 = \sqrt{\frac{2}{m\omega}}\,\sqrt{P_1}\,\sin Q^1, \qquad q^2 = \sqrt{\frac{2}{m\omega}}\,\sqrt{P_2}\,\sin Q^2,$$

$$p_1 = \sqrt{2m\omega}\,\sqrt{P_1}\,\cos Q^1, \qquad p_2 = \sqrt{2m\omega}\,\sqrt{P_2}\,\cos Q^2.$$

(b) Show that the transformation is canonical by verifying that $dp_i \wedge dq^i = dP_i \wedge dQ^i$.

(c) Show that

$$\frac{1}{2}\,p_i dq^i + \frac{1}{2}\,q^i dp_i = p_i dq^i - P_i dQ^i.$$

Integrate the left-hand side of the above equation and show that a generating function of the first kind can be obtained as

$$F_1(q^i, Q^i) = \frac{1}{2}\,m\omega\,\left[(q^1)^2 \cot Q^1 + (q^2)^2 \cot Q^2\right].$$

Verify that this generating function generates the correct canonical transformation.

(d) Show that the new Hamiltonian is given by

$$H' = \omega\,(P_1 + P_2).$$

This form of the Hamiltonian immediately shows that P_1 and P_2 are constants of motion.

(e) Solve Hamiltonian's equations of motion in the canonical coordinates (Q^i, P_i) for Q^i and transform back to the old coordinates. Write explicit expressions for the time-dependences of $q^1(t), q^2(t), p_1(t)$ and $p_2(t)$. Interpret the meanings of the new coordinates Q^i and P_i.

Problem 37.6 Consider the problem of a charged particle moving in a constant magnetic field (solved in Chapter 13 by elementary methods). Here we will take the Hamiltonian approach. It was shown in Chapter 33 [cf. (33.70)] that the *minimal coupling* Hamiltonian can be written as (with the scalar potential ϕ set equal to zero):

$$H = \frac{\left(\boldsymbol{p} - \frac{e}{c}\boldsymbol{A}\right)^2}{2m},$$

where $\boldsymbol{A}$ is the vector potential corresponding to the magnetic field.

(a) Let the constant magnetic field be given by $\boldsymbol{B} = (mc\omega/e)\,\boldsymbol{e}_3$, where ω is a constant frequency. Use the relationship $\boldsymbol{B} = \nabla \times \boldsymbol{A}$ to show that the vector potential (in a certain gauge) is given by

$$\boldsymbol{A} = \frac{mc\omega}{2e}\,(-y\,\boldsymbol{e}_1 + x\,\boldsymbol{e}_2),$$

where x, y are the Cartesian coordinates of the charged particle.

(b) Show that the Hamiltonian is given by

$$H = \frac{1}{2m}\left(p_x + \frac{1}{2}m\omega y\right)^2 + \frac{1}{2m}\left(p_y - \frac{1}{2}m\omega x\right)^2 .$$

(c) Show that the transformation of canonical coordinates

$$x = \sqrt{\frac{1}{m\omega}}\left(\sqrt{2P_1}\,\sin Q^1 + P_2\right), \qquad y = \sqrt{\frac{1}{m\omega}}\left(\sqrt{2P_1}\,\cos Q^1 + Q^2\right),$$

$$p_x = \frac{1}{2}\sqrt{m\omega}\left(\sqrt{2P_1}\,\cos Q^1 - Q^2\right), \quad p_y = \frac{1}{2}\sqrt{m\omega}\left(-\sqrt{2P_1}\,\sin Q^1 + P_2\right),$$

is canonical by verifying that $dp_x \wedge dx + dp_y \wedge dy = dP_1 \wedge dQ^1 + dP_2 \wedge dQ^2$.

(d) Show that the new Hamiltonian is given by

$$H' = \omega P_1 .$$

This immediately implies that $Q^1(t) = \omega t + \varphi_1$, and that Q^2, P_1 and P_2 are constants of motion. Write explicit expressions for $x(t), y(t), p_x(t)$ and $p_y(t)$, and interpret the meanings of Q^2, P_1 and P_2.

Problem 37.7 Consider the time-dependent Hamiltonian for the damped undriven harmonic oscillator introduced in Problem 33.5. We will introduce the following time-dependent transformation of canonical coordinates:

$$Q = e^{bt/m}\,q\,, \qquad P = e^{-bt/m}\,p\,.$$

(a) Show that the function

$$F_2(q, P, t) = e^{bt/(2m)}\,qP$$

serves as the correct generating function for the above transformation, which shows that the transformation is canonical.

(b) Show that the new Hamiltonian H' is time-independent and is given by

$$H'(Q, P) = \frac{P^2}{2m} + \frac{k}{2}Q^2 + \frac{b}{2m}QP\,.$$

If it were possible to have a canonical transformation $\{q^i, p_i\} \to \{Q^i, P_i\}$ such that the new Hamiltonian function $H'(Q^i, P_i, t)$ vanishes identically (in the new canonical variables and t), then, according to Hamilton's equations of motion in the new variables, the new canonical coordinates Q^i and P_i will all be constants of motion. If, in addition, the generating function F_α for such a transformation is known, the solution of Hamilton's equations in the old variables $q^i(t)$ and $p_i(t)$ is achieved by expressing them in terms of the constants Q^i and P_i with the help of the generating function. It is a very consequential fact that such a canonical transformation always exists, as guaranteed by the following theorem.

Theorem 37.3. *Corresponding to a given set of Hamiltonian's equations of motion [cf. (33.16)] in the canonical coordinates q^i and p_i there always exists a canonical transformation $\{q^i, p_i\} \to \{Q^i, P_i\}$ such that the new Hamiltonian function $H'(Q^i, P_i, t)$ is identically zero.*

Proof. Consider the set of Hamiltonian's equations of motion (in the old variables) with the Hamiltonian function $H(q^i, p_i, t)$:

$$\frac{dq^i}{dt} = \frac{\partial H}{\partial p_i}, \qquad \frac{dp_i}{dt} = -\frac{\partial H}{\partial q^i} \qquad (i = 1, \dots, n). \tag{37.64}$$

Note that we have assumed the general case where the Hamiltonian may be explicitly time-dependent. It will be convenient to rewrite this set of $2n$ equations as the following single set [cf. (36.28)]:

$$\frac{dz_k}{dt} = (\mathcal{J}\partial_z H)_k \equiv g_k(z_l, t) \qquad (k = 1, \dots, m = 2n), \tag{37.65}$$

where $z_1 = q^1, \dots, z_n = q^n$; $z_{n+1} = p_1, \dots, z_{2n} = p_n$, and $\partial_z H$ is the column matrix whose elements are $\partial H/\partial z_r$. By the theory of ordinary differential equations, there exists a unique solution $z_k = z_k(\zeta_1, \dots, \zeta_m, t)$ corresponding to the initial condition $z_k(t = \tau) = \zeta_k$ for a suitable range of values $|t - \tau|$. For fixed τ we can consider ζ_l, t as independent variables in the functions $z_k = z_k(\zeta_l, t)$ and rewrite the set of ordinary differential equations (37.65) as the following set of partial differential equations:

$$\frac{\partial z_k}{\partial t} = g_k(z_l, t). \tag{37.66}$$

Thus

$$\frac{\partial^2 z_k}{\partial \zeta_l\, \partial t} = \frac{\partial^2 z_k}{\partial t\, \partial \zeta_l} = \sum_{r=1}^{m=2n} \frac{\partial g_k}{\partial z_r} \frac{\partial z_r}{\partial \zeta_l}. \tag{37.67}$$

Defining the $(2n \times 2n)$ matrices $\mathcal{M}$ and $\mathcal{G}$ by

$$\mathcal{M}_{rl} \equiv \frac{\partial z_r}{\partial \zeta_l}, \qquad \mathcal{G}_{kr} \equiv \frac{\partial g_k}{\partial z_r}, \tag{37.68}$$

Eq. (37.67) can be written as the matrix differential equation

$$\frac{\partial \mathcal{M}}{\partial t} = \mathcal{G}\mathcal{M}, \tag{37.69}$$

with the initial condition $\mathcal{M}(\tau) = I$ (identity matrix), since $z_k(\zeta_1, \dots, \zeta_m, \tau) = \zeta_k$.

Defining the column matrix $g = (g_1, \dots, g_m)^T$, the second equality of (37.65) can be written as the matrix equation $\mathcal{J}\partial_z H = g$. This implies $\partial_z H = \mathcal{J}^{-1} g = -\mathcal{J} g$. Thus $\partial_z \partial_z H = -\mathcal{J}\partial_z g = -\mathcal{J}\mathcal{G}$. Since $\partial_z \partial_z H$ is obviously a symmetric matrix, $\mathcal{J}\mathcal{G}$ must be symmetric, which in turn implies that $\mathcal{M}^T \mathcal{J}\mathcal{G}\mathcal{M}$

is symmetric. But from (37.69), $\mathcal{M}^T\mathcal{J}\mathcal{G}\mathcal{M} = \mathcal{M}^T\mathcal{J}(\partial\mathcal{M}/\partial t)$. Therefore, $\mathcal{M}^T\mathcal{J}(\partial\mathcal{M}/\partial t)$ is symmetric, and it follows that

$$\mathcal{M}^T\mathcal{J}\frac{\partial\mathcal{M}}{\partial t} = \left(\mathcal{M}^T\mathcal{J}\frac{\partial\mathcal{M}}{\partial t}\right)^T = -\frac{\partial\mathcal{M}^T}{\partial t}\mathcal{J}\mathcal{M}\,, \tag{37.70}$$

or

$$\mathcal{M}^T\mathcal{J}\frac{\partial\mathcal{M}}{\partial t} + \frac{\partial\mathcal{M}^T}{\partial t}\mathcal{J}\mathcal{M} = \frac{\partial}{\partial t}\left(\mathcal{M}^T\mathcal{J}\mathcal{M}\right) = 0\,. \tag{37.71}$$

Consequently $\mathcal{M}^T\mathcal{J}\mathcal{M}$ is independent of t. (Note that in the partial derivative with respect to t above the other variables are $\zeta_1,\dots,\zeta_m$, which are independent of t.) But $\mathcal{M}(\tau)=I$, so $\mathcal{M}^T(\tau)\mathcal{J}\mathcal{M}=\mathcal{J}$, and we conclude that $\mathcal{M}^T\mathcal{J}\mathcal{M}=\mathcal{J}$. Thus $\mathcal{M}$ is a symplectic matrix and represents a canonical transformation.

Comparing the definition of $\mathcal{M}$ [as given by the first equation of (37.68)] with (36.15) and (36.16), we see that $\mathcal{M}$ represents the canonical transformation

$$\{z_1=q^1,\dots,z_n=q^n,z_{n+1}=p_1,\dots,z_{2n}=p_n\}$$
$$\longrightarrow \{\zeta_1=Q^1,\dots,\zeta_n=Q^n,\zeta_{n+1}=P_1,\dots,\zeta_{2n}=P_n\}\,,$$

where the new variables are the (constant) initial values of $z_1(t),\dots,z_{2n}(t)$ at some fixed time $t=\tau$. The new canonical coordinates ζ_k must satisfy the Hamilton's equations

$$\frac{d\zeta_k}{dt} = (\mathcal{J}\partial_\zeta H)_k\,. \tag{37.72}$$

In order for the solution $\zeta_k(t)$ to be all constants, $H(\zeta_k)$ must be independent of any ζ_k, and can be taken to be zero. The theorem is thus proved. □

A consequence of this theorem is that *the mapping of a point at a time t on a trajectory in the phase space of a Hamiltonian system to a point at another time (either earlier or later) along the same trajectory is always achieved by a canonical transformation.*

We saw in the proof of Theorem 37.3 that the Jacobian matrix $\mathcal{M}(t)$ for the canonical transformation $\{q^i(t),p_i(t)\}\to\{q^i(\tau),p_i(\tau)\}$ satisfies $\mathcal{M}(\tau)=I$. Writing $\mathcal{M}$ in the form (36.15), we see that $\mathcal{A}(\tau)=I$. By reasons of continuity the determinant $|\mathcal{A}(t)|$ will not vanish for t sufficiently close to τ. Thus the canonical transformation under consideration is of the second kind in this neighborhood. (It is actually also of the third kind.) According to our earlier discussion, then, we can use *Hamiltonian's principal function* $S(q^i,P_i)$ [denoted by $F_2(q^i,P_i)$ earlier] as the generating function, where the P_i are now the initial values of $p_i(t)$. From the first equation of (37.23) we have

$$p_i = \frac{\partial S(q^j,P_j,t)}{\partial q^i}\,. \tag{37.73}$$

Since the new Hamiltonian $H'(Q^i,P_it)$ vanishes, (37.25) implies that

$$\boxed{H\left(q^i,\frac{\partial S}{\partial q^i},t\right)+\frac{\partial S}{\partial t}=0}\,. \tag{37.74}$$

This nonlinear first-order *partial* differential equation, to be solved for the function $S(q^i, P_i, t)$ parametrized by n constants $P_1, \ldots, P_n$ and subject to the condition $|\partial^2 S/\partial q^j \partial P_i| \neq 0$, is known as the **Hamilton-Jacobi equation**. Solving this equation is equivalent to solving Hamilton's equations of motion, which is a system of first-order *ordinary* differential equations.

Note that if the old Hamiltonian function $H(q^i, p_i)$ does not depend explicitly on time, then the energy is conserved: $H(q^i, p_i) = E$ for a particular set of initial conditions for the canonical coordinates q^i and p_i. In this case a general solution for Hamiltonian's principal function (which separates out the time-dependence) is

$$S(q^i, P_i, t) = W(q^i, P_i) - Et \,, \tag{37.75}$$

and the Hamilton-Jacobi equation above reduces to

$$H\left(q^i, \frac{\partial W}{\partial q^i}\right) = E \,. \tag{37.76}$$

This equation is also called the Hamilton-Jacobi equation, and the time-independent function $W(q^i, P_i)$ is called **Hamilton's characteristic function**. Note that in either form of this equation, there is no explicit reference to the exact identities of the n constants of motion on which Hamilton's principal (or characteristic) function may depend. They may be the initial values $p_i(\tau)$ of the canonical momenta, or any other set of n independent constants of motion. As long as one can find a solution parametrized by such a set, the existence of a corresponding canonical transformation is guaranteed. The example below will illustrate this situation.

Even though it appears that the Hamilton-Jacobi equation is more difficult to solve than the Hamilton's equations, its solution (if found) turns out to be the most powerful method known for exact integration of a mechanical problem. Furthermore many problems which were solved by the Hamilton-Jacobi approach cannot be solved by other methods. Historically, it was also important for the formulation of the Schrödinger equation in non-relativistic quantum mechanics (see, for example, Goldstein 1965).

We will now illustrate a method of solution of the Hamilton-Jacobi equation with a relatively simple but non-trivial example – the problem of gravitational attraction by two fixed force centers. (For more examples, see Problems 37.8, 37.9, and 37.10.) Consider the planar problem of gravitational attraction by two fixed point masses of equal mass, separated from each other by a fixed distance $2D$. Let a moving mass (assumed to be unity) be subjected to the gravitational pulls by the fixed masses. Suppose the distances of the moving mass from the fixed masses are r_1 and r_2. We will use **elliptical coordinates** $\xi = q^1$ and $\eta = q^2$ as the canonical coordinates, defined by $\xi \equiv r_1 + r_2$ and $\eta \equiv r_1 - r_2$. The lines $\xi = constant$ are ellipses with foci at the two fixed points, while those with $\eta = constant$ are hyperbolas with the same foci. These two families of curves are always orthogonal to each other, so we can write the infinitesimal line element ds as

$$ds^2 = a(\xi, \eta)\, d\xi^2 + b(\xi, \eta)\, d\eta^2 \,. \tag{37.77}$$

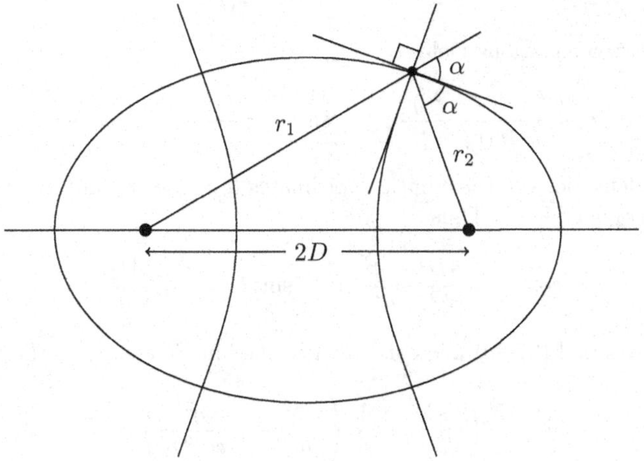

Fig. 37.1

To find the quantities a and b we refer to Fig. 37.1. Along an ellipse ($d\xi = 0$), we have

$$dr_1 = ds\,\cos\alpha\,, \qquad dr_2 = -ds\,\cos\alpha\,. \tag{37.78}$$

Thus

$$d\eta = d(r_1 - r_2) = dr_1 - dr_2 = 2\cos\alpha\, ds \qquad (d\xi = 0)\,. \tag{37.79}$$

Along a hyperbola ($d\eta = 0$),

$$dr_1 = dr_2 = ds\,\sin\alpha\,. \tag{37.80}$$

This implies

$$d\xi = d(r_1 + r_2) = dr_1 + dr_2 = 2\sin\alpha\, ds \qquad (d\eta = 0)\,. \tag{37.81}$$

Therefore, (37.75) becomes

$$ds^2 = \frac{d\xi^2}{4\sin^2\alpha} + \frac{d\eta^2}{4\cos^2\alpha} = \frac{(dq^1)^2}{4\sin^2\alpha} + \frac{(dq^2)^2}{4\cos^2\alpha}\,. \tag{37.82}$$

The angle α can be expressed in terms of ξ and η by referring to Fig. 37.1 again. We see that $r_1^2 + r_2^2 + 2r_1r_2\cos 2\alpha = 4D^2$, which implies

$$\cos^2\alpha - \sin^2\alpha = \frac{4D^2 - r_1^2 - r_2^2}{2r_1r_2}\,. \tag{37.83}$$

On the other hand, we can write

$$\cos^2\alpha + \sin^2\alpha = \frac{2r_1r_2}{2r_1r_2}\,. \tag{37.84}$$

The above two equations yield

$$\cos^2\alpha = \frac{4D^2 - (r_1 - r_2)^2}{4r_1r_2}\,, \qquad \sin^2\alpha = \frac{(r_1 + r_2)^2 - 4D^2}{4r_1r_2}\,. \tag{37.85}$$

From the definitions of the elliptic coordinates ($\xi = r_1 + r_2$ and $\eta = r_1 - r_2$) we obtain $4r_1r_2 = \xi^2 - \eta^2$. Thus

$$\cos^2\alpha = \frac{4D^2 - \eta^2}{\xi^2 - \eta^2}\,, \qquad \sin^2\alpha = \frac{\xi^2 - 4D^2}{\xi^2 - \eta^2}\,. \tag{37.86}$$

These results and (37.80) allow us to write the kinetic energy in the form

$$\begin{aligned} T &= \frac{1}{2}\left(\frac{ds}{dt}\right)^2 = \frac{1}{8}\left(\frac{\dot\xi^2}{\sin^2\alpha} + \frac{\dot\eta^2}{\cos^2\alpha}\right) \\ &= \frac{1}{8}(\xi^2 - \eta^2)\left(\frac{\dot\xi^2}{\xi^2 - 4D^2} + \frac{\dot\eta^2}{4D^2 - \eta^2}\right), \end{aligned} \tag{37.87}$$

while the potential energy is given by

$$U = -k\left(\frac{1}{r_1} + \frac{1}{r_2}\right) = -\frac{4k(r_1 + r_2)}{4r_1r_2} = -\frac{4k\xi}{\xi^2 - \eta^2}\,. \tag{37.88}$$

The Lagrangian is given by $L = T - U$. From (37.85) we have

$$p_\xi = p_1 = \frac{\partial L}{\partial\dot\xi} = \frac{\partial T}{\partial\dot\xi} = \frac{\dot\xi(\xi^2 - \eta^2)}{4(\xi^2 - 4D^2)}\,, \tag{37.89a}$$

$$p_\eta = p_2 = \frac{\partial L}{\partial\dot\eta} = \frac{\partial T}{\partial\dot\eta} = \frac{\dot\eta(\xi^2 - \eta^2)}{4(4D^2 - \eta^2)}\,. \tag{37.89b}$$

These equations yield

$$\dot\xi = \frac{4p_\xi(\xi^2 - 4D^2)}{\xi^2 - \eta^2}\,, \qquad \dot\eta = \frac{4p_\eta(4D^2 - \eta^2)}{\xi^2 - \eta^2}\,. \tag{37.90}$$

Substituting these into $H = p_i\dot q^i - L(q^j, \dot q^j)$ where $\dot q^i$ occurs we obtain the Hamiltonian function for our problem in the following form [with the canonical coordinates written in usual notation (q^i, p_i)]:

$$H(q^1, q^2, p_1, p_2) = 2p_1^2\frac{(q^1)^2 - 4D^2}{(q^1)^2 - (q^2)^2} + 2p_2^2\frac{4D^2 - (q^2)^2}{(q^1)^2 - (q^2)^2} - \frac{4kq^1}{(q^1)^2 - (q^2)^2}\,, \tag{37.91}$$

where $q^1 = \xi$, $q^2 = \eta$, $p_1 = p_\xi$, $p_2 = p_\eta$.

This Hamiltonian is explicitly time-independent, so we can use the form of the Hamilton-Jacobi equation given by (37.76), which then appears, for the present problem, as

$$\frac{2[(q^1)^2 - 4D^2]}{(q^1)^2 - (q^2)^2}\left(\frac{\partial W}{\partial q^1}\right)^2 + \frac{2[4D^2 - (q^2)^2]}{(q^1)^2 - (q^2)^2}\left(\frac{\partial W}{\partial q^2}\right)^2 - \frac{4kq^1}{(q^1)^2 - (q^2)^2} = E\,. \tag{37.92}$$

We will rewrite this as follows:

$$[(q^1)^2 - 4D^2]\left(\frac{\partial W}{\partial q^1}\right)^2 + [4D^2 - (q^2)^2]\left(\frac{\partial W}{\partial q^2}\right)^2 = E\,[(q^1)^2 - (q^2)^2] + 4k(q^1)^2\,. \tag{37.93}$$

This equation allows the following separation of variables procedure. We can set

$$[(q^1)^2 - 4D^2]\left(\frac{\partial W}{\partial q^1}\right)^2 - 4kq^1 - E(q^1)^2 = C\,, \tag{37.94}$$

$$[4D^2 - (q^2)^2]\left(\frac{\partial W}{\partial q^2}\right)^2 + E(q^2)^2 = -C\,, \tag{37.95}$$

where C is a constant. From these equations, W can be found by *quadratures* in terms of the following *elliptic integrals*:

$$W(q^1, q^2, E, C) = \int \sqrt{\frac{C + E(q^1)^2 + 4kq^1}{(q^1)^2 - 4D^2}}\, dq^1 + \int \sqrt{\frac{-C - E(q^2)^2}{4D^2 - (q^2)^2}}\, dq^2\,. \tag{37.96}$$

As discussed above the constants of motion E and C can be identified with the new canonical momenta $P_1 = E$ and $P_2 = C$ (for example), with the corresponding new (time-constant) canonical positions $Q^1 = Q^E$ and $Q^2 = Q^C$ given, according to the second equation of (37.23), by

$$Q^i = \frac{\partial S(q^i, E, C, t)}{\partial P_i}\,; \tag{37.97}$$

or, by (37.75),

$$Q^1 = Q^E = \frac{\partial W(q^1, q^2, C, E)}{\partial E} - t\,, \qquad Q^2 = Q^C = \frac{\partial W(q^1, q^2, C, E)}{\partial C}\,. \tag{37.98}$$

These equations allow the old canonical positions q^i to be solved as functions $q^i = q^i(Q^E, Q^C, E, C, t)$ $(i = 1, 2)$. The rest of the canonical coordinates, the canonical momenta p_i, can then be obtained through the first equation of (37.23):

$$p_i = \frac{\partial S(q^i, E, C, t)}{\partial q^i} = \frac{\partial W(q^1, q^2, E, C)}{\partial q^i}\,. \tag{37.99}$$

Thus the problem is completely solved.

Problem 37.8 Consider a Hamiltonian of a one-dimensional system with a time-dependent potential function:

$$H(q,p,t) = \frac{p^2}{2m} - kqt ,$$

where $k > 0$ is a constant. The Hamilton-Jacobi equation is

$$\frac{1}{2m}\left(\frac{\partial S}{\partial q}\right)^2 - kqt + \frac{\partial S}{\partial t} = 0 .$$

(a) Write $S(q,t)$ in the form

$$S(q,t) = f(t)q + h(t) ,$$

and substitute this expression in the Hamilton-Jacobi equation above. Solve for $f(t)$ and $h(t)$ and show that the results lead to the solution

$$S(q,P,t) = \left(\frac{kt^2}{2} + P\right) q - \left(\frac{k^2}{40m}t^5 + \frac{kP}{6m}t^3 + \frac{P^2}{2m}t\right) .$$

(b) By setting $Q = \partial S/\partial P$ and solving the equation for q show that

$$q(t) = Q + \frac{k}{6m}t^3 + \frac{P}{m}t ,$$

This verifies that Q is the initial value of $q(t)$.

(c) Set $p = \partial S/\partial q$ and show that

$$p(t) = \frac{k}{2}t^2 + P .$$

This verifies that P is the initial value of $p(t)$.

(d) Solve this problem by elementary means: Find the force $F = -\partial U/\partial q$. Use Newton's Second Law $F = m\ddot{q}$ and integrate to find $q(t)$. Verify that this leads to the same result as obtained above through the Hamilton-Jacobi equation.

Problem 37.9 Consider the time-independent Hamiltonian for the damped harmonic oscillator given in Problem 37.7. We will rewrite it here as [using (q,p) instead of (Q,P) and H instead of H']:

$$H(q,p) = \frac{p^2}{2m} + \frac{k}{2}q^2 + \frac{b}{2m}qp .$$

Since H is explicitly time-independent we can use the Hamilton-Jacobi equation for Hamilton's characteristic function $W(q,P)$ [(37.76)], which for the present problem appears as

$$\frac{1}{2m}\left(\frac{\partial W}{\partial q}\right)^2 + \frac{k}{2}q^2 + \frac{b}{2m}q\left(\frac{\partial W}{\partial q}\right) = P ,$$

where $P = H(q(t),p(t))$ is a constant of motion. [Note that by (37.75), Hamilton's principal function $S(q,P,t)$ is given by $S(q,P,t) = W(q,P) - Pt$.] Solve this equation with the following steps.

(a) Show that

$$\frac{\partial W}{\partial q} = -\frac{b}{2}\,q \pm \frac{1}{2}\sqrt{(b^2 - 4mk)\,q^2 + 8mP}\,.$$

(b) By using [cf. second equation of (37.23)]

$$Q = \frac{\partial S}{\partial P} = \frac{\partial W}{\partial P} - t = \textit{constant}\,,$$

show that

$$t+Q = \pm 2m \int \frac{dq}{\sqrt{(b^2-4mk)\,q^2+8mP}} = \pm\frac{2m}{\sqrt{4mk-b^2}}\,\sin^{-1}\left(\sqrt{\frac{4mk-b^2}{8mP}}\,q\right)\,.$$

(c) Assuming the underdamped case $[k/m - (b/2m)^2 \equiv \omega^2 > 0]$ show that the result in (b) reduces to

$$q(t) = \pm\frac{1}{\omega}\sqrt{\frac{2P}{m}}\,\textit{sin}\,[\omega(t+Q)]\,.$$

Recalling the transformation $q(t) = x(t)\,e^{bt/m}$ introduced in Problem 37.7, we have the expected result for the solution $x(t)$ for the physical displacement of the damped oscillator.

Problem 37.10 Consider the Kepler problem for a point particle of mass m with an added constant force F along the z-direction, so that the potential energy function is given by

$$U(x,y,z) = -\frac{k}{r} - Fz \qquad (k>0)\,.$$

Use the **parabolic coordinates** (ξ, η, φ) defined by

$$x = \sqrt{\xi\eta}\,\cos\varphi\,, \qquad y = \sqrt{\xi\eta}\,\sin\varphi\,, \qquad z = \frac{1}{2}\,(\xi - \eta)\,.$$

Let $\rho \equiv \sqrt{x^2+y^2} = \sqrt{\xi\eta}$. Thus

$$r^2 = x^2 + y^2 + z^2 = \rho^2 + z^2 = \frac{1}{4}\,(\xi+\eta)^2\,,$$

or $r = (\xi+\eta)/2$. The kinetic energy can be written as

$$T = \frac{m}{2}\,(\dot\rho^2 + \rho^2\dot\varphi^2 + \dot z^2)\,.$$

(a) Show that, in terms of parabolic coordinates, the kinetic energy appears as

$$T = \frac{m}{8}\,(\xi+\eta)\left(\frac{\dot\xi^2}{\xi} + \frac{\dot\eta^2}{\eta}\right) + \frac{m}{2}\,\xi\eta\,\dot\varphi^2$$

and thus that the Lagrangian is given by

$$L(\xi,\eta,\varphi,\dot\xi,\dot\eta,\dot\varphi) = \frac{m}{8}\,(\xi+\eta)\left(\frac{\dot\xi^2}{\xi} + \frac{\dot\eta^2}{\eta}\right) + \frac{m}{2}\,\xi\eta\,\dot\varphi^2 + \frac{2k}{\xi+\eta} + \frac{F}{2}\,(\xi-\eta)\,.$$

(b) Show that

$$p_\xi = \frac{m}{4}(\xi+\eta)\frac{\dot{\xi}^2}{\xi}\,, \quad p_\eta = \frac{m}{4}(\xi+\eta)\frac{\dot{\eta}^2}{\eta}\,, \quad p_\varphi = m\,\xi\eta\,\dot{\varphi}\,.$$

Hence show that the Hamiltonian is given by

$$H(\xi,\eta,\varphi,p_\xi,p_\eta,p_\varphi) = \frac{2}{m}\left(\frac{\xi\,p_\xi^2+\eta\,p_\eta^2}{\xi+\eta}\right) + \frac{p_\varphi^2}{2m\,\xi\eta} - \frac{2k}{\xi+\eta} - \frac{F}{2}(\xi-\eta)\,.$$

We immediately see that $H = constant = E \equiv P_1$. Also, since $\partial H/\partial\varphi = 0$, $p_\varphi \equiv P_2 = constant$.

(c) Write Hamilton's principal function as

$$S(\xi,\eta,\varphi,P_1,P_2,P_3,t) = W(\xi,\eta,\varphi,P_1,P_2,P_3) - Et\,.$$

Then Hamilton's characteristic function W satisfies the following Hamilton-Jacobi equation

$$\frac{2}{m\,(\xi+\eta)}\left\{\xi\left(\frac{\partial W}{\partial\xi}\right)^2 + \eta\left(\frac{\partial W}{\partial\eta}\right)^2\right\} + \frac{1}{2m\,\xi\eta}\left(\frac{\partial W}{\partial\varphi}\right)^2 - \frac{2k}{\xi+\eta} - \frac{F}{2}(\xi-\eta) = E\,.$$

Show that this equation is separable by using the following ansatz:

$$W(\xi,\eta,\varphi,P_1,P_2,P_3) = W_\xi(\xi,P_1,P_2,P_3) + W_\eta(\eta,P_1,P_2,P_3) + P_2\varphi\,.$$

Do this by substituting into the Hamilton-Jacobi equation above and verify that one then obtains (with $P_1 = E, P_2 = p_\varphi$)

$$2\xi\left(\frac{\partial W_\xi}{\partial\xi}\right)^2 + \frac{p_\varphi^2}{2\xi} - 2mk - \frac{mF}{2}\xi^2 - mE\,\xi$$
$$= -2\eta\left(\frac{\partial W_\eta}{\partial\eta}\right)^2 - \frac{p_\varphi^2}{2\eta} - \frac{mF}{2}\eta^2 + mE\,\eta\,.$$

Since the left-hand side only depends on ξ while the right-hand side only depends on η, we can set both sides equal to a constant. Denote this constant by P_3.

(d) Show that

$$S(\xi,\eta,\varphi,P_1,P_2,P_3,t) = \int d\xi\sqrt{\frac{mP_1}{2} + \frac{P_3+2mk}{2\xi} + \frac{mF}{4}\xi - \frac{P_2^2}{4\xi^2}}$$
$$+\int d\eta\sqrt{\frac{mP_1}{2} - \frac{P_3}{2\eta} - \frac{mF}{4}\eta - \frac{P_2^2}{4\eta^2}} + P_2\varphi - P_1 t\,.$$

(e) From the requirements that [cf. (37.23)]

$$Q^1 = \frac{\partial S}{\partial P_1}\,, \quad Q^2 = \frac{\partial S}{\partial P_2}\,, \quad Q^3 = \frac{\partial S}{\partial P_3}\,; \quad p_\xi = \frac{\partial S}{\partial\xi}\,, \quad p_\eta = \frac{\partial S}{\partial\eta}\,, \quad p_\varphi = \frac{\partial S}{\partial\varphi}\,,$$

and that Q^1, Q^2, Q^3 (as well as P_1, P_2, P_3) are constants, discuss how one can obtain, at least in principle, the solutions for

$$\xi(Q^i,P_i,t)\,, \qquad \eta(Q^1,P_i,t)\,, \qquad \varphi(Q^i,P_i,t)\,.$$

It is immediately clear that $p_\varphi = P_2 = constant$. Write expressions for $p_\xi(Q^i,P_i,t)$ in terms of $\xi(Q^i,P_i,t)$ and P_i, and $p_\eta(Q^i,P_i,t)$ in terms of $\eta(Q^i,P_i,t)$ and P_i.

Chapter 38

Integrability, Invariant Tori, Action-Angle Variables

The Hamiltonian formulation of classical mechanics lends itself to the use of powerful geometric techniques (in *symplectic geometry*) in its analysis and an interpretation of the dynamics in terms of beautiful and intuitive geometric pictures. In this chapter we will begin to explore the latter aspect with the fundamental notion of *integrability*. The next chapter will introduce some basic elements of symplectic geometry and explain their relevance in Hamiltonian mechanics. The basic ideas and mathematical techniques introduced in these two chapters will form the basis for further developments in the rest of the book.

We will be concerned with Hamiltonian functions which do not depend explicitly on the time t. Then

$$\frac{dH}{dt} = \frac{\partial H}{\partial t} = 0\,, \tag{38.1}$$

and the energy E is conserved (E is a constant of motion).

Recall (from Chapter 33) that a necessary condition for a system with N degrees of freedom to be *integrable* is that there exists N functionally independent constants of motion (*first integrals*). This means the existence N independent functions of the canonical coordinates (q^i, p_i):

$$F_n(q^i, p_i) \qquad (n = 1, \dots, N)$$

that are constant along each trajectory of the Hamiltonian phase flow, where functional independence can be characterized most economically by the condition

$$dF_1 \wedge \dots \wedge dF_N \neq 0\,, \tag{38.2}$$

with

$$dF_i = \sum_{j=1}^{N} \left(\frac{\partial F_i}{\partial q^j}\, dq^j + \frac{\partial F_i}{\partial p_j}\, dp_j \right)\,.$$

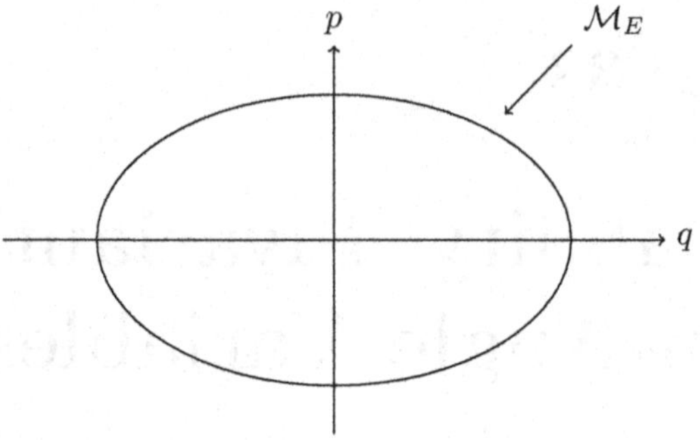

Fig. 38.1

For an integrable system, then, the N equations

$$F_1(q^i, p_i) = \alpha_1 \; ,, \; \dots \; , \; F_N(q^i, p_i) = \alpha_N \; , \tag{38.3}$$

where $\alpha_1, \dots, \alpha_N$ are constants, determine an N-dimensional submanifold $\mathcal{M}$ of *phase space* (which is a $2N$-dimensional manifold). For example, the Hamiltonian of a 1-dimensional harmonic oscillator is given by

$$H(p, q) = \frac{p^2}{2m} + \frac{kq^2}{2} \; . \tag{38.4}$$

The phase space, with canonical coordinates q and p, is 2-dimensional. The submanifold $\mathcal{M}_E$, specified by the curve

$$\frac{p^2}{2m} + \frac{kq^2}{2} = E \; , \tag{38.5}$$

where E is the conserved total energy, is a 1-dimensional submanifold of phase space (see Fig. 38.1).

We will now recall the important concept of *canonical transformations*, to set further conditions for the N first integrals, which, together with the constancy of the integrals along phase trajectories, will constitute necessary and sufficient conditions for integrability. We will use Hamilton's principal function $S(q^i, P_i, t)$ to generate the desired canonical transformation. This generating function satisfies

$$p_i = \frac{\partial S}{\partial q^i} \; , \quad Q^i = \frac{\partial S}{\partial P_i} \; , \tag{38.6}$$

which implies

$$dS = \sum_{i=1}^{N} \left(\frac{\partial S}{\partial q^i} dq^i + \frac{\partial S}{\partial P_i} dP_i \right) + \frac{\partial S}{\partial t} dt = p_i dq^i + Q^i dP_i + \frac{\partial S}{\partial t} dt \,. \tag{38.7}$$

In the present case where $\partial H/\partial t = 0$ ($H(q^i, p_i) = E = constant$), we can set $S(q^i, P_i, t) = W(q^i, P_i) - Et$ [cf. (37.75)], where $W(q^i, P_i)$, Hamilton's characteristic function, also satisfies a set of equations similar to (38.6):

$$p_i = \frac{\partial W}{\partial q^i} \,, \quad Q^i = \frac{\partial W}{\partial P_i} \,, \tag{38.8}$$

so that

$$dW = p_i \, dq^i + Q^i \, dP_i \,.$$

$W(q^i, P_i)$, as distinct from $S(q^i, P_i, t)$, can be viewed as a generating function of the second kind in its own right, and generates a different canonical transformation from that generated by $S(q^i, P_i, t)$, the latter being such that all the new canonical coordinates Q^i and P_i are constants of motion. As discussed in the last chapter, it satisfies the (time-independent) Hamilton-Jacobi equation in the form (37.76). In both cases, the condition $dp_i \wedge dq^i = dP_i \wedge dQ^i$ [cf. (37.59)] gives a necessary and sufficient condition for the transformation $\{q^i, p_i\} \to \{Q^i, P_i\}$ to be canonical. The significance of the invariant 2-form $dp_i \wedge dq^i$, called the *symplectic form*, will be explained more fully in the next chapter.

The first equation of (38.8) implies that the generating function W can be written as the following integral (called an **action integral**):

$$W(q^i, P_i) = \int_{(q^i)_0}^{q^i} p_i(q^j, P_j) \, dq^i \,, \tag{38.9}$$

where the integrand p_i is expressed explicitly as functions of (q^j, P_j), by solution of the set of equations

$$P_1(q^i, p_i) = P_1, \ldots, P_N(q^i, p_i) = P_N \,, \tag{38.10}$$

with the right-hand sides being constants. Thus if a Hamiltonian system is integrable, it is possible to make a canonical transformation

$$\{q^i, p_i\} \mapsto \{Q^i, P_i\} \,, \qquad H'(Q^i, P_i) = H(q^i(Q^j, P_j), p_i(Q^j, P_j)) \,,$$

by using an appropriate generating function $W(q^i, P_i)$ such that all new N canonical momenta P_i are constants of motion. (Note that the above form of the new Hamiltonian applies since $\partial W/\partial t = 0$.) Then the second equation of Hamilton's equations ($\dot{P}_i = -\partial H'/\partial Q^i$) implies that the Hamiltonian $H'(Q^i, P_i)$ must be a function of the P_i's only (and none of the Q^i). The Hamilton's equations $\dot{Q}^i = \partial H'/\partial P_i$ then imply that the new canonical coordinates Q^i are all linear functions of time:

$$Q^i(t) = \left(\frac{\partial H'}{\partial P_i} \right) t + Q^i(0) \,, \tag{38.11}$$

where the $\partial H'/\partial P_i$ are again constants of motion, since the P_i's are, and $H' = H'(P_i)$. In principle, then, the problem is solved by relating the old canonical coordinates (q^i, p_i) to the new ones (Q^i, P_i) by first inverting the second equation of (38.8) to find $q^i(Q^j, P_j)$, and then using the second equation of (38.8) to find $p_i(q^j(Q^k, P_k), P_j)$.

To gain more mathematical insight into the Hamiltonian formulation of classical mechanics, we will introduce an analytical construct known as the *Poisson bracket.*

Definition 38.1. *The* ***Poisson bracket*** $\{u, v\}$ *of any two functions* $u(q^i, p_i)$ *and* $v(q^i, p_i)$ *on phase space is defined by:*

$$\{u, v\} \equiv \sum_{k=1}^{N} \left(\frac{\partial u}{\partial q^k} \frac{\partial v}{\partial p_k} - \frac{\partial u}{\partial p_k} \frac{\partial v}{\partial q^k} \right) . \tag{38.12}$$

It is easily verified that Poisson brackets satisfy the following relations:

$$\{u, v\} = -\{v, u\} , \tag{38.13}$$

$$\{u, \{v, w\}\} + \{v, \{w, u\}\} + \{w, \{u, v\}\} = 0 . \tag{38.14}$$

The first says that the Poisson bracket as a multiplication rule is anti-commutative, while the second is recognized to be the *Jacobi identity* [cf. (1.46)]. Loosely speaking, a set of objects satisfying the above relations constitute a *Lie algebra.* It is also easily seen that

$$\{p_i, p_j\} = \{q^i, q^j\} = 0 , \tag{38.15}$$

$$\{q^i, p_j\} = \delta_{ij} , \tag{38.16}$$

$$\{u, q^i\} = -\frac{\partial u}{\partial p_i} , \tag{38.17}$$

$$\{u, p_i\} = \frac{\partial u}{\partial q^i} . \tag{38.18}$$

The last two equations imply that

$$\{q^i, H\} = \frac{\partial H}{\partial p_i} = \dot{q}^i , \tag{38.19}$$

$$\{p_i, H\} = -\frac{\partial H}{\partial q^i} = \dot{p}_i , \tag{38.20}$$

where H is the Hamiltonian. In the last two equations, the second equality follows from Hamilton's equations of motion. The Poisson bracket also has the important property that *it is invariant under canonical transformations*:

$$\{u, v\}_{q,p} = \{u, v\}_{Q,P} . \tag{38.21}$$

This result, together with (38.19) and (38.20), corroborate the fact that Hamilton's equations of motion are invariant under canonical transformations.

Problem 38.1 Use the definition of the Poisson bracket given by Def. 38.1 to verify the properties (38.13) to (38.16), and the fact that Poisson brackets are invariant under canonical transformations [(38.21)].

Problem 38.2 Show that the Poisson brackets for the components L^i $(i = 1, 2, 3)$ of the angular momentum vector $\boldsymbol{L} = \boldsymbol{r} \times \boldsymbol{p}$ satisfy the commutation relations:

$$\{L^i, L^j\} = \varepsilon^{ij}{}_k L^k \,,$$

where $\varepsilon^{ij}{}_k$ is the Levi-Civita tensor.

The time dependence of an arbitrary function $u(q^i, p_i)$ is then obtained as follows:

$$\frac{du}{dt} = \frac{\partial u}{\partial q^i}\dot{q}^i + \frac{\partial u}{\partial p_i}\dot{p}_i + \frac{\partial u}{\partial t} = \frac{\partial u}{\partial q^i}\frac{\partial H}{\partial p_i} - \frac{\partial u}{\partial p_i}\frac{\partial H}{\partial q^i} + \frac{\partial u}{\partial t} \,. \tag{38.22}$$

Thus

$$\frac{du}{dt} = \{u, H\} + \frac{\partial u}{\partial t} \,. \tag{38.23}$$

In particular, Eq. (38.1) follows, since $\{H, H\} = 0$. Note that since the N constants of motion P_i are the new canonical momenta, (38.15) implies that

$$\{P_i, P_j\} = 0 \qquad \text{for any } i, j \,. \tag{38.24}$$

We also say that the N constants of motion are **in involution** with each other.

The N equations (38.10), with the right-hand sides all constants, determine the N-dimensional submanifold $\mathcal{M}$ of the integrable system. We can then state and prove the following critical theorem for integrable systems

Theorem 38.1. *Suppose $P_i(q^j, p_j)$ $(i = 1, \dots, N)$ are N constants of motion of an integrable system that are in involution ($\{P_i, P_j\} = 0$ for all $i, j = 1, \dots, N$) and $\mathcal{M}$ is the corresponding N-dimensional integral submanifold of the phase space of the system. Then the N vector fields*

$$\boldsymbol{T}_m = \left(\frac{\partial P_m}{\partial p_1}, \dots, \frac{\partial P_m}{\partial p_N}, -\frac{\partial P_m}{\partial q^1}, \dots, -\frac{\partial P_m}{\partial q^N}\right) \qquad (m = 1, \dots, N) \tag{38.25}$$

(each with the indicated $2N$ components) are all tangent to $\mathcal{M}$.

Proof. The N independent normal vector fields to $\mathcal{M}$ are given by

$$\nabla P_n = \left(\frac{\partial P_n}{\partial q^i}, \frac{\partial P_n}{\partial p_i}\right) \qquad (n = 1, \dots, N) \,. \tag{38.26}$$

We then have

$$\boldsymbol{T}_m \cdot \nabla P_n = \sum_i \left(\frac{\partial P_m}{\partial p_i}\frac{\partial P_n}{\partial q^i} - \frac{\partial P_m}{\partial q^i}\frac{\partial P_n}{\partial p_i}\right) = -\{P_n, P_m\} = 0 \,, \tag{38.27}$$

where the last equality follows from the assumption that all the P_i are in involution. We can thus conclude that the vector fields $\boldsymbol{T}_m$ must all be tangent to $\mathcal{M}$. □

Since the $\boldsymbol{T}_m$ are globally defined on $\mathcal{M}$, they form a *global frame field* on $\mathcal{M}$. In other words the manifold $\mathcal{M}$ is **parallelizable** by the N independent tangent vector fields $\boldsymbol{T}_m$. Colloquially, we say that if an N-dimensional manifold $\mathcal{M}$ is parallelizable, we can comb it in N different ways without running into singularities. This leads to a very important consequence for the topological structure of the integral submanifold $\mathcal{M}$, due to the following well-known theorem in topology (stated here without proof; for a proof, see, for example, Arnold 1978):

Theorem 38.2. *Any N-dimensional compact manifold parallelizable by N independent vector fields is* ***homeomorphic*** *(topologically equivalent) to an N-torus.*

We note that

$$N\text{-torus} \sim \underbrace{S^1 \times S^1 \times \cdots \times S^1}_{N \text{ times}}\,, \tag{38.28}$$

where S^1 is the 1-sphere (a circle). An example of a compact manifold that is not parallelizable is the 2-sphere S^2 (any vector field has at least one singularity on S^2). We will consider bound motions so that the $(2N-1)$-dimensional energy hypersurface $\mathcal{E}$ is compact, which implies that the N-dimensional integral submanifold $\mathcal{M} \subset \mathcal{E}$ must also be compact. It follows from the above theorem that $\mathcal{M}$ must be an N-torus. Recall (from Chapter 33) that phase space is the cotangent bundle T^*Q of the configuration space Q. Each integral submanifold $\mathcal{M} \subset T^*Q$ specified by the *level set*

$$\mathcal{M}_{\alpha_1,\ldots,\alpha_N} = \{(q^i,p_i) \in T^*Q \,|\, F_1(q^i,p_i) = \alpha_1, \ldots, F_N(q^i,p_i) = \alpha_N\}\,, \tag{38.29}$$

and characterized by the N constants $\alpha_1, \ldots, \alpha_N$ is called an **invariant torus** of the integrable system. They are so-called because a phase trajectory induced by the Hamiltonian flow of an integrable system never leaves a particular torus.

Now consider the action integral

$$\oint p_i(q^j, P_j)\, dq^i \tag{38.30}$$

calculated over closed loops in an invariant torus. A well-known topological result is that *an N-torus has N homotopy classes of closed loops*, where, roughly speaking, two loops γ and γ' in $\mathcal{M}$ belong to the same **homotopy class** if one can be continuously deformed into the other without obstructions (and without leaving $\mathcal{M}$). Let γ and γ' be homotopic to each other in an invariant torus, then $\gamma - \gamma'$ is the boundary of some region $\sigma \subset \mathcal{M}$. We write $\gamma - \gamma' = \partial\sigma$. Figure 38.2 illustrates the two classes of homotopic loops for a 2-torus.

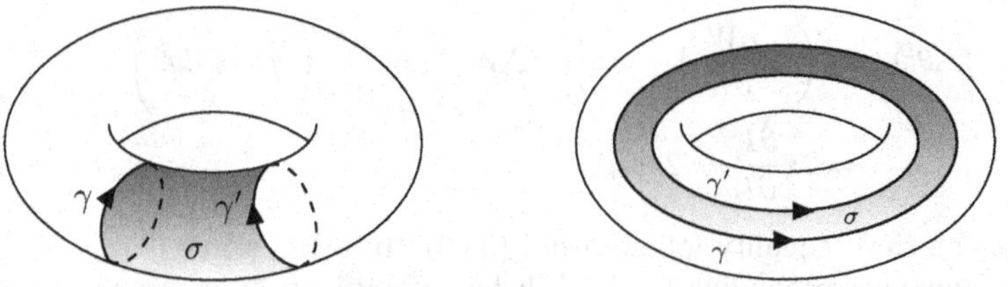

Fig. 38.2

It is easily seen that the action integral (38.30) over a closed loop only depends on the homotopy class of the loop. Indeed,

$$\begin{aligned} &\oint_{\gamma} p_i\, dq^i - \oint_{\gamma'} p_i\, dq^i = \oint_{\gamma-\gamma'=\partial\sigma} p_i\, dq^i = \int_{\sigma} d(p_i\, dq^i) \\ &= \int_{\sigma} dp_i \wedge dq^i = \int_{\sigma} dP_i \wedge dQ^i = 0\,, \end{aligned} \tag{38.31}$$

where the second equality follows from Stokes' Theorem [(8.33)], the fourth equality from the fact that $(q^i, p_i) \to (Q^i, P_i)$ is a canonical transformation [cf. (37.59)], and the last equality from the fact that $dP_i = 0$ on $\mathcal{M}$ (all P_i's are constants on an invariant torus). We also say that the integral submanifold $\mathcal{M}$ of an integrable system is a **null manifold** of the symplectic form $dp_i \wedge dq^i$.

We can then use the loop action integrals over N loops $\gamma_1, \dots, \gamma_N$ belonging to the N different homotopy classes of the N-torus to construct a set of N special canonical momenta I_n, the so-called **action variables**:

$$I_n(P_m) \equiv \frac{1}{2\pi} \oint_{\gamma_n} p_j\, dq^j\,, \quad n = 1, \dots N\,. \tag{38.32}$$

These action variables, being constants on an invariant torus, can also serve as the N constants of motion. The Hamiltonian can then be written as a function of just the action variables:

$$H = H(I_1, \dots, I_N)\,. \tag{38.33}$$

From the second equation of (38.8) we can define the so-called **angle variables** (which are the canonical coordinates conjugate to the action variables):

$$\theta^i \equiv \frac{\partial W(q^i, I_i)}{\partial I_i} . \tag{38.34}$$

These variables are so called because on traversing a loop γ_j, the change in value of θ^i is given by

$$\begin{aligned}(\Delta\theta^i)_{\gamma_j} &= \left(\Delta \frac{\partial W}{\partial I_i}\right)_{\gamma_j} = \frac{\partial}{\partial I_i}(\Delta W)_{\gamma_j} = \frac{\partial}{\partial I_i}\left(\oint_{\gamma_j} p_k \, dq^k\right) \\ &= 2\pi \frac{\partial I_j}{\partial I_i} = 2\pi\delta_{ij} ,\end{aligned} \tag{38.35}$$

where the third equality follows from (38.9). In other words θ^i $(i = 1, \ldots, N)$ is the angle corresponding to the i-th homotopy loop in an N-torus, since on traversing the loop once its value changes by 2π. On the other hand, on traversing the loop γ_j, where $j \neq i$, θ^i does not change in value at all.

When any set of canonical variables in phase space q^i and p_i are expressed as functions of the action-angle variables (I_j, θ^j), the functions must all be periodic in all the θ^j's, with periods equal to 2π. Thus q^i and p_i can all be expressed as multiple Fourier series:

$$q^i(I_1, \ldots, I_N; \theta^1, \ldots, \theta^N) = \sum_{m_1 \ldots m_N = -\infty}^{\infty} \tilde{q}^i_{m_1 \ldots m_N}(I_1, \ldots, I_N)\, e^{i(m_1\theta^1 + \cdots + m_N\theta^N)} , \tag{38.36}$$

$$p_i(I_1, \ldots, I_N; \theta^1, \ldots, \theta^N) = \sum_{m_1 \ldots m_N = -\infty}^{\infty} \tilde{p}_{i;\, m_1 \ldots m_N}(I_1, \ldots, I_N)\, e^{i(m_1\theta^1 + \cdots + m_N\theta^N)} . \tag{38.37}$$

The angle variables all depend linearly on time:

$$\theta^i(t) = \omega_i t + \delta^i , \tag{38.38}$$

where δ^i is the initial value $\theta^i(0)$, and

$$\dot{\theta}^i = \frac{\partial H}{\partial I_i} \equiv \omega_i = \text{constant} . \tag{38.39}$$

Setting

$$\boldsymbol{\delta} = (\delta_1, \ldots, \delta_N) , \quad \boldsymbol{m} = (m_1, \ldots, m_N) , \quad \boldsymbol{\omega} = (\omega_1, \ldots, \omega_N) , \tag{38.40}$$

we can write

$$q^i(t) = \sum_{m_1 \ldots m_N} \tilde{q}^i_{\boldsymbol{m}}(I_1, \ldots, I_N)\, e^{i\boldsymbol{m}\cdot(\boldsymbol{\omega} t + \boldsymbol{\delta})} , \tag{38.41}$$

$$p_i(t) = \sum_{m_1 \ldots m_N} \tilde{p}_{i\boldsymbol{m}}(I_1, \ldots, I_N)\, e^{i\boldsymbol{m}\cdot(\boldsymbol{\omega} t + \boldsymbol{\delta})} . \tag{38.42}$$

These are called **multiply (conditionally) periodic functions**.

A trajectory in an invariant torus is *periodic* if *closure* occurs after n_1 circuits of θ^1, n_2 circuits of $\theta^2, \dots, n_N$ circuits of θ^N, that is, if $\boldsymbol{\omega} = \boldsymbol{n}\omega_0$ for some ω_0, where $\boldsymbol{n} = (n_1, \dots, n_N)$; or $\boldsymbol{\omega}\tau = 2\pi\boldsymbol{n}$, where $\tau = 2\pi/\omega_0$ is the period of the orbit. An equivalent condition is that there exist $N-1$ linearly independent relations of the form $\boldsymbol{\omega} \cdot \boldsymbol{m} = 0$, where $\boldsymbol{m}$ is a non-zero integer vector (for example, $3\omega_1 = 2\omega_2$ for $N = 2$, in which case $n_1 = 2$ and $n_2 = 3$). We then say that $\boldsymbol{\omega}$ is **commensurate**. If $\boldsymbol{\omega}$ is not commensurate, the trajectory will cover the invariant torus *uniformly* and *densely* in time and is sometimes loosely referred to as being **ergodic**, that is, it explores all regions of the invariant torus after a sufficiently long time. In this case the trajectory never repeats itself in the torus. We will henceforth refer to this kind of trajectory as being **quasiperiodic**. Strictly speaking the concept of ergodicity refers to the trajectory covering the entire energy hypersurface (of dimension $2N-1$) uniformly and densely in a sufficiently long time, and is a much stronger condition than that of commensurability of $\boldsymbol{\omega}$. We will not go into a detailed discussion of the important concept of ergodicity here, or the closely related one of *mixing*, and the significance of these in the general subject of *chaos* in dynamical systems (beyond Hamiltonian ones) and the foundations of statistical mechanics; the interested reader is referred to the literature on these topics [see, for example the comprehensive monograph by Katok and Hasselblatt (Katok and Hasselblatt 1995), or the much shorter but very informative one by Ruelle (Ruelle 1989) on dynamical systems theory, and the classic monograph by Khinchin on statistical mechanics (Khinchin 1949)].

Problem 38.3 In the course of one day Jupiter moves along its orbit by an amount $\theta_J = 299.1$" and Saturn by an amount $\theta_S = 120.5$". Find (small) integers n_J and n_S such that $n_J\omega_J + n_S\omega_S \approx 0$, where ω_J and ω_S are angular frequencies of Jupiter and Saturn in their orbits around the sun, respectively.

Problem 38.4 Consider an integrable system with three degrees of freedom and an invariant 3-torus with a commensurate $\boldsymbol{\omega}$, so that $\boldsymbol{\omega} = \boldsymbol{n}\omega_0$, where $\boldsymbol{n} = (n_1, n_2, n_3)$ is a 3-vector with non-zero integer components. Suppose there are two linearly independent relations of the form

$$l_1\omega_1 + l_2\omega_2 + l_3\omega_3 = 0 \;, \qquad m_1\omega_1 + m_2\omega_2 + m_3\omega_3 = 0 \;,$$

where both $\boldsymbol{l} = (l_1, l_2, l_3)$ and $\boldsymbol{m} = (m_1, m_2, m_3)$ are non-zero 3-vectors with integer components. Show that the above equations uniquely determine the integers n_1, n_2 and n_3. Verify that

$$n_1 = m_2 l_3 - m_3 l_2 \;, \quad n_2 = m_3 l_1 - m_1 l_3 \;, \quad n_3 = m_1 l_2 - m_2 l_1 \;.$$

The system of two equations to be solved for integer values of (n_1, n_2, n_3) have integer coefficients, and are known as *diophantine equations.*

The main results of this chapter can now be summarized by the following theorem:

Theorem 38.3. *Suppose a mechanical system with configuration space Q has N degrees of freedom ($\dim(Q) = N$), and there exist N functionally independent constants of motion $F_n(q^i, p_i)$ on the $2N$-dimensional phase space T^*Q (cotangent bundle of Q) that are in involution with each other ($\{P_i, P_j\} = 0$, for all $i, j = 1, \ldots, N$). Then the mechanical system is integrable by quadratures, which means:*

(1) The level set $\mathcal{M}_{\alpha_1,\ldots,\alpha_N}$ as defined by (38.29) is a smooth N-dimensional submanifold of T^*Q that is invariant under the phase flow of the Hamiltonian function.

(2) If an integral submanifold $\mathcal{M}_{\alpha_1,\ldots,\alpha_N}$ is compact and connected, then it is diffeomorphic to the N-torus. Each such integral submanifold is called an invariant torus.

(3) On each invariant torus there exists a special set of canonical coordinates, called the action-angle variables, where the N action variables $I_n(q^i, p_i)$ are all constants of motion so that $I_n = I_n(\alpha_1, \ldots, \alpha_N)$, and canonically conjugate to them are the angle-variables $\theta^1, \ldots, \theta^N$ specifying the position on the torus, which all range from 0 to 2π. In terms of these canonical variables the Hamiltonian function $H(I_1, \ldots, I_N)$ in general determines multiply (conditionally) periodic phase trajectories on the invariant torus described by

$$\frac{d\theta^i}{dt} = \omega_i = \frac{\partial H}{\partial I_i} \, .$$

Note that a system with one degree of freedom, for example, the simple harmonic oscillator with Hamiltonian given by (38.4), is always integrable, since the Hamiltonian $H(p, q)$ is a constant of motion. The invariant tori for bound motions, each characterized by a constant value of the energy $E = H(p, q)$, are all 1-tori (ellipses in the case of the simple harmonic oscillator) in the 2-dimensional (p, q)-plane, which is the phase space. The single angle variable $0 < \theta < 2\pi$ designates the position around the torus, while the single action variable $I(E)$ is the area in the (p, q)-plane bounded by the curve $H(p, q) = E$. Likewise a system with two degrees of freedom with a constant of motion F that is independent of the Hamiltonian H is also necessarily integrable, since, by (38.23), $\{F, H\} = 0$ and F and H are thus in involution. In this case, the invariant tori are each 2-tori in 4-dimensional phase space characterized by the constraints $H(p, q) = E$ and $F(p, q) = \alpha$. This is the case that is directly visualizable geometrically. Each constant-energy hypersurface is a 3-dimensional submanifold of a 4-dimensional phase space, and can be imagined as being fibered by 2-dimensional tori. These can be pictured in ordinary 3-dimensional space as a family of concentric tori lying inside one another, each characterized by a different set of values for (I_1, I_2) all yielding the same energy value (see Fig. 38.3). In the general case of N degrees of freedom, we speak of

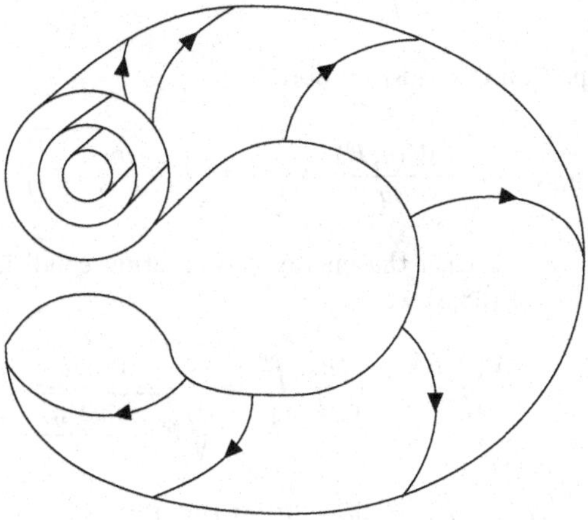

Fig. 38.3

the $2N$-dimensional phase space of an integrable system as being **foliated** into an N-parameter family of invariant tori.

We will illustrate integrability with a few examples. Consider first the simple harmonic oscillator, a system with one degree of freedom. Its Hamiltonian function is given by

$$H(p,q) = \frac{p^2}{2m} + \frac{m\omega^2}{2}\, q^2 \,. \tag{38.43}$$

This is a constant of motion itself. So

$$H(p,q) = E = \text{constant} \tag{38.44}$$

along any integral curve of the Hamiltonian flow, and the (one-dimensional) integral submanifold $\mathcal{M}_E \subset T^*Q$ characterized by a particular value of E is given by

$$\mathcal{M}_E = \{(q,p) \in T^*Q \,|\, H(p,q) = E\} \,. \tag{38.45}$$

Taking E to be the new canonical momentum of a canonical transformation $(q,p) \to (Q,E)$, the generating function $W(q,E)$ (Hamilton's characteristic function) satisfies the following Hamilton-Jacobi equation [(37.76)]:

$$\frac{1}{2m}\left(\frac{\partial W}{\partial q}\right)^2 + \frac{m\omega^2}{2}\, q^2 = E \,, \tag{38.46}$$

with the solution

$$W(q,E) = \int_{q_0}^{q} dq \sqrt{2m\left(E - \frac{m\omega^2}{2}q^2\right)} \,. \tag{38.47}$$

From the first equation of (38.8) we obtain

$$p(q,E) = \frac{\partial W(q,E)}{\partial q} = \sqrt{2m\left(E - \frac{m\omega^2}{2}q^2\right)} \,, \tag{38.48}$$

which, of course, agrees with the energy conservation condition (38.5). From the second equation of (38.8) we have

$$Q(q,E) = \frac{\partial W(q,E)}{\partial E} = \sqrt{\frac{m}{2}} \int_{q_0}^{q} dq \frac{1}{\sqrt{E - \dfrac{m\omega^2 q^2}{2}}} \,. \tag{38.49}$$

Since $H(p,q) = H'(E) = E$, we have, from Hamilton's equations (in the new coordinates),

$$\dot{Q} = \frac{\partial H'}{\partial E} = 1 \,, \tag{38.50}$$

and so

$$Q(t) = Q_0 + t \,. \tag{38.51}$$

To express q directly in terms of Q we have to "turn (38.49) inside out" by first evaluating the integral in that equation. Doing so we find that

$$Q = \frac{1}{\omega} \int_{q_0}^{q} \frac{dq}{\sqrt{\dfrac{2E}{m\omega^2} - q^2}} = \frac{1}{\omega} \sin^{-1}\left(\sqrt{\frac{m\omega^2}{2E}}\, q\right)\Bigg|_{q_0}^{q} \,. \tag{38.52}$$

Choosing $q_0 = \sqrt{2E/(m\omega^2)}$ we have

$$\omega Q = \sin^{-1}\left(\sqrt{\frac{m\omega^2}{2E}}\, q\right) - \frac{\pi}{2} \,. \tag{38.53}$$

Using (38.51) for Q we finally obtain the expected result for the explicit time dependence of q:

$$q(t) = \sqrt{\frac{2E}{m\omega^2}} \cos(\omega t + \omega Q_0) \,. \tag{38.54}$$

This result agrees with the one obtained by much more elementary means. The explicit time dependence of p is then obtained by substituting this result in (38.48).

It is clear that in this example the single action variable $I(E)$ [calculated by using (38.32) with the function $p(q,E)$ given by (38.48)] is the area bounded by the curve $H(p,q) = E$ in the (q,p)-plane divided by 2π; and the corresponding action variable θ is simply the polar angle of a point on that curve.

Problem 38.5 For the simple harmonic oscillator problem defined by the Hamiltonian $H(p,q)$ given by (38.43) calculate the action variable

$$I(E) = \oint_{\gamma(E)} p(q,E)dq ,$$

where $\gamma(E)$ is the closed curve $H(p,q) = E$ (an ellipse) in the (p,q)-plane. Then express the Hamiltonian $H'(I)$ in terms of I. Verify that the frequency of the angular motion $\theta(t)$ as calculated by $\partial H'(I)/\partial I$ is precisely the quantity ω in (38.43).

Next we will revisit the Kepler (two-body) problem, with $N = dim(Q) = 2$, given by the Hamiltonian [cf. (33.35)]:

$$H(p_r, p_\theta; r, \theta) = \frac{p_r^2}{2m} + \frac{p_\theta^2}{2mr^2} - \frac{\alpha}{r} \qquad (\alpha > 0) , \tag{38.55}$$

where $q^1 = r, q^2 = \theta, p_1 = p_r, p_2 = p_\theta$. As discussed in Chapter 33, besides H, the angular momentum $L = p_\theta = mr^2\dot{\theta}$ is also a constant of motion. Since L is necessarily in involution with H ($\{H, L\} = 0$) the problem is integrable (in the sense of Theorem 38.3). Consider bound orbits only, so that each integral submanifold $\mathcal{M}_{E,L}$ is an invariant 2-torus. We have the following expressions for the two action variables:

$$I_1 = \frac{1}{2\pi}\oint_{\gamma_1} p_r dr , \tag{38.56}$$

$$I_2 = \frac{1}{2\pi}\oint_{\gamma_2} p_\theta d\theta = \frac{L}{2\pi}\int_0^{2\pi} d\theta = L . \tag{38.57}$$

Consider a particular bound orbit with total energy $E\,(< 0)$ such that the turning points r_1 and r_2 satisfy $0 < r_1 < r < r_2$. The closed path of integration γ_1 in (38.56) for the calculation of I_1 then consists of a path from r_1 to r_2, with p_r positive (outward motion), and then from r_2 back to r_1, with p_r negative (inward motion). Although the evaluation of the integral is amenable to elementary methods it is instructive to do it by the elegant method of contour integration. So we will digress to present the mathematical procedure in this case, which involves an appreciation of the concept of *Riemann surfaces* in complex analysis.

Equation (38.55), with $H = E$, implies

$$p_r(r) = \mp\sqrt{2m\left(E + \frac{\alpha}{r}\right) - \frac{L^2}{r^2}} = \mp\sqrt{\frac{-2m|E|}{r^2}\left(r^2 - \frac{\alpha}{|E|}r + \frac{L^2}{2m|E|}\right)}$$
$$= \mp\frac{\sqrt{2m|E|}}{r}\sqrt{(r-r_1)(r_2-r)} , \tag{38.58}$$

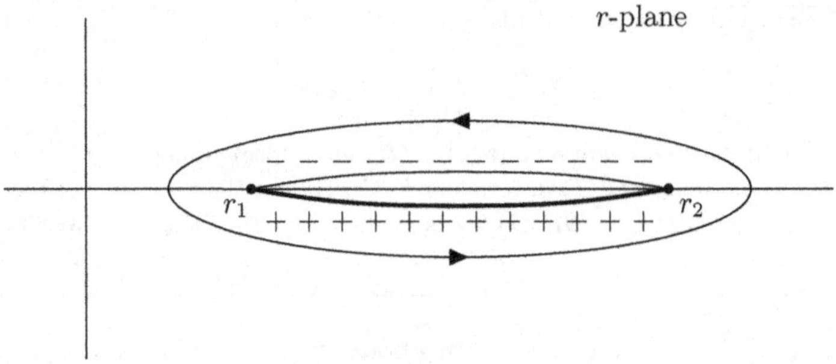

Fig. 38.4

where

$$r_{\frac{1}{2}} = \frac{\alpha}{2|E|} \mp \frac{1}{2}\sqrt{\frac{\alpha^2}{E^2} - \frac{2L^2}{m|E|}}\,. \tag{38.59}$$

When analytically continued to the complex r-plane, the function $p_r(r)$ as given by (38.58) has a square root branch structure with *branch points* at $r = r_1$ and $r = r_2$ (where $p_r = 0$). To construct the Riemann surface for the function one can make a *branch cut* along the real axis from r_1 to r_2, such that at the lower (upper) bank of the cut p_r is positive (negative). The contour of integration γ_1 is then shown in Fig. 38.4, which also depicts the first sheet of the Riemann surface. The second sheet can be pictured similarly but with p_r assuming positive (negative) values at the upper (lower) bank of the cut. The upper (lower) bank of the cut on the second sheet is imagined to be joined to the lower (upper) bank of the first sheet to form the 2-sheeted Riemann surface of the *meromorphic function* $p_r(r)$.

The *topological structure* of the 2-sheeted Riemann surface described above can be better understood if we perform a 1-point *compactification* of each of the sheets to include *the point at infinity* (denoted ∞), so that each sheet is topologically a 2-sphere (*Riemann sphere*). When the two Riemann spheres with cuts are appropriately joined, positive bank to positive bank and negative bank to negative bank, as shown in Fig. 38.5, the resulting surface is again topologically a 2-sphere.

To do the integral (36.64) for I_1 we have to investigate the analytical proper-

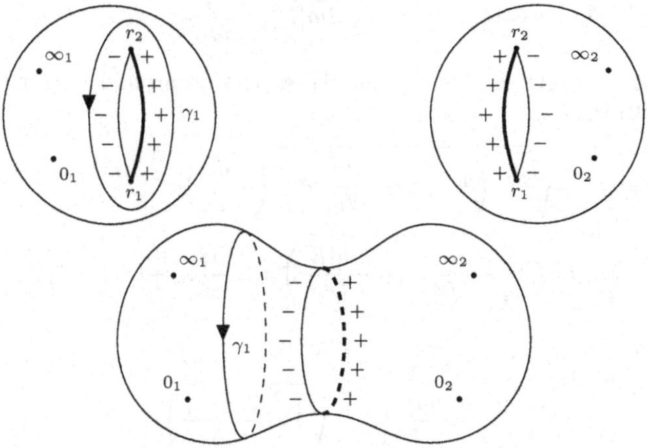

Fig. 38.5

ties of the function $p_r(r)$. We claim that the differential $p_r(r)dr$ has first order poles at $r = 0_1$ and $r = 0_2$, and second order poles at $r = \infty_1$ and $r = \infty_2$, where the subscripts 1 and 2 refer to locations of r on the first and second Riemann sheets, respectively. Note that the negative (positive) sign in front of the square root sign in (36.66) refers to values of $p_r(r)$ on the first (second) sheet. This can be seen by expanding $p_r(r)$ around $r = 0$ and $p_r(r)dr = p_r(1/z)(-dz/z^2)$ $(z \equiv 1/r)$ around $z = 0$ as *Laurent series*. Writing $C \equiv \sqrt{2m|E|} > 0$, we have from (38.58), around $r = 0$,

$$\begin{aligned} p_r(r) &= \mp\frac{C}{r}\sqrt{(r-r_1)(r_2-r)} = \mp\frac{C}{r}\sqrt{r_1r_2\left(\frac{r}{r_1}-1\right)\left(1-\frac{r}{r_2}\right)} \\ &= \mp\frac{C}{r}\sqrt{-r_1r_2}\left(1-\frac{r}{r_1}\right)^{1/2}\left(1-\frac{r}{r_2}\right)^{1/2} \\ &= \mp\frac{C}{r}\sqrt{-r_1r_2}\left(1-\frac{r}{2r_1}+\dots\right)\left(1-\frac{r}{2r_2}+\dots\right) \\ &= \mp\frac{C}{r}\sqrt{-r_1r_2}\left\{1-\frac{r}{2}\left(\frac{1}{r_1}+\frac{1}{r_2}\right)+\dots\right\}. \end{aligned} \tag{38.60}$$

From (38.59) we have

$$r_1r_2 = \frac{\alpha^2}{4E^2} - \frac{1}{4}\left(\frac{\alpha^2}{E^2} - \frac{2L^2}{m|E|}\right) = \frac{L^2}{2m|E|}\,. \tag{38.61}$$

The residues of $p_r(r)$ at the first order poles $r = 0_1$ and $r = 0_2$ are then seen from (38.60) to be

$$Res\left(0_{\frac{1}{2}}\right) = \mp C\sqrt{-r_1 r_2} = \mp\sqrt{2m|E|}\sqrt{\frac{-L^2}{2m|E|}} = \mp\sqrt{-L^2}\,. \tag{38.62}$$

Similarly, on setting $z \equiv 1/r$ for the expansion around $r = \infty$, it follows from (38.60) that

$$\begin{aligned} p_r(r)dr &= \mp\sqrt{2m\left(E + \alpha z - \frac{L^2}{2m}z^2\right)}\left(-\frac{dz}{z^2}\right) \\ &= \pm\frac{dz}{z^2}\sqrt{-L^2\left(z^2 - \frac{2m\alpha}{L^2}z + \frac{2m|E|}{L^2}\right)} = \pm\frac{dz}{z^2}\sqrt{-L^2(z-z_1)(z-z_2)}\,, \end{aligned} \tag{38.63}$$

where

$$z_{\frac{1}{2}} = \frac{m\alpha}{L^2}\left(1 \mp \sqrt{1 - \frac{2L^2|E|}{m\alpha^2}}\right), \tag{38.64}$$

from which it follows that

$$z_1 + z_2 = \frac{2m\alpha}{L^2}\,, \qquad z_1 z_2 = \frac{2m|E|}{L^2}\,. \tag{38.65}$$

Continuing to expand (38.63) around $z = 0$ ($r = \infty$) we have

$$\begin{aligned} p_r(r)dr &= \pm\frac{dz}{z^2}\sqrt{-L^2 z_1 z_2\left(1 - \frac{z}{z_1}\right)\left(1 - \frac{z}{z_2}\right)} \\ &= \pm\frac{dz}{z^2}\sqrt{-2m|E|}\left(1 - \frac{z}{2z_1} + \dots\right)\left(1 - \frac{z}{z_2} + \dots\right) \\ &= \pm\frac{dz}{z^2}\sqrt{-2m|E|}\left\{1 - \frac{z}{2}\left(\frac{1}{z_1} + \frac{1}{z_2}\right) + \dots\right\}. \end{aligned} \tag{38.66}$$

We see that the poles at $z = 0_1$ ($r = \infty_1$) and $z = 0_2$ ($r = \infty_2$) are second order and the residues are given by

$$\begin{aligned} Res\left(r = \infty_{\frac{1}{2}}\right) &= \mp\frac{1}{2}\sqrt{2mE}\left(\frac{1}{z_1} + \frac{1}{z_2}\right) = \mp\frac{1}{2}\sqrt{2mE}\left(\frac{z_1 + z_2}{z_1 z_2}\right) \\ &= \mp\alpha\sqrt{\frac{m}{2E}}\,, \end{aligned} \tag{38.67}$$

where the last equality follows directly from (38.65).

From the orientation of the closed loop of integration γ_1 shown in Fig. 36.4, we finally have, by using the **residue theorem**,

$$\begin{aligned} I_1 &= \frac{1}{2\pi}2\pi i\left(Res\,(0_2) + Res\,(\infty_2)\right) = -\frac{1}{2\pi}2\pi i\left(Res\,(0_1) + Res\,(\infty_1)\right) \\ &= i\left(\sqrt{-L^2} + \alpha\sqrt{\frac{m}{2E}}\right) = -L + \alpha\sqrt{-\frac{m}{2E}}\,. \end{aligned} \tag{38.68}$$

Recalling from (38.57) that $L = I_2$, we have

$$(I_1 + I_2)^2 = -\frac{\alpha^2 m}{2E} . \tag{38.69}$$

We finally obtain the following expression for the Hamiltonian function $H'(I_1, I_2)$ in terms of the action variables:

$$H'(I_1, I_2) = -\frac{\alpha^2 m}{2(I_1 + I_2)^2} . \tag{38.70}$$

A historically significant development from the above equation beyond classical mechanics follows immediately from the imposition of the so-called **Bohr-Sommerfeld quantization rules** on the action integrals I_i:

$$I_i = \frac{1}{2\pi} \oint_{\gamma_i} p_j dq^j = n_i \hbar , \qquad n_i = 1, 2, 3, \ldots , \tag{38.71}$$

where $\hbar = h/(2\pi)$ and h is **Planck's constant**. This yields, on setting $\alpha = e^2$ (e being the electronic charge), the Bohr energy formula for the hydrogen atom:

$$E_n = -\frac{me^4}{2\hbar^2 n^2} , \qquad n = n_1 + n_2 = 1, 2, 3, \ldots . \tag{38.72}$$

Before leaving this problem we would like to highlight some remarkable mathematical facts concerning compact Riemann surfaces that are manifested by our method to obtain I_1. First we note that in our example, the sum of the residues at all the singularities $r = 0_1, 0_2, \infty_1$ and ∞_2 of the differential $p_r(r)dr$ vanishes. This confirms a general statement of the *residue theorem*: *The sum of the residues of an* ***Abelian differential*** *on a compact Riemann surface is always zero.* An Abelian differential is a differential of the form $w(z)dz$ where $w(z)$ is an **algebraic function**, namely, one which satisfies an algebraic equation of the form

$$w^N + a_1(z)w^{N-1} + \cdots + a_N(z) = 0 , \tag{38.73}$$

with the coefficients $a_i(z)$ being *rational functions* of the complex variable z. It is a fundamental theorem of classical function theory that *the Riemann surface of an algebraic function satisfying an algebraic equation of degree N is compact and homeomorphic to a sphere with g handles, $g = 0, 1, 2, \ldots$ being known as the* ***genus*** *of the surface.* Furthermore, the genus of the Riemann surface of an algebraic function is determined in a simple manner by the so-called **Riemann-Hurwitz formula**, which states that:

$$g = \frac{B}{2} - N + 1 , \tag{38.74}$$

where B is the number of branch points of the Riemann surface, and N is the degree of the algebraic equation satisfied by the algebraic function. This formula immediately implies that the number of branch points must be even.

In our example, $B = N = 2$, so $g = 0$. This confirms the fact that the Riemann surface of $p_r(r)$ is topologically a 2-sphere, for which $g = 0$.

Problem 38.6 Consider a billiard ball moving on a tabletop with a rectangular shape, and suffering only *elastic* collisions on hitting the sides, of lengths a and b. Assume that the tabletop is frictionless and horizontal.

(a) Show that $|p_x|$ and $|p_y|$ are both constants of motion, and justify that the system is integrable. (The x and y axes are along the directions of the sides of the table.)

(b) Express the action variables

$$I_1 = \frac{1}{2\pi}\oint p_x dx\,, \qquad I_2 = \frac{1}{2\pi}\oint p_y dy$$

in terms of $|p_x|$ and $|p_y|$.

(c) Express the frequencies of the angle variables in terms of I_1 and I_2.

(d) Sketch graphs for $x(t)$ and $p_x(t)$.

(e) Write Fourier series expansions for the two-dimensional vector quantities $\boldsymbol{x}(t)$ and $\boldsymbol{p}(t)$.

Problem 38.7 Consider the so-called **Toda lattice**, a system with three degrees of freedom described by the Hamiltonian

$$H(q^i, p_i) = \frac{1}{2}\left((p_1)^2 + (p_2)^2 + (p_3)^2\right) + \left\{e^{-(q^1-q^2)} + e^{-(q^2-q^3)} + e^{-(q^3-q^1)} - 3\right\}. \tag{38.75}$$

We will go through several steps to show that this system is integrable. First make a (non-canonical) transformation to the new variables $\{a_i, b_i\}$ $(i = 1, 2, 3)$:

$$a_i = \frac{1}{2}\, e^{-(q^i - q^{i+1})/2} \quad (q^{i+3} = q^i)\,, \qquad b_i = \frac{p_i}{2}\,. \tag{38.76}$$

Then define the symmetric matrix $A(t)$ and the anti-symmetric matrix $B(t)$ as follows:

$$A(t) \equiv \begin{pmatrix} b_1 & a_1 & a_3 \\ a_1 & b_2 & a_2 \\ a_3 & a_2 & b_3 \end{pmatrix} \qquad B(t) \equiv \begin{pmatrix} 0 & a_1 & -a_3 \\ -a_1 & 0 & a_2 \\ a_3 & -a_2 & 0 \end{pmatrix}. \tag{38.77}$$

(a) Show that the Hamiltonian is given in terms of the new coordinates by

$$H = 2\,Tr(A^2) - 3\,, \tag{38.78}$$

where Tr designates the trace of a matrix.

(b) Show that Hamilton's equations of motion in the (q^i, p_i) coordinates are equivalent to the following matrix equation:

$$\frac{dA(t)}{dt} = B(t)A(t) - A(t)B(t)\,. \tag{38.79}$$

Now introduce another matrix $O(t)$ which satisfies the differential equation:

$$\frac{dO(t)}{dt} = B(t)O(t)\,. \tag{38.80}$$

(c) Show that

$$\frac{dO^{-1}}{dt} = -O^{-1}(t)B(t)\,. \tag{38.81}$$

(d) Noting that $B(t)$ is anti-symmetric, show that $O(t)$ is orthogonal, that is, $O^{-1} = O^T$.

(e) Show that (38.79), (38.80) and (38.81) together imply

$$\frac{dA}{dt} = \frac{dO}{dt}\mathcal{L}O^{-1} + O\mathcal{L}\frac{dO^{-1}}{dt}\,, \tag{38.82}$$

where

$$\mathcal{L} \equiv O^{-1}(t)A(t)O(t)\,. \tag{38.83}$$

Show that $\mathcal{L}$ is time-independent.

The matrix $O(t)$ can be interpreted as an evolution operator which propagates $A(t)$ in time such that $A(0) = \mathcal{L}$. Introduce the eigenvalue equation

$$A(t)\phi(t) = \lambda(t)\phi(t)\,, \tag{38.84}$$

where $\lambda(t)$ and $\phi(t)$ are the eigenvalue and eigenvector, respectively, of $A(t)$.

(f) Show from (38.83) that

$$\mathcal{L}O^{-1}(t)\phi(t) = \lambda(t)\,O^{-1}(t)\phi(t)\,. \tag{38.85}$$

Hence $\lambda(t)$ is also an eigenvalue of $\mathcal{L}$. The crucial point is that, since $\mathcal{L}$ is time-independent, as shown above in (e), the eigenvalues must also be time-independent. The three time-independent eigenvalues λ_1, λ_2 and λ_3 are thus three independent constants of motion of the Toda-lattice. We have thus shown that the Toda-lattice is integrable.

Problem 38.8 Consider the following Hamiltonian with a $(2, 2)$ resonance:

$$H(\phi_1, \phi_2, J_1, J_2) = H_0(J_1, J_2) + \alpha J_1 J_2 \cos(2\phi_1 - 2\phi_2)\,, \tag{38.86}$$

where

$$H_0(J_1, J_2) = J_1 + J_2 - (J_1)^2 - 3J_1J_2 + (J_2)^2\,. \tag{38.87}$$

$\{J_i, \phi_i\}$ are the action-angle variables of the integrable unperturbed Hamiltonian H_0 and α is a small parameter characterizing the strength of the perturbation.

(a) Show that, besides H, $I_1 \equiv J_1 + J_2$ is also a constant of motion. Thus the full Hamiltonian, including the perturbation, is also integrable.

Make the transformation of variables $\{J_i, \phi_i\} \to \{I_i, \theta_i\}$ given by:

$$I_1 = J_1 + J_2\,, \qquad \theta_1 = \phi_1\,, \tag{38.88a}$$

$$I_2 = J_2\,, \qquad \theta_2 = \phi_2 - \phi_1\,. \tag{38.88b}$$

(b) Show that the above transformation is canonical by showing that

$$\sum_i dI_i \wedge d\theta_i = \sum_i dJ_i \wedge d\phi_i \,.$$

(c) Show that in the new coordinates the Hamiltonian takes the form

$$H(I_1, I_2, \theta_1, \theta_2) = I_1 - (I_1)^2 - I_1 I_2 + 3(I_2)^2 + \alpha I_2(I_1 - I_2)\cos 2\theta_2 \,. \tag{38.89}$$

Hamilton's equations of motion in terms of I_i and θ_i are thus

$$\frac{dI_1}{dt} = -\frac{\partial H}{\partial \theta_1} = 0 \,, \tag{38.90a}$$

$$\frac{d\theta_1}{dt} = \frac{\partial H}{\partial I_1} = 1 - 2I_1 - I_2 + \alpha I_2 \cos 2\theta_2 \,, \tag{38.90b}$$

$$\frac{dI_2}{dt} = -\frac{\partial H}{\partial \theta_2} = 2\alpha I_2(I_1 - I_2)\sin 2\theta_2 \,, \tag{38.90c}$$

$$\frac{d\theta_2}{dt} = \frac{\partial H}{\partial I_2} = -I_1 + 6I_2 + \alpha(I_1 - 2I_2)\cos 2\theta_2 \,. \tag{38.90d}$$

Thus I_1 is a constant of motion and the last two equations (those for I_2 and θ_2) are decoupled from the first two. We will use the methods for *linear dynamical systems* discussed in Chapter 18 to investigate the solutions for the system of two coupled equations (38.90c) and (38.90d).

(d) Find the *fixed points* on the (I_2, θ_2) *surface of section*. By definition these are obtained from solutions to the equations $dI_2/dt = 0$ and $d\theta_2/dt = 0$. Show that these fixed points are given by I_0 and $\theta_2 = n\pi/2$, where I_0 is a solution of the equation

$$-I_1 + 6I_0 + 2\alpha\cos(n\pi)(I_1 - 2I_0) = 0 \,. \tag{38.91}$$

(e) To determine the nature of the fixed points (*elliptic* or *hyperbolic*), we linearize Hamilton's equations of motion for I_2 and θ_2 about the fixed points. Thus we write

$$I_2(t) \approx I_0 + \Delta I_2(t) \,, \qquad \theta_2(t) \approx \frac{n\pi}{2} + \Delta\theta_2(t) \,. \tag{38.92}$$

Show that ΔI_2 and $\Delta\theta_2$ satisfy the linear equations

$$\begin{pmatrix} \dfrac{d(\Delta I_2)}{dt} \\ \dfrac{d(\Delta\theta_2)}{dt} \end{pmatrix} = \begin{pmatrix} 0 & 4\alpha\cos(n\pi)(I_1 - I_0) \\ 6 - 2\alpha\cos(n\pi) & 0 \end{pmatrix} \begin{pmatrix} \Delta I_2 \\ \Delta\theta_2 \end{pmatrix} . \tag{38.93}$$

(f) Solve the above linearized equations by finding the eigenvalues of the 2×2 matrix in (38.93). Show that for small perturbations ($\alpha \ll 1$) the eigenvalues are given by

$$\lambda_\pm = \pm\sqrt{\frac{10}{3}\alpha(I_1)^2\cos(n\pi)} \,. \tag{38.94}$$

Then show that

$$\begin{pmatrix} \Delta I_2(t) \\ \Delta\theta_2(t) \end{pmatrix} = a_+ e^{\lambda_+ t} \begin{pmatrix} \dfrac{b}{\lambda_+} \\ 1 \end{pmatrix} + a_- e^{\lambda_- t} \begin{pmatrix} \dfrac{b}{\lambda_-} \\ 1 \end{pmatrix} , \tag{38.95}$$

where $b \equiv 5\alpha(I_1)^2/9$ and a_+, a_- are arbitrary constants depending on initial conditions.

(g) Out of the fixed points found in (d), identify which are elliptic (trajectories oscillate about these) and which are hyperbolic (trajectories approach or recede from these exponentially). Note that they alternate.

(h) Show that for $\alpha \ll 1$, fixed points exist only for $E < 3/13$, where E is the constant energy.

Chapter 39

Symplectic Geometry in Hamiltonian Dynamics, Hamiltonian Flows, and Poincaré-Cartan Integral Invariants

In this chapter we will begin to give an introductory account of symplectic geometry, a kind of geometry that forms the mathematical basis of the Hamiltonian formulation of classical mechanics. We will first discuss this geometry within the context of Hamiltonian mechanics using a coordinatized approach in terms of the canonical positions and momenta q^i and p_i, and then in the next chapter give a more general and abstract account, eventually leading up to a fundamental theorem of symplectic geometry, known as Darboux's Theorem.

As explained in Chapter 33, given a configuration manifold Q of a mechanical system of dimension n, the Hamiltonian $H(q^1, \dots, q^n; p_1, \dots, p_n)$ is a function on the cotangent bundle $\pi : T^*Q \to Q$ (phase space), where (q^i, p_i) are the local (canonical) coordinates, with $(q^i) \in Q$ being the base space coordinates and (p_i) being the fiber coordinates. In Chapter 37 we saw that on T^*Q there is a special two-form

$$\omega = dp_i \wedge dq^i \,, \tag{39.1}$$

defined globally on T^*Q, which is invariant under canonical transformations of the canonical coordinates [cf. (37.59)]. This special two-form is known as the **symplectic form** on the cotangent bundle T^*Q, and endows the latter with a so-called **symplectic structure**, analogous to the situation in Riemannian geometry, where a symmetric, non-degenerate metric two-form $g_{ij}\, dx^i \otimes dx^j$ endows the underlying manifold with a Riemannian structure. The canonical coordinates (q^i, p_i) are also called **symplectic coordinates** of T^*Q, and canonical transformations also called **symplectic transformations**. We say that the cotangent bundle T^*Q, being endowed with the special two-form ω, is naturally a **symplectic manifold**.

As can be seen readily from (39.1), the symplectic form ω is closed: $d\omega = 0$. We denote its action by $\omega(\boldsymbol{\xi}, \boldsymbol{\eta})$, where $\boldsymbol{\xi}$ and $\boldsymbol{\eta}$ are arbitrary tangent vector fields on T^*Q. Since each such vector field is the assignment of a vector on the tangent space $T_{(p,q)}(T^*Q)$ at each point $(p,q) \in T^*Q$, we write $\boldsymbol{\xi}, \boldsymbol{\eta} \in \Gamma(T_*(T^*Q))$ (the space of smooth tangent vector fields on the cotangent bundle T^*Q), following the notation established immediately preceding Def. 22.4. It is also readily seen that the symplectic form ω is *non-degenerate*, in other words, $\omega(\boldsymbol{\xi}, \boldsymbol{\eta}) = 0$ for all $\boldsymbol{\eta}$ implies $\boldsymbol{\xi} = 0$ (cf. discussion at the end of Chapter 1). Indeed, for any $(p,q) \in T^*Q$, let the tangent vectors $\boldsymbol{\xi}_{(p,q)}, \boldsymbol{\eta}_{(p,q)} \in T_{(p,q)}(T^*Q)$ be expressed in terms of their components with respect to the natural frame $\{\partial/\partial q^i, \partial/\partial p_i\}$ [at the point (p,q)] by

$$\boldsymbol{\xi}_{(p,q)} = \xi_{q^i} \frac{\partial}{\partial q^i} + \xi_{p_i} \frac{\partial}{\partial p_i}, \qquad \boldsymbol{\eta}_{(p,q)} = \eta_{q^i} \frac{\partial}{\partial q^i} + \eta_{p_i} \frac{\partial}{\partial p_i}, \tag{39.2}$$

where the components $\xi_{p_i}, \xi_{q^i}, \eta_{p_i}, \eta_{q^i}$ are all real numbers. Then

$$\begin{aligned} &\omega_{(p,q)}(\boldsymbol{\xi}_{(p,q)}, \boldsymbol{\eta}_{(p,q)}) = (dp_i \wedge dq^i)_{(p,q)}(\boldsymbol{\xi}_{(p,q)}, \boldsymbol{\eta}_{(p,q)}) \\ &= \frac{1}{2}\sum_{i=1}^{n} \begin{vmatrix} dp_i(\boldsymbol{\xi}_{(p,q)}) & dp_i(\boldsymbol{\eta}_{(p,q)}) \\ dq^i(\boldsymbol{\xi}_{(p,q)}) & dq^i(\boldsymbol{\eta}_{(p,q)}) \end{vmatrix} = \frac{1}{2} \begin{vmatrix} \xi_{p_i} & \eta_{p_i} \\ \xi_{q^i} & \eta_{q^i} \end{vmatrix} = \frac{1}{2}\sum_{i=1}^{n} \left(\xi_{p_i}\eta_{q^i} - \xi_{q^i}\eta_{p_i}\right), \end{aligned} \tag{39.3}$$

where the second equality follows from the evaluation formula of exterior forms given by (2.29). The supposition that the right-hand side vanishes for all possible values of η_{q^i} and η_{p_i} then entails the vanishing of ξ_{p_i} and ξ_{q^i}. This shows the non-degeneracy of ω.

From the discussion at the end of Chapter 1, we see that the symplectic form ω, due to its non-degeneracy, induces an isomorphism between the tangent space $T_{(p,q)}(T^*Q)$ and its dual space, the cotangent space $T^*_{(p,q)}(T^*Q)$, for any point $(p,q) \in T^*Q$. We specify this isomorphism $\mathcal{I}_{(p,q)} : T_{(p,q)}(T^*Q) \to T^*_{(p,q)}(T^*Q)$ by [cf. (1.66)]

$$\mathcal{I}_{(p,q)}(\boldsymbol{\xi}_{(p,q)})(\boldsymbol{\eta}_{(p,q)}) = 2\omega_{(p,q)}(\boldsymbol{\eta}_{(p,q)}, \boldsymbol{\xi}_{(p,q)}), \tag{39.4}$$

where $\mathcal{I}_{(p,q)}(\boldsymbol{\xi}_{(p,q)}) \in T^*_{(p,q)}(T^*Q)$ is a real linear function on $T_{(p,q)}(T^*Q)$ and hence acts on any tangent vector $\boldsymbol{\eta}_{(p,q)} \in T_{(p,q)}(T^*Q)$ to produce a real number. Note the order of the vector fields on the right-hand side of the above equation in this specification. The isomorphism $\mathcal{I}_{(p,q)}$, given above for each point $(p,q) \in T^*Q$, gives rise in an obvious pointwise fashion to the corresponding isomorphism between the spaces of tangent vector fields and one-forms on T^*Q,

$$\mathcal{I} : \Gamma(T_*(T^*Q)) \longrightarrow \Gamma(T^*(T^*Q)), \qquad \boldsymbol{\xi} \mapsto \mathcal{I}(\boldsymbol{\xi}) \equiv \omega_{\boldsymbol{\xi}}.$$

So we have, analogous to (39.4),

$$\mathcal{I}(\boldsymbol{\xi})(\boldsymbol{\eta}) = \omega_{\boldsymbol{\xi}}(\boldsymbol{\eta}) = 2\omega(\boldsymbol{\eta}, \boldsymbol{\xi}) = \sum_{i=1}^{n} \left(\eta_{p_i}(p,q)\,\xi_{q^i}(p,q) - \eta_{q^i}(p,q)\,\xi_{p_i}(p,q)\right) , \tag{39.5}$$

where all the components of the tangent vector fields are now functions of the symplectic coordinates (p, q). It follows that the one-form $\omega_{\boldsymbol{\xi}}$ can be expressed in terms of the symplectic (canonical) coordinates as

$$\omega_{\boldsymbol{\xi}} = \sum_{i=1}^{n} \left((\omega_{\boldsymbol{\xi}})_{q^i}\, dq^i + (\omega_{\boldsymbol{\xi}})_{p_i}\, dp_i\right) = \sum_{i=1}^{n} \left(-\xi_{p_i}(p,q)\, dq^i + \xi_{q^i}(p,q)\, dp_i\right) . \tag{39.6}$$

In matrix notation,

$$\begin{pmatrix} (\omega_{\boldsymbol{\xi}})_{q^i} \\ (\omega_{\boldsymbol{\xi}})_{p_i} \end{pmatrix} = \begin{pmatrix} -\xi_{p_i} \\ \xi_{q^i} \end{pmatrix} = \begin{pmatrix} 0 & -1 \\ 1 & 0 \end{pmatrix} \begin{pmatrix} \xi_{q^i} \\ \xi_{p_i} \end{pmatrix} , \tag{39.7}$$

where each entry in the column matrices is an $n \times 1$ column matrix, the entry 1 represents the $n \times n$ identity matrix, and the entry 0 represents the $n \times n$ zero matrix. Thus the matrix representations of the isomorphism maps $\mathcal{I}$ and its inverse $(\mathcal{I})^{-1}$, with respect to the natural frame fields $(\partial/\partial q^i, \partial/\partial p_i)$ and (dq^i, dp_i), respectively, are given by

$$\mathcal{I} = \begin{pmatrix} 0 & -1 \\ 1 & 0 \end{pmatrix} , \qquad \mathcal{I}^{-1} = \begin{pmatrix} 0 & 1 \\ -1 & 0 \end{pmatrix} = \mathcal{J} , \tag{39.8}$$

where, again, each entry is an $n \times n$ matrix. We note that the matrix $\mathcal{I}^{-1}$ was denoted by $\mathcal{J}$ in Chapter 36 [cf. (36.23)], and was there introduced as the matrix behind the definition for the symplectic group and canonical transformations.

Suppose now $H(q^i, p_i)$ is a smooth function on T^*Q. Then $dH \in \Gamma(T^*(T^*Q))$ is a one-form on T^*Q; so that $\mathcal{I}^{-1}(dH) \in \Gamma(T_*(T^*Q))$ is a tangent vector field on T^*Q. If H is the Hamiltonian function of a mechanical system, then we call $\mathcal{I}^{-1}(dH)$ the **Hamiltonian vector field** of the mechanical system. Since

$$dH = \sum_{i=1}^{n} \left(\frac{\partial H}{\partial q^i}\, dq^i + \frac{\partial H}{\partial p_i}\, dp_i\right) , \tag{39.9}$$

the corresponding differential equations, for $(q^i, p_i) \in T^*Q$, are thus

$$\begin{pmatrix} \dfrac{dq^i}{dt} \\ \dfrac{dp_i}{dt} \end{pmatrix} = \mathcal{I}^{-1}(dH) = \begin{pmatrix} 0 & 1 \\ -1 & 0 \end{pmatrix} \begin{pmatrix} \dfrac{\partial H}{\partial q^i} \\ \dfrac{\partial H}{\partial p_i} \end{pmatrix} = \begin{pmatrix} \dfrac{\partial H}{\partial p_i} \\ -\dfrac{\partial H}{\partial q^i} \end{pmatrix} \tag{39.10}$$

These are precisely *Hamilton's canonical equations of motion* (33.16). We can also write

$$\mathcal{I}^{-1}(dH) = \sum_{i=1}^{n} \left(\frac{dq^i}{dt}\frac{\partial}{\partial q^i} + \frac{dp_i}{dt}\frac{\partial}{\partial p_i} \right) . \tag{39.11}$$

The Poisson bracket introduced in Def. 38.1 can be defined alternatively in a coordinate-free manner in terms of the symplectic isomorphism $\mathcal{I}$ as follows.

Definition 39.1. *Let F and G be smooth functions on the cotangent bundle T^*Q of a configuration space Q endowed naturally with the symplectic form $\omega = dp_i \wedge dq^i$. Then the smooth function $\{F\,,\,G\}$ defined by*

$$\boxed{\{F\,,\,G\} \equiv 2\,\omega\left(\mathcal{I}^{-1}(dG)\,,\,\mathcal{I}^{-1}(dF)\right)} \tag{39.12}$$

*is called the **Poisson bracket** of the functions F and G.*

The following theorem gives the main properties of the Poisson bracket.

Theorem 39.1. *The Poisson bracket $\{\,,\,\}$ is bilinear and non-degenerate. Non-degeneracy here means that if $(p,q) \in T^*Q$ is not a critical point of a function F on T^*Q, that is, if $F(p,q) \neq 0$, then there exists a smooth non-zero function G on T^*Q such that $\{F\,,\,G\} \neq 0$. Furthermore, $\{\,,\,\}$ satisfies the following properties:*

(1) $\{F\,,\,G\} = -\{G\,,\,F\}$ *(skew-symmetry)*, (39.13)

(2) $\{F_1F_2\,,\,G\} = F_1\,\{F_2\,,\,G\} + F_2\,\{F_1\,,\,G\}$ *(Leibnitz rule)*, (39.14)

(3) $\{\{H\,,\,F\}\,,\,G\} + \{\{F\,,\,G\}\,,\,H\} + \{\{G\,,\,H\}\,,\,F\} = 0$ *(Jacobi identity)*. (39.15)

Thus the set of smooth functions on the cotangent bundle T^*Q constitute a Lie algebra under the Poisson bracket (as the internal multiplication).

Problem 39.1 Use (39.5) and (39.8) to verify that the above coordinate-free definition for the Poisson bracket is equivalent to Def. 38.1, which is given in terms of partial derivatives with respect to the symplectic coordinates (q^i, p_i).

Problem 39.2 Verify the three Lie algebra properties of the Poisson bracket given in Theorem 39.1 by using Def. 39.1.

If $H(q^i, p_i)$ is the Hamiltonian function and $F(q^i, p_i)$ is an arbitrary smooth function on T^*Q, the Poisson bracket as defined by Def. 39.1 allows us to write

$$\begin{aligned} \{F\,,\,H\} &= 2\omega\left(\mathcal{I}^{-1}(dH)\,,\,\mathcal{I}^{-1}(dF)\right) = \mathcal{I}\left(\mathcal{I}^{-1}(dF)\right)\left(\mathcal{I}^{-1}(dH)\right) \\ &= (dF)\left(\mathcal{I}^{-1}(dH)\right) = \frac{\partial F}{\partial q^i}\frac{dq^i}{dt} + \frac{\partial F}{\partial p_i}\frac{dp_i}{dt}\,, \end{aligned} \tag{39.16}$$

where the second equality follows from (39.5) and the last equality follows from (39.10). Thus the equation of motion for an arbitrary function F on T^*Q can be expressed in terms of Poisson brackets as

$$\frac{dF}{dt} = \{F\,,\,H\}\,, \tag{39.17}$$

where H is the Hamiltonian of the system. This is in conformity with (38.23). Replacing F by q^i and p_i, and using Def. 38.1 of the Poisson bracket (in terms of partial derivatives), one regains again Hamilton's canonical equations of motion (38.19) and (38.20) [or (33.16)].

To appreciate better the geometrical implications of Hamilton's equations of motion it is useful to consider them in light of the notion of a *one-parameter group of diffeomorphisms* on a manifold M, introduced in Chapter 11. According to Theorem 11.1, the tangent vector field $\mathcal{I}^{-1}(dH)$ on T^*Q corresponding to a Hamiltonian function H on T^*Q induces a (local) one-parameter group of diffeomorphisms $\varphi_t : T^*Q \to T^*Q$ defined by

$$\left.\frac{d}{dt}\varphi_t(x)\right|_{t=0} = \left(\mathcal{I}^{-1}(dH)\right)(x)\,, \qquad x \in T^*Q\,. \tag{39.18}$$

The one-parameter group φ_t (parametrized by the time t) is called the **Hamiltonian phase flow** of the Hamiltonian function H. By Def. 11.1 and Theorem 11.2 we can rephrase the content of (39.18) as follows. *The tangent vector field $\mathcal{I}^{-1}(dH)$ on T^*Q is invariant under the Hamiltonian phase flow φ_t* [cf. (11.11)]:

$$(\varphi_t)_*\left(\mathcal{I}^{-1}(dH)\right) = \mathcal{I}^{-1}(dH)\,. \tag{39.19}$$

The group element φ_t for each t is in fact a symplectic (or canonical) transformation on T^*Q. A consequence of Theorem 37.2 can be stated as the following theorem:

Theorem 39.2. *The Hamiltonian phase flow φ_t preserves the symplectic structure $\omega = dp_i \wedge dq^i$ on phase space (the cotangent bundle T^*Q of a configuration manifold Q): $(\varphi_t)^*\omega = \omega$; or equivalently, the symplectic structure ω is invariant under the Hamiltonian phase flow.*

Certain integrals on regions of phase space, called *Poincaré-Cartan invariants,* manifest themselves as invariants under Hamiltonian phase flows. To demonstrate this we first present two definitions.

Definition 39.2. *A differential k-form α on T^*Q (phase space) is called a* ***relative integral invariant*** *of the phase flow φ_t if*

$$\int_{\varphi_t C} \alpha = \int_C \alpha \tag{39.20}$$

for every closed k-dimensional region (region without boundary, also called a k-cycle) C in T^*Q. (We express the fact that C is without boundary by writing $\partial C = \emptyset$, cf. Def. 8.2.)

At this point it will be useful to recall the topological concepts of *homology* and *cohomology* as introduced in Chapter 21.

Definition 39.3. *A differential k-form α on T^*Q (phase space) is called an* ***absolute integral invariant*** *of the phase flow φ_t if (39.20) is satisfied for every k-dimensional region (also called a k-chain) C in T^*Q. An absolute integral invariant is also referred to just as an* ***integral invariant.***

Note that in the above definitions the phase flow φ_t is not necessarily the Hamiltonian phase flow. We will now present a series of theorems involving relative and absolute integral invariants.

Theorem 39.3. *If α is a relative integral invariant of a phase flow φ_t in phase space, then $d\alpha$ is an (absolute) integral invariant of φ_t.*

Proof. Suppose a k-form α on T^*Q is a relative integral invariant of a phase flow φ_t on T^*Q. Let C be a $k+1$-dimensional region in T^*Q. Then

$$\int_C d\alpha = \int_{\partial C} \alpha = \int_{\varphi_t(\partial C)} \alpha = \int_{\partial(\varphi_t C)} \alpha = \int_{\varphi_t C} d\alpha \,, \tag{39.21}$$

where the first and last equalities follow from Theorem 8.1 (Stokes' Theorem), and the second equality is a consequence of the supposition that α is a relative integral invariant and the fact that $\partial(\partial C) = \emptyset$ (the boundary of a boundary always vanishes). □

Theorem 39.4. *A k-form α on phase space (T^*Q) is an (absolute) integral invariant of a phase flow φ_t if and only if α is invariant under φ_t, that is, if $\varphi_t^*\alpha = \alpha$.*

Proof. Suppose $\varphi_t^*\alpha = \alpha$, and C is an arbitrary k-dimensional region in T^*Q. Then

$$\int_{\varphi_t C} \alpha = \int_C \varphi_t^* \alpha = \int_C \alpha \,, \tag{39.22}$$

where the first equality follows from (8.32). Thus α is an integral invariant of φ_t. Conversely, suppose α is an integral invariant of φ_t. Then, for any k-dimensional region C in T^*Q,

$$\int_C \varphi_t^*\alpha = \int_{\varphi_t C} \alpha = \int_C \alpha \,, \tag{39.23}$$

where the first equality again follows from (8.32) and the second from the supposition. Since C is an arbitrary k-dimensional region in T^*Q, it must be true that $\varphi_t^*\alpha = \alpha$. □

As a consequence of Theorem 39.2, the above theorem immediately implies the following important fact.

Theorem 39.5. *The symplectic form $\omega = dp_i \wedge dq^i$ on phase space (T^*Q) is an (absolute) integral invariant of the Hamiltonian phase flow φ_t, that is, for any 2-dimensional region (2-chain) C in T^*Q,*

$$\int_{\varphi_t C} dp_i \wedge dq^i = \int_C dp_i \wedge dq^i \, . \tag{39.24}$$

The global one-form

$$\gamma = p_i \, dq^i \tag{39.25}$$

on T^*Q, whose exterior derivative is obviously the symplectic form ω:

$$d\gamma = \omega = dp_i \wedge dq^i \, , \tag{39.26}$$

is called the **canonical form** on T^*Q. We have already seen the importance of the integral of this one-form in relation to the *action integral* [cf. (38.9)] and the *action variables* in integrable systems [cf. (38.32)]. We will now demonstrate another one of its useful properties, namely, *the canonical one-form γ is a relative integral invariant of a Hamiltonian phase flow when it is integrated over a closed loop (a 1-cycle) C situated entirely in a simply-connected region M of phase space*, that is, when C is contractible to a point in the region. In particular, γ is always a relative integral invariant of a Hamiltonian phase flow when phase space has the topological structure of $\mathbb{R}^{2n}$, since the latter is everywhere simply-connected. The important topological fact to realize about such loops is that C is always the boundary of some 2-dimensional region D in the simply-connected region M: $C = \partial D$. Suppose now that φ_t is a Hamiltonian phase flow, considered to be restricted to the simply-connected region M. We have

$$\begin{aligned} &\int_{\varphi_t C} \gamma = \int_{\varphi_t(\partial D)} \gamma = \int_{\partial(\varphi_t D)} \gamma = \int_{\varphi_t D} d\gamma = \int_{\varphi_t D} \omega = \int_D \omega \\ &= \int_D d\gamma = \int_{\partial D} \gamma = \int_C \gamma \, , \end{aligned} \tag{39.27}$$

where the third and seventh equalities follow from Stokes' Theorem, and the fifth from the fact that the symplectic form ω is an (absolute) integral invariant of the Hamiltonian phase flow, according to Theorem 39.5.

The following useful theorem will be stated without proof.

Theorem 39.6. *If α and β are both integral invariants of a phase flow φ_t, then $\alpha \wedge \beta$ is also an integral invariant of φ_t.*

Using this theorem one can immediately construct a series of so-called **Poincaré-Cartan integral invariants** of the Hamiltonian phase flow, beginning with the symplectic form ω. If $\dim(Q) = n$, one has

$$\omega, \, \omega \wedge \omega, \, \omega \wedge \omega \wedge \omega, \, \ldots, \underbrace{\omega \wedge \cdots \wedge \omega}_{n \text{ times}} \, .$$

The last form, being a non-zero $2n$-form, is a top form on T^*Q and must be proportional to the volume element $(dp_1 \wedge dq^1) \wedge \cdots \wedge (dp_n \wedge dq^n)$ in T^*Q. In fact, it can be readily shown, using the rules of exterior algebra presented in Chapter 2, that

$$\underbrace{\omega \wedge \cdots \wedge \omega}_{n \text{ times}} = (n!)\,(dp_1 \wedge dq^1) \wedge \cdots \wedge (dp_n \wedge dq^n) \,. \tag{39.28}$$

By Theorem 39.6, then, we obtain the following important theorem.

Theorem 39.7. ***(Liouville's Theorem)*** *The Hamiltonian phase flow preserves volumes in phase space. In other words, if $V \subset T^*Q$ is a certain $2n$-dimensional region in phase space (of dimension $2n$) and φ_t is the Hamiltonian phase flow, then*

$$\int_V (dp_1 \wedge dq^1) \wedge \cdots \wedge (dp_n \wedge dq^n) = \int_{\varphi_t V} (dp_1 \wedge dq^1) \wedge \cdots \wedge (dp_n \wedge dq^n) \,. \tag{39.29}$$

Problem 39.3 Show that the *canonical 1-form* γ [cf. (39.25)] and the symplectic 2-form ω on T^*Q can be defined globally in a coordinate-free manner as follows. First define the projection map

$$\pi : T^*Q \to Q \,, \qquad (p,q) \mapsto q \quad \text{for each } x = (p,q) \in T^*Q \,.$$

Then define a global 1-form γ on T^*Q by requiring that

$$\langle \boldsymbol{X}, \gamma \rangle = \langle \pi_* \boldsymbol{X}, \boldsymbol{p} \rangle \tag{39.30}$$

for all tangent vector fields $\boldsymbol{X}$ on T^*Q and all cotangent vector fields $\boldsymbol{p}$ on Q. The symplectic form is then defined globally without reference to local coordinates by $\omega = d\gamma$. Show that in local coordinates γ is given by $\gamma = p_i\, dq^i$, in agreement with (39.25).

When the Hamiltonian $H(q^i, p_i, t)$ is explicitly time-dependent it will be useful to introduce the so-called **extended phase space** Σ coordinatized by $q^i, p_i, t\ (i = 1, \ldots, n)$, with $2n$ being the dimension of the phase space. We can write the extended phase space as $\Sigma = T^*Q \times I$, where I is a closed interval in $\mathbb{R}$. Introduce the one-form ω^1 on Σ defined by

$$\omega^1 = p_i\, dq^i - H\, dt \,. \tag{39.31}$$

We see that the canonical 1-form $\gamma = p_i\, dq^i$ introduced in (39.25) is the restriction of ω^1 to T^*Q. We would like to show that just as γ, ω^1 is a distinguished 1-form possessing a special invariance property which is actually a generalization of (39.27). To explain this we refer to Fig. 39.1 and rely on a hydrodynamical

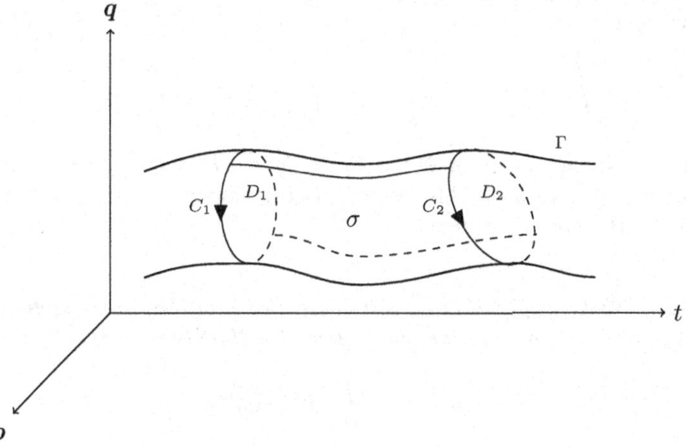

Fig. 39.1

interpretation of the situation. In this figure, let C_1 be a closed oriented path in Σ. Points on this path can be regarded as initial positions in phase space (at various times) for trajectories of the Hamiltonian phase flow. The various trajectories emanating from points on C_1 then form a smooth surface Γ in extended phase space. We can refer to this as a *tube of trajectories*, akin to fluid flowing through a pipe so that Γ can be viewed as the wall of the pipe. Let C_2 be another closed oriented curve in Σ lying on the surface of the tube and encircling it so that $C_1 - C_2$ is the boundary of a segment of the tube, which we will denote by σ. In other words, $\partial\sigma = C_1 - C_2$.

The special property that we referred to is stated as the following theorem.

Theorem 39.8. *Suppose two arbitrary curves C_1 and C_2 encircle the same tube of trajectories of a Hamiltonian phase flow. Then*

$$\oint_{C_1} p_i\, dq^i - H\, dt = \oint_{C_2} p_i\, dq^i - H\, dt\,. \tag{39.32}$$

The distinguished 1-form ω^1 is often called the primary **Poincaré-Cartan integral invariant**. We will shortly indicate the main ideas behind the proof of this theorem but will not go into details. Before doing so we will mention two consequences of the theorem. First, suppose both loops C_1 and C_2 are oriented so that C_1 corresponds to the constant time t_1 and C_2 to the constant time t_2.

Then in both line integrals in (39.32) $dt = 0$, and so ω^1 reduces to the canonical one-form $\gamma = p_i\, dq^i$ and (39.32) reduces to (39.27), that is, γ is a relative integral invariant of the phase flow φ_t of the Hamiltonian. Second, by using the Stokes theorem (Theorem 8.1) and letting $C_1 = \partial D_1$ and $C_2 = \partial D_2$ (see Fig. 39.1), we have

$$\int_{D_1} d\omega^1 = \int_{\partial D_1 = C_1} \omega^1 = \int_{\partial D_2 = C_2} \omega^1 = \int_{D_2} d\omega^1 \,. \tag{39.33}$$

We can regard the two-form $d\omega^1 = dp_i \wedge dq^i - dH \wedge dt$ as another Poincaré-Cartan integral invariant. This result is useful enough that we will state it as a corollary to the above theorem.

Corollary 39.1. *Suppose D_1 and D_2 are two arbitrary cross sections of the same tube of trajectories of the phase flow of a Hamiltonian H. Then*

$$\int_{D_1} dp_i \wedge dq^i - dH \wedge dt = \int_{D_2} dp_i \wedge dq^i - dH \wedge dt \,. \tag{39.34}$$

Let us now return to an outline of the proof of Theorem 39.8, following the approach by Arnold (Arnold 1978). First we need to introduce the notion of *vortex lines* of a 1-form ω^1. At any point $x = (p_i, q^i, t)$ in the extended phase space Σ, the differential 2-form $d\omega^1$ gives rise to an exterior algebraic 2-form $d\omega^1\big|_x$ which acts on two vectors in the tangent space $T_x\Sigma$. The crucial fact is that Σ is odd-dimensional ($dim\, T_x\Sigma = 2n+1$). We have the following lemma.

Lemma 39.1. *Let ω^2 be an algebraic 2-form on an odd-dimensional vector space $\mathbb{V}$ ($\dim \mathbb{V} = 2n+1$). Then there exists a vector $\boldsymbol{u} \in \mathbb{V}$ such that*

$$\omega^2(\boldsymbol{u}, \boldsymbol{v}) = 0 \,, \qquad \textit{for all } \boldsymbol{v} \in \mathbb{V} \,. \tag{39.35}$$

Proof. The exterior 2-form ω^2 can be represented by a skew-symmetric $(2n+1) \times (2n+1)$ matrix A ($A^T = -A$) , so that

$$\omega^2(\boldsymbol{u}, \boldsymbol{v}) = (A\boldsymbol{u}, \boldsymbol{v}) \,, \tag{39.36}$$

where the bracket on the right-hand side represents the usual Euclidean scalar product of two vectors. The determinant of A must vanish, since

$$det\,(A) = det\,(A^T) = det\,(-A) = (-1)^{2n+1}\, det\,(A) = -det\,(A) \,. \tag{39.37}$$

Therefore A must possess a non-zero eigenvector $\boldsymbol{u}$ with eigenvalue 0: $A\boldsymbol{u} = \boldsymbol{0}$. So the lemma is proved. □

A vector $\boldsymbol{u} \in \mathbb{V}$ for which $\omega^2(\boldsymbol{u}, \boldsymbol{v}) = 0$ for all $\boldsymbol{v} \in \mathbb{V}$ is called a **null vector** of the form ω^2. The totality of null vectors clearly forms a subspace of $\mathbb{V}$. The 2-form ω^2 is said to be **nonsingular** if the dimension of this subspace is one for odd-dimensional $\mathbb{V}$ and zero for even-dimensional $\mathbb{V}$. Now let $\mathbb{V} = T_x\Sigma$. It is not hard to show that $d\omega^1\big|_x$ [where ω^1 is defined by (39.31)] is nonsingular for all $x \in \Sigma$. We then say that the differential 2-form $d\omega^1$ is nonsingular. Thus for every $x \in \Sigma$ there exists a vector $\boldsymbol{u}_x \in T_x\Sigma$ having the property that

$$d\omega^1|_x(\boldsymbol{u}_x, \boldsymbol{v}_x) = 0\,, \qquad \text{for all } \boldsymbol{v}_x \in T_x\Sigma\,. \tag{39.38}$$

The direction of $\boldsymbol{u}_x$ is uniquely determined by ω^1 for every $x \in \Sigma$. We call this direction field the **vortex direction field** of the 1-form ω^1. We then have the following definition.

Definition 39.4. *The integral curves of the vortex direction field of the* 1*-form* ω^1 *are called the* ***vortex lines*** *(or characteristic lines) of* ω^1.

The proof of Theorem 39.8 depends now on the following lemma, which will be stated without proof (for a proof see Arnold 1978).

Lemma 39.2. *The integral curves of the Hamiltonian vector field (the trajectories of the phase flow) are the same as the vortex lines of the* 1*-form* $\omega^1 = p_i\, dq^i - H\, dt$ *on the extended phase space.*

Theorem 39.8 now follows from the Stokes Theorem. Indeed, let σ be the surface on the tube of trajectories bounded by the loops C_1 and C_2, that is, $\partial\sigma = C_1 - C_2$ (see Fig. 39.1), then the above lemma implies that the flow lines on σ are the vortex lines of ω^1, so (39.38) applies with $\boldsymbol{u}_x$ along the flow lines of the Hamiltonian for all $x \in \sigma$, and we have

$$\int_\sigma d\omega^1 = 0\,. \tag{39.39}$$

On the other hand, by the Stokes theorem,

$$\int_\sigma d\omega^1 = \int_{\partial\sigma} \omega^1 = \int_{C_1 - C_2} \omega^1 = \int_{C_1} \omega^1 - \int_{C_2} \omega^1\,. \tag{39.40}$$

Thus (39.32) follows.

Problem 39.4 Show that the 2-form $d\omega^1 = dp_i \wedge dq^i - dH \wedge dt$ on the extended phase space is nonsingular. (Refer to the definition of nonsingularity of a 2-form in the discussion immediately above (39.38).)

Problem 39.5 In this problem we will demonstrate that (39.32) is the generalization of the description of fluid flow in Euclidean space $\mathbb{R}^3$. Figure 39.1 can be used again, with the three axes now denoting the x, y and z axes. Let $\boldsymbol{v}$ be a vector field in $\mathbb{R}^3$. The integral curves of $\nabla \times \boldsymbol{v}$ are called the vortex lines of $\boldsymbol{v}$. Let C_1 and C_2 be oriented closed curves in $\mathbb{R}^3$ encircling a vortex tube. Show that

$$\int_{C_1} \boldsymbol{v} \cdot d\boldsymbol{l} = \int_{C_2} \boldsymbol{v} \cdot d\boldsymbol{l} \,. \tag{39.41}$$

Hint: Use the Stokes theorem of vector calculus [cf. (8.14)].

Chapter 40

Darboux's Theorem in Symplectic Geometry

This chapter will be devoted to the explanation and proof of a central theorem in symplectic geometry, Darboux's Theorem, which essentially states that every symplectic manifold is locally like a tangent or cotangent space of some smooth manifold. This is analogous to the more general fact that, locally, every differentiable manifold is a Euclidean space. Our presentation will be adapted mainly from that by Woodhouse (Woodhouse 1992) and Abraham and Marsden (Abraham and Marsden 1982). For authoritative and comprehensive accounts on the deep relationship between symplectic geometry and physics (in particular classical mechanics), the reader may consult the monographs by Souriau (Souriau 1997) and Guillemin and Sternberg (Guillemin and Sternberg 1984).

Based on the more specialized discussion of symplectic geometry within the context of Hamiltonian dynamics in the last chapter, we can now give a general and more abstract definition of a symplectic manifold.

Definition 40.1. *Let M be an even-dimensional differentiable manifold. A **symplectic structure** on M is a closed, non-degenerate 2-form ω, called a **symplectic form**, that is defined everywhere on M. In other words, ω satisfies the following properties:*

(1) $d\omega = 0$, $\quad$ (ω is closed) .

(2) Let $\boldsymbol{X}$ be a smooth vector field on M [$\boldsymbol{X} \in \Gamma(TM)$]. If $\omega(\boldsymbol{X}, \boldsymbol{Y}) = 0$ for all smooth vector fields $\boldsymbol{Y} \in \Gamma(TM)$, then $\boldsymbol{X} = 0$ (non-degeneracy of ω).

*An even-dimensional manifold endowed with a symplectic structure is called a **symplectic manifold**.*

From the discussion surrounding (39.4) we see that the condition of non-degeneracy of the symplectic form ω is equivalent to the fact that the map

$$\boldsymbol{X}_x \mapsto \boldsymbol{X}_x \lrcorner\, \omega_x \in T_x^* M \tag{40.1}$$

is a linear isomorphism at each point $x \in M$. In the above equation, $\boldsymbol{X}_x$ is a tangent vector in the tangent space T_xM and ω_x is the symplectic form evaluated at x. The meaning of the *contraction map* $\lrcorner$ has been given in (11.26).

By means of the symplectic structure of a symplectic manifold, each tangent space T_xM becomes a *symplectic vector space.* This notion is defined in general as follows.

Definition 40.2. *A* ***symplectic vector space****, denoted by the pair* (V, ω)*, is a vector space* V *endowed with an antisymmetric, non-degenerate bilinear form* ω *on* V*. Antisymmetry means* $\omega(\boldsymbol{x}, \boldsymbol{y}) = -\omega(\boldsymbol{y}, \boldsymbol{x})$ *for all* $\boldsymbol{x}, \boldsymbol{y} \in V$*. Non-degeneracy means* $\boldsymbol{x} \lrcorner \omega = \boldsymbol{0}$ *(the zero linear form on* V*) only if* $\boldsymbol{x} = \boldsymbol{0}$*, where the contraction operation* $\lrcorner$ *is here defined by* $(\boldsymbol{x} \lrcorner \omega)(\boldsymbol{y}) \equiv \omega(\boldsymbol{x}, \boldsymbol{y})$ *[cf. (1.66)].*

Conversely, a symplectic vector space can be made into a symplectic manifold by using the components of vectors in V with respect to some basis as local coordinates of the manifold. The bilinear form ω then becomes a closed 2-form. We will make use of this fact in the proof of Darboux's Theorem later.

Associated with a symplectic vector space there are several important kinds of subspaces. To identify these we have to introduce the notion of a *symplectic complement.*

Definition 40.3. *Let* (V, ω) *be a symplectic vector space and* $W \subset V$ *be a subspace. The* ***symplectic complement*** *of* W *is the subspace* $W^\perp$ *defined by*

$$W^\perp \equiv \{\boldsymbol{x} \in V \,|\, \omega(\boldsymbol{x}, \boldsymbol{y}) = 0 \textit{ for all } \boldsymbol{y} \in W\} . \tag{40.2}$$

The symplectic complement is also referred to as the ω*-****orthogonal complement****.*

Problem 40.1 Show that the symplectic complement $W^\perp$ of a subspace W of a symplectic vector space V is also a subspace of V.

Problem 40.2 Let S and T be subspaces of a symplectic vector space (V, ω). Prove the following properties of symplectic complements.

$$S \subset T \implies T^\perp \subset S^\perp , \tag{40.3}$$

$$(S + T)^\perp = S^\perp \cap T^\perp . \tag{40.4}$$

The notation $S + T$ means the set of all $\boldsymbol{s} + \boldsymbol{t}$ for $\boldsymbol{s} \in S$ and $\boldsymbol{t} \in T$.

A very useful fact relating the dimensions of a subspace and its symplectic complement is the following.

Theorem 40.1. *Let $W \subset V$ be a subspace of a symplectic vector space (V, ω). Then*

$$dim(V) = dim(W) + dim(W^{\perp}) . \tag{40.5}$$

Proof. The non-degenerate bilinear form ω induces the *conjugate isomorphism* $G_\omega : V \to V^*$ given by [cf. (1.66)]

$$G_\omega(\boldsymbol{x})(\boldsymbol{y}) = (\boldsymbol{x} \lrcorner \omega)(\boldsymbol{y}) = \omega(\boldsymbol{x}, \boldsymbol{y}) \quad \text{for all } \boldsymbol{x}, \boldsymbol{y} \in V . \tag{40.6}$$

Let the restriction of G_ω to W be written as

$$G_\omega^W : W \to V^* , \quad \boldsymbol{w} \mapsto \boldsymbol{w} \lrcorner \omega , \quad \boldsymbol{w} \in W . \tag{40.7}$$

Consider the linear form $G_\omega^W(\boldsymbol{w}) \in V^*$ for $\boldsymbol{w} \in W$. This is a linear function on V. For $\boldsymbol{v}_1, \boldsymbol{v}_2 \in V$ such that $\boldsymbol{v}_1 - \boldsymbol{v}_2 \in W^{\perp}$, we have

$$\begin{aligned} &G_\omega^W(\boldsymbol{w})(\boldsymbol{v}_1) - G_\omega^W(\boldsymbol{w})(\boldsymbol{v}_2) = G_\omega^W(\boldsymbol{w})(\boldsymbol{v}_1 - \boldsymbol{v}_2) = (\boldsymbol{w} \lrcorner \omega)(\boldsymbol{v}_1 - \boldsymbol{v}_2) \\ &= \omega(\boldsymbol{w} , \boldsymbol{v}_1 - \boldsymbol{v}_2) = 0 . \end{aligned} \tag{40.8}$$

We can then define a new induced map (induced from ω)

$$\mathcal{G}_\omega^W : W \longrightarrow \left(\frac{V}{W^{\perp}} \right)^* , \tag{40.9}$$

where the *quotient space* $V/W^{\perp}$ is the vector space of *equivalence classes* $\{\boldsymbol{v}\}$, with the equivalence relationship $\sim$ defined by $\boldsymbol{v}_1 \sim \boldsymbol{v}_2$ if $\boldsymbol{v}_1 - \boldsymbol{v}_2 \in W^{\perp}$ (cf. Problem 1.13). Since G_ω is an isomorphism, G_ω^W is an injection, and so must be $\mathcal{G}_\omega^W$. We then have

$$dim(W) \leq dim((V/W^{\perp})^*) = dim(V/W^{\perp}) = dim(V) - dim(W^{\perp}) , \tag{40.10}$$

where the last equality follows from (1.69). Now consider the map $\tilde{G}_\omega : V \to W^*$ defined by

$$\tilde{G}_\omega(\boldsymbol{v}) = \text{restriction of } G_\omega(\boldsymbol{v}) \text{ to } W , \tag{40.11}$$

so that for $\boldsymbol{w} \in W$, $\tilde{G}_\omega(\boldsymbol{v})(\boldsymbol{w}) = \omega(\boldsymbol{v}, \boldsymbol{w})$. We see that the kernel of this map is precisely the subspace $W^{\perp}$, and

$$dim(W) = dim(W^*) \geq dim(Im\, \tilde{G}_\omega) . \tag{40.12}$$

By applying (1.68) to the map $\tilde{G}_\omega$ we have

$$dim(V) = dim(Ker\, \tilde{G}_\omega) + dim(Im\, \tilde{G}_\omega) = dim(W^{\perp}) + dim(Im\, \tilde{G}_\omega) . \tag{40.13}$$

It follows from (40.12) that

$$dim(W) \geq dim(V) - dim(W^{\perp}) . \tag{40.14}$$

This equation and (40.10) then imply the theorem. □

Problem 40.3 Use Theorem 40.1 and the result (40.4) in Problem 40.2 to prove that

$$(S^{\perp})^{\perp} = S \,, \tag{40.15}$$

$$(S \cap T)^{\perp} = S^{\perp} + T^{\perp} \,. \tag{40.16}$$

We will now define several important kinds of subspaces of a symplectic vector space.

Definition 40.4. *A subspace W of a symplectic vector space is said to be*

(a) ***isotropic*** *if $W \subset W^{\perp}$;*

(b) ***coisotropic*** *if $W^{\perp} \subset W$;*

(c) ***symplectic*** *if $W \cap W^{\perp} = \{\mathbf{0}\}$ (the zero vector) ;*

(d) ***Lagrangian*** *if $W = W^{\perp}$.*

The concept of a Lagrangian subspace plays an important and special role in symplectic geometry, due to the following theorems.

Theorem 40.2. *Every finite dimensional symplectic vector space contains a Lagrangian subspace and is even-dimensional.*

Proof. Let (V, ω) be a finite-dimensional symplectic vector space. Consider any 1-dimensional subspace W_1 of V. Any two vectors $\boldsymbol{x}, \boldsymbol{y} \in W_1$ must be related by $\boldsymbol{x} = a\boldsymbol{y}$ for some real number a. Thus $\omega(\boldsymbol{x}, \boldsymbol{y}) = a\,\omega(\boldsymbol{y}, \boldsymbol{y}) = 0$, since ω is antisymmetric. This implies that $W_1 \subset W_1^{\perp}$, so W_1 is isotropic. Now suppose $W \subset V$ is an isotropic subspace, and $dim(W) = k$. We have just seen that $k \geq 1$. On the other hand, since $W \subset W^{\perp}$, we must have

$$k \leq dim(W^{\perp}) = dim(V) - k \,, \tag{40.17}$$

where the second equality follows from Theorem 40.1. So the possible values k of the dimension of an isotropic subspace are given by

$$1 \leq k \leq \frac{1}{2} dim(V) \,. \tag{40.18}$$

If $k < dim(V)/2$ we can generate a bigger isotropic subspace W' from W, of dimension $k+1$, by letting $W' = span\,\{W, \boldsymbol{v}\}$, where $\boldsymbol{v}$ is some vector in $W^{\perp} - W$. We can then always start with a one-dimensional isotropic subspace, and generate a series of isotropic subspaces, each of dimension one larger than the previous one, until we reach an isotropic subspace $L \subset V$ of dimension $dim(V)/2$. By Theorem 40.1, $dim(L^{\perp})$ is also $dim(V)/2$. Since $L \subset L^{\perp}$ (L being isotropic), we must have $L = L^{\perp}$. Thus L is Lagrangian. The fact that $dim(L) = dim(V)/2$ necessarily implies that V is even-dimensional. □

Problem 40.4 Show that another definition of a Lagrangian subspace (alternative to that given in Def. 40.4) is the following. A subspace $W \subset V$ of a symplectic vector space V is Lagrangian if W is isotropic and has an isotropic complement, that is, if $V = W \oplus W'$, where both W and W' are isotropic.

The following theorem will be crucial later in the proof of Darboux's theorem.

Theorem 40.3. *Let (V, ω) be a $2n$-dimensional symplectic vector space. Then there exists a basis $\{\boldsymbol{x}_1, \ldots, \boldsymbol{x}_n, \boldsymbol{y}^1, \ldots, \boldsymbol{y}^n\}$ of V, called a* ***symplectic frame****, such that*

$$\omega(\boldsymbol{x}_i, \boldsymbol{x}_j) = 0\,, \qquad 2\omega(\boldsymbol{y}^j, \boldsymbol{x}_i) = \delta_i^j\,, \qquad \omega(\boldsymbol{y}^i, \boldsymbol{y}^j) = 0\,, \tag{40.19}$$

where $i, j = 1, \ldots, n$. (Note: The positioning of the indices here is made in accordance with the positioning of the indices of the canonical coordinates (q^i, p_i) in phase space. For example, a vector in $\boldsymbol{v} \in V$ may be expressed as $\boldsymbol{v} = q^i\, \boldsymbol{x}_i + p_i\, \boldsymbol{y}^i$ using Einstein's summation convention.)

Proof. By Theorem 40.2 there always exists a Lagrangian subspace of V. Call one such subspace L, and let $V = L \oplus W$ (the *direct sum* of L and W), where W is some n-dimensional subspace. W can be identified with L^* by the ω-induced isomorphism $G_\omega^{WL^*} : W \to L^*$ defined by

$$G_\omega^{WL^*}(\boldsymbol{w})(\boldsymbol{l}) = 2\omega(\boldsymbol{w}, \boldsymbol{l}) \qquad (\boldsymbol{w} \in W,\, \boldsymbol{l} \in L)\,. \tag{40.20}$$

The non-degeneracy of ω guarantees that $G_\omega^{WL^*}$ is injective (one-to-one), while the fact that L^* has the same dimension as W, being n in each case (due to the assumption that L is Lagrangian), ensures that it is surjective also. Now let $\{\boldsymbol{x}_1, \ldots, \boldsymbol{x}_n\}$ be a basis of L, and let $\{\boldsymbol{z}^1, \ldots, \boldsymbol{z}^n\}$ be the *dual basis* (to $\{\boldsymbol{x}_i\}$) for $W \sim L^*$, such that

$$2\omega(\boldsymbol{z}^j, \boldsymbol{x}_i) = \delta_i^j\,. \tag{40.21}$$

Since $L = L^\perp$, we have $\omega(\boldsymbol{x}_i, \boldsymbol{x}_j) = 0$. Define $\lambda^{ij} \equiv \omega(\boldsymbol{z}^i, \boldsymbol{z}^j) = -\lambda^{ji}$, and let

$$\boldsymbol{y}^i = \boldsymbol{z}^i + \lambda^{ij}\boldsymbol{x}_j \qquad (i = 1, \ldots, n)\,. \tag{40.22}$$

It is clear that the $\boldsymbol{y}^i$ are linearly independent of each other and of the $\boldsymbol{x}_i$. The set $\{\boldsymbol{x}_1, \ldots, \boldsymbol{x}_n, \boldsymbol{y}^1, \ldots, \boldsymbol{y}^n\}$ is therefore a basis for V. Furthermore, we have

$$\begin{aligned} \omega(\boldsymbol{y}^i, \boldsymbol{y}^j) &= \omega(\boldsymbol{z}^i + \lambda^{ik}\boldsymbol{x}_k\,,\, \boldsymbol{z}^j + \lambda^{jm}\boldsymbol{x}_m) \\ &= \omega(\boldsymbol{z}^i, \boldsymbol{z}^j) + \lambda^{ik}\,\omega(\boldsymbol{x}_k, \boldsymbol{z}^j) + \lambda^{jm}\,\omega(\boldsymbol{z}^i, \boldsymbol{x}_m) + \lambda^{ik}\lambda^{jm}\,\omega(\boldsymbol{x}_k, \boldsymbol{x}_m) \\ &= \lambda^{ij} - \frac{1}{2}\lambda^{ik}\delta_k^j + \frac{1}{2}\lambda^{jm}\delta_m^i = \lambda^{ij} - \frac{1}{2}\lambda^{ij} + \frac{1}{2}\lambda^{ji} = \frac{1}{2}\left(\lambda^{ij} + \lambda^{ji}\right) = 0\,; \end{aligned} \tag{40.23}$$

and

$$2\omega(\boldsymbol{y}^i, \boldsymbol{x}_j) = 2\omega(\boldsymbol{z}^i + \lambda^{ik}\boldsymbol{x}_k, \boldsymbol{x}_j) = 2\omega(\boldsymbol{z}^i, \boldsymbol{x}_j) = \delta_j^i\,. \tag{40.24}$$

Therefore, $\{\boldsymbol{x}_1, \ldots, \boldsymbol{x}_n, \boldsymbol{y}^1, \ldots, \boldsymbol{y}^n\}$ is the sought-for symplectic frame, and the theorem is proved. □

We will now return to the consideration of symplectic manifolds. It is important to keep in mind that although the most commonly encountered example of symplectic manifolds is the cotangent bundle T^*Q, not all symplectic manifolds have the structure of cotangent bundles globally. However, all symplectic manifolds have this structure locally, as asserted by a basic theorem in symplectic geometry – Darboux's Theorem. Before proving this theorem we will state and prove the following key lemma.

Lemma 40.1. *Suppose ω and ω' are (different) symplectic structures on a symplectic manifold M. If $\omega(x) = \omega'(x)$ at some point $x \in M$, then there exist neighborhoods U and V of x and a diffeomorphism $\rho : U \to V$ such that $\rho(x) = x$ (x is a fixed point of ρ) and $\rho^*(\omega') = \omega$.*

Proof. Since all symplectic forms are by definition closed, $d\omega' = d\omega = 0$. So $d(\omega' - \omega) = 0$. By Poincaré's lemma, $\omega' - \omega$ must be locally exact. Then there exists a local one-form α in a neighborhood W' of x in M such that

$$d\alpha = \omega' - \omega \,. \tag{40.25}$$

α is only determined up to the gradient df of a scalar function f. By adding a suitable gradient we can ensure that $\alpha(x) = 0$. Now define a time-dependent two-form (or a one-parameter set of two-forms) Ω_t on W':

$$\Omega_t \equiv \omega + t(\omega' - \omega) \,. \tag{40.26}$$

This two-form, however, may not be non-degenerate throughout W'. But since, by assumption, $\omega'(x) = \omega(x)$, $\Omega_t(x) = \omega(x)$. So Ω_t must be non-degenerate in some (possibly smaller) neighborhood $W \subseteq W'$, since ω is non-degenerate everywhere in M. The two-form Ω_t then gives rise to a well-defined time-dependent vector field $\boldsymbol{X}_t$ on W determined by

$$\boldsymbol{X}_t \lrcorner \Omega_t + \alpha = 0 \,. \tag{40.27}$$

By a generalization of the *Cartan formula* (11.27) for *Lie derivatives* (for time-dependent forms) we have

$$L_{\boldsymbol{X}_t}\Omega_t = d(\boldsymbol{X}_t \lrcorner \Omega_t) + \boldsymbol{X}_t \lrcorner d\Omega_t + \frac{\partial \Omega_t}{\partial t} \,. \tag{40.28}$$

Since $d\Omega_t = 0$ and $\partial\Omega_t/\partial t = \omega' - \omega$, (40.25) and (40.27) imply

$$L_{\boldsymbol{X}_t}\Omega_t = -d\alpha + \omega' - \omega = 0 \,. \tag{40.29}$$

Let φ_t be the local group of diffeomorphisms (cf. discussion in Chapter 11), or the flow, corresponding to the vector field $\boldsymbol{X}_t$. The vanishing of the Lie derivative $L_{\boldsymbol{X}_t}\Omega_t$ then implies, according to (11.23),

$$\varphi_t^*\Omega_t = \Omega_0 \,, \tag{40.30}$$

or $\varphi_1^*\Omega_1 = \Omega_0$. Since, from (40.26), $\Omega_0 = \omega$ and $\Omega_1 = \omega'$, we have, on defining $\rho \equiv \varphi_1$,

$$\rho^*\omega' = \omega \, . \tag{40.31}$$

Also, by (40.27), the vector field $\boldsymbol{X}_t$ has a *singularity* at x $[\boldsymbol{X}_t(x) = 0]$ since $\alpha(x) = 0$. So $\rho(x) = \varphi_1(x) = x$. Finally, even though $\boldsymbol{X}_t$ is well defined on W, $\rho(= \varphi_1)$ may not be, since the integral curves of $\boldsymbol{X}_t$ may leave this region at $t = 1$. But since $\rho(x) = x$, there must exist a neighborhood $U \subset W$ such that ρ is well defined on U and $\rho(U) \subset W$. The proof is concluded by taking $V = \rho(U)$. □

With the aid of Lemma 40.1, we are now ready to prove Darboux's Theorem, stated as follows:

Theorem 40.4. ***(Darboux's Theorem)*** *Let M be a $2n$-dimensional symplectic manifold with a (global) symplectic form ω. At every point $x \in M$ there is a neighborhood of x with local coordinates (q^i, p_i), $i = 1, \ldots, n$, such that in this neighborhood ω can be expressed locally as $\omega = dp_i \wedge dq^i$.*

Proof. Consider a point $x \in M$. Then T_xM is a symplectic vector space whose symplectic form is ω_x [cf. (40.1)]. By Theorem 40.3, there exists a symplectic frame $\{\boldsymbol{x}_1, \ldots, \boldsymbol{x}_n, \boldsymbol{y}^1, \ldots, \boldsymbol{y}^n\}$ in T_xM. Let the components of a vector $\boldsymbol{v} \in T_xM$ with respect to this frame be (x^i, y_i):

$$\boldsymbol{v} = x^i\,\boldsymbol{x}_i + y_i\,\boldsymbol{y}^i \, . \tag{40.32}$$

One can use (x^i, y_i) as local coordinates for M in some neighborhood U' of x, and construct a local symplectic form ω' in this neighborhood given by

$$\omega' = dy_i \wedge dx^i \, . \tag{40.33}$$

Obviously,

$$\omega'(x) = \omega_x = \omega(x) \, . \tag{40.34}$$

By Lemma 40.1, there exist neighborhoods U'' and V'' of x and a diffeomorphism $\rho : U'' \to V''$ such that

$$\rho(x) = x \qquad \text{and} \qquad \rho^*(\omega') = \omega \, . \tag{40.35}$$

Choose U to be the smaller of U' and U''. Then we also have a well-defined diffeomorphism $\rho : U \to V$, where both U and V are neighborhoods of x, such that the two equations in (40.35) are still valid. Now we define local coordinates (q^i, p_i) in the neighborhood U of x by

$$q^i = x^i \circ \rho \, , \qquad p_i = y_i \circ \rho \, . \tag{40.36}$$

The second equation in (40.35) then implies $\omega = dp_i \wedge dq^i$ in the neighborhood U. □

Chapter 41

The Kolmogorov-Arnold-Moser (KAM) Theorem

Hamiltonian dynamical systems are more often than not non-integrable, although in the historical development of the subject prior to the beginning of the 20th century, this was not thought to be so. Indeed, the whole formalism of *generating functions* and the *Hamilton-Jacobi theory* (cf. Chapter 37) was developed with the explicit purpose of achieving integrability. Within the context of the n-body problem in celestial mechanics, where n massive objects interact solely through Newtonian gravitation, this program was carried out exhaustively by many physicists and mathematicians, culminating in the ground-breaking work of Poincaré as presented in his celebrated three-volume magnum opus: "Les Méthodes Nouvelles de la Méchanique Céleste" (Poincaré, 1892, 1893, 1899). Through the pioneering work of Poincaré it was realized that integrability was the exception rather than the rule, and that the n-body problem was non-integrable for $n > 2$.

Even so, the concepts of *invariant tori* and *action-angle variables* central to integrable systems (cf. Chapter 38) remain instrumental in the analysis of a certain class of non-integrable problems. In this chapter we will present an introductory treatment of this class of problems, which can be represented by writing the Hamiltonian in the form

$$H(\theta^i, I_i, \mu) = H_0(I_i) + \mu H_1(\theta^i, I_i) + O(\mu^2) , \tag{41.1}$$

where $H_0(I_i)$ is the Hamiltonian function for an integrable system with action-angle variables (θ^i, I_i), μ is a small real parameter, and $O(\mu^2)$ represents the sum of higher order terms in μ than the first. We will further require the full Hamiltonian $H(\theta^i, I_i, \mu)$ to be analytic in the $2N$ variables (θ^i, I_i) and periodic in $\theta^1, \ldots, \theta^N$ with period 2π. Thus we can view this class of problems as those involving a certain kind of small perturbations to an integrable system, of which the n-body problem in celestial mechanics is a prime example. Since H_0 is integrable, we know from Chapter 38 that the $2N$-dimensional phase space [with canonical coordinates (θ^i, I_i)] is *foliated* into an N-parameter family of invariant tori $I_i = c_i = constant$ $(i = 1, \dots, N)$, on which the flow is given by

$$\frac{d\theta^i}{dt} = \left.\frac{\partial H_0}{\partial I_i}\right|_{I_i = c_i} \equiv \omega_i(c_j) = constant\,. \tag{41.2}$$

The main question is the following: What happens to this foliation under small perturbations (for small values of μ)? Stated in another way: Do the invariant tori of the unperturbed system persist or will they break up under small perturbations? The definitive answer to this question is provided by the celebrated *Kolmogorov-Arnold-Moser (KAM) Theorem*, which represents the most significant theoretical advance in the study of dynamical systems since the achievements of Poincaré.

We will first approach the problem by standard perturbation techniques. In the course of doing so, it will be shown that one will run into the so-called *small-divisors problem*, which leads to either divergent or very poorly converging infinite series and thus throws into doubt the general validity of the perturbation approach.

In order for invariant tori to exist for the perturbed system, one should be able to find a canonical transformation $\{\theta^i, I_i\} \to \{\theta'^i, I'_i\}$, where the transformed coordinates will be the new action-angle variables. The Hamilton's characteristic function $W(\theta^i, I'_i, \mu)$ generating this transformation will then be a solution of the (time-independent) Hamilton-Jacobi equation [cf. (37.76)]

$$H\left(\theta^i, \frac{\partial W}{\partial \theta^i}, \mu\right) = H'(I'_i)\,, \tag{41.3}$$

and satisfy [cf. (38.8)]

$$I_i = \frac{\partial W}{\partial \theta^i}\,, \qquad \theta'^i = \frac{\partial W}{\partial I'_i}\,. \tag{41.4}$$

We will express W as a formal power series in the small parameter μ. For $\mu = 0$, W should generate the identity transformation ($I_i = I'_i$, $\theta'^i = \theta^i$), and so is given by

$$W(\theta^i, I'_i, 0) = \theta^i I'_i\,, \tag{41.5}$$

as confirmed by (41.4). So we write

$$W(\theta^i, I'_i, \mu) = \theta^i I'_i + \sum_{k=1}^{\infty} \mu^k\, W_k(\theta^i, I'_i)\,. \tag{41.6}$$

From the first equation of (41.4) we then have

$$I_i = I'_i + \sum_{k=1}^{\infty} \mu^k\, \frac{\partial W_k}{\partial \theta^i}\,. \tag{41.7}$$

It follows from (41.1) that the Hamilton-Jacobi equation (41.3) then appears as

$$H_0\left(I'_i + \sum_{k=1}^{\infty} \mu^k \frac{\partial W_k}{\partial \theta^i}\right) + \mu\, H_1\left(\theta^i,\, I'_i + \sum_{k=1}^{\infty} \mu^k \frac{\partial W_k}{\partial \theta^i}\right) + \dots = H'(I'_i)\,. \quad (41.8)$$

The hope is that on comparing terms for like powers of μ on both sides of the equation, one can obtain a set of equations from which both the generating function W and the new Hamiltonian H' can be determined.

Since, by assumption, $H_1(\theta^i, I_i)$ is periodic in $\theta^1, \dots, \theta^N$, we can expand it as a Fourier series [cf. (38.41) and (38.42)]:

$$H_1(\theta^i, I_i) = \sum_{\{m_i\}\in\mathbb{Z}^N} H_{1,\{m_i\}}(I_i)\, e^{i m_j \theta^j}\,, \quad (41.9)$$

where $\mathbb{Z}^N$ is the set of N-tuples of integers. We also have, by (41.2),

$$\theta^i = \theta^i_0 + \omega_i t\,. \quad (41.10)$$

Likewise, within the expansion for $W(\theta^i, I'_i, \mu)$ in (41.6), we can expand

$$W_k(\theta^i, I'_i) = \sum_{\{m_i\}\,\in\,\mathbb{Z}^N - \{0\}} W_{k,\{m_i\}}(I'_i)\, e^{i m_j \theta^j}\,. \quad (41.11)$$

Note that the term in the Fourier sum for $m_i = 0$ can be excluded since only $\partial W_k/\partial\theta^i$ enters into (41.8). The terms on the left-hand side of (41.8) can be expanded as formal power series in μ as follows (the expansions will be given explicitly only up to order μ^2):

$$\begin{aligned}
&H_0\left(I'_i + \sum_{k=1}^{\infty}\mu^k \frac{\partial W_k}{\partial\theta^i}\right) = H_0\left(I'_i + \mu\frac{\partial W_1}{\partial\theta^i} + \mu^2\frac{\partial W_2}{\partial\theta^i} + \dots\right)\\
&= H_0(\boldsymbol{I}') + \sum_i \left(\mu\frac{\partial W_1}{\partial\theta^i} + \mu^2\frac{\partial W_2}{\partial\theta^i} + \dots\right) \left.\frac{\partial H_0}{\partial I_i}\right|_{\boldsymbol{I}=\boldsymbol{I}'}\\
&\quad + \frac{1}{2}\sum_{i,j}\left(\mu\frac{\partial W_1}{\partial\theta^i} + \mu^2\frac{\partial W_2}{\partial\theta^i} + \dots\right)\left(\mu\frac{\partial W_1}{\partial\theta^j} + \mu^2\frac{\partial W_2}{\partial\theta^j}\right)\left.\frac{\partial^2 H_0}{\partial I_i \partial I_j}\right|_{\boldsymbol{I}=\boldsymbol{I}'}\\
&\quad + \dots\dots\\
&= H_0(\boldsymbol{I}') + \mu\sum_i \frac{\partial W_1}{\partial\theta^i}\cdot\left.\frac{\partial H_0}{\partial I_i}\right|_{\boldsymbol{I}=\boldsymbol{I}'}\\
&\quad + \mu^2\left(\sum_i \frac{\partial W_2}{\partial\theta^i}\cdot\left.\frac{\partial H_0}{\partial I_i}\right|_{\boldsymbol{I}=\boldsymbol{I}'} + \frac{1}{2}\sum_{i,j}\left.\frac{\partial^2 H_0}{\partial I_i\partial I_j}\right|_{\boldsymbol{I}=\boldsymbol{I}'}\cdot\frac{\partial W_1}{\partial\theta^i}\frac{\partial W_1}{\partial\theta^j}\right) + O(\mu^3)\,,
\end{aligned} \quad (41.12)$$

$$\mu H_1\left(\theta^i,\, I'_i + \mu\frac{\partial W_1}{\partial\theta^i} + \dots\right) = \mu\left\{H_1(\boldsymbol{\theta}, \boldsymbol{I}') + \mu\sum_i \frac{\partial W_1}{\partial\theta^i}\left.\frac{\partial H_1}{\partial I_i}\right|_{\boldsymbol{I}=\boldsymbol{I}'}\right\} + O(\mu^3)\,. \quad (41.13)$$

In the last two equations the vector notation $\boldsymbol{I} = (I_1, \ldots, I_N)$, $\boldsymbol{I}' = (I'_1, \ldots, I'_N)$ and $\boldsymbol{\theta} = (\theta^1, \ldots, \theta^N)$ has been used. On recalling (41.2) and further expressing as vectors the quantities $\boldsymbol{\omega} = (\omega_1, \ldots, \omega_N)$, $\nabla_{\boldsymbol{\theta}} W_{1,2} = (\partial W_{1,2}/\partial \theta^i)$ and $\nabla_{\boldsymbol{I}} H_1 = (\partial H_1/\partial I_i)$, Eq. (41.8) then becomes

$$\begin{aligned} &H_0(\boldsymbol{I}') + \mu \left\{\boldsymbol{\omega}(\boldsymbol{I}') \cdot \nabla_{\boldsymbol{\theta}} W_1 + H_1(\boldsymbol{\theta}, \boldsymbol{I}')\right\} \\ &+ \mu^2 \left\{ \boldsymbol{\omega}(\boldsymbol{I}') \cdot \nabla_{\boldsymbol{\theta}} W_2 + \nabla_{\boldsymbol{I}} H_1 \big|_{\boldsymbol{I}=\boldsymbol{I}'} \cdot \nabla_{\boldsymbol{\theta}} W_1 + \frac{1}{2} \sum_{i,j} \left. \frac{\partial^2 H_0}{\partial I_i \partial I_j} \right|_{\boldsymbol{I}=\boldsymbol{I}'} \cdot \frac{\partial W_1}{\partial \theta^i} \frac{\partial W_1}{\partial \theta^j} \right\} \\ &+ O(\mu^3) = H'(\boldsymbol{I}') \,. \end{aligned} \tag{41.14}$$

We will assume that the *Hessian* $(\partial^2 H_0/\partial I_i \partial I_j)$ satisfies

$$\det \left(\frac{\partial^2 H_0}{\partial I_i \partial I_j} \right) \neq 0 \qquad \text{for } I_i = c_i \,, \tag{41.15}$$

that is, for values of I_i on invariant tori [cf. (41.2)]. From (41.9) and (41.11) we have the following Fourier series:

$$H_1(\boldsymbol{\theta}, \boldsymbol{I}) = H_{1,\boldsymbol{0}}(\boldsymbol{I}) + \sum_{\boldsymbol{m} \neq \boldsymbol{0}} H_{1,\boldsymbol{m}}(\boldsymbol{I})\, e^{i\boldsymbol{m}\cdot\boldsymbol{\theta}} \,, \tag{41.16}$$

$$W_1(\boldsymbol{\theta}, \boldsymbol{I}') = \sum_{\boldsymbol{m} \neq \boldsymbol{0}} W_{1,\boldsymbol{m}}(\boldsymbol{I}')\, e^{i\boldsymbol{m}\cdot\boldsymbol{\theta}} \,, \tag{41.17}$$

$$W_2(\boldsymbol{\theta}, \boldsymbol{I}') = \sum_{\boldsymbol{m} \neq \boldsymbol{0}} W_{2,\boldsymbol{m}}(\boldsymbol{I}')\, e^{i\boldsymbol{m}\cdot\boldsymbol{\theta}} \,, \tag{41.18}$$

where $\boldsymbol{m} = (m_1, \ldots, m_N) \in \mathbb{Z}^N$ are vectors whose components are all integers and $\boldsymbol{0} = (0, \ldots, 0)$. So we can write the terms on the left-hand side of (41.14) involving Fourier components as follows.

$$\begin{aligned} &\mu \left\{\boldsymbol{\omega}(\boldsymbol{I}') \cdot \nabla_{\boldsymbol{\theta}} W_1 + H_1(\boldsymbol{\theta}, \boldsymbol{I}')\right\} \\ &= \mu H_{1,\boldsymbol{0}}(\boldsymbol{I}') + \mu \sum_{\boldsymbol{m} \neq \boldsymbol{0}} \left\{ (i\boldsymbol{\omega}(\boldsymbol{I}') \cdot \boldsymbol{m})\, W_{1,\boldsymbol{m}}(\boldsymbol{I}') + H_{1,\boldsymbol{m}}(\boldsymbol{I}') \right\} e^{i\boldsymbol{m}\cdot\boldsymbol{\theta}} \,, \end{aligned} \tag{41.19}$$

$$\begin{aligned} &\mu^2 \left\{ \boldsymbol{\omega}(\boldsymbol{I}') \cdot \nabla_{\boldsymbol{\theta}} W_2 + \nabla_{\boldsymbol{I}} H_1 \big|_{\boldsymbol{I}=\boldsymbol{I}'} \cdot \nabla_{\boldsymbol{\theta}} W_1 \right\} = \mu^2 \sum_{\boldsymbol{m} \neq \boldsymbol{0}} (i\boldsymbol{\omega}(\boldsymbol{I}') \cdot \boldsymbol{m})\, W_{2,\boldsymbol{m}}(\boldsymbol{I}')\, e^{i\boldsymbol{m}\cdot\boldsymbol{\theta}} \\ &+ \mu^2 \sum_{\boldsymbol{l}} \sum_{\boldsymbol{n} \neq \boldsymbol{0}} \left(i\boldsymbol{n} \cdot \nabla_{\boldsymbol{I}} H_{1,\boldsymbol{l}} \big|_{\boldsymbol{I}=\boldsymbol{I}'} \right) W_{1,\boldsymbol{n}}(\boldsymbol{I}')\, e^{i(\boldsymbol{n}+\boldsymbol{l})\cdot\boldsymbol{\theta}} \,, \end{aligned} \tag{41.20}$$

$$\begin{aligned} &\frac{\mu^2}{2} \sum_{i,j} \left. \frac{\partial^2 H_0}{\partial I_i \partial I_j} \right|_{\boldsymbol{I}=\boldsymbol{I}'} \cdot \frac{\partial W_1}{\partial \theta^i} \frac{\partial W_1}{\partial \theta^j} \\ &= -\frac{\mu^2}{2} \sum_{\boldsymbol{l},\boldsymbol{n} \neq \boldsymbol{0}} \left(\sum_{i,j} l_i n_j \left. \frac{\partial^2 H_0}{\partial I_i \partial I_j} \right|_{\boldsymbol{I}=\boldsymbol{I}'} \right) W_{1,\boldsymbol{l}}(\boldsymbol{I}') W_{1,\boldsymbol{n}}(\boldsymbol{I}')\, e^{i(\boldsymbol{l}+\boldsymbol{n})\cdot\boldsymbol{\theta}} \,. \end{aligned} \tag{41.21}$$

We will first identify the Fourier components in the above three expansions corresponding to terms independent of $\boldsymbol{\theta}$. In the double sums in (41.20) and (41.21), this requires $\boldsymbol{n}+\boldsymbol{l}=\boldsymbol{0}$ and hence $\boldsymbol{n}=-\boldsymbol{l}$. Since in both sums $\boldsymbol{n}\neq\boldsymbol{0}$, the double sums can be reduced to the single sum $\sum_{\boldsymbol{l}\neq\boldsymbol{0}}$, or, on replacing $\boldsymbol{l}$ by $\boldsymbol{m}$, $\sum_{\boldsymbol{m}\neq\boldsymbol{0}}$. Equating the terms independent of $\boldsymbol{\theta}$ on both sides of (41.14) we then have

$$H'(\boldsymbol{I}') = H_0(\boldsymbol{I}') + \mu H_{1,\boldsymbol{0}}(\boldsymbol{I}') - \mu^2 \left\{ i \sum_{\boldsymbol{m}\neq\boldsymbol{0}} \left(\boldsymbol{m}\cdot\nabla_{\boldsymbol{I}} H_{1,\boldsymbol{m}}\big|_{\boldsymbol{I}=\boldsymbol{I}'}\right) W_{1,-\boldsymbol{m}}(\boldsymbol{I}') \right.$$
$$\left. + \frac{1}{2}\sum_{\boldsymbol{m}\neq\boldsymbol{0}} \left(\sum_{i,j} m_i m_j \left.\frac{\partial \omega_i}{\partial I_j}\right|_{\boldsymbol{I}=\boldsymbol{I}'} \right) W_{1,\boldsymbol{m}}(\boldsymbol{I}')\, W_{1,-\boldsymbol{m}}(\boldsymbol{I}') \right\} + O(\mu^3)\,. \tag{41.22}$$

Since the right-hand side of (41.14) is independent of μ, we can now set each term corresponding to a non-zero power of μ equal to zero on the left-hand side; and for each such term, separately set equal to zero a particular Fourier coefficient of $\exp(i\boldsymbol{m}\cdot\boldsymbol{\theta})$ for $\boldsymbol{m}\neq\boldsymbol{0}$. Doing this for the first-order term in μ we obtain, on using (41.9),

$$W_{1,\boldsymbol{m}}(\boldsymbol{I}') = \frac{iH_{1,\boldsymbol{m}}(\boldsymbol{I}')}{\boldsymbol{\omega}(\boldsymbol{I}')\cdot\boldsymbol{m}} \qquad (\boldsymbol{m}\neq\boldsymbol{0})\,. \tag{41.23}$$

For the second order term in μ we will need to rewrite the double sums on the right-hand sides of (41.20) and (41.21) by setting $\boldsymbol{n}+\boldsymbol{l}=\boldsymbol{m}$ and requiring $\boldsymbol{m}\neq\boldsymbol{0}$. These then become, respectively,

$$\sum_{\boldsymbol{m}\neq\boldsymbol{0}}\sum_{\boldsymbol{l}\neq\boldsymbol{m}} i(\boldsymbol{m}-\boldsymbol{l})\cdot\nabla_{\boldsymbol{I}} H_{1,\boldsymbol{l}}\big|_{\boldsymbol{I}=\boldsymbol{I}'}\; W_{1,\boldsymbol{m}-\boldsymbol{l}}(\boldsymbol{I}')\, e^{i\boldsymbol{m}\cdot\boldsymbol{\theta}}\,, \tag{41.24}$$

and

$$\sum_{\boldsymbol{m}\neq\boldsymbol{0}}\sum_{\boldsymbol{l}\neq\boldsymbol{m}} \left(\sum_{i,j} l_i(m_j-l_j) \left.\frac{\partial^2 H_0}{\partial I_i \partial I_j}\right|_{\boldsymbol{I}=\boldsymbol{I}'} \right) W_{1,\boldsymbol{l}}(\boldsymbol{I}')\, W_{1,\boldsymbol{m}-\boldsymbol{l}}(\boldsymbol{I}')\, e^{i\boldsymbol{m}\cdot\boldsymbol{\theta}}\,. \tag{41.25}$$

On substituting the result (41.23) for $W_{1,\boldsymbol{m}}$ into the above two expressions, setting the sum of the right-hand sides of (41.20) and (41.21) equal to zero, and then separately setting equal to zero a particular Fourier coefficient for $\boldsymbol{m}\neq\boldsymbol{0}$, we obtain the following second-order result:

$$W_{2,\boldsymbol{m}}(\boldsymbol{I}') = -\frac{i}{\boldsymbol{m}\cdot\boldsymbol{\omega}(\boldsymbol{I}')}\sum_{\boldsymbol{l}\neq\boldsymbol{m}} \frac{H_{1,\boldsymbol{m}-\boldsymbol{l}}(\boldsymbol{I}')}{(\boldsymbol{m}-\boldsymbol{l})\cdot\boldsymbol{\omega}(\boldsymbol{I}')}$$
$$\times\left\{ (\boldsymbol{m}-\boldsymbol{l})\cdot\nabla_{\boldsymbol{I}} H_{1,\boldsymbol{l}}\big|_{\boldsymbol{I}=\boldsymbol{I}'} - \frac{1}{2}\frac{H_{1,\boldsymbol{l}}(\boldsymbol{I}')}{\boldsymbol{l}\cdot\boldsymbol{\omega}(\boldsymbol{I}')} \sum_{i,j} l_i(m_j-l_j) \left.\frac{\partial^2 H_0}{\partial I_i \partial I_j}\right|_{\boldsymbol{I}=\boldsymbol{I}'} \right\} \tag{41.26}$$

$(\boldsymbol{m}\neq\boldsymbol{0})\,.$

Higher order (and increasingly more complicated) expressions for $W_{k,\boldsymbol{m}}$ ($k > 2$) can be similarly generated, but we will not present them here.

The result (41.23) for $W_{1,\boldsymbol{m}}$ would right away, through (41.22), yield a formal power series expansion for the new Hamiltonian $H'(\boldsymbol{I}')$; it and the result (41.26) for $W_{2,\boldsymbol{m}}$ [as well as the higher order results for $W_{2,\boldsymbol{m}}$ ($k = 2$)] would provide, through the Fourier expansion (41.6), a formal power series expansion for the full generating function $W(\boldsymbol{\theta}, \boldsymbol{I}', \mu)$. It would appear, then, that the perturbation approach described above solves the problem specified by (41.1). On a closer examination of (41.23) and (41.26), however, a serious problem is revealed. In both equations, powers of quantities of the form $\boldsymbol{\omega}(\boldsymbol{I}') \cdot \boldsymbol{m}$ appear in denominators, where $\boldsymbol{m}$ is a vector with non-zero integer components to be summed over. If the frequency components $\omega_i(\boldsymbol{I}')$ of $\boldsymbol{\omega}$ are such that there exists a non-zero set of N rational numbers $r_1, \dots, r_N$ satisfying $\omega_1 r_1 + \cdots + \omega_N r_N = 0$, in which case $(\omega_1, \dots, \omega_N)$ is said to be **rationally degenerate**, there will exist a non-zero set of integers $m_1, \dots, m_N$ such that $m_1\omega_1 + \cdots + m_N\omega_N = 0$, and the perturtbative expansions would have no meaning. Moreover, even if $\boldsymbol{\omega}$ is not rationally degenerate, there will still be infinitely many terms in the sum over $\boldsymbol{m}$ for which $\boldsymbol{\omega} \cdot \boldsymbol{m}$ will be arbitrarily close to zero, and the Fourier expansions would diverge. This is known as the **small-divisor (small-denominator) problem**, as mentioned earlier. One should note, however, that the perturbative approach may still be of value in the prediction of long-time behavior of Hamiltonian systems in certain situations provided that the Fourier expansions are suitably truncated. The important point is that it fails for *arbitrarily* long times, and thus cannot be used to answer questions pertaining to asymptotic stability.

The KAM Theorem addresses this problem in two respects. First, it establishes the convergence of the perturbation series for certain "sufficiently irrational" sets of frequencies $(\omega_1, \dots, \omega_N)$, and thereby guarantees that the invariant tori corresponding to these frequencies survive with *quasi-periodic* flows on them. Second, it employs a generalization of *Newton's Method* for the solution of equations of the form $f(x) = 0$ to effect a fast convergence of the iterative scheme for an infinite series of canonical transformations which will eventually yield the invariant surviving tori under small perturbations. Furthermore, it leads to quantitative measures on the size of the domains of surviving and destroyed tori in a constant energy hypersurface in the phase space of the dynamical system. In what follows we will first present a rigorous statement of the theorem itself, and then provide a qualitative discussion of the main ideas behind the proof, the details of which, involving a plethora of technical considerations, are far beyond the scope of this text. The interested reader is referred to authoritative accounts by the discoverers of the Theorem themselves (Kolmogorov 1954, Arnold 1963, and Moser 1973). Our presentation will follow the treatment by Moser. The statement of the KAM Theorem will be followed by that of a theorem (also presented without proof) on the size of the regions of phase space covered by the preserved tori.

Theorem 41.1. *(The **KAM Theorem**) Let T^N be the N-torus and $Y \subset \mathbb{R}^N$ be an open set. Suppose $H(\boldsymbol{\theta}, \boldsymbol{I}, \mu)$ is a real analytic function for all $\boldsymbol{\theta} \in T^N$ and $\boldsymbol{I} \in Y$, and near $\mu = 0$; and $H_0(\boldsymbol{I}) = H(\boldsymbol{\theta}, \boldsymbol{I}, 0)$ is independent of $\boldsymbol{\theta}$. Let $\boldsymbol{I}_0$ be chosen so that*

$$\det \left(\frac{\partial^2 H_0}{\partial I_i \partial I_j} \right) \bigg|_{\boldsymbol{I}_0} \tag{41.27}$$

does not vanish, and that the freqiencies

$$\omega_i \equiv \frac{\partial H_0}{\partial I_i} \bigg|_{\boldsymbol{I}_0} \tag{41.28}$$

satisfy the inequality

$$\boxed{\, |\, \boldsymbol{m} \cdot \boldsymbol{\omega} \,| \geq \frac{\gamma(\mu)}{||\boldsymbol{m}||^{\tau}} \,} \tag{41.29}$$

for all non-zero vectors $\boldsymbol{m}$ with integer components, with $||\boldsymbol{m}|| \equiv \sum_{i=1}^{N} |\, m_i \,|$ (≥ 1), and $\gamma(\mu), \tau$ being some positive constants (γ depending on μ). Then for sufficiently small $|\mu|$, there exists an invariant torus

$$\boldsymbol{\theta} = \boldsymbol{\xi} + \boldsymbol{u}_0(\boldsymbol{\xi}, \mu) \,, \qquad \boldsymbol{I} = \boldsymbol{I}_0 + \boldsymbol{v}_0(\boldsymbol{\xi}, \mu) \,, \tag{41.30}$$

where $\boldsymbol{u}_0(\boldsymbol{\xi}, \mu)$ and $\boldsymbol{v}_0(\boldsymbol{\xi}, \mu)$ are real analytic functions of μ and of $\boldsymbol{\xi} = (\xi^1, \dots, \xi^N)$, periodic in $\xi^1, \dots, \xi^N$ with period 2π, and vanish for $\mu = 0$. Furthermore the flow on this torus is given by

$$\dot{\xi}^i = \omega_i \qquad (i = 1, \dots, N) \,, \tag{41.31}$$

that is, the flow is quasiperiodic with the same frequencies as those on the unperturbed torus specified by $\boldsymbol{I}_0$.

We will refer to the inequality (41.29), which quantifies the condition for $\boldsymbol{\omega}$ of being "sufficiently irrational", as the KAM condition.

Theorem 41.2. *Suppose the conditions of the KAM Theorem (Theorem 41.1) hold, and in addition, assume that Y has a compact closure. Let δ be a given positive number, then there exists a positive number $\mu_1(\delta) > 0$ such that for $\mu < \mu_1$ the set $\Sigma(\mu) \subset T^N \times Y$ corresponding to the surviving invariant tori satisfies*

$$m_L \left(T^N \times Y - \Sigma(\mu) \right) < \delta\, m_L(T^N \times Y) \,, \tag{41.32}$$

where m_L denotes the Lesbegue measure *in $T^N \times Y$.*

This Theorem essentially states that, under a small perturbation (small μ), invariant tori "almost always" survive: most of phase space trajectories remain for all times on tori T^N and do not migrate into the entire $2N-1$-dimensional energy hypersurface. On the other hand, "almost all" unperturbed tori found in the proximity of tori with rationally degenerate frequencies [those not satisfying the KAM condition (41.29)] are destroyed.

The discussion of the proof outline of the KAM Theorem given below (in most of the remainder of this chapter) will be for the case of *reversible systems*, which presents many technical simplifications but yet preserves the crucial features. Our presentation will be largely adapted from the approach by Moser (for details see Moser 1973). Consider a dynamical system given by

$$\frac{d\boldsymbol{\theta}}{dt} = \boldsymbol{f}(\boldsymbol{\theta}, \boldsymbol{I}) \,, \qquad \frac{d\boldsymbol{I}}{dt} = \boldsymbol{g}(\boldsymbol{\theta}, \boldsymbol{I}) \,. \tag{41.33}$$

Reversibility requires this system to be invariant under the reflection $\rho : (\boldsymbol{\theta}, \boldsymbol{I}) \to (-\boldsymbol{\theta}, \boldsymbol{I})$ and the time reversal $t \to -t$. This imposes the following condition on the vector field of the system:

$$\boldsymbol{f}(-\boldsymbol{\theta}, \boldsymbol{I}) = \boldsymbol{f}(\boldsymbol{\theta}, \boldsymbol{I}) \,, \qquad \boldsymbol{g}(-\boldsymbol{\theta}, \boldsymbol{I}) = -\boldsymbol{g}(\boldsymbol{\theta}, \boldsymbol{I}) \,. \tag{41.34}$$

Under the coordinate transformation $(\boldsymbol{\theta}, \boldsymbol{I}) \to (\boldsymbol{\xi}, \boldsymbol{\eta})$ given by

$$\boldsymbol{\theta} = \boldsymbol{u}(\boldsymbol{\xi}, \boldsymbol{\eta}) \,, \qquad \boldsymbol{I} = \boldsymbol{v}(\boldsymbol{\xi}, \boldsymbol{\eta}) \,, \tag{41.35}$$

where $\boldsymbol{\xi} = (\xi^1, \dots, \xi^N)$ are the new angle variables, reversibility then requires that (see Problem 41.1 below)

$$\boldsymbol{u}(-\boldsymbol{\xi}, \boldsymbol{\eta}) = -\boldsymbol{u}(\boldsymbol{\xi}, \boldsymbol{\eta}) \,, \qquad \boldsymbol{v}(-\boldsymbol{\xi}, \boldsymbol{\eta}) = \boldsymbol{v}(\boldsymbol{\xi}, \boldsymbol{\eta}) \,. \tag{41.36}$$

We expect that as θ^i changes by 2π, ξ^i changes by 2π also. Thus $\boldsymbol{u}(\boldsymbol{\xi}, \boldsymbol{\eta}) - \boldsymbol{\xi}$ and $\boldsymbol{v}(\boldsymbol{\xi}, \boldsymbol{\eta})$ must be periodic in each ξ^i with period 2π.

We will denote by G_R the group of local coordinate transformations of the form (41.35), with $\boldsymbol{u}$ and $\boldsymbol{v}$ being real analytic functions in all the indicated variables, that leaves the reversible dynamical system (41.33) invariant and reversible. In the new coordinates the dynamical system is described by the equations

$$\frac{d\boldsymbol{\xi}}{dt} = \boldsymbol{\phi}(\boldsymbol{\xi}, \boldsymbol{\eta}) \,, \qquad \frac{d\boldsymbol{\eta}}{dt} = \boldsymbol{\psi}(\boldsymbol{\xi}, \boldsymbol{\eta}) \,, \tag{41.37}$$

where, analogous to (41.34),

$$\boldsymbol{\phi}(-\boldsymbol{\xi}, \boldsymbol{\eta}) = \boldsymbol{\phi}(\boldsymbol{\xi}, \boldsymbol{\eta}) \,, \qquad \boldsymbol{\psi}(-\boldsymbol{\xi}, \boldsymbol{\eta}) = -\boldsymbol{\psi}(\boldsymbol{\xi}, \boldsymbol{\eta}) \,. \tag{41.38}$$

Furthermore we will require that all elements $g \in G_R$ are close to the identity transformation. Note that the group G_R will be replaced by the symplectic group of canonical transformations (cf. Chapter 36) when general Hamiltonian systems are considered (instead of just reversible ones).

Problem 41.1 Show that (41.36) follows from the requirement that the transformed equations for the dynamical system (41.33) in the new variables $\boldsymbol{\xi}$ and $\boldsymbol{\eta}$ are invariant under the reflection $(\boldsymbol{\xi} \to -\boldsymbol{\xi}, \boldsymbol{\eta} \to \boldsymbol{\eta})$.

The unperturbed system corresponding to (41.33) is

$$\frac{d\boldsymbol{\theta}}{dt} = \boldsymbol{F}(\boldsymbol{I}) , \qquad \frac{d\boldsymbol{I}}{dt} = \boldsymbol{0} , \tag{41.39}$$

where

$$\det\left(\frac{\partial F^i}{\partial I_j}\right) \neq 0 . \tag{41.40}$$

We will take

$$\boldsymbol{F}(\boldsymbol{I}) = \boldsymbol{\omega} + \boldsymbol{I} , \tag{41.41}$$

where $\boldsymbol{\omega}$ is a constant frequency-vector that satisfies the KAM condition (41.29). This choice certainly satisfies (41.40). Moreover it is such that $\boldsymbol{F}(\boldsymbol{0}) = \boldsymbol{\omega}$. More generally we require

$$\boldsymbol{F}(\boldsymbol{I}_0) = \left.\frac{\partial H_0}{\partial I_i}\right|_{\boldsymbol{I}_0} = \omega_i , \tag{41.42}$$

so that the special choice for $\boldsymbol{F}(\boldsymbol{I})$ given by (41.41) just amounts to choosing $\boldsymbol{I}_0 = \boldsymbol{0}$, which can always be done. Thus

$$\boldsymbol{\theta}(t) = \boldsymbol{\omega} t + \boldsymbol{\theta}(0) , \qquad \boldsymbol{I} = \boldsymbol{0} \tag{41.43}$$

is a family of quasiperiodic solutions for the unperturbed system (41.39).

To show that the general formula for $\boldsymbol{F}(\boldsymbol{I})$ can be reduced to (41.41) (with $\boldsymbol{\omega}$ satisfying the KAM condition), we first make the following observation. Suppose $\boldsymbol{F}(\boldsymbol{I})$ is defined in a domain $D \subset \mathbb{R}^N$. Since the set of all $\boldsymbol{\omega}$ satisfying the KAM condition for some $\gamma > 0$ and $\tau > 0$ is *dense* in $\mathbb{R}^N$, the non-singularity condition

$$\det\left(\left.\frac{\partial F^i}{\partial I_j}\right|_{\boldsymbol{I}_0}\right) \neq 0 , \tag{41.44}$$

where $\boldsymbol{F}(\boldsymbol{I}_0) = \boldsymbol{\omega}$, basically ensures, by the *Inverse Function Theorem*, that there exists a neighborhood $U \subset D$ of $\boldsymbol{I}_0$ such that $V = \boldsymbol{F}(U)$ is a neighborhood of $\boldsymbol{F}(\boldsymbol{I}_0)$ and the inverse function $\boldsymbol{F}^{-1}$ is defined for every $\boldsymbol{\omega}' \in \boldsymbol{F}(U)$ satisfying the KAM condition: $\boldsymbol{F}^{-1}(\boldsymbol{\omega}') = \boldsymbol{I} \in U$. Now write

$$\boldsymbol{F}(\boldsymbol{I}) - \boldsymbol{F}(\boldsymbol{I}_0) = Q\,(\boldsymbol{I} - \boldsymbol{I}_0) + \dots , \tag{41.45}$$

where the matrix

$$Q \equiv \left.\frac{\partial F^i}{\partial I_j}\right|_{\boldsymbol{I}_0} \tag{41.46}$$

is, as stipulated by (41.44), non-singular. Define

$$\boldsymbol{z} \equiv Q\,(\boldsymbol{I} - \boldsymbol{I}_0) \approx \boldsymbol{F}(\boldsymbol{I}) - \boldsymbol{F}(\boldsymbol{I}_0) = \boldsymbol{\omega}' - \boldsymbol{\omega} , \tag{41.47}$$

so that

$$\boldsymbol{I} = \boldsymbol{I}_0 + Q^{-1}\boldsymbol{z} . \tag{41.48}$$

Then

$$\begin{aligned} \boldsymbol{F}(\boldsymbol{I}) &= \boldsymbol{F}(\boldsymbol{I}_0) + \left(\left.\frac{\partial F^i}{\partial I_j}\right|_{\boldsymbol{I}_0} \right) (Q^{-1}\boldsymbol{z}) + \dots = \boldsymbol{\omega} + QQ^{-1}\boldsymbol{z} + \dots \\ &= \boldsymbol{\omega} + \boldsymbol{z} + O(z^2) \,. \end{aligned} \tag{41.49}$$

If $|\boldsymbol{z}|$ is sufficiently small (or equivalently, $\boldsymbol{\omega}'$ is sufficiently close to $\boldsymbol{\omega}$), $\boldsymbol{I} = \boldsymbol{F}^{-1}(\boldsymbol{\omega}')$ will be in U. We can then rescale $\boldsymbol{z}$ by the constant matrix Q^{-1}, set $\boldsymbol{I}_0 = \boldsymbol{0}$, and replace $\boldsymbol{z}$ by $\boldsymbol{I}$ on the right-hand side of the last equality in (41.49), thus obtaining (41.41) in the limit $|\boldsymbol{z}| \to 0$.

For a sufficiently non-resonant torus to survive under small perturbations, we will need a coordinate transformation in G_R (for reversible systems) such that a solution to the system (41.37), in analogy to (41.43), is

$$\boldsymbol{\xi}(t) = \boldsymbol{\omega} t + \boldsymbol{\xi}(0) \,, \qquad \boldsymbol{\eta} = \boldsymbol{0} \,, \tag{41.50}$$

which yields the desired family of quasiperiodic solutions in the variables $\boldsymbol{\theta}$ and $\boldsymbol{I}$ according to (41.35):

$$\boldsymbol{\theta}(t) = \boldsymbol{u}(\boldsymbol{\omega} t + \boldsymbol{\xi}(0), 0) \,, \qquad \boldsymbol{I}(t) = \boldsymbol{v}(\boldsymbol{\omega} t + \boldsymbol{\xi}(0), 0) \,. \tag{41.51}$$

For this to occur we require, in analogy to (41.39) and (41.49) [cf. (41.37)],

$$\boldsymbol{\phi}(\boldsymbol{\xi}, \boldsymbol{\eta}) = \boldsymbol{\omega} + \boldsymbol{\eta} + O(\eta^2) \,, \qquad \boldsymbol{\psi}(\boldsymbol{\xi}, \boldsymbol{\eta}) = O(\eta^2) \,. \tag{41.52}$$

Note that this satisfies the reversibility requirement (41.38), and that on the new distorted torus (whose existence the KAM Theorem asserts) the frequencies of the quasiperiodic motion as the same as those on the unperturbed torus.

The mathematical basis of the statements made in the above paragraph, and hence of the KAM Theorem, is the following key Lemma, which we will state for reference but without proof. Note that this Lemma requires for consideration the complexification of the functions $\boldsymbol{f}(\boldsymbol{\theta}, \boldsymbol{I})$ and $\boldsymbol{g}(\boldsymbol{\theta}, \boldsymbol{I})$ in (41.33), that is, the variables $\theta^1, \dots, \theta^N$ and $I_1, \dots, I_N$ are to be considered as complex variables.

Lemma 41.1. *Suppose an invariant torus of the unperturbed system corresponding to the reversible syatem (41.33) with a $2N$-dimensional phase space has frequencies $\boldsymbol{\omega}$ satisfying the KAM condition (41.29) for some $\gamma > 0$ and $\tau > 0$. Let $0 < \alpha < 1$ and*

$$0 < \beta < min \left(\frac{\alpha}{4N + 4\tau + 1}, 1 - \alpha \right) \,. \tag{41.53}$$

Then there exists a positive number $\delta_0 = \delta_0(\alpha, \beta, \gamma, \tau, N)$ such that if the complexified system

$$\dot{\boldsymbol{\theta}} = \boldsymbol{f}(\boldsymbol{\theta}, \boldsymbol{I}, \mu) \,, \qquad \dot{\boldsymbol{I}} = \boldsymbol{g}(\boldsymbol{\theta}, \boldsymbol{I}, \mu) \tag{41.54}$$

satisfies the condition

$$|\, \boldsymbol{f} - \boldsymbol{\omega} - \boldsymbol{I} \,| + |\, \boldsymbol{g} \,| < \delta^{1+\alpha} \tag{41.55}$$

in the complex domain

$$|\,Im\,(\theta^i)\,| < \delta^\beta\,, \qquad |\,I_i\,| < \delta \tag{41.56}$$

for some δ in $0 < \delta < \delta_0$ as well as the reversibility condition (41.34), it can be transformed by a coordinate transformation

$$\boldsymbol{\theta} = \boldsymbol{u}(\boldsymbol{\xi}, \boldsymbol{\eta})\,, \qquad \boldsymbol{I} = \boldsymbol{v}(\boldsymbol{\xi}, \boldsymbol{\eta}) \tag{41.57}$$

into the system

$$\dot{\boldsymbol{\xi}} = \boldsymbol{\phi}(\boldsymbol{\xi}, \boldsymbol{\eta})\,, \qquad \dot{\boldsymbol{\eta}} = \boldsymbol{\psi}(\boldsymbol{\xi}, \boldsymbol{\eta})\,, \tag{41.58}$$

with

$$\boldsymbol{\phi} = \boldsymbol{\omega} + \boldsymbol{\eta} + O(\eta^2)\,, \qquad \boldsymbol{\psi} = O(\eta^2)\,. \tag{41.59}$$

Moreover, both $\boldsymbol{u}(\boldsymbol{\xi}, \boldsymbol{\eta})$ and $\boldsymbol{v}(\boldsymbol{\xi}, \boldsymbol{\eta})$ are linear in $\boldsymbol{\eta}$ and satisfy

$$|\,\boldsymbol{u} - \boldsymbol{\xi}\,| + |\,\boldsymbol{v}\,| < \delta \tag{41.60}$$

in the complex domain

$$|\,Im\,(\xi^i)\,| < \frac{\delta^\beta}{2}\,, \qquad |\,\eta_i\,| < \frac{\delta}{2}\,. \tag{41.61}$$

In the above $|\,\boldsymbol{z}\,| \equiv \sqrt{|z_1|^2 + \cdots + |z_N|^2}$ (where on the right-hand side, $|z_i|^2$ means the square of the magnitude of the complex quantity z_i).

The proof of this Lemma will show that the functions $\boldsymbol{u}$ and $\boldsymbol{v}$ will depend on any parameters on which $\boldsymbol{f}$ and $\boldsymbol{g}$ depend analytically, in particular the small perturbation parameter μ. A particular solution for the system (41.58) is given by (41.50). Thus the perturbed surviving torus corresponding to $\boldsymbol{\omega}$ is given by (41.51). In order to show that this is of the form (41.30) in the KAM Theorem (Theorem 41.1), we will have to solve for the functional forms for $\boldsymbol{u}(\boldsymbol{\xi}, \boldsymbol{\eta})$ and $\boldsymbol{v}(\boldsymbol{\xi}, \boldsymbol{\eta})$, subject to the reversibility conditions (41.36), and the conditions (41.59) for the vector field $(\boldsymbol{\phi}, \boldsymbol{\psi})$ in the above Lemma.

We will now proceed to derive the partial differential equations satisfied by $\boldsymbol{u}(\boldsymbol{\xi}, \boldsymbol{\eta})$ and $\boldsymbol{v}(\boldsymbol{\xi}, \boldsymbol{\eta})$. We have, from (41.57),

$$\dot{\boldsymbol{\theta}} = u_\xi\,\dot{\boldsymbol{\xi}} + u_\eta\,\dot{\boldsymbol{\eta}}\,, \qquad \dot{\boldsymbol{I}} = v_\xi\,\dot{\boldsymbol{\xi}} + v_\eta\,\dot{\boldsymbol{\eta}}\,, \tag{41.62}$$

where u_ξ denotes the matrix $\partial u^i/\partial \xi^j$, etc. It follows from (41.58) that

$$\dot{\boldsymbol{\theta}} = u_\xi\,\boldsymbol{\phi}(\boldsymbol{\xi}, \boldsymbol{\eta}) + u_\eta\,\boldsymbol{\psi}(\boldsymbol{\xi}, \boldsymbol{\eta})\,, \qquad \dot{\boldsymbol{I}} = v_\xi\,\boldsymbol{\phi}(\boldsymbol{\xi}, \boldsymbol{\eta}) + v_\eta\,\boldsymbol{\psi}(\boldsymbol{\xi}, \boldsymbol{\eta})\,. \tag{41.63}$$

By assumption (41.55) we have

$$\boldsymbol{f} - \boldsymbol{\omega} - \boldsymbol{I} \approx O(\delta^2)\,, \qquad \boldsymbol{g} \approx O(\delta^2)\,. \tag{41.64}$$

So we write, on replacing $\boldsymbol{I}$ by $\boldsymbol{I} - \boldsymbol{I}_0$,

$$\boldsymbol{f} = \boldsymbol{\omega} + \boldsymbol{I} - \boldsymbol{I}_0 + \hat{\boldsymbol{f}}\,, \qquad \boldsymbol{g} = \hat{\boldsymbol{g}}\,, \tag{41.65}$$

where $\hat{\boldsymbol{f}}$ and $\hat{\boldsymbol{g}}$ are small. Lemma 41.1 requires us to seek solutions of the form [cf. (41.60) and the condition on $\boldsymbol{\eta}$-linearity stated immediately above the equation]

$$\boldsymbol{u}(\boldsymbol{\xi},\boldsymbol{\eta}) = \boldsymbol{\xi} + \boldsymbol{u}_0(\boldsymbol{\xi}) + U_1(\boldsymbol{\xi})\,\boldsymbol{\eta}\,, \qquad \boldsymbol{v}(\boldsymbol{\xi},\boldsymbol{\eta}) = \boldsymbol{I}_0 + \boldsymbol{v}_0(\boldsymbol{\xi}) + V_1(\boldsymbol{\xi})\,\boldsymbol{\eta}\,, \quad (41.66)$$

where $\boldsymbol{u}_0, \boldsymbol{v}_0$ are N-vectors and U_1, V_1 are $N \times N$ matrices, and all are periodic in $\xi^1, \dots, \xi^N$ with period 2π. The above equations can be written in component form as follows (with repeated indices summed over):

$$u^i = \xi^i + (u_0)^i + (U_1)^i_k\,\eta_k\,, \qquad v_i = (I_0)_i + \eta_i + (v_0)_i + (V_1)_{ik}\,\eta_k\,. \quad (41.67)$$

The derivative matrices can now be calculated:

$$(u_\xi)^i_j = \frac{\partial u^i}{\partial \xi^j} = \delta^i_j + \frac{\partial (u_0)^i}{\partial \xi^j} + \frac{\partial (U_1)^i_k}{\partial \xi^j}\,\eta_k\,. \quad (41.68)$$

$$(u_\eta)^i_j = \frac{\partial u^i}{\partial \eta_j} = (U_1)^i_j\,, \quad (41.69)$$

$$(v_\xi)_{ij} = \frac{\partial v_i}{\partial \xi^j} = \frac{\partial (v_0)_i}{\partial \xi^j} + \frac{\partial (V_1)_{ik}}{\partial \xi^j}\,\eta_k\,, \quad (41.70)$$

$$(v_\eta)_{ij} = \frac{\partial v_i}{\partial \eta_j} = \delta_{ij} + (V_1)_{ij}\,; \quad (41.71)$$

or, in matrix forms:

$$u_\xi = 1 + (u_0)_\xi + (U_1)_\xi\,\eta\,, \quad (41.72)$$

$$u_\eta = U_1\,, \quad (41.73)$$

$$v_\xi = (v_0)_\xi + (V_1)_\xi\,\eta\,, \quad (41.74)$$

$$v_\eta = 1 + V_1\,, \quad (41.75)$$

where 1 on the right-hand sides denotes the $N \times N$ identity matrix, and

$$((U_1)_\xi\,\eta)^i_j \equiv \frac{\partial (U_1)^i_k}{\partial \xi^j}\,\eta_k\,, \quad (41.76)$$

with a similar equation for the matrix elements of $(V_1)_\xi\,\eta$.

Using (41.65) for $\boldsymbol{f}$ and $\boldsymbol{g}$ [on the right-hand side of (41.33)], (41.59) for $\boldsymbol{\phi}$ and $\boldsymbol{\psi}$ and the matrix equations (41.72) to (41.75) for $u_\xi, u_\eta, v_\xi, v_\eta$ [on the right-hand side of (41.63)], and the second equation of (41.66) for $\boldsymbol{v}$ [to replace $\boldsymbol{I}$ in the right-hand side of (41.65)], the two equations in (41.63) can be recast as (with terms to different orders in η appearing explicitly):

$$\begin{aligned}&(1 + (u_0)_\xi + (U_1)_\xi\,\eta)\,(\boldsymbol{\omega} + \boldsymbol{\eta} + O(\eta^2)) + U_1(O(\eta^2))\\ &= \boldsymbol{\omega} + \boldsymbol{\eta} + \boldsymbol{v}_0 + V_1\,\boldsymbol{\eta} + \hat{\boldsymbol{f}}\,,\end{aligned} \quad (41.77)$$

$$((v_0)_\xi + (V_1)_\xi\,\eta)\,(\boldsymbol{\omega} + \boldsymbol{\eta} + O(\eta^2)) + (1 + V_1)(O(\eta^2)) = \hat{\boldsymbol{g}}\,. \quad (41.78)$$

On the left-hand sides of the above equations, a matrix product of the form $A\boldsymbol{a}$ means an $N \times N$ matrix A multiplied (on the right) by the $N \times 1$ column matrix (or vector) $\boldsymbol{a}$, yielding an $N \times 1$ column matrix (or vector). Gathering terms to zeroth order in (or independent of) η_i on both sides of these equations and equating, we obtain

$$(\boldsymbol{u}_0)_{\boldsymbol{\xi}}\boldsymbol{\omega} = \boldsymbol{v}_0 + \hat{\boldsymbol{f}}(\boldsymbol{\xi} + \boldsymbol{u}_0, \boldsymbol{I}_0 + \boldsymbol{v}_0)\,, \tag{41.79}$$

$$(\boldsymbol{v}_0)_{\boldsymbol{\xi}}\boldsymbol{\omega} = \hat{\boldsymbol{g}}(\boldsymbol{\xi} + \boldsymbol{u}_0, \boldsymbol{I}_0 + \boldsymbol{v}_0)\,, \tag{41.80}$$

where in the arguments for $\hat{\boldsymbol{f}}$ and $\hat{\boldsymbol{g}}$, we have used (41.66) with $\boldsymbol{\eta} = \boldsymbol{0}$. These two equations can be rewritten as the following pair of coupled non-linear partial differential equations:

$$\boxed{\partial \boldsymbol{u}_0 - \boldsymbol{v}_0 = \hat{\boldsymbol{f}}(\boldsymbol{\xi} + \boldsymbol{u}_0,\, \boldsymbol{I}_0 + \boldsymbol{v}_0)\,, \qquad \partial \boldsymbol{v}_0 = \hat{\boldsymbol{g}}(\boldsymbol{\xi} + \boldsymbol{u}_0,\, \boldsymbol{I}_0 + \boldsymbol{v}_0)}\,, \tag{41.81}$$

where the partial differential operator ∂ is defined by

$$\partial \equiv \sum_k \omega_k \frac{\partial}{\partial \xi^k} = \boldsymbol{\omega} \cdot \partial_{\boldsymbol{\xi}}\,. \tag{41.82}$$

These equations will in principle determine the functions $\boldsymbol{u}_0$ and $\boldsymbol{v}_0$. In order to obtain the equations satisfied by U_1 and V_1 [cf. (41.66)] we have to work with the first-order terms in $\boldsymbol{\eta}$ in (41.77) and (41.78). We first expand the functions $\hat{\boldsymbol{f}}$ and $\hat{\boldsymbol{g}}$ in powers of $\boldsymbol{\eta}$. To begin,

$$\begin{aligned}
&\hat{\boldsymbol{f}}(\boldsymbol{\theta} = \boldsymbol{u}(\boldsymbol{\xi}, \boldsymbol{\eta}),\, \boldsymbol{I} = \boldsymbol{v}(\boldsymbol{\xi}, \boldsymbol{\eta})) \\
&= \hat{\boldsymbol{f}}(\boldsymbol{\theta} = \boldsymbol{\xi} + \boldsymbol{u}_0 + U_1\boldsymbol{\eta},\, \boldsymbol{I} = \boldsymbol{I}_0 + \boldsymbol{\eta} + \boldsymbol{v}_0 + V_1\boldsymbol{\eta}) \\
&= \hat{\boldsymbol{f}}(\boldsymbol{\xi} + \boldsymbol{u}_0,\, \boldsymbol{I}_0 + \boldsymbol{v}_0) + \hat{\boldsymbol{f}}_{\boldsymbol{\theta}}\Big|_{\boldsymbol{\xi}+\boldsymbol{u}_0,\, \boldsymbol{I}_0+\boldsymbol{v}_0} U_1\boldsymbol{\eta} + \hat{\boldsymbol{f}}_{\boldsymbol{I}}\Big|_{\boldsymbol{\xi}+\boldsymbol{u}_0,\, \boldsymbol{I}_0+\boldsymbol{v}_0} (1 + V_1)\boldsymbol{\eta}\,.
\end{aligned} \tag{41.83}$$

On defining

$$\hat{\boldsymbol{f}}_{\boldsymbol{\theta}}(0) \equiv \hat{\boldsymbol{f}}_{\boldsymbol{\theta}}\Big|_{\boldsymbol{\xi}+\boldsymbol{u}_0,\, \boldsymbol{I}_0+\boldsymbol{v}_0}\,, \qquad \hat{\boldsymbol{f}}_{\boldsymbol{I}}(0) \equiv \hat{\boldsymbol{f}}_{\boldsymbol{I}}\Big|_{\boldsymbol{\xi}+\boldsymbol{u}_0,\, \boldsymbol{I}_0+\boldsymbol{v}_0}\,, \tag{41.84}$$

Eq. (41.83) can be rewritten as

$$\hat{\boldsymbol{f}}(\boldsymbol{\xi}, \boldsymbol{\eta}) = \hat{\boldsymbol{f}}(\boldsymbol{\xi} + \boldsymbol{u}_0(\boldsymbol{\xi}),\, \boldsymbol{I}_0 + \boldsymbol{v}_0(\boldsymbol{\xi})) + \hat{\boldsymbol{f}}_{\boldsymbol{\theta}}(0)\, U_1\boldsymbol{\eta} + \hat{\boldsymbol{f}}_{\boldsymbol{I}}(0)\,(1 + V_1)\boldsymbol{\eta}\,. \tag{41.85}$$

Similarly, on defining

$$\hat{\boldsymbol{g}}_{\boldsymbol{\theta}}(0) \equiv \hat{\boldsymbol{g}}_{\boldsymbol{\theta}}|_{\boldsymbol{\xi}+\boldsymbol{u}_0,\, \boldsymbol{I}_0+\boldsymbol{v}_0}\,, \qquad \hat{\boldsymbol{g}}_{\boldsymbol{I}}(0) \equiv \hat{\boldsymbol{g}}_{\boldsymbol{I}}|_{\boldsymbol{\xi}+\boldsymbol{u}_0,\, \boldsymbol{I}_0+\boldsymbol{v}_0}\,, \tag{41.86}$$

we have, analogous to (41.85),

$$\hat{\boldsymbol{g}}(\boldsymbol{\xi}, \boldsymbol{\eta}) = \hat{\boldsymbol{g}}(\boldsymbol{\xi} + \boldsymbol{u}_0(\boldsymbol{\xi}),\, \boldsymbol{I}_0 + \boldsymbol{v}_0(\boldsymbol{\xi})) + \hat{\boldsymbol{g}}_{\boldsymbol{\theta}}(0)\, U_1\boldsymbol{\eta} + \hat{\boldsymbol{g}}_{\boldsymbol{I}}(0)\,(1 + V_1)\boldsymbol{\eta}\,. \tag{41.87}$$

Substituting the above results for $\hat{\boldsymbol{f}}$ and $\hat{\boldsymbol{g}}$ into (41.77) and (41.78), and collecting and equating first-order terms in $\boldsymbol{\eta}$, we obtain

$$(u_0)_\xi\boldsymbol{\eta} + (U_1)_\xi\eta\,\boldsymbol{\omega} = V_1\boldsymbol{\eta} + \hat{\boldsymbol{f}}_{\boldsymbol{\theta}}(0)\,U_1\boldsymbol{\eta} + \hat{\boldsymbol{f}}_{\boldsymbol{I}}(0)\,(1+V_1)\boldsymbol{\eta}\,, \tag{41.88}$$
$$(v_0)_\xi\boldsymbol{\eta} + (V_1)_\xi\eta\,\boldsymbol{\omega} = \hat{\boldsymbol{g}}_{\boldsymbol{\theta}}(0)\,U_1\boldsymbol{\eta} + \hat{\boldsymbol{g}}_{\boldsymbol{I}}(0)\,(1+V_1)\boldsymbol{\eta}\,. \tag{41.89}$$

The above two equations can be read as matrix equations, with each term being an $N\times 1$ matrix (or column vector). For example (with repeated indices summed over),

$$((u_0)_\xi\,\boldsymbol{\eta})^i = \frac{\partial(u_0)^i}{\partial\xi^j}\,\eta_j\,, \tag{41.90}$$
$$(V_1\boldsymbol{\eta})_i = (V_1)_{ij}\,\eta_j\,, \tag{41.91}$$
$$((U_1)_\xi\eta\,\boldsymbol{\omega})_i = ((U_1)_\xi\eta)_{ik}\,\omega_k = \frac{\partial(U_1)_{ij}}{\partial\xi^k}\,\eta_j\omega_k\,, \tag{41.92}$$
$$\left(\hat{\boldsymbol{f}}_{\boldsymbol{\theta}}(0)U_1\,\boldsymbol{\eta}\right)_i = \left(\hat{\boldsymbol{f}}_{\boldsymbol{\theta}}(0)U_1\right)_{ij}\eta_j\,, \tag{41.93}$$

where in the last equation we recall that both $\hat{\boldsymbol{f}}_{\boldsymbol{\theta}}(0)$ and U_1 are $N\times N$ matrices, and in the second equality of (41.92) we have used (41.76). Equating the coefficients of $\boldsymbol{\eta}$ on both sides of (41.88) and (41.89) we get the matrix equations

$$(u_0)_\xi + (U_1)_\xi\,\omega = V_1 + \hat{\boldsymbol{f}}_{\boldsymbol{\theta}}(0)\,U_1 + \hat{\boldsymbol{f}}_{\boldsymbol{I}}(0)\,(1+V_1)\,, \tag{41.94}$$
$$(v_0)_\xi + (V_1)_\xi\,\omega = \hat{\boldsymbol{g}}_{\boldsymbol{\theta}}(0)\,U_1 + \hat{\boldsymbol{g}}_{\boldsymbol{I}}(0)\,(1+V_1)\,, \tag{41.95}$$

where every term is an $N\times N$ matrix, and the matrices $(U_1)_\xi\,\omega$ and $(V_1)_\xi\,\omega$ are defined similarly as in (41.76). Using the partial differential operator ∂ as defined in (41.82) the above coupled differential matrix equations can be recast as

$$\begin{aligned}\partial U_1 - V_1 &= -(u_0)_\xi + \hat{\boldsymbol{f}}_{\boldsymbol{\theta}}(0)\,U_1 + \hat{\boldsymbol{f}}_{\boldsymbol{I}}(0)\,(1+V_1)\,,\\ \partial V_1 &= -(v_0)_\xi + \hat{\boldsymbol{g}}_{\boldsymbol{\theta}}(0)\,U_1 + \hat{\boldsymbol{g}}_{\boldsymbol{I}}(0)\,(1+V_1)\,.\end{aligned} \tag{41.96}$$

To prove Lemma 41.1 we need to be able to solve the coupled nonlinear partial differential equations (41.81) and (41.96) for $\boldsymbol{u}_0(\boldsymbol{\xi}), \boldsymbol{v}_0(\boldsymbol{\xi}), U_1(\boldsymbol{\xi})$ and $V_1(\boldsymbol{\xi})$, which are all periodic in $\xi^1,\dots,\xi^N$ with period 2π and satisfy the reversibility conditions (41.36). One can imagine using an iterative scheme. Start by replacing the arguments of $\hat{\boldsymbol{f}}$ and $\hat{\boldsymbol{g}}$ on the right-hand sides of (41.81) by $(\boldsymbol{\xi}, \boldsymbol{I}_0)$ (setting $\boldsymbol{u}_0 = \boldsymbol{v}_0 = 0$), and also dropping all products of U_1, V_1 with the derivative terms $\hat{\boldsymbol{f}}_{\boldsymbol{\theta}}, \hat{\boldsymbol{g}}_{\boldsymbol{\theta}}, \hat{\boldsymbol{f}}_{\boldsymbol{I}}, \hat{\boldsymbol{g}}_{\boldsymbol{I}}$ on the right-hand sides of (41.96), to obtain the following (much simpler) system of linear partial differential equations with constant coefficients:

$$\partial\boldsymbol{u}_0 - \boldsymbol{v}_0 = \hat{\boldsymbol{f}}(\boldsymbol{\xi}, \boldsymbol{I}_0)\,, \qquad \partial\boldsymbol{v}_0 = \hat{\boldsymbol{g}}(\boldsymbol{\xi}, \boldsymbol{I}_0)\,, \tag{41.97}$$
$$\partial U_1 - V_1 = -(u_0)_\xi + \hat{\boldsymbol{f}}_{\boldsymbol{I}}(0)\,, \qquad \partial V_1 = -(v_0)_\xi + \hat{\boldsymbol{g}}_{\boldsymbol{I}}(0)\,. \tag{41.98}$$

The hope is to be able to first solve these equations and then effect a transformation of the original dynamical system (41.57) by using the coordinate transformation defined by (41.66) and the solutions of the above linear system of equations, with the result that the transformed dynamical system has smaller error terms than the original one. Specifically, this means that in the transformed system [cf. (41.58)]

$$\dot{\boldsymbol{\xi}} = \boldsymbol{\phi}(\boldsymbol{\xi},\boldsymbol{\eta}) = \boldsymbol{\omega} + \boldsymbol{\eta} + \hat{\boldsymbol{\phi}}\,, \qquad \dot{\boldsymbol{\eta}} = \boldsymbol{\psi}(\boldsymbol{\xi},\boldsymbol{\eta}) = \hat{\boldsymbol{\psi}}\,, \tag{41.99}$$

the errors incurred by $\hat{\boldsymbol{\phi}}$ and $\hat{\boldsymbol{\psi}}$ will be smaller than those by $\hat{\boldsymbol{f}}$ and $\hat{\boldsymbol{g}}$. The transformation can then be repeated indefinitely, with each iteration step yielding smaller errors than the previous one, until, finally, the composition of the infinitely many coordinate transformations converges to the one specified in Lemma 41.1 [defined by (41.57) to (41.61)]. We will see later that the construction of the desired infinite sequence of coordinate transformations follows the procedure for a refinement of Newton's method for solving algebraic equations of the form $f(x) = 0$.

To examine this process in more detail we consider the second equation in (41.97). It can be formally solved by expanding $\hat{\boldsymbol{g}}(\boldsymbol{\xi}, \boldsymbol{I}_0)$ in a Fourier series:

$$\hat{\boldsymbol{g}}(\boldsymbol{\xi}, \boldsymbol{I}_0) = \sum_{\boldsymbol{m}\neq \boldsymbol{0}} \hat{\boldsymbol{g}}_{\boldsymbol{m}}(\boldsymbol{I}_0)\, e^{i\boldsymbol{m}\cdot\boldsymbol{\xi}}\,, \tag{41.100}$$

where $\boldsymbol{m}$ is an N-vector with integer components. The constant term in the sum (for $\boldsymbol{m} = \boldsymbol{0}$) does not appear due to the requirement that $\hat{\boldsymbol{g}}$ is an odd function of $\boldsymbol{\xi}$. Thus

$$\partial \boldsymbol{v}_0 = \boldsymbol{\omega}\cdot\partial_{\boldsymbol{\xi}}\boldsymbol{v}_0 = \sum_{\boldsymbol{m}\neq \boldsymbol{0}} \hat{\boldsymbol{g}}_{\boldsymbol{m}}(\boldsymbol{I}_0)\, e^{i\boldsymbol{m}\cdot\boldsymbol{\xi}}\,. \tag{41.101}$$

The formal solution is

$$\boldsymbol{v}_0(\boldsymbol{\xi}) = \sum_{\boldsymbol{m}\neq \boldsymbol{0}} \frac{\hat{\boldsymbol{g}}_{\boldsymbol{m}}(\boldsymbol{I}_0)}{i\boldsymbol{m}\cdot\boldsymbol{\omega}}\, e^{i\boldsymbol{m}\cdot\boldsymbol{\xi}} \equiv \Lambda\hat{\boldsymbol{g}}(\boldsymbol{\xi}, \boldsymbol{I}_0)\,, \tag{41.102}$$

where in the last equality the operator Λ is defined. We see here the emergence of the small-denominators problem again! But this result will also allow us to see how the KAM condition (41.29) overcomes the threat of divergence of perturbation series, as will be demonstrated below.

Revisiting (41.97) and (41.98) we see that in these equations, the partial differential operator ∂ acts componentwise on the vectors $\boldsymbol{u}_0, \boldsymbol{v}_0$ and the matrices U_1, V_1. Letting $\alpha(\boldsymbol{\xi})$ and $\beta(\boldsymbol{\xi})$ be particular components of $\boldsymbol{u}_0, \boldsymbol{v}_0, U_1$ or V_1, we obtain the following pair of scalar equations in place of (41.97) and (41.98):

$$\partial\alpha(\boldsymbol{\xi}) - \beta(\boldsymbol{\xi}) = a(\boldsymbol{\xi})\,, \qquad \partial\beta(\boldsymbol{\xi}) = b(\boldsymbol{\xi})\,, \tag{41.103}$$

where $a(\boldsymbol{\xi})$ and $b(\boldsymbol{\xi})$ are scalar functions periodic in $\xi^1, \dots, \xi^N$ with period 2π, and satisfy the reversibility conditions

$$a(-\boldsymbol{\xi}) = a(\boldsymbol{\xi})\,, \qquad b(-\boldsymbol{\xi}) = -b(\boldsymbol{\xi})\,. \tag{41.104}$$

As stipulated before [cf. (41.36)] we seek solutions $\alpha(\boldsymbol{\xi})$ and $\beta(\boldsymbol{\xi})$ that have the same periodicity conditions and satisfy the reversibility conditions

$$\alpha(-\boldsymbol{\xi}) = -\alpha(\boldsymbol{\xi}) , \qquad \beta(-\boldsymbol{\xi}) = \beta(\boldsymbol{\xi}) . \tag{41.105}$$

To obtain the formal solutions we write

$$\beta = \Lambda b + \beta_0 , \tag{41.106}$$

where the operator Λ is defined as in (41.102), and β_0 is as yet an arbitrary constant; that is,

$$\Lambda b = \sum_{\boldsymbol{m}\neq 0} \frac{b_{\boldsymbol{m}}}{i\boldsymbol{m}\cdot\boldsymbol{\omega}} e^{i\boldsymbol{m}\cdot\boldsymbol{\xi}} , \tag{41.107}$$

with the Fourier expansion of $b(\boldsymbol{\xi})$ given by

$$b(\boldsymbol{\xi}) = \sum_{\boldsymbol{m}\neq 0} b_{\boldsymbol{m}}\, e^{i\boldsymbol{m}\cdot\boldsymbol{\xi}} , \tag{41.108}$$

where, again, the constant term (for $\boldsymbol{m} = \boldsymbol{0}$) in the expansion does not occur because $b(\boldsymbol{\xi})$ is odd in $\boldsymbol{\xi}$. We also Fourier expand $a(\boldsymbol{\xi})$ as follows

$$a(\boldsymbol{\xi}) = a_0 + \sum_{\boldsymbol{m}\neq 0} a_{\boldsymbol{m}}\, e^{i\boldsymbol{m}\cdot\boldsymbol{\xi}} \tag{41.109}$$

and choose $\beta_0 = -a_0$. Thus we can write

$$\alpha = \Lambda(a+\beta) = \Lambda(a - a_0 + a_0 + \Lambda b + \beta_0) = \Lambda((a-a_0) + \Lambda b) , \tag{41.110}$$

where the second equality follows from (41.106); or

$$\alpha = \Lambda(a-a_0) + \Lambda^2 b . \tag{41.111}$$

We will now show the convergence of $\alpha(\boldsymbol{\xi})$ and $\beta(\boldsymbol{\xi})$ despite the appearance of small denominators when these functions are expressed as infinite series like (41.107), if the KAM condition is satisfied. The conclusion will be stated more precisely in the form of the following lemma. Note that for the purpose of precise mathematical formulation, as in Lemma 41.1, the functions $\alpha(\boldsymbol{\xi})$ and $\beta(\boldsymbol{\xi})$ are to be considered as functions of the complex variables $\xi^1, \dots, \xi^N$. First we define a *functional analytic* concept, the r-norm $||a||_r$ of the function $a(\boldsymbol{\xi})$, by

$$||a||_r \equiv \sup_{|Im\,\xi^k|<r} |a(\boldsymbol{\xi})| . \tag{41.112}$$

The r-norms of the other relevant functions $b(\boldsymbol{\xi}), \alpha(\boldsymbol{\xi})$ and $\beta(\boldsymbol{\xi})$ are defined similarly.

Lemma 41.2. *If $0<\rho<r<1$, then the solutions for the functions $\alpha(\boldsymbol{\xi})$ and $\beta(\boldsymbol{\xi})$ of (41.101) are real analytic in $Im\,\xi^k<\rho$ and satisfy*

$$||\alpha||_\rho + ||\beta||_\rho \le \frac{c_1}{(r-\rho)^{2(\tau+N)}} \left(||a||_r + ||b||_r\right) , \tag{41.113}$$

where $c_1 = c_1(\gamma,\tau,N)$ is a positive constant, and γ,τ are constants appearing in the KAM condition (41.29).

Proof. We have, by (41.107),

$$\Lambda(a - a_0) = \sum_{\boldsymbol{m} \neq 0} \frac{a_{\boldsymbol{m}}}{i\boldsymbol{m} \cdot \boldsymbol{\omega}} \, e^{i\boldsymbol{m} \cdot \boldsymbol{\xi}} \,. \tag{41.114}$$

Hence, by (41.112),

$$||\Lambda(a - a_0)||_\rho = \sup_{|Im\, \xi^k| < \rho} \sum_{\boldsymbol{m} \neq 0} \frac{|a_{\boldsymbol{m}}| \, |e^{i\boldsymbol{m} \cdot \boldsymbol{\xi}}|}{|\boldsymbol{m} \cdot \boldsymbol{\omega}|} = \sum_{\boldsymbol{m} \neq 0} \frac{|a_{\boldsymbol{m}}| \, e^{||\boldsymbol{m}||\rho}}{|\boldsymbol{m} \cdot \boldsymbol{\omega}|} \,, \tag{41.115}$$

where $||\boldsymbol{m}||$ is defined immediately following (41.29). The magnitude of the Fourier coefficient $a_{\boldsymbol{m}}$ can be estimated by using its r-norm. We have

$$||a||_r = \sup_{|Im\, \xi^k| < r} \sum_{\boldsymbol{m}} |a_{\boldsymbol{m}}| \, \left| e^{i\boldsymbol{m} \cdot \boldsymbol{\xi}} \right| = \sum_{\boldsymbol{m}} |a_{\boldsymbol{m}}| \, e^{||\boldsymbol{m}||r} \,. \tag{41.116}$$

Thus $||a||_r \geq |a_{\boldsymbol{m}}| \exp(||\boldsymbol{m}||r)$ for any $\boldsymbol{m}$, and

$$|a_{\boldsymbol{m}}| \leq ||a||_r \, e^{-||\boldsymbol{m}||r} \,, \tag{41.117}$$

which expresses the exponential decay of Fourier coefficients of a real analytic function. It follows from (41.115) that

$$||\Lambda(a - a_0)||_\rho \leq \frac{||a||_r}{\gamma} \sum_{\boldsymbol{m} \neq 0} ||\boldsymbol{m}||^\tau \, e^{-||\boldsymbol{m}||(r-\rho)} \,, \tag{41.118}$$

where we have assumed the KAM condition (41.29), and γ and τ are the positive constants therein. The right-hand side is clearly an exponentially decaying quantity when $0 < \rho < r < 1$, which we assume to be the case. We will estimate the sum as follows. Let $n(||\boldsymbol{m}||, N)$ be the number of distinct non-zero $\boldsymbol{m}$ (N-vectors with integer components) having the same norm $||\boldsymbol{m}||(\geq 1)$, which we estimate by

$$n(||\boldsymbol{m}||, N) \leq A''(N) \, ||\boldsymbol{m}||^{N-1} \qquad (A''(N) > 2N) \,. \tag{41.119}$$

The sum in (41.118) can be first written as the left-hand side of the following inequality and then estimated according to the above inequality:

$$\sum_{m=1}^{\infty} n(m, N) \, m^\tau e^{-mx} \leq A''(N) \sum_{m=1}^{\infty} m^{N+\tau-1} e^{-mx} \qquad (x \equiv r - \rho, \, 0 < x < 1) \,. \tag{41.120}$$

An appropriate function $A(N) \geq A''(N)$ can always be chosen so that the sum on the right-hand side is bounded by the integral

$$A(N) \int_0^\infty dt' \, (t')^{N+\tau-1} e^{-t'x} \,. \tag{41.121}$$

Changing the integration variable to $t = t'x$ we obtain the following bound on the sum in (41.118):

$$\sum_{m=1}^{\infty} n(m,N)\, m^{\tau} e^{-mx} \leq \frac{A(N)}{x} \int_0^{\infty} \left(\frac{t}{x}\right)^{N+\tau-1} e^{-t} = \frac{A(N)}{x^{N+\tau}}\, \Gamma(N+\tau)\,, \tag{41.122}$$

where $\Gamma(z)$ is the *gamma function* with the integral representation

$$\Gamma(z) = \int_0^{\infty} t^{z-1} e^{-1}\, dz\,. \tag{41.123}$$

We can thus rewrite the inequality (41.118) as

$$||\Lambda(a - a_0)||_{\rho} \leq \frac{c(N,\tau,\gamma)}{(r-\rho)^{N+\tau}}\, ||a||_r \qquad (0 < \rho < r < 1)\,, \tag{41.124}$$

where

$$c(N,\tau,\gamma) \equiv \frac{A(N)}{\gamma}\, \Gamma(N+\tau)\,. \tag{41.125}$$

Similarly, since $\beta = \Lambda b + \beta_0$, we have

$$||\beta||_{\rho} \leq \frac{c'(N,\tau,\gamma)}{(r-\rho)^{N+\tau}}\, ||b||_r \qquad (0 < \rho < r < 1)\,, \tag{41.126}$$

where

$$c'(N,\tau,\gamma) \equiv \frac{A'(N)}{\gamma}\, \Gamma(N+\tau)\,, \tag{41.127}$$

with an appropriate function $A'(N)$, similar to $A''(N)$, introduced in (41.119). Finally we estimate $||\alpha||_{\rho}$. Recalling (41.111) for $\alpha(\boldsymbol{\xi})$ we have

$$||\alpha||_{\rho} \leq ||\Lambda(a - a_0)||_{\rho} + ||\Lambda^2 b||_{\rho}\,. \tag{41.128}$$

The first term on the right-hand side has already been estimated by (41.124). To estimate the second term we first write, from (41.107),

$$\Lambda^2 b = \sum_{\boldsymbol{m}\neq 0} \left(-\frac{b_{\boldsymbol{m}}}{(\boldsymbol{m}\cdot\boldsymbol{\omega})^2}\right) e^{i\boldsymbol{m}\cdot\boldsymbol{\xi}}\,. \tag{41.129}$$

Hence, following the same reasoning leading from (41.118) to (41.122), we have

$$
\begin{aligned}
\|\Lambda^2 b\|_\rho &\le \frac{\|b\|_r}{\gamma^2} \sum_{\boldsymbol{m}\neq 0} \|\boldsymbol{m}\|^{2\tau}\, e^{-\|\boldsymbol{m}\|(r-\rho)} \\
&\le \frac{\|b\|_r}{\gamma^2} A''(N) \sum_{m=1}^{\infty} m^{N+2\tau-1} e^{-mx} \qquad (0 < x \equiv r-\rho < 1) \\
&\le \frac{\|b\|_r}{\gamma^2} A''(N) \sum_{m=1}^{\infty} m^{2(N+\tau)-1} e^{-mx} \\
&\le \frac{\|b\|_r}{\gamma^2} A(N) \int_0^\infty dt'\, (t')^{2(N+\tau)-1} e^{-t'x} \\
&= \frac{A(N)}{\gamma^2} \int_0^\infty dt\, t^{2(N+\tau)-1} e^{-t}\, \frac{\|b\|_r}{x^{2(N+\tau)}} \\
&= \frac{A(N)}{\gamma^2} \Gamma(2(N+\tau)) \frac{\|b\|_r}{x^{2(N+\tau)}} = \frac{c''(N,\tau,\gamma)}{x^{2(N+\tau)}} \|b\|_r \,,
\end{aligned}
\tag{41.130}
$$

where

$$
c''(N,\tau,\gamma) \equiv \frac{A(N)}{\gamma^2} \Gamma(2(N+\tau)) \,. \tag{41.131}
$$

Collecting the results (41.124), (41.126), (41.128) and (41.130), we have

$$
\begin{aligned}
&\|\alpha\|_\rho + \|\beta\|_\rho \le \frac{c}{(r-\rho)^{N+\tau}} \|a\|_r + \frac{c''}{(r-\rho)^{2(N+\tau)}} \|b\|_r + \frac{c'}{(r-\rho)^{N+\tau}} \|b\|_r \\
&\le \frac{1}{(r-\rho)^{2(N+\tau)}} \left[c\, \|a\|_r + (c'' + c')\, \|b\|_r \right] .
\end{aligned}
\tag{41.132}
$$

Equation (41.113) is then obtained by choosing $c_1 = \max(c, c'' + c')$, and the lemma is proved. □

In order to state succinctly the improvement in errors achieved by the transformation (41.66) and (41.99) [that is, estimates for the small quantities $\hat{\phi}$ and $\hat{\psi}$ in (41.99)], we first compactify our notation and define the following quantities [with reference to the dynamical system (41.33) and the coordinate transformation (41.66)]:

$$
\boldsymbol{h}(\boldsymbol{\xi},\boldsymbol{\eta}) \equiv \begin{pmatrix} \boldsymbol{f} \\ \boldsymbol{g} \end{pmatrix} , \qquad \hat{\boldsymbol{h}} \equiv \begin{pmatrix} \boldsymbol{f} - \boldsymbol{\omega} - (\boldsymbol{I}(\boldsymbol{\xi},\boldsymbol{\eta}) - \boldsymbol{I}_0) \\ \boldsymbol{g} \end{pmatrix} = \begin{pmatrix} \hat{\boldsymbol{f}} \\ \hat{\boldsymbol{g}} \end{pmatrix} , \tag{41.133}
$$

$$
\boldsymbol{w} \equiv \begin{pmatrix} \boldsymbol{u} \\ \boldsymbol{v} \end{pmatrix} = \begin{pmatrix} \boldsymbol{\xi} + \boldsymbol{u}_0 + U_1 \boldsymbol{\eta} \\ \boldsymbol{I}_0 + \boldsymbol{v}_0 + V_1 \boldsymbol{\eta} \end{pmatrix} , \tag{41.134}
$$

$$\hat{\boldsymbol{w}} \equiv \boldsymbol{w} - \begin{pmatrix} \boldsymbol{\xi} \\ \boldsymbol{I}_0 \end{pmatrix} = \begin{pmatrix} \boldsymbol{u}_0 + U_1\boldsymbol{\eta} \\ \boldsymbol{v}_0 + V_1\boldsymbol{\eta} \end{pmatrix} . \tag{41.135}$$

With $\boldsymbol{\zeta} \equiv (\boldsymbol{\xi}, \boldsymbol{\eta})^T$, the transformed dynamical system (41.99) can be written as

$$\dot{\boldsymbol{\zeta}} = \boldsymbol{\chi}(\boldsymbol{\zeta}) \equiv \begin{pmatrix} \boldsymbol{\phi}(\boldsymbol{\xi}, \boldsymbol{\eta}) \\ \boldsymbol{\psi}(\boldsymbol{\xi}, \boldsymbol{\eta}) \end{pmatrix} = \begin{pmatrix} \boldsymbol{\omega} + \boldsymbol{\eta} \\ \boldsymbol{0} \end{pmatrix} + \hat{\boldsymbol{\chi}} , \tag{41.136}$$

where

$$\hat{\boldsymbol{\chi}} \equiv \begin{pmatrix} \hat{\boldsymbol{\phi}} \\ \hat{\boldsymbol{\psi}} \end{pmatrix} . \tag{41.137}$$

We also define the (r, s)-norm of the function $\hat{\boldsymbol{h}}(\boldsymbol{\xi}, \boldsymbol{\eta})$ [analogously to (41.112)] as follows:

$$||\hat{\boldsymbol{h}}||_{r,s} \equiv \sup_{\substack{|Im\,\xi^k|<r \\ |\eta_k|<s}} |\hat{\boldsymbol{h}}(\boldsymbol{\xi}, \boldsymbol{\eta})| , \tag{41.138}$$

where the supremum is taken over all the components of the vector $\hat{\boldsymbol{h}}$. The norm for $\hat{\boldsymbol{w}}$ is similarly defined.

The first important result deducible from Lemma 41.2 is the following (the steps will not be shown in detail here)

$$||\hat{\boldsymbol{w}}||_{\rho,s} \leq \frac{c_2(N,\tau,\gamma)}{(r-\rho)^{\lambda}} ||\hat{\boldsymbol{h}}||_{r,s} \qquad (0 < \rho < r < 1) , \tag{41.139}$$

where c_2 is a positive constant with the indicated dependencies and

$$\lambda \equiv 4N + 4\tau + 1 . \tag{41.140}$$

One can also establish the following more comprehensive lemma (stated without proof) concerning the estimates for various norms:

Lemma 41.3. *Let $0 < \alpha < 1$, and β, λ be given by (41.53) and (41.140), respectively. Assume that the positive numbers r, ρ, s, σ satisfy*

$$0 < \rho < r < 1 , \qquad 0 < 2\sigma < s < (r-\rho)^{\frac{1}{1-\alpha}} < 1 , \tag{41.141}$$

and that, with the constant c_2 as appearing in (41.139),

$$\frac{2c_2}{(r-\rho)^{\lambda}}\, s^{\alpha} < 1 . \tag{41.142}$$

If the given dynamical system (41.33) satisfies

$$||\hat{\boldsymbol{h}}||_{r,s} < s^{1+\alpha} , \tag{41.143}$$

then the coordinate transformation (41.66) will transform the system into (41.99), with

$$||\hat{\chi}||_{\rho,\sigma} < c_3(N,\tau,\gamma)\, s^{1+\alpha} \left[\frac{s^\alpha}{(r-\rho)^\lambda} + \left(\frac{\sigma}{s}\right)^2 \right] , \tag{41.144}$$

where c_3 is a constant (related to c_2) with the indicated dependencies. Moreover [cf. (41.139)],

$$||\hat{w}||_{\rho,\sigma} < \frac{c_2(N,\tau,\gamma)}{(r-\rho)^\lambda}\, s^{1+\alpha} . \tag{41.145}$$

The above lemma will lead to the proof of Lemma 41.1 by allowing us to construct an infinite sequence of coordinate transformations of the form (41.66), each giving rise to smaller errors $||\hat{\chi}||$ than the previous, which converges to the one specified in that lemma. We will outline the reasoning behind this claim as follows.

If we choose σ so that

$$2c_3 \frac{s^\alpha}{(r-\rho)^\lambda} = \left(\frac{\sigma}{s}\right)^{1+\alpha} , \tag{41.146}$$

or

$$\sigma = c_4(N,\tau,\gamma)\,(r-\rho)^{-\frac{\lambda}{1+\alpha}}\, s^{1+\frac{\alpha}{1+\alpha}} , \tag{41.147}$$

where

$$c_4 \equiv (2c_3)^{\frac{1}{1+\alpha}} , \tag{41.148}$$

then it follows from (41.144) that

$$\begin{aligned} ||\hat{\chi}||_{\rho,\sigma} &< s^{1+\alpha} \left[\frac{1}{2}\left(\frac{\sigma}{s}\right)^{1+\alpha} + c_3 \left(\frac{\sigma}{s}\right)^2 \right] \\ &< s^{1+\alpha} \left[\frac{1}{2}\left(\frac{\sigma}{s}\right)^{1+\alpha} + \frac{1}{2}\left(\frac{\sigma}{s}\right)^{1+\alpha} \right] = \sigma^{1+\alpha} , \end{aligned} \tag{41.149}$$

provided σ/s is made small enough, which can be done by assuring that c_2 in (41.142) is made large enough. This estimate for $||\hat{\chi}||$ is of the form (41.143) for $||\hat{h}||$, and is therefore suitable for our sought-for iteration process. To carry out the iteration we construct two infinite sequences r_k, s_k as follows:

$$r_k = \left(\frac{1}{2} + \frac{1}{2^{k+1}}\right) r_0 , \tag{41.150a}$$

$$s_{k+1} = c_4\,(r_k - r_{k+1})^{-\frac{\lambda}{1+\alpha}}\, s_k^{1+\frac{\alpha}{1+\alpha}} \qquad (k=0,1,2,\dots) , \tag{41.150b}$$

where r_0 and s_0 will be chosen subsequently. Using the recursion relation for r_k we see that the recursion relation for s_k can be expressed as

$$s_{k+1} = c_4 \left(\frac{1}{2}\right)^{-\frac{\lambda}{1+\alpha}} \left(\frac{1}{2}\right)^{-\frac{\lambda}{1+\alpha}(k+1)} r_0^{-\frac{\lambda}{1+\alpha}}\, s_k^{1+\frac{\alpha}{1+\alpha}} . \tag{41.151}$$

A constant $c_5(N,\tau,\gamma)$ can then be introduced to write

$$s_{k+1} \le c_5^{k+1}\, r_0^{-\frac{\lambda}{1+\alpha}}\, s_k^{1+\frac{\alpha}{1+\alpha}} , \tag{41.152}$$

or, on multiplying by $r_0^{-\frac{\lambda}{\alpha}}$,

$$r_0^{-\frac{\lambda}{\alpha}} s_{k+1} \leq c_5^{k+1} \left(r_0^{-\frac{\lambda}{\alpha}} s_k \right)^{1+\frac{\alpha}{1+\alpha}} . \tag{41.153}$$

Define the sequence ε_k by

$$\varepsilon_k \equiv r_0^{-\frac{\lambda}{\alpha}} s_k , \tag{41.154}$$

the above inequality becomes

$$\varepsilon_{k+1} \leq c_5^{k+1} \, \varepsilon_k^\kappa \qquad \left(1 < \kappa \equiv 1 + \frac{\alpha}{1+\alpha} < \frac{3}{2} \right) . \tag{41.155}$$

Clearly, with the *convergence exponent* $\kappa > 1$ as indicated, $\varepsilon_k \xrightarrow[k\to\infty]{} 0$ if $\varepsilon_0 = r_0^{-\frac{\lambda}{\alpha}} s_0$ is small. In this case $s_k \xrightarrow[k\to\infty]{} 0$ also. Note that the sequences r_k, s_k defined above satisfy the conditions (41.141) and (41.142) in Lemma 41.3 if we set $r = r_k, \rho = r_{k+1}, s = s_k, \sigma = s_{k+1}$ and $r_0^{-\frac{\lambda}{\alpha}} s_0$ is chosen small enough. Thus at each step of the iteration of coordinate transformations these conditions are satisfied and the lemma can be applied.

Now return to consider the parameters α, β and δ specified in Lemma 41.1, and set

$$s_0 = \delta , \qquad r_0 = \delta^\beta . \tag{41.156}$$

Since, by (41.53), $\beta < \alpha/\lambda$, we see that the quantity

$$r_0^{-\frac{\lambda}{\alpha}} s_0 = (\delta^\beta)^{-\frac{\lambda}{\alpha}} \delta = \delta^{1-\frac{\beta\lambda}{\alpha}} \tag{41.157}$$

can be made small by making δ small. We can then apply Lemma 41.3 to each step of the iteration, beginning by setting $r = r_0, s = s_0, \rho = r_1, \sigma = s_1$. At the k-th step of the iteration we obtain the dynamical system

$$\dot{\boldsymbol{\zeta}}_{(k)} = \boldsymbol{\chi}_{(k)}(\boldsymbol{\zeta}_{(k)}) , \tag{41.158}$$

with the error given by, as follows from (41.149),

$$||\hat{\boldsymbol{\chi}}_{(k)}||_{r_k, s_k} < s_k^{1+\alpha} . \tag{41.159}$$

This implies that $\hat{\boldsymbol{\chi}}_{(k)}$, that is, the error at the k-th step [cf. (41.136)], converges to zero as $k \to \infty$. Furthermore, by **Cauchy's inequality** and the above equation, the derivatives with respect to $\boldsymbol{\eta}$ of the error terms at $\boldsymbol{\eta} = \mathbf{0}$ satisfy

$$\left| \frac{\partial}{\partial \eta_i} \hat{\boldsymbol{\chi}}_{(k)}(\boldsymbol{\xi}, \boldsymbol{\eta} = \mathbf{0}) \right| \leq \frac{1}{s_k} \, ||\hat{\boldsymbol{\chi}}_{(k)}||_{r_k, s_k} < s_k^\alpha . \tag{41.160}$$

Hence the $\boldsymbol{\eta}$-derivatives of $\hat{\boldsymbol{\chi}}_{(k)}(\boldsymbol{\xi}, \boldsymbol{\eta})$ at $\boldsymbol{\eta} = \mathbf{0}$ converge as well to zero as $k \to \infty$. It follows from (41.136) and (41.137) that the iterations of coordinate transformations converge to one of the forms (41.59) in Lemma 41.1 [with the $O(\eta^2)$ dependence specified therein], the convergence being guaranteed by (41.145).

Since $\lim_{k\to\infty} r_k = r_0/2$ [cf. (41.150(a))], the composition of the iterated transformations converge in the complex domain $|Im\,\xi^i| < r_0/2 = \delta^\beta/2$, thereby satisfying the first condition in (41.61). Also, since each iterated transformation is linear in $\boldsymbol{\eta}$ [cf. (41.66)], we have convergence for the composition in, say, $|\eta_i| < s_0/2 = \delta/2$, thereby satisfying the second condition in (41.61). Finally, we have, by putting $r = r_0, \rho = r_0/2, s = s_0$ and $\sigma = s_0/2$ in (41.145),

$$\begin{aligned} ||\hat{\boldsymbol{w}}||_{\frac{r_0}{2},\frac{s_0}{2}} &< c_6 r_0^{-\lambda} s_0^{1+\alpha} = c_6\,(r_0^{-\lambda} s_0^{\alpha})\, s_0 \\ &= c_6\,(r_0^{-\frac{\lambda}{\alpha}} s_0)^{\alpha} s_0 < \delta\,, \end{aligned} \tag{41.161}$$

where $c_6 \equiv 2^\lambda c_2$, and we have used the facts that $s_0 = \delta$ and $r_0^{-\frac{\lambda}{\alpha}} s_0$ can be made arbitrarily small by decreasing δ. This implies (41.60). Thus Lemma 41.1 is proved, and so is the KAM Theorem (Theorem 41.1) (for reversible dynamical systems).

As mentioned earlier, it turns out that the iteration procedure discussed at length above for the solution of the system of nonlinear partial differential equations (41.81) and (41.96), a procedure that is key to the proof of the KAM Theorem, is derived from the well-known **Newton's Method** for finding roots of algebraic equations of the form $f(x) = 0$ (see Problem 41.4 below). In order to help the reader appreciate this remarkable fact, we will devote the remainder of this chapter to demonstrate how a particular formulation of the latter can be generalized for use in the KAM Theorem. Our discussion is again adapted from the treatment by Moser (Moser 1973).

Consider an algebraic equation $f(x) = 0$ to be solved for the unknown x. Let $\xi = f(x)$ and $x = v(\xi)$ be inverse functions of each other: $x \xrightarrow{f} \xi$, $\xi \xrightarrow{v} x$, so that the desired solution is $x = v(\xi = 0)$. Start with $x \approx x_0 = 0$ as the initial estimate (zeroth approximation) for the root. Write

$$f(x) = x + \hat{f}(x)\,, \tag{41.162}$$

and define the function $\phi(\xi)$ by

$$\phi(\xi) \equiv f(v(\xi))\,. \tag{41.163}$$

The function $\hat{f}(x)$ is isolated as a small error function for $f(x)$ when $x = 0$ is a good approximation to the desired root. We will similarly isolate a function $\hat{v}(\xi)$ by writing

$$v(\xi) = \xi + \hat{v}(\xi)\,, \tag{41.164}$$

so that $\hat{v}(0)\,(= v(0))$ is the exact root. To approximate $\hat{v}(0)$ we will further define an function $\hat{\phi}(\xi)$ by

$$\phi(\xi) = \xi + \hat{\phi}(\xi)\,. \tag{41.165}$$

The three "small" functions $\hat{f}(v(\xi)), \hat{v}(\xi)$ and $\hat{\phi}(\xi)$ can be related by observing that

$$\phi(\xi) = f(v(\xi)) = v(\xi) + \hat{f}(v(\xi)) = \xi + \hat{v}(\xi) + \hat{f}(v(\xi)) = \xi + \hat{\phi}(\xi)\,. \tag{41.166}$$

Hence

$$\hat{\phi}(\xi) = \hat{f}(v(\xi)) + \hat{v}(\xi) . \tag{41.167}$$

We will attempt to generate better and better approximations to the root of $f(x) = 0$ by making a sequence of coordinate transformations of the form $x \to \xi$, and at each step requiring that $\hat{\phi}(\xi) = 0$. Start by letting

$$f_0(x) \equiv f(x) , \qquad v_0(\xi) \equiv v(\xi) , \qquad \xi_0 \equiv \xi . \tag{41.168}$$

The first two equations also imply

$$\hat{f}_0(v(\xi)) = \hat{f}(v(\xi)) , \qquad \hat{v}_0(\xi) = \hat{v}(\xi) . \tag{41.169}$$

With $x_0 = 0$ as the zeroth approximation to the root, the first approximation x_1 is obtained by requiring, as mentioned above, $\hat{\phi}(\xi) = 0$. From (41.167) we obtain

$$\hat{v}_0(\xi) = -\hat{f}_0(v_0(\xi)) = -\hat{f}_0(\xi + \hat{v}_0(\xi)) \approx -\hat{f}_0(\xi) . \tag{41.170}$$

We further approximate $\hat{f}_0(\xi)$ by its linear part:

$$\hat{f}_0(\xi) \approx \hat{f}_0(0) + \xi \hat{f}_0'(0) \equiv \hat{f}_{0(l)}(\xi) . \tag{41.171}$$

Thus the first approximation to the root, x_1, is given by

$$v_0(\xi = 0) = \hat{v}_0(\xi = 0) \approx -\hat{f}_{0(l)}(0) = -\hat{f}_0(0) = x_1 , \tag{41.172}$$

or

$$x_1 = -\hat{f}(0) = -f(0) . \tag{41.173}$$

This result already anticipates the following recursion relation for the root:

$$x_{k+1} = x_k - a_k f(x_k) \qquad (k = 0, 1, 2, \dots) , \tag{41.174}$$

on setting $x_0 = 0$ and $a_0 = 1$. We will justify this result subsequently, and, of course, have to derive a recursion relation for the sequence a_k.

To obtain higher order approximations to the root we envision a sequence of coordinate transformations as follows:

$$x \xleftarrow[v_0(\equiv v)]{} \xi_0(\equiv \xi) \xleftarrow[v_1]{} \xi_1 \xleftarrow[v_2]{} \xi_2 \leftarrow \dots \xleftarrow[v_k]{} \xi_k . \tag{41.175}$$

We can then write

$$\begin{aligned} f(x) &\equiv f_0(x) = f_0(v_0(\xi_0)) = f_0(v_0 \circ v_1(\xi_1)) = \dots \\ &= f_0(v_0 \circ v_1 \circ \dots \circ v_{k-1}(\xi_{k-1})) = f_0(u_k(\xi_{k-1})) \quad (k = 1, 2, \dots) , \end{aligned} \tag{41.176}$$

where

$$u_k \equiv v_0 \circ v_1 \circ \dots \circ v_{k-1} \qquad (k = 1, 2, \dots) . \tag{41.177}$$

Define the functions $f_k(\xi_{k-1})$ by

$$f_k(\xi_{k-1}) \equiv f_0(u_k(\xi_{k-1})) \qquad (k = 1, 2, \dots) , \tag{41.178}$$

or equivalently,

$$f_{k+1}(\xi_k) \equiv f_0(u_{k+1}(\xi_k)) \qquad (k = 0, 1, 2, \dots) . \tag{41.179}$$

By (41.177) we can also write

$$\begin{aligned} f_{k+1}(\xi_k) &= f_0 \circ v_0 \circ v_1 \circ v_2 \circ \cdots \circ v_k(\xi_k) \\ &= f_1 \circ v_1 \circ v_2 \circ \cdots \circ v_k(\xi_k) \\ &= f_2 \circ v_2 \circ \cdots \circ v_k(\xi_k) = \dots\dots \\ &= (f_k \circ v_k)(\xi_k) . \end{aligned} \tag{41.180}$$

Thus the f_k satisfy the recursion relation

$$f_{k+1} = f_k \circ v_k \qquad (k = 0, 1, 2, \dots) . \tag{41.181}$$

Analogous to (41.163) and (41.165) we also define

$$\phi_k(\xi_k) \equiv f_{k+1}(\xi_k) = \xi_k + \hat{\phi}_k(\xi_k) \qquad (k = 0, 1, 2, \dots) . \tag{41.182}$$

We also have, analogous to (41.164),

$$\xi_{k-1} = v_k(\xi_k) = \xi_k + \hat{v}_k(\xi_k) \qquad (k = 1, 2, \dots) . \tag{41.183}$$

At this point, it will be useful to lay further groundwork for the general problem of extracting roots of algebraic equations by reviewing the following elementary considerations in calculus. Suppose we want to find the roots of the equation $F(x) = 0$. Initially we guess that there is a root near $x = x_0$. Defining $x' \equiv x - x_0$ and setting $f(x') = F(x)$, the zeroth-order root for $f(x) = 0$ is, as before, $x = 0$. Using the Taylor series for $f(x)$ near $x = 0$ we define

$$f^{(2)}(x) \equiv f(0) + x f'(0) + \frac{x^2}{2} f''(0) . \tag{41.184}$$

It follows that

$$|f^{(2)}(x)| \le |f(0)| + |x|\,|f'(0)| + |x|^2 |f''(0)| , \tag{41.185}$$

or

$$|f^{(2)}(x)| \le |f(0)| + r\,|f'(0)| + r^2\,|f''(0)| \qquad \text{for} \quad |x| < r . \tag{41.186}$$

If we then require $|f(0)| + r^2\,|f''(0)| \le r\,|f'(0)|$, then $|f''(x)| \le 2r\,|f'(0)|$. So by choosing r small $|f^{(2)}(x)|$ for $|x| < r$ can be made as small as we please, as long as the condition

$$\frac{1}{r}\,|f(0)| + r\,|f''(0)| \le |f'(0)| \tag{41.187}$$

is fulfilled. Recalling the definition for $\hat{f}(x)$ given by (41.162), we see that $f'(x) = 1 + \hat{f}'(x)$ and $f''(x) = \hat{f}''(x)$. Hence $f'(0) = 1 + \hat{f}'(0)$, which implies

$|f'(0)| \leq 1 + |\hat{f}'(0)|$. Since $f(0) = \hat{f}(0)$ and $f''(0) = \hat{f}''(0)$, the inequality (41.187) is equivalent to

$$\frac{1}{r}\,|\hat{f}(0)| + r\,|\hat{f}''(0)| \leq 1 + |\hat{f}'(0)|\,. \tag{41.188}$$

Guided by this inequality we define the r-norm $||\hat{f}||_r$ of the function $\hat{f}(x)$ by

$$||\hat{f}||_r \equiv \sup_{|x|<r}\left\{\frac{1}{r}\,|\hat{f}(x)| + r|\hat{f}''(x)|\right\}\,. \tag{41.189}$$

The criterion for $f(x) = 0$ to have a root x in the range $|x| < r$, then, is that the r-norm of $\hat{f}$ be small for small r.

The following useful lemma (stated without proof) will allow us to compare the norm for the "small" function $\hat{\phi}_0(\xi_0)$ [cf. (41.182)] with that for the "small" function $\hat{f}_0(x)$. This lemma can be regarded as the precursor to Lemma 41.3.

Lemma 41.4. *If* $||\hat{f}_0||_r = \varepsilon < 1/4$ *and* ρ *is chosen such that* $\varepsilon < \dfrac{\rho}{r} < 1 - 2\varepsilon$, *then*

$$||\hat{\phi}_0||_\rho \leq 6\left(\varepsilon\,\frac{\rho}{r} + \varepsilon^2\,\frac{r}{\rho}\right)\,. \tag{41.190}$$

The minimum value for $||\hat{\phi}_0||_\rho$ *is achieved when* $\rho = r\sqrt{\varepsilon}$, *in which case*

$$||\hat{\phi}_0||_\rho \leq c\,\varepsilon^{3/2}\,, \tag{41.191}$$

where $c = 12$.

Suppose we choose ε such that $0 < \varepsilon < 1/c^2 < 1/4$, and construct the sequence:

$$\varepsilon_0 = \varepsilon,\ \varepsilon_1 = c\,\varepsilon_0^{3/2},\ \varepsilon_2 = c\,\varepsilon_1^{3/2}, \dots, \varepsilon_k = c\,\varepsilon_{k-1}^{3/2}, \dots\quad (k = 1, 2, \dots)\,. \tag{41.192}$$

Then $0 < c\sqrt{\varepsilon} < 1$, and it is not hard to see that

$$\varepsilon_k = (c\,\sqrt{\varepsilon_{k-1}})\,\varepsilon_{k-1} < (c\,\sqrt{\varepsilon_0})\,\varepsilon_{k-1} < \varepsilon_{k-1}\,. \tag{41.193}$$

Thus $\lim_{k\to\infty}\varepsilon_k = 0$ and the convergence is rapid, with exponent 3/2. Now construct the sequence

$$r_0 = r,\ r_1 = r_0\,\varepsilon_0^{1/2},\ r_2 = r_1\,\varepsilon_1^{1/2}, \dots, r_k = r_{k-1}\,\varepsilon_{k-1}^{1/2}, \dots\,. \tag{41.194}$$

Recalling (41.182) we will apply Lemma 41.4 to estimate the following sequence of norms:

$$||\hat{\phi}_{k-1}(\xi_{k-1})||_{r_k} = ||f_k(\xi_{k-1}) - \xi_{k-1}||_{r_k} \qquad (k = 1, 2, \dots)\,. \tag{41.195}$$

Carrying out the replacements $r \to r_{k-1}, \rho \to r_k$ and $\varepsilon \to \varepsilon_{k-1}$ in (41.190) we get

$$||f_k(\xi_{k-1}) - \xi_{k-1}||_{r_k} \leq \frac{c}{2}\left(\varepsilon_{k-1}\,\frac{r_k}{r_{k-1}} + (\varepsilon_{k-1})^2\,\frac{r_{k-1}}{r_k}\right)\,. \tag{41.196}$$

Since $r_k/r_{k-1} = \varepsilon_{k-1}^{1/2}$ [by (41.194)] we conclude that

$$||f_k(\xi_{k-1}) - \xi_{k-1}||_{r_k} \leq c\,\varepsilon_{k-1}^{3/2} = \varepsilon_k \xrightarrow[k\to\infty]{} 0\,, \tag{41.197}$$

or, on using (41.178),

$$||f(u_k(\xi_{k-1})) - \xi_{k-1}||_{r_k} \xrightarrow[k\to\infty]{} 0\,. \tag{41.198}$$

Letting ξ_{k-1} approach 0, the above equation for norms translate to

$$\lim_{k\to\infty} f(u_k(0)) = 0\,. \tag{41.199}$$

It can be shown that (the details will be omitted here) the functions u_k converge to a function u: $\lim_{k\to\infty} u_k = u$, and that each u_k is a linear function of its argument (ξ_{k-1}). Hence the limit function u is a linear function and

$$f(u(0)) = 0\,, \tag{41.200}$$

indicating that $x_\infty \equiv u(0)$ is the desired root. We will also express this by writing

$$x_\infty = \lim_{k\to\infty} u_k(0)\,. \tag{41.201}$$

More explicitly,

$$f(u(0)) = \lim_{k\to\infty} f(u_k(0)) = \lim_{k\to\infty} f_k(0) = 0\,. \tag{41.202}$$

Using Taylor's expansion it can be shown quite straightforwardly that, for any twice differentiable function f, $|f'(x)| \leq 2||f||_r$ for $|x| < r$ (see Problem 41.2). Thus, on differentiating (41.197) with respect to ξ_{k-1}, we have $\lim_{k\to\infty}(f_k'(0) - 1) = 0$, or

$$\lim_{k\to\infty} f_k'(0) = 1\,. \tag{41.203}$$

Now consider the function

$$f_\infty(\xi_\infty) \equiv f(u(\xi_\infty)) = \lim_{k\to\infty} f(u_k(\xi_{k-1}))\,. \tag{41.204}$$

Its derivative can be calculated by the chain rule as follows:

$$\left.\frac{df_\infty}{d\xi_\infty}\right|_0 = f'(u(0))\,u'(0) = \lim_{k\to\infty} f'(u_k(0))\,u_k'(0) = \lim_{k\to\infty} f_k'(0) = 1\,, \tag{41.205}$$

where we have made use of (41.203) in the last equality. Thus

$$u'(0) = \frac{1}{f'(u(0))}\,. \tag{41.206}$$

From the above discussion we arrive at the fact that

$$u(\xi_\infty) = x_\infty + a_\infty\,\xi_\infty\,, \qquad a_\infty \equiv \frac{1}{f'(x_\infty)}\,. \tag{41.207}$$

The relationship between the successive iterations x_k for the root and the coordinate transformations $u_k(\xi_{k-1})$ can be delineated as follows:

$$\begin{aligned} x_0 = 0 \rightarrow x_1 = u_1(0) \rightarrow x_2 = u_2(0) \rightarrow \cdots \rightarrow x_k = u_k(0) \rightarrow \dots \\ \rightarrow x_\infty = u_\infty(0) = u(0) \,, \end{aligned} \tag{41.208}$$

where

$$u_k(\xi_{k-1}) = x_k + a_k \, \xi_{k-1} \,. \tag{41.209}$$

Problem 41.2 Show that for a twice differential function $f(x)$, its first derivative is dominated by the norm: $|f'(x)| \leq ||f||_r$ in $|x| < r$. [Hints: Let $x, y \in (-r, r)$. Use Taylor's expansion to write

$$f'(x) = \frac{f(y) - f(x)}{y - x} - \frac{1}{2} f''(x')(y - x) \,,$$

where x' is some intermediate point between x and y. Then let $y = x + r$ or $x - r$ and use the norm definition (41.189).]

We can now construct the iteration scheme for x_k [already conjectured in (41.174)] and a_k, with $x_0 = 0$ and $a_0 = 1$. Recalling (41.182) and that $f_0 \equiv f$, we can use (41.162) and (41.179) to write

$$\xi_k + \hat{\phi}(\xi_k) = u_{k+1}(\xi_k) + \hat{f}(u_{k+1}(\xi_k)) \,. \tag{41.210}$$

Guided by the limit equation (41.198) we set $\hat{\phi}(\xi_k)$ to be zero for every k and obtain

$$u_{k+1}(\xi_k) = \xi_k - \hat{f}(u_{k+1}(\xi_k)) \qquad (k = 0, 1, 2, \dots) \,. \tag{41.211}$$

This is the algebraic equation that needs to be solved for $u_{k+1}(\xi_k)$ to find the k-th iteration for the root $x_k = u_{k+1}(\xi_k = 0)$. Its counterparts in the KAM iteration scheme presented earlier in this chapter are the coupled partial differential equations (41.81) and (41.96). Using (41.209) and (41.162) the above equation appears as

$$x_{k+1} + a_{k+1}\xi_k = \xi_k - [f(x_{k+1} + a_{k+1}\xi_k) - (x_{k+1} + a_{k+1}\xi_k)] \,. \tag{41.212}$$

From this equation we obtain

$$\begin{aligned} 0 &= \xi_k - f(x_{k+1} + a_{k+1}\xi_k) = \xi_k - f(x_k + (x_{k+1} - x_k) + a_{k+1}\xi_k) \\ &\approx \xi_k - [f(x_k) + f'(x_k) \{(x_{k+1} - x_k) + a_{k+1}\xi_k\}] \\ &= -f(x_k) - f'(x_k)(x_{k+1} - x_k) + \xi_k \{1 - f'(x_k) a_{k+1}\} \,. \end{aligned} \tag{41.213}$$

Setting $\xi_k = 0$ we obtain

$$x_{k+1} - x_k \approx -\frac{f(x_k)}{f'(x_k)} . \tag{41.214}$$

Using (41.211) in the form

$$u_k(\xi_{k-1}) = \xi_{k-1} - \hat{f}(u_k(\xi_{k-1})) , \tag{41.215}$$

we have, analogous to (41.212),

$$x_k + a_k\, \xi_{k-1} = \xi_{k-1} - [f(x_k + a_k\xi_{k-1}) - (x_k + a_k\xi_{k-1})] , \tag{41.216}$$

which implies

$$\begin{aligned} 0 &= \xi_{k-1} - f(x_k + a_k\xi_{k-1}) \approx \xi_{k-1} - [f(x_k) + f'(x_k)a_k\xi_{k-1}] \\ &= -f(x_k) + \xi_{k-1}\,[1 - f'(x_k)\, a_k] . \end{aligned} \tag{41.217}$$

Setting the coefficient of ξ_{k-1} on the right-hand side equal to zero we see that a_k is an approximation for $1/f'(x_k)$:

$$a_k \approx \frac{1}{f'(x_k)} . \tag{41.218}$$

Putting this result and (41.214) together we obtain the recursion relation (41.174) for the approximate root.

To obtain the recursion relation for a_k we use (41.218) to write

$$\frac{1}{a_k} - \frac{1}{a_{k+1}} \approx f'(x_k) - f'(x_{k+1}) = f'(x_k)\left[1 - \frac{f'(x_{k+1})}{f'(x_k)}\right] . \tag{41.219}$$

This implies

$$\frac{a_{k+1} - a_k}{a_k a_{k+1}} \approx \frac{1}{a_k}\,[1 - a_k f'(x_{k+1})] , \tag{41.220}$$

or

$$a_{k+1} - a_k \approx a_{k+1}\,[1 - a_k f'(x_{k+1})] . \tag{41.221}$$

We will replace the a_{k+1} on the right-hand side by a_k and finally present the two recursion relations together for x_k and a_k below:

$$x_{k+1} = x_k - a_k\, f(x_k) , \qquad a_{k+1} = a_k + a_k\,[1 - a_k\, f'(x_{k+1})] \qquad (x_0 = 0,\, a_0 = 1) . \tag{41.222}$$

Note that the advantage of this iteration procedure is that the derivative $f'(x_k)$ never appears in the denominator.

These iteration sequences will be shown to converge rapidly, with a convergence exponent $\kappa = 2$ [cf. (41.155)]. Define

$$z_k \equiv x_k - x_\infty , \qquad y_k \equiv 1 - a_k\, f'(x_k) . \tag{41.223}$$

From (41.208) and the second equation of (41.207) we see that both sequences converge to 0 as $k \to \infty$. Using Taylor's expansion we have

$$0 = f(x_\infty) = f(x_k) + f'(x_k)(x_\infty - x_k) + O[(x_\infty - x_k)^2] , \tag{41.224}$$

or

$$0 = f(x_k) - f'(x_k) z_k + O(z_k^2) . \tag{41.225}$$

Using the first equation of (41.222) we then have

$$z_{k+1} - z_k = x_{k+1} - x_k = -a_k [f'(x_k) z_k + O(z_k^2)] , \tag{41.226}$$

or

$$z_{k+1} = [1 - a_k\, f'(x_k)]\, z_k + O(z_k^2) , \tag{41.227}$$

that is,

$$z_{k+1} = y_k z_k + O(z_k^2) . \tag{41.228}$$

On the other hand, we can use the second equation of (41.222) to get

$$\begin{aligned} y_{k+1} &= 1 - a_{k+1} f'(x_{k+1}) \\ &= 1 - [a_k + a_k(1 - a_k f'(x_{k+1}))]\, f'(x_{k+1}) \\ &= 1 - a_k [2 - a_k f'(x_{k+1})]\, f'(x_{k+1}) , \end{aligned} \tag{41.229}$$

or

$$y_{k+1} = [1 - a_k\, f'(x_{k+1})]^2 . \tag{41.230}$$

But, on using the definition of y_k [(41.223)], we can write

$$\begin{aligned} 1 - a_k\, f'(x_{k+1}) &= y_k - a_k\, [f'(x_{k+1}) - f'(x_k)] \\ = y_k - a_k (x_{k+1} - x_k)\, f'' &= y_k + a_k^2 f'(x_k) f''\, z_k + O(z_k^2) , \end{aligned} \tag{41.231}$$

where the second derivative f'' is calculated at some intermediate value between x_k and x_{k+1}; for the second equality we have used the *mean-value theorem*, and for the third equality we have used (41.226). From (41.228) we obtain

$$|z_{k+1}| \le |z_k y_k| + C_1 |z_k|^2 , \tag{41.232}$$

and from (41.230) and (41.231) we obtain

$$|y_{k+1}| \le C_2 (z_k^2 + y_k^2) , \tag{41.233}$$

where in the above two equations, C_1 and C_2 are both positive constants. These equations in turn yield

$$|z_{k+1}|^2 + |y_{k+1}|^2 \le C_3 (z_k^2 + y_k^2)^2 , \tag{41.234}$$

where C_3 is another positive constant. We have thus demonstrated that the convergence of the iteration sequences in (41.222) is quadratic (convergence exponent $\kappa = 2$). Stated more explicitly, suppose we define the total error at each step by $\varepsilon_k \equiv z_k^2 + y_k^2$, then [cf. (41.155)]

$$\varepsilon_{k+1} \leq C_3\, \varepsilon_k^\kappa \qquad (\kappa = 2)\,, \tag{41.235}$$

so that beginning with the initial error ε_0, we arrive at the k-th iteration step an error $\varepsilon_k \leq (\varepsilon_0)^{2^k}$.

Problem 41.3 Consider finding the estimates for the roots, and the corresponding errors, of an algebraic equation $f(x) = 0$ by a straightforward application of Taylor's expansion. Use the following steps:

(a) Let x_0 be the zeroth-order approximation of a root. Write, for $x \sim x_0$,

$$f(x) = f(x_0) + f'(x_0)(x - x_0) + \dots\,. \tag{41.236}$$

Let the first approximation x_1 be defined by

$$f(x_0) + f'(x_0)(x_1 - x_0) = 0\,, \tag{41.237}$$

so that

$$x_1 = x_0 - \frac{f(x_0)}{f'(x_0)}\,. \tag{41.238}$$

Let the exact root be x, so that the error for x_1 is $e_1 \equiv x - x_1$. Define $\varepsilon \equiv x - x_0$. We then have the following two equations:

$$0 \approx f(x_0) + f'(x_0)\,\varepsilon + \frac{1}{2} f''(x_0)\,\varepsilon^2 \tag{41.239}$$

$$0 = f(x_0) + f'(x_0)(x_1 - x_0)\,. \tag{41.240}$$

Show, by subtracting one equation from the other, that the error for x_1 is estimated by

$$e_1 \equiv x - x_1 \approx -\frac{1}{2}\frac{f''(x_0)}{f'(x_0)}\,\varepsilon^2\,. \tag{41.241}$$

(b) Truncate the Taylor expansion (41.236) to the n-th order term in $x - x_0$ and write

$$P^{(n)}(x) \equiv \sum_{m=0}^{n} \frac{f^{(m)}(x_0)}{m!}(x - x_0)^m = 0\,, \tag{41.242}$$

where $P^{(n)}(x)$ denotes an n-th order polynomial. Note that this equation reduces to (41.237) when $n = 1$. The n solutions, $x_1, x_2, \dots, x_n$, to this algebraic equation can then be considered as the first n approximations to the root. Analogous to (41.239) and (41.240) we have

$$0 \approx f(x_0) + f'(x_0)(x - x_0) + \cdots + \frac{1}{n!} f^{(n)}(x - x_0)^n + \frac{1}{(n+1)!} f^{(n+1)}(x - x_0)^{n+1}\,, \tag{41.243}$$

$$0 = f(x_0) + f'(x_0)(x_n - x_0) + \cdots + \frac{1}{n!} f^{(n)}(x_0)(x - x_0)^n\,. \tag{41.244}$$

Show, by subtracting one equation from the other, and setting

$$(x-x_0)^2-(x_n-x_0)^2=\cdots=(x-x_0)^n-(x_n-x_0)^n=0$$

in the result, that the error for x_n, namely $e_n \equiv x - x_n$, can be estimated by

$$e_n \approx -\frac{1}{(n+1)!}\frac{f^{(n+1)}(x_0)}{f'(x_0)}\,\varepsilon^{n+1}\,. \tag{41.245}$$

Note the slow geometric convergence of the errors.

Problem 41.4 Consider the conventional *Newton's Method* for finding roots for an algebraic equation $f(x)=0$. The first step is the same as in the previous problem, and we obtain

$$x_1 = x_0 - \frac{f(x_0)}{f'(x_0)}\,,\qquad e_1 \equiv x - x_1 \approx \alpha(x_0)\,\varepsilon^2\,, \tag{41.246}$$

where

$$\alpha(x) \equiv -\frac{1}{2}\frac{f''(x)}{f'(x)}\,. \tag{41.247}$$

(a) To obtain the second-order approximation x_2 we expand $f(x)$ around x_1 and write

$$f(x) = f(x_1) + f'(x_1)(x-x_1) + \cdots = 0\,. \tag{41.248}$$

Then define x_2 by

$$f(x_1) + f'(x_1)(x_2 - x_1) = 0\,. \tag{41.249}$$

This yields the second-order approximation to the root

$$x_2 = x_1 - \frac{f(x_1)}{f'(x_1)}\,. \tag{41.250}$$

Subtract (41.249) from the equation

$$f(x_1) + f'(x_1)(x-x_1) + \frac{1}{2}f''(x_1)(x-x_1)^2 \approx 0 \tag{41.251}$$

to obtain the following error estimate for x_2:

$$e_2 \equiv x - x_2 \approx \alpha(x_1)(\alpha(x_0))^2\varepsilon^4\,. \tag{41.252}$$

(b) Show that for the third-order approximation to the root we have

$$x_3 = x_2 - \frac{f(x_2)}{f'(x_2)}\,,\qquad e_3 \equiv x - x_3 \approx \alpha(x_2)(\alpha(x_1))^2(\alpha(x_0))^4\,\varepsilon^8\,. \tag{41.253}$$

(c) Show that the n-th order results are given by, for $n=1,2,\ldots,$

$$x_n = x_{n-1} - \frac{f(x_{n-1})}{f'(x_{n-1})}\,,\qquad e_n \equiv x - x_n \approx \left(\prod_{i=1}^{n}(\alpha(x_{i-1}))^{2^{n-i}}\right)\varepsilon^{2^n}\,. \tag{41.254}$$

Clearly, the convergence of errors is much faster than in (41.245).

Problem 41.5 Consider a Hamiltonian system of the form (41.1) with two degrees of freedom ($N = 2$). Suppose the two unperturbed frequencies corresponding to invariant tori of H_0 are ω_1 and ω_2.

(a) Show that one way to re-express the KAM condition (41.29) is

$$\left| \frac{\omega_1}{\omega_2} - \frac{|m_2|}{|m_1|} \right| > \frac{K(\mu)}{|m_1|^{\tau+1}} , \tag{41.255}$$

where $K(\mu)$ is a constant depending on $\gamma(\mu)$.

(b) Consider the interval $I = 0 \le \omega_1/\omega_2 \le 1$ on the real line. Show that the sum of the lengths of all subintervals corresponding to values of ω_1/ω_2 for which the KAM condition (41.255) is *not* true is given by

$$2 \sum_{|m_1|=1}^{\infty} |m_1| \frac{K(\mu)}{|m_1|^{1+\tau}} = 2K(\mu)\,\zeta(\tau) , \tag{41.256}$$

where $\zeta(\tau)$ is the **Riemann zeta function** defined by the *Dirichlet series*

$$\zeta(s) = \sum_{n=1}^{\infty} \frac{1}{n^s} \qquad (Re\, s > 1) . \tag{41.257}$$

Hints: Each subinterval in I for which the KAM condition (41.255) is not true is of the form $|\,\omega_1/\omega_2 - |m_2|/|m_1|\,| \le K(\mu)/|m_1|^{\tau+1}$ and thus has length $2K(\mu)/|m_1|^{\tau+1}$. The integer $|m_1|$ can be thought of as the number of $|m_2|$ values for which $|m_2|/|m_1| < 1$, for a particular $|m_1|$.

Problem 41.6 In this problem we will study a simple application of the KAM Theorem. Revisit the "toy" three-body problem first introduced in Problem 33.4, but now in the context of an asteroid (of mass m) moving mainly under the gravitational pull of the sun (of mass M) and perturbed by the gravitational pull of Jupiter (of mass μ). We have $m \ll \mu \ll M$. Assume that all three bodies move in a space-fixed plane, and that Jupiter moves in a circular orbit centered at the center of mass of the sun-Jupiter system with constant angular speed Ω. As in Problem 33.4 we will use a rotating frame $\{\boldsymbol{e}_1, \boldsymbol{e}_2\}$ with origin at the center of mass of the sun and Jupiter, and co-rotating with Jupiter (with angular speed Ω).

(a) Use the result in part (c) of Problem 33.4 to show that, with respect to the rotating frame $\{\boldsymbol{e}_1, \boldsymbol{e}_2\}$, the Hamiltonian for the asteroid is

$$H(\boldsymbol{p}, \boldsymbol{q}) = \frac{\boldsymbol{p}^2}{2m} - \Omega p_\theta - \frac{GMm}{r} - \frac{Gm\mu}{|\boldsymbol{q} - \boldsymbol{r}_\mu|} , \tag{41.258}$$

where $\boldsymbol{q}$ is the position vector of the asteroid with respect to $\{\boldsymbol{e}_1, \boldsymbol{e}_2\}$, p_θ is the conserved angular momentum of the asteroid in the absence of Jupiter, r is the instantaneous distance between the asteroid and the sun, and $\boldsymbol{r}_\mu$ is the (fixed) position vector of Jupiter with respect to the rotating frame $\{\boldsymbol{e}_1, \boldsymbol{e}_2\}$.

Express the Hamiltonian in part (a) as [cf. (41.1)]

$$H(\boldsymbol{p}, \boldsymbol{q}; \mu) = H_0 + \mu H_1 , \tag{41.259}$$

where

$$H_0(\boldsymbol{p},\boldsymbol{q}) = H(\boldsymbol{p},\boldsymbol{q}=\boldsymbol{r};\mu=0) = \frac{1}{2m}\left(p_r^2 + \frac{p_\theta^2}{r^2}\right) - \Omega p_\theta - \frac{GMm}{r}\,, \tag{41.260}$$

and

$$\mu H_1 = -\mu\left(\frac{Gm}{|\boldsymbol{q}-\boldsymbol{r}_\mu|}\right)\,. \tag{41.261}$$

In (41.260), $\boldsymbol{q}$ is set equal to $\boldsymbol{r}$ since, as $\mu \to 0$, the CM of the Jupiter-sun system approaches the position of the sun. In (41.261), the mass of Jupiter, μ, actually serves as the small parameter in the perturbation Hamiltonian. Let the action variables for H_0 be $I_1 \equiv I_r$ and $I_2 \equiv I_\theta$. From (38.57) we have

$$I_2 = \frac{1}{2\pi}\int_0^{2\pi} p_\theta\, d\theta = p_\theta \qquad \text{(conserved angular momentum)}\,. \tag{41.262}$$

(b) Use (38.68) to show that

$$I_1 = -I_2 + GMm^2\sqrt{-\frac{1}{2m(H_0+\Omega I_2)}}\,. \tag{41.263}$$

(c) From the above equation show that

$$H_0 = -\Omega I_2 - \frac{G^2M^2m^3}{2(I_1+I_2)^2}\,. \tag{41.264}$$

(d) Show that the frequencies corresponding to H_0 are

$$\omega_1 = \omega_r = \frac{\partial H_0}{\partial I_1} = \frac{G^2M^2m^3}{(I_1+I_2)^3}\,, \tag{41.265a}$$

$$\omega_2 = \omega_\theta = \frac{\partial H_0}{\partial I_2} = -\Omega + \frac{G^2M^2m^3}{(I_1+I_2)^3} = -\Omega + \omega_1\,. \tag{41.265b}$$

From the above two equations we conclude that

$$\frac{\omega_2}{\omega_1} = 1 - \frac{\Omega}{\omega_1}\,. \tag{41.266}$$

Ω and ω_1 can be interpreted as the rotational frequencies of Jupiter and the asteroid around the sun, respectively, with ω_1 being the unperturbed rotational frequency of the asteroid. According to the KAM Theorem and the above equation, the ratio of these frequencies determine the stability of motion of the asteroid (whether invariant tori are preserved under perturbation). The invariant tori for frequency ratios Ω/ω_1 sufficiently near a rational number less than 1 would be destroyed under the perturbation due to Jupiter. The so-called **Kirkwood gaps** in the asteroid belt between Mars and Jupiter illustrate this situation. The gaps (marked by a scarcity of asteroids) appear prominently at $\Omega/\omega_1 = 1/2, 1/3, 1/4$, and less so at $\Omega/\omega_1 = 2/3, 2/5, 2/7$, as evidenced by empirical observations relating the number of asteroids in the belt as a function of Ω/ω_1.

Chapter 42

The Homoclinic Tangle and Instability, Shifts as Subsystems

The KAM Theorem discussed in the previous chapter gives vital and precise information on distinguishing between regions in the energy level submanifold of phase space where invariant tori are preserved or destroyed under small perturbations, but does not yield information on what happens in regions where the tori are destroyed. In this chapter we will give an introductory survey on some important mathematical developments that illuminate this problem. Our discussion will be restricted to the simplest case (beyond the trivial) where the dynamical system has $N = 2$ degrees of freedom (so the dimension of phase space is 4). We will see that even in this relatively "simple" case, which has the distinct advantage that the geometric picture can be visualized directly (a constant energy submanifold of phase space being 3-dimensional), the time evolution of phase orbits is already anything but simple, involving as it does a rich and dazzling panorama of bewilderingly complex patterns, matched only by an equally intricate web of mathematical concepts and theorems advanced for its elucidation. As opposed to the traditional analytic and geometric methods used most often in classical mechanics, these often borrowed heavily from modern topology. The big strides in the mathematical development began at the dawn of the 20th century with the pioneering work of H. Poincaré, which was further solidified by G. D. Birkhoff and enriched by S. Smale, among a host of other mathematicians and a few physicists. In the course of our discussion we will state precisely and discuss the main mathematical theorems underlying the crucial results presented, but as usual for weighty theorems abstain from giving any proofs, which involve too many technical details and are consequently beyond the scope of this text. The significant theorems are the *Poincaré-Birkhoff Fixed Point Theorem*, *Moser's Twist Theorem*, the *Smale-Birkhoff Theorem*, and *Zehnder's Theorem*.

We will begin by recalling the concepts of the *Poincaré section* and *Poincaré map* introduced in Chapter 18 (cf. Defs. 18.8 and 18.9), and apply them to the $N = 2$ case. The first important fact about the Poincaré maps is that they are **area preserving**. This can be seen as follows. Suppose a set of canonical coordinates $\{q^1, q^2, p_1, p_2\}$ has been chosen and the Poincaré surface of section, denoted by S, is chosen to be some neighborhood of a *periodic point* of the flow in the q^1, p_1 plane. So the Poincaré map, as an induced map by the phase flow φ_t of the Hamiltonian system and denoted by P, maps S into S via a canonical transformation. Consider a closed two-dimensional region $D \subset S$ and its image $P(D)$ under P. We will show that D is mapped to $P(D)$ by a set of reduced Hamilton's equations on a reduced 2-dimensional phase space (with local canonical coordinates q^1 and p_1), and hence, according to Liouville's Theorem (Theorem 39.7), D has the same area as $P(D)$. The full set of Hamilton's equations is given by

$$\frac{dq^1}{dt} = \frac{\partial H}{\partial p_1}, \qquad \frac{dq^2}{dt} = \frac{\partial H}{\partial p_2}, \qquad \frac{dp_1}{dt} = -\frac{\partial H}{\partial q^1}, \qquad \frac{dp_2}{dt} = -\frac{\partial H}{\partial q^2}. \tag{42.1}$$

Assuming that the Hamiltonian H is explicitly time-independent, $H(q^1, q^2, p_1, p_2)$ is constant along a flow line. So we define a 3-dimensional level set

$$H(q^1, q^2, p_1, p_2) = h = constant\,, \tag{42.2}$$

and invert this equation to obtain

$$p_2 = P_h(q^1, p_1\,;\, q^2)\,, \tag{42.3}$$

where q^2 is to be treated as a continuous parameter. Within the constant-energy level set we obtain, by differentiating (42.2) and using (42.3),

$$\frac{dH}{dq^1} = 0 = \frac{\partial H}{\partial q^1} + \frac{\partial H}{\partial p_2}\frac{\partial P_h}{\partial q^1}, \qquad \frac{dH}{dp_1} = 0 = \frac{\partial H}{\partial p_1} + \frac{\partial H}{\partial p_2}\frac{\partial P_h}{\partial p_1}. \tag{42.4}$$

These equations imply

$$\frac{\partial H}{\partial q^1} = -\frac{\partial H}{\partial p_2}\frac{\partial P_h}{\partial q^1}, \qquad \frac{\partial H}{\partial p_2} = -\left(\frac{\partial H}{\partial p_1}\right)\left(\frac{\partial P_h}{\partial p_1}\right)^{-1}. \tag{42.5}$$

Choosing coordinates in some subset of the constant-energy level surface such that

$$\frac{dq^2}{dt} = \frac{\partial H}{\partial p_2} \neq 0\,, \tag{42.6}$$

we can then eliminate the t-dependence from the full Hamilton's equations (42.1) and write

$$\frac{dq^1}{dq^2} = \frac{\dot{q}^1}{\dot{q}^2} = \frac{\partial H/\partial p_1}{\partial H/\partial p_2} = -\frac{\partial P_h}{\partial p_1}, \tag{42.7}$$

$$\frac{dp_1}{dq^2} = \frac{\dot{p}_1}{\dot{q}^2} = -\frac{\partial H/\partial q^1}{\partial H/\partial p_2} = \frac{\dfrac{\partial H}{\partial p_2}\dfrac{\partial P_h}{\partial q^1}}{\dfrac{\partial H}{\partial p_2}} = \frac{\partial P_h}{\partial q^1}. \tag{42.8}$$

Rewriting these results as

$$\frac{dq^1}{dq^2} = -\frac{\partial P_h}{\partial p_1}\,, \qquad \frac{dp_1}{dq^2} = \frac{\partial P_h}{\partial q^1}\,, \tag{42.9}$$

we obtain a system of Hamilton's equations of motion in the Poincaré section S [a 2-dimensional phase space with canonical coordinates (q^1, p_1)], with the parameter q^2 playing the role of a "time" variable and the "time"-dependent Hamiltonian given by $-P_h(q^1, p_1\,;\, q^2)$. The above set of equations can be regarded as a set of *reduced* Hamilton's equations (reduced from the original set on a 4-dimensional phase space). If q^2 is the angle variable θ^2, the Poincaré map P is precisely the diffeomorphism $\varphi_{2\pi}$ generated by the reduced Hamiltonian $-P_h$. By Liouville's Theorem (Theorem 39.7) (applied to the reduced 2-dimensional phase space, that is, the Poincaré section S) we conclude that P is area preserving.

On the Poincaré surface S we first consider dynamics due to an unperturbed (integrable) Hamiltonian of the form $H_0(I_i)$ in (41.1). The invariant tori will each be characterized by a specific set of action variables (I_1, I_2), and positions on each torus will be described by the corresponding angle variables (θ^1, θ^2). We assume that each invariant torus intersects S in a circle of a certain radius. Since a phase point on one such circle always returns to the same circle the next time the orbit meets S (assuming the orbit under consideration intersects S transversally), P has the effect of simply rotating points on a circle. Let the center of a family of concentric circles formed by the intersection of a family of invariant tori with S be the origin of a polar coordinate system with coordinates (r, θ), and suppose P maps a point (r_0, θ_0) to another point (r_1, θ_1). The Poincaré map P can then be represented by

$$P\begin{pmatrix} r_0 \\ \theta_0 \end{pmatrix} = \begin{pmatrix} r_1 = r_0 \\ \theta_1 = \theta_0 + 2\pi\alpha(r_0) \end{pmatrix}, \tag{42.10}$$

where

$$2\pi\alpha(r_0) = \omega_1 \tau = \omega_1 \left(\frac{2\pi}{\omega_2}\right). \tag{42.11}$$

So

$$\alpha(r_0) = \frac{\omega_1}{\omega_2}\,. \tag{42.12}$$

The action of the map P is represented in Fig. 42.1. Note that the two frequencies ω_1 and ω_2 for the time-dependences of the two angle variables θ^1 and θ^2 correspond to a particular invariant torus, and thus the above ratio depends only on r_0, the radius of the circle of intersection between the torus and S. In general $\alpha(r)$ is an increasing function of r. An area-preserving map characterized by (42.10) is called a **twist map**. It maps each circle to itself and radii to curved lines in such a way that areas are preserved (see Fig. 42.2, in which the shaded areas are equal to each other).

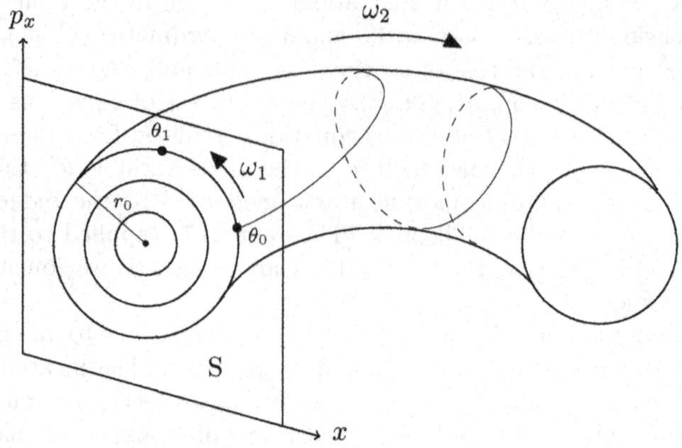

Fig. 42.1

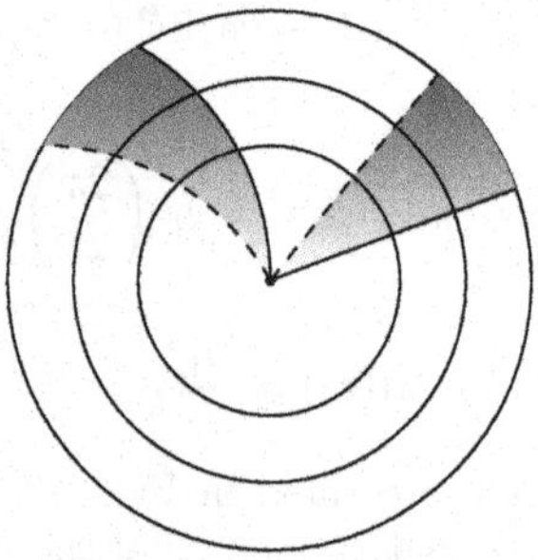

Fig. 42.2

Let us now turn on a perturbation to the Hamiltonian H_0, with μ as the small parameter [cf. (41.1)]. The Poincaré map is now denoted by P_μ and its action is expressed in the following form:

$$P_\mu \begin{pmatrix} r_0 \\ \theta_0 \end{pmatrix} = \begin{pmatrix} r_1 = r_0 + \mu f(r_0, \theta_0) \\ \theta_1 = \theta_0 + 2\pi\alpha(r_0) + \mu g(r_0, \theta_0) \end{pmatrix}, \tag{42.13}$$

where the functions f and g are required to (1) make P_μ still area-preserving, (2) satisfy the condition that $f(0, \theta_0) = g(0, \theta_0) = 0$, and (3) be sufficiently smooth but not necessarily analytic. The KAM Theorem asserts that tori will be destroyed in annular regions of S where $\alpha(r) = \omega_1/\omega_2 \approx m/n$ (sufficiently close to a rational number). The main question is then: What can one deduce about these regions of destroyed tori using the above stated properties for P_μ?

Suppose C_+, C and C_- are three concentric circles on S, with radii r_+, r and r_-, respectively, and $r_+ > r > r_-$. Let $r = m/n, r_+ > m/n$ and $r_- < m/n$, where m, n are positive integers relatively prime to each other. Consider the action of the n-th iterate of the Poincaré map P, P^n, on C: $P^n(r_0, \theta_0) = (r_1, \theta_1)$, $(r_0, \theta_0) \in C$. We see that $r_1 = r_0$, and

$$\theta_1 = \theta_0 + \omega_1 n \left(\frac{2\pi}{\omega_2}\right) = \theta_0 + 2m\pi\,. \tag{42.14}$$

Hence every point on the circle C is a *fixed point* of P^n. But it is quite obvious that P^n maps C_+ and C_- in opposite directions: C_- is rotated clockwise and C_+ counter-clockwise. Thus P^n is an example of an *annulus map*, which we will define more formally as follows.

Definition 42.1. *An* ***annulus map*** $\phi : A \to A$ *is an area-preserving homeomorphism defined on an annulus* $A \in \mathbb{R}^2$ *of the plane that maps the boundaries in opposite directions. If the boundaries of* A *are two concentric circles of radii* a *and* b, *where* $0 \neq a < b$, *then the action of* ϕ *can be expressed as*

$$r_1 = f(r_0, \theta_0)\,, \qquad \theta_1 = \theta_0 + g(r_0, \theta_0)\,, \tag{42.15}$$

where (r_0, θ_0) *are the polar coordinates of a point in* A, f *and* g *are continuous functions in* A *and both are periodic in* θ_0 *with period* 2π, *and*

$$g(a, \theta)\, g(b, \theta) < 0\,. \tag{42.16}$$

The last condition reflects the requirement that the boundaries are mapped in opposite directions.

Note that in this definition the boundaries of the annulus are not necessarily preserved by the map.

Problem 42.1 Show explicitly that P^n maps the circles C_+ and C_- in opposite directions.

We will now state the first significant theorem of this chapter (without proof). This theorem was first conjectured by Poincaré and subsequently proved by Birkhoff. For details on the proofs, the reader is referred to Guckenheimer and Holmes 1986, or Moser 1971.

Theorem 42.1. ***(Poincaré-Birkhoff)*** *An annulus map possesses at least one fixed point in the interior of the annulus.*

Let us now use the n-th iterate of the (perturbed) Poincaré map P_μ, that is, P_μ^n, to act on the annulus bounded by the circles C_- and C_+ introduced earlier. We see that, under the conditions for P_μ stated following (42.13), P_μ^n, like P^n, is still an annulus map for some annulus. According to the Poincaré-Birkhoff Theorem above, there is then a fixed point of P_μ^n inside the annulus. We will denote this fixed point by X. Such points are of great interest because they obviously imply periodic trajectories. The first pertinent question to ask is: Under what conditions would C_+ and C_- survive under small perturbations of an integrable Hamiltonian? This is answered by the following theorem on annulus maps, the second main theorem of this chapter, which, as expected, is a counterpart to the KAM Theorem and asserts that, under certain conditions, not only fixed points but also *invariant curves* exist for such maps. To state this theorem, which we will do shortly without proof, we need to introduce the concept of the l-norm of a C^l function $h(r,\theta)$ on an annulus A (C^l means continuously differentiable to order l) [cf. a similar function-theoretic concept introduced in (41.112) in the last chapter for the purpose of the proof of the KAM Theorem].

Definition 42.2. *The l-norm of an annulus function $h(r,\theta) \in C^l(A)$, denoted by $||h||_l$, is defined by the following supremum:*

$$||h||_l \equiv sup_{m+n\leq l,A} \left| \frac{\partial^{m+n} h}{\partial r^m \partial \theta^n} \right| , \tag{42.17}$$

where m, n are non-negative integers.

Theorem 42.2. ***(Moser Twist Theorem)*** *Suppose ϕ is an annulus map defined on the annulus A with boundaries $a < b$. Let $\alpha(r)$ be a C^l function and $|\alpha'(r)| \geq \nu \geq 0$ in $a \leq r \leq b$, for some $l \geq 5$. Let ε be a positive number and suppose also f, g [cf. (42.15)] are C^l functions. Then there exists a positive number $\delta(\varepsilon, l, \alpha(r))$ such that*

$$||f - r||_l + ||g - \alpha(r)||_l < \nu\delta , \tag{42.18}$$

and ϕ possesses an invariant curve C of the form (see Fig. 42.3)

$$r = c + u(\xi) , \qquad \theta = \xi + v(\xi) , \tag{42.19}$$

in the interior of the annulus A, where $u(\xi)$ and $v(\xi)$ are continuously differentiable, periodic in ξ with period 2π, and satisfy

$$||u||_1 + ||v||_1 < \varepsilon , \tag{42.20}$$

with $a < c < b$. Moreover, the induced mapping on this invariant curve is given by

$$\xi \longrightarrow \xi + \omega , \tag{42.21}$$

where ω is incommensurate with 2π and satisfies infinitely many conditions of the form

$$\left| \frac{\omega}{2\pi} - \frac{m}{n} \right| \geq \frac{\gamma}{n^\tau} \tag{42.22}$$

for some positive numbers γ and τ, and for all integers $n > 0$ and m. In fact, each choice of ω in the range of $\alpha(r)$ and satisfying (42.22) gives rise to such an invariant curve.

We note that the area-preserving property of ϕ is essential for this theorem.

Corollary 42.1. *An invariant curve of an annulus map ϕ partitions the annulus domain into two annuli each of which is an invariant set of ϕ.*

Returning to the annulus formed between the boundary circles C_+ and C_- on the Poincaré section S, we will construct a closed curve $R_\mu^{(\phi)}$ corresponding to an annulus map ϕ in the interior of the annulus by the following procedure. Take any radial line joining the common center of the circles and C_+, for example, the line 012 (see Fig. 42.4). ϕ maps point 1 clockwise and point 2 counterclockwise. Hence there must be a unique point somewhere between 1 and 2 on the radial line at which θ remains invariant under ϕ (say the point $1'$ in Fig. 42.4). Let $\phi = P_\mu^n$. The locus of such radially mapped points will then form the sought-for closed curve $R_\mu^{(\phi)}$ lying between C_+ and C_-.

By applying Birkhoff's Fixed Point Theorem (Theorem 42.1) to the map P_μ^n and the annulus bounded by C_+ and C_-, we see that there must be at least one fixed point in the interior of the annulus. We will denote one such fixed point by X. It is clear that X must lie on the closed curve $R_\mu^{(\phi)}$. But the fact that X is a fixed point of P_μ^n, that is, $P_\mu^n X = X$, implies that the n iterated points $P_\mu X, P_\mu^2 X, \dots, P_\mu^{n-1} X$ are also fixed points of P_μ^n, since

$$P_\mu^n(P_\mu^i X) = P_\mu^i(P_\mu^n X) = P_\mu^i X \qquad (i = 1, \dots, n-1) . \tag{42.23}$$

So the existence of one fixed point of P_μ^n necessarily implies the existence of n fixed points of the map, each of which must also lie on the closed curve $R_\mu^{(\phi)}$. Let us now construct another closed curve from $R_\mu^{(\phi)}$ by applying P_μ^n to it. This

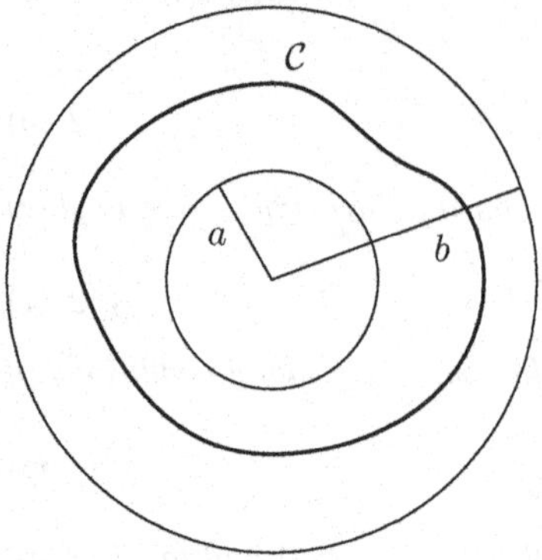

Fig. 42.3

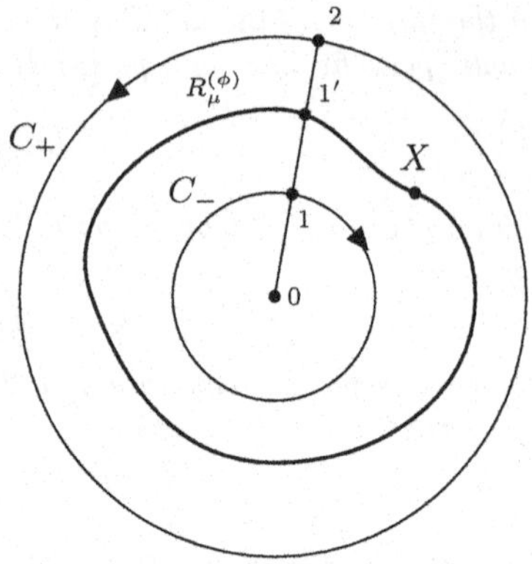

Fig. 42.4

curve will be denoted by $P_\mu^n R_\mu^{(\phi)}$. Since P_μ^n is area-preserving, $P_\mu^n R_\mu^{(\phi)}$ must intersect $R_\mu^{(\phi)}$ transversally, and at points other than the intersections, $R_\mu^{(\phi)}$ is either pushed radially inwards or outwards, so as to preserve the area bounded by it. It is also clear that the intersections must be fixed points of P_μ^n. By looking at Fig. 42.5, which depicts the role played by the area-preserving property of the map P_μ^n, we see that there must be an even number of intersection points between the curves $P_\mu^n R_\mu^{(\phi)}$ and $R_\mu^{(\phi)}$. So each of the n fixed points in (42.23) gives rise to an even number ($2k$, say) of fixed points. Hence we have $2kn$ fixed points of the area-preserving map P_μ^n (for some fixed positive integer k) in a region of destroyed tori in the annulus between C_- and C_+, which, we recall, contains C, the circular section of the destroyed torus corresponding to $\alpha = m/n$. Among these (even number of) fixed points, it is not hard to see that half of them are so-called *elliptic* fixed points, and the other half are *hyperbolic* fixed points (cf. Def. 18.4), and these two types of fixed points must alternate in position.

We recall here briefly how the fixed points are classified according to the eigenvalues λ_1 and λ_2 of the Jacobian matrix $D\phi$ at a fixed point (cf. Chapter 18). Since ϕ is area-preserving, we have $\lambda_1\lambda_2 = 1$. Since the mapping is real, we have either $\lambda_1 = \overline{\lambda}_1, \lambda_2 = \overline{\lambda}_2$ (both eigenvalues real), in which case the fixed point is hyperbolic or *parabolic*, or $\overline{\lambda}_1 = \lambda_2 = \lambda_1^{-1}, \lambda_1 \neq \lambda_2$ (both eigenvalues complex and lie on the unit circle), in which case the fixed point is **elliptic**. In the first case, when $\lambda_1 \neq \lambda_2$ are both real and also not equal to ± 1, the fixed

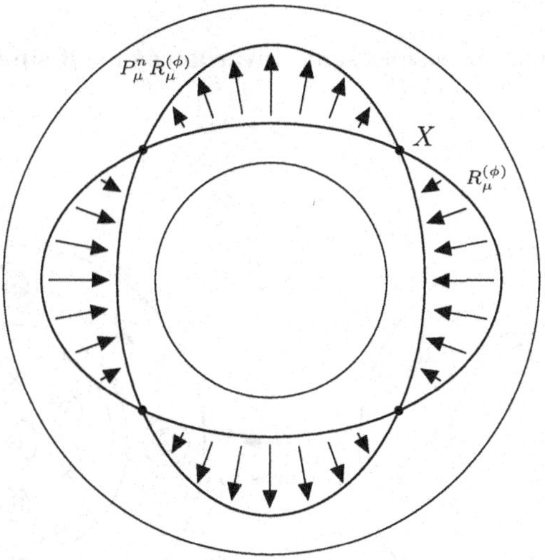

Fig. 42.5

point is **hyperbolic**. When $\lambda_1 = \lambda_2 = \pm 1$, the fixed point is an exceptional one known as a **parabolic** fixed point.

As a result of our discussions so far, we begin to see a mind-boggling heirarchy of patterns that appear in the so-called *stochastic* regions (where invariant tori are destroyed), as depicted in Fig. 42.6. The size of these regions grow with the strength of the perturbation. In the figure, the solid circles represent intersections of surviving tori with the Poincaré section S, while the dashed circles indicate positions of destroyed tori. The crosses and dots represent alternate positions of hyperbolic and elliptic fixed points, respectively, in the stochastic regions. Around each elliptic fixed point are also shown smaller solid circles representing more surviving tori, with the annuli regions between them representing yet more stochastic regions. This pattern repeats itself ad infinitum as one moves to smaller and smaller scales. This is shown by a blow-up picture of a neighborhood of an elliptic point, which looks exactly the same as the entirety of the previous picture. Perhaps one can summarize the state of affairs by the following whimsical description: As the "measles" break out around the elliptic fixed points in regions where tori are destroyed upon the introduction of a perturbation to an integrable Hamiltonian, they proliferate around more elliptic points in smaller and smaller stochastic regions, ad infinitum. The following lines by Jonathan Swift come to mind:

Great fleas have lesser fleas

Upon their backs to bite 'em
And lesser fleas have lesser still
And so ad infinitum.

We refer to this phenomenon as **scale invariance** or **self similarity**.

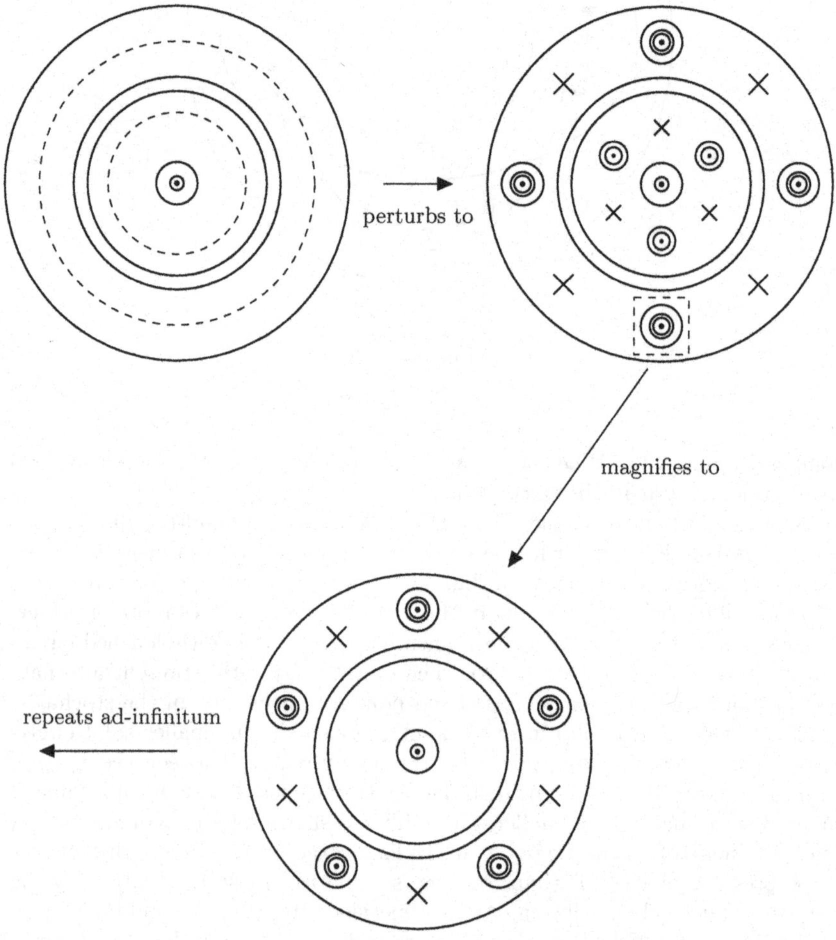

Fig. 42.6

Even more intricate and striking patterns are revealed when one examines the neighborhood of a hyperbolic fixed point p (Fig. 42.7). We first distinguish between two point sets corresponding to each hyperbolic fixed point of an area-preserving map ϕ: the **stable manifold** W^+ and the **unstable manifold** W^-, defined by

$$W^+ \equiv= \{x \mid \lim_{k\to\infty} \phi^k(x) = p\} , \qquad W^- \equiv= \{x \mid \lim_{k\to-\infty} \phi^k(x) = p\} . \quad (42.24)$$

In Fig. 42.8, these are represented by the directed lines meeting at the point p.

We will cite two main facts (without proofs) concerning the stable and unstable manifolds of a hyperbolic fixed-point: (1) neither W^+ nor W^- can intersect itself; (2) W^+ and W^- can, and will, intersect each other. The following definitions will set the stage for further discussions.

Definition 42.3. *Let p and q be two different fixed points of an area-preserving map ϕ and $r \neq p, q$ be a point such that $\phi^k(r)$ is defined for all integers $k = \pm1, \pm2, \pm3, \ldots,$ and that*

$$\lim_{k\to-\infty} \phi^k(r) = p , \qquad \lim_{k\to\infty} \phi^k(r) = q . \qquad (42.25)$$

Then r is called a **heteroclinic point** *(see Fig. 42.8). More generally, a point r is called* heteroclinic *if $\lim_{k\to-\infty} \phi^k(r)$ approaches one periodic orbit $\{p, \phi(p), \ldots, \phi^m(p) = p\}$ and $\lim_{k\to\infty} \phi^k(r)$ approaches another periodic orbit $\{q, \phi(q), \ldots, \phi^m(q) = q\}$. If these two orbits approach the same periodic orbit, then r is said to be a* **homoclinic point**.

We will focus on the special case where $p = q$ is a hyperbolic fixed point. The set

$$W_p^- \equiv \{x \mid \lim_{k\to-\infty} \phi^k(x) = p\} \qquad (42.26)$$

then constitutes the *unstable manifold* of p and the set

$$W_p^+ \equiv \{x \mid \lim_{k\to\infty} \phi^k(x) = p\} \qquad (42.27)$$

constitutes the *stable manifold* of p. A point $r \neq p$ of intersection between W_p^- and W_p^+ is then a *homoclinic point* (see Fig. 42.9).

Problem 42.2 Justify with qualitative arguments the fact that neither W^+ nor W^- can intersect itself.

Three salient facts concerning homoclinic points will next be stated, for which mathematical justifications in the form of relevant theorems will be provided later. These are (1) Homoclinic points exist, (2) Continuity of the area-preserving map ϕ implies that the existence of one homoclinic point implies the

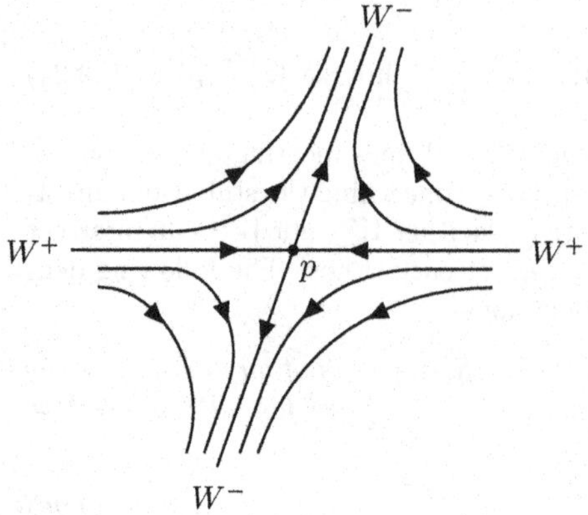

Fig. 42.7

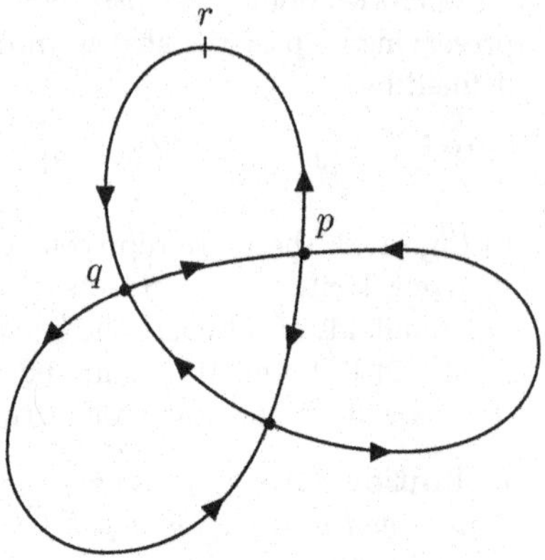

Fig. 42.8

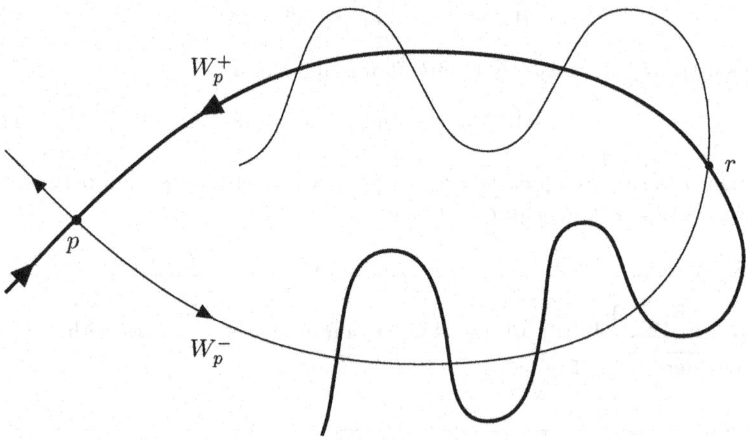

Fig. 42.9

existence of infinitely many others, and (3) Area-preservation of ϕ implies that homoclinic points form a *dense* set in any segment between any two homoclinic points on $W_p^{\pm}$. These lead to a highly intricate web of intersections between the stable and unstable manifolds W_p^- and W_p^+ emanating from a hyperbolic fixed point p, in which both of these manifolds must wind in and out so as to intersect with each other infinitely many times. This complicated behavior of homoclinic points corresponding to a hyperbolic fixed point is known as the **homoclinic tangle**, a representation of which is shown in Fig. 42.10. In this figure points in the stable vs unstable manifolds are drawn as lines with different thicknesses for clarity. We immediately note a typical property bestowed by a homoclinic tangle. Consider three points A, B and C in W_p^- in the figure that are initially close to each other. Following the directions of the arrows we see that after a sufficient number of steps ϕ^k (k large), the resulting corresponding points A', B' and C' will be arbitrarily far apart. This indicates, as expected, *stochastic behavior* or *randomness*. Different degrees of randomness, such as *ergodicity, mixing* etc. can be classified rigorously. But we will not enter into a discussion of the details here. The reader is referred to, for example, Arnold and Avez 1968.

Indeed, Poincaré already recognized the complexity of the flow within homoclinic tangles, and referred to it explicitly in descriptive prose (Section 397, Part 3, Poincaré 1899, English translation 1993):

> *When we try to represent the figure formed by these two curves and their infinitely many intersections, each corresponding to a doubly asymptotic solution, these intersections form a type of trellis, tissue, or grid with infinitely fine mesh. Neither of the two curves must ever cut across itself again, but it must bend back upon itself in a very complex manner in order to cut across all the meshes in the grid an infinite number of times.*
>
> *The complexity of this figure is striking, and I shall not even try to draw it. Nothing is more suitable for providing us with an idea of the complex nature of the three-body problem, and of all the problems of dynamics in general, where there is no uniform integral and where the Bohlin series are divergent.*

As a historical note, Arnold finally drew the figure for the first time, which appeared on p. 91 of Arnold and Avez 1968.

For the remainder of this chapter, we will give an introductory account of what is called **symbolic dynamics**, an immensely powerful and elegant mathematical method for the study of area-preserving maps ϕ in two dimensions that will shed much light on the main results discussed earlier. This method derives from the properties of sets S of *doubly-infinite* (infinite in both directions) or finite sequences of a finite or infinite number of symbols. The original advances were already found in the work of Birkhoff, but it was Smale who gave the method a rigorous foundation by focusing on the topological properties of ϕ (Smale 1967). The main idea is that a one-to-one correspondence can be

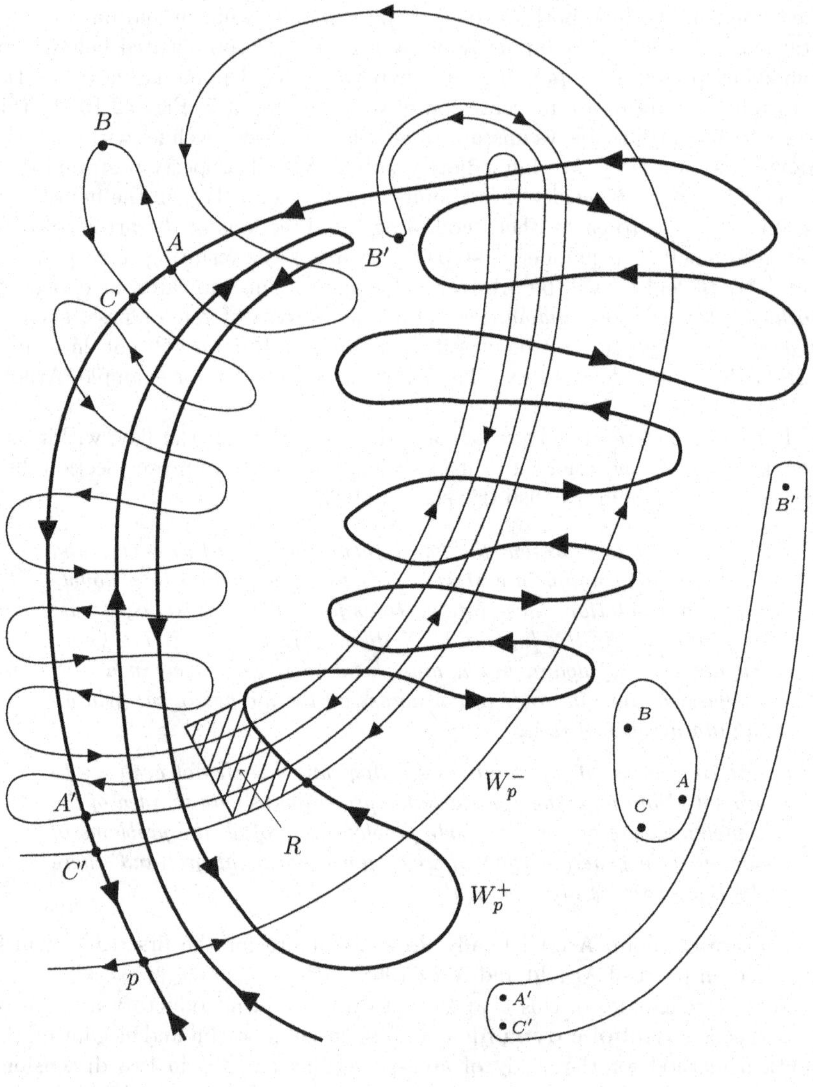

Fig. 42.10

established between a sequence set S and the points of an invariant (*Cantor*) set I of all iterates of an area-preserving map ϕ, in such a way that ϕ can be associated with the operation of a **shift** of the sequence (moving the symbols in the sequence one slot to the right, for example). We then speak of the *shift as a subsystem of the mapping* ϕ. In this way, the topological properties of orbits of a dynamical system can be associated with topological properties of S, considered as a topological space. We will naturally not be able to enter into the details of formulations and proofs of this large subject, but we hope to be able to convey some of the key elements.

The general concept of the shift as a subsystem can be defined as follows.

Definition 42.4. *Let $\phi : I \to I$ be a topological mapping of the topological space I into itself and $\sigma : S \to S$ be a second such mapping of the topological space S into itself. σ is said to be a subsystem of ϕ if there exists a homeomorphism $\tau : S \to \tau(S) \subset I$ such that the following diagram commutes:*

$$\begin{array}{ccc} S & \xrightarrow{\ \sigma\ } & S \\ {\scriptstyle\tau}\downarrow & & \downarrow{\scriptstyle\tau} \\ I & \xrightarrow{\ \phi\ } & I \end{array} \quad , \tag{42.28}$$

in other words,

$$\tau\sigma = \phi\tau \ , \qquad \text{or} \qquad \phi = \tau\sigma\tau^{-1} \ . \tag{42.29}$$

One of the main goals of our development is to be able to appreciate the *Smale-Birkhoff Theorem*, a result that underlies the instabilities of orbits near hyperbolic fixed points. This theorem will be stated more precisely later. Loosely speaking, it asserts that the presence of a homoclinic point belonging to a hyperbolic fixed point on a Poincaré section S implies that the shift σ on infinitely many symbols is a subsystem of a diffeomorphism ϕ on S.

To set the stage for an understanding of shift maps, we will consider the classic construction of a continuous map (topological map) known as the **Smale's horseshoe**. Let $\phi : Q \to \mathbb{R}^2$ be a continuous map from the closed square Q into the plane, so that $\phi(Q) \cap Q$ is a pair of horizontal strips H_1 and H_2 in Q (Fig. 42.11). This can be visually realized by the following successive operations: (1) stretching the square Q along the horizontal direction into a long horizontal strip, (2) bending the resulting strip into the shape of a horseshoe, and (3) placing the horseshoe back into the region originally occupied by Q and allowing the horseshoe to overlap the square in 2 horizontal strips. So we have

$$\phi(Q) \cap Q = H_1 \cup H_2 \equiv H^{(1)} \ . \tag{42.30}$$

The figure also indicates that

$$\phi^{-1}(H_1) = V_1 \ , \qquad \phi^{-1}(H_2) = V_2 \ , \tag{42.31}$$

where V_1 and V_2 are a pair of vertical strips in Q. We can also define

$$V^{(1)} \equiv V_1 \cup V_2 = \phi^{-1}(H^{(1)}) \ . \tag{42.32}$$

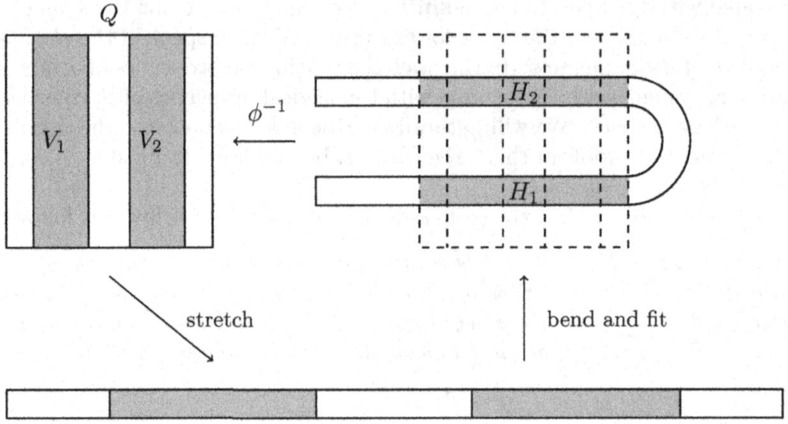

Fig. 42.11

Note that the intersection $H^{(1)} \cap V^{(1)}$ consists of four disjoint quadrilaterals in Q. Now take the square Q with the two horizontal strips H_1 and H_2 marked and apply ϕ a second time. Proceeding as we did the first time with the successive stretching, folding and overlapping the square, we see that the image of H_1 and H_2 under ϕ will intersect Q in four (narrower) horizontal strips (Fig. 42.12). These four horizontal strips can be identified by

$$\begin{aligned} H_{11} &\equiv H_1 \cap \phi(H_1)\,, \quad & H_{12} &\equiv H_1 \cap \phi(H_2)\,, \\ H_{21} &\equiv H_2 \cap \phi(H_1)\,, \quad & H_{22} &\equiv H_2 \cap \phi(H_2)\,. \end{aligned} \tag{42.33}$$

The union of these can be denoted $H^{(2)}$ and we can write

$$H^{(2)} = H_{11} \cup H_{12} \cup H_{21} \cup H_{22} = \phi^2(Q) \cap Q\,. \tag{42.34}$$

Similar to (42.31) we see that the second pre-image of $H^{(2)}$ (under ϕ) consists of four vertical strips in Q. Figure 42.13 illustrates how this situation arises through two operations of the inverse map ϕ^{-1}. (In this figure, the top picture shows a strip that is obtained from unbending the horseshoe in Fig. 42.12, and the second picture from the top represents the result of one operation of ϕ^{-1}.) The union of these vertical strips is denoted by $V^{(2)}$ and we can write, analogous to (42.32),

$$V^{(2)} = \phi^{-2}(H^{(2)})\,. \tag{42.35}$$

It is clear from Fig. 42.12 and Fig. 42.13 that the intersection $H^{(2)} \cap V^{(2)}$ consists of 16 small quadrilaterals, and $H^{(2)} \cap V^{(2)} \subset H^{(1)} \cap V^{(1)}$.

Problem 42.3 Draw figures for (a) horizontal strips $H^{(3)}$, (b) the vertical strips $V^{(3)}$ for the Smale's horse, and (c) the intersection of these two sets. Justify your results for the forward map ϕ^3 and the inverse map ϕ^{-3} by following procedures similar to those shown in Figs. 42.12 and 42.13.

Continuing inductively, we can generate the sets

$$H^{(n)} \equiv \phi^n(Q) \cap Q\,, \qquad V^{(n)} \equiv \phi^{-n}(H^{(n)}) \quad (n = 1, 2, \ldots)\,, \tag{42.36}$$

with the property that

$$H^{(n)} \cap V^{(n)} \subset H^{(n-1)} \cap V^{(n-1)} \qquad (n = 2, 3, \ldots)\,. \tag{42.37}$$

One then obtains in the limit the following invariant set I in which all iterates ϕ^k are defined:

$$I = \lim_{n\to\infty} H^{(n)} \cap V^{(n)}\,, \tag{42.38}$$

and can show that it is not empty and is in fact a **Cantor set**. A point $p \in I$ can be identified with a doubly infinite sequence

$$\{\ldots, s_{-2}, s_{-1}, s_0, s_1, s_2, \ldots\} \tag{42.39}$$

of two symbols (1 and 2) in the following manner. For any $k \in \mathbb{Z}$ (the set of positive and negative integers including zero), ϕ^{-k} is defined for p. Obviously $p \in H^{(k)}$. So by (42.36)

$$\phi^{-k}(p) \in V^{(k)} \subset V^{(1)} \subset V_1 \cup V_2\,. \tag{42.40}$$

This implies that $\phi^{-k}(p)$ is either in V_1 or V_2. If it is in V_1, s_k is assigned the value of 1; if it is in V_2, s_k is assigned the value of 2. We can summarize this rule by writing

$$\phi^{-k}(p) \in V_{s_k}\,. \tag{42.41}$$

This provides a mapping of the invariant set I into the sequence space S (of doubly infinite sequences of two symbols). Under suitable conditions for ϕ (to be specified subsequently) this map has an inverse $\tau : S \to I$ and τ is a homeomorphism. Now it is evident from (42.41) that

$$\phi^{-k}(\phi^{-1}(p)) \in V_{s_{k+1}}\,, \tag{42.42}$$

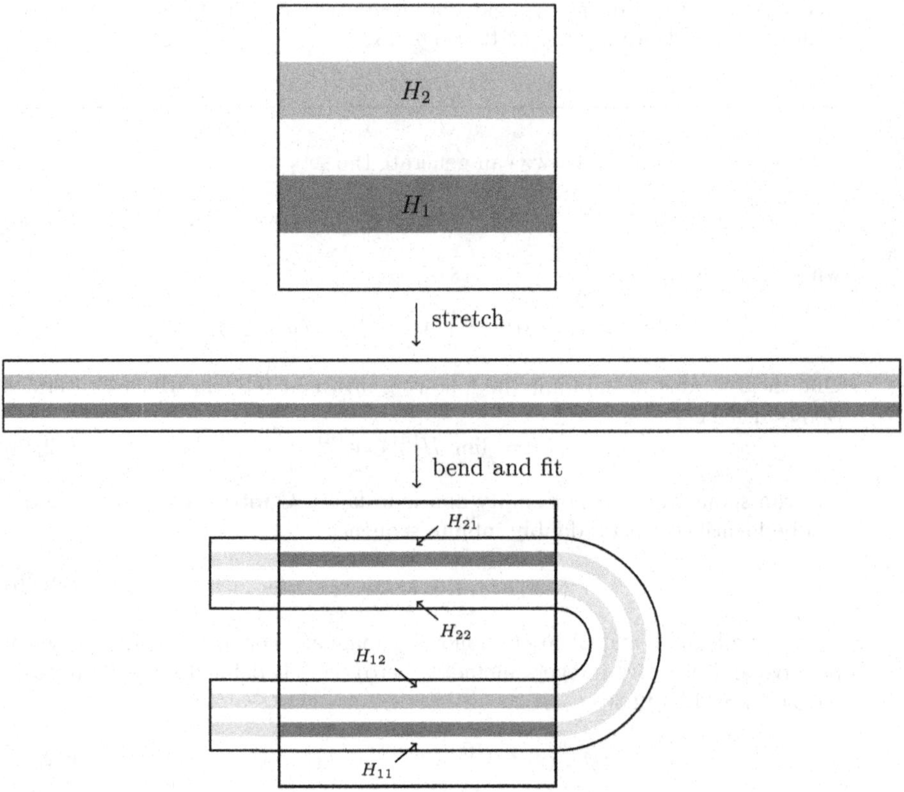

Fig. 42.12

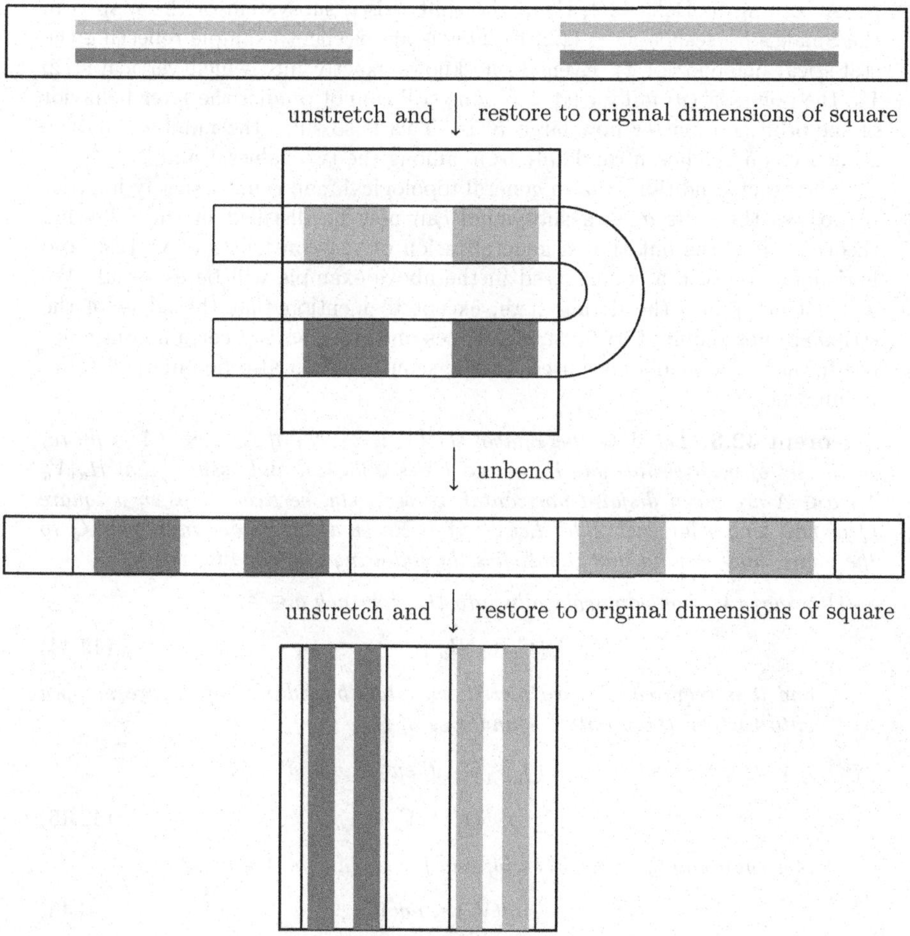
unstretch and
restore to original dimensions of square
unbend
unstretch and
restore to original dimensions of square

Fig. 42.13

which implies that if, under the above rule, a sequence s is assigned to a point $p \in I$, then to the point $\phi(p)$, the sequence s' is assigned, where s' is s right-shifted by one place. This result is equivalent to the statement that

$$\phi|_I = \tau\sigma\tau^{-1} , \tag{42.43}$$

where $\sigma : S \to S$ is the shift map one place to the right on the sequence set S. According to Def. 42.4, then, the shift σ is a subsystem of the map ϕ in the Smale's horseshoe [cf. (42.29)]. This is an excellent example reflecting the statistical behavior of ϕ: Even if one knows exactly into which vertical strip V_{s_k} the points $\phi^k(p)$ fall for $|k| < K$, one still cannot predict the later behavior of the orbit, no matter how large K is. This is so since the numbers s_k in a sequence can be chosen randomly from among the two values 1 and 2.

The exact conditions that a general topological map ϕ must satisfy in order to possess the shift σ as a subsystem can now be distilled in the following theorem. It turns out that a generalization of the constructs of vertical and horizontal strips in a square used in the above example will be essential. We will not enter into the details here, except to mention that the edges of the strips are not required to be straight lines and must satisfy certain continuity conditions. The symbol set used for the sequences can also be enlarged to an infinite set.

Theorem 42.3. *Let A be the symbol set $\{1, 2, \dots, N\}$ if $N < \infty$ (A is finite) or the set of positive integers if $N = \infty$ (A is infinite), and assume that H_a, V_a for $a \in A$ are given disjoint horizontal, respectively, vertical strips in a square Q of unit length for each side. Let $\phi : Q \to \mathbb{R}^2$ be a continuous map from Q to the plane, and assume that it satisfies the following two conditions:*

(i) ϕ maps V_a homeomorphically into H_a, for each $a \in A$:

$$\phi(V_a) = H_a , \qquad a \in A , \tag{42.44}$$

and it is required that vertical (horizontal) boundaries of V_a are mapped onto vertical (horizontal) boundaries of H_a.

(ii) If V is a vertical strip in $\bigcup_{a\in A} V_a$, then for any $a \in A$,

$$\phi^{-1}(V) \cap V_a = \tilde{V}_a \tag{42.45}$$

is a (non-empty) vertical strip; and for some ν in $0 < \nu < 1$,

$$d(\tilde{V}_a) \le \nu d(V_a) , \tag{42.46}$$

where $d(V)$ denotes the diameter *(maximum width) of the strip V. Similarly, if H is a horizontal strip in $\bigcup_{a\in A} H_a$, then*

$$\phi(H) \cap H_a = \tilde{H}_a \tag{42.47}$$

is a (non-empty) horizontal strip with

$$d(\tilde{H}_a) \le \nu d(H_a) , \tag{42.48}$$

for some ν in $0 < \nu < 1$.

Then ϕ possesses the shift σ on the set S of sequences of the elements of A as a substem: There exists a homeomorphism τ of S into Q such that

$$\phi\tau = \tau\sigma \,. \tag{42.49}$$

In particular, if A consists of a finite set of symbols, then $\tau(S)$ is a closed and invariant set [under any ϕ^k] in Q, which, like S, is a Cantor set.

Problem 42.4 Verify that the Smale's horseshoe map satisfies the two conditions stated in Theorem 42.3.

Problem 42.5 In this problem we wish to understand the main ideas behind the proof of Theorem 42.3 (see Mosher 1973 for details). For this purpose we need to identify the invariant set consisting of points $p \in Q$ for which $\phi^{-k}(p) \in V_{s_k}$ $(k = 0, \pm 1, \pm 2, \dots)$.

(a) Define inductively for $n \geq 1$ the following sets:

$$V_{s_0 s_{-1} \dots s_{-n}} \equiv V_{s_0} \cap \phi^{-1}(V_{s_{-1} \dots s_{-n}}) \tag{42.50}$$

and verify, inductively, by using condition (ii) of the theorem, that these are all vertical strips.

(b) Verify, by using condition (ii) again, that the following relationship holds for the diameters of the vertical strips in (a):

$$d(V_{s_0 s_{-1} \dots s_{-n}}) \leq \nu\, d(V_{s_{-1} \dots s_{-n}}) \leq \nu^n\, d(V_{s_{-n}}) \leq \nu^n \,. \tag{42.51}$$

Hence the diameters of these vertical strips tend to zero as $n \to \infty$.

(c) Show that

$$\begin{aligned} V_{s_0 s_{-1} \dots s_{-n}} &= \{p \in Q \,|\, \phi^k(p) \in V_{s_{-k}} \; (k = 0, 1, \dots, n)\} \\ &= \{p \in Q \,|\, \phi^{-k}(p) \in V_{s_k} \; (k = 0, -1, \dots, -n)\} \,. \end{aligned} \tag{42.52}$$

Hence

$$V_{s_0 s_{-1} \dots s_{-n}} \subset V_{s_0 s_{-1} \dots s_{-n+1}} \,, \tag{42.53}$$

so these vertical strips form a nested sequence.

(d) For a given sequence $s \in S$, convince yourself, without actually showing the details of the proof, that the intersection

$$V(s) \equiv \bigcap_{n=0}^{\infty} V_{s_0 s_{-1} \dots s_n} = \{p \in Q \,|\, \phi^{-k}(p) \in V_{s_k} \; (k = 0, -1, -2, \dots)\} \tag{42.54}$$

defines a vertical curve, depending uniquely on the left half of the sequence s.

(e) Define inductively, for $n \geq 2$, the following sets:

$$H_{s_1 s_2 \dots s_n} \equiv H_{s_1} \cap \phi(H_{s_2 \dots s_n}) \,. \tag{42.55}$$

Show that, as for the vertical strips (42.50) in (a), these are nested horizontal strips with diameters $\leq \nu^{n-1}$.

(f) Similar to part (d), convince yourself that the intersection

$$H(s) \equiv \bigcap_{n=1}^{\infty} H_{s_1 s_2 \dots s_n} = \{p \in Q \,|\, \phi^{-k+1}(p) \in H_{s_k} \; (k \geq 1)\} \\ = \{p \in Q \,|\, \phi^{-k} \in V_{s_k} \; (k \geq 1)\} \tag{42.56}$$

is a horizontal curve, depending uniquely on the right half of the sequence s.

The idea is that the two curves $V(s)$ and $H(s)$ (one vertical and one horizontal) intersect at exactly one point in Q, which will be identified as the image of τ [$\tau(s) = V(s) \cap H(s)$], If $\tau(s) = p$, then the shifted sequence $\sigma(s)$ is mapped to $\phi(p)$, that is, $\tau\sigma = \phi\tau$. We will not be concerned here with the proof that τ defined above is injective and bi-continuous.

The half-infinite sequence $\{\dots, s_{-2}, s_{-1}, s_0\}$ describes the random behavior of the iterates $\phi^{-k}(p)\,(k = 1, 2, \dots)$ or the trajectory of the dynamical system indefinitely backwards in time, while the half-infinite sequence $\{s_1, s_2, \dots\}$ describes that of the iterates $\phi^k(p)\,(k = 1, 2, \dots)$, or the trajectory indefinitely forward in time. Note that if the symbol set A is finite ($N < \infty$), the sequence set is compact. But if $N = \infty$, S will have to be compactified to a larger sequence set $\overline{S}$, which includes as a subset S, and also finite sequences with the symbol ∞ at both ends [of the form $(\infty, s_{\kappa+1}, \dots, s_{\lambda-1}, \infty)$, where $\kappa \leq 0, \lambda \geq 1$], or half-infinite sequences ending with ∞ on the right [such as $(\dots, s_{-2}, s_{-1}, \infty)$, where $\kappa = -\infty, \lambda = 2$], or on the left [such as $(\infty, s_{-2}, s_{-1}, s_0, s_1, s_2, \dots)$, where $\kappa = -3, \lambda = \infty$]. The new sequences correspond to solutions of the dynamical system which "escape" for positive or negative times. (For a physical example and more discussion on these see the following chapter.) The shift map σ will have to be changed to $\overline{\sigma}$, whose domain is given by $D(\overline{\sigma}) = \{s \in \overline{S} \,|\, s_0 \neq \infty\}$, and whose range is $R(\overline{\sigma}) = \{s \in \overline{S} \,|\, s_1 \neq \infty\}$. The homeomorphism τ will also have to be expanded to $\overline{\tau}$, mapping $\overline{S}$ into Q, such that

$$\overline{\tau}\,\overline{\sigma} = \phi\,\overline{\tau}|_{D(\overline{\sigma})} \;. \tag{42.57}$$

$\overline{\tau}(\overline{S})$ is compact but not invariant under ϕ^k for arbitrary k.

Theorem 42.4 applies to continuous maps ϕ. In case of continuously differentiable (C^1) maps, condition (ii) in the Theorem can be replaced by a more easily verifiable one. We will call this condition (iii), and state it as follows.

(iii) Suppose ϕ is a C^1 function on $\mathbb{R}^2$ represented in coordinates by:

$$x_1 = f(x_0, y_0) \,, \qquad y_1 = g(x_0, y_0) \,, \tag{42.58}$$

where (x_1, y_1) is the image point of (x_0, y_0); and the tangent map $d\phi$ takes the tangent vector (ξ_0, η_0) at (x_0, y_0) to the tangent vector (ξ_1, η_1) at (x_1, y_1), where

$$\xi_1 = f_x \xi_0 + f_y \eta_0 \,, \qquad \eta_1 = g_x \xi_0 + g_y \eta_0 \,. \tag{42.59}$$

For some μ in $0 < \mu < 1$, the bundle of vectors

$$S^+ \equiv \{(\xi, \eta) \mid |\eta| \le \mu |\xi|\} \tag{42.60}$$

defined over the vertical strips $\bigcup_{a \in A} V_a$ is mapped by $d\phi$ into itself, that is,

$$d\phi(S^+) \subset S^+ . \tag{42.61}$$

Moreover, if $(\xi_0, \eta_0) \in S^+$ and (ξ_1, η_1) its image point, then

$$|\xi_1| \ge \mu^{-1} |\xi_0| . \tag{42.62}$$

Similarly the bundle of vectors S^- defined over the horizontal strips $\bigcup_{a \in A} H_a$ by

$$S^- \equiv \{(\xi, \eta) \mid |\xi| \le \mu |\eta|\} \tag{42.63}$$

is mapped into itself by $d\phi^{-1}$, that is,

$$d\phi^{-1}(S^-) \subset S^- . \tag{42.64}$$

Moreover, if $(\xi_1, \eta_1) \in S^-$ and (ξ_0, η_0) its pre-image, then

$$|\eta_0| \ge \mu^{-1} |\eta_1| . \tag{42.65}$$

This condition expresses the *instability* of the C^1 mapping ϕ under iteration in metrical fashion: Eq. (42.55) implies that the horizontal components of tangent vectors in S^+ (those with longer x-components) are amplified by a factor of at least μ^{-1} under each "forward" iteration; while (42.58) implies that the vertical components of tangent vectors in S^- (those with longer vertical components) are amplified by a factor of at least μ^{-1} under each "backward" iteration. With this condition in mind, we have the following counterpart theorem to Theorem 42.3 for C^1 mappings (stated without proof).

Theorem 42.4. *If ϕ is a C^1 mapping that satisfies condition (i) of Theorem 42.3 and condition (iii) above with $0 < \mu < 1/2$, then condition (ii) of Theorem 42.3 is also satisfied with $\nu = \mu(1-\mu)^{-1}$, and hence the conclusion of that theorem as expressed in its last sentence – that ϕ possesses the shift σ as a subsystem – holds.*

The invariant set $I = \tau(S)$ for a C^1 map ϕ that satisfies conditions (i) and (iii) possesses a special structure called *hyperbolicity* (I is a *hyperbolic set*) if the parameter μ in (iii) satisfies a certain condition. The notion of a hyperbolic set is defined as follows.

Definition 42.5. *A set I is said to be a* **hyperbolic set** *if there exist two line bundles L^+, L^- with I as the base manifold that satisfy the following conditions:*

(a) Each of L^+ and L^- is a one-dimensional smooth distribution on I (a one-dimensional tangent subspace field on I) [cf. (10.9)].

(b) For each point $p \in I$, L_p^+ and L_p^- are invariant under ϕ, that is

$$d\phi\,(L_p^{\pm}) = L_{\phi(p)}^{\pm}\,, \tag{42.66}$$

where $L_p^{\pm}$ are the fibers of $L^{\pm}$ over p.

(c) Define the norm $||\zeta||$ of tangent vector $\zeta = (\xi, \eta)$ by $||\zeta|| \equiv max\,(|\xi|, |\eta|)$. For some constant $\lambda > 1$,

$$||\,d\phi\,(\zeta)\,|| \geq \lambda\,||\zeta|| \qquad \textit{for} \quad \zeta \in L_p^+\,, \tag{42.67a}$$

$$||\,d\phi^{-1}\,(\zeta)\,|| \geq \lambda\,||\zeta|| \qquad \textit{for} \quad \zeta \in L_p^-\,. \tag{42.67b}$$

The sufficient condition for the invariant set I being a hyperbolic set is given by the following theorem (stated without proof).

Theorem 42.5. *Suppose ϕ is a C^1 map on the square Q that satisfies conditions (i) of Theorem 42.3 and condition (iii) above, and I is the invariant set of ϕ^k $(k = \pm 1, \pm 2, \ldots)$. Let the action of ϕ be represented by (42.58) in coordinates, so that the Jacobian of the tangent map $d\phi$ at a point $p \in I$ can be written*

$$\begin{pmatrix} \left.\dfrac{\partial f}{\partial x}\right|_p & \left.\dfrac{\partial f}{\partial y}\right|_p \\ \left.\dfrac{\partial g}{\partial x}\right|_p & \left.\dfrac{\partial g}{\partial y}\right|_p \end{pmatrix} = \begin{pmatrix} a & b \\ c & d \end{pmatrix}\,, \tag{42.68}$$

and define

$$\Delta(p) \equiv ad - bc\,. \tag{42.69}$$

If the parameter μ in condition (iii) satisfies

$$\Delta(p)\,,\,(\Delta(p))^{-1} \leq \frac{1}{2}\mu^{-2} \qquad \textit{for all } p \in I\,, \tag{42.70}$$

then I is a hyperbolic set. Furthermore, if, for all $p \in I$,

$$0 < \mu \leq \frac{1}{2}\,min\,(|\Delta(p)|^{1/2}\,,\,|\Delta(p)|^{-1/2})\,, \tag{42.71}$$

then the horizontal line $H(s)$ and the vertical line $V(s)$ in Q (cf. Problem 42.5) are continuously differentiable curves whose tangents at each point $p \in I$ coincide with the lines L_p^+ and L_p^-, respectively, of the hyperbolic structure of I.

Now we are finally ready to make some exact statements concerning the properties of homoclinic points. One additional concept needs to be introduced, that of the *transversal map*. Suppose we construct a quadrilateral R near a homoclinic point r, two of its sides consisting of parts of W_p^+ and W_p^-, and the others can be straight lines parallel to the tangents of W_p^+ and W_p^- at r (see Fig. 42.10). For a point $q \in R$, we let $k = k(q)$ be the smallest positive integer for which $\phi^k(p) \in R$, if it exists. Denote the set of $q \in R$ for which such a $k > 0$ exists by $D(\tilde{\phi})$, where the **transversal map** $\tilde{\phi}$ is defined by

$$\tilde{\phi}(q) \equiv \phi^k(q) \qquad \text{for} \qquad q \in D(\tilde{\phi}) \,. \tag{42.72}$$

The main result on the properties of homoclinic points is then given by the following theorem, which is the central theorem of this chapter.

Theorem 42.6. *(Smale-Birkhoff) If a C^∞-(smooth) diffeomorphism ϕ possesses a homoclinic point r at which the curves W_p^+ and W_p^- of a hyperbolic fixed point p intersect, then in any neighborhood of r the transversal map $\tilde{\phi}$ of a quadrilateral possesses an invariant subset I homeomorphic to the set S of sequences of N $(< \infty)$ symbols via a homeomorphism $\tau : S \to I$ such that*

$$\tilde{\phi}\tau = \tau\sigma \,, \tag{42.73}$$

where σ is the shift mapping on S. Moreover, if $N = \infty$, τ can be extended to $\overline{\tau} : \overline{S} \to \overline{I}$ such that

$$\tilde{\phi}\overline{\tau} = \overline{\tau}\,\overline{\sigma} \,, \tag{42.74}$$

if both sides are restricted to $D(\overline{\sigma})$ [cf. (42.57)].

This powerful theorem at once entails the result that the existence of one homoclinic point implies the existence of infinitely many other homoclinic and heteroclinic points, as was pointed out before. Indeed, all sequences in $\overline{S}$ ending on both sides with the symbol ∞ correspond, via $\overline{\tau}$, to homoclinic points belonging to p, and these sequences are *dense* in $\overline{S}$. So the Smale-Birkhoff theorem at once yields the following important result.

Corollary 42.2. *If a C^∞-diffeomorphism ϕ possesses one homoclinic point r belonging to a fixed hyperbolic point p, then in any neighborhood of r it possesses a dense set of homoclinic points in a subset of that neighborhood which is an invariant set of the transversal map $\tilde{\phi}$.*

In fact other homoclinic points are obtained by using $\overline{\tau}$ to map sequences that end with the same repeated periodic blocks on both sides, and heteroclinic points are obtained by mapping sequences that end with two different repeating blocks on the right and left sides.

One can also ask the question: Can homoclinic points exist near elliptic fixed points? An affirmative answer is provided by *Zehnder's Theorem*, which attacks the problem by imposing a topology on sets of area-preserving real analytic maps, so that the notion of a neighborhood of a certain area-preserving map makes sense. We will state this theorem without proof.

Theorem 42.7. *(Zehnder) If ϕ is an area-preserving real analytic map with $(0,0)$ as an elliptic fixed point, then any neighborhood of ϕ (with respect to a certain topology introduced into the set of area-preserving real analytic maps) contains an area-preserving map ψ with the properties: (a) $(0,0)$ is an elliptic fixed point with an eigenvalue that is not a root of unity, and (b) every neighborhood of $(0,0)$ contains a homoclinic point.*

Theorems 42.6 and 42.7 provide the rigorous mathematical justifications for the complicated nature of the self-similar "measle" like patterns (Fig. 42.6) and the homoclinic tangle (Fig. 42.10) for $N = 2$ systems that appear in regions of destroyed tori described earlier. The behavior of trajectories for systems with number of degrees of freedom > 2 involves even more layers of complication and intricacy, such as the process of *Arnold diffusion*.

Finally, we mention a result concerning the nonexistence of a certain kind of integrals, in particular, real analytic integrals. A real-valued, non-constant function f in a planar domain D is said to be an integral of a diffeomorphism ϕ on D if $f(\phi(p)) = f(p)$, for any $p \in D$. The result follows.

Theorem 42.8. *Any diffeomorphism ϕ on a square Q in a planar region satisfying conditions (i) of Theorem 42.3 and condition (iii) above does not possess a real analytic integral in Q.*

This theorem has a direct application to the so-called *restricted three-body problem.*

Chapter 43

The Restricted Three-Body Problem

In this final chapter we will provide a glimpse of the power and elegance of the method of symbolic dynamics introduced in the last chapter by applying it on a version of the restricted three-body problem initially studied by Sitnikov in the early 1960's, whose results were subsequently extended by Alekseev, and by Moser, among others (see Moser 1973, and the references therein on original works by Sitnilov and Alekseev). This problem has a deceptively simple appearance, but is an excellent example of what is called **deterministic chaos**, that is, the appearance of complicated and random trajectories governed by exact equations of motion. In this introductory account, we will follow the authoritative exposition given by Moser, but will leave out most of the technical mathematical details. The main result is that a certain equivalence can be established between the set of possible trajectories $z(t)$ and the set $\overline{S}$ of doubly-infinite, infinite, or finite sequences of an infinite number of symbols [see Theorem 43.1 below and recall the discussion just before (42.57)], and thus the nature of the possible solutions to the problem can be "read off" from the corresponding sequences. As we will see, a number of rather surprising solutions are possible, such as unbounded, oscillatory, but non-escaping solutions.

The problem can be quite simply stated as follows. Consider two equal point masses $m_1 = m_2 > 0$ (known as *primaries*) moving in elliptic orbits on a plane under Newtonian attraction by each other [cf. the discussion leading up to (13.14)]. Assume that the center of mass (CM) of this two-body system is at rest. A third vanishing mass ($m_3 = 0$) moves in a straight line L perpendicular to the plane of motion of the primaries and passing through the CM of the primaries. This is always possible by choosing the initial position and velocity of m_3 appropriately (with initial position on L and initial velocity along the direction of L). The geometry is depicted in Fig. 43.1. The vanishing of the third mass m_3 assures that the motion of the primaries is not influenced by the presence of this mass at all, while the equality of the masses of the primaries

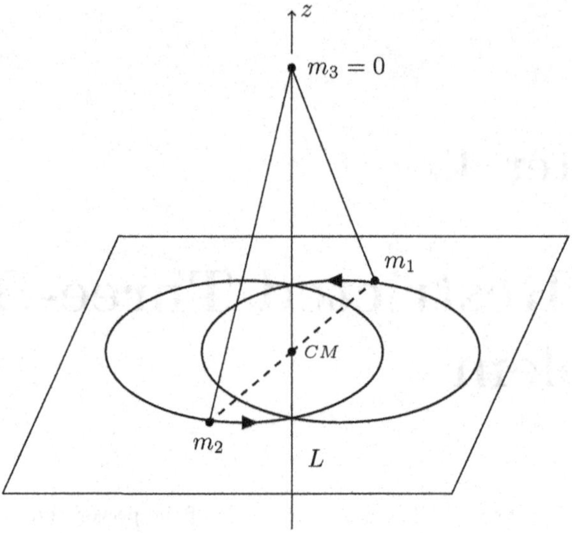

Fig. 43.1

provides the symmetry that makes the gravitational tugs due to the primaries on m_3 cancel each other along the plane of the ellipses. The problem then is to determine the position of m_3 for all times on L, given its initial position (on L) and its initial velocity (along L).

We will normalize units so that the (common) period of the primaries is $T_1 = T_2 = 2\pi$, $m_1 = m_2 = 1/2$, and the gravitational (Newton's) constant is $G = 1$. Suppose the position on the line L is measured by the variable z, and let the CM have coordinate $z_{CM} = 0$. Then, using Newton's Law of universal gravitation on the vanishing mass, one obtains the following second-order, non-linear differential equation:

$$\frac{d^2 z}{dt^2} = -\frac{z}{\{z^2 + r^2(t)\}^{3/2}}, \tag{43.1}$$

where t is the time, and $r(t)$ is the instantaneous distance of each primary from the CM. This equation is sometimes known as **Sitnikov's equation**. When the eccentricity $\varepsilon > 0$ of the elliptic orbits of the primaries is small, one readily obtains

$$r(t) = \frac{1}{2}(1 - \varepsilon \cos t) + O(\varepsilon^2). \tag{43.2}$$

Obviously $r(t)$ is an even function: $r(t) = r(-t)$, and periodic with period 2π: $r(t + 2\pi) = r(t)$.

Problem 43.1 Derive Sitnikov's equation (43.1), using Newton's law of gravitation and the conditions stated above the equation.

Intuitively, it is reasonable to expect that all solutions with small $|z| + |\dot{z}|$ at $t = 0$ will remain bounded for all real t. This is indeed the case, but is already non-trivial to prove. On the other hand, if $|\dot{z}(0)|$ is sufficiently large the solution will tend to infinity (the mass m_3 will escape). One of the big surprises, as we shall see, is that for initial velocities smaller than the escape velocity, infinitely many *oscillatory, unbounded* solutions exist. For these solutions, the mass m_3 behaves in a rather strange fashion by oscillating above and below the plane of the primary orbits with larger and larger amplitudes! In order to see how this can come about we would like to study solutions near (and below) the critical escape velocity and show that there is a set of orbits to each of which one can associate a sequence of integers. In other words, we will use the theory of *symbolic dynamics* developed in the last chapter, which will actually yield a number of other interesting and important results. We will discuss these towards the end of this chapter.

The integers referred to above will be chosen in the following manner. For each oscillatory solution $z(t)$ there will be infinitely many zeroes [t values at which $z(t) = 0$]. (Note that a zero always exists for any solution besides the trivial solution for which $z(t)$ is identically zero, see Problem 43.2 below.) We will denote these by t_k ($k = 0, \pm 1, \pm 2, \dots$) and assume that they are arranged in increasing order, so that $t_k < t_{k+1}$. An infinite set of positive integers s_k is then defined by

$$s_k \equiv \left[\frac{t_{k+1} - t_k}{2\pi} \right] , \tag{43.3}$$

where $[x]$ is the largest integer less than x. Each of these integers obviously gives the number of complete revolutions of the primaries between two consecutive times that m_3 passes through their orbital plane. In this way we can associate with every oscillatory solution a doubly infinite sequence of integers.

Problem 43.2 Show that a non-trivial solution $z(t)$ of the Sitnikov equation always has a real zero. [*Hint*: If no zero exists for a solution $z(t)$, it must be true that either $z(t) > 0$ for all t or $z(t) < 0$ for all t, depending on the initial position. Suppose $z(t) > 0$. Then show that $z(t)$ must also be *concave* ($\ddot{z} < 0$) for all t. Argue that it is impossible for $z(t)$ to be *both* positive *and* concave for all t.]

As a consequence of the assertion stated in Problem 43.2 and the uniqueness theorem for differential equations, we can characterize any non-trivial solution $z(t) \not\equiv 0$ of (43.1) by giving a time t_0 at which $z(t_0) = 0$ and the velocity at this time, $\dot{z}(t_0)$. Furthermore, since (43.1) is invariant under $z \to -z$ and $t \to 2n\pi$ (n an integer), only the magnitude of the initial velocity $v_0 \equiv |\dot{z}(t_0)|$ is important and it is sufficient to give the zero by $t_0\ (mod\, 2\pi)$. The value $v_0 = 0$ obviously corresponds to the trivial solution $z(t) \equiv 0$. So (v_0, t_0) can be regarded as polar coordinates on the plane (with v_0 the radius and t_0 the angle), and we have the convenient situation that any solution can be characterized by these coordinates.

We would now like to define a map $\phi : \mathbb{R}^2 \to \mathbb{R}^2$ on the plane that is analogous to the Poincaré return map used extensively in the last chapter. For a point (v_0, t_0) on the plane, that is, for a set of initial conditions $z(t_0) = 0, \dot{z}(t_0) = v_0$, one follows the trajectory in time until the *next* time, say $t_1 > t_0$, if it exists, when $z(t_1) = 0$, that is, when the solution reaches its next zero, and set $v_1 = |\dot{z}(t_1)|$. Then we define the map ϕ by

$$\phi(v_0, t_0) = (v_1, t_1) . \tag{43.4}$$

The question naturally arises as to whether or not the trajectory will hit $z = 0$ again, or more comprehensively, what the domain of the map ϕ is. This is answered by the following lemma, which is stated without proof.

Lemma 43.1. *There exists a real analytic simple curve in the plane $\mathbb{R}^2$ in whose interior D_0 the map ϕ of (43.4) corresponding to Sitnikov's equation (43.1) is defined. For initial conditions (v_0, t_0) outside D_0, the solutions escape.*

We next observe that, since the differential equation (43.1) is invariant under the time reflection $(z, t) \to (z, -t)$, the map ϕ satisfies some special properties given in the following result:

Lemma 43.2. *The map ϕ of (43.4) maps $D_0 \subset \mathbb{R}^2$ onto a domain $D_1 \subset \mathbb{R}^2$, which is the same as the reflected domain:*

$$D_1 = \phi(D_0) = \rho(D_0) , \tag{43.5}$$

where ρ is the time-reflection map

$$\rho : (v, t) \mapsto (v, -t) . \tag{43.6}$$

Moreover, ϕ preserves the area element $vdv \wedge dt$ in D_0, and

$$\phi^{-1} = \rho^{-1}\phi\rho . \tag{43.7}$$

Proof. Let $z(t;\, v_0, t_0)$ be the solution of (43.1) that satisfies

$$z(t_0;\, v_0, t_0) = 0\,, \qquad \dot{z}(t_0;\, v_0, t_0) = v_0\,. \tag{43.8}$$

From the property that (43.1) is invariant under the reflection $(z, \dot{z}, t) \to (z, -\dot{z}, -t)$ we have

$$z(-t;\, -v_0, -t_0) = z(t;\, v_0, t_0)\,. \tag{43.9}$$

We wish now to determine the domain of ϕ^{-1}, which, of course, is the range of ϕ. Suppose

$$\phi^{-1} : (v_0, t_0) \mapsto (v_{-1}, t_{-1})\,, \tag{43.10}$$

where t_{-1} $(< t_0)$, *if it exists*, is the nearest zero to t_0 that is less than t_0, so

$$z(t_{-1};\, v_0, t_0) = 0\,, \tag{43.11}$$

and

$$v_{-1} = |\,\dot{z}(t_{-1};\, v_0, t_0)\,|\,. \tag{43.12}$$

It follows from (43.9) and (43.11) that

$$z(-t_{-1};\, -v_0, -t_0) = 0\,. \tag{43.13}$$

We observe that, since $t_{-1} < t_0$, $-t_{-1} > -t_0$. Also, the initial conditions $(-v_0, -t_0)$ for the solution given by the above equation corresponds to the point (v_0, t_0) (as polar coordinates) in the domain D_0 of ϕ. So (43.13) implies, since ϕ maps in the forward time direction,

$$\phi(v_0, -t_0) = (v_{-1}, -t_{-1})\,. \tag{43.14}$$

Hence

$$\begin{aligned}(\rho^{-1}\phi\rho)(v_0, t_0) &= (\rho^{-1}\phi)(v_0, -t_0) = \rho^{-1}(v_{-1}, -t_{-1}) = \rho(v_{-1}, -t_{-1}) \\ &= (v_{-1}, t_{-1})\,,\end{aligned} \tag{43.15}$$

where in the second equality we have made use of (43.14), and in the third we have used the fact that, since $\rho^2 = 1$, $\rho^{-1} = \rho$. Comparing this result with (43.10), (43.7) follows.

Denote the domain of ϕ^{-1} by D^{-1}. From (43.7), D^{-1} must be such that $\rho(D^{-1}) = D_0$, since D_0 is the domain of ϕ. It follows that

$$D^{-1} = \rho^{-1}(D_0) = \rho(D_0)\,, \tag{43.16}$$

since $\rho^{-1} = \rho$. But, as we observed above, the domain of ϕ^{-1} must be $\phi(D_0)$, so (43.5) is established.

It remains to establish the area-preserving property of ϕ. For this we note that (43.1) can be expressed in Hamiltonian form:

$$\dot{z} = \frac{\partial H}{\partial v}\,, \qquad \dot{v} = -\frac{\partial H}{\partial z}\,, \tag{43.17}$$

where the time-dependent Hamiltonian can be written as

$$H(z,v,t) = \frac{v^2}{2} - \frac{1}{\sqrt{z^2 + r^2(t)}} . \tag{43.18}$$

(The reader should verify this.) Recalling (39.34) in Corollary 39.1 (with p_1 identified as v and q^1 as z in the present context) we see that the 2-form

$$dv \wedge dz - dH \wedge dt \tag{43.19}$$

on the extended phase space coordinatized by $(z, v, t(mod\, 2\pi))$ is a Poincaré-Cartan integral invariant. On both D_0 and D_1, $z = 0 = constant$ and thus $dz = 0$. Restricting the above 2-form to D_0, then, we see that

$$-dH \wedge dt = -v\, dv \wedge dt \tag{43.20}$$

is invariant under ϕ, which proves the area-preservation property of ϕ. □

At this point it is useful to examine what happens when the eccentricity vanishes ($\varepsilon = 0$). In this case $r = const = 1/2$ [by (43.2)]. This makes the Hamiltonian (43.18) explicitly time-independent and the total energy E of m_3 an integral (constant of motion). We have

$$E = \frac{v^2}{2} - \frac{1}{\sqrt{z^2 + 1/4}} . \tag{43.21}$$

For $E < 0$ this equation yields closed level curves in the (v, z) plane; while for $E \geq 0$, it gives rise to curves extending to infinity. In the former case ($E < 0$) the closed curves intersect $z = 0$ at values of v given by

$$v^2 = 2(E + 2) < 4 . \tag{43.22}$$

Thus the bound solutions for $\varepsilon = 0$ correspond to initial conditions represented by points in the disc $v < 2$ on the (v, t) plane, that is, D_0 is a disc with radius < 2. The mapping ϕ can be expressed as

$$\phi : (v_0, t_0) \mapsto (v_1 = v_0, t_1 = t_0 + T(v_0)) \qquad (v_0 < 2) , \tag{43.23}$$

where $T(v_0)$, the time of return to $z = 0$, is independent of t_0. The action of the map ϕ, for which $D_1 = \phi(D_0) = D_0$, is thus the rotation of concentric circles $v_0 = const < 2$ by an angle $T(v_0)$. As v_0 approaches 2 from below the rotation angle $T(v_0) \to \infty$. Another way of describing this is that the image of any radius of a circle with radius < 2 is a curve spiraling infinitely about the origin as it approaches the boundary circle.

We now return to the general case of $\varepsilon > 0$, and continue to state (without proofs) some crucial facts leading to the final result of this chapter (Theorem 43.1).

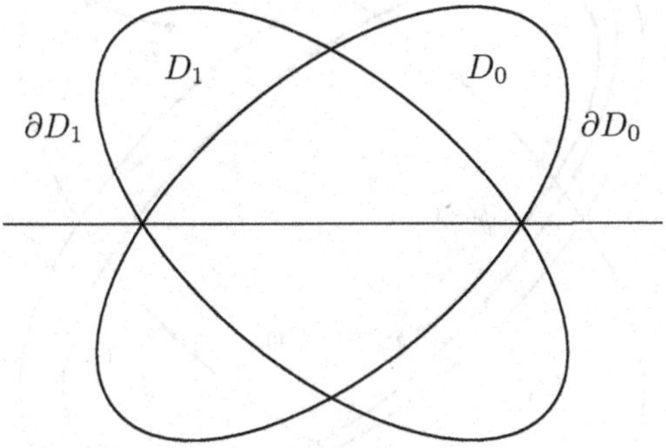

Fig. 43.2

Lemma 43.3. *If ε is sufficiently small and positive then $D_1 \neq D_0$, and the boundary curves ∂D_0 and ∂D_1 intersect on the symmetry line of time-reflection nontangentially (see Fig. 43.2).*

The twisting and spiraling property of a radius by the map ϕ for $\varepsilon = 0$ is retained when $\varepsilon > 0$. This is described by the following lemma.

Lemma 43.4. *Let $\gamma(\lambda) = (v_0(\lambda), t_0(\lambda))$ $(0 \leq \lambda \leq 1)$ be a C^1 curve in D_0 such that γ meets ∂D_0 non-tangentially at the endpoint p when $\lambda = 0$, and γ/p (γ with the point p removed) lies entirely in D_0. Then the image curve*

$$\phi(\gamma) = (v_1(\lambda), t_1(\lambda)) \tag{43.24}$$

approaches the boundary ∂D_1 in a spiraling fashion, in such a way that (see Fig. 43.3)

$$\lim_{\lambda \to 0} t_1(\lambda) = \infty\,, \tag{43.25}$$

that is, the curve $\phi(\gamma)$ spirals indefinitely towards the boundary ∂D_1 as $\lambda \to 0$.

The above lemma does not indicate in which direction the image curve $\phi(\gamma)$ approaches the boundary ∂D_1. In fact, and this is crucially important, *it approaches the boundary in such a way that its tangent direction approaches the tangent direction of the nearest point on the boundary,* as shown schematically

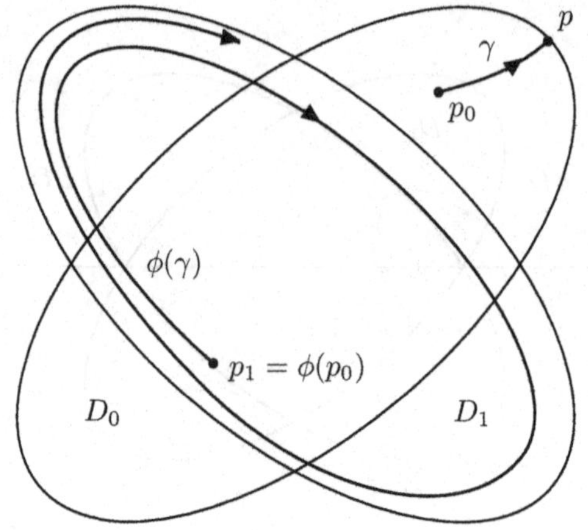

Fig. 43.3

in Fig. 43.3. Some more groundwork need to be laid in order for us to be able to state the lemma allowing this fact to be established.

For a sufficiently small number $\delta > 0$ we can define the set $D_0(\delta) \subset D_0$ to consist of points in D_0 whose distance from ∂D_0 is less than δ. Since, by Lemma 43.1, ∂D_0 is continuously differentiable (it being real analytic by that lemma), we can associate with any point $p \in D_0(\delta)$ a unique closest point $q \in \partial D_0$ (see Fig. 43.4). Over $D_0(\delta)$, then, we can define two bundles of sectors [two sectors associated with each point in $D_0(\delta)$]. The first bundle $\Sigma_0(\delta^{1/3})$ assigns to every point $p \in D_0(\delta)$ the sector (set of lines) passing through p which form an angle $\leq \delta^{1/3}$ with the line through p parallel to the tangent of the boundary curve ∂D_0 at q (see Fig. 43.4 again). The second bundle Σ_0' assigns to every point $p \in D_0(\delta)$ the set of lines passing through p complementary (not belonging) to Σ_0. One can similarly define the two bundles of sectors Σ_1 and Σ_1' over D_1. We can now state the following lemma, which implies the property of the tangent directions of $\phi(\gamma)$ stated in the last paragraph.

Lemma 43.5. *For a sufficiently small $\delta > 0$ there exists a number β ($0 < \beta < 1$) such that the mapping ϕ of (43.4) takes $D_0(\delta)$ into $D_1(\delta^\beta)$, and the tangent mapping $d\phi$ takes the bundle $\Sigma_0' = \Sigma_0'(\delta^{1/3})$ into the bundle $\Sigma_1 = \Sigma_1(\delta^{\beta/3})$. Moreover, if $\zeta_0 \in \Sigma_0'$, then $\zeta_1 = d\phi(\zeta_0) \in \Sigma_1$; and if ξ_0 and ξ_1 are the orthogonal projections of ζ_0 and ζ_1 into the center lines of Σ_0' and Σ_1, respectively (see Fig.*

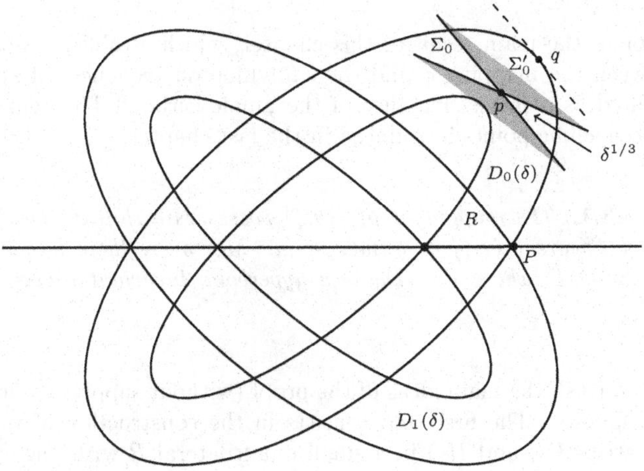

Fig. 43.4

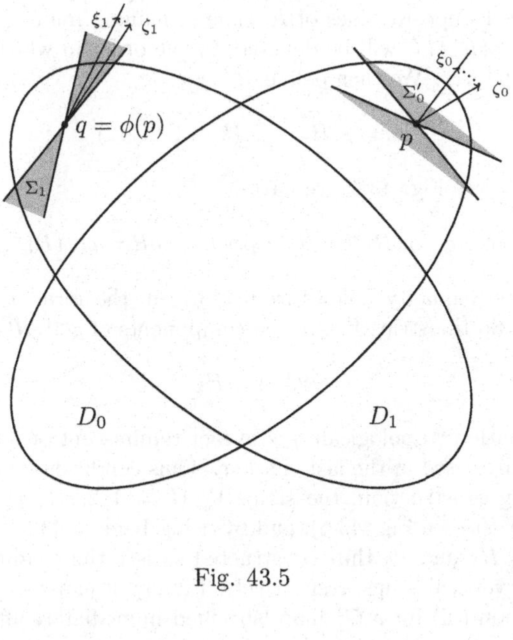

Fig. 43.5

43.5), then

$$|\xi_1| \le \delta^{-1/3} |\xi_0| \,. \tag{43.26}$$

Now comes the main result of this chapter, which establishes a correspondence between the mapping ϕ and the shift map on sequences of symbols. It can be viewed as a special instance of the Smale-Birkhoff Theorem (Theorem 42.6) on homoclinic points introduced in the last chapter.

Theorem 43.1. *The mapping ϕ of (43.4) possesses a shift $\overline{\sigma} : D(\overline{\sigma}) \to \overline{S}$ [cf. (42.57) and Theorem 42.6] as a subsystem. Moreover, there is a homeomorphism $\tau : S \to I$, where $I \subset D_0$ is a hyperbolic invariant subset, satisfying $\phi = \tau\sigma\tau^{-1}$.*

We will furnish the main ideas of the proof (without supplying the technical details) as follows. The first step consists in the construction of vertical and horizontal strips (V_k and H_k) in a small quadrilateral R with one corner at a point P of intersection between D_0 and D_1. For sufficiently small $\delta > 0$, the quadrilateral is given by $D_0(\delta) \cap D_1(\delta)$, which contains in its closure the point P (see Fig. 43.4). In fact, R will play the role of the square Q in Theorem 42.3. Since two opposite sides of R are curves meeting ∂D_0 non-tangentially, Lemma 43.4 implies that the image curves of these sides under ϕ will spiral towards ∂D_1 indefinitely, and $\phi(R)$ will be a spiraling strip that intersects R in infinitely many components. Aside from finitely many of these components they will connect opposite sides of R. The remaining infinite number of these components in $\phi(R) \cap R$ will be denoted, in the order in which they appear in R, by $H_1, H_2, H_3, \ldots$. We then define

$$V_k = \rho(H_k) \qquad (k = 1, 2, \ldots) \,. \tag{43.27}$$

Since $\rho(R) = R$ (see Fig. 43.4) we have

$$\phi^{-1}(R) \cap R = \phi^{-1}\rho(R) \cap \rho(R) = \rho\phi(R) \cap \rho R = \rho(\phi(R) \cap R) \,, \tag{43.28}$$

where the second equality follows from (43.7) in the form $\phi^{-1}\rho = \rho\phi$ (recall that $\rho^{-1} = \rho$). So the strips V_k are the components of $\phi^{-1}(R) \cap R$, and

$$\phi(V_k) = H_k \,. \tag{43.29}$$

The whole situation (topologically) is in fact reminiscent of the Smale's horseshoe example discussed in the last chapter. Thus condition (i) of Theorem 42.3 is satisfied. By construction, the strips V_k ($k = 1, 2, \ldots$) are disjoint closed sets, as are the H_k (see Fig. 43.6); and by using Lemma 43.5, it can be verified that the strips H_k and V_k thus constructed satisfy the conditions for general horizontal and vertical strips, respectively. Finally, it can also be shown that ϕ satisfies condition (iii) for a C^1 map [specified immediately after (42.57)] with

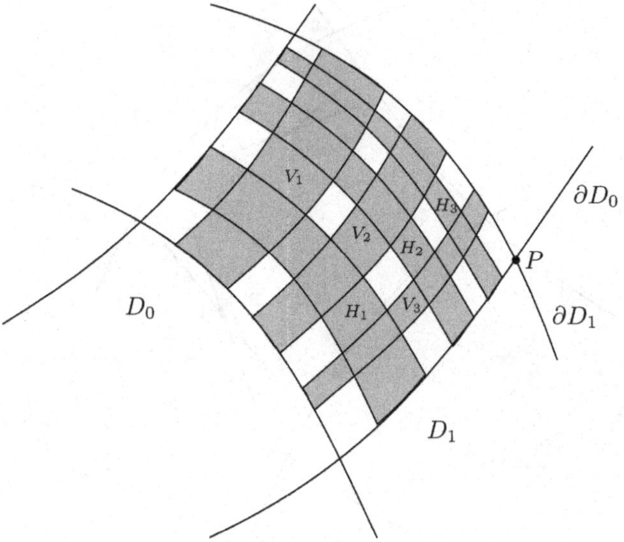

Fig. 43.6

$\mu = c\delta^{1/3}$, by making use of the properties of the formerly defined bundles Σ_i, Σ_i' ($i = 0, 1$). The crucial result which allows one to establish this fact is that in the quadrilateral R,

$$\Sigma_1(\delta^{1/3}) \subset \Sigma_0'(\delta^{\beta/3}) \,, \tag{43.30}$$

which follows from Lemma 43.5 (see Fig. 43.7). Thus Theorem 42.3 is applicable and Theorem 43.1 is proved.

A number of important and highly interesting consequences illustrating the richness and complexity of the set of solutions for the restricted three-body problem follow immediately from Theorem 43.1. We will conclude this chapter by discussing a few of these below. They mostly follow from the fact that the sequence of integers $\{s_k\}$ can be chosen completely randomly.

(1) *The existence of infinitely many capture, escape, and transitory orbits.* Consider a half-infinite sequence of the form

$$s = \{\ldots, s_{-1}, s_0.s_1, \ldots, s_{\lambda-1}, \infty\} \qquad (\lambda \geq 1) \,. \tag{43.31}$$

This sequence corresponds to a point $p \in \overline{\tau}(s)$ for which $\phi^{\lambda-1} \in \partial D_0$, so that p is not in the domain of ϕ^λ. The corresponding solution $z(t)$ never returns to $z = 0$ after one final encounter. If the s_k are bounded for $k \leq 0$, the above sequence represents an **escape orbit**, for which the

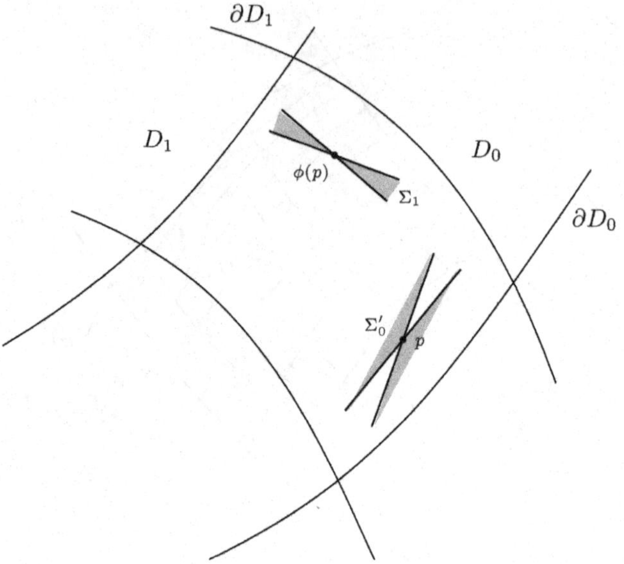

Fig. 43.7

solution escapes for positive times. Similarly, a half-infinite sequence of the form

$$\{\infty, s_{\kappa+1}, \ldots, s_0, s_1, \ldots\} \qquad (\kappa \leq 0) \tag{43.32}$$

represents a **capture orbit**. If the s_k are bounded for $k \geq 1$, the solution begins to encounter $z = 0$ plane at a certain time, and then performs infinitely many oscillations about this plane as $t \to \infty$. Finally, a finite sequence of the form

$$\{\infty, s_{\kappa+1}, \ldots, s_{\lambda-1}, \infty\} \qquad (\kappa \leq 0,\, \lambda \geq 1) \tag{43.33}$$

represents a **transitory orbit**, corresponding to a solution $z(t)$ which began to encounter the plane $z = 0$ at a certain past time, oscillates for a finite number of times about the plane, encounters the plane one last time in the future, and then escapes.

The existence of these types of orbits is a consequence of the following theorem, which itself is a consequence of Theorem 43.1.

Theorem 43.2. *For a sufficiently small eccentricity $\varepsilon > 0$ there exists an integer $m = m(\varepsilon)$ such that any sequence with $s_k \geq m$ for all k corresponds to a solution of the Sitnikov equation (43.1).*

(2) *The existence of infinitely many periodic orbits.* In the present context a periodic orbit is one which returns to $z = 0$ after a fixed period of time, indefinitely. Thus the sequence corresponding to this type of solution must have constant terms: $s_k = m$, for all k. Such a sequence, of which there are infinitely many, is definitely in S, which, according to Theorem 43.1, is homeomorphic to an invariant hyperbolic subset $I \subset D_0$ through the homeomorphism $\tau : S \to I$. The points $\tau(s)$ in I, where s is any sequence with constant terms, then correspond to periodic solutions.

(3) *The existence of infinitely many solutions corresponding to periodic points of ϕ.* A periodic point of ϕ is an invariant point of ϕ^n, for some positive integer n. Such a point obviously corresponds to a periodic sequence $s \in S$, such as $\{\dots, 2, 3, 5, 2, 3, 5, 2, 3, 5, \dots\}$, for which $n = 3$. There are infinitely many such sequences in S, so by Theorem 43.1, there are infinitely many points of ϕ with the corresponding solutions.

(4) *The existence of infinitely many oscillating and unbound solutions.* One can clearly have infinitely many unbounded sequences in S, for each term s_k can be chosen to be as large as one pleases, as long as it is finite. By Theorem 43.1, the points $\tau(s)$ for such unbounded sequences then correspond to solutions $z(t)$ having infinitely many zeroes, that is, oscillating indefinitely about the plane $z = 0$, but yet can have arbitrarily long intervals between consecutive encounters with the plane.

(5) *Existence of invariant curves.* The map ϕ of (43.4) in this chapter satisfies the conditions of Moser's Twist Theorem (Theorem 42.2) (we will not verify this in detail). The theorem then implies the existence of invariant curves in $0 < v_0 \leq r_0$ with $r_0 < 2$, if ε is small enough. The solutions corresponding to these curves are then bounded for all real t.

Problem 43.3 Show that fixed points of the mapping ϕ (which guarantee the existence of periodic solutions of the Sitnikov equation) can be constructed in the following manner. Let γ be the diagonal line of the quadrilateral R (refer to Fig. 43.4) through the intersection point P between ∂D_0 and ∂D_1, so γ lies on the time-reflection symmetry line. Since it meets ∂D_0 non-tangentially, Lemma 43.4 implies that $\phi(\gamma)$ spirals towards ∂D_1 indefinitely, and so it intersects γ at infinitely many points p_k ($k = 1, 2, \dots$). If these points are ordered according to the order of intersection, show that they are all fixed points of ϕ, and that

$$\phi(p_k) = p_k \,. \tag{43.34}$$

Hints: Write $\bigcup_{k=1}^{\infty} p_k = \gamma \cap \phi^{-1}(\gamma)$ and show that this set is invariant under ϕ. Use (43.7) and the fact that $\rho(\gamma) = \gamma$.

References

R. Abraham and J. E. Marsden, *Foundations of Mechanics*, Second edition, revised, enlarged, reset, Benjamin/Cummings (1982).

V. I. Arnold, *Small Denominators and Problems of Stability of Motion in Classical and Celestial Mechanics*, in Russian Mathematical Surveys, Vol. 18:6, pp. 85 – 191 (1963).

V. I. Arnold, *Mathematical Methods of Classical Mechanics*, Springer-Verlag (1978).

V. I. Arnold and A. Avez, *Ergodic Problems of Classical Mechanics*, Benjamin (1968).

P. Bamberg and S. Sternberg, *A Course in Mathematics for Students of Physics*, Vol. 2, Cambridge University Press (1991).

E. Cartan, *La Théorie des Groupes Finis et Continus et la Géométrie Differentielle Traitées par la Méthode du Repére Mobile*, Gauthier-Villars (1937).

S. S. Chern, *Curves and Surfaces in Euclidean Space*, in *MAA Studies in Mathematics, Vol. 4: Studies in Global Geometry and Analysis*, pp. 16 – 56, edited by S. S. Chern, The Mathematical Association of America, distributed by Prentice-Hall, Inc. (1967).

S. S. Chern, Soc. Math. de France, Astérisque, 67 (1985). Reprinted in S. S. Chern, *Shiing-Shen Chern Selected Papers*, Vol. IV, pp. 139 – 149, Springer-Verlag (1989).

S. S. Chern, *Historical Remarks on Gauss-Bonnet*, in *Analysis, etc., Volume in Honor of Jürgen Moser*, pp. 209 – 217, Academic Press (1990). Reprinted in S. Y. Cheng, P. Li, and G. Tian (ed), *A Mathematician and His Mathematical Work: Selected Papers of S. S. Chern*, pp. 539 – 547, World Scientific (1996).

S. S. Chern, W. H. Chen and K. S. Lam, *Lectures on Differential Geometry*, World Scientific (1999).

D. Chruściński and A. Jamiołkowski, *Geometric Phases in Classical and Quantum Mechanics*, Birkhäuser (2004).

M. P. do Carmo, *Riemannian Geometry*, translated by F. Flaherty, Birkhäuser (1992).

H. Flanders, *Differential Forms with Applications to the Physical Sciences*, Academic Press (1963).

H. Goldstein, *Classical Mechanics*, Addison-Wesley (1965).

J. Guckenheimer and P. Holmes, *Nonlinear Oscillations, Dynamical Systems, and Bifurcations of Vector Fields*, Springer-Verlag (Second Printing, Revised and Corrected 1986).

V. Guillemin and S. Sternberg, *Symplectic Techniques in Physics*, Cambridge University Press (reprinted with corrections 1990).

D. Hilbert and S. Cohn-Vossen, translated by P. Nemenyi, *Geometry and the Imagination*, Chelsea Publishing Company (1952).

M. W. Hirsch and S. Smale, *Differential Equations, Dynamical Systems, and Linear Algebra*, Academic Press (1974).

A. Katok and B. Hasselblatt, *Introduction to the Modern Theory of Dynamical Systems*, Cambridge University Press (1995).

A. I. Khinchin, *Mathematical Foundations of Statistical Mechanics*, Dover Publications (1949).

A. N. Kolmogorov, *The General Theory of Dynamical Systems and Classical Mechanics*, in Russian, in Proc. of the 1954 International Congress of Mathematicians, pp. 315 – 333, North Holland (1957). An English translation of this article appears in the Appendix of Abraham and Marsden 1982 (see above).

K. S. Lam, *Topics in Contemporary Mathematical Physics*, World Scientific (2006).

K. S. Lam, *Non-relativistic Quantum Theory: Dynamics, Symmetry, and Geometry*, World Scientific (2009).

L. D. Landau and E. M. Lifshitz, *Fluid Mechanics, Course of Theoretical Physics, Vol. 6*, Pergamon Press (1979).

R. G. Littlejohn and M. Reinsch, Reviews of Modern Physics, Vol. 69, No. 1, pp. 213 – 275 (1997).

J. Moser, *Stable ans Random Motions in Dynamical Systems*, Princeton University Press (1973).

H. Poincaré, *Les Méthodes Nouvelles de la Méchanique Céleste*, Vols. 1 - 3, Gauthier-Villars (1892, 1893, 1899); reprinted, Dover (1957); in English translation, edited and introduced by D. Goroff, *New Methods of Celestial Mechanics*, American Institute of Physics (1993).

D. Ruelle, *Chaotic Evolution and Strange Attractors: The Statistical Analysis of Time Series for Deterministic Nonlinear Systems*, Cambridge University Press (1989).

A. Shapere and F. Wilczek, *Geometric Phases in Physics*, World Scientific (1989).

C. L. Siegel and J. K. Moser, *Lectures on Celestial Mechanics*, Springer-Verlag (1971).

I. M. Singer and J. A. Thorpe, *Lecture Notes on Elementary Topology and Geometry*, Undergraduate Texts in Mathematics, Springer-Verlag (1976).

S. Smale, *Differentiable Dynamical Systems*, Bulletin of the American Mathematical Society, Vol. 73 (1967). Reprinted in *The Mathematics of Time*, by S. Smale, Springer-Verlag (1980).

J.-M. Souriau (translated by C. H. Cushman-de Vries), *Structure of Dynamical Systems – A Symplectic View of Physics*, Birkhäuser, (1997).

M. Spivak, *A Comprehensive Introduction to Differential Geometry*, Vols. I - V, Publish or Perish, Inc. (1979).

S. Sternberg, *Group Theory and Physcs*, Cambridge University Press (1994).

E. T. Whittaker, *A Treatise on the Analytical Dynamics of Particles and Rigid Bodies*, Fourth Edition, Cambridge University Press (1965).

N. M. J. Woodhouse, *Geometric Quantization*, Second Edition, Clarendon Press, Oxford (1992).

Index

Printed in the United States
By Bookmasters